Geology of a transpressional orogen developed during ridge-trench interaction along the North Pacific margin

Edited by

Virginia B. Sisson
Department of Earth Science
Rice University
6100 Main Street, MS-126
Houston, Texas 77005-1892
USA

and

Sarah M. Roeske
Department of Geology
University of California
One Shields Avenue
Davis, California 95616
USA

and

Terry L. Pavlis
Department of Geology and Geophysics
University of New Orleans
New Orleans, Louisiana 70148
USA

THE
GEOLOGICAL
SOCIETY
OF AMERICA

Special Paper 371

3300 Penrose Place, P.O. Box 9140 ▪ Boulder, Colorado 80301-9140 USA

2003

Published by The Geological Society of America, Inc.
3300 Penrose Place, P.O. Box 9140, Boulder, Colorado 80301
www.geosociety.org

Printed in U.S.A.

GSA Books Science Editor: Abhijit Basu

Library of Congress Cataloging-in-Publication Data

Geology of a transpressional orogen developed during ridge-trench interaction along the North Pacific margin / edited by Virginia B. Sisson, Sarah M. Roeske, Terry L. Pavlis.
p. cm. — (Special paper ; 371)
Includes bibliographic references and index.
ISBN 0-8137-2371-X
1. Geology, Structural--North Pacific Ocean. 2. Continental margins--North America. I. Sisson, Virginia Baker. II. Roeske, Sarah Melissa. III Pavlis, Terry L., 1953- IV. Special papers (Geological Society of America) ; 371.

QE601.G445 2003
551.8'2'091643—dc21

2003049093

Cover:

Backgound: Andalusite-bearing quartz veins in the Chugach metamorphic complex from the Tana River section discussed by Bowman et al. in Chapter 10. Vein is approximately 6 cm across. These show the voluminous fluid flow that is associated with low-pressure metamorphism and gold deposition synchronous with ridge subduction.

Top and Middle: Unnamed mountain at northern terminus of Quintino Sello Glacier, Canada. This shows near trench intrusives (gray felsic granites) intruding into the Chugach metamorphic complex (brown rocks).

Bottom: Hummingbird Ridge on Mount Logan, Canada, showing Border Ranges Fault zone. This is the contact between Wrangellia Composite terrane (light gray granitic rocks) with Chugach metamorphic complex (dark rocks).

10 9 8 7 6 5 4 3 2 1

Contents

Geological Society of America
Special Paper 371
2003

Introduction: An overview of ridge-trench interactions in modern and ancient settings

Virginia B. Sisson*
Department of Earth Science, MS-126, Rice University, Houston, Texas 77005-1892, USA

Terry L. Pavlis
Department of Geology and Geophysics, University of New Orleans, New Orleans, Louisiana 70148, USA

Sarah M. Roeske
Department of Geology, One Shields Avenue, University of California, Davis, California 95616-8605, USA

Derek J. Thorkelson
Department of Earth Sciences, Simon Fraser University, Burnaby, British Columbia V5A 1S6, Canada

ABSTRACT

Virtually all subduction zones eventually interact with a spreading ridge, and this interaction leads to a great diversity of tectonic processes in the vicinity of the triple junction. In the present-day Pacific basin, there are seven examples of active or recently extinct spreading ridges and transforms interacting with trenches. In contrast, there are only a few well-documented cases of spreading ridge interactions in the ancient geologic record, which indicates this process is grossly underrepresented in tectonic syntheses of plate margins. Analogies with modern systems can identify some distinctive processes associated with triple junction interactions, yet an incomplete understanding of those processes, and their effects, remains. Additional insights can be gained from well-documented examples of ancient ridge subduction because exhumation has revealed deeper levels of the tectonic system and such systems provide a temporal record of complex structural, metamorphic, igneous, and sedimentary events. This volume focuses on ridge-trench interactions in the Paleogene forearc record of the northern Cordillera (north of the 49th parallel). Insights from this system and modern analogs suggest that there is no single unique signature of ridge subduction events, but a combination of processes (e.g., igneous associations, changes in kinematics, motion of forearc slivers, thermal events, etc.) can be diagnostic, especially when they are time-transgressive along a plate margin. Understanding these processes in both modern and ancient systems is critical to our understanding of the creation and evolution of continental crust and provides a new framework for evaluating the evolution of the onshore and offshore tectonic history of the northern North American Cordillera.

Keywords: Neogene, Pacific plate, Cordillera, triple junction interaction, ridge subduction, Antarctica, Chile, Japan, Spain, accretionary prism, forearc magmatism.

INTRODUCTION

The typical view of subduction of oceanic lithosphere is a two-dimensional process dominated by underflow of oceanic lithosphere, refrigeration of the forearc, and development of an associated linear chain of calc-alkaline volcanoes situated 70–150 km above the downgoing plate. As first noted by DeLong and Fox (1977), all trenches must ultimately interact with a spreading center, yet there are very few ridge subduction events described in the geologic record. It is doubtful that this is an accurate reflection of the tectonic history of Earth.

*jinnys@rice.edu

Sisson, V.B., Pavlis, T.L., Roeske, S.M., and Thorkelson, D.J., 2003, Introduction: An overview of ridge-trench interactions in modern and ancient settings, *in* Sisson, V.B., Roeske, S.M., and Pavlis, T.L., eds., Geology of a transpressional orogen developed during ridge-trench interaction along the North Pacific margin: Boulder, Colorado, Geological Society of America Special Paper 371, p. 1–18.

A triple junction that develops when a spreading center interacts with a convergent margin is a combination of two or more of the following types of boundaries: an active spreading ridge, a transform fault, or a trench. In this volume, we do not consider subduction of aseismic ridges, though some consider this as ridge subduction. Specifically, we consider the following triple junction interactions: ridge-trench-trench (e.g., Chile triple junction), ridge-trench-transform (e.g., Rivera triple junction), and trench-transform-transform (e.g., Mendocino triple junction). Ridge subduction can be either a ridge-trench-trench or a ridge-trench-transform fault triple junction as both involve subduction of an active spreading ridge (Fig. 1). As a margin evolves, it can interact with different boundaries, which makes interpretation of geologic events difficult. For example, a triple junction can switch from ridge-trench-trench to ridge-trench-transform, as we speculate occurred in southern Alaska (Sisson et al., this volume, Chapter 13), or from trench-ridge-transform to trench-transform-transform as has occurred along the California margin at ca. 26 Ma.

A difficulty faced by the geoscience community in assessing the role of triple junction interactions is the diversity of possible geometries, all of them characterized by four-dimensional processes. Not only are there dozens of possible plate configurations (e.g., Cronin, 1992; Farrar and Dixon, 1993), but the actual geologic consequences are strongly influenced by plate geometry and velocity. For example, ridge-trench interaction may generate a "slab window" (Dickinson and Snyder, 1979), an asthenosphere-filled gap beneath the continental margin. A slab window can produce significant forearc thermal anomalies and geochemically unusual magmatic events that migrate with the triple junction. A simple scenario of a slab window migrating steadily in one direction may become geometrically and geologically complicated as subduction of ridge segments alternates with subduction of transform faults (Thorkelson, 1996) and plate motions change (this volume, Chapters 1 and 8).

Figure 1. Lambert projection of circum-Pacific basin showing locations of modern ridge subduction events and the region in the northern Cordillera affected by Cretaceous-Tertiary subduction of the Kula-Farallon ridge. Double thin lines indicate spreading ridges, thin lines indicate strike-slip boundaries, and thick lines with teeth are for subduction zones. Map created using projections available at http://www.aquarius.geomar.de/omc/omc_intro.html.

Two of us (VBS and TLP) convened a Penrose conference in 1994 on the general problem of triple junction interactions at convergent margins. Two conference summaries (Sisson et al., 1994; Pavlis et al., 1995) arose from that meeting. One result of that conference was a general acceptance that the Paleocene and Eocene geologic record of the northern Cordillera of North America preserves an example of the geologic consequences of ridge-trench interaction and subsequent ridge subduction (Fig. 2). Although there are several modern examples that provide excellent analogs for ancient systems, the northern Cordillera provides an important perspective on the long-term temporal evolution of a ridge subduction event, its regional manifestations, and something no modern system can provide, exposure of mid-crustal rocks that reflect some of the deep processes that accompany the event.

This volume presents results of a number of studies of the processes and consequences of Paleogene ridge-trench interaction beneath the northern Cordilleran margin. The main emphasis of this volume is on the effects of the ridge-trench interaction in the accretionary prism and forearc regions, but a number of the papers also look at consequences of terrane translation. Ridge subduction took place during a time of oblique convergence and large-scale northward translation of blocks along much of the margin. The combination of the relative motions of both the triple junction migration and the inboard terranes make the reconstruction of the paleogeography particularly challenging.

In this introductory paper, we first summarize existing information on modern triple junction interactions, and then summarize available data on ancient triple junction interactions. This summary sets up the framework for understanding triple junction interactions in general and the specific implications of these processes to the evolution of the northern Cordillera. We end with a brief summary of the results reported in this volume.

Triple Junction Processes: Insights from Modern Analogs

There is only modest consensus on what the effects of triple junction interactions are when integrated over time in the geologic record. However, modern examples of triple junction interactions in the circum-Pacific region (Fig. 1) do provide insight into processes that may be indicators of triple junction interactions along convergent margins. A summary of key indicator features appears in Table 1.

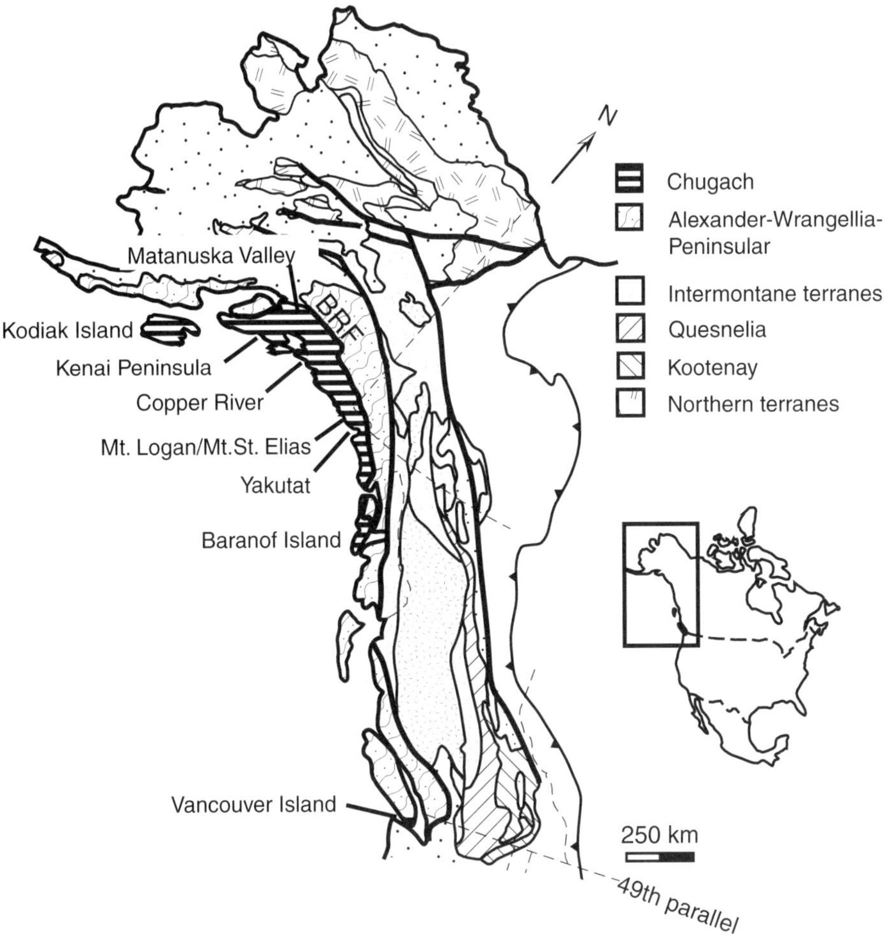

Figure 2. Tectonostratigraphic terrane map of the northern Cordillera showing geographic locations discussed in this volume. Most of the ridge subduction processes discussed in this volume occur in the Chugach terrane even though the effects are observed far from the Eocene Cordilleran margin. For simplicity, we have grouped together many of the terranes in the Intermontane belt as well as the Ruby Mountains, Seward Peninsula, and Brooks Range in northern Alaska. The stippled and unpatterned regions represent sedimentary sequences that were deposited since terrane amalgamation. BRF—Border Ranges fault system. Adapted from McClelland (2001, personal commun.).

Modern Triple Junctions of Western North and Central America: Mendocino, Rivera, British Columbia, and Panama Examples

The present-day triple junction interactions along western North and Central America involve ridge-trench-trench (e.g., British Columbia triple junction), ridge-trench-transform (Rivera triple junction) and trench-transform-transform (Mendocino and Panama triple junctions). Along this margin, the configuration of the various components has changed through time. This has led to considerable debate about the geological consequences of their interactions and whether or not a slab window forms in these complicated tectonic settings.

Mendocino

The Mendocino triple junction and its companions along the western margin of North America originated in the Oligocene (Atwater and Stock, 1998). These modern triple junctions began to evolve when the easternmost promontory of the Farallon-Pacific spreading ridge first encountered the subduction zone of western North America, and the Farallon plate was broken into several smaller microplates (McKenzie and Morgan, 1969; Atwater, 1970; Atwater and Stock, 1998). This interaction set up an unusual plate configuration in which the relative motion of two of the plates, North America and Pacific, was within a few degrees of parallel to the plate margin. The unusual plate configuration resulted in the formation of the San Andreas transform system between the two triple junctions (McKenzie and Morgan, 1969). Migration of the triple junctions, particularly the Mendocino triple junction, produced a lengthening of the San Andreas transform with time, and produced numerous broader effects throughout the southwestern edge of North America (Atwater, 1970; and numerous more recent studies).

Recent studies have focused particular attention on the Mendocino triple junction where two very different scenarios are called upon to explain observations in the vicinity of the triple junction: (1) a "stalled plate" scenario in which either microplates are still present as underplated slabs, or that subducted slab stretched to fill the gap (ten Brink et al., 1999; van Wijk et al., 2001); or (2) a "slab window" scenario, in which an asthenosphere-filled gap opened inboard of the triple

TABLE 1. FEATURES USED TO EVALUATE CONVERGENT MARGIN TRIPLE JUNCTION INTERACTIONS

Geologic process	Characteristics	Examples in coastal regions of the Northern Cordillera
Kinematics	Temporal and spatial variations in structural history; rapid shifts in kinematics in one region.	Change in regional kinematics (Chapters 1 and 8) Strike-slip faults (Chapters 6, 7, and 8) Emplacement of plutons (Chapters 12 and 14) Brittle deformation (Chapter 5)
Magmatism	Forearc magmatism that migrates through time; magma source from mantle and/or crust. Magmatic arc ceases or increases in activity; arc may develop alkaline signature. Backarc magmatism with alkalic and ocean-island signatures; slab-melt magmas in arc and back arc.	Sanak-Baranof forearc magmatic belt (Chapters 12 and 13) Magmatism in southern Vancouver Island (Chapter 14)
Fluid circulation	High temperature fluid interactions in forearc and accretionary prism; hydrothermal activity with stratiform sulfide deposits; gold mineralization.	Regional gold deposits (Chapter 5) High temperature fluids (Chapters 7, 9, and 10) Volcanogenic massive sulfide deposits (Crowe et al., 1992)
Metamorphism	Low-pressure/high-temperature metamorphism of forearc; possibly regional-scale contact metamorphism; Buchan series metamorphism.	High-temperature metamorphism (Chapters 9, 10, 11, and 14)
Basin evolution	Rapid shifting of depocenters and uplift within the forearc; destruction of basins by forearc strike-slip systems.	Transpressional evolution of forearc basins (Chapter 4)
Accretionary prism	Punctuated periods of tectonic erosion during ridge interactions, systematic younging of accreted sediment culminating in zero age pelagic sediments associated with near-trench MORBs accreted near the trench.	Uplift of accretionary margin (Chapters 2, 3, and 14)
Suprasubduction zone ophiolite	Accretion of forearc ophiolites with unusual chemistry.	Resurrection Peninsula ophiolite (Chapter 1)

Note: Both temporal and spatial variations with progressive younging along a margin can be used as a guide for evaluating the likelihood of triple junction interactions.

junction as the Cascadia subducted slab was carried northward with the triple junction (Dickinson and Snyder, 1979; Zandt and Furlong, 1982; Atwater and Stock, 1998). Both scenarios are consistent with higher heat flow (Lachenbruch and Sass, 1980; Underwood et al., 1999) and northward progression of volcanism (Dickinson and Snyder, 1979; Fox et al., 1985; Dickinson, 1997). The transient thermal effects have also been linked to ephemeral crustal thickening through viscous coupling between the Gorda plate and North America (Furlong and Govers, 1999). The northward migration of the triple junction is also attributed to variations in forearc sedimentation and deformation, particularly northward tilting and reduced deposition in the forearc basin (McCrory, 2000; Gulick et al., 2002). In the Gorda plate, immediately north of the present-day triple junction, there has been counterclockwise rotation of preexisting structures, uplift in the Eel River Basin, channel incision, shift in sediment source, and decrease in sediment preservation (Gulick and Meltzer, 2002; Burger et al., 2002).

The Mendocino triple junction is widely used as an instructional example of the effects of triple junction interactions. Unfortunately, from the broader perspective of ridge subduction processes, the Mendocino triple junction is not an ideal type example as its stability is highly dependent on maintenance of a specific geometry through time. As more data have accumulated, the broader effects of this unusual geometry have become increasingly clear (e.g., see Bohannon and Parson's [1995] reappraisal of Atwater's [1970] hypothesis).

Rivera

To the south, the Neogene breakup of the East Pacific rise also generated the Rivera triple junction. This ridge-transform-trench triple junction geometry resulted in complex microplate interactions that produced multiple triple junctions, many of which were unstable (e.g., Bandy and Pardo, 1994; Nicholson et al., 1994; Bourgois and Michaud, 2002). As a consequence, broad areas of deformation surround these triple junctions, both on the continental and oceanic plates (Bandy and Pardo, 1994; Nicholson et al., 1994; Atwater and Stock, 1998). The genesis of this deformation is difficult to assess within a broader context, however, because two ridge jumps occurred during the history of the Rivera triple junction (Atwater and Stock, 1998). For example, Dickinson and Snyder (1979) proposed that the Rivera triple junction caused rifting and formation of the California Borderlands. However, Fletcher et al. (2000) found that extension in Baja California postdates southward passage of the triple junction and Miller (2002) attributed Miocene extension and mafic underplating in the California Borderlands to a ridge-transform jump during southward migration. Several papers in this volume also examine the association of microplate formation with triple junction interactions (see Chapters 6, 7, and 8).

British Columbia

Prior to development of the Mendocino and Rivera triple junctions, the northern end of the Pacific-Farallon spreading ridge intersected the North American trench along the coastline of British Columbia and southeast Alaska. The triple junction first formed after the death of the Kula plate at ca. 43 Ma (Lonsdale, 1988) and migrated southward from southeast Alaska to its present position north of Vancouver Island (Riddihough, 1982). Currently, the plate margin north of the triple junction is dominated by the dextral Queen Charlotte transform fault. South of the triple junction, the Juan de Fuca plate is subducting eastward beneath British Columbia and Washington. In detail, the plate geometry in the region near the triple junction is complex, involving intraplate deformation and microplate formation (Riddihough, 1984; Rohr and Furlong, 1995). High heat flow (up to 100 mW/m^2) in northern British Columbia and the Yukon Territory near the present triple junction suggests that the ridge has been intersecting the trench in approximately the same location for several m.y. (Lewis and Hyndman, 1998; Hyndman and Lewis, 1999). Neogene volcanic rocks on northern Vancouver Island (Armstrong et al., 1985) and the Queen Charlotte Islands (Hamilton and Dostal, 2001) have been interpreted as products of "descending plate edge" and slab window environments, respectively. Throughout British Columbia and the Yukon, Miocene to Recent volcanic rocks range from small monogenetic centers to large polygenetic successions with alkaline to tholeiitic compositions and intraplate trace element signatures (Bevier et al., 1979; Hart and Villeneuve, 1999; Edwards and Russell, 2000). The boundaries of this intraplate volcanic field correspond closely with the margins of the proposed slab window.

According to Thorkelson and Taylor (1989), the combined effects of northward subduction of the Pacific plate and eastward subduction of the Farallon (Juan de Fuca) plate yielded a large, long-lived slab window beneath much of British Columbia, Yukon Territory, and southeast Alaska. In their model, the window is bounded to the northwest by a subducted slab of the Pacific plate and to the south by the Juan de Fuca slab. Using seismic tomography, Frederiksen et al. (1998) imaged the subducted eastern edge of the Pacific plate near the Yukon-Alaska border and identified anomalously hot mantle to the east and south, which they attributed to upwelling of asthenosphere through the slab window.

Panama

Another North American triple junction, the Cocos-Nazca-Caribbean triple junction in Panama is a poorly understood but potentially significant modern analog for ancient triple junction interactions. Johnston and Thorkelson (1997) made the case that although this triple junction is now a trench-trench-transform triple junction, a slab window is probably present beneath Panama due to relatively recent subduction of a ridge-segment. This section of the Middle American arc has ocean-island-like geochemical and isotopic characteristics, interpreted as a result of contamination of the mantle wedge by enriched sub-slab asthenosphere, and by localized "adakitic" volcanic assemblages interpreted as slab melts (Drummond et al., 1995; Johnston and Thorkelson, 1997). The absence of sublithospheric seismicity beneath the region is also consistent with a slab window between the Cocos and Nazca plates.

Modern Southern Pacific Triple Junctions: Woodlark, Chilean, and Antarctic Examples

All three triple junctions in the southern Pacific involve ridge subduction. The Chilean triple junction and the Woodlark Basin (Fig. 1) represent triple junction configurations that are probably the best modern examples illustrating the range of processes associated with ridge subduction. These systems display active subduction of ridge segments with different plate configurations: The ridge enters the trench at a high angle in the Woodlark system, whereas the ridge segments of the Chilean margin are nearly parallel to the trench. This variation is important for studies reported in this volume, as both configurations probably occurred during the Paleogene ridge-trench interactions in the northern Cordillera (e.g., Sisson and Pavlis, 1993).

Woodlark Basin

The Woodlark ridge-trench interaction developed entirely within oceanic microplates where the New Britain–South Solomon trench system is now consuming the Woodlark spreading system (Fig. 1). Ridge-trench interaction began when the Ontong Java Plateau collided with the Solomon Island arc (Taylor and Exon, 1987), causing the subduction zone to flip and initiating consumption of the Woodlark spreading center. The system is characterized by very high convergence rates (10.7 cm per year) as well as high spreading rates (7.2 cm per year) for the subducting ridge. The plate geometry is the closest modern analog for ridge-trench interactions in which oceanic plates on both sides of a ridge meet a trench at a moderate angle and are subducted. Different angles of incidence of the ridge with the trench would produce different triple junction migration rates along the trench. The sense of offset of transforms along the ridge would further control the rate of migration along the trench, possibly even causing a temporary reversal of migration direction.

In the Woodlark triple junction, only a small segment of the forearc is subjected to the effects of ridge subduction because of the configuration of the ridge-transform segments and the resultant back tracking of migration of the triple junction (Taylor and Exon, 1987). Some of the key signatures of this interaction are (1) a gap in deep seismicity in the vicinity of the triple junction, (2) the lack of a bathymetrically distinct trench, (3) rapid rates of uplift of the forearc near the site of the triple junction that produce local emergence and exhumation of the forearc and arc, and perhaps most significant, (4) voluminous forearc magmatism, including formation of large forearc volcanoes and volcanoes outboard of the triple junction on the ocean floor (Johnson et al., 1987; Perfit et al., 1987; Staudigel et al., 1987; Taylor and Exon, 1987; Crook and Taylor, 1994). The Paleogene of the northern Cordillera has similar forearc uplift and magmatism, which are well documented (this volume, Chapters 1, 3, 13, and 14).

Chilean Triple Junction

The Chilean triple junction, where the Nazca-Antarctic ridge is being consumed in the Andean subduction zone, provides a second example that may be similar to the Paleogene of the northern Cordillera (Forsythe and Nelson, 1985; Cande and Leslie, 1986; Tebbens et al., 1997; Lagabrielle et al., 2000). In this system, subducting ridge segments are subparallel to the trench. The Nazca plate subducts at about 70 mm per year compared to only 20 mm per year for the Antarctic plate, and the resultant plate geometry produces rapid triple junction migration during subduction of ridge segments and minor migration during transform subduction. Interestingly, the spreading rate for the ridge apparently slows down when the ridge segments collide and increases when transforms are being subducted (Tebbens et al., 1997). The earliest documented ridge collision occurred in southern Patagonia (54°S) at ca. 18 Ma (Forsythe and Nelson, 1985). The locus of ridge subduction then migrated northward through time, with major segments subducted at 14–10 Ma, 6 Ma, 3 Ma, and 100 ka to present (Forsythe and Nelson, 1985; Cande and Leslie, 1986; Bourgois et al., 2000).

Behrmann et al. (1994) summarized the results of ODP Leg 141 drilling and noted heat flow anomalies in the forearc, changes in forearc sedimentation, and hydrothermal alteration and mineral deposition, all consistent with subduction of a ridge. Other recent studies have identified many other features attributable to ridge subduction. The most prominent of these are magmatic effects, including the development of geochemically unusual mid-ocean ridge basalts on ridge-segments near the triple junction (Klein and Karsten, 1995; Karsten et al., 1996; Sherman et al., 1997; Sturm et al., 1999), anomalous forearc magmatism and a gap in the volcanic arc (Forsythe et al., 1986), the local occurrence of adakitic volcanic rocks interpreted as slab melts (Kay et al., 1993), and the presence of high-Mg backarc plateau basalts with ocean-island affinity (Gorring et al., 1997; Gorring and Kay, 2001; D'Orazio et al., 2000, 2001). In addition, the Taitao ridge and Bahai Barrientos ophiolite were obducted in the forearc during ridge subduction periods (Kaeding et al., 1990; Nelson et al., 1993; Lemoigne et al., 1996; Moores, 1998; Guivel et al., 1999; Bourgois et al., 2000; Shervais, 2001). At the same time, a forearc sliver developed north of the triple junction due to oblique subduction, and a small rift basin developed at the southern edge of the forearc sliver (Nelson et al., 1993; Flint et al., 1994; Behrmann and Kopf, 2001; Bourgois et al., 2000). Similar effects are seen in the Paleogene of the northern Cordillera (Chapters 1, 4, 5, 6, 7, and 8, this volume).

The region near the subducted ridge in Chile also has a complicated upper mantle structure which suggests significant orogen-perpendicular mantle flow (Murdie and Russo, 1999). The long history of convergence between the Chile rise and South America and the absence of significant microplate development implies that the ridge-transform system subducted beneath the continent and progressively formed a slab window (Forsythe and Nelson, 1985; Kay et al., 1993; Daniel et al., 2001). Unlike the Woodlark example, there has not been uplift in the forearc or backarc (Thomson et al., 2001).

At the Chilean triple junction, three large plates have interacted for an extended period of time, in contrast to the complex, short-lived microplate interactions at Woodlark. A configuration similar to the Chilean triple junction probably characterized part of the Paleogene north Pacific interactions discussed in this volume (e.g., Sisson and Pavlis, 1993), and thus, is particularly important as a modern analog. In particular, the Chilean system has a forearc ophiolite as well as forearc volcanism, both of which are observed in the north Pacific geologic record (Chapters 1, 12, 13, and 14). In addition, both the Chilean and Woodlark triple junction have anomalously high thermal gradients in the forearc during ridge subduction, a process that is represented in the northern Cordillera by extensive Paleogene hydrothermal activity in southern Alaska (Chapters 5, 9, and 10) as well as areas of high-grade metamorphism and plutonism (Chapters 1 and 8–14).

Antarctic Peninsula

The Aluk-Antarctic spreading center subducted beneath the Antarctic Peninsula in Miocene and Pliocene time (Barker et al., 1984; Larter and Barker, 1991; Henriet et al., 1992; Maldonado et al., 1994; Birkenmajer, 1994). Timing of ridge and transform fault subduction is constrained by collisional unconformities between deformed and undeformed horizontal strata, seen in seismic profiles of the shelf sediments (Larter and Barker, 1991; Bart, 1993). As the ridge moved northward along the Peninsula, there was increased uplift and erosion of the continental shelf (Bart, 1993). In addition, eruption of a small volume of alkalic basalt, geochemically indistinguishable from ocean island and continental alkali basalts, occurred in Recent time (Hole and Larter, 1993; Hole et al., 1994; Hole and Saunders, 1996). Farther south along the Antarctic Peninsula, near Alexander Island, there was high-Mg magmatism in the forearc that migrated northward (McCarron and Millar, 1997). Also in this region, the accretionary prism was uplifted and high mountains formed (Storey et al., 1996). This region is complicated by subduction of fracture zones that break up the margin into fragments with different structural and sedimentologic histories (Johnson and Swain, 1995; Johnson, 1997). Some speculate that there were two previous ridge subduction events in the Mid-to-Late Jurassic and Late Cretaceous (Scarrow et al., 1998; McCarron and Larter, 1998). Although important in the general context of ridge subduction processes, the plate configuration of this system is distinct from the Paleogene northern Cordillera. A few processes, such as uplift of the accretionary prism, are seen in the northern Cordillera (this volume, Chapter 1).

Synopsis of Recent Triple Junction Interactions

Significant signatures of triple junction interactions observed in recent convergent margins include: a gap in arc magmatism; anomalous locations and compositions of arc magmatism; forearc magmatism ranging from felsic (including peraluminous) to mafic (largely of MORB affinity); arc and backarc volcanics with characteristics ranging from slab melts to ocean island basalts; lateral migration of sedimentary sources and basins; ophiolite emplacement, high geothermal gradients extending from forearc to backarc; abrupt changes in plate kinematics that migrate with time; formation and juxtaposition of microplates both offshore as well as in forearc slivers driven by oblique subduction; and diachroneity of these geologic events (Table 1). Not all these features occur at all sites of triple junction interactions and subsequent ridge subduction. The different geometry and relative rates of plate interactions lead to a range of possible signatures. Although some of these geologic signatures are not unique to ridge subduction, several taken together can help identify regions where ridge subduction and/or triple junction interactions have occurred in the past.

INSIGHTS FROM ANCIENT SYSTEMS

Although modern triple junction interactions provide important constraints on variations in the driving processes and short-term effects of triple junction interaction, ancient examples provide additional insights to the process because (1) they provide a temporal record of the long-term evolution of processes such as magmatism, kinematic changes, and thermal perturbations, and (2) later exhumation can provide important views of mid-crustal processes. In addition to the northern Cordilleran system described in this volume, other areas have been suggested as sites of past triple junction interactions and serve as important guidelines for features characteristic of triple junction interactions. Here we present examples from the Precambrian to Cenozoic.

Precambrian Examples: Slave, Arunta Inlier, and Hemlo

A long-standing controversy has been the origin of Archean granite-greenstone terranes, and Kusky and Polat (1999) proposed that these terranes represent Archean ridge subduction episodes. Typically, granite-greenstone terranes are similar to large accretionary complexes with terrigenous sediment intruded by a tonalite-trondhjemite-granitoid (TTG) suite. The TTG suite was likely generated by melting of hot young subducted slabs. These rocks are commonly associated with a sanukitoid suite formed from the mantle wedge. Two aspects of Archean plate tectonics are particularly relevant. First, many more ridges probably occurred in the oceanic crust, thus increasing the chance of ridge-trench interaction. Secondly, oceanic slabs were hotter and thus more buoyant, so that both normal subduction and ridge-trench triple junction interactions produced potentially different effects compared to the Phanerozoic (Abbott and Menke, 1990).

The presence of a TTG suite may be an indicator of ridge subduction as it is also common in southern Alaska (see Chapters 1, 12, 13, and 14), but this association has not been extensively explored. The abundance of this igneous association in the Archean and Abbott and Menke's (1990) analysis of the probability of ridge subduction events in the Archean together suggest that forearc metamorphism and felsic magmatism related to ridge subduction may have an important, but generally overlooked, role in continental growth.

Kusky (1990) proposed that ridge subduction occurred during Late Archean evolution of the Slave province of northern Canada and ascribed low-pressure metamorphism in association with anatectic migmatites as a consequence of ridge subduction. However, others have interpreted these *P-T* conditions as reflecting crustal thinning and rifting (Thompson, 1989). Another possible example of Precambrian ridge subduction is in the northern Arunta Inlier of central Australia, which contains two separate episodes of low-pressure/high-temperature metamorphism related to granitoid emplacement (Collins and Williams, 1995; Myers et al., 1996). A third possible example is the late Archean Hemlo-Schreiber and White River–Dayohessarah greenstone belts of the Superior province, Canada (Polat et al., 1998).

Paleozoic Example: Spain

Castro et al. (1996, 1999) proposed that the Aracena metamorphic belt in the Paleozoic Hercynian region of Spain is related to ridge subduction. They also describe a linear belt of low-pressure/high-temperature metamorphism with steep lateral thermal gradients associated with norites that are geochemically similar to boninites. There is also a belt of low-pressure/high-temperature amphibolites with an inverted metamorphic gradient. Recently, Castro et al. (1999) documented a west to east linear progression of Rb-Sr ages from 343 to 328 Ma. This temporal evolution is suggestive of migration of a triple junction. Also, the magma geochemistry and high-temperature conditions suggest a ridge subduction event. There is not an associated ophiolite, although the belt of amphibolites has a mid-ocean ridge affinity. Thus, these amphibolites may be a high-temperature equivalent of suprasubduction ophiolites in other regions.

Superimposed Mesozoic and Cenozoic Example: Japan

There have been several episodes of triple junction interactions with Japan (see Brown, 1998a), from a relatively well-documented Miocene event to two more speculative Cretaceous and Tertiary events (Uyeda and Miyashiro, 1974; Taira et al., 1988; Osozawa et al., 1990; Osozawa, 1992; Tokunaga, 1992; Kiminami et al., 1993; Underwood et al., 1993a; Sakaguchi, 1996).

The Miocene subduction of the Kyushu ridge between the Pacific and Philippine Sea plates (ca. 15 Ma) resulted in anomalous near-trench magmatism, cessation of backarc spreading in the Shikoku Basin, opening of the Sea of Japan, rotation of crustal blocks, and relatively high geothermal gradients in the forearc (Underwood et al., 1993b). Another signature of this ridge subduction event is that the forearc accretionary complex records accretion of sediments with a successively narrower age range until sediments and associated basalts are indistinguishable in age, implying subduction of successively younger lithosphere as the triple junction approached (Osozawa et al., 1990). The magmatism in the forearc was mostly felsic with minor mafic bodies (Hibbard and Karig, 1990). The high geothermal gradients resulted in anthracite coals with paleotemperatures of 180 to 315 °C and up to 500 °C adjacent to intrusives (Tokunaga, 1992; DiTullio et al., 1993; Hibbard et al., 1993; Laughland and Underwood, 1993; Underwood et al., 1993b). Sakaguchi (1999a, 1999b) also recognized high-temperature fluids in association with these high paleogeothermal gradients. Another feature is a set of normal faults that parallel the trend of the subducting ridge in the Shimanto accretionary sequence (Osozawa, 1992).

Evidence for a Cretaceous ridge-trench encounter in Japan is seen in the migration of igneous activity along the Shimanto accretionary complex in southwest Japan (Kinoshita and Ito, 1986; Kiminami et al., 1993). This event has been correlated with eastward migration of the Kula-Pacific ridge as it was subducted underneath Eurasia. Coeval "in situ basalts" in the accretionary prism have a large compositional range from alkalic and high-alumina to tholeiitic (Osozawa and Yoshida, 1997). Sakaguchi (1996) noted high geothermal gradients in the accretionary prism coincide with this magmatic activity, which is consistent with the ridge subduction model. Osozawa (1993) also identified a series of normal faults associated with this event.

To the northwest of the Shimanto belt is the Ryoke belt, a long linear belt with a high-temperature/moderate-pressure metamorphic history. This metamorphic belt is historically significant as it was the type example of the high-temperature half of Miyashiro's (1959) paired metamorphic belts. For years, this belt was virtually synonymous with an ancient arc terrane. Work by Brown (1998b), however, questioned that association, and it now appears this classic "arc terrane" is actually the product of ridge-trench interactions. The metamorphic assemblages record clockwise pressure-temperature paths that parallel the temperature axis for both heating and cooling (also known as hairpin pressure-temperature paths; Brown, 1998b; Ikeda, 1998). The metamorphic isograds are independent of the location of voluminous granitoids. Based on K-Ar biotite ages of granitoids, Kinoshita and Ito (1988, 1990) and Kinoshita (1995, 1999) found a west to east decrease in cooling ages from 96 to 63 Ma. They postulated that these cooling ages were associated with subduction of the Farallon-Izanagi ridge. Geochronologic data on monazite show, however, that the cooling trend is not the same as the emplacement history (Suzuki and Adachi, 1998). Recently, Kinoshita (2002) reevaluated all geochronologic data on granitoids and adakitic intrusives in this area and determined that ridge subduction produced a V-shaped slab window oriented at an angle to the Japanese margin, based on the geometry of the trends in the geochronologic data both along and across the entire belt.

A problem in the ridge subduction interpretation of this belt is the occurrence of the adjacent high-pressure/low-temperature blueschist belt that was the basis of Miyashiro's (1959) original interpretation of the belt. Brown (1998b) suggested that strike-slip offset has juxtaposed these two coeval belts to form the present paired belts, a suggestion consistent with well-known evidence for strike-slip rejuvenation of the Median tectonic line, the major structure separating the Ryoke and Sambagawa belts. Alternatively, new thermal models suggest that it is possible to develop the low-pressure metamorphism simultaneously with blueschist facies metamorphism during ridge subduction (Iwamori, 2000).

In northeast Japan, the Abukuma belt is an andalusite-sillimanite terrane, which was originally correlated with the Ryoke belt (Miyashiro, 1959). Subsequent work has shown that this correlation is incorrect as there was an initial low-pressure metamorphism followed by high-pressure metamorphism and subsequent isothermal decompression; all of these events occurred 10 m.y. before peak metamorphism in the Ryoke belt (Hiroi et al., 1998). Brown (1998a) speculated that the initial low-pressure metamorphism may relate to the subduction of the Farallon-Izanagi ridge and that subsequent plate reorganization caused the high-pressure event. The terrain was then displaced northward along arc-parallel strike-slip faults.

Many of the features seen in the Japanese ridge subduction events are also recorded in the Cenozoic northern Cordillera. These include low-pressure/high-temperature metamorphism, normal faults, and age progression in forearc magmatism that includes both felsic and mafic magmas.

Mesozoic Example: North American Cordillera South of the 49th Parallel

Pavlis (1997) speculated that ridge subduction occurred during the Jurassic beneath the Klamath Mountains of northern

California. This region has voluminous granitoid plutons, calc-alkaline volcanic rocks, and low-pressure/high-temperature metamorphism superimposed on forearc assemblages. Most previous workers have interpreted these relationships in the context of a magmatic arc constructed on previously accreted forearc assemblages (e.g., Wright and Fahan, 1988; Saleeby et al., 1992), and associated ophiolites generally have been interpreted as backarc basin systems (e.g., Harper and Wright, 1984). Alternatively, ophiolites like the Josephine ophiolite (Harper and Wright, 1984) and Coast Ranges ophiolite (e.g., Saleeby et al., 1992) may represent forearc ophiolites accreted during ridge subduction. The calc-alkaline volcanic associations and oblique spreading axes documented for these ophiolites have direct analogs in the modern Chilean triple junction. Thus, these ophiolite complexes do not require a backarc origin, and a forearc origin is possible. The best candidate for an entirely forearc event, however, is the Middle Jurassic history of the Marble Mountain and Hayfork terranes. In this area, most of the pre-Jurassic rocks were accreted as forearc accretionary complexes that were subsequently invaded by Jurassic intrusive complexes (e.g., Saleeby et al., 1992). In the Marble Mountains terrane, for example, low-pressure amphibolite-facies metamorphism is superimposed on older mélange (Donato, 1989). Also, a complex history of high-T metamorphism, deformation, and intrusion affects the Hayfork terrane, much of which is composed of forearc accretionary complexes (Wright and Wyld, 1994). These now exist only as thin thrust sheets atop younger forearc accretionary assemblages, and in most cases the present exposures are too small to comprise an entire oceanic arc assemblage. Comparison between this belt and parts of the Paleogene forearc of southern Alaska are striking, and it is conceivable that much, or all, of the mid-Mesozoic high-T metamorphism in the Klamath province is the product of a ridge subduction event.

The Kula-Farallon spreading center may also have had effects on the southern Cordillera that predate the events described in the volume. From ca. 83 to 65 Ma, the Kula-Farallon spreading ridge is likely to have intersected the continent somewhere along the Mexico-Alaska coastline (Atwater, 1970). Engebretson et al. (1985) suggested that the triple junction in Late Cretaceous time was located alongside Mexico or California and traveled northward toward British Columbia by the Eocene. Thorkelson and Taylor (1989) and Thorkelson (1995) used this model to propose the existence of a long-lived northerly migrating slab window beneath the western United States, and suggested that the Laramide magmatic gap of Late Cretaceous through Paleocene age was generated by the absence of subducted slab beneath much of the western United States. Babcock et al. (1992), Breitsprecher et al. (2003), and Haeussler et al. (2003) suggested that a slab window was positioned alongside Oregon, Washington, and Vancouver Island in Paleocene to Eocene time. However, the location of the Kula–Farallon–North America spreading ridge remains controversial and strong arguments for a more northerly (Alaskan) triple junction, particularly in the early Cenozoic, are presented in some articles of this volume and are highlighted below.

Cenozoic Example: North American Cordillera North of the 49th Parallel

Overview

The first geologic mapping in the Gulf of Alaska investigated reports of copper mineralization in the McCarthy region near the Copper River (Alaskan Military Expedition, 1898; Scharder and Spencer, 1901; Fig. 3). The expedition geologists recognized an association of greenstone and low-grade metamorphic rocks even though they had a difficult time traversing the glaciated terrane on foot and by boat. Sixty years later, Brabb and Miller (1962) were the first geologists to traverse this glaciated region from the coast to the interior. Later, Hudson and Plafker (1982) systematically described the metamorphic rocks of the Chugach metamorphic complex and hypothesized several different scenarios to account for their presence in a paleoaccretionary margin. On a larger scale, Marshak and Karig (1977) recognized the occurrence of anomalous near-trench felsic plutons in the southern Alaskan complex, and proposed the then revolutionary idea that ridge subduction could be an appropriate explanation for the high heat flow and related magmatism. Inspired by the Marshak and Karig (1977) article and their own studies of anomalous processes in the Kodiak accretionary complex, faculty and students at University of California, Santa Cruz, began to interpret the Paleogene history of the southern accretionary complex in the context of a ridge subduction event (Byrne, 1979; Hill et al., 1981; Moore et al., 1983). These early works set the stage for recent geologic studies that investigate whether or not Cenozoic ridge subduction is a plausible tectonic scenario for southern Alaska. In the 1980s, the U.S. Geological Survey Trans-Alaska Crustal Transect program also began to investigate various portions of southern Alaska related to ridge subduction. For example, James et al. (1989) investigated thermal models related to subduction of young oceanic crust. Many other studies were published from these studies (see references in various chapters).

Highlights of This Volume

Almost all the chapters in this volume examine the latest Cretaceous and Cenozoic evolution of the margin of the northern Cordillera, focusing on regions where ridge subduction processes may have occurred between Kodiak and Vancouver Islands (Fig. 2). These chapters are organized approximately by location from west to east along the margin as well as by topic ranging from sedimentology, structural evolution, fluid history, metamorphism, and magmatism. A summary of some of the central themes of the papers follows.

Dwight Bradley, Tim Kusky, Peter Haeussler, Dave Rowley, Rich Goldfarb, and Steve Nelson present an overview paper with regional data as well as new data that pertain to the Paleogene events of the northern Cordillera with particular

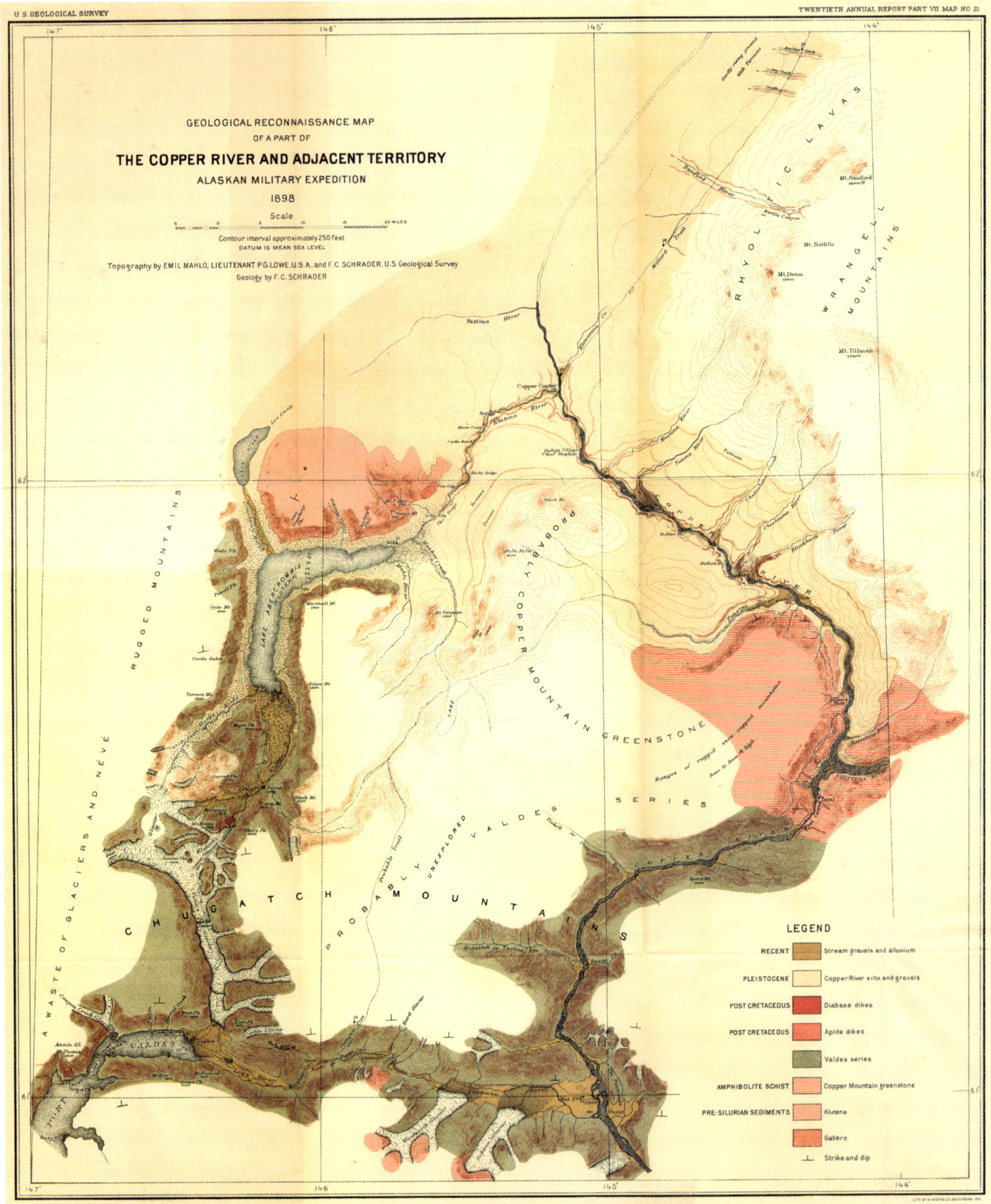

Figure 3. First geologic map of the Prince William Sound region, Alaska (Alaskan Military Expedition, 1898). Some of the geologic units are not correctly identified in terms of stratigraphic age, but all units are important components related to Eocene ridge subduction.

emphasis on Alaska (this volume, Chapter 1). For readers unfamiliar with this region, this paper is an ideal introduction to the general features associated with ridge-trench interactions. This is particularly important because the Eocene was a period of many complex interactions in the North American Cordillera, many of which are probably only indirectly related to a ridge subduction event—e.g., the close of the Laramide orogeny, the transition to Eocene extensional terranes of the southern Canadian Cordillera, exhumation in the Coast Mountains of western Canada and southeast Alaska, Eocene magmatism in the Coast Mountains synchronous with coast-parallel strike-slip faulting (Hollister and Andronicos, 1997; Stowell and McLelland, 2000), and the accretion of an oceanic plateau/seamount assemblage in what is now the Cascadia subduction zone. Thus, this overview paper sets the stage for what events are likely to be key processes associated with ridge subduction and triple junction interactions. The authors also evaluate alternative models and emphasize that the ridge subduction hypothesis is the only available hypothesis that satisfactorily explains all of the available data. They also address specific issues for the early Cenozoic northern Cordilleran region, including (1) implications for north Pacific plate reconstruction, including the provocative hypothesis of a fourth north Pacific plate (Resurrection plate); (2) uncertainties from poorly constrained reconstructions of the northern Cordilleran—problems of margin-parallel strike slip, early Cenozoic formation of the Alaskan orocline, and reference frames that moved in time; and (3) the broader effects of ridge subduction ranging from magmatism, deformation, and uplift within the forearc to possible far-field effects.

James Sample and Mary Reid reexamine the tectonic implications of the phenomenal thickness of the clastic member of the Chugach accretionary prism, known as the Kodiak Formation on Kodiak Island (this volume, Chapter 2). New faunal data and calculations of the size of the turbidite complex indicate it was deposited as an immense turbidite fan in a very short time period and thus records very large-scale erosional processes in an adjacent mountain belt. Nd isotope signatures and petrography of the sandstones lead them to conclude that the turbidites originated from the Coast Plutonic Complex of southeast Alaska and British Columbia and were deposited in a near-trench and within-trench environment, not far from the source as first postulated by Hollister (1979). The short time between deposition and accretion leads them to interpret the current location of the Kodiak Formation as due to margin-parallel tectonic transport of the accretionary complex.

Bill Clendenen, Don Fisher, and Tim Byrne discuss thermochronology of the Kodiak Islands (this volume, Chapter 3). Their data show the interplay between exhumation and changing thermal gradients associated with ridge subduction. Early Cretaceous low-temperature cooling ages from rocks on either side of the Border Ranges fault, the backstop to the accretionary prism, suggest that no significant slip occurred along this fault since that time and that there was no additional heating related to ridge subduction. The authors note a temporal correlation between subduction accretion with uplift in the adjacent Cook Inlet forearc (Trop et al., this volume, Chapter 4). Within the accretionary prism, zircon fission track cooling ages correlate with location of a regional antiform, whereas apatite data show a post-40 Ma cooling history. Thus, cooling from the ridge subduction event on the Kodiak Islands is regionally dispersed or did not produce significant thermal manifestations because the rocks had already been uplifted to relatively shallow levels. This is a similar to the present-day subduction of the Chile Ridge (Thomson et al., 2001).

Jeff Trop, Ken Ridgway, and Terry Spell report detailed sedimentological and stratigraphic studies of the early Cenozoic deposits exposed in the Matanuska Valley and southern Talkeetna Mountains (this volume, Chapter 4). Their study documents paleogeography accompanying early Cenozoic ridge-trench interactions in a large forearc basin complex. They identify sources from the arc to the north and the accretionary complex to the south within a Paleogene coarsening-upward megasequence. The authors give two alternative interpretations of this sequence that depend on the magnitude of strike-slip displacement on the Border Ranges fault. In a minimum slip scenario, the depositional sequence would record basinward progradation and syndepositional deformation and uplift, whereas in a maximum slip scenario there would be basin wide uplift forming a major unconformity. Interestingly, the authors note that either or both uplift events recorded in the basin (latest Cretaceous and middle Eocene) could be tied to regional effects of ridge subduction, and both uplift events have corresponding cooling histories recorded in the accretionary prism to the southwest (Clendenen et al., this volume, Chapter 3).

Peter Haeussler, Dwight Bradley, and Rich Goldfarb present details on the brittle deformation during and after the ridge subduction episode (this volume, Chapter 5). This deformation is commonly associated with hydrothermal activity that deposited gold-quartz veins and dikes. There is a diffuse network of faults that host small lode gold deposits in central southern Alaska, quite distinct from the significant dextral faults with large gold deposits in southeastern Alaska. The geometry of the faults is similar to brittle faults near the Chile triple junction, fault orientations determined from earthquake focal mechanisms associated with ridge subduction in the Woodlark basin, and extensional faults in the Shimanto accretionary prism. The authors speculate that both extensional and strike-slip deformation may be a common occurrence above slab windows.

Sarah Roeske, Larry Snee, and Terry Pavlis investigated the Border Ranges fault zone east of the Copper River in southern Alaska using a combination of detailed field mapping and structural analysis with Ar/Ar geochronology (this volume, Chapter 6). This fault system originated as a subduction zone megathrust between Wrangellia and the Chugach terrane. Subsequently, it was reactivated as a dextral fault zone with some oblique slip in the Late Cretaceous, prior to or synchronous with accretion of the coarse clastic unit of the Chugach accretionary complex, the Valdez Group, and continued through the Early Eocene. The long-lived (>20 Ma) history of oblique convergence occurred before and after the ridge subduction event and constrains plate geometries during this time. The zone of dextral slip may extend to Glacier Bay and Baranof Island in southeast Alaska, a distance of ~700 km. A 170 Ma diorite provides a possible piercing point for the Border Ranges fault system, indicating a minimum of 700 km of displacement, and possibly much more, since the Late Cretaceous.

In a pair of companion papers, Terry Pavlis, Kevin Marty, and Jinny Sisson (this volume, Chapter 7) and Terry Pavlis and Jinny Sisson (this volume, Chapter 8) describe the structural evolution in the vicinity of the western termination of the Chugach metamorphic complex. In these papers, the authors use field observations, detailed mapping, finite strain analyses, metamorphic petrology, and fluid inclusion data to develop a perspective on this down-plunge view of the western termination of the Chugach metamorphic complex. A transect to the west of the Chugach metamorphic complex crosses a significant dextral shear zone near Stuart Creek that developed during the ridge subduction event. Observed constrictional finite strains suggest that the bulk flow was constrictional—lengthening shear or lengthening-narrowing shear. The constrictional strain probably reflects vertical shortening associated with ridge subduction occurring at the same time as dextral shear produced subhorizontal shortening and orogen-parallel elongation.

In Chapter 8, this structural theme is carried into the high-grade core of the Chugach metamorphic complex, where there is a lithologically homogeneous section from greenschist facies phyllite to upper amphibolite facies gneiss. Within this sequence is a petrologic transition from schist to gneiss coincident with a regional, low-angle structural boundary that separates rocks displaying grossly different tectonic responses. Above this contact, the latest deformation is a system of dextral shear zones separated by relatively undeformed domains. As these shear zones are traced downward, however, they disappear into homogeneously deformed gneissic rocks below the schist-gneiss transition. Pavlis and Sisson propose that this boundary represents a decoupling horizon developed during dextral shearing that occurred at the same time (ca. 54 Ma) as the Kula-Farallon ridge was subducted along the margin. They hypothesize that the gneiss transition is an important rheological boundary between gneisses characterized by ~Newtonian viscous flow and overlying schists with a power-law, shear-thinning rheology, a conclusion that is also consistent with experimental and theoretical studies of rock deformation. If this rheological stratification is typical of schist-gneiss transitions, this conclusion has implications well beyond regions associated with ridge subduction.

Jill Weinberger and Jinny Sisson discuss higher-than-normal geothermal gradients resulting in greenschist facies metamorphism and a significant volume of quartz veins associated with Cenozoic subduction of the Kula-Farallon ridge (this volume, Chapter 9). Some of these veins record significant pressure drops of up to 140 MPa resulting from a change from lithostatic to hydrostatic conditions. In addition, one vein system was emplaced close to the brittle/ductile transition: Its appearance changes from ductile to brecciated. At some point after the ridge subduction event, these veins trapped late stage hydrocarbons that may indicate long-distance, post-metamorphic, fluid flow through this region.

John Bowman, Jinny Sisson, John Valley, and Terry Pavlis present an oxygen stable isotope study for a transect through the low-pressure/high-temperature Chugach metamorphic complex (this volume, Chapter 10). There is a systematic decrease in $\delta^{18}O$ values of quartz from both rock matrix and associated veins that is interpreted to result from a geometrically complex fluid flow system with at least two major periods of fluid infiltration. This infiltration occurred prior to the peak metamorphism with fluids that had exchanged with magmatic intrusions. Thus, magmatic heating associated with ridge subduction is most likely responsible for this episode of high-temperature/low-pressure Buchan metamorphism.

Cathy Zumsteg, Glenn Himmelberg, Sue Karl, and Peter Haeussler discuss low-pressure/high-temperature metamorphism on Baranof Island, the southern end of the Sanak-Baranof plutonic belt (this volume, Chapter 11). This region experienced four periods of metamorphism; only two, M1 and M4, are discussed in this paper. The first metamorphism, M1, was a Mesozoic moderate pressure amphibolite facies event. The final event, M4 at ca. 50 Ma, began at moderate pressure in sillimanite facies and then the region was exhumed as evidenced by the presence of andalusite. They suggest that the extensive (~25 km wide) biotite zone may be associated with a slab window formed during ridge subduction. They point out that the metamorphic history of this region is different from the Chugach metamorphic complex to the northwest. Instead, this region has a similar metamorphic history to the Hidaka belt of northwestern Japan, which records rapid changes in metamorphic pressure during a ridge subduction event (Hiroi et al., 1998).

Tim Kusky, Dwight Bradley, Thomas Donley, Dave Rowley, and Peter Haeussler used plutons of the Sanak-Baranof belt as time and strain markers during the ridge subduction event (this volume, Chapter 12). The early structures in the accretionary prism are typical of convergence followed by margin-parallel strike-slip faulting. An orthorhombic set of late faults may be the conduit for magma emplacement and may have also facilitated exhumation. A younger plutonic suite (ca. 35 Ma) that intruded after the ridge subduction event records a significant component of dextral transpression.

Jinny Sisson, Anne Poole, Holly Cooper Burner, Nancy Harris, Terry Pavlis, Peter Copeland, Ray Donelick, and Bill McClelland investigated geochemistry and geochronology of felsic plutons and associated rocks between the Copper River and Johns Hopkins Inlet in Glacier Bay National Park and Preserve (this volume, Chapter 13). The data indicate that this predominantly tonalite-trondhjemite-granitoid suite with minor mafic intrusives was derived from melted accretionary wedge and underplated mafic material during subduction of the Kula-

Farallon ridge. The amount of mafic material incorporated into these felsic magmas appears to increase toward the east. Previous geochronologic studies imply a linear age progression younging from west to east. In detail, this pattern is not linear as there are identical ages in the western Chugach metamorphic complex and Nunatak Fjord, 130 km to the east. The authors propose that this apparent age jump is related to either subduction of a transform fault and formation of a microplate sliver or a period of northward migration of the ridge.

Wes Groome, Derek Thorkelson, Richard Friedman, Jim Mortensen, Nick Massey, Daniel Marshal, and Paul Layer studied the Leech River complex, a fault bounded sliver on southern Vancouver Island (this volume, Chapter 14). New geochronologic data identify the age of deposition (103–88 Ma) of the metasediments as well as two felsic intrusive events at 88 Ma and 51 Ma. Both intrusive events occurred at depths of approximately 10 km. There are also undated metagabbros and greenschist facies volcanics, both with MORB geochemical signatures. Rapid cooling and exhumation in the Eocene was followed by deposition of Oligocene sediments. This assemblage is more than 1000 km to the south of coeval metamorphic and plutonic suites of the Sanak-Baranof belt and supports the concept of large-magnitude strike-slip displacement of the Alaskan segment northward (Sample and Reid, Chapter 2; Roeske et al., Chapter 6, this volume); however, it may also indicate rapid margin-parallel ridge migration (Pavlis and Sisson, Chapter 8, this volume) or the presence of a missing microplate (Bradley et al., Chapter 1, this volume).

SUMMARY

Figure 4 is a schematic diagram of our vision of the ridge subduction processes that affected the Paleogene northern Cordillera margin. The diagram shows two oceanic plates that are growing along a mid-ocean spreading ridge and are concurrently subducting beneath an overriding plate (shown partially transparent and cut away for clarity). As the spreading ridge enters the trench, it "unzips" and widens into a slab window (geometric principles and assumptions provided in Thorkelson, 1996). The ridge-trench-trench triple junction is migrating to the left, but jumps either left or right as transform faults are subducted. Mantle that was previously beneath the slab upwells through the slab window toward the base of the overriding plate. Mantle metasomatism from the downgoing slabs is absent in the slab window region, and arc volcanism is interrupted and replaced by magmatism of differing character, extending from forearc to backarc. Forearc magmatism occurs where the subducting ridge has opened into a narrow window, and MORB magmas rise into the forearc, triggering crustal anatexis, a process aptly named the "blowtorch effect" by DeLong et al. (1979). This effect also

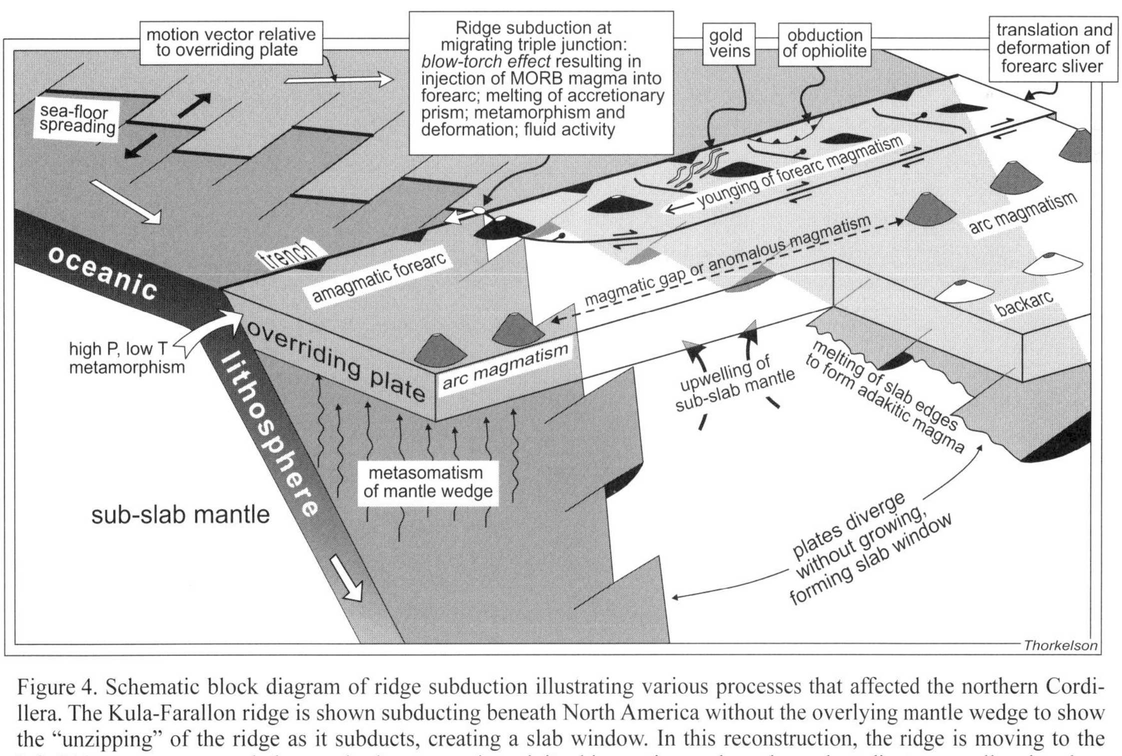

Figure 4. Schematic block diagram of ridge subduction illustrating various processes that affected the northern Cordillera. The Kula-Farallon ridge is shown subducting beneath North America without the overlying mantle wedge to show the "unzipping" of the ridge as it subducts, creating a slab window. In this reconstruction, the ridge is moving to the left. This creates temporal changes in the magmatic activity, kinematics, and geothermal gradients, as well as involves creation of a microplate or forearc sliver and obduction of an ophiolite. These processes are described more fully in the text. Modified from Thorkelson (1996).

generates thermal metamorphism and hydrothermal activity, supplanting the high-*P*/low-*T* metamorphism typical of most forearcs. If this process is localized for extended times due to slow triple junction migration velocity or repeated interactions, the entire forearc could be rapidly heated to high temperature to form high-*T*, low-*P* assemblages like those observed in the Chugach metamorphic complex (Chapter 10, this volume) or in the Cretaceous Ryoke belt of Japan (see above). Inboard, partial fusion of the slab edges results in adakitic magmas, and low-percentage melts of the upwelling sub-slab asthenosphere generate alkalic magmas. Deformation and exhumation in the forearc are variably affected by subduction of ridge segments and transform faults. All of these effects are time-transgressive and controlled by the migration of the triple junction, resulting in a general younging of triple junction–slab window features to the left (or south, in present-day coordinates). Basal drag from lateral movement of the triple junction is imparted to the overriding plate, and a trench-parallel strike-slip fault forms. The resultant forearc sliver is partially coupled to one of the descending plates and moves dextrally relative to the rest of the overriding plate.

We hope that this volume will stimulate researchers to re-examine the established tectonic models and consider the potential role that triple junction processes have played in various convergent margins. If the classic "arc terrane" of Japan can be reinterpreted as a forearc metamorphic belt generated by ridge subduction (Brown, 1998a, 1998b), then many other "arc terranes" may need re-evaluation. The papers here, and citations within, should serve as a template for evaluating other systems.

ACKNOWLEDGMENTS

This work would not have started without the pioneering efforts of George Plafker. His gracious invitation for three of us (VBS, TLP, SMR) to join the U.S. Geological Survey Trans-Alaska Crustal Transect field-mapping program in 1984 and 1985 was the beginning of our studies in this region. J. Casey Moore introduced a large cadre of students to the geology of Kodiak Island. We thank both George and Casey and many others for assistance in our fieldwork. Lincoln Hollister, Gary Winkler, Travis Hudson, and Warren Nokleberg provided thought-provoking ideas that helped our interpretations. Lincoln Hollister, Sean Gulick, and Will Maze provided helpful reviews that improved this preface. We thank Janice Fong for a superb job of scanning the antique map for Figure 3. Danny Rosencrans of the Wrangell-St. Elias National Park and Preserve aided with our logistics, as did Ken Bunch of Gulkana Air Service and Walt Woodrow and Jan Jorgensen of ERA Helicopters, Valdez, Alaska.

We are indebted to the following reviewers for their thorough reviews of the manuscripts; all of their efforts have greatly improved the content of the volume: Hans G. Avé Lallemant, Rice University; James Beard, Virginia Museum of Natural History; Steven C. Bergman, University of Texas at Dallas; Mark T. Brandon, Yale University; Ronald B. Cole, Allegheny College; Maria Luisa Crawford, Bryn Mawr College; J. Steven Davis, ExxonMobil Exploration Company; Susan DeBari, Western Washington University; Steve Dunn, Mount Holyoke College; George Gehrels, University of Arizona; Sean P.S. Gulick, University of Texas Institute for Geophysics; James P. Hibbard, North Carolina State University; Mary S. Hubbard, Kansas State University; Lincoln S. Hollister, Princeton University; Raymond Ingersoll, University of California, Los Angeles; Jonathon Lewis, University of California, Davis (now at University of Massachusetts); Kathleen Marsaglia, California State University, Northridge; William B. Maze, ExxonMobil Exploration Company; William C. McClelland, University of Idaho; Peter I. Nabelek, University of Missouri, Columbia; Daniel L. Orange, Monterey Bay Aquarium Research Institute; Terry L. Pavlis, University of New Orleans; James Ryan, Geological Survey of Canada, Vancouver; Sarah M. Roeske, University of California, Davis; Margaret E. Rusmore, Occidental College; Virginia B. Sisson, Rice University; John A. Stamatakos, Southwest Research Institute; Harold H. Stowell, University of Alabama; Christian Teyssier, University of Minnesota; Michael J. Underwood, University of Missouri, Columbia; Cees T. van Staal, Geological Survey of Canada, Ottawa; Aaron S. Yoshinobu, Texas Tech University.

REFERENCES CITED

Abbott, D., and Menke, W., 1990, Length of the global plate boundaries at 2.4 Ga: Geology, v. 18, p. 58–61.

Alaskan Military Expedition, 1898, The Copper River and adjacent territory, 20th Annual Report Part VII, Map no. 21. Topography by Emil Mahlo, Lt. P.G. Lowe, and F.C. Schrader. Geology by F.C. Schrader. 1 Plate, scale 1:380,160.

Armstrong, R.L., Muller, J.E., Harakal, J.E., and Muehlenbachs, K., 1985, The Neogene Alert Bay volcanic belt of northern Vancouver Island, Canada: Descending-plate-edge volcanism in the arc-trench gap: Journal of Volcanology and Geothermal Research, v. 26, p. 75–97.

Atwater, T., 1970, Implications of plate tectonics for the Cenozoic tectonic evolution of western North America: Geological Society of America Bulletin, v. 81, p. 3515–3536.

Atwater, T., and Stock, J., 1998, Pacific North America plate tectonics of the Neogene southwestern United States: An update, *in* Ernst, W.G., and Nelson, C.A., eds., Integrated Earth and environmental evolution of the southwestern United States: The Clarence A. Hall, Jr., volume: Columbia, Maryland, Bellwether Publishing, p. 393–420.

Babcock, S., Burmester, R.F., Engebretson, D.C., Warnock, A., and Clark, K.P., 1992, A rifted margin origin for the Crescent basalts and related rocks in the northern Coast Range volcanic province, Washington and British Columbia: Journal of Geophysical Research, v. 97, p. 6799–6821.

Bandy, W., and Pardo, M., 1994, Statistical examination of the existence and relative motion of the Jalisco and southern Mexico blocks: Tectonics, v. 13, p. 755–768.

Barker, P.F., Barber, P.L., and King, E.C., 1984, An Early Miocene ridge crest trench collision on the South Scotia Ridge near 36-degrees-W: Tectonophysics, v. 102, p. 315–332.

Bart, P., 1993, Glacial and tectonic development of the Antarctic Peninsula [M.A. thesis]: Houston, Texas, Rice University, 119 p.

Behrmann, J.H., and Kopf, A., 2001, Balance of tectonically accreted and subducted sediment at the Chile triple junction: International Journal of Earth Sciences, v. 90, p. 753–768.

Behrmann, J.H., Lewis, S.D., Cande, S.C., Musgrave, R., Bangs, N., Boden, P., Brown, K., Collombat, H., Didenko, A.N., Didyk, B.M., Froelich, P.N., Golovchenko, X., Forsythe, R., Kurnosov, V., Lindsley-Griffin, N., Marsaglia, K., Osozawa, S., Prior, D., Sawyer, D., Scholl, D., Spiegler, D., Strand, K., Takahashi, K., Torres, M., Vegafaundez, M., Vergara, H., and Wasada, A., 1994, Tectonics and geology of spreading ridge subduction at the Chile triple junction: A synthesis of results from Leg 141 of the Ocean Drilling Program: Geologische Rundschau, v. 83, p. 832–852.

Bevier, M.L., Armstrong, R.L., and Souther, J.G., 1979, Miocene peralkaline volcanism in west-central British Columbia: Its temporal and plate-tectonics setting: Geology, v. 7, p. 389–392.

Birkenmajer, K., 1994, Evolution of the pacific margin of the northern Antarctic Peninsula: An overview: Geologische Rundschau, v. 83, p. 309–321.

Bohannon, R.G., and Parsons, T., 1995, Tectonic implications of post-30 Ma Pacific and North American relative plate motions: Geological Society of America Bulletin, v. 107, p. 937–959.

Bourgois, J., and Michaud, F., 2002, Comparison between the Chile and Mexico triple junction areas substantiates slab window development beneath northwestern Mexico during the past 12–10 Myr: Earth and Planetary Science Letters, v. 201, p. 35–44.

Bourgois, J., Guivel, C., Lagabrielle, Y., Calmus, T., Boulegue, J., and Daux, V., 2000, Glacial-interglacial trench supply variation, spreading-ridge subduction, and feedback controls on the Andean margin development at the Chile triple junction area (45–48 degrees S): Journal of Geophysical Research, v. 105, p. 8355–8386.

Brabb, E.E., and Miller, D.J., 1962, Reconnaissance traverse across the eastern Chugach Mountains, Alaska: U.S. Geological Survey Miscellaneous Geologic Investigations Map I-341, scale 1:96,000, 1 sheet.

Breitsprecher, K., Thorkelson, D.J., Groome, W.G., and Dostal, J., 2003, Geochemical confirmation of the Kula-Farallon slab window beneath the Pacific Northwest in Eocene time: Geology, v. 31, p. 351–354.

Brown, M., 1998a, Ridge-trench interactions and high-*T*–low-*P* metamorphism with particular reference to the Cretaceous evolution of the Japanese Islands, *in* Treloar, P.J., and O'Brien, P.J., eds., What drives metamorphism and metamorphic reactions?: Geological Society [London] Special Publication No. 138, p. 137–169.

Brown, M., 1998b, Unpairing metamorphic belts: *P-T* paths and a tectonic model for the Ryoke belt, southwest Japan: Journal of Metamorphic Geology, v. 16, p. 3–22.

Burger, R.L., Fulthrope, C.S., Austin, J.A., Jr., and Gulick, S.P.S., 2002, Lower Pleistocene to present structural deformation and sequence stratigraphy of the continental shelf, offshore Eel River basin, northern California: Marine Geology, v. 185, p. 249–281.

Byrne, T., 1979, Late Paleocene demise of the Kula-Pacific spreading center: Geology, v. 7, p. 341–344.

Cande, S., and Leslie, R., 1986, Late Cenozoic tectonics of the southern Chile Trench, Journal of Geophysical Research, v. 91, p. 471–496.

Castro, A., Fernandez, C., de la Rosa, J.D., Moreno-Ventas, I., El-Hmidi, H., El-Biad, M., Bergamin, J.F., and Sanchez, N., 1996, Triple-junction migration during Paleozoic plate convergence: The Aracena metamorphic belt, Hercynian Massif, Spain: Geologische Rundschau, v. 85, p. 180–185.

Castro, A., Fernandez, C., El-Hmidi, H., El-Biad, M., Diaz, M., de la Rosa, J., and Stuart, F., 1999, Age constraints to the relationships between magmatism, metamorphism and tectonism in the Aracena metamorphic belt, southern Spain: International Journal of Earth Sciences, v. 88, p. 26–37.

Collins, W.J., and Williams, I.S., 1995, SHRIMP ion probe dating of short lived Proterozoic tectonic cycles in the Northern Arunta Inlier, central Australia: Precambrian Research, v. 71, p. 69–89.

Cronin, V.S., 1992, Types and kinematic stability of triple junctions: Tectonophysics, v. 207, p. 287–301.

Crook, K.A.W., and Taylor, B., 1994, Structure and Quaternary tectonic history of the Woodlark triple junction region, Solomon Islands: Marine Geophysical Researches, v. 16, p. 65–89.

Crowe, D.E., Nelson, S.W., Brown, P.E., Shanks, W.C., and Valley, J.W., 1992, Geology and Geochemistry of volcanogenic massive sulfide deposits and related igneous rocks, Prince-William Sound, south-central Alaska: Economic Geology, v. 87, p. 1722–1746.

Daniel, A.J., Kusznir, N.J., and Styles, P., 2001, Thermal and dynamic modeling of deep subduction of a spreading center: Implications for the fate of the subducted Chile Rise, southern Chile: Journal of Geophysical Research, v. 106, p. 4293–4304.

DeLong, S.E., and Fox, P., 1977, Geological consequences of ridge subduction, *in* Talwani, M., and Pittman, W.C., III, eds., Island arcs, deep sea trenches and back-arc basins: American Geophysical Union Maurice Ewing Series, no. 1, p. 221–228.

DeLong, S.E., Schwarz, W.M., and Anderson, R.N., 1979, Thermal effects of ridge subduction: Earth and Planetary Science Letters, v. 44, p. 239–246.

Dickinson, W.R., 1997, Tectonic implications of Cenozoic volcanism in coastal California: Geological Society of America Bulletin, v. 109, p. 936–954.

Dickinson, W.R., and Snyder, W.S., 1979, Geometry of triple junctions related to San Andreas transform: Journal of Geophysical Research, v. 84, p. 561–572.

DiTullio, L., Laughland, M.M., and Byrne, T., 1993, Thermal maturity and constraints on deformation from illite cyrstallinity and vitrinite reflectance in the shallow levels of an accretionary prism, Eocene-Oligocene Shimanto belt, southwest Japan, *in* Underwood, M.B., ed., Thermal evolution of the Tertiary Shimanto belt, southwest Japan: An example of ridge-trench interaction: Boulder, Colorado, Geological Society of America Special Paper 273, p. 63–82.

Donato, M.M., 1989, Metamorphism of an ophiolitic tectonic melange, Northern California Klamath Mountains, USA: Journal of Metamorphic Geology, v. 7, p. 515–528.

D'Orazio, M., Agostini, S., Mazzarini, F., Innocenti, F., Manetti, P., Haller, M.J., and Lahsen, A., 2000, The Pali Aike Volcanic Field, Patagonia: Slab-window magmatism near the tip of South America: Tectonophysics, v. 321, p. 407–427.

D'Orazio, M., Agostini, S., Innocenti, F., Haller, M.J., Manetti, P., and Mazzarini, F., 2001, Slab window-related magmatism from southernmost South America: The late Miocene mafic volcanics from the Estancia Glencross area (~52° S, Argentina-Chile): Lithos, v. 57, p. 67–89.

Drummond, M.S., Bordelon, M., Deboer, J.Z., Defant, M.J., Bellon, H., and Feigenson, M.D., 1995, Igneous petrogenesis and tectonic setting of plutonic and volcanic rocks of the Cordillera de Talamanca, Costa Rica Panama, Central American arc: American Journal of Science, v. 295, p. 875–919.

Edwards, B.R., and Russell, J.K., 2000, Distribution, nature, and origin of Neogene-Quaternary magmatism in the northern Cordilleran volcanic province, Canada: Geological Society of America Bulletin, v. 112, p. 1280–1295.

Engebretson, D.C., Cox, A., and Gordon, R.G., 1985, Relative motions between oceanic and continental plates in the Pacific Basin: Boulder, Colorado, Geological Society of America Special Paper 206, 59 p.

Farrar, E., and Dixon, J.M., 1993, Ridge subduction: Kinematics and implications for the nature of mantle upwelling: Canadian Journal of Earth Sciences, v. 30, p. 893–907.

Fletcher, J.M., Kohn, B.P., Foster, D.A., and Gleadow, A.J.W., 2000, Heterogeneous Neogene cooling and exhumation of the Los Cabos Block, southern Baja California: Evidence from fission-track thermochronology: Geology, v. 28, p. 107–110.

Flint, S.S., Prior, D.J., Agar, S.M., and Turner, P., 1994, Stratigraphic and structural evolution of the Tertiary Cosmelli Basin and its relationship to the Chile triple junction: Geological Society [London] Journal, v. 151, p. 251–268.

Forsythe, R.E., and Nelson, E., 1985, Geological manifestations of ridge collision: Evidence from the Golfo de Penas–Taitao Basin, southern Chile: Tectonics, v. 4, p. 477–495.

Forsythe, R.E., Nelson, E., Carr, M., Kaeding, M., Herve, M., Modozis, C., Soffia, J., and Harambour, S., 1986, Pliocene near-trench magmatism in southern Chile: Possible manifestation of ridge collision: Geology, v. 14, p. 23–27.

Fox, K.F., Jr., Fleck, R.J., Curtis, G.H., and Meyer, C.E., 1985, Implications of the northwestwardly younger age of the volcanic rocks of west-central California: Geological Society of America Bulletin, v. 96, p. 647–654.

Frederiksen, A.W., Bostock, M.G., VanDecar, J.C., and Cassidy, J.F., 1998, Seismic structure of the upper mantle beneath the northern Canadian Cordillera from teleseismic travel-time inversion: Tectonophysics, v. 294, p. 43–55.

Furlong, K.P., and Govers, R., 1999, Ephemeral crustal thickening at a triple junction: The Mendocino crustal conveyor: Geology, v. 27, p. 127–130.

Gorring, M.L., and Kay, S.M., 2001, Mantle processes and sources of Neogene slab window magmas from southern Patagonia, Argentina: Journal of Petrology, v. 42, p. 1067–1094.

Gorring, M.L., Kay, S.M., Zeitler, P.K., Ramos, V.A., Rubiolo, D., Fernandez, M.I., and Panza, J.L., 1997, Neogene Patagonian plateau lavas: Continental magmas associated with ridge collision at the Chile triple junction: Tectonics, v. 16, p. 1–17.

Guivel, C., Lagabrielle, Y., Bourgois, J., Maury, R.C., Fourcade, S., Martin, G., and Arnoud, N., 1999, New geochemical constraints for the origin of ridge subduction-related plutonic and volcanic suites from the Chile triple junction (Taitao Peninsula and Site 862, LEG ODP141 on the Taitao Ridge): Tectonophysics, v. 311, p. 83–111.

Gulick, S.P.S., and Meltzer, A.S., 2002, Effect of the northward migrating Mendocino triple junction on the Eel River forearc basin, California:

Structural evolution: Geological Society of America Bulletin, v. 114, p. 1505–1519.

Gulick, S.P.S., Meltzer, A.S., and Clarke, S.H., Jr., 2002, Effect of the northward-migrating Mendocino triple junction on the Eel River forearc basin, California: Stratigraphic development: Geological Society of America Bulletin, v. 114, p. 178–191.

Haeussler, P.J., Bradley, D.C., Wells, R.E., and Miller, M.L., 2003, Life and death of the Resurrection plate: Evidence for its existence and subduction in the northeastern Pacific in Paleocene-Eocene time: Geological Society of America Bulletin, v. 115, no. 7, p. 867–880.

Hamilton, T.S., and Dostal, J., 2001, Melting of heterogeneous mantle in a slab window environment: Examples from the middle Tertiary Masset basalts, Queen Charlotte Islands, British Columbia: Canadian Journal of Earth Sciences, v. 38, p. 825–838.

Harper, G.D., and Wright, J.E., 1984, Middle to Late Jurassic tectonic evolution of the Klamath Mountains, California-Oregon: Tectonics, v. 3, p. 759–772.

Hart, C.J.R., and Villeneuve, M., 1999, Geochronology of Neogene alkaline volcanic rocks (Miles Canyon Basalt), southern Yukon Territory, Canada: The relative effectiveness of laser $^{40}Ar/^{39}Ar$ and K-Ar geochronology: Canadian Journal of Earth Sciences, v. 36, p. 1495–1507.

Henriet, J.P., Meissner, R., Miller, H., Bialas, J., Bittner, R., DeBatist, M., Fluh, E., Hedrich, K., Jokat, W., Kaul, N., Moons, A., Sorensen, G., Wever, T., and Versteeg, W., 1992, Active margin processes along the Antarctic Peninsula: Tectonophysics, v. 201, p. 229–253.

Hibbard, J.P., and Karig, D.E., 1990, Structural and magmatic responses to spreading ridge subduction: An example from southwest Japan: Tectonics, v. 9, p. 207–230.

Hibbard, J.P., Laughland, M.M., Kang, S.M., and Karig, D., 1993, The thermal imprint of spreading ridge subduction on the upper structural levels of an accretionary prism, southwest Japan, *in* Underwood, M.D., ed., Thermal Evolution of the Tertiary Shimanto Belt, Southwest Japan: An example of ridge-trench interaction: Geological Society of America Special Paper 273, p. 83–101.

Hill, M.D., Morris, J., and Whelan, J., 1981, Hybrid granodiorites intruding the accretionary prism, Kodiak, Shumagin, and Sanak Islands, southwest Alaska: Journal of Geophysical Research, v. 86, p. 10,569–10,590.

Hiroi, Y., Kishi, S., Nohara, T., Sato, K., and Goto, J., 1998, Cretaceous high-temperature rapid loading and unloading in the Abukuma metamorphic terrane, Japan: Journal of Metamorphic Geology, v. 16, p. 67–81.

Hole, M.J., and Larter, R.D., 1993, Trench-proximal volcanism following ridge crest-trench collision along the Antarctic Peninsula: Tectonics, v. 12, p. 897–910.

Hole, M.J., and Saunders, A.D., 1996, The generation of small melt-fractions in truncated melt columns: Constraints from magmas erupted above slab windows and implications for MORB genesis: Mineralogical Magazine, v. 60, p. 173–189.

Hole, M.J., Saunders, A.D., Rogers, G., and Sykes, M.A., 1994, The relationship between alkaline magmatism, lithospheric extension and slab window formation along continental destructive plate margins, *in* Smellie, J.L., ed., Volcanism associated with extension at consuming plate margins: Geological Society [London] Special Publication No. 81, p. 265–285.

Hollister, L.S., 1979, Metamorphism and crustal displacements: New insights: Episodes, v. 3, p. 3–8.

Hollister, L.S., and Andronicos, C.L., 1997, A candidate for the Baja British Columbia fault system in the Coast Plutonic Complex: GSA Today, v. 7, p. 1–7.

Hudson, T., and Plafker, G., 1982, Paleogene metamorphism of an accretionary flysch terrane, eastern Gulf of Alaska: Geological Society of America Bulletin, v. 93, p. 1280–1290.

Hyndman, R.D., and Lewis, T.J., 1999, Geophysical consequences of the Cordillera-Craton thermal transition in southwestern Canada, *in* Christoph, C., et al., eds., Thermal regimes in the continental and oceanic lithosphere: Tectonophysics, v. 306, p. 397–422.

Ikeda, T., 1998, Phase equilibria and the pressure-temperature path of the highest-grade Ryoke metamorphic rocks in the Yanai district, SW Japan: Contributions to Mineralogy and Petrology, v. 132, p. 321–335.

Iwamori, H., 2000, Thermal effects of ridge subduction and its implications for the origin of granitic batholith and paired metamorphic belts: Earth and Planetary Science Letters, v. 181, p. 131–144.

James, T.S., Hollister, L.S., and Morgan, W.J., 1989, Thermal modeling of the Chugach metamorphic complex: Journal of Geophysical Research, v. 94, p. 4392–4410.

Johnson, A.C., 1997, Cenozoic tectonic evolution of the Marguerite Bay area, Antarctic Peninsula, interpreted from geophysical data: Antarctic Science, v. 9, p. 268–280.

Johnson, A.C., and Swain, C.J., 1995, Further evidence of fracture-zone induced tectonic segmentation of the Antarctic Peninsula from detailed aeromagnetic anomalies: Geophysical Research Letters, v. 22, p. 1917–1920.

Johnson, R.W., Jaques, A.L., Langmuir, C.H., Perfit, M.R., Staudigel, H., Dunkley, P.N., Chappell, B.W., Taylor, S.R., and Baekisapa, M., 1987, Ridge subduction and forearc volcanism: Petrology and geochemistry of rocks dredged from the western Solomon Arc and Woodlark Basin, *in* Taylor, B., and Exon, N.F., eds., Marine geology, geophysics, and geochemistry of the Woodlark Basin–Solomon Islands: Houston, Texas, Circum-Pacific Council for Energy and Mineral Resources, Earth Science Series, v. 7, p. 155–226.

Johnston, S.T., and Thorkelson, D.J., 1997, Cocos-Nazca slab window beneath Central America: Earth and Planetary Science Letters, v. 146, p. 465–474.

Kaeding, M., Forsythe, R.D., and Nelson, E.P., 1990, Geochemistry of the Taito ophiolite and near-trench intrusions from the Chile margin triple junction: Journal of South American Earth Science, v. 4., p. 161–177.

Karsten, J.L., Klein, E.M., and Sherman, S.B., 1996, Subduction zone geochemical characteristics in ocean ridge basalts from the southern Chile Ridge: Implications of modern ridge subduction systems for the Archean: Lithos, v. 37, p. 143–161.

Kay, S.M., Ramos, V.A., and Marquez, M., 1993, Evidence in Cerro Pampa volcanic rocks for slab-melting prior to ridge-trench collision in southern South America: Journal of Geology, v. 101, p. 703–714.

Kiminami, K., Miyashita, S., and Kawabata, K., 1993, Active ridge-forearc collision and its geological consequences: An example from Late Cretaceous southwest Japan, *in* Komatsu, M., ed., Evolution of the arc crust in southwest Japan: Geological Society of Japan Memoir 42, p. 167–182 (in Japanese with an English abstract).

Kinoshita, O., 1995, Migration of igneous activities related to ridge subduction in southwest Japan and the East Asian continental margin from the Mesozoic to the Paleogene: Tectonophysics, v. 245, p. 25–35.

Kinoshita, O., 1999, A migration model of magmatism explaining ridge subduction and its details on a statistical analysis of the granite ages in Cretaceous southwest Japan: The Island Arc, v. 8, p. 181–189.

Kinoshita, O., 2002, Possible manifestations of slab window magmatisms in Cretaceous southwest Japan: Tectonophysics, v. 344, p. 1–13.

Kinoshita, O., and Ito, H., 1986, Migration of Cretaceous igneous activity in southwest Japan: Journal of the Geological Society of Japan, v. 92, p. 732–736 (in Japanese with an English abstract).

Kinoshita, O., and Ito, H., 1988, Cretaceous magmatism in southwest and northeast Japan related to two ridge subduction and Mesozoic magmatism along the East Asia continental margin: Journal of the Geological Society of Japan, v. 94, p. 925–944 (in Japanese with an English abstract).

Kinoshita, O., and Ito, H., 1990, Reconstruction of southwest and northeast Japan based on trends of Mesozoic igneous rock ages: Journal of the Geological Society of Japan, v. 96, p. 821–838 (in Japanese with an English abstract).

Klein, E.M., and Karsten, J.L., 1995, Ocean-ridge basalts with convergent margin geochemical affinities from the Chile Ridge: Nature, v. 374, p. 52–57.

Kusky, T.M., 1990, Evidence for Archean ocean opening and closing in the southern Slave province: Tectonics, v. 9., p. 1533–1563.

Kusky, T.M., and Polat, A., 1999, Growth of granite-greenstone terranes at convergent margins, and stabilization of Archean cratons: Tectonophysics, v. 305, p. 43–73.

Lachenbruch, A.H., and Sass, J.H., 1980, Heat flow and energetics of the San Andreas fault zone: Journal of Geophysical Research, v. 85, p. 6185–6222.

Lagabrielle, Y., Guivel, Y., Maury, R.C., Bourgois, J., Fourcade, S., and Martin, H., 2000, Magmatic-tectonic effects of high thermal regime at the site of active ridge subduction: The Chile triple junction model: Tectonophysics, v. 326, p. 255–268.

Larter, R.D., and Barker, P.F., 1991, Effects of ridge crest trench interaction on Antarctic Phoenix spreading: Forces on a young subducting plate: Journal of Geophysical Research, v. 96, p. 19,583–19,607.

Laughland, M.M., and Underwood, M.B., 1993, Vitrinite reflectance and estimates of paleotemperature within the Upper Shimanto Group, Muroto Peninsula, Shikoku, Japan, *in* Underwood, M.B., ed., Thermal evolution of the Tertiary Shimanto belt, southwest Japan: An example of ridge-trench interaction: Boulder, Colorado, Geological Society of America Special Paper 273, p. 25–44.

Lemoigne, J., Lagabrielle, Y., Whitechurch, H., Girardeau, J., Bourgois, J., and Maury, R., 1996, Petrology and geochemistry of the ophiolitic and volcanic suites of the Taitao peninsula Chile triple junction area: Journal of South American Earth Sciences, v. 9., p. 43–58.

Lewis, T.J., and Hyndman, R.D., 1998, Heat flow transitions along the SNORCLE transect: Asthenospheric flow, *in* Cook, F., and Erdmer, P., compilers, 1998, Slave-Northern Cordillera Lithospheric Evolution (SNORCLE) Transect and Cordilleran Tectonics Workshop Meeting: Lithoprobe Report No. 64, p. 110–119.

Lonsdale, P., 1988, Paleogene history of the Kula plate: Offshore evidence and onshore implications: Geological Society of America Bulletin, v. 100, p. 733–754.

Maldonado, A., Larter, R.D., and Aldaya, F., 1994, Forearc tectonic evolution of the South Shetland margin, Antarctic Peninsula: Tectonics, v. 13, p. 1345–1370.

Marshak, R.S., and Karig, D.E., 1977, Triple junctions as a cause for anomalous near-trench activity between the trench and volcanic arc: Geology, v. 5, p. 233–236.

McCarron, J.J., and Larter, R.D., 1998, Late Cretaceous to early Tertiary subduction history of the Antarctic Peninsula: Journal of the Geological Society [London], v. 155, p. 255–268.

McCarron, J.J., and Millar, I.L., 1997, The age and stratigraphy of forearc magmatism on Alexander Island, Antarctica: Geological Magazine, v. 134, p. 507–522.

McCrory, P.A., 2000, Upper plate contraction north of the migrating Mendocino triple junction, Northern California: Implications for partitioning of strain: Tectonics, v. 19, p. 1144–1160.

McKenzie, D.P., and Morgan, W.J., 1969, Evolution of triple junctions: Nature, v. 224, p. 125–133.

Miller, K.C., 2002, Geophysical evidence for Miocene extension and mafic magmatic addition in the California Continental Borderland: Geological Society of America Bulletin, v. 114, p. 497–512.

Miyashiro, A., 1959, Abukuma, Ryoke, and Sambagawa metamorphic belts: Journal of the Geological Society of Japan, v. 65, p. 624–637.

Moore, J.C., Byrne, T., Plumley, P.W., Reid, M., Gibbons, H., and Coe, R.S., 1983, Paleogene evolution of the Kodiak Islands, Alaska: Consequences of ridge-trench interaction in a more southerly latitude: Tectonics, v. 2, p. 265–293.

Moores, E.M., 1998, Ophiolites, the Sierra Nevada, "Cordilleria," and orogeny along the Pacific and Caribbean margins of North and South America: International Review of Geology, v. 40, p. 40–54.

Murdie, R.E., and Russo, R.M., 1999, Seismic anisotropy in the region of the Chile margin triple junction: Journal of South American Earth Sciences, v. 12, p. 261–270.

Myers, J.S., Shaw, R.D., and Tyler, I.M., 1996, Tectonic evolution of Proterozoic Australia: Tectonics, v. 15, p. 1431–1446.

Nelson, E., Forsythe, R., Diemer, J., Allen, M., and Urbina, O., 1993, Taitao ophiolite: A ridge collision ophiolite in the forearc of southern Chile (46 degrees S): Revista Geologia de Chile, v. 20, p. 137.

Nicholson, C., Sorlien, C.C., Atwater, T., Crowell, J.C., and Luyendyk, B.P., 1994, Microplate capture, rotation of the western Transverse Ranges, and initiation of the San Andreas transform as a low-angle fault system: Geology, v. 22, p. 491–495.

Osozawa, S., 1992, Double ridge subduction recorded in the Shimanto accretionary wedge complex, Japan, and plate reconstruction: Geology, v. 20, p. 939–942.

Osozawa, S., 1993, Normal faults in an accretionary complex formed at trench-trench-ridge triple junction, as an indicator of angle between the trench and subducted ridge: The Island Arc, v. 3, p. 142–151.

Osozawa, S., and Yoshida, T., 1997, Arc-type and intraplate-type ridge basalts formed at the trench-trench-ridge triple junction: Implication for the extensive sub-ridge mantle heterogeneity: The Island Arc, v. 6, p. 197–212.

Osozawa, S., Sakai, T., and Naito, T., 1990, Miocene subduction of an active midocean ridge and origin of the Setogawa ophiolite, central Japan: Journal of Geology, v. 98, p. 763–771.

Pavlis, T.L., 1997, The Jurassic history of the Klamath Mountains province: A product of ridge subduction?: Geological Society of America Abstracts with Programs, v. 29, no. 5, p. 56.

Pavlis, T.L., Underwood, M., Sisson, V.B., Serpa, L.F., Prior, D., Marsaglia, K.M., Lewis, S.D., and Byrne, T., 1995, The effects of triple junction interactions at convergent plate margins: Report on the results of the joint JOI/USSAC and GSA Penrose conference: Washington, D.C., Joint Oceanographic Institutions, 44 p.

Perfit, M.R., Langmuir, C.H., Baekisapa, M., Chappell, B.W., Johnson, R.W., Staudigel, H., and Taylor, S.R., 1987, Geochemistry and petrology of volcanic rocks from the Woodlark Basin addressing questions of ridge subduction, *in* Taylor, B., and Exon, N.F., eds., Marine geology, geophysics, and geochemistry of the Woodlark Basin–Solomon Islands: Houston, Texas, Circum-Pacific Council for Energy and Mineral Resources, Earth Science Series, v. 7, p. 113–154.

Polat, A., Kerrich, R., and Wyman, D.A., 1998, The late Archean Schreiber-Hemlo and White River Dayohessarah greenstone belts, Superior province: Collages of oceanic plateaus, oceanic arcs, and subduction-accretion complexes: Tectonophysics, v. 289, p. 295–326.

Riddihough, R.P., 1982, One hundred million years of plate tectonics in western Canada: Geoscience Canada, v. 9, p. 28–34.

Riddihough, R.P., 1984, Recent movements of the Juan de Fuca plate system: Journal of Geophysical Research, v. 89, p. 6980–6994.

Rohr, K.M.M., and Furlong, K.P., 1995, Ephemeral plate tectonics at the Queen Charlotte triple junction: Geology, v. 23, p. 1035–1038.

Sakaguchi, A., 1996, High paleothermal gradient with ridge subduction beneath the Cretaceous Shimanto accretionary prism, southwest Japan: Geology, v. 24, p. 795–798.

Sakaguchi, A., 1999a, Thermal maturity in the Shimanto accretionary prism, southwest Japan, with the thermal change of the subducting slab: Fluid inclusion and vitrinite reflectance study: Earth and Planetary Science Letters, v. 173, p. 61–74.

Sakaguchi, A., 1999b, Thermal structure and paleo-heat flow in the Shimanto accretionary prism, southwest Japan: The Island Arc, v. 8, p. 350–372.

Saleeby, J.B., Busby-Spera, C., Oldow, J.S., Dunne, G.C., Wright, J.E., Cowan, D.S., Walker, N.W., and Allmendinger, R.W., 1992, Early Mesozoic tectonic evolution of the western U.S. Cordillera, *in* Burchfiel, B.C., et al., eds., The Cordilleran Orogen: Conterminous U.S.: Boulder, Colorado, Geological Society of America, Geology of North America, v. G-3, p. 107–168.

Scarrow, J.H., Leat, P.T., Wareham, C.D., and Millar, I.L., 1998, Geochemistry of mafic dykes in the Antarctic Peninsula continental-margin batholith: A record of arc evolution: Contributions to Mineralogy and Petrology, v. 131, p. 289–305.

Scharder, F.C., and Spencer, A.C., 1901, The geology and mineral resources of a portion of the Copper River District, Alaska: U.S. Geological Survey Report to 56th Congress, Document No. 546, 94 p.

Sherman, S.B., Karsten, J.L., and Klein, E.M., 1997, Petrogenesis of axial lavas from the southern Chile Ridge: Major element constraints: Journal of Geophysical Research, v. 102, p. 14,963–14,990.

Shervais, J.W., 2001, Birth, death, and resurrection: The life cycle of supra-subduction zone ophiolites: Geochemistry, Geophysics, Geosystems, v. 2, p. U1–U45.

Sisson, V.B., 1994, Effects of triple junction interactions discussed: Joint Oceanographic Institutions/U.S. Science Advisory Committee Newsletter, v. 7, no. 2, p. 5.

Sisson, V.B., and Pavlis, T.L., 1993, Geologic consequences of plate reorganization: An example from the Eocene southern Alaska forearc: Geology, v. 21, p. 913–916.

Staudigel, H., McCulloch, M.T., Zindler, A., and Perfit, M.R., 1987, Complex ridge subduction and island arc magmatism; an isotopic study of the New Georgia forearc and the Woodlark Basin, *in* Taylor, B., and Exon, N.F., eds., Marine geology, geophysics and geochemistry of the Woodlark Basin-Solomon Islands: Houston, Texas, Circum-Pacific Council for Energy and Mineral Resources, Earth Science Series, v. 7, p. 227–240.

Storey, B.C., Brown, R.W., Carter, A., Doubleday, P.A., Hurford, A.J., MacDonald, D.I.M., and Nell, P.A.R., 1996, Fission-track evidence for the thermotectonic evolution of a Mesozoic-Cenozoic forearc, Antarctica: Journal of the Geological Society [London], v. 153, p. 65–82.

Stowell, H.H., and McClelland, W.C., 2000, Tectonics of the Coast Mountains, southeastern Alaska and British Columbia: Boulder, Colorado, Geological Society of America Special Paper 343, 289 p.

Sturm, M.E., Klein, E.M., Graham, D.W., and Karsten, J., 1999, Age constraints on crustal recycling to the mantle beneath the southern Chile Ridge: He-Pb-Sr-Nd isotope systematics: Journal of Geophysical Research, v. 104, p. 5097–5114.

Suzuki, K., and Adachi, M., 1998, Denudation history of the high T/P Ryoke metamorphic belt, southwest Japan: Constraints from CHIME monazite ages of gneisses and granitoids: Journal of Metamorphic Geology, v. 16, p. 23–37.

Taira, A., Katto, J., Tashiro, M., Okamura, M., and Kodama, K., 1988, The Shimanto belt in Shikoku, Japan; evolution of Cretaceous to Miocene accretionary prism: Modern Geology, v. 12, p. 5–46.

Taylor, B., and Exon, N.F., 1987, An investigation of ridge subduction in the Woodlark-Solomons region: Introduction and overview, *in* Taylor, B., and Exon, N.F., eds., Marine geology, geophysics, and geochemistry of the Woodlark Basin-Solomon Islands: Houston, Texas, Circum-Pacific Council for Energy and Mineral Resources, Earth Science Series, v. 7, p. 1–24.

Tebbens, S.F., Cande, S.C., Kovacs, L., Parra, J.C., La Brecque, J.L., and Vergara, H., 1977, The Chile Ridge: A tectonic framework: Journal of Geophysical Research, v. 102, p. 12,035–12,059.

ten Brink, U.S., Shimizu, N., and Molzer, P.C., 1999, Plate deformation at depth under northern California: Slab gap or stretched slab?: Tectonics, v. 18, p. 1084–1098.

Thompson, P.H., 1989, Moderate overthickening of thinned sialic crust and the origin of granitic magmatism and regional metamorphism in low-*P*–high-*T* terrains: Geology, v. 17, p. 520–523.

Thomson, S.N., Herve, F., and Stöckhert, B., 2001, Mesozoic-Cenozoic denudation history of the Patagonian Andes (southern Chile) and its correlation to different subduction processes: Tectonics, v. 20, p. 693–711.

Thorkelson, D.J., 1995, The Paleogene Kula-Farallon slab window: Geological Association of Canada, Program with Abstracts, v. 20, p. A-105.

Thorkelson, D.J., 1996, Subduction of diverging plates and the principles of slab window formation: Tectonophysics, v. 255, p. 47–63.

Thorkelson, D.J., and Taylor, R.P., 1989, Cordilleran slab windows: Geology, v. 17, p. 833–836.

Tokunaga, T., 1992, Duplexing and intraprism deformation of the Paleogene Shimanto Supergroup in western Shikoku, southwest Japan: Tectonics, v. 11, p. 1168–1179.

Underwood, M.B., Hibbard, J.P., and DiTullio, L., 1993a, Geologic summary and conceptual framework for the study of thermal maturity within the Eocene-Miocene Shimanto belt, Shikoku, Japan, *in* Underwood, M.B., ed., Thermal evolution of the Tertiary Shimanto belt, southwest Japan: An example of ridge-trench interaction: Boulder, Colorado, Geological Society of America Special Paper 273, p. 1–24.

Underwood, M.B., Laughland, M.M., and Kang, S.M., 1993b, A comparison among organic and inorganic indicators of diagenesis and low-temperature metamorphism, Tertiary Shimanto belt, Shikoku, Japan, *in* Underwood, M.B., ed., Thermal evolution of the Tertiary Shimanto belt, southwest Japan: An example of ridge-trench interaction: Boulder, Colorado, Geological Society of America Special Paper 273, p. 45–62.

Underwood, M.B., Shelton, K.L., McLaughlin, R.J., Laughland, M.M., and Solomon, R.M., 1999, Middle Miocene paleotemperature anomalies within the Franciscan Complex of northern California: Thermo-tectonic responses near the Mendocino triple junction: Geological Society of America Bulletin, v. 111, p. 1448–1467.

Uyeda, S., and Miyashiro, A., 1974, Plate tectonics and the Japanese Islands: A synthesis: Geological Society of America Bulletin, v. 85, p. 1159–1170.

van Wijk, J.W., Govers, R., and Furlong, K.P., 2001, Three-dimensional thermal modeling of the California upper mantle: A slab window vs. stalled slab: Earth and Planetary Science Letters, v. 186, p. 175–186.

Wright, J.E., and Fahan, M.R., 1988, An expanded view of Jurassic orogenesis in the western United States Cordillera: Middle Jurassic (pre-Nevadan) regional metamorphism and thrust faulting within an active arc environment, Klamath Mountains, California: Geological Society of America Bulletin, v. 100, p. 859–876.

Wright, J.E., and Wyld, S.J., 1994, The Rattlesnake Creek terrane, Klamath Mountains, California: An early Mesozoic volcanic arc and its basement of tectonically disrupted oceanic crust: Geological Society of America Bulletin, v. 106, p. 1033–1056.

Zandt, G., and Furlong, K.P., 1982, Evolution and thickness of the lithosphere beneath coastal California: Geology, v. 10, p. 376–381.

Manuscript Accepted by the Society February 5, 2003

Printed in the USA

Geological Society of America
Special Paper 371
2003

Geologic signature of early Tertiary ridge subduction in Alaska

Dwight Bradley*
U.S. Geological Survey, 4200 University Drive, Anchorage, Alaska 99508, USA

Tim Kusky
Department of Earth and Atmospheric Sciences, St. Louis University, St. Louis, Missouri 63103, USA

Peter Haeussler
U.S. Geological Survey, 4200 University Drive, Anchorage, Alaska 99508, USA

Rich Goldfarb
U.S. Geological Survey, Box 25046, MS 973, Denver Federal Center, Denver, Colorado 90225, USA

Marti Miller
Julie Dumoulin
Steven W. Nelson
Sue Karl
U.S. Geological Survey, 4200 University Drive, Anchorage, Alaska 99508, USA

ABSTRACT

A mid-Paleocene to early Eocene encounter between an oceanic spreading center and a subduction zone produced a wide range of geologic features in Alaska. The most striking effects are seen in the accretionary prism (Chugach–Prince William terrane), where 61 to 50 Ma near-trench granitic to gabbroic plutons were intruded into accreted trench sediments that had been deposited only a few million years earlier. This short time interval also saw the genesis of ophiolites, some of which contain syngenetic massive sulfide deposits; the rapid burial of these ophiolites beneath trench turbidites, followed immediately by obduction; anomalous high-*T*, low-*P*, near-trench metamorphism; intense ductile deformation; motion on transverse strike-slip and normal faults; gold mineralization; and uplift of the accretionary prism above sea level. The magmatic arc experienced a brief flare-up followed by quiescence. In the Alaskan interior, 100 to 600 km landward of the paleotrench, several Paleocene to Eocene sedimentary basins underwent episodes of extensional subsidence, accompanied by bimodal volcanism. Even as far as 1000 km inboard of the paleotrench, the ancestral Brooks Range and its foreland basin experienced a pulse of uplift that followed about 40 million years of quiescence.

All of these events—but most especially those in the accretionary prism—can be attributed with varying degrees of confidence to the subduction of an oceanic spreading center. In this model, the ophiolites and allied ore deposits were produced at the soon-to-be subducted ridge. Near-trench magmatism, metamorphism, deformation, and gold mineralization took place in the accretionary prism above a slab window, where hot asthenosphere welled up into the gap between the two subducted, but still diverging, plates. Deformation took place as the critically tapered accretionary prism

*dbradley@usgs.gov

Bradley, D., Kusky, T., Haeussler, P., Goldfarb, R., Miller, M., Dumoulin, J., Nelson, S.W., and Karl, S., 2003, Geologic signature of early Tertiary ridge subduction in Alaska, *in* Sisson, V.B., Roeske, S.M., and Pavlis, T.L., eds., Geology of a transpressional orogen developed during ridge-trench interaction along the North Pacific margin: Boulder, Colorado, Geological Society of America Special Paper 371, p. 19–49.

adjusted its shape to changes in the bathymetry of the incoming plate, changes in the convergence direction before and after ridge subduction, and changes in the strength of the prism as it was heated and then cooled. In this model, events in the Alaskan interior would have taken place above more distal, deeper parts of the slab window. Extensional (or transtensional) basin subsidence was driven by the two subducting plates that each exerted different tractions on the upper plate. The magmatic lull along the arc presumably marks a time when hydrated lithosphere was not being subducted beneath the arc axis. The absence of a subducting slab also may explain uplift of the Brooks Range and North Slope: Geodynamic models predict that long-wavelength uplift of this magnitude will take place far inboard from Andean-type margins when a subducting slab is absent. Precise correlations between events in the accretionary prism and the Alaskan interior are hampered, however, by palinspastic problems. During and since the early Tertiary, margin-parallel strike-slip faulting has offset the near-trench plutonic belt—i.e., the very basis for locating the triple junction and slab window—from its backstop, by an amount that remains controversial.

Near-trench magmatism began at 61 Ma at Sanak Island in the west but not until 51 Ma at Baranof Island, 2200 km to the east. A west-to-east age progression suggests migration of a trench-ridge-trench triple junction, which we term the Sanak-Baranof triple junction. Most workers have held that the subducted ridge separated the Kula and Farallon plates. As a possible alternative, we suggest that the ridge may have separated the Kula plate from another oceanic plate to the east, which we have termed the Resurrection plate.

Keywords: massive sulfide deposits, gold mineralization, brittle faulting, accretionary prism, ridge subduction, Chugach–Prince William terrane, Sanak Island, triple junction.

INTRODUCTION

Southern Alaska has a long history as a convergent margin, where various plates of the Pacific Ocean have been subducted. During much of the time since the Jurassic, it is possible to recognize the same standard elements of a subduction zone that are seen today: backarc, magmatic arc, forearc basin, and accretionary prism. Each of these elements underwent major changes in the early Tertiary. The most striking changes were in the accretionary prism, where near-trench granodiorite plutons—some of batholithic proportions—were intruded into accreted trench deposits that had been deposited only a few million years earlier (Moore et al., 1983) (Figs. 1 and 2). At this same time, the accretionary prism was metamorphosed at high-T, low-P conditions (Sisson et al., 1989), ductilely deformed over large regions (Pavlis and Sisson, 1995), cut by strike-slip and normal faults (Kusky et al., 1997a), and mineralized by gold-bearing hydrothermal systems (Haeussler et al., 1995). Meanwhile, very young ophiolites were accreted onto the leading edge of the subduction complex (Kusky and Young, 1999). It was an unusual time in the history of the convergent margin, presumably due to special circumstances.

One important clue to the cause of early Tertiary events in Alaska can be found offshore, in the magnetic anomalies of the northern Pacific plate. A ~70-degree bend in the anomaly pattern, known as the Great Magnetic Bight, implies that three oceanic plates existed here in the early Tertiary, separated by three spreading ridges (Atwater, 1989). One ridge, between the Kula and Farallon plates, headed in the general direction of North America—but exactly where it intersected the margin cannot be deciphered from marine magnetic anomalies, because all of the evidence has been subducted (Page and Engebretson, 1984; Atwater, 1989). The track of a migrating trench-ridge-trench triple junction is suggested, however, by the near-trench plutons in Alaska. As will be shown, ridge subduction provides the most reasonable explanation for these plutons, which define the Sanak-Baranof plutonic belt of Hudson et al. (1979). If the Great Magnetic Bight is a smoking gun, the near-trench plutons would then be the bullet holes. In fact, ridge subduction provides such an elegant rationale for the age, location, and geochemistry of these plutons that since being first proposed a quarter-century ago (Marshak and Karig, 1977), it has become the leading hypothesis (e.g., Helwig and Emmet, 1981; Hill et al., 1981; Moore et al., 1983; Barker et al., 1992; Sisson and Pavlis, 1993; Bradley et al., 1993; Pavlis and Sisson, 1995; Haeussler et al., 1995; Poole, 1996; Harris et al., 1996).

The case for early Tertiary ridge subduction in Alaska has not been universally embraced. For example, Engebretson's (1985) widely cited plate reconstructions showed the Kula-Farallon ridge being subducted at the latitude of Washington, and no ridge-trench interactions in Alaska. In Plafker and Berg's 1994 *Decade of North American Geology* synthesis of Alaskan tectonic evolution, they likewise showed no ridge intersecting the Alaskan margin during the time in question. Various alternative explanations for the near-trench plutons—published and unpublished—have been put forth: They were emplaced along an arc that had jumped to an extreme trenchward position (Kienle and Turner, 1976); they record anatectic melting of a thick pile of recently subducted flysch (Hudson et al., 1979); the trench overrode a hotspot; very

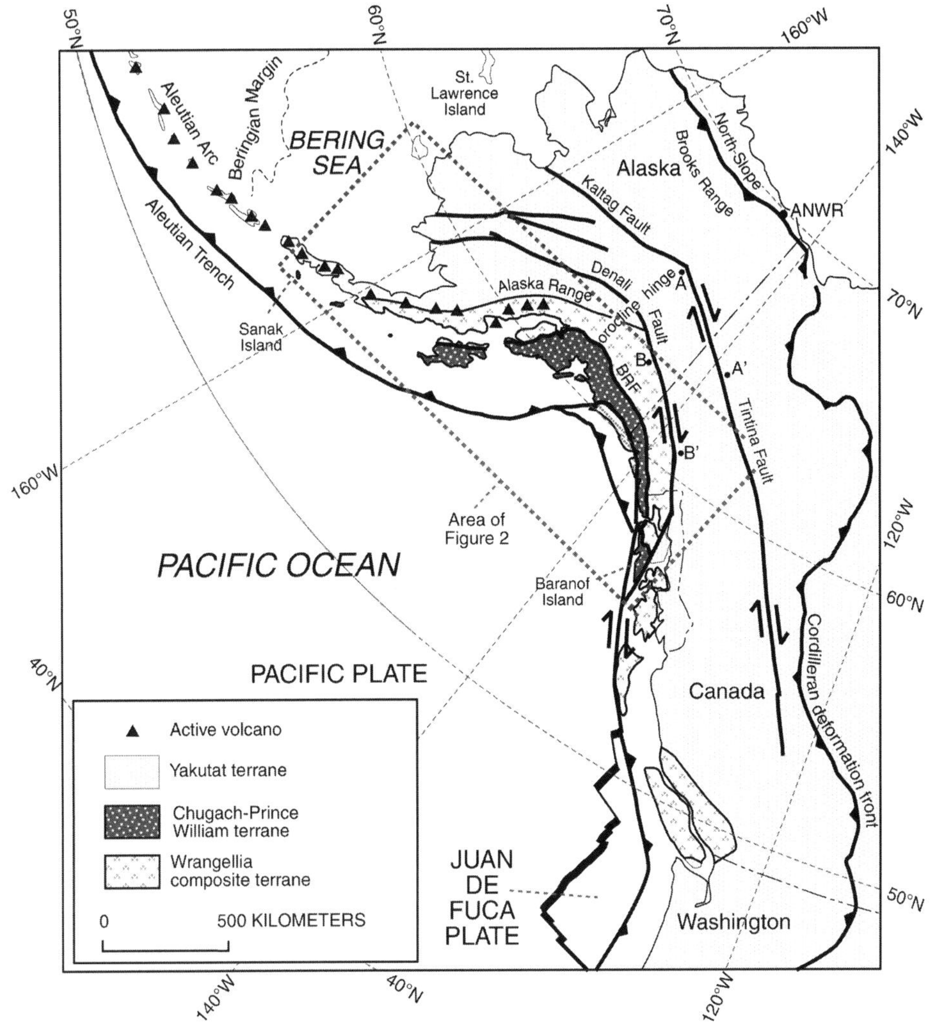

Figure 1. Map of Alaska and surrounding areas showing geologic features mentioned in text. A and A′ mark offset across Tintina fault system (after Dover, 1994); B and B′ mark offset across Denali fault (Lowey, 1998). Wrangellia composite terrane includes Wrangellia, Alexander, and Peninsular terranes.

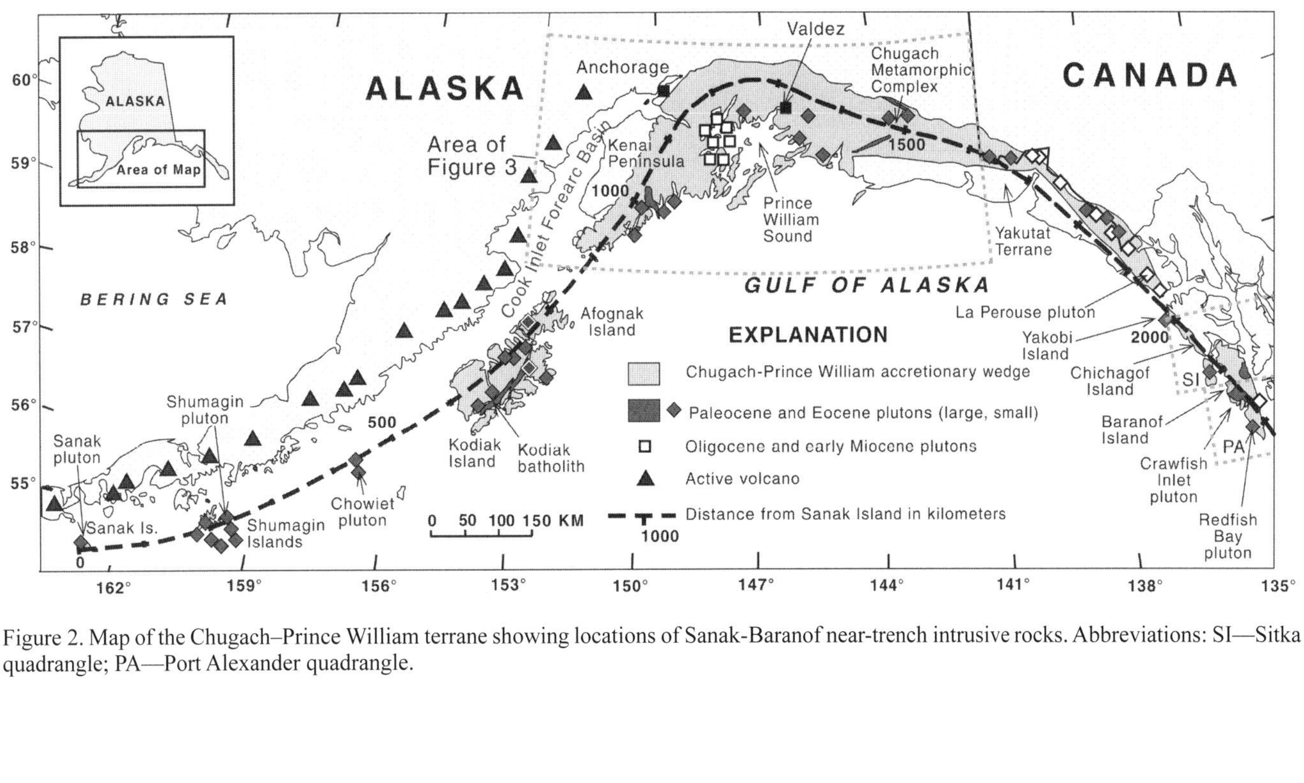

Figure 2. Map of the Chugach–Prince William terrane showing locations of Sanak-Baranof near-trench intrusive rocks. Abbreviations: SI—Sitka quadrangle; PA—Port Alexander quadrangle.

young, hot, oceanic lithosphere was subducted; a "leaky transform" entered the trench (Tysdal et al., 1977); near-trench magmatism was the product of flexural extension of a single downgoing plate (Tysdal et al., 1977); and a subducted slab broke off, exposing the overriding plate to a slab window. As will be discussed at greater length below, the evidence now in hand is best explained by the ridge subduction hypothesis. Accordingly, as this volume itself makes clear, the emphasis of current research has shifted from testing alternative tectonic scenarios, to refining the ridge subduction scenario and studying processes of ridge subduction. By analogy, studies of the Franciscan Complex have long since advanced beyond the question of whether or not it is the product of subduction, to questions about subduction zones. Still, it behooves us to not forget alternative explanations that are no longer in the forefront. This philosophy guides the first part of the present paper, a non-interpretive discussion of Alaska's early Tertiary geologic record.

Our descriptive overview of the geology is based on our own experience gained through field work during the past two decades, as well as on published findings. Most of our work was done in connection with 1:250,000-scale reconnaissance bedrock mapping and resource assessment by the U.S. Geological Survey in the Seldovia, Seward, Cordova, Valdez, Sitka, Port Alexander, Talkeetna, and Talkeetna Mountains quadrangles (Figs. 2 and 3). We emphasize geology of the accretionary prism, rocks most pertinent to the subduction history. Specific aspects of the early Tertiary geology of the accretionary prism have been discussed in several dozen publications—but a comprehensive synthesis has been lacking. The early Tertiary record of interior Alaska is covered here in less detail, because new research has just been started that will quickly eclipse our present knowledge. In the Alaskan interior, various episodes of early Tertiary basin subsidence, uplift, magmatism, and mineralization attest to significant tectonic activity that presumably was related to events along the continental margin—but reliable correlations must await tighter dating and new information on the amount of margin-parallel strike slip.

The second part of the paper is interpretive. Here we present the case that ridge subduction provides a unifying rationale for the broad range of evidence, and we consider the various competing, but to us less satisfactory, hypotheses. We conclude with some general observations about ridge subduction, and a discussion of some outstanding problems in Alaska.

GEOLOGIC SETTING

Alaska consists of a collage of terranes that were accreted to the western margin of North America as a result of complex plate interactions through most of the Phanerozoic. Comprehensive reviews and bibliographies of the regional geology are given by Plafker et al. (1994) and Nokleberg et al. (1994). Those terranes of particular significance to the present study are introduced here along a south-to-north transect along the hinge of the southern Alaska orocline (Fig. 1). Alaska's Pacific margin is underlain by a Mesozoic and Cenozoic accretionary prism, referred to as the Chugach–Prince William terrane (or composite terrane), which is discussed in more detail below. It was built in piecemeal fashion by subduction-accretion and lies seaward of the Wrangellia composite terrane (Fig. 3). Between the two, and resting on both, lies the Cook Inlet forearc basin. The modern forearc basin had an ancestry in the Triassic, Jurassic, and parts of the Cretaceous when it received many kilometers of sediment from the nearby magmatic arc. The early Tertiary history of this basin bears on the ridge subduction hypothesis (Trop et al., this volume, Chapter 4). The active volcanoes of the modern arc are built largely on basement of the Wrangellia composite terrane, which includes arc batholiths and associated volcanic and volcaniclastic rocks of Jurassic, Cretaceous, and early Tertiary age. The inboard suture of the Wrangellia composite terrane is marked by a belt of deformed Jurassic-Cretaceous flysch, host to a suite of 59–55 Ma granitic plutons that will be mentioned in a later section on possible inboard effects of the slab window. The complex region between the Alaska Range and Brooks Range (Fig. 1), underlain largely by the Yukon-Tanana continental margin terrane, is of only peripheral interest to the present study. In northern Alaska, the Brooks Range orogen records a Jurassic to Early Cretaceous collision between the Yukon-Koyukuk magmatic arc and a south-facing (present coordinates) passive margin of the Arctic Alaska terrane. Tectonic transport was to the north and the resulting orogenic load formed a foreland basin on Alaska's North Slope (Fig. 1). The Alaskan terrane collage is cut by margin-parallel strike-slip faults such as the Border Ranges, Denali, and Tintina (Fig. 1). As will be discussed, the sense, amount, and timing of motion bears on correlations between events in the Alaskan interior and those in the Chugach–Prince William terrane.

The Chugach–Prince William accretionary prism is subdivided into belts of accreted ocean-floor rocks that young in a seaward direction. The most landward belt consists of high-pressure metamorphic rocks (Fig. 3) that were metamorphosed during the Early Jurassic, as summarized, for example, by Dusel-Bacon et al. (1993). The next belt is an argillite-matrix melange containing blocks and fault slices of Triassic to middle Cretaceous chert and basalt, undated graywacke, rare Permian limestone, and a few Triassic-Jurassic ultramafic-mafic complexes (Uyak Complex, McHugh Complex, and Kelp Bay Group; Connelly, 1978; Bradley and Kusky, 1992; Decker, 1980). Small near-trench plutons and numerous dikes intrude these rocks. The next outboard belt contains Upper Cretaceous flysch that has been assigned to the Shumagin Formation, Kodiak Formation, Valdez Group, and part of the Sitka Graywacke (Moore, 1973; Nilsen and Moore, 1979; Nilsen and Zuffa, 1982; Decker, 1980). The biggest near-trench plutons and largest areas of high-temperature, *low*-pressure metamorphism occur in the Upper Cretaceous flysch belt. In Prince William Sound, mafic volcanic rocks are interbedded with flysch assigned to the Valdez Group. Farther outboard lie belts of flysch assigned to the Ghost Rocks Formation and Orca Group (Moore et al., 1983; Moore and Allwardt, 1980; Helwig and

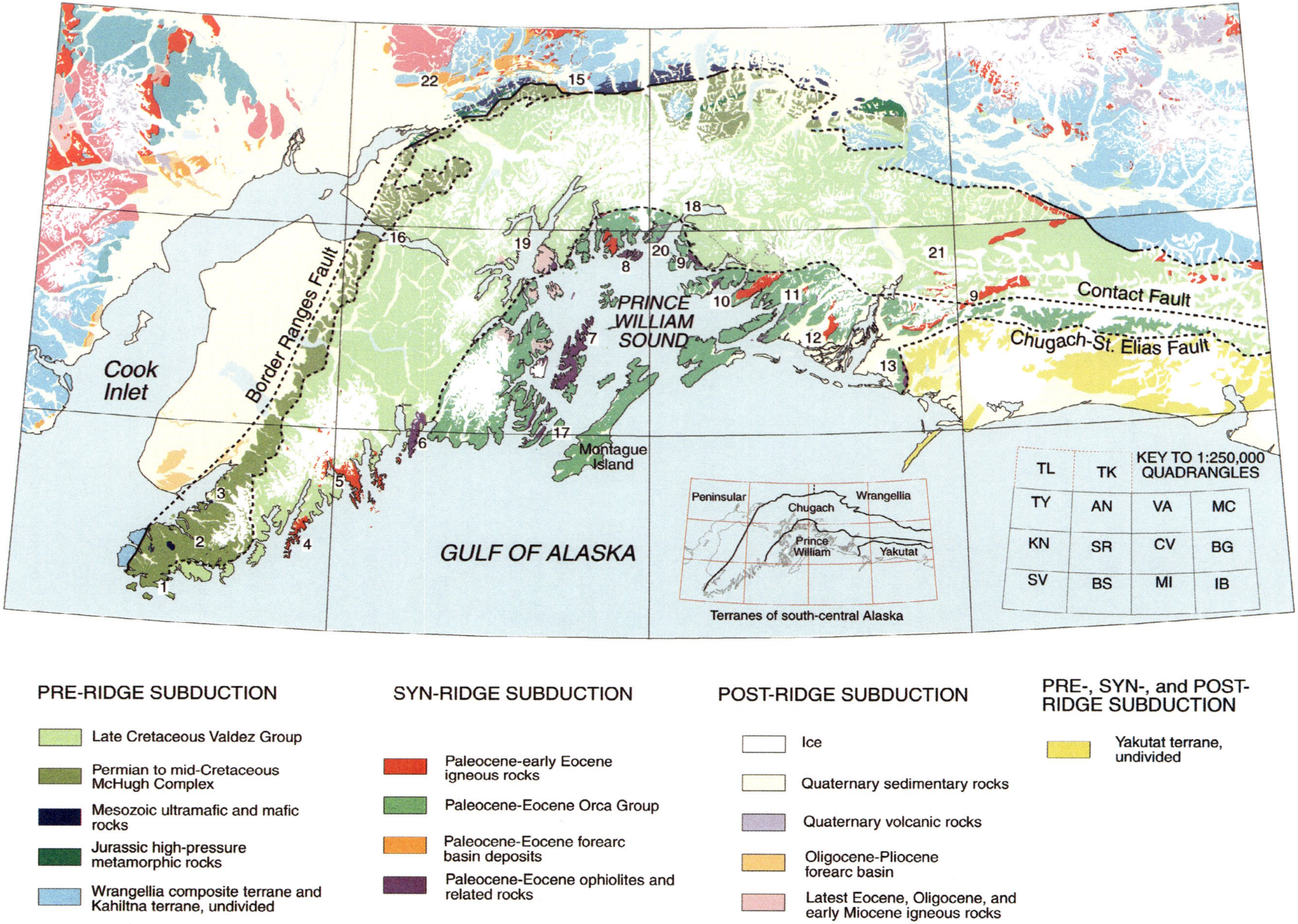

Figure 3. Geologic map of south-central Alaska showing belts of accreted rocks, near-trench igneous rocks, and localities mentioned in text. Abbreviations for quadrangles: AN—Anchorage; BG—Bering Glacier; BS—Blying Sound; CV—Cordova; IB—Icy Bay; KN—Kenai; MC—McCarthy; MI—Middleton Island; SR—Seward; SV—Seldovia; TK—Talkeetna Mountains; TL—Talkeetna; TY—Tyonek; VA—Valdez. (Talkeetna and Talkeetna Mountains quadrangles are outside the map area.)

Emmet, 1981). These units include mafic and ultramafic rocks that, we will argue, formed near the trench-ridge-trench triple junction. Penetrative deformation and regional metamorphism occurred in the accretionary prism during and shortly after subduction-accretion in the Jurassic, Cretaceous, and Paleocene (Dusel-Bacon et al., 1993). Near-trench plutons were emplaced into the accretionary prism after this deformation. More deep-sea turbidites (Eocene Sitkalidak Formation and the outboard part of the Orca Group; Moore and Allwardt, 1980; Helwig and Emmet, 1981) occur outboard of the Sanak-Baranof belt; these younger turbidites are not cut by the plutons, and were deposited and accreted after near-trench magmatism.

The Yakutat terrane, which occupies a belt along the coast where the Alaskan panhandle joins the mainland (Figs. 2 and 3), is regarded as a segment of the Chugach–Prince William terrane, now located some 600 km north of its original position (Plafker et al., 1994). Like the Chugach–Prince William terrane, the Yakutat includes Mesozoic melange and flysch, and Tertiary near-trench plutons (Plafker et al., 1994). A 3 km thick succession of late Paleocene to Eocene marine clastic rocks, basalt, and minor coal, which has no obvious counterpart in the Chugach–Prince William terrane, spans the age of near-trench magmatism—but connections between particular rock units and tectonic regime have not yet been studied.

NEAR-TRENCH INTRUSIVE ROCKS

Near-trench intrusive rocks along the Gulf of Alaska margin (Fig. 2) are of two broad age groups: Paleocene to early Eocene, and latest Eocene to early Miocene. The latter present an intriguing problem, but are younger than events of interest here. The Paleocene to early Eocene plutons include granodiorite, granite, tonalite, and gabbro (Hudson et al., 1979; Hudson, 1983). Swarms of basaltic to rhyolitic dikes, known or inferred to be of the same range of ages as the Paleocene to early Eocene plutons, have been mapped along some sectors of the margin.

The mapped limits of the Sanak-Baranof plutonic belt and the Chugach–Prince William accretionary prism are nearly identical. To the west of Sanak Island, any offshore continuation of the near-trench intrusive belt would presumably follow the Beringian margin (Fig. 1) where a pre-Eocene accretionary prism is inferred in the subsurface (Worrall, 1991). The Sanak-Baranof belt is not recognized to the southeast of Baranof Island, but, subject to restoration of 600 km of margin-parallel dextral displacement, the near-trench plutons of the Yakutat terrane would appear to represent the former southeastward continuation of the belt.

Geochronology

Near-trench magmatism progressed from west to east. This is shown in two plots, one for U/Pb and $^{40}Ar/^{39}Ar$ ages and another for conventional K/Ar and Rb/Sr isochron ages (Fig. 4). The conventional K/Ar data (Fig. 4B) have considerable scatter and large error bars, and the black-box nature of the method sometimes makes it difficult if not impossible to tell a "good" date from one that reflects excess argon, partial argon loss, or complete resetting. Two decades ago, when only K/Ar and Rb/Sr data were available, some researchers held that the near-trench intrusives fell into two age groups, ca. 60 Ma in the west and ca. 50 Ma in the east (e.g., Hudson et al., 1979). Others (e.g., Helwig and Emmet, 1981; Moore et al., 1983) interpreted a more gradual younging trend. Following W. Clendenen (1989, written commun.), Bradley et al. (1993) compiled available geochronology and, by means of early versions of Figures 4A and 4B, showed a clear west-to-east younging trend. Since 1993, new $^{40}Ar/^{39}Ar$ and U/Pb dating has confirmed and refined the age progression (Fig. 4A) (Bradley et al., 1993, 2000; Haeussler et al., 1995). Most convincing is the trend shown by the concordant U/Pb TIMS data from zircon and monazite—the least equivocal way to date granitic rocks of this age. These were deliberately shown with the boldest symbols in Figure 4A. The other data in Figure 4A show the same overall west-to-east younging trend, but are not as bombproof—the discordant U/Pb ages because they rest on a judgment of whether inheritance or lead loss was responsible for discordance, and the $^{40}Ar/^{39}Ar$ ages because they date cooling rather than intrusion.

The age-distance plots in Figure 4 show several other tectonically significant features. (1) Other fields. Most of the conventional K/Ar data shown (Fig. 4B) clearly correspond to the diachronous cluster of U/Pb and $^{40}Ar/^{39}Ar$ data (Fig. 4A)—the "main Sanak-Baranof trend" as used below. Some of the K/Ar ages, however, require other explanations. Fields 2 and 3 are younger than the main Sanak-Baranof trend, and are seen in both Figures 4A and 4B. Field 2 corresponds to the late Eocene to Oligocene plutons of Prince William Sound. Field 3 likely includes a combination of emplacement ages in the range 35–45 Ma, and reset or cooling ages of older plutons. Field 1, older than the main Sanak-Baranof trend, is a group of conventional K/Ar amphibole ages suspected of reflecting excess argon (Onstott et al., 1989). Bradley et al. (1993) provided some additional discussion of these fields. (2) Onset of magmatism. Magmatism began ca. 61 Ma on Sanak Island but not until 51 Ma on Baranof Island, some 2200 km along strike to the east (Fig. 4). The 61 Ma age quoted here for the onset of magmatism at the western end of the belt is based on a new concordant U/Pb zircon age of the Shumagin pluton (Bradley et al., 2000) and is younger than the 63–66 Ma age range quoted by Bradley et al. (1993), which was based on the K/Ar and Rb/Sr ages shown in Figure 4B. The 51 Ma age quoted here for the *onset* of magmatism at the eastern end of the belt (Bradley et al., 1993) is based on one new concordant U/Pb and two new $^{40}Ar/^{39}Ar$ plateau ages, and is older than the 50 Ma age quoted by Bradley et al. (1993). (3) Duration of magmatism at a given place. At face value, the dates in Figure 4A would seem to suggest that magmatism locally spanned 3–5 million years, at least along certain sectors of the margin. Conversely, near-trench magmatism would appear to have taken place simultaneously along a considerable length (~500 km) of continental margin.

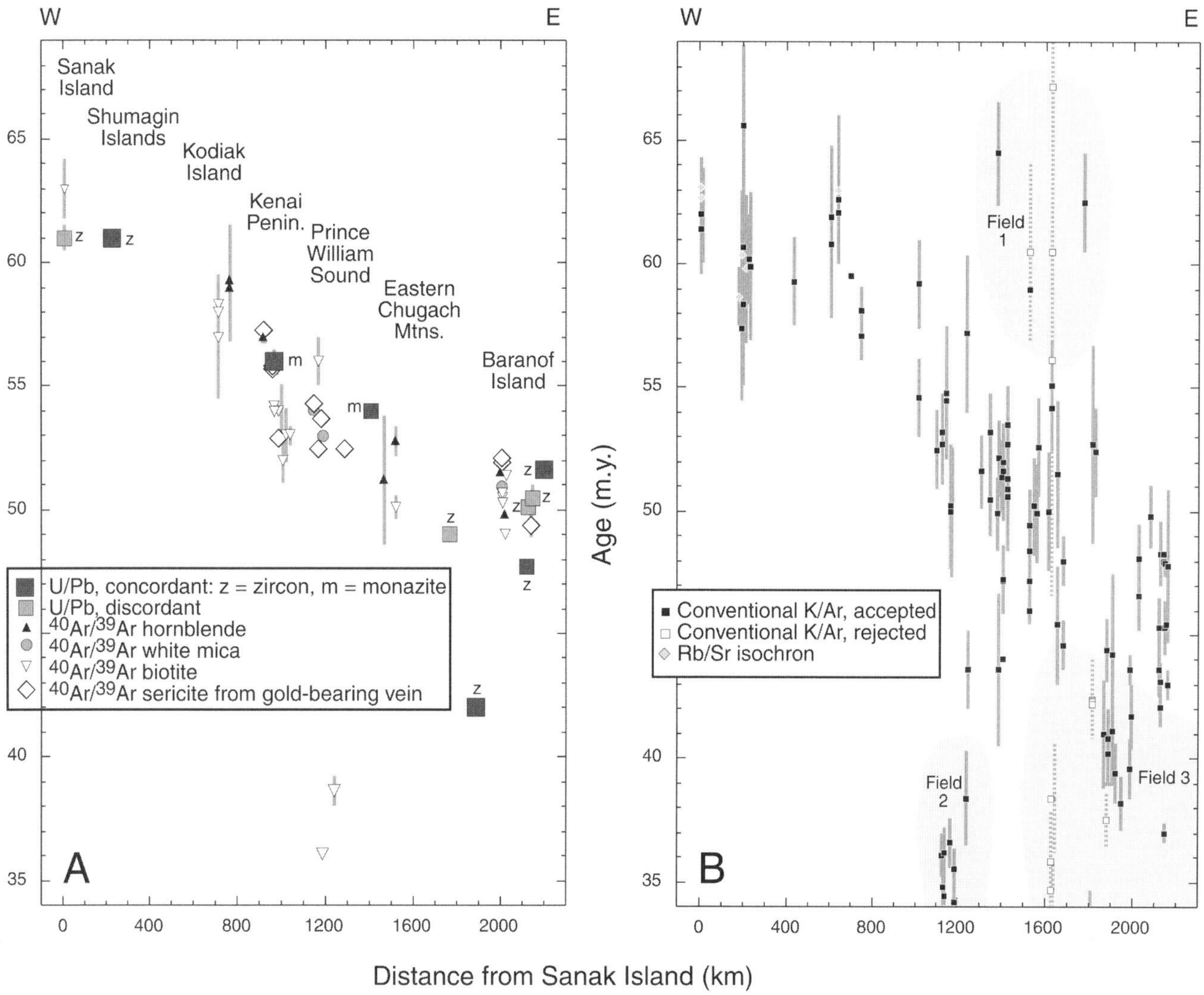

Figure 4. Isotopic ages of near-trench intrusive rocks of the Sanak-Baranof belt, plotted against distance along strike. Distances along the x-axis were measured from Sanak Island on Beikman's (1980) geologic map of Alaska; the map projection was responsible for errors up to about +5% in some cases. Data in this plot are mainly from the tabulation by Bradley et al. (1993) and two sequential updates, by Haeussler et al. (1995) and by Bradley et al. (2000). Table 1 summarizes new results since the latter paper went to press, and was current as of September 2001. Other papers in this volume may contain new dates that are not plotted here or listed in Table 1. A: U/Pb and $^{40}Ar/^{39}Ar$ ages. Short error bars are obscured behind the symbols for most dates. B: Conventional K/Ar and Rb/Sr isochron ages. "Rejected" K/Ar ages are shown for the sake of completeness, but have errors >10% or are significantly younger than other K/Ar ages from the same sample.

Only on Baranof Island, however, has a significant (~4 m.y.) age span been confirmed by concordant U/Pb ages. Some of the apparent age span is probably an artifact of cooling ages.

Petrology and Geochemistry

The predominant plutonic rock of the Sanak-Baranof belt is granodiorite (Fig. 5). Some plutons also include granite, tonalite, and quartz diorite phases (Fig. 5), which mainly differ from the granodiorites in the proportion of K-feldspar. Tonalite and quartz diorite are more common in the eastern than the central or western parts of the belt. All of these show calc-alkaline affinities (Fig. 6).

Point-count data from a variety of plutons indicate plagioclase, quartz, biotite and K-feldspar as the major minerals listed in order of decreasing abundance. Hornblende is present in a few plutons, muscovite in others. Zircon and apatite appear to be ubiquitous; monazite has been noted in two plutons. Magmatic andalusite occurs in one pluton in the Chugach metamorphic complex (J. Sisson, 2001, written commun.). The Nuka pluton (Fig. 3) appears to be the source of corundum in pan concentrates from a number of drainages entirely within the pluton (D. Tripp, U.S. Geological Survey, 1993, written commun.). Kyanite xenocrysts, discussed later, have been reported from at least five plutons (Hill et al., 1981; Kusky et al., this volume, Chapter 12).

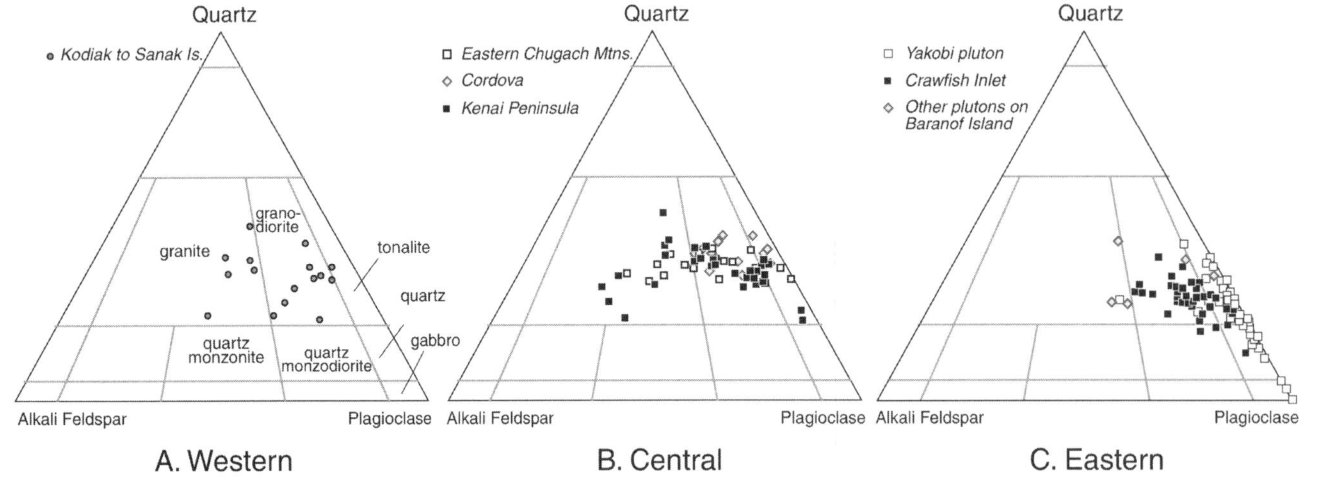

Figure 5. Ternary diagrams of point counts of plutonic rocks from three sectors of the Sanak-Baranof belt. Rock classification fields from Streckeisen (1973). Data are from Hudson et al. (1979), Tysdal and Case (1979), Reifenstuhl (1986), and our data from Seldovia quadrangle, Prince William Sound, and Sitka quadrangles.

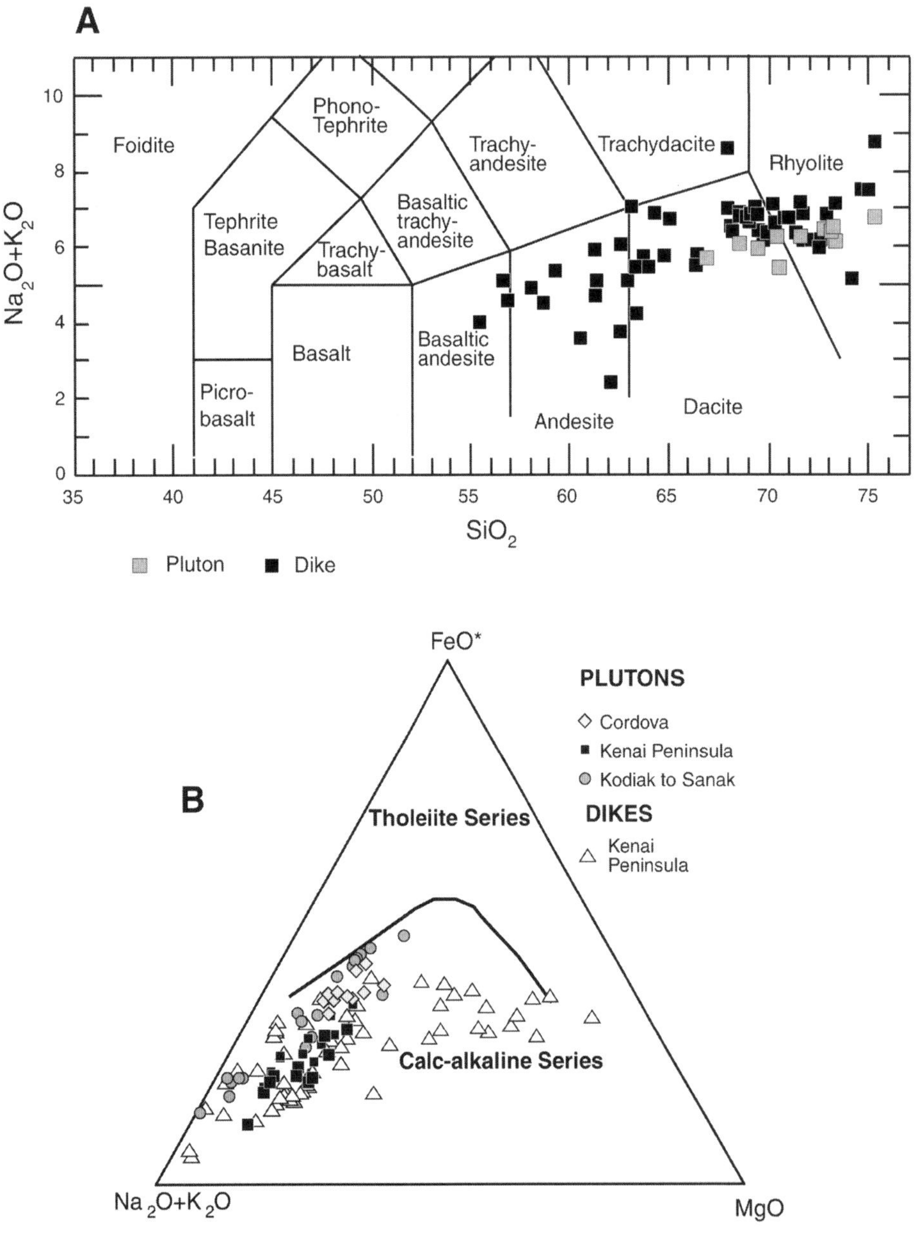

Figure 6. A: $K_2O + Na_2O$ plotted against SiO_2 for the least altered dikes and plutonic rocks from the Kenai Peninsula. Petrographic fields are from LeBas et al. (1986). B: AFM ternary diagram showing calc-alkaline affinity of the same samples. A—alkalis as $K_2O + Na_2O$; F—iron as $FeO + Fe_2O_3$; M—magnesium as MgO.

Gabbroic plutons have been described from as far west as Kodiak (Moore et al., 1983), but appear to be much more abundant in the eastern half of the Sanak-Baranof belt. Until very recently, the geochronological basis for including these mafic plutons in the Sanak-Baranof belt was problematic and merits a brief digression. Conventional K/Ar ages on these rocks range from 19 to 64 Ma (Bradley et al., 1993); none, however fall exactly within the 61–50 Ma range that has been established for Sanak-Baranof magmatism by modern geochronological methods. When Bradley et al. compiled the available geochronology in 1993, the most compelling evidence for including any of the mafic plutons in the Sanak-Baranof belt was from the Rude River pluton (location 10, Fig. 3), which includes a gabbroic phase that had yielded a conventional K/Ar hornblende age of 64.5 Ma (probably too old due to excess argon), and a granodiorite phase that had yielded a K/Ar biotite age of 52.3 Ma. Better evidence is now available. Gabbroic bodies on Yakobi and Chichagof Islands (Fig. 2) grade into and are comagmatic with tonalite (Himmelberg et al., 1987) that recently has yielded $^{40}Ar/^{39}Ar$ biotite ages of 50.3 and 50.5 Ma (Table 1). In this volume (Chapter 13), Sisson et al. report a $^{40}Ar/^{39}Ar$ hornblende plateau age of 53.8 ± 0.4 Ma from a diorite in the Yakutat terrane near Yakutat. Himmelberg et al. (1987, their Figs. 3–5) showed that the mafic rocks on Yakobi and Chichagof Islands range from hornblende pyroxenite to hornblende-pyroxene gabbronorite to quartz-bearing gabbronorite and gabbronorite. Local magmatic sulfide deposits (e.g., Bohemia Basin), which presumably formed from an immiscible sulfide melt in the silicate magma, contain pyrite, chalcopyrite, and pentlandite (Himmelberg et al., 1987). Another well-studied pluton, associated with the much larger Brady Glacier magmatic sulfide deposit, is the funnel-shaped La Perouse layered mafic-ultramafic intrusion (Fig. 2) (Loney and Himmelberg, 1983). Unfortunately, it has not been reliably dated, having yielded a number of difficult-to-interpret K/Ar and $^{40}Ar/^{39}Ar$ ages on biotite and hornblende ranging from 44 to 28 Ma (Bradley et al., 1993).

Dikes of the Sanak-Baranof belt have been studied most thoroughly in the Seldovia quadrangle (Fig. 3) (Bradley et al., 1999; Lytwyn et al., 2000). Almost all are dacite, rhyolite, or andesite (Fig. 6A). The rhyolite dikes are porphyritic, containing 10–20% phenocrysts (.5 to 2 mm) in a very fine groundmass. Phenocrysts include plagioclase ± resorbed quartz ± biotite. The microcrystalline to granular groundmass has plagioclase > quartz > potassium feldspar, and secondary white mica and locally, minor epidote. The dacite dikes, where porphyritic, contain phenocrysts of plagioclase (commonly zoned), lesser to no mafic minerals (altered biotite and locally amphibole), and rarely potassium feldspar and garnet. Groundmass consists of plagioclase > quartz > potassium feldspar, and locally biotite altered to chlorite. Secondary white mica and calcite are ubiquitous. The andesite dikes are primarily fine-grained seriate texture. Locally, the andesite is porphyritic, containing up to 10% phenocrysts of brown hornblende in a groundmass of plagioclase, brown hornblende, chlorite after an unknown mafic mineral, minor quartz, and little to no potassium feldspar. Hornblende is locally altered to actinolite + chlorite. Some samples have secondary white mica and prehnite. The basaltic andesite is generally fine-grained seriate in texture consisting of plagioclase, clinopyroxene, and lesser quartz. The clinopyroxene is partly replaced by brown hornblende and later chlorite. Secondary prehnite, sphene, and minor epidote occur between grains; calcite alteration is present locally. Alteration is pervasive, as might be expected of dikes emplaced into a submarine accretionary prism.

The dikes are correlated on geochronological grounds with the Sanak-Baranof plutons. A dacite dike yielded a $^{40}Ar/^{39}Ar$ hornblende age of 57.2 ± 0.8 Ma (Bradley et al., 2000); an altered, weakly mineralized dike yielded a $^{40}Ar/^{39}Ar$ sericite age of 57.3 ± 0.1 Ma (Haeussler et al., 1995). Comparable ages have been obtained from nearby plutons and all of these dates fall on the Sanak-Baranof age trend (Fig. 4). One other dike from the Seldovia quadrangle yielded a surprisingly old $^{40}Ar/^{39}Ar$ hornblende age of 115.0 ± 1.7 Ma (Bradley et al., 2000) (location 2, Fig. 3). This hornblende-phyric basaltic andesite, which cuts the McHugh Complex, is petrographically distinct—but its existence does make us suspect that at least a few other dikes in the McHugh Complex are older than the Sanak-Baranof intrusive rocks.

Major- and trace-element geochemical data have been reported for a number of plutons and dikes in the Sanak-Baranof belt (e.g., Hill et al., 1981; Barker et al., 1992; Lytwyn et al., 2000). Probably the most comprehensive data set available—11 plutons and 74 dikes—is from the Seldovia quadrangle (analytical data are in the Appendix[1]). Although the dikes as a group have a much wider range of SiO_2 than the plutons, major- and trace-element data overlap. Both plot in the calc-alkaline field of an AFM diagram (Fig. 6B). Chondrite-normalized REE abundances vary with silica composition (Fig. 7A–E). The least-altered mafic rock, a basaltic andesite, has a flat, MORB-like pattern; the intermediate and siliceous dikes and the granodiorite plutons are enriched in light REE. Some rhyolites and granites show a pronounced negative Eu anomaly. Trace element spidergrams from the same samples (Fig. 7F–J), normalized to chondrite after Sun (1980), show similar variations with silica composition. The basaltic andesite sample lacks any strong anomalies and broadly resembles the E-MORB pattern illustrated by Sun and McDonough (1989). The intermediate and siliceous dikes and the granodiorite plutons show negative anomalies in Ta, P, and Ti and a strong positive Pb anomaly. Mafic dikes from the Seldovia area are transdicriminant on various tectonic discriminant plots (Lytwyn et al., 2000).

Recent workers have differed only in detail on the petrogenesis of the Sanak-Baranof intrusives. On the basis of

[1]GSA Data Repository item 2003123, Appendix 1, geochemical analytical data for near-trench intrusive rocks, Seldovia quadrangle, Alaska (Table DR1, "Geologic signature of early Tertiary ridge subduction in Alaska"), is available on request from Documents Secretary, GSA, P.O. Box 9140, Boulder, CO 80301-9140, USA, editing@geosociety.org, at www.geosociety.org/pubs/ft2003.htm, or on the CD-ROM accompanying this volume.

TABLE 1. RECENTLY DETERMINED ISOTOPIC AGES OF INTRUSIVE ROCKS FROM THE SANAK-BARANOF BELT

Igneous body	Quad	Field number or location	Latitude (N.)	Longitude (W.)	Age (Ma)	Error (m.y.)	Method and mineral	Distance from Sanak Island (km)	Reference or geochronologist
Lost Cove pluton	SI	93CH-47	57°51'33"	136°24'36"	51.5	0.1	$^{40}Ar/^{39}Ar$ hornblende	2003	C. Taylor, R. Goldfarb, L. Snee, written communication, 2001
Lost Cove pluton	SI	93CH-47	57°51'33"	136°24'36"	52.0	0.1	$^{40}Ar/^{39}Ar$ biotite	2003	C. Taylor, R. Goldfarb, L. Snee, written communication, 2001
Mirror Harbor pluton	SI	92RG-15	57°46'16"	136°18'12"	50.3	0.1	$^{40}Ar/^{39}Ar$ biotite	2007	Taylor et al. (1994)
Mirror Harbor pluton	SI	92RG-14A	57°46'50"	136°18'10"	50.5	0.1	$^{40}Ar/^{39}Ar$ biotite	2007	Taylor et al. (1994)
Lake Elfendahl pluton	SI	93CH-46	57°50'10"	136°18'05"	50.7	0.1	$^{40}Ar/^{39}Ar$ biotite	2005	C. Taylor, R. Goldfarb, L. Snee, written communication, 2001
Lake Elfendahl pluton	SI	79RB062A	57°49'45"	136°17'05"	50.9	0.1	$^{40}Ar/^{39}Ar$ muscovite	2005	L. Snee & S. Karl, unpublished data
White Sisters–Granite Islands pluton at White Sisters	SI	93CH-29	57°38'00"	136°15'10"	49.0	0.1	$^{40}Ar/^{39}Ar$ biotite	2020	C. Taylor, R. Goldfarb, L. Snee, written communication, 2001
White Sisters–Granite Islands pluton at Outer Rocks	SI	93CH-27B	57°34'12"	136°09'10"	49.8	0.1	$^{40}Ar/^{39}Ar$ hornblende	2022	C. Taylor, R. Goldfarb, L. Snee, written communication, 2001
White Sisters–Granite Islands pluton at Granite Islands	SI	D152a	57°35'58"	136°11'51"	46.6	1.4	K/Ar	2025	Decker (1980)
Khaz Bay pluton	SI	D281a	57°34'07"	136°04'40"	48.1	1.4	K/Ar	2030	Decker (1980)
Khaz Bay pluton	SI	92RG-09	57°34'09"	136°04'42"	51.4	0.1	$^{40}Ar/^{39}Ar$ biotite	2030	Taylor et al. (1994)
Crawfish Inlet pluton	PA	99SK121A	56°44'27"	135°04'30"	50.5	0.5	U/Pb zircon	2150	S. Karl, P. Haeussler, R. Friedman, unpublished data
Redfish Bay pluton	PA	99PH567A	56°19'42"	134°47'48"	51.6	0.3	U/Pb zircon	2198	S. Karl, P. Haeussler, R. Friedman, unpublished data

Note: This summary of the most recent isotopic ages of intrusive rocks from the Sanak-Baranof belt, southern Alaska, is plotted in Figure 4. SI—Sitka 1:250,000 quadrangle. PA—1:250,000 quadrangle. Distance along strike was measured from the western tip of Sanak Island on Beikman's (1980) geologic map of Alaska; the map projection causes some distortion. This table updates the original compilation by Bradley et al. (1993), and the more recent supplements by Haeussler et al. (1995) and Bradley et al. (2000).

Sr-isotopes and trace elements, Hill et al. (1981) interpreted the Kodiak, Shumagin, and Sanak granodiorite plutons as the products of mixing between MORB from a subducted spreading center and melted flysch of the overlying accretionary prism. Kyanite and garnet xenocrysts in several of the plutons (Hill et al., 1981) imply that the granodiorite magmas originated at depths of at least 20 km in aluminous metasediments of the accretionary prism. Barker et al. (1992), working with the most complete geochemical and isotopic data set, interpreted three granodiorite plutons near Cordova as the product of partial melting of arc-derived flysch at depth in the accretionary prism. These plutons have the following isotopic compositions: $^{87}Sr/^{86}Sr$ = 0.7051–0.7067, $^{206}Pb/^{204}Pb$ = 19.04–19.20, $^{207}Pb/^{204}Pb$ = 15.60–15.66, $^{208}Pb/^{204}Pb$ = 38.59–38.85, and ε_{Nd} = +2.1 to −3.3—values that overlap with isotopic compositions from the Orca Group flysch and provide evidence that the granodiorites formed by melting of flysch. Barker et al. (1992) suggested that the most likely source of heat, and of mantle melts that formed the gabbroic plutons, was a ridge-transform complex similar to the one presently bounding the Explorer plate off the Cascadia margin. They interpreted dacite dikes in their study area as the products of partial melting of basalts at depth in the accretionary prism. Harris et al. (1996), on the basis of trace-element and Sr- and Nd-isotopic abundances, similarly interpreted felsic dikes in the Chugach metamorphic complex in terms of mixing between melted flysch and a mafic source associated with subduction of the Kula-Farallon ridge. Based on trace elements, Lytwyn et al. (2000) interpreted dikes from the Seldovia quadrangle as representing a spectrum from I-type basalts and andesites to S-type dacites and rhyolites, the more silicic

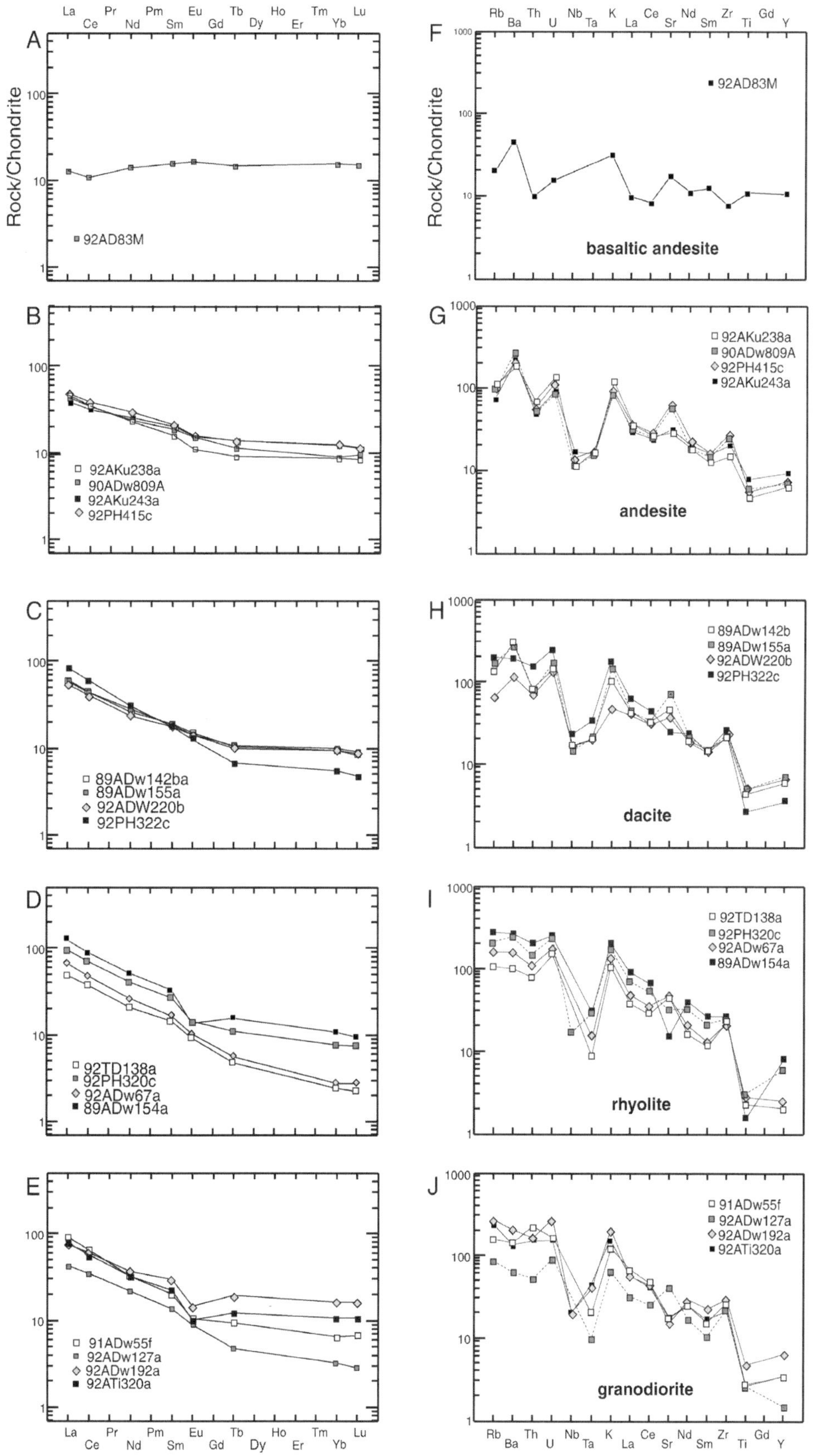

Figure 7. A–E: REE diagrams for dikes and plutons of the Sanak-Baranof belt, Seldovia quadrangle, normalized to chondrite after Sun and McDonough (1989). F–J: Spidergrams of the same samples, normalized to chondrite after Sun (1980). Samples plotted are the least altered representatives of each rock type.

compositions reflecting assimilation of sedimentary protoliths at the base of the accretionary prism.

Structure of the Intrusive Rocks

Intrusive bodies of the Sanak-Baranof belt range from dikes to plugs to batholiths. The larger plutons vary in shape from concordant, elongate sill-like bodies (Kodiak batholith) to discordant, irregularly shaped masses (Aialik pluton). The plutons and dikes were intruded after most of the regional deformation, but some penetrative deformation was coeval with magmatism, and some deformation happened even later. In the Seldovia quadrangle, dikes typically crosscut accretion-related structures (Fig. 8C), showing that near-trench magmatism occurred late in the structural sequence. Most are unfoliated, but a few display a weak tectonic foliation. Similarly, the Nuka and Aialik plutons (locations 4 and 5, Fig. 3) are discordant in most places, truncating folds, cleavage, and melange fabrics, which are all attributed to subduction-accretion (Kusky et al., 1997b). The interiors of these plutons are unfoliated but a tectonic foliation is well developed along the Nuka pluton's concordant landward contact (Fig. 8B) (Kusky et al., this volume, Chapter 12). Contact-metamorphic textures in the adjacent Valdez Group provide additional evidence for syn-emplacement deformation (Fig. 9). In the Chugach metamorphic complex (Fig. 2), dikes, sills, and plutons of the Sanak-Baranof belt have been assigned to four generations, of which the second and third were emplaced syntectonically (Pavlis and Sisson, 1995; Harris et al., 1996).

Dikes are abundant in parts of the Sanak-Baranof belt. A regional compilation map of dike orientations (Fig. 10; from Haeussler et al., this volume, Chapter 5) shows several important features. (1) In southeastern Alaska, dikes strike NW, subparallel to the margin. (2) Dikes in the broad hinge area of the southern Alaska orocline have quite uniform strikes of approximately N-S, at a high angle to the margin but at a varying angle with respect to the local structural grain. (3) In the Seldovia quadrangle, dike strikes are roughly E-W, nearly perpendicular to the structural grain. Separate rose diagrams for mafic to intermediate and felsic dikes in Seldovia quadrangle show slightly different orientations. Except in southeastern Alaska, the net effect of dike emplacement was orogen-parallel extension. The cumulative thickness of dikes measured in the Seldovia quadrangle suggests as much as 1% extension due to dike injection alone.

OBDUCTED EARLY TERTIARY OPHIOLITES AND ASSOCIATED MASSIVE SULFIDE DEPOSITS

Geology and Geochemistry

In Prince William Sound and on Kodiak Island, Paleocene to Eocene turbidites are interbedded with mafic pillow lavas and associated subvolcanic diabase and gabbro, an association that has been interpreted in terms of a ridge-trench encounter. Three of the larger belts include pillow lavas, sheeted dikes, and gabbro and are categorized as ophiolites: the Resurrection Peninsula (Nelson et al., 1989; Kusky and Young, 1999), Knight Island (Richter, 1965; Nelson and Nelson, 1993), and Glacier Island ophiolites. Among worldwide examples of ophiolites, these are important because their geologic relations leave little doubt that they formed along an oceanic ridge and *not* in a supra-subduction setting.

The ophiolites are known or inferred to be late Paleocene in age. Plagiogranite from the Resurrection Peninsula ophiolite has been dated at 57 ± 1 Ma (U/Pb zircon; Nelson et al., 1989). The ophiolites at Knight Island and Glacier Island, although undated, are believed to be about the same age as the Resurrection Peninsula body because they are petrologically similar, and crop out amongst the same turbiditic formations in the same strike belt. In the Resurrection Peninsula ophiolite (Tysdal and Case, 1979; Miller, 1984; Nelson et al., 1985; Bol et al., 1992; Kusky and Young, 1999), the pillow lava succession is interbedded with siliciclastic turbidites, suggesting proximity to a continental sediment source during seafloor spreading. An upward transition from thinly bedded to thickly bedded turbidites has been interpreted as a distal-to-proximal facies change recording the migration of the ophiolite toward the continent (Kusky and Young, 1999). The ophiolite is juxtaposed against the Valdez Group along a thrust (Fox Island shear zone of Kusky and Young, 1999) along which a 53.4 ± 0.9 Ma granodiorite pluton was emplaced (Bradley et al., 2000). Thus, only a few million years elapsed between formation of the ophiolite and its incorporation into the accretionary prism. Trace-element abundances from basalts of the Resurrection and Knight Island ophiolites are broadly MORB-like but show some enrichment in incompatible elements (Lytwyn et al., 1997; Nelson and Nelson, 1993; Crowe et al., 1992). These trends can be explained in terms of contamination of normal MORB with sediment in a near-trench setting.

Basalt flows are interbedded with deep-sea fan deposits that have been assigned to the Upper Cretaceous Valdez Group, the Paleocene-Eocene Orca Group, and the Paleocene Ghost Rocks Formation (Fig. 3). Volcanics of the Orca Group (Nelson et al., 1985; Crowe et al., 1992) are thought to be broadly comparable to the ophiolites described above. Volcanics in the Ghost Rocks Formation of Kodiak Island include andesite and basalt flows interbedded with Paleocene turbidites. Volcanic rocks of the Valdez Group (Nelson et al., 1985; Lull and Plafker, 1990) have not been directly dated but if their nominal Late Cretaceous age assignment is indeed correct, they would appear to be older than events of interest here.

Massive Sulfide Deposits

More than 600 massive sulfide deposits and prospects are known from the Chugach–Prince William terrane in Prince William Sound (Crowe et al., 1992). These deposits are inferred to have formed syngenetically along a spreading ridge that at times was sediment-covered, but at other times, or perhaps along

Figure 8. Photographs; all locations are keyed to Figure 3. A: Kyanite xenocryst, Harris Bay, Aialik pluton (location 5). B: Dike cutting melange that marks the McHugh-Valdez contact in the Seldovia quadrangle, showing that most of the accretion-related deformation of the Maastrichtian Valdez Group predated near-trench magmatism (location 1). C: Dike cutting melange fabrics in the McHugh Complex at Grewingk Glacier, Seldovia quadrangle (location 3) (Bradley and Kusky, 1992). D: Foliation cutting both granodiorite of the Aialik pluton and a late-stage aplite dike, showing that the foliation is of tectonic origin (location 5). E: Dikes cutting McHugh Complex, Seldovia quadrangle (location 2). F: Pillow basalt, Resurrection Peninsula ophiolite (location 6). G: Sheeted dikes, Resurrection Peninsula ophiolite (location 6). H: Interbedded basalt and black shale, Orca Group, Prince William Sound (location 20). I: Migmatite (protolith was Valdez Group), Cordova quadrangle (location 21).

certain sectors, was sediment-starved. Crowe et al. (1992) identified two main deposit types: (1) Cyprus-type deposits hosted by pillow basalts, massive basaltic flows, and tuffs (e.g., Rua Cove deposit; location 7, Fig. 3); and (2) Besshi-type deposits hosted by slate and graywacke, which are interbedded with the basaltic rocks (e.g., Ellamar and Beatson deposits; locations 9 and 17, Fig. 3). The largest deposits, up to 5 million tons, are sediment-hosted. Sulfide assemblages typically include pyrite + arsenopyrite + pyrrhotite + chalcopyrite ± sphalerite. Fluid inclusions indicate temperatures of 170–290 °C, consistent with mineralization in sea-floor hydrothermal systems. Sulfur and oxygen isotopes reported by Crowe et al. (1992) are also consistent with submarine mineralization. Mineralization has not been directly dated but most of the deposits, namely those

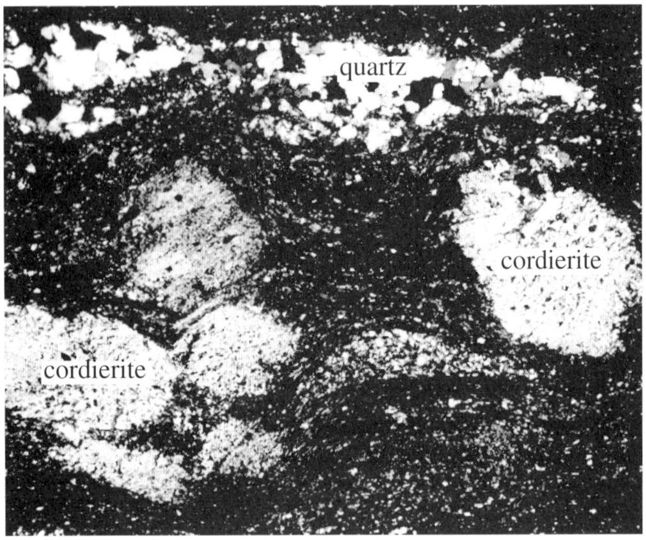

Figure 9. Photomicrograph from the aureole of Nuka pluton showing textural evidence for syntectonic contact metamorphism. Phenocrysts of cordierite surrounded by asymmetric strain shadows of biotite indicate that metamorphism was coeval with deformation. Note weak inclusion trails in cordierite grains that record a layering—either bedding or foliation—that existed prior to contact metamorphism ca. 56 Ma. Regional metamorphic grade is chlorite; cordierite and biotite are found only within the aureoles of the Nuka and allied plutons.

associated with the Knight Island and Glacier Island ophiolites or with volcanic rocks assigned to the Paleocene-Eocene Orca Group, would appear to be roughly the same age as the 57 Ma Resurrection Peninsula ophiolite. The origin of massive sulfide deposits associated with volcanic rocks that have been mapped as part of the Valdez Group remains problematic.

LATE CRETACEOUS–EARLY TERTIARY DEEP-MARINE SEDIMENTATION

Late Cretaceous through Eocene deposition of turbidites that were offscraped or underplated to build the Chugach–Prince William accretionary prism was underway well before, and continued long after, emplacement of the Sanak-Baranof plutons. In the following we will refer to "stage 1" as the ~10 m.y. interval preceding near-trench magmatism at a given location, "stage 2" as the time of near-trench magmatism at that location, and "stage 3" as the ~10 m.y. interval that followed.

Turbidites of stage 1, assigned to the Shumagin and Kodiak Formations, Valdez Group, and Sitka Graywacke, have been interpreted as trench-fill deposits (Moore, 1973; Nilsen and Zuffa, 1982). The Maastrichtian fossil age of these deposits shows that they were deposited 4 to 11 million years before the onset of near-trench magmatism at the western end of the

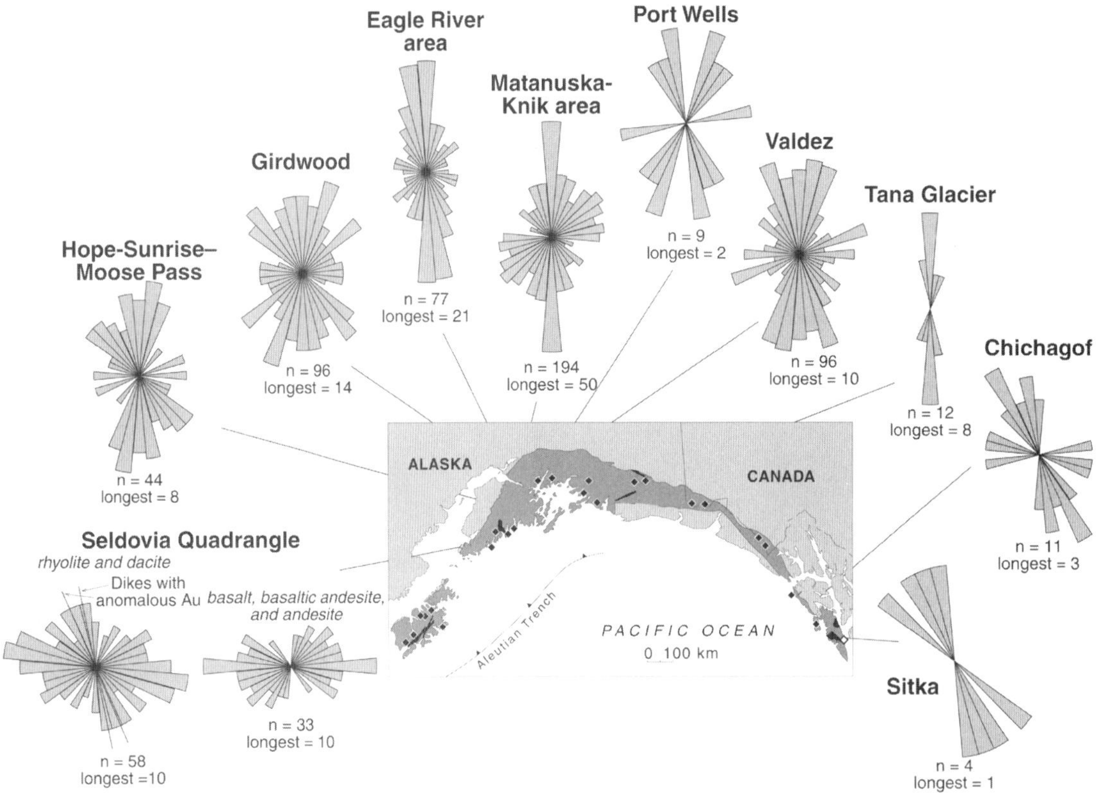

Figure 10. Map showing the strike of dikes in the Sanak-Baranof belt (from Haeussler et al., this volume).

belt. Paleoflow was along the trench axis, mainly toward the western end of the belt (Moore, 1973; Nilsen and Moore, 1979) (Fig. 11).

Sedimentary rocks that are interbedded with volcanic rocks in the Orca Group, the Knight Island and Resurrection Peninsula ophiolites, and the Ghost Rocks Formation correspond to stage 2. Paleocurrent data from the Ghost Rocks Formation indicate turbidity flows toward virtually every point of the compass, in contrast with strong axial paleocurrents of the slightly older Kodiak Formation (Nilsen and Moore, 1979). Some paleocurrent data are also available for the Orca Group (Winkler, 1976), which show flow directions toward the north, west, and south, but not toward the east.

Accreted sedimentary rocks related to stage 3 lie immediately outboard of the Sanak-Baranof near-trench intrusives, the ophiolites, and the volcanic rocks. They are interpreted as closely post-dating the near-trench magmatism episode. On Kodiak Island and adjacent islands, Eocene deep-sea turbidites of this outboard belt have been named the Sitkalidak Formation. Paleocurrent directions reported by Nilsen and Moore (1979) are predominantly toward the southeast (i.e., seaward, perpendicular to the margin).

Sandstone compositions show progressive changes through the time interval of interest, as well as some anomalies corresponding to stage 2 (Fig. 12). In Prince William Sound (Dumoulin, 1987, 1988) sandstones of Stage 1 (Valdez Group and inboard part of the Orca Group) contain abundant volcanic lithic clasts, locally abundant grains of pyroxene and amphibole, and little quartz. They were derived from a progressively unroofing magmatic arc with increasing input from subduction-complex sources through time. Similar compositions have been reported for the Kodiak Formation to the west, and the Sitka Graywacke to the east (e.g., Zuffa et al., 1980). Stage 2 sandstones interbedded with volcanic rocks (locations 6, 7, and 8, Fig. 3) are richer in feldspar and quartz and generally finer grained and better sorted than sandstones in adjacent areas. These differences could be due to barrier effects of local topography. Stage 3 sandstones of the Orca Group on Montague Island (Fig. 3) and the Ragged Mountain area (location 13, Fig. 3) are relatively rich in quartz and sedimentary lithic clasts and correlate well with the Sitkalidak Formation on Kodiak Island to the west (Nilsen and Moore, 1979; Moore et al., 1983). Provenance of these younger rocks reflects increased contributions from subduction complex sources and waning input from a magmatic arc.

GOLD VEINS AND HYDROTHERMAL FLUID-FLOW IN THE ACCRETIONARY PRISM

The Chugach–Prince William terrane is host to dozens of small turbidite-hosted lode-gold deposits. Similarities in structural setting, vein mineralogy, alteration assemblages, stable isotope geochemistry, and fluid inclusion *P-T-*X data for a distance of >1000 km along the continental margin suggest a common origin for these deposits (Goldfarb et al., 1986, 1997). In south-central Alaska, most of the deposits are hosted by rocks of the Valdez Group or the landward part of the Orca Group. In southeastern Alaska, the gold deposits are hosted by the Sitka Graywacke, and by older metamorphic and igneous rocks of the Wrangellia composite terrane within about 5–10 km of the Border Ranges fault system. Many occupy faults, as discussed under the next heading. $^{40}Ar/^{39}Ar$ ages on micas from ten of the gold veins fall along the same age trend as the Sanak-Baranof plutons (Haeussler et al., 1995) (Fig. 4A), requiring a common tectonic setting for magmatism and the regional hydrothermal event.

The gold-bearing veins throughout the Alaskan margin are typically discordant, steeply dipping, high-grade (many 30 g/t Au), and relatively small. The quartz veins contain gold, arsenopyrite, pyrite, and hydrothermal muscovite/sericite. The small size of vein systems suggests that, on the scale of the accretionary prism as a whole, fluid flow was relatively diffuse. The most important producers in south-central Alaska were the Cliff mine (location 18, Fig. 3), which yielded about 1.5 tonnes Au from ore averaging 53 g/t, and the Granite mine (location 19, Fig. 3), which yielded about 0.75 tonnes Au. None of the other small mines in south-central Alaska yielded more than 200,000 g Au. In contrast, the Chichagof/Hirst-Chichagof vein systems in southeastern Alaska contain about 30 tonnes Au (production plus reserves) from ore averaging slightly more than 30 g/t Au. These larger, more productive veins indicate a more voluminous concentration of hydrothermal fluid. Goldfarb et al. (1997) suggested that the onset of transcurrent motion along the southern part of the Border Ranges fault system may have favored establishment of a large second-order fault system that was capable of the more extensive fluid focussing.

The gold deposits are regionally restricted to areas of greenschist facies metamorphic rocks and are only rarely present in lower- or higher-grade metamorphic rocks (Goldfarb et al., 1986, 1997). $\delta^{18}O$ values for quartz from more than 30 deposits range between 13 and 17.5‰; fluid-inclusion microthermometry suggests that veins were deposited from low-salinity aqueous-dominant fluids containing about 5–10 mol% gas at 210–360 °C and 100–300 MPa (1–3 kbar). Quantitative mass spectrometry indicates that the non-aqueous volatiles in the fluids include CO_2, CH_4, N_2, and H_2S. These data are together interpreted to indicate that vein-forming fluids were produced from devolatilization reactions within the previously accreted flysch. Whereas quartz veins (and thus fluid flow) are present throughout the Chugach–Prince William terrane, the restriction of most gold and arsenopyrite to veins in greenschist regions is consistent with desulfidization reactions predominating in these moderate *P-T* regions (e.g., Ferry, 1981). The H_2S made available to crustal fluids during conversion of sedimentary pyrite to pyrrhotite is critical for complexing gold and arsenic in hydrothermal solutions. The H_2S is additionally likely to complex mercury, and transport it into relatively low-*T* crustal regimes. Cinnabar, which has been found in pan concentrates in many drainages in the Seldovia quadrangle (Fig. 2), probably was derived from some of the most shallowly emplaced vein systems.

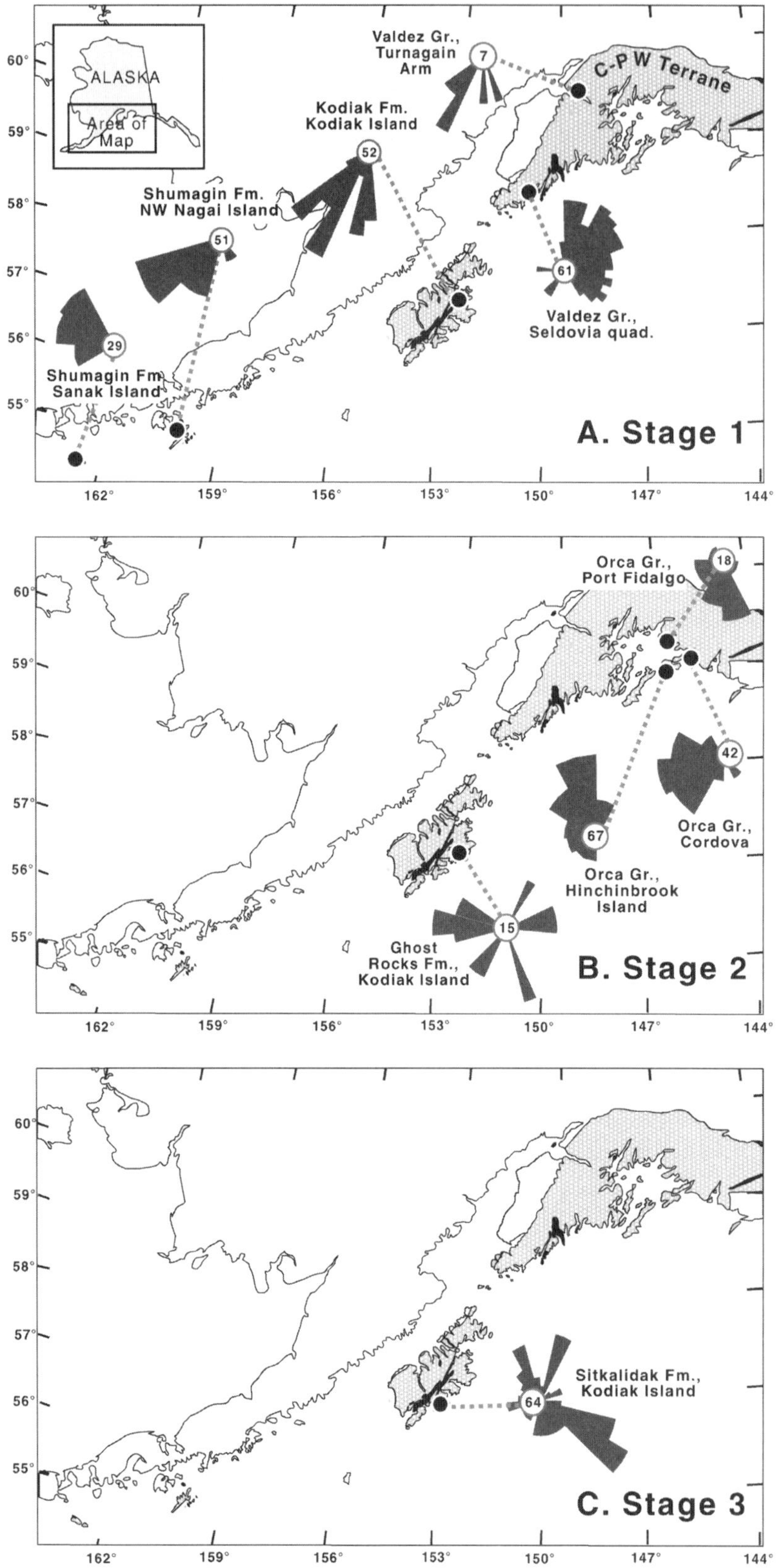

Figure 11. Map of the central portion of the Chugach–Prince William terrane, showing paleocurrent directions from unimodal indicators, mainly turbidite cross laminae and flute casts. A: Stage 1, before near-trench magmatism. B: Stage 2, during near-trench magmatism. C: Stage 3, after near-trench magmatism. Number in each rose diagram corresponds to number of measurements. Rose diagrams for the Shumagin Formation are from Moore (1973); rose diagrams for the Kodiak, Ghost Rocks, and Sitkalidak Formations are derived from Nilsen and Moore (1979); rose diagrams for the Orca Group are from Winkler (1976); rose diagrams for the Valdez Group are based on new data.

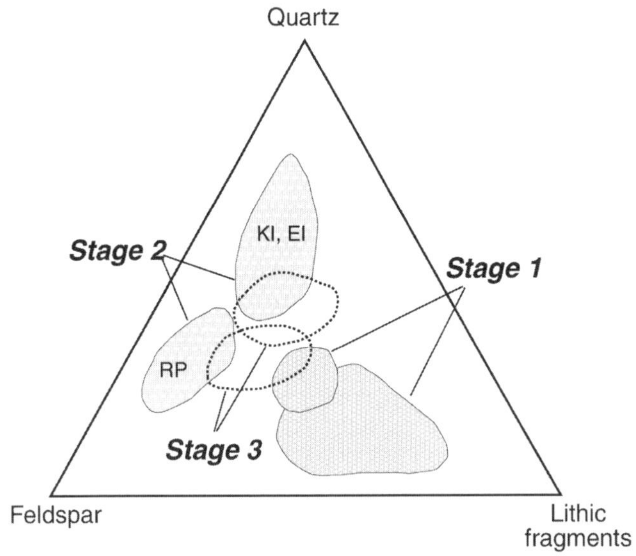

Figure 12. QFL (quartz, feldspar, lithic clasts) diagrams of sandstones deposited before (Stage 1), during (Stage 2), and after (Stage 3) ridge subduction. Abbreviations: RP—Resurrection Peninsula; KI—Knight Island; EI—Ellamar Island. From Dumoulin (1987).

BRITTLE FAULTING

Late brittle faults are common throughout many parts of the Chugach–Prince William terrane, and some of these, at least, were active at the time of near-trench magmatism. Most of the gold-bearing veins described above were developed synkinematically along high-angle faults. $^{40}Ar/^{39}Ar$ sericite ages from these gold lodes thus link fault motion to the time of near-trench magmatism. A structural synthesis of lode-gold deposits along the length of the Chugach–Prince William terrane (Haeussler and Bradley, 1993; Haeussler et al., this volume, Chapter 5) has revealed both common themes and differences in the brittle faulting history. In southeastern Alaska, gold mineralization was coeval with movement on northwest-striking, margin-parallel dextral faults. In south-central Alaska, gold mineralization was associated with normal and strike-slip faults that strike at a high angle to the continental margin and facilitated orogen-parallel extension.

Some of the nonmineralized brittle faults from other parts of the Chugach–Prince William terrane are likely the same age as those associated with gold. Brittle faults are seen at nearly every outcrop along Turnagain Arm near Anchorage (location 16, Fig. 3); they have been described by Bradley and Kusky (1990) and Kusky et al. (1997a), who recognized: (1) a conjugate set of normal faults that strike perpendicular to the structural grain of accreted rocks; (2) a conjugate set of sinistral and dextral strike-slip faults, the acute bisector of which lies perpendicular to regional strike; and (3) a conjugate set of synthetic and antithetic thrusts, parallel to regional strike. Kusky et al. (1997a) interpreted the three sets as components of a single orthorhombic fault population. Curved slickenlines occur on faults of each type. Populations of minor brittle faults that are broadly similar to those along Turnagain Arm have also been reported from Kodiak Island (Byrne, 1984), Afognak Island (Sample and Moore, 1987), and the Seldovia quadrangle (Bradley and Kusky, 1992), as well as the various gold districts (Haeussler et al., this volume, Chapter 5).

The abundance of brittle faults at many places along the length of the Chugach–Prince William terrane, together with evidence that many, at least, are broadly coeval with Sanak-Baranof magmatism, indicates that the accretionary prism underwent bulk, but non-penetrative deformation at the time of near-trench magmatism. In Prince William Sound and the Kenai Peninsula, brittle faulting resulted in a combination of orogen-perpendicular shortening and orogen-parallel extension. In southeastern Alaska, the structural regime was one of dextral strike slip.

METAMORPHISM OF THE ACCRETIONARY PRISM AND ASSOCIATED DUCTILE DEFORMATION

The Chugach–Prince William terrane, unlike most accretionary prisms, is characterized by anomalous areas of relatively high-temperature, low-pressure metamorphism. Typical background metamorphism, attributed to "normal" processes of subduction-accretion, is the low-grade greenschist facies, which in graywacke is recognized by the assemblage chlorite + epidote ± white mica (Miller, in Goldfarb et al., 1986). Fluid-inclusion studies of synmetamorphic veins on Kodiak Island suggest pressures of 200–300 MPa (2–3 kbar) (Vrolijk, 1987). Chlorite geothermometry from quartz veins in the Chugach metamorphic complex suggests 325–350 °C during the greenschist-facies event (Weinberger and Sisson, this volume, Chapter 9).

The largest area of high-grade rocks is the Chugach metamorphic complex (Hudson and Plafker, 1982; Sisson et al., 1989), a 25 × 180 km belt of amphibolite-facies metamorphism affecting the Valdez Group. Mineral assemblages in the center of the complex include biotite + garnet + staurolite + sillimanite (Sisson et al., 1989). Metamorphic temperatures ranged from 600 to 650 °C and pressures ranged from 250 to 350 MPa (2.5–3.5 kbar), corresponding to depths of 7.5 to 10.5 km (Sisson et al., 1989). The high-temperature metamorphism was superimposed on, and took place after the regional low-grade event.

Smaller areas of anomalous metamorphism, also superimposed on the regional low-grade greenschist facies event, have been mapped on Baranof Island (Loney and Brew, 1987; Karl et al., 1999) (Fig. 2), Prince William Sound (Miller, in Goldfarb et al., 1986), the Kenai Peninsula (Bradley et al., 1999), and the Matanuska Glacier area (Little and Naeser, 1989) (Fig. 13). Several of the hotter areas reached amphibolite facies, which is recognized in graywacke protoliths by the relatively low-pressure assemblage biotite > muscovite + andalusite ± cordierite ± garnet (Miller, in Goldfarb et al., 1986). Areas of amphibolite-facies metamorphism in the southern Kenai Peninsula (Seldovia quadrangle; Bradley et al., 1999) are concentrically zoned around ca. 56 Ma plutons and clearly are contact aureoles; metamorphic textures indicate that pluton emplacement was syntectonic (Fig. 10).

Ductile deformation that accompanied high-grade metamorphism has been best documented in the Chugach metamorphic complex, where Pavlis and Sisson (1995) recognized the following structural sequence. The first deformation consisted of imbrication and shortening of Valdez Group turbidites. When traced beyond the Chugach metamorphic complex, this deformation can be seen to have taken place at chlorite grade and to correspond to the main accretion-related deformation that is characteristic of Upper Cretaceous flysch units of the Chugach–Prince William terrane (e.g., Sample and Moore, 1987; Nokleberg et al., 1989). The two subsequent deformations (D2 and D3) were synchronous with the high-*T* metamorphism discussed above. D2 resulted in ductile, orogen-parallel extension, and in vertical shortening, which is indicated by recumbent folding of previously subvertical fabrics. D3 involved dextral transpression. Pavlis and Sisson (1995) identified four generations of Sanak-Baranof intrusives, the oldest of which predated D2 and accompanying metamorphism, and the youngest of which, north-striking dikes, are undeformed and were emplaced after D3.

EARLY TERTIARY EVENTS IN INTERIOR ALASKA

To this point we have focused on the accretionary prism. We now briefly discuss early Tertiary tectonism in areas located 150 to as much as 1000 km inboard of the paleotrench (Fig. 14). Episodes of basin subsidence, magmatism, and uplift took place at about the same time as near-trench magmatism, suggesting the possibility of genetic connections. Precise correlations between events in the accretionary prism and the interior, however, are hampered by uncertainties involving age control and the timing and magnitude of displacements on margin-parallel strike-slip faults (Miller et al., 2003).

Basin Subsidence

Several sedimentary basins in interior Alaska are approximately coeval with near-trench magmatism along the adjacent sector of the Sanak-Baranof belt. The Matanuska subbasin—one of two arms of Cook Inlet forearc basin at its northern end (location 15, Fig. 3)—underwent an episode of extension in early Tertiary time (Little and Naeser, 1989). The basin lay about 150 km landward of the paleotrench during the early Tertiary. Just prior to the time of interest, Upper Cretaceous shales and turbidites of the Matanuska Formation were being deposited in the forearc basin. These marine strata are disconformably overlain by nonmarine conglomerate, sandstone, shale, and coal of the Paleocene to Eocene Chickaloon Formation. Alluvial-fan conglomerates along the southern flank of the basin contain clasts of the McHugh Complex (Little, 1988); the Chugach–Prince William accretionary prism must therefore have risen above sea level by this time. Little and Naeser (1989) showed that north-dipping normal faults were active during deposition of the Chickaloon Formation along the southern basin margin. Three volcanic ash horizons interbedded with coal in the upper part of the Chickaloon Formation have yielded ages of 53.3 ± 1.5 to 55.8 ± 1.7 Ma (plagioclase K/Ar and zircon fission track, respectively; Triplehorn et al., 1984). The Talkeetna Mountains, which flank the Matanuska subbasin on the north, also rose during the early Tertiary. This is recorded by northerly derived boulder conglomerates of the Arkose Ridge Formation (Trop and Ridgway, 1999). The Arkose Ridge is assigned a late Paleocene through early Eocene age based on fossil plants (Silberman and Grantz, 1984) and on conventional K/Ar ages from intercalated basalt (50.0 ± 2.5, 51.8 ± 1.6, and 56.2 ± 1.7 Ma—all whole rock) and rhyolite (50.5 ± 1.5 whole rock, 45.5 ± 1.8 whole rock, and 51.4 ± 1 K-feldspar) (Winkler, 1992). A belt of Jurassic(?) mica schist (Winkler, 1992) just to the north has yielded $^{40}Ar/^{39}Ar$ white-mica ages of 59.7 ± 0.3 Ma (near-plateau) and 56.6 ± 0.2 Ma (plateau) (S. Harland and L. Snee, 2001, written commun.). Simultaneous extension and bimodal volcanism in the Matanuska subbasin and exhumation of metamorphic rocks in flanking highlands (location 22, Fig. 3) suggests that the southern Talkeetna Mountains may represent a core complex. Subject to the limitations of the geochronology, extension in the forearc basin would appear to have been coeval with near-trench magmatism along the adjacent part of the Sanak-Baranof belt (e.g., 54 Ma Crow Pass pluton) (Fig. 4).

Approximately 400 km inboard of the paleotrench, in the Alaska Range, the Cantwell Basin (Fig. 14) entered its second of two evolutionary stages at about the same time. During the first stage in Late Cretaceous time, the Cantwell Basin had been the site of fluvial deposition; it has been interpreted as a thrust-top basin formed in response to northward convergence between the Wrangellia composite terrane and previously accreted terranes to the north (Ridgway et al., 1997). The second stage, during the late Paleocene to earliest Eocene, was marked by mixed mafic and felsic volcanism and nonmarine sedimentation. The volcanic rocks range in age from 59 Ma at the base to 55 Ma at the top ($^{40}Ar/^{39}Ar$ biotite ages; Cole et al., 1999). This second phase of Cantwell Basin evolution broadly correlates with near-trench magmatism, and may be related to an episode of dextral motion along the Denali fault system (Figs. 1 and 14) (Miller et al., 2003).

Approximately 600 km inboard of the paleotrench, the Ruby-Rampart basin (Kirschner, 1994) (Fig. 14) is located along the Victoria Creek fault, a strand of the Kaltag strike-slip fault system. Its fill consists of nonmarine sandstone, conglomerate, coal, basalt, and rhyolite. Several of the volcanic rocks have yielded Paleocene $^{40}Ar/^{39}Ar$ ages: A rhyolite has been dated at 56 Ma (whole rock); a basalt at 60 Ma (whole rock); two mafic dikes at 58 and 59 Ma (whole rock); and two quartz-syenite dikes at 57.2 (biotite) and 57.8 Ma (whole rock) (P. Layer, quoted in Reifenstuhl et al., 1997). The basin is flanked

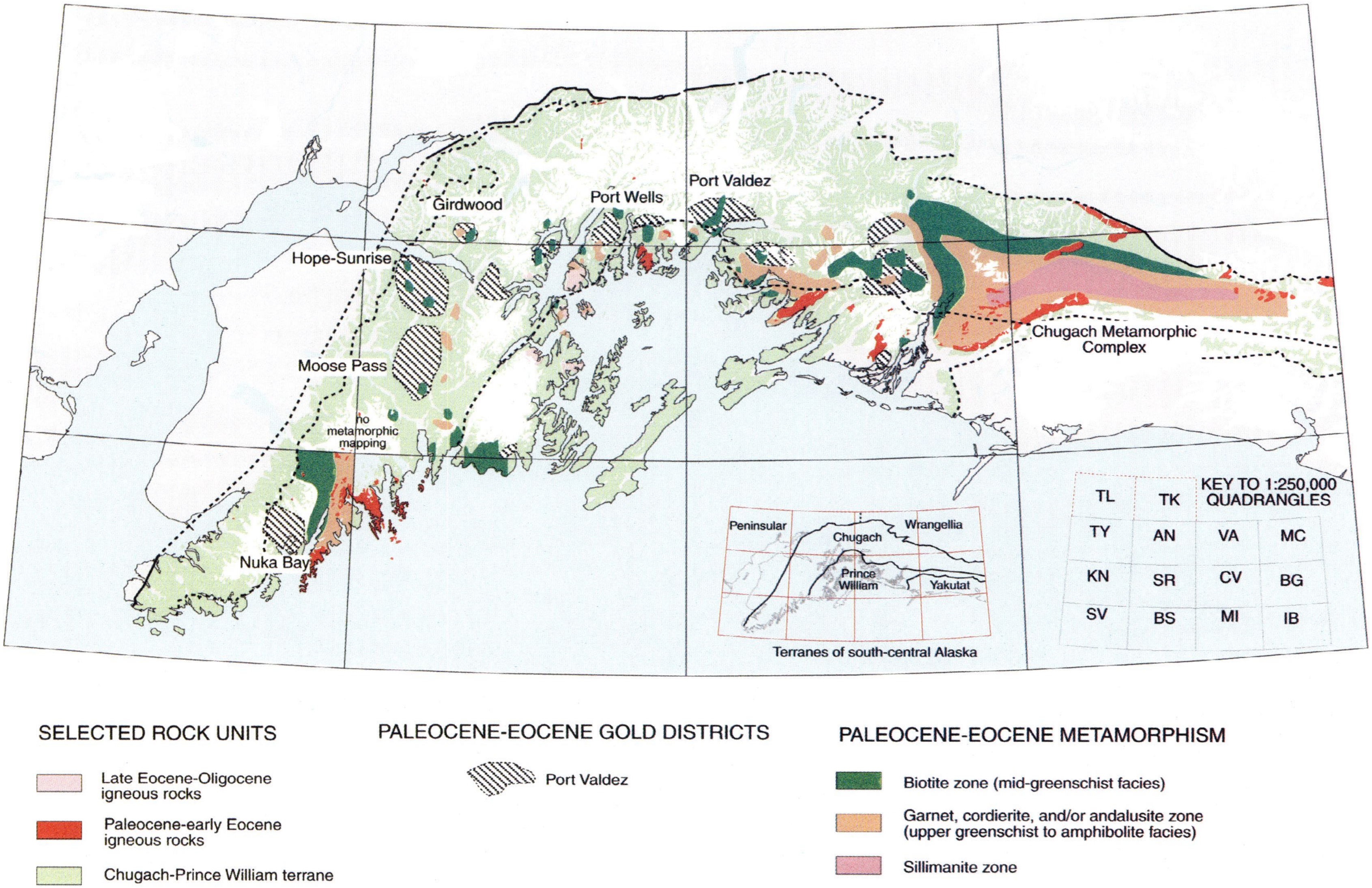

Figure 13. Map of the Chugach–Prince William terrane showing areas of anomalous, high-*T* metamorphism (from Sisson et al., 1989; Miller in Goldfarb et al., 1986; and Bradley et al., 1999), ca. 55 Ma gold districts (from Goldfarb et al., 1986), and near-trench plutonic rocks. The spotty map pattern of isograds results, in part, from the reconnaissance nature of metamorphic mapping.

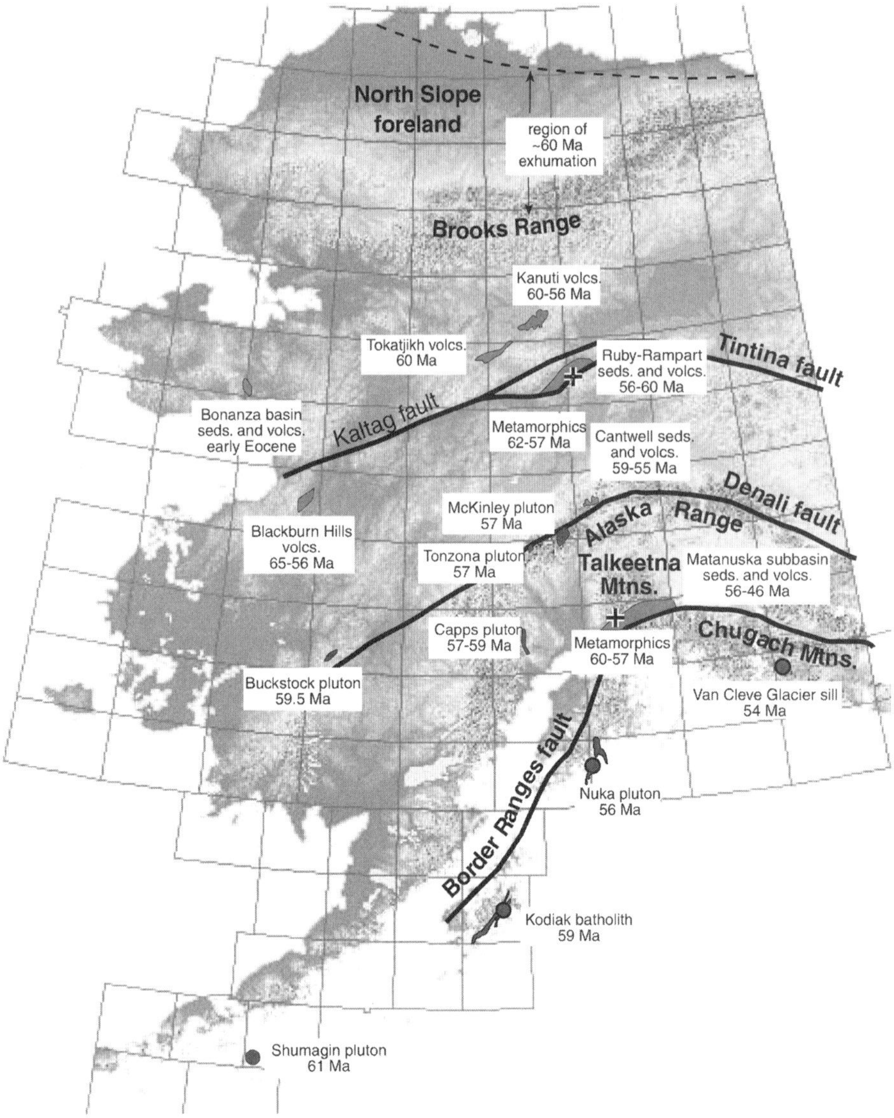

Figure 14. Map of Alaska showing selected Paleocene-Eocene plutons, volcanic centers, basins, metamorphic rocks, and other features. The four best-dated plutons of this portion of the Sanak-Baranof belt show the timing of near-trench magmatism. Note synchronism of various events in interior Alaska and near-trench magmatism. Detailed interpretation of interior events is hampered, however, by the still-controversial offsets across the margin-parallel strike-slip faults. Large (many hundreds of kilometers) dextral offsets since ca. 55 Ma would require comparing a particular location in the backstop with a no-longer adjacent part of the Sanak-Baranof belt.

to the southeast by a fault-bounded belt of anomalously young metamorphic rocks (mica schist, amphibolite, marble, and quartzite), which have yielded $^{40}Ar/^{39}Ar$ ages of 62 Ma (white mica) and 57 Ma (hornblende). Reifenstuhl et al. (1997) interpreted this unit as Ruby terrane or Yukon-Tanana terrane basement (Paleozoic protolith, metamorphosed in Jurassic to Early Cretaceous) that was rapidly exhumed during early Tertiary extension. Simultaneous basin subsidence, bimodal volcanism, and exhumation of metamorphic rocks within a strike-slip fault system suggests a transtensional core complex.

Magmatism

The interval from ca. 75 Ma to 58 Ma was characterized by widespread subduction-generated magmatism in the Alaska Range and backarc magmatism in the Kuskokwim region (Moll-Stalcup, 1994). In the Alaska Range (Fig. 1), the interval from 58 to 55 Ma was marked by emplacement of the McKinley series of plutons—peraluminous granites that Lanphere and Reed (1985) interpreted as products of partial melting of Mesozoic flysch in the unnamed collisional orogen north of the Wrangellia composite terrane. A magmatic lull then ensued, which was to last from ca. 55 to 40 Ma (Wallace and Engebretson, 1984). North of the Alaska Range, the 66–47 Ma Yukon-Kanuti magmatic belt includes several volcanic fields that overlap the age of near-trench magmatism along the adjacent Sanak-Baranof belt. The following description is from Moll-Stalcup (1994) and the rock units discussed are shown in Figure 14. The Kanuti field consists of dacite, andesite, and rhyodacite flows, domes, and tuffs erupted between 59.5 and 55.9 Ma. Similarly, the Tokatjikh field consists of andesite and dacite dated at 59.6 ± 0.6 Ma. The Blackburn Hills field consists of andesite flows dated at 65 Ma near the base, which toward the top are interbedded with rhyolite domes and basalt flows, and intruded by 56 Ma granodiorite. The upper part of the Blackburn Hills succession shows a change from calc-alkalic ("arc-type") to mildly alkalic ("post-arc-type") compositions ca. 56 Ma (Moll-Stalcup, 1994).

Uplift of the Brooks Range and North Slope

Apatite fission track studies reveal that 1000 km inboard of the trench, the Brooks Range and the North Slope foreland basin (Fig. 1) were exhumed during Paleocene time (O'Sullivan et al., 1997). Rocks in the central Brooks Range cooled rapidly through 110–50 °C during discrete episodes at ca. 100 ± 5, ca. 60 ± 4, and ca. 24 ± 3 Ma. Foreland-basin strata just to the north record cooling pulses at ca. 60 ± 4, 46 ± 3, 35 ± 2, and ca. 24 ± 3 Ma. The 60 Ma event came after 40 million years of quiescence that had followed after collision-related Brookian shortening, and is notable because both the orogen and its proximal foreland were affected—a pairing that requires a mechanism other than thrust loading.

EARLY TERTIARY TECTONIC MODELS FOR ALASKA

We are now in a position to more thoroughly consider the various tectonic scenarios mentioned in the Introduction. Any viable model will need to account for the full range of coeval early Tertiary events in the accretionary prism: near-trench magmatism; low-pressure, high-temperature metamorphism; brittle and ductile deformation; ophiolite genesis; and trench sedimentation. Events in the Alaskan interior are less constraining for tectonic models because they can only be loosely tied to events along the coast; a successful tectonic model nonetheless should account for these events, at least qualitatively.

Eight possible explanations for near-trench magmatism—all involving subduction along the Gulf of Alaska margin—are explored below. The final three are variants of each other.

Arc magmatism. Kienle and Turner (1976) interpreted the near-trench plutons as the roots of an arc that had migrated to an extreme near-trench position. Even if the near-trench location might be explained in terms of very steep subduction, the arc model does not account for the diachronous age trend, nor for a coeval magmatic belt ~200–300 km farther inland along what had been the arc axis in latest Cretaceous time (Wallace and Engebretson, 1984). For a brief time, the seaward and landward belts coexisted, then magmatism tapered off in both.

Spontaneous melting of flysch. Hudson et al. (1979) regarded the Sanak-Baranof granitoids as anatectic melts of recently accreted trench sediments. At best, this is only a partial explanation, because the gabbroic plutons and basaltic dikes of the Sanak-Baranof belt—volumetrically minor though they are—require a mantle source. Moreover, spontaneous near-trench melting is not happening at any of the thickly sedimented Neogene accretionary prisms of the modern world (e.g., Makran, Cascadia, Barbados). The spontaneous anatectic melting model also fails to account for the fact that the latest Cretaceous flysch is the same age along the 2200 km length of the belt, yet there is a ten million–year difference between magmatism at the either end of the belt. Finally, this model provides no driving mechanism for deformation of the accretionary prism.

Trench-hotspot encounter. Another possible cause of near-trench magmatism is overriding a sea-floor hot spot. Two lines of evidence argue against this model. The first involves the age and location of magmatism. Overriding a hotspot would be expected to cause magmatism in only one place in an accretionary prism, not along its entire length. Only in the contrived case where the trend of the subduction zone was parallel to the relative motion of the overriding plate in a hot-spot reference frame would near-trench magmatism be diachronous along strike. Second, the distinctive alkalic signature of hotspot magmas are lacking.

Slab breakoff. This idea would entail the failure of what had been a single, subducting oceanic plate. The more deeply subducted part of the plate would have broken away and descended into the deep mantle, leaving, in its wake, a slab window. Slab breakoff has been suggested in a number of collisional orogens—the result of jamming a subduction zone with buoyant lithosphere—but to our knowledge has not been demonstrated for any oceanic subduction zones.

Flexural extension of a subducting plate. Tysdal et al. (1977) suggested that the Resurrection and Knight Island ophiolites formed by a variant of seafloor spreading when a single subducting plate was flexed upon entering a trench. Flexural extension, however, only affects the upper crust of a bending plate, so a clear mechanism for partial melting of the mantle would have been lacking.

Ridge subduction. Subduction of an active ridge-transform system, our preferred model, provides a single, unifying explanation for near-trench magmatism, low-pressure metamorphism, deformation, and hydrothermal activity, geology of the ophiolites, and the history of trench sedimentation and volcanism.

Subduction of very young seafloor. Melting of the accretionary prism might perhaps have been a consequence of subduction of very young ocean floor, but here the distinction blurs with the ridge subduction model, because very young ocean floor implies the nearby presence of even younger (zero-age) ocean floor at a spreading ridge.

Subduction of a "leaky transform." This mechanism, proposed by Tysdal et al. (1977) to explain the Resurrection and Knight Island ophiolites, also is merely a variant on the ridge subduction model because all spreading ridges have transform offsets and fracture-zone extensions.

RIDGE SUBDUCTION–SLAB WINDOW MODEL

On a global mosaic of two dozen large and small plates, divergent and convergent plate boundaries inevitably will intersect. The term "ridge subduction" describes the entry of an oceanic spreading center into a subduction zone; it is not to be confused with "aseismic ridge subduction," which applies to such things as seamount chains that are embedded in a single subducting plate. A spreading center, as such, ceases to exist once it enters a subduction zone. In its place, an ever-widening "slab window" opens downdip of the trench (e.g., Thorkelson, 1996). This window is the interface between the base of the overriding plate and hot asthenosphere that wells up from beneath the two subducted, but still diverging plates. Several triple junction types can result from the intersection between broadly divergent and broadly convergent plate boundaries (Thorkelson, 1996). The simplest and most fundamental type, associated with subduction of ridge segments along a spreading system, is the trench-ridge-trench triple junction (e.g., Figs. 15A and 15C). The Chile triple junction, where South America is overriding the spreading center between the Nazca and Antarctic plates, is the best modern example of the trench-ridge-trench geometry (e.g., Cande and Leslie, 1986). Ridges are commonly offset by transforms, and when one of these enters a trench, the result is a trench-transform-trench triple junction (Fig. 15B). This still qualifies as "ridge subduction" in the loose sense that we favor. Two triple junctions along the western margin of North America represent special cases of "ridge subduction" in which one of the oceanic plates—in this case the Pacific—is moving parallel to the subduction zone, giving rise to trench-ridge-transform and trench-transform-transform triple junctions. The principles of ridge subduction have been examined by DeLong and Fox (1977), Dickinson and Snyder (1979), Thorkelson and Taylor (1989), Hole et al. (1991) and most thoroughly, by Thorkelson (1996).

Subduction of a ridge-transform system seems to account for the wide range of early Tertiary events in Alaska, particularly in the accretionary prism (Fig. 16). Ridge subduction elegantly accounts for the existence and diachronous age progression of near-trench magmatism (Bradley et al., 1993). Because triple junctions migrate, any effects of ridge subduction *should* be diachronous along the strike of a subduction zone, except in the two special cases where the ridge is either exactly parallel or exactly perpendicular to the subduction zone. A subducting ridge accounts for interbedded flysch and pillow lavas in the early Tertiary Resurrection ophiolite (Nelson et al., 1989; Bol et al., 1992; Kusky and Young, 1999), and for submarine volcanics interbedded with the Orca and Ghost Rocks flysch sequences (Moore et al., 1983). Such a ridge could have provided the heat source for the anomalous low-pressure, high-temperature metamorphism in the Chugach metamorphic complex (Sisson and Pavlis, 1993), and a way of generating and moving gold-bearing hydrothermal fluids in the accretionary prism (Haeussler et al., 1995). A ridge-trench encounter is likely to result in deformation, due to indentation of the accretionary prism, to changes in upper-plate rheology, and to changes in subduction direction. Extension and magmatism in the Alaskan interior are entirely consistent with subduction of a ridge and attendant opening of a slab window.

IMPLICATIONS OF THE SANAK-BARANOF TRIPLE JUNCTION FOR PLATE RECONSTRUCTIONS

Identity of Subducted Plates

Although tectonism along the Pacific rim of Alaska since the Cretaceous has undoubtedly been related to plate motions in the Pacific, the identity of the oceanic plates that have bordered Alaska is well established only since about the beginning of the Oligocene. Subduction of a vast tract of oceanic crust has obliterated the older marine magnetic anomalies that would provide the basis for unique plate reconstructions for the Late Cretaceous, Paleocene, and Eocene (Atwater, 1989). All Late Cretaceous to early Tertiary marine-based plate reconstructions show a Kula-Farallon spreading center that must have intersected the western margin North America somewhere—the question is where. Page and Engebretson (1984) handled this uncertainty by showing possible end-member locations for this trench-ridge-trench triple junction, a northern option in Washington (Fig. 17A), and a southern option in Mexico (the choice does not affect Alaska one way or the other, because both are so far south).

The near-trench plutons in Alaska provide onshore geologic evidence that a trench-ridge-trench triple junction existed along the continental margin—but in rocks now located some 3000 km along strike to the north of Page and Engebretson's (1984) northern option. Most workers have nonetheless interpreted the Sanak-Baranof near-trench magmatic belt as the consequence of subduction of the Kula-Farallon ridge (e.g., Marshak and Karig, 1977; Hill et al., 1981; Helwig and Emmett, 1981; Bradley et al., 1993; Haeussler et al., 1995; Sisson and Pavlis, 1993) (Fig. 17B). This is the most obvious choice of ridge as it is the only one that is required to have then existed in the northern Pacific but whose location is subject to conjecture. An alternative—permitted by the vanished magnetic anomaly record—is that yet another spreading ridge was responsible (Bradley et al., 1993; Miller et al., 2003; Haeussler et al., 2000, 2003); such a ridge would have separated the Kula plate from a hypothetical plate to the east (Fig. 17C). We suggest the name Resurrection plate for this hypothesized plate, for the Resurrection Ophiolite, which would be a remnant of it. In the "Resurrection plate" model, evidence for Paleocene-Eocene ridge subduction in Washington would be explained in terms of the Farallon-Resurrection ridge, whereas the Sanak-Baranof belt would be attributed to the Kula-Resurrection ridge. We now lean toward a configuration resembling that in Figure 17C, but acknowledge that a geometry like that in Figure 17B cannot be ruled out. In light of this ambiguity, it is probably best to refer to the site of inferred ridge subduction in Alaska as the "Sanak-Baranof triple junction." It bears repeating that some widely cited reconstructions show no ridge-trench encounter in Alaska between 61 and 50 Ma (e.g., Engebretson et al., 1984, 1985; Wallace and Engebretson, 1984; Plafker and Berg, 1994, their Fig. 5).

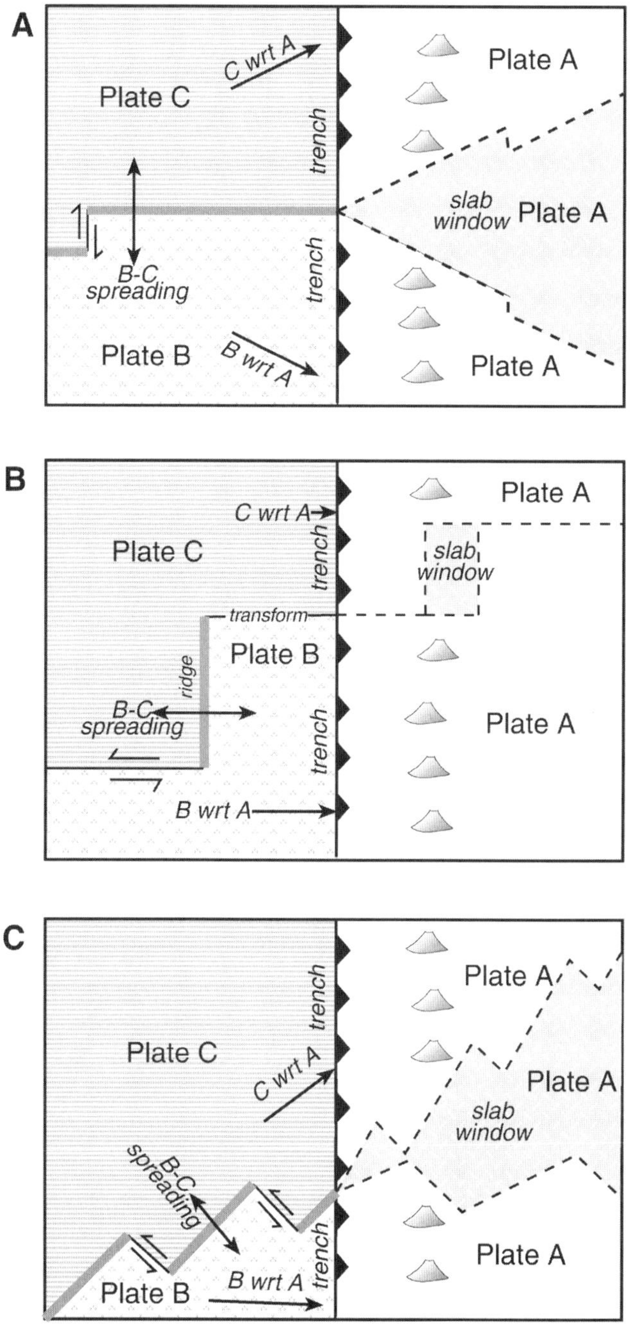

Figure 15. Schematic maps showing hypothetical interactions between ridge-transform systems and a convergent plate boundary. A: Ridge segments are perpendicular to the trench. A slab window opens between the subducted parts of the diverging plates. Were this gap at Earth's surface, new seafloor would have filled it; instead it is filled with hot asthenosphere that wells up to fill the vacancy. The trench-ridge-trench triple junction is stationary but will suddenly jump to a new position when the next transform reaches the trench. B: Ridge segments are parallel to the trench. A transform is being subducted; divergence across a previously subducted ridge segment has formed an isolated slab window at depth. The trench-transform-trench triple junction is stationary but will suddenly jump to a new position when the next ridge segment reaches the trench. C: Ridge segments are at an acute angle to the trench. The irregular shape of the slab window resulted from subduction of transform segments. The trench-ridge-trench triple junction will migrate southward until the next transform enters the trench, whereupon the triple junction will move back in the other direction. *C wrt A* and *B wrt A* are relative motions vectors. Adapted, in part, from Thorkelson (1996).

Palinspastic Complication #1: Oroclinal Bending

Formation of the southern Alaska orocline must also be accounted for in relating the Sanak-Baranof triple junction to plate reconstructions. Paleomagnetic data from a number of sites suggest that during the early Tertiary, southwestern Alaska rotated about 44° ± 11° counterclockwise with respect to cratonic North America (Coe et al., 1985, 1989). The time of oroclinal bending is bracketed between 66 and 44 Ma (Hillhouse and Coe, 1994)—but it is not known how much, if any, of the bending took place before, during, and after ridge subduction. Oroclinal rotation affected about half of the Chugach–Prince William terrane, along with its near-trench plutons that record the position of the Sanak-Baranof triple junction. Two quite different mechanisms for bending have been suggested. One possibility, implied by Coe et al. (1989), is that the hinge and eastern limb remained nearly fixed with respect to stable North America, while the western limb rotated southward. In this scenario, but depending also on the age of northward translation of the Chugach–Prince William terrane, the Sanak pluton could have moved as much as 1000 km *southward* during oroclinal rotation. An alternative possibility (Scholl et al., 1992) is that the orocline formed by indentation of an originally straighter margin, the hinge region corresponding to the indented area. In this case, the Sanak end of the belt would *not* have moved south with respect to North America. This choice between mechanisms is important to early Tertiary tectonic reconstructions because some plutons that track the Sanak-Baranof triple junction might or might not have moved considerable distances—before, during, or after ridge subduction. There are so many possible combinations that the chances of guessing correctly are poor.

Palinspastic Complication #2: Northward Terrane Transport

Paleomagnetic and geologic data each provide evidence, some of it conflicting, on the amount of northward movement of outboard terranes in Alaska. Some paleomagnetic studies have been interpreted to show very large (1000–4000 km) northward translations that were presumably taken up on margin-parallel strike-slip faults (e.g., Irving et al., 1996; Cowan et al., 1997). Most important to the present discussion is the amount of coastwise translation of the Chugach–Prince William

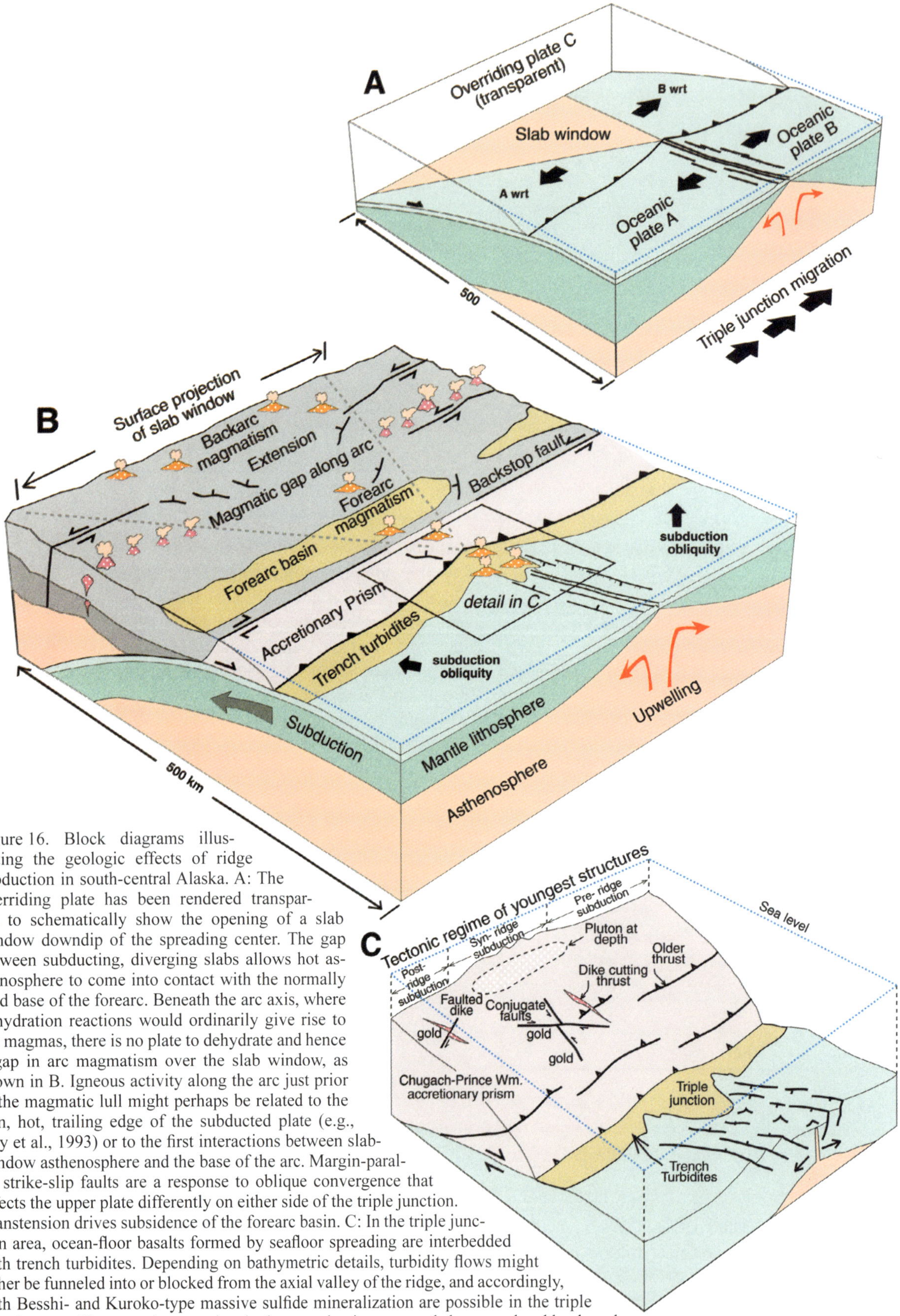

Figure 16. Block diagrams illustrating the geologic effects of ridge subduction in south-central Alaska. A: The overriding plate has been rendered transparent to schematically show the opening of a slab window downdip of the spreading center. The gap between subducting, diverging slabs allows hot asthenosphere to come into contact with the normally cold base of the forearc. Beneath the arc axis, where dehydration reactions would ordinarily give rise to arc magmas, there is no plate to dehydrate and hence a gap in arc magmatism over the slab window, as shown in B. Igneous activity along the arc just prior to the magmatic lull might perhaps be related to the thin, hot, trailing edge of the subducted plate (e.g., Kay et al., 1993) or to the first interactions between slab-window asthenosphere and the base of the arc. Margin-parallel strike-slip faults are a response to oblique convergence that affects the upper plate differently on either side of the triple junction. Transtension drives subsidence of the forearc basin. C: In the triple junction area, ocean-floor basalts formed by seafloor spreading are interbedded with trench turbidites. Depending on bathymetric details, turbidity flows might either be funneled into or blocked from the axial valley of the ridge, and accordingly, both Besshi- and Kuroko-type massive sulfide mineralization are possible in the triple junction area. The accretionary prism is shown as having emerged above sea level landward of the triple junction. The presence of near-trench, forearc volcanoes is inferred by analogy with the Woodlark triple junction; only near-trench plutons and dikes remain at the crustal levels now exposed in Alaska. Brittle faults, some of them mineralized by gold-quartz veins, facilitate orogen-normal shortening (i.e., the narrowing of the prism) and orogen-parallel extension.

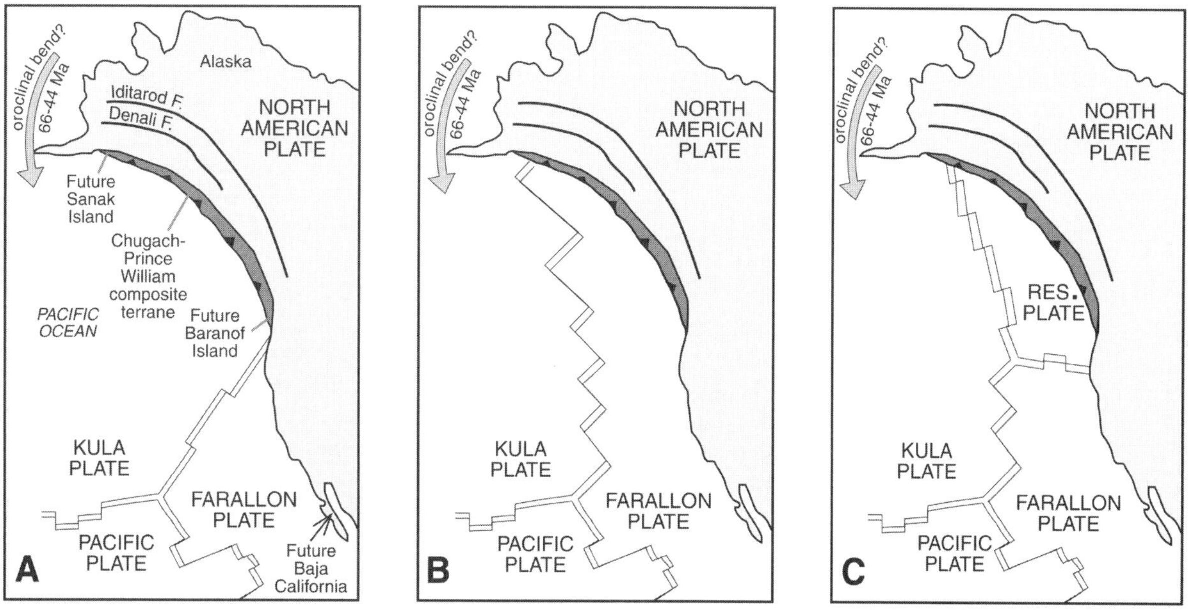

Figure 17. Alternative plate reconstructions. In each, the southern Alaska orocline has been somewhat straightened and the Chugach–Prince William terrane has been restored ~820 km southward along the margin. A: Kula-Farallon ridge intersects North America at the position of Engebretson et al.'s (1985) "northern option." B: Kula-Farallon ridge intersects the Chugach–Prince William terrane (modified by Bradley et al., 1993, from a diagram by Bol et al., 1992). C: Preferred model: a variant on B in which the ridge intersecting the Chugach–Prince William terrane separated the Kula plate from a previously unrecognized plate to the east, which we have named the Resurrection plate (Haeussler et al., 2000; Miller et al., 2003).

terrane (Fig. 1), and particularly its near-trench plutons, which record the position of the Sanak-Baranof trench-ridge-trench triple junction. For reconstructing the global plate mosaic, the key issue is restoration of the Chugach–Prince William terrane with respect to cratonic North America—the craton being the starting point of a multiplate reconstruction circuit. In order to link inboard and outboard events in Alaska, however, a more intricate puzzle must be solved, involving timing and amount of motion on all the significant margin-parallel faults. Two paleomagnetic results are particularly relevant to repositioning the Sanak-Baranof triple junction. Paleomagnetic data have been interpreted to indicate 13° ± 9° of northward displacement of the Chugach–Prince William terrane since 57 Ma (Bol et al., 1992). Allowing for the 325° strike of the margin, this equates to a coast-parallel translation of ~1745 ± 1205 km, or a range of about 540 to 2950 km. It bears noting, however, that Bol et al.'s (1992) paleomagnetic data are from the Resurrection Peninsula ophiolite (Fig. 1), which is interpreted to have been formed at a spreading ridge some distance offshore from earlier accreted parts of the Chugach–Prince William terrane. An unknown, but conceivably large, fraction of the northward travel of the Resurrection Peninsula thus took place before it was even part of the accretionary wedge (Kusky and Young, 1999). Paleomagnetic data from the more inland Wrangellia composite terrane in south-central Alaska show no latitudinal discrepancy with respect to North America as far back as 65–55 Ma (Hillhouse and Coe, 1994).

Geologic evidence from western Canada and eastern Alaska suggests approximately 450 km of dextral displacement across the Tintina fault (Dover, 1994) and approximately 370 km across the Denali fault east of the oroclinal hinge (Lowey, 1998) (Fig. 1). These distances are consistent with the paleomagnetic data from the Chugach–Prince William terrane, but at odds with the paleomagnetic data from the Wrangellia composite terrane, which also lies outboard of the Tintina and Denali faults. Total offset across the Border Ranges fault (i.e., the inboard boundary of the Chugach–Prince William terrane) is debatable, but available evidence allows only minor offset since the Sanak-Baranof belt formed. In southeastern Alaska, mapping by Johnson and Karl (1985) showed that a near-trench intrusion, the ca. 51 Ma Lake Elfendahl pluton (Table 1), crosscuts the fault, sealing this segment since early Eocene. Similarly, Little and Naeser (1989) identified a loose sedimentary linkage across the Border Ranges fault in south-central Alaska, which suggests at most only a few tens of kilometers of dextral offset between the two since the early Eocene. On the other hand, Roeske et al. (this volume, Chapter 6) have reported evidence for perhaps 700–1000 km dextral motion between Late Cretaceous and middle Eocene, and Smart et al. (1996) suggested as much as 600 km dextral offset based on the notion that 120 Ma plutonic rocks found on either side of the fault restore to a single cluster. Acknowledging that no interpretation can satisfy all the geologic and paleomagnetic data, in the following discussion, we use geologic offsets on all faults totaling 820 km.

Plate Geometry and Kinematics

Subject to straightening of the orocline and restoration of 820 km of dextral offset along the margin, Sanak Island would have lain in a position near the present Kenai Peninsula at 61 Ma, the onset of near-trench magmatism. If the Sanak-Baranof triple junction involved the Kula plate and the hypothesized Resurrection plate to the east (Fig. 17C), it would imply that the Resurrection plate would have been subducted before 61 Ma, the Kula plate from 61 to 43 Ma, and the Pacific plate since then. Alternatively, if the Sanak-Baranof triple junction involved the Kula and Farallon plates (Fig. 17B), the Farallon plate would have been subducted beneath Sanak Island until 61 Ma, the Kula plate subducted from 60 to 43 Ma, and the Pacific plate since then. As the ridge-transform system approached the subduction zone, minor plates would undoubtedly have formed; these would have been akin to today's Juan de Fuca and Gorda plates—small remnants of the once-vast Farallon plate.

No direct evidence exists for the configuration of ridges and transforms along the subducted ridge but transform segments almost certainly would have existed. Paleomagnetic declinations from sheeted dikes of the Resurrection Peninsula suggest that the ridge trended 026° ± 12° (Bol et al., 1992); transforms would thus have trended 116° ± 12°. The Sanak-Baranof belt includes "hotter" sectors within which intrusive rocks and/or high-temperature metamorphic rocks are abundant, separated by relatively "colder" areas. This might reflect, at least in part, the subduction of ridge segments versus transform segments. Sisson and Pavlis (1993), Pavlis and Sisson (1995), and Poole (1996) have suggested that some of the complicated history of the Chugach metamorphic complex might have been due to backtracking of the triple junction along the continental margin in connection with ridge-transform complexities.

The nature of tectonism along the margin just before and just after near-trench magmatism suggests that the Sanak-Baranof triple junction was the trench-ridge-trench type, as opposed to, for example, a trench-ridge-transform type. This is best documented on Kodiak Island and adjacent islands. In the latest Cretaceous and earliest Paleocene, deposition of thick turbidite sequences (Kodiak and Ghost Rocks Formations) was followed within a few million years by penetrative deformation, including melange formation, and then by near-trench plutonism. Immediately outboard of the belts containing near-trench intrusions are accreted Eocene turbidites (Sitkalidak Formation) that are themselves deformed and unconformably overlain by undeformed Oligocene strata (Moore and Allwardt, 1980). The style and sequence of deformations of the Kodiak, Ghost Rocks, and Sitkalidak Formations has been widely interpreted in terms of subduction-accretion along a convergent margin (Byrne, 1984; Sample and Moore, 1987).

As a subducting ridge passes a point along a convergent margin, the direction of subduction abruptly changes. Depending on the angle between the ridge and margin, the sense of subduction obliquity might also change. Such is the case at the present-day Chile Triple Junction, which marks a divide between dextral and sinistral margin-parallel strike-slip faults in the forearc (Kay et al., 1993). In Alaska, however, dextral strike slip would appear to have taken place before, during, and after the inferred ridge subduction event. Roeske et al. (this volume, Chapter 6) have described evidence for dextral motion on the Hanagita strand of the Border Ranges fault system between ca. 85 and ca. 51 Ma, which spans the time when the triple junction is inferred to have migrated past this part of the margin. Passage of the Sanak-Baranof triple junction coincided with episodes of margin-parallel strike slip elsewhere, as well. In the Matanuska subbasin (location 15, Fig. 3), Little (1988) has shown that dextral strike slip on the Border Ranges fault system was coeval with emplacement of dacite dikes that he correlated with similar dikes of the Sanak-Baranof belt. In the Chugach metamorphic complex (Fig. 2), D3 deformation took place relatively late in the near-trench high-T sequence, and involved dextral transpression (Pavlis and Sisson, 1995). As discussed earlier, subsidence and volcanism in the Cantwell and Ruby-Rampart basins are consistent with episodes of strike slip on the Denali and Kaltag faults, respectively (Fig. 1). Finally, in southeastern Alaska, gold mineralization on margin-parallel dextral faults was coeval with near-trench magmatism, and all strike-slip faults that postdate gold-quartz veins are dextral. To the extent that margin-parallel strike slip faithfully records subduction obliquity, subduction of both the leading and trailing oceanic plates would appear to have had a dextral-oblique component.

Some recent attempts have been made to fine-tune the plate reconstructions of Engebretson et al. (1984, 1985) in light of information from the Sanak-Baranof belt (e.g., Sisson and Pavlis, 1993; Pavlis and Sisson, 1995; Kusky and Young, 1999). The Engebretson et al. (1985) reconstructions were a major landmark but they are now in need of revision on several counts, irrespective of any new onshore geologic constraints: (1) mounting evidence has called into question the fixed hotspot assumption on which these reconstructions were based (e.g., Molnar and Stock, 1987; Tarduno and Cottrell, 1997); (2) new magnetic anomaly data and picks are available for virtually the entire world ocean, which will support more accurate North America–Africa–Antarctica–Pacific–Kula–Farallon plate circuits; and (3) significant revisions have been made to the magnetic polarity time scale (e.g., Gradstein et al., 1994; Berggren et al., 1995)—the basis for correlation between seafloor spreading history and onshore geology. Until a new generation of oceanic-plate reconstructions is available, improvements suggested by the Sanak-Baranof belt and triple junction are probably best left as hypothetical scenarios (Fig. 17).

IMPLICATIONS FOR RIDGE-SUBDUCTION PROCESSES

Ridge Subduction and Forearc Tectonics

Submarine accretionary prisms and subaerial contractional orogenic belts owe their critical taper to a balance between the strength of the wedge, basal friction, and basal and surface slopes (Davis et al., 1983; Platt, 1986). As a subducting ridge migrates along a convergent margin, several mechanical effects might be expected in an accretionary prism. One effect is simply topographic: At 2.7 km below sea level, a spreading ridge is a high-standing, buoyant object rising far above normal ocean floor. The trench would be expected to shallow, then deepen, as a subducting ridge passes a particular location (DeLong and Fox, 1977). This effect can be seen in the Chile Trench, which systematically shallows toward the Chile triple junction (e.g., Thornburg and Kulm, 1987, their Fig. 2). Judging from the Chilean example, the bathymetric anomaly associated with the subducting ridge in Alaska could have been perhaps 3 km high and several thousand kilometers long. Geologic confirmation of this long-wavelength effect has not been specifically sought in Alaska, but anecdotal observations are at least consistent with what might be expected. Rise of the accretionary prism above sea level during late Paleocene or Eocene is recorded by detritus shed landward into the forearc basin (Little and Naeser, 1989). The paleocurrent patterns of turbidites deposited along or near the continental margin show a profound change at about the time of near-trench magmatism, which could have been due to bathymetric changes associated with the passage of the triple junction.

On a shorter wavelength, the subducting ridge crest itself is likely to induce shortening and uplift in the immediate vicinity of the triple junction. At the Chile triple junction, the ridge indents the inner trench wall (Forsythe and Prior, 1992), and a similar effect can be seen elsewhere in the world where aseismic ridges are presently being subducted (e.g., Cocos Ridge at the Middle American Trench; Corrigan et al., 1990). The structural evidence summarized above shows that margin-perpendicular shortening did take place in the accretionary prism at the time of ridge subduction in Alaska.

Another likely effect of ridge subduction involves a weakening of the accretionary prism as a result of heating, both advectively by rising magmas and conductively from hot asthenosphere of the slab window. The fact that the Chugach–Prince William accretionary prism was heated is self-evident from the extent of near-trench plutons (Fig. 2) and of areas of mid-greenschist and higher metamorphism (Fig. 13). In the Chugach metamorphic complex, D2 recumbent folds that Pavlis and Sisson (1995) attributed to vertical shortening might be attributed to a loss of strength of the prism. There is no evidence that the weakened prism extended in a seaward direction, but it did extend parallel to the margin in areas of both ductile and brittle deformation (Fig. 16). Eventually, as the triple junction moved on, it would have left in its wake an accretionary prism that cooled with time. The final outcome of ridge subduction in Alaska was an accretionary prism that was riddled with isotropic plutons armored by hornfels rims, undoubtedly stronger—and hence capable of sustaining a steeper critical taper—than before.

Magmatism Related to Ridge Subduction

Five igneous provinces in Alaska can be temporally linked to ridge subduction (Fig. 16B). The following list starts at the seaward end of a north-south transect in the vicinity of the oroclinal axis: (1) pillow basalts and andesites that are interbedded with trench deposits; (2) the near-trench plutons and dikes that are the main basis for the ridge-subduction model; (3) volcanics in the forearc basin; (4) the youngest plutons along the arc axis in the Alaska Range that predate a magmatic lull of about 15 million years; and (5) widespread volcanics in a continental backarc region. Each of these has potential analogues in the modern world.

The Knight Island and Resurrection Peninsula ophiolites are presumed to be analogous to ocean floor that is presently being produced at spreading centers that are about to be subducted, such as the Chile Rise and the Gorda Ridge. Sheeted dike complexes attest to an extensional tectonic setting. Interbedded turbidites within the pillow sequence suggest proximity to a trench. Trace-element data can be explained by sediment contamination of N-MORB-like magmas to yield arc-like trace element signatures (Crowe et al., 1992; Nelson and Nelson, 1993; Lytwyn et al., 1997). Off-axis igneous activity in the trench is perhaps represented by basaltic and andesitic flows that are interbedded with flysch sequences and are not associated with sheeted dike complexes. Such anomalous volcanism *on the downgoing plate* is the hallmark feature of ridge subduction east of New Guinea, where the Woodlark spreading center has encountered the Solomon Trench (Taylor and Exon, 1987).

The near-trench intrusive rocks provide the strongest evidence for ridge subduction. Trace-element and isotopic data are best interpreted in terms of melting of flysch at depth in the accretionary prism, the heat having been supplied by upwelling asthenosphere in the slab window. The forearc volcanoes shown in Figure 15 would be the extrusive equivalents of intrusive rocks seen today at the present level of exposure. Modern analogues are known from the Woodlark triple junction, where active volcanoes dot the trench and forearc (Taylor and Exon, 1987), and the Chile triple junction, where a near-trench granodiorite pluton has been dated at 3.6 Ma (Forsythe and Nelson, 1985; Forsythe et al., 1986). Paleocene to Eocene volcanic horizons in the Matanuska subbasin (location 15, Fig. 3) may represent the landward continuation of the Sanak-Baranof igneous province.

The age of arc magmatism in Alaska also can be explained in terms of ridge subduction. Igneous activity along the 74 to 55 Ma Alaska Range magmatic belt was followed by a magmatic lull from 55 to 40 Ma (Wallace and Engebretson, 1984), after which magmatism resumed. This sequence might be interpreted in terms of arc magmatism related to subduction of the leading plate (prior to ridge subduction) followed by passage of a slab window beneath the arc. By analogy, a volcanic gap in the Andean arc landward of the Chile triple junction corresponds closely to the predicted position of the slab window (Kay et al., 1993; Cande and Leslie, 1986). Plutons in the Alaska Range that just preceded the lull (e.g., McKinley Series of granitic rocks of Lanphere and Reed, 1985) would thus correlate with passage of the thin, hot, trailing edge of the subducted slab beneath the arc. These plutons would be analogous to the most southerly volcano in the Southern Volcanic Belt in Chile (Kay et al., 1993), which is expected to be the next volcano to go extinct as the slab window migrates into a position beneath it. An alternative view is that the entire 80–45 Ma history of arc magmatism in Alaska was due to subduction of a single plate, the Kula (Bergman and Hudson, 1987); in our view, this does not account for near-trench magmatism.

Early Tertiary volcanic rocks north of the Alaska Range (e.g., Kanuti, Tokatjikh, and Blackburn Hills fields) correlate in part with magmatism in the adjacent sector of the Sanak-Baranof belt. Kay et al. (1993) and d'Orazio et al. (2000, 2001) have described Neogene continental backarc volcanics in Patagonia above the modern slab window associated with the Chile triple junction. The Patagonian volcanics, which occur as far as ~650 km from the trench, are potential analogs for the Alaskan volcanics.

Far-Field Uplift

Brief mention has been made of a Paleocene episode of exhumation in the Brooks Range and its proximal foreland, which permissibly overlaps the age of Sanak-Baranof magmatism in south-central Alaska. The question is whether or not a slab window could conceivably have affected an area located as far as 1000 km from the paleotrench. Fission track studies from Patagonia, landward of the Chile triple junction, show that Neogene ridge subduction has not caused denudation of the upper plate (Thomson et al., 2001). On the other hand, far-field uplift is predicted by geodynamic models of sublithospheric flow that inevitably takes place in the wedge of asthenosphere above a subducted slab and below an overriding plate (Mitrovica et al., 1989). Modeling revealed that protracted subduction of ocean floor in the Pacific would have caused long-wavelength (>1000 km), high-amplitude (1000 m) depression of both the Canadian Rockies and the associated foreland. Conversely, upon cessation of subduction (e.g., as a result of collision, or opening of a slab window), both orogen and foreland rebounded (Mitrovica et al., 1989).

CLOSING COMMENTS

Ridge subduction is not uncommon. On the present plate mosaic, there are six locations—each with its own geometry—where a ridge-transform system intersects a convergent margin. Moreover, because triple junctions migrate, a single ridge-transform system can eventually work its way along many hundreds to thousands of kilometers of convergent margin—the "moving blowtorch" effect. Ridge subduction was probably even more common in the deep past. For example, when the number of plates was about twice that at present near the end of the Archean (Abbott and Menke, 1990), ridge subduction would have been twice as common. Nonetheless, surprisingly few hypothesized ancient examples have yet been recognized, presumably for want of clear-cut criteria. Alaska is among the most prominent of the Tertiary examples and warrants continued research in the effort to recognize the geologic effects of ridge subduction.

ACKNOWLEDGMENTS

Much of the research reported here was conducted under the Alaska Mineral Resource Assessment Program of the U.S. Geological Survey. Haeussler was supported by a U.S. Geological Survey post-doctoral fellowship from 1992 to 1994. Kusky drew support from NSF Grant 9304647 for studies of near-trench plutonism. Collaborations with Randy Parrish, Larry Snee, Paul Layer, Dan Lux, Matt Heizler, and Bill Clendenen yielded many of the dates on which our interpretations now rest. Figure 3 was derived from digital geology provided in part by Nora Shew and Ric Wilson. Aaron Otteman assisted with geochronological compilations. Mark Brandon, Steve Bergman, Cynthia Dusel-Bacon, Don Richter, and Jinny Sisson provided insightful reviews. We thank editors Jinny Sisson, Sarah Roeske, and Terry Pavlis, for putting together this volume.

REFERENCES CITED

Atwater, T., 1989, Plate tectonic history of the northeast Pacific and western North America, *in* Winterer, E.L., et al., eds., The geology of North America: The eastern Pacific Ocean and Hawaii: Boulder, Colorado, Geological Society of America, Geology of North America, v. N, p. 21–72.

Abbott, D., and Menke, W., 1990, Length of the global plate boundary at 2.4 Ga: Geology, v. 18, p. 58–61.

Barker, F., Farmer, G.L., Ayuso, R.A., Plafker, G., and Lull, J.S., 1992, The 50 Ma granodiorite of the eastern Gulf of Alaska: Melting in an accretionary prism in the forearc: Journal of Geophysical Research, v. 97, p. 6757–6778.

Beikman, H., 1980, Geologic map of Alaska: U.S. Geological Survey, scale 1: 2,500,000.

Berggren, W.A., Kent, D.V., Swisher, C.C., III, and Aubry, M., 1995, A revised Cenozoic geochronology and chronostratigraphy: SEPM (Society for Sedimentary Geology) Special Publication No. 54, p. 144–139.

Bergman, S.C., and Hudson, T.L., 1987, Magmatic rock evidence for a Paleocene change in the tectonic setting of Alaska: Geological Society of America Abstracts with Programs, v. 19, no. 7, p. 586–587.

Bol, A.J., Coe, R.S., Gromme, C.S., and Hillhouse, J.W., 1992, Paleomagnetism of the Resurrection Peninsula, Alaska: Implications for the tectonics of southern Alaska and the Kula-Farallon Ridge: Journal of Geophysical Research, v. 97, p. 17,213–17,232.

Bradley, D.C., and Kusky, T.M., 1990, Kinematics of late faults along Turnagain Arm, Mesozoic accretionary complex, south-central Alaska: U.S. Geological Survey Bulletin 1946, p. 3–10.

Bradley, D.C., and Kusky, T.M., 1992, Deformation history of the McHugh Complex, Seldovia quadrangle, south-central Alaska: U.S. Geological Survey Bulletin 1999, p. 17–32.

Bradley, D.C., Haeussler, P., and Kusky, T.M., 1993, Timing of Early Tertiary ridge subduction in southern Alaska: U.S. Geological Survey Bulletin 2068, p. 163–177.

Bradley, D.C., Kusky, T., Haeussler, P., Karl, S., and Donley, D.T., 1999, Geologic map of the Seldovia quadrangle, Alaska: U.S. Geological Survey Open-File Report 99-18, scale 1:250,000.

Bradley, D.C., Parrish, R., Clendenen, W., Lux, D., Layer, P., Heizler, M., and Donley, D.T., 2000, New geochronological evidence for the timing of early Tertiary ridge subduction in southern Alaska: U.S. Geological Survey Professional Paper 1615, p. 5–21.

Byrne, T., 1984, Early deformation in mélange terranes of the Ghost Rocks Formation, Kodiak Islands, Alaska, *in* Raymond, L.A., ed., Melanges: Their nature, origin, and significance: Boulder, Colorado, Geological Society of America Special Paper 198, p. 21–51.

Cande, S.C., and Leslie, R.B., 1986, Late Cenozoic tectonics of the southern Chile Trench: Journal of Geophysical Research, v. 91, p. 471–496.

Coe, R.S., Globerman, B.R., Plumley, P.R., and Thrupp, G.A., 1985, Paleomagnetic results from tectonic implications, *in* Howell, D.G., ed., Tectonostratigraphic terranes of the circum-Pacific region: Houston, Texas, Circum-Pacific Council for Energy and Mineral Resources, Earth Science Series, v. 1, p. 85–108.

Coe, R.S., Globerman, B.R., Plumley, P.R., and Thrupp, G.A., 1989, Rotation of central and southern Alaska in the Early Tertiary: Oroclinal bending by megakinking?, *in* Kissel, C., and Laj, C., eds., Paleomagnetic rotations and continental deformation: Boston, Kluwer Academic Publishers, NATO-ASI series, p. 327–339.

Cole, R.B., Ridgway, K.D., Layer, P.W., and Drake, J., 1999, Kinematics of basin development during the transition from terrane accretion to strike-slip tectonics, Late Cretaceous–early Tertiary Cantwell Formation, south-central Alaska: Tectonics, v. 18, p. 1224–1244.

Connelly, W., 1978, Uyak Complex, Kodiak Islands, Alaska: A Cretaceous subduction complex: Geological Society of America Bulletin, v. 89, p. 755–769.

Corrigan, J., Mann, P., and Ingle, J.C., Jr., 1990, Forearc response to subduction of the Cocos Ridge, Panama–Costa Rica: Geological Society of America Bulletin, v. 102, p. 628–652.

Cowan, D.S., Brandon, M.T., and Garver, J.I., 1997, Geologic tests of hypotheses for large coastwise displacements—A critique illustrated by the Baja British Columbia hypothesis: American Journal of Science, v. 297, p. 117–173.

Crowe, D.E., Nelson, S.W., Brown, P.E., Shanks, W.C., III, and Valley, J.W., 1992, Geology and geochemistry of volcanogenic massive sulfide deposits and related igneous rocks, Prince William Sound, southern Alaska: Economic Geology, v. 87, p. 1722–1746.

Davis, D., Suppe, J., and Dahlen, F.A., 1983, Mechanics of fold-and-thrust belts and accretionary wedges: Journal of Geophysical Research, v. 88, p. 1153–1172.

Decker, J.E., 1980, Geology of a Cretaceous subduction complex, western Chichagof Island, southeastern Alaska [Ph.D. thesis]: Stanford, California, Stanford University, 135 p.

DeLong, S.E., and Fox, P.J., 1977, Geologic consequences of ridge subduction: American Geophysical Union, Maurice Ewing Series, v. 1, p. 221–228.

Dickinson, W.R., and Snyder, W.S., 1979, Geometry of subducted slabs related to the San Andreas transform: Journal of Geology, v. 87, p. 609–627.

d'Orazio, M., Agostini, S., Mazzarini, F., Innocenti, F., Manetti, P., Haller, M.J., and Lahsen, A., 2000, The Pali Aike Volcanic Field, Patagonia: Slab-window magmatism near the tip of South America: Tectonophysics, v. 321, p. 407–427.

d'Orazio, M., Agostini, S., Innocenti, F., Haller, M.J., Manetti, P., and Mazzarini, F., 2001, Slab window–related magmatism from southernmost South America: The Late Miocene mafic volcanics from the Estancia Glencross area (~52° S, Argentina-Chile): Lithos, v. 57, p. 67–89.

Dover, J.H., 1994, Geology of part of east-central Alaska, *in* Plafker, G., and Berg, H.C., eds., The geology of Alaska: Boulder, Colorado, Geological Society of America, Geology of North America, v. G-1, p. 153–204.

Dumoulin, J.A., 1987, Sandstone composition of the Valdez and Orca Groups, Prince William Sound, Alaska: U.S. Geological Survey Bulletin 1774, 37 p.

Dumoulin, J.A., 1988, Sandstone petrographic evidence and the Chugach–Prince William terrane boundary in southern Alaska: Geology, v. 16, p. 456–460.

Dusel-Bacon, C., Csejtey, B., Jr., Foster, H.L., Doyle, E.O., Nokleberg, W.J., and Plafker, G., 1993, Distribution, facies, ages, and proposed tectonic associations of regionally metamorphosed rocks in east- and south-central Alaska: U.S. Geological Survey Professional Paper 1497-C, p. C1–C72.

Engebretson, D.C., Cox, A., and Gordon, R.G., 1984, Relative motions between oceanic plates of the Pacific Basin: Journal of Geophysical Research, v. 89, p. 10,291–10,310.

Engebretson, D.C., Cox, A., and Gordon, R.G., 1985, Relative motions between oceanic and continental plates in the Pacific Basin: Boulder, Colorado, Geological Society of America Special Paper 206, 59 p.

Ferry, J.M., 1981, Petrology of graphitic sulfide-rich schists from south-central Maine: An example of desulfidation during prograde regional metamorphism: American Mineralogist, v. 66, p. 908–930.

Forsythe, R.D., and Nelson, E.P., 1985, Geological manifestations of ridge collision: Evidence from the Golfo de Penas-Taitao basin, southern Chile: Tectonics, v. 4, p. 477–495.

Forsythe, R.D., Nelson, E.P., Carr, M.J., Kaeding, M.E., Herve, M., Mpodozis, C., Soffia, J.M., and Harambour, S., 1986, Pliocene near-trench magmatism in southern Chile: A possible manifestation of ridge collision: Geology, v. 14, p. 23–27.

Forsythe, R., and Prior, D., 1992, Cenozoic continental geology of South America and its relations to the evolution of the Chile Triple Junction: Proceedings of the Ocean Drilling Program, Initial reports, Volume 141: College Station, Texas, Ocean Drilling Program, p. 23–31.

Goldfarb, R.J., Leach, D.L., Miller, M.L., and Pickthorn, W.J., 1986, Geology, metamorphic setting, and genetic constraints of epigenetic lode-gold mineralization within the Cretaceous Valdez Group, south-central Alaska: Geological Association of Canada Special Paper 32, p. 87–105.

Goldfarb, R.J., Miller, L.D., Leach, D.L., and Snee, L.W., 1997, Gold deposits in metamorphic rocks in Alaska: Economic Geology Monograph 9, p. 151–190.

Gradstein, F.M., Agterberg, F.P., Ogg, J.G., Hardenbol, J., van Veen, P., Thierry, J., and Huang, Zehui, 1994, A Mesozoic time scale: Journal of Geophysical Research, v. 99, p. 24,051–24,074.

Haeussler, P.J., and Bradley, D.C., 1993, Map and compilation of structural data from lode-gold mineral occurrences in the Chugach–Prince William terrane of southern Alaska: U.S. Geological Survey Open-File Report 93-325, 53 p., 1 plate.

Haeussler, P.J., Bradley, D.C., Goldfarb, R., and Snee, L., 1995, Link between ridge subduction and gold mineralization in southern Alaska: Geology, v. 23, p. 995–998.

Haeussler, P.J., Bradley, D.C., Miller, M.L., and Wells, R., 2000, Life and death of the Resurrection plate: Evidence for an additional plate in the NE Pacific in Paleocene-Eocene time: Geological Society of America Abstracts with Programs, v. 32, no. 7, p. A-382.

Haeussler, P.J., Bradley, D.C., Wells, R., and Miller, M.L., 2003, Life and death of the Resurrection plate: Evidence for an additional plate in the northeastern Pacific in Paleocene-Eocene time: Geological Society of America Bulletin, v. 115, p. 867–880.

Haeussler, P.J., Bradley, D.C., and Goldfarb, R.J., 2003, Brittle deformation along the Gulf of Alaska margin in response to Paleocene-Eocene triple junction migration, *in* Sisson, V.B., Roeske, S.M., and Pavlis, T.L., eds, Geology of a transpressional orogen developed during ridge-trench interaction along the North Pacific margin: Boulder, Colorado, Geological Society of America Special Paper 371, p. 119–140 (this volume).

Harris, N.R., Sisson, V.B., Wright, J.E., and Pavlis, T.E., 1996, Evidence for Eocene mafic underplating during forearc intrusive activity, eastern Chugach Mountains, Alaska: Geology, v. 24, p. 263–266.

Helwig, J., and Emmet, P., 1981, Structure of the Early Tertiary Orca Group in Prince William Sound and some implications for the plate tectonic history of southern Alaska: Journal of the Alaska Geological Society, v. 1, p. 12–35.

Hill, M., Morris, J., and Whelan, J., 1981, Hybrid granodiorites intruding the accretionary prism, Kodiak, Shumagin, and Sanak Islands, southwest Alaska: Journal of Geophysical Research, v. 86, p. 10,569–10,590.

Hillhouse, J.W., and Coe, R.S., 1994, Paleomagnetic data from Alaska, *in* Plafker, G., and Berg, H.C., eds., The geology of Alaska: Boulder, Colorado, Geological Society of America, Geology of North America, v. G-1, p. 797–812.

Himmelberg, G.R., Loney, R.A., and Nabelek, P.I., 1987, Petrogenesis of gabbronorite at Yakobi and northwest Chichagof Islands, Alaska: Geological Society of America Bulletin, v. 98, p. 265–279.

Hole, M.J., Rogers, G., Saunders, A.D., and Storey, M., 1991, Relation between alkalic volcanism and slab window formation: Geology, v. 19, p. 657–660.

Hudson, T., 1983, Calc-alkaline plutonism along the Pacific rim of southern Alaska, *in* Roddick, J.A., ed., Circum-Pacific plutonic terranes: Boulder, Colorado, Geological Society of America Memoir 159, p. 159–169.

Hudson, T., and Plafker, G., 1982, Paleogene metamorphism in an accretionary flysch terrane, eastern Gulf of Alaska: Geological Society of America Bulletin, v. 93, p. 1280–1290.

Hudson, T., Plafker, G., and Peterman, Z.E., 1979, Paleogene anatexis along the Gulf of Alaska margin: Geology, v. 7, p. 573–577.

Irving, E., Wynne, P.J., Thorkelson, D.J., and Schiarizza, P., 1996, Large (1000 to 4000 km) northward movements of tectonic domains in the Northern Cordillera, 83 to 45 Ma: Journal of Geophysical Research, v. 101, p. 17,901–17,916.

Johnson, B.R., and Karl, S.M., 1985, Geologic map of western Chichagof and Yakobi Islands, southeastern Alaska: U.S. Geological Survey Map MI-1506, scale 1:125,000, 15 p.

Karl, S.M., Zumsteg, C., Haeussler, P.J., and Himmelberg, G.R., 1999, Slab window metamorphism of the Chugach accretionary complex, Baranof

Island, southeastern Alaska: Eos (Transactions, American Geophysical Union), v. 80, p. F1025.

Kay, S.M., Ramos, V.A., and Márquez, 1993, Evidence in Cerro Pampa volcanic rocks for slab-melting prior to ridge-collision in southern South America: Journal of Geology, v. 101, p. 703–714.

Kienle, J., and Turner, D.L., 1976, The Shumagin-Kodiak batholith—A Paleocene magmatic arc?, *in* Short notes on Alaskan geology, 1976: Alaska Division of Geological and Geophysical Surveys Geologic Report 51, p. 9–11.

Kirschner, C.E., 1994, Interior basins of Alaska, *in* Plafker, G., and Berg, H.C., eds., The geology of Alaska: Boulder, Colorado, Geological Society of America, Geology of North America, v. G-1, p. 469–493.

Kusky, T.M., and Young, C.M., 1999, Emplacement of the Resurrection Peninsula ophiolite in the southern Alaska forearc during a ridge-trench encounter: Journal of Geophysical Research, v. 104, p. 29,025–29,054.

Kusky, T.M., Bradley, D.C., and Haeussler, P., 1997a, Progressive deformation of the Chugach accretionary complex, Alaska, during a Paleogene ridge-trench encounter: Journal of Structural Geology, v. 19, p. 139–157.

Kusky, T.M., Bradley, D.C., Haeussler, P., and Karl, S.J., 1997b, Controls on accretion of flysch and melange belts at accretionary margins: Evidence from the Chugach Bay thrust and Iceworm melange, Chugach accretionary wedge, Alaska: Tectonics, v. 16, p. 855–878.

Kusky, T.M., Bradley, D., Donley, D.T., Rowley, D., and Haeussler, P., 2003, Controls on intrusion of near-trench magmas of the Sanak-Baranof Belt, Alaska, during Paleogene ridge subduction, and consequences for forearc evolution, *in* Sisson, V.B., Roeske, S.M., and Pavlis, T.L., eds, Geology of a transpressional orogen developed during ridge-trench interaction along the North Pacific margin: Boulder, Colorado, Geological Society of America Special Paper 371, p. 269–292 (this volume).

Lanphere, M.A., and Reed, B.L., 1985, The McKinley sequence of granitic rocks: A key element in the accretionary history of southern Alaska: Journal of Geophysical Research, v. 90, p. 11,413–11,430.

LeBas, M.J., LeMaitre, R.W., Streckeisen, A., and Zanettin, B., 1986, A chemical classification of volcanic rocks based on the total alkali silica diagram: Journal of Petrology, v. 27, p. 745–750.

Little, T.A., 1988, Tertiary tectonics of the Border Ranges fault system, north-central Chugach Mountains, Alaska: Sedimentation, deformation, and uplift along the inboard edge of a subduction complex [Ph.D. thesis]: Stanford, California, Stanford University, 343 p.

Little, T.A., and Naeser, C.W., 1989, Tertiary tectonics of the Border Ranges fault system, Chugach Mountains, Alaska: Deformation and uplift in a forearc setting: Journal of Geophysical Research, v. 94, p. 4333–4359.

Loney, R.A., and Brew, D.A., 1987, Regional thermal metamorphism and deformation of the Sitka Graywacke, southern Baranof Island, southeastern Alaska: U.S. Geological Survey Bulletin 1779, 17 p.

Loney, R.A., and Himmelberg, G.R., 1983, Structure and petrology of the La Perouse gabbro intrusion, Fairweather Range, southeastern, Alaska: Journal of Petrology, v. 24, p. 377–423.

Lowey, G.W., 1998, A new estimate of the amount of displacement on the Denali fault system based on the occurrence of carbonate megaboulders in the Dezadeash Formation (Jura-Cretaceous), Yukon, and the Nutzotin Mountains sequence (Jura-Cretaceous), Alaska: Bulletin of Canadian Petroleum Geology, v. 46, p. 379–386.

Lull, J.S., and Plafker, G., 1990, Geochemistry and paleotectonic implications of metabasaltic rocks in the Valdez Group, southern Alaska: U.S. Geological Survey Bulletin 1946, p. 29–38.

Lytwyn, J., Casey, J., Gilbert, S., and Kusky, T., 1997, Arc-like mid-ocean ridge basalt formed seaward of a trench-forearc system just prior to ridge subduction: An example from subaccreted ophiolites in southern Alaska: Journal of Geophysical Research, v. 102, p. 10,225–10,243.

Lytwyn, J., Casey, J., Gilbert, S., and Kusky, T., 2000, Geochemistry of near-trench intrusive rocks associated with ridge subduction, Seldovia quadrangle, southern Alaska: Journal of Geophysical Research, v. 105, p. 27,957–27,978.

Marshak, R.S., and Karig, D.E., 1977, Triple junctions as a cause for anomalously near-trench igneous activity between the trench and volcanic arc: Geology, v. 5, p. 233–236.

Miller, M.L., 1984, Geology of the Resurrection Peninsula, *in* Winkler, G.L., et al., eds., Guide to the bedrock geology of a traverse of the Chugach Mountains from Anchorage to Cape Resurrection: Alaska Geological Society Guidebook, p. 25–34.

Miller, M.L., Bradley, D.C., and Bundtzen, T.K., and McClelland, W., 2002, Late Cretaceous through Cenozoic strike-slip tectonics of southwestern Alaska: Journal of Geology, v. 110, p. 247–270.

Mitrovica, J.X., Beaumont, C., and Jarvis, G.T., 1989, Tilting of continental interiors by the dynamical effects of subduction: Tectonics, v. 8, p. 1079–1094.

Moll-Stalcup, E.J., 1994, Latest Cretaceous and Cenozoic magmatism in mainland Alaska, *in* Plafker, G., and Berg, H.C., eds., The geology of Alaska: Boulder, Colorado, Geological Society of America, Geology of North America, v. G-1, p. 589–619.

Molnar, P., and Stock, J.M., 1987, Relative motions of hotspots in the Pacific, Atlantic, and Indian Oceans since Late Cretaceous time: Nature, v. 327, p. 587–591.

Moore, J.C., 1973, Cretaceous continental margin sedimentation, southwestern Alaska: Geological Society of America Bulletin, v. 84, p. 595–614.

Moore, J.C., and Allwardt, A., 1980, Progressive deformation of a Tertiary trench slope, Kodiak islands, Alaska: Journal of Geophysical Research, v. 85, p. 4741–4756.

Moore, J.C., Byrne, T., Plumley, P.W., Reid, M., Gibbons, H., and Coe, R.S., 1983, Paleogene evolution of the Kodiak Islands, Alaska: Consequences of ridge-trench interaction in a more southerly latitude: Tectonics, v. 2, p. 265–293.

Nelson, S.W., and Nelson, M.S., 1993, Geochemistry of ophiolitic rocks from Knight Island, Prince William Sound, Alaska: U.S. Geological Survey Bulletin 2068, p. 130–142.

Nelson, S.W., Dumoulin, J.A., and Miller, M.L., 1985, Geologic map of the Chugach National Forest, Alaska: U.S. Geological Survey Miscellaneous Field Studies Map MF-1645-B, 14 p., 1 sheet, scale 1:250,000.

Nelson, S.W., Miller, M.L., and Dumoulin, J.A., 1989, Resurrection Peninsula ophiolite, *in* Nelson, S.W., and Hamilton, T.W., eds., Guide to the geology of Resurrection Bay, eastern Kenai Fiords area, Alaska: Anchorage, Geological Society of Alaska, p. 10–20.

Nilsen, T.H., and Moore, G.W., 1979, Reconnaissance study of Upper Cretaceous to Miocene stratigraphic units and sedimentary facies, Kodiak and adjacent islands, Alaska: U.S. Geological Survey Professional Paper 1093, 34 p.

Nilsen, T.H., and Zuffa, G.G., 1982, The Chugach terrane, a Cretaceous trench-fill deposit, southern Alaska: Geological Society [London] Special Publication No. 10, p. 213–227.

Nokleberg, W.J., Plafker, G., Lull, J.S., Wallace, W.K., and Winkler, G.R., 1989, Structural analysis of the southern Peninsular, southern Wrangellia, and northern Chugach terranes along the Trans-Alaska Crustal Transect, northern Chugach Mountains, Alaska: Journal of Geophysical Research, v. 94, p. 4297–4320.

Nokleberg, W.J., Plafker, G., and Wilson, F.H., 1994, Geology of south-central Alaska, *in* Plafker, G., and Berg, H.C., eds., The geology of Alaska: Boulder, Colorado, Geological Society of America, Geology of North America, v. G-1, p. 311–366.

Onstott, T.C., Sisson, V.B., and Turner, D.L., 1989, Initial argon in amphiboles from the Chugach Mountains, southern Alaska: Journal of Geophysical Research, v. 94, p. 4361–4372.

O'Sullivan, P.B., Murphy, J.M., and Blythe, A.E., 1997, Late Mesozoic and Cenozoic thermotectonic evolution of the central Brooks Range and adjacent North Slope foreland basin, Alaska: Including fission-track results from the Trans-Alaska Crustal Transect (TACT): Journal of Geophysical Research, v. 102, p. 20,821–20,845.

Page, B.M., and Engebretson, D.C., 1984, Correlation between the geologic record and computed plate motions for central California: Tectonics, v. 3, p. 133–155.

Pavlis, T.L., and Sisson, V.B., 1995, Structural history of the Chugach metamorphic complex in the Tana River region, eastern Alaska: A record of Eocene ridge subduction: Geological Society of America Bulletin, v. 107, p. 1333–1355.

Plafker, G., and Berg, H.C., 1994, Overview of the geology and tectonic evolution of Alaska, *in* Plafker, G., and Berg, H.C., eds., The geology of Alaska: Boulder, Colorado, Geological Society of America, Geology of North America, v. G-1, p. 989–1021.

Plafker, G., Moore, J.C., and Winkler, G.R., 1994, Geology of the southern Alaska margin, *in* Plafker, G., and Berg, H., eds., The geology of Alaska: Boulder, Colorado, Geological Society of America, Geology of North America, v. G-1, p. 389–449.

Platt, J.P., 1986, Dynamics of orogenic wedges and the uplift of high-pressure metamorphic rocks: Geological Society of America Bulletin, v. 97, p. 1037–1053.

Poole, A.R., 1996, Age and geochemical constraints on ridge subduction for igenous rocks of the eastern Chugach Mountains, Alaska [master's thesis]: Houston, Texas, Rice University, 83 p.

Reifenstuhl, R.R., 1986, Geology of the Goddard Hot Springs area, Baranof Island, southeastern Alaska: Alaska Division of Geological and Geophysical Surveys, Public-Data File 86-2, 82 p., 1 plate, scale 1:40,000.

Reifenstuhl, R.R., Dover, J.H., Newberry, R.J., Clautice, K.H., Liss, S.A., Blodgett, R.B., Bundtzen, T.K., and Weber, F.R., 1997, Interpretive bedrock geologic map of the Tanana B1 quadrangle, central Alaska: Alaska Division of Geological and Geophysical Surveys, Report of Investigations 97-15B, 15 p., scale 1:63,360.

Richter, D.H., 1965, Geology and mineral deposits of central Knight Island, Prince William Sound, Alaska: Alaska Division of Mines and Minerals, Geologic Report 16, 37 p.

Ridgway, K.D., Trop, J.M., and Sweet, A.R., 1997, Thrust-top basin formation along a suture zone, Cantwell basin, Alaska Range: Implications for development of the Denali fault system: Geological Society of America Bulletin, v. 109, p. 505–523.

Roeske, S.M., Snee, L.W., Pavlis, T.L., 2003, Dextral-slip reactivation of an arc-forearc boundary during Late Cretaceous–Early Eocene oblique convergence in the northern Cordillera, *in* Sisson, V.B., Roeske, S.M., and Pavlis, T.L., eds, Geology of a transpressional orogen developed during ridge-trench interaction along the North Pacific margin: Boulder, Colorado, Geological Society of America Special Paper 371, p. 141–169 (this volume).

Sample, J.C., and Moore, J.C., 1987, Structural style and kinematics of an underplated slate belt, Kodiak and adjacent islands, Alaska: Geological Society of America Bulletin, v. 99, p. 7–20.

Scholl, D.W., Stevenson, A.J., Mueller, S., Geist, E., Engebretson, D.C., and Vallier, T.L., 1992, Exploring the notion that southeast-Asian-type escape tectonics and trench clogging are involved in regional-scale deformation of Alaska and the formation of the Aleutian-Bering Sea region, *in* Southeast Asia structure, tectonics and magmatism: Geodynamics Research Institute Symposium, College Station, Texas, Program and Abstracts, Texas A&M University, p. 57–61.

Silberman, M.L., and Grantz, A., 1984, Paleogene volcanic rocks of the Matanuska Valley and the displacement history of the Castle Mountain fault: U.S. Geological Survey Circular 868, p. 82–86.

Sisson, V.B., and Pavlis, T.L., 1993, Geologic consequences of plate reorganization: An example from the Eocene southern Alaska forearc: Geology, v. 21, p. 913–916.

Sisson, V.B., Hollister, L.S., and Onstott, T.C., 1989, Petrologic and age constraints on the origin of a low-pressure/high-temperature metamorphic complex, southern Alaska: Journal of Geophysical research, v. 94, p. 4392–4410.

Smart, K.J., Pavlis, T.L., Sisson, V.B., Roeske, S.M., and Snee, L.W., 1996, The Border Ranges fault system in Glacier Bay National Park, Alaska: Evidence for major early Cenozoic dextral strike-slip motion: Canadian Journal of Earth Sciences, v. 33, p. 1268–1282.

Sisson, V.B., Poole, A.R., Harris, N.R., Cooper Burner, H., Pavlis, T.L., Copeland, P., Donelick, R.A., and McLelland, W.C., 2003, Geochemical and geochronologic constraints for genesis of a tonalite-trondhjemite suite and associated mafic intrusive rocks in the eastern Chugach Mountains, Alaska: A record of ridge-transform subduction, *in* Sisson, V.B., Roeske, S.M., and Pavlis, T.L., eds, Geology of a transpressional orogen developed during ridge-trench interaction along the North Pacific margin: Boulder, Colorado, Geological Society of America Special Paper 371, p. 293–326 (this volume).

Streckeisen, A.L., 1973, Plutonic rocks; classification and nomenclature recommended by the IUGS subcommision on the systematics of igneous rocks: Geotimes, v. 18, p. 26–30.

Sun, S.-S., 1980, Lead isotopic study of young volcanic rocks from mid-ocean ridges, ocean islands, and volcanic arcs: Royal Society of London Philosophical Transactions, v. 297, p. 409–445.

Sun, S.-S., and McDonough, W.F., 1989, Chemical and isotopic systematics of oceanic basalts: Implications for mantle compositions and processes, *in* Saunders, A.D., and Norry, M.J., eds., Magmatism in the ocean basins: Geological Society [London] Special Publication No. 42, p. 313–345.

Tarduno, J.A., and Cottrell, R.D., 1997, Paleomagnetic evidence for motion of the Hawaiian hotspot during formation of the Emperor seamounts: Earth and Planetary Science Letters, v. 153, p. 171–180.

Taylor, B., and Exon, N.F., 1987, An investigation of ridge subduction in the Woodlark-Solomons region: Introduction and overview, *in* Taylor, B., and Exon, N., eds., Marine geology, geophysics, and geochemistry of the Woodlark Basin-Solomon Islands: Houston, Texas, Circum-Pacific Council for Energy and Mineral Resources, Earth Science Series, v. 7, p. 1–24.

Taylor, C.D., Goldfarb, R.J., Snee, L.W., Gent, C.A., Karl, S.M., and Haeussler, P.J., 1994, New age data for gold deposits and granites, Chichagof Mining District, SE Alaska: Evidence for a common origin: Geological Society of America Abstracts with Programs, v. 26, no. 7, p. A-140.

Thomson, S.N., Hervé, F., and Stöckert, B., 2001, Mesozoic-Cenozoic denudation history of the Patagonian Andes (southern Chile) and its correlation to different tectonic processes: Tectonics, v. 20, p. 693–711.

Thorkelson, D.J., 1996, Subduction of diverging plates and principles of slab window formation: Tectonophysics, v. 255, p. 47–63.

Thorkelson, D.J., and Taylor, R.P., 1989, Cordilleran slab windows: Geology, v. 17, p. 833–836.

Thornburg, T.M., and Kulm, L.D., 1987, Sedimentation in the Chile Trench: Depositional morphologies, lithofacies, and stratigraphy: Geological Society of America Bulletin, v. 98, p. 33–52.

Triplehorn, D.M., Turner, D.L., and Naeser, C.W., 1984, Radiometric age of the Chickaloon Formation of south-central Alaska: Location of the Paleocene-Eocene boundary: Geological Society of America Bulletin, v. 95, p. 740–742.

Trop, J.M., and Ridgway, K.D., 1999, Sedimentology and provenance of the Paleocene-Eocene Arkose Ridge Formation, Cook Inlet–Matanuska Valley forearc basin, southern Alaska: Alaska Division of Geological and Geophysical Surveys, Professional Report 119, p. 129–144.

Trop, J.M., Ridgway, K.D., and Spell, T.L., 2003, Sedimentary record of transpressional tectonics and ridge subduction in the Tertiary Matanuska Valley–Talkeetna Mountains forearc basin, southern Alaska, *in* Sisson, V.B., Roeske, S.M., and Pavlis, T.L., eds, Geology of a transpressional orogen developed during ridge-trench interaction along the North Pacific margin: Boulder, Colorado, Geological Society of America Special Paper 371, p. 89–118 (this volume).

Tysdal, R.G., and Case, J.E., 1979, Geologic map of the Seward and Blying Sound quadrangles, Alaska: U.S. Geological Survey Miscellaneous Investigations Series I-1150, 12 p., 1 sheet, scale 1:250,000.

Tysdal, R.G., Case, J.E., Winkler, G.R., and Clark, S.H.B., 1977, Sheeted dikes, gabbro, and pillow basalt in flysch of coastal southern Alaska: Geology, v. 5, p. 377–383.

Vrolijk, P., 1987, Tectonically driven fluid flow in the Kodiak accretionary complex, Alaska: Geology, v. 15, p. 466–469.

Wallace, W.K., and Engebretson, D.C., 1984, Relationships between plate motions and Late Cretaceous to Paleogene magmatism in southwestern Alaska: Tectonics, v. 3, p. 295–315.

Weinberger, J., and Sisson, V.B., 2003, Pressure and temperature conditions of brittle/ductile vein emplacement in the greenschist facies, Chugach metamorphic complex, Alaska: Evidence from fluid inclusions, *in* Sisson, V.B., Roeske, S.M., and Pavlis, T.L., eds, Geology of a transpressional orogen developed during ridge-trench interaction along the North Pacific margin: Boulder, Colorado, Geological Society of America Special Paper 371, p. 217–235 (this volume).

Winkler, G.R., 1976, Deep-sea fan deposition of the lower Tertariy Orca Group, eastern Prince William Sound, Alaska, *in* Miller, T., ed., Recent and ancient sedimentary environments in Alaska: Alaska Geological Society Symposium Proceedings, p. R1–R20.

Winkler, G.R., 1992, Geologic map and summary geochronology of the Anchorage 1° × 3° quadrangle, southern Alaska: U.S. Geological Survey Map I-2283, scale 1:250,000.

Worrall, D.M., 1991, Tectonic history of the Bering Sea and the evolution of Tertiary strike-slip basins of the Bering shelf: Boulder, Colorado, Geological Society of America Special Paper 257, 120 p.

Zuffa, G.G., Nilsen, T.H., and Winkler, G.R., 1980, Rock-fragment petrography of the Upper Cretaceous Chugach terrane, southern Alaska: U.S. Geological Survey Open-File Report 80-713, 28 p.

Manuscript Accepted by the Society February 5, 2003

Printed in the USA

Geological Society of America
Special Paper 371
2003

Large-scale, latest Cretaceous uplift along the northeast Pacific Rim: Evidence from sediment volume, sandstone petrography, and Nd isotope signatures of the Kodiak Formation, Kodiak Islands, Alaska

James C. Sample*
Department of Geology, Northern Arizona University, P.O. Box 4099, Flagstaff, Arizona 86011-4099, USA

Mary R. Reid*
Department of Earth and Space Sciences, University of California, Los Angeles, California 90095-1567, USA

ABSTRACT

The Chugach terrane largely comprises a voluminous flysch belt that extends for over 2000 km along the southern margin of Alaska. Sandstone petrography, paleocurrent patterns, lithofacies, and Nd isotopic characteristics of the flysch belt in Kodiak and adjacent islands constrain the predominant source terrane of the Chugach flysch as a rapidly uplifting recycled orogen with a lesser component of a magmatic arc source.

The Kodiak Formation composes the portion of the Chugach flysch that crops out in Kodiak and adjacent islands. The Kodiak Formation was added to the North American continental margin by underplating in a subduction zone during latest Cretaceous time. It is a thick sequence of deep-water turbidites, most of which are coarse-grained, thickly bedded sandstone-shale sequences interleaved with units of thinly bedded sandstone and shale or nearly massive shale. Several occurrences of *Inoceramus* species, including new finds reported here, confirm that the Kodiak Formation was deposited during early Maastrichtian time, probably in less than four million years. The short absolute time period of deposition, coupled with estimates of the volume of the Chugach flysch, suggests that the original turbidite fan system represented by this unit was immense. Its depositional rate was probably within a factor of two of the rate for the Bengal-Nicobar fan system composed of erosional products of the Himalayan uplift. 118 absolute paleocurrent indicators show two components of sediment transport: a dominant component to the south, highly oblique to the northeast-trending structural grain, with a smaller component oriented north. Petrography of 23 sandstones spanning the width of the formation fall into two distinct categories, suggesting a mixture of provenances. The dominant provenance is a recycled orogen relatively high in quartz ($Q_{49}F_{15}L_{36}$). The lesser provenance is an uplifting magmatic arc rich in lithic volcaniclastic grains ($Q_{10}F_{22}L_{68}$). Nd isotopic signatures of volcaniclastic sandstones (ε_{Nd} = +2 to +5) are significantly more radiogenic than those of quartzose sandstones (ε_{Nd} = 0 to –2). Nd model ages suggest that the source terrane progressed from a recycled orogen, composed in part of Proterozoic materials, to a continental margin volcanic arc.

The combination of sedimentological data indicates that the Kodiak Formation was deposited in a large submarine fan system near its source, filling the trench and spilling out onto the downgoing oceanic plate. The magnitude of uplift in the source region must have been large. Our data alone do not give a unique solution to the source terrane, but

*E-mail, Sample: james.sample@nau.edu. Present address, Reid: Department of Geology, Northern Arizona University, P.O. Box 4099, Flagstaff, Arizona 86011-4099, USA.

Sample, J.C., and Reid, M.R., 2003, Large-scale, latest Cretaceous uplift along the northeast Pacific Rim: Evidence from sediment volume, sandstone petrography, and Nd isotope signatures of the Kodiak Formation, Kodiak Islands, Alaska, *in* Sisson, V.B., Roeske, S.M., and Pavlis, T.L., eds., Geology of a transpressional orogen developed during ridge-trench interaction along the North Pacific margin: Boulder, Colorado, Geological Society of America Special Paper 371, p. 51–70.

considering many aspects of the geology of the North American margin, the most likely candidate is the Coast Mountains belt of western British Columbia, including the Coast Plutonic Complex. This orogen satisfies constraints of sandstone composition, uplift history, and microplate reconstructions, and suggests a major convergent-margin event caused uplift of the westernmost margin of North America to begin during latest Cretaceous time. Deposition and accretion of the Chugach flysch may have occurred substantially south of its current latitude along a convergent margin with significant dextral shear.

Keywords: Kodiak Island, Alaska, tectonics, terranes, Maastrichtian, turbidite, paleocurrents, sedimentary petrology, sedimentology, neodymium isotopes, Kodiak Formation.

INTRODUCTION

The Chugach terrane is a belt of accreted deep-marine sedimentary rocks, which extends for 2000 km across southern Alaska (Fig. 1). The landwardmost portion of this terrane is a mélange comprising pelagic sedimentary rocks and volcanic rocks, but an Upper Cretaceous slate and graywacke belt composes the seawardmost 90% of the terrane, hereafter referred to as the Chugach flysch (the "Chugach flysch terrane" of Nilsen and Zuffa, 1982). Structures in the Chugach flysch evolved by underplating at a subduction zone (Moore, 1973; Plafker et al., 1977; Sample and Fisher, 1986; Sample and Moore, 1987; Fisher and Byrne, 1987), but the location of this subduction-accretion along the North American margin is in dispute. Lithofacies patterns, petrographic data, Nd isotope data, and paleocurrent data have been interpreted to show a source region for the slate belt in south-central British Columbia, with detritus transported hundreds to thousands of kilometers along a continuous trench before final subduction-accretion at the southern Alaska margin (Nilsen and Moore, 1979; Farmer et al., 1993). One implication of this model is that accretion occurred at the Chugach terrane's current position relative to North America. An alternative view is that accretion of the Chugach flysch occurred approximately 2000 km south of its expected position

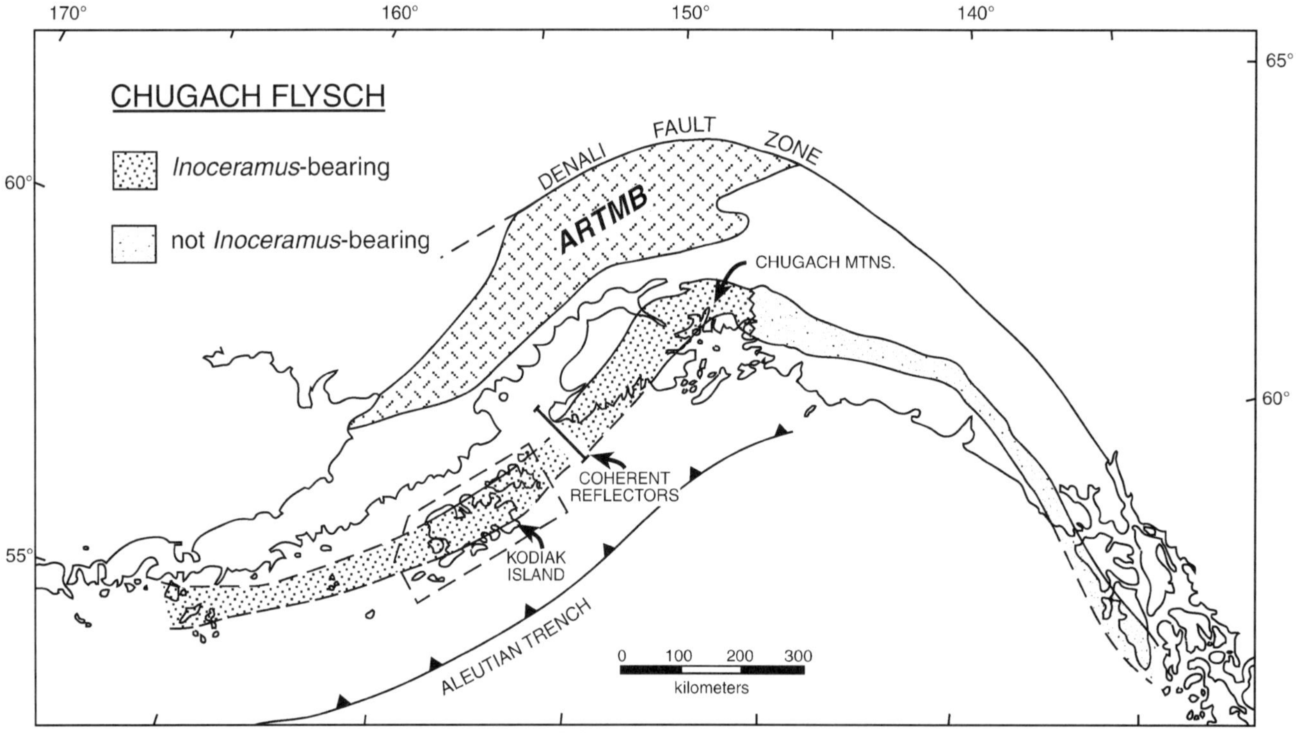

Figure 1. Location of Chugach terrane and study area in southern Alaska. The Chugach flysch composes approximately the southern 80%–90% (in outcrop width) of the Chugach terrane. Only the western half of the slate and graywacke belt contains *Inoceramus kusiroensis* and *Inoceramus balticus* diagnostic of the early Maastrichtian age. Location of Figure 2 shown by box around Kodiak and adjacent islands. Location of seismic reflection line with coherent reflectors (Event A) used to constrain geometry of flysch in Figure 4 is also shown. ARTMB denotes belt of Upper Cretaceous to Paleocene magmatic and volcanic rocks (modified from Hudson et al., 1979; Wallace and Engebretson, 1984).

relative to North America, based on paleomagnetic data from the Paleocene volcanic rocks on Kodiak Island and along strike in Prince William Sound (Moore et al., 1983; Plumley et al., 1983; Bol et al., 1992; Coe et al., 1985). By virtue of a plutonic tie between the Paleocene volcanic rocks and Chugach terrane to the northwest, the Chugach terrane on Kodiak Island also must have been at lower latitudes during its deposition and accretion (Davies and Moore, 1984).

To help resolve the provenance of the Chugach flysch and the location of its accretion, this study focuses on sedimentology and Nd isotopic characteristics of the Kodiak Formation, the part of the Chugach flysch located on Kodiak and adjacent islands (hereafter referred to as the "Kodiak islands"). Field data presented here were collected during mapping of a transect through the Kodiak Formation at a scale of 1:25,000 (Fig. 2). The large volume of sediment and apparently short depositional time interval of this slate belt suggests very high sedimentation rates that were the result of special circumstances in western North American paleogeography during latest Cretaceous time. We use the available data to discuss possible depositional settings and the tectonics of possible source terranes that could have combined to form such a voluminous flysch belt in a very short period of time.

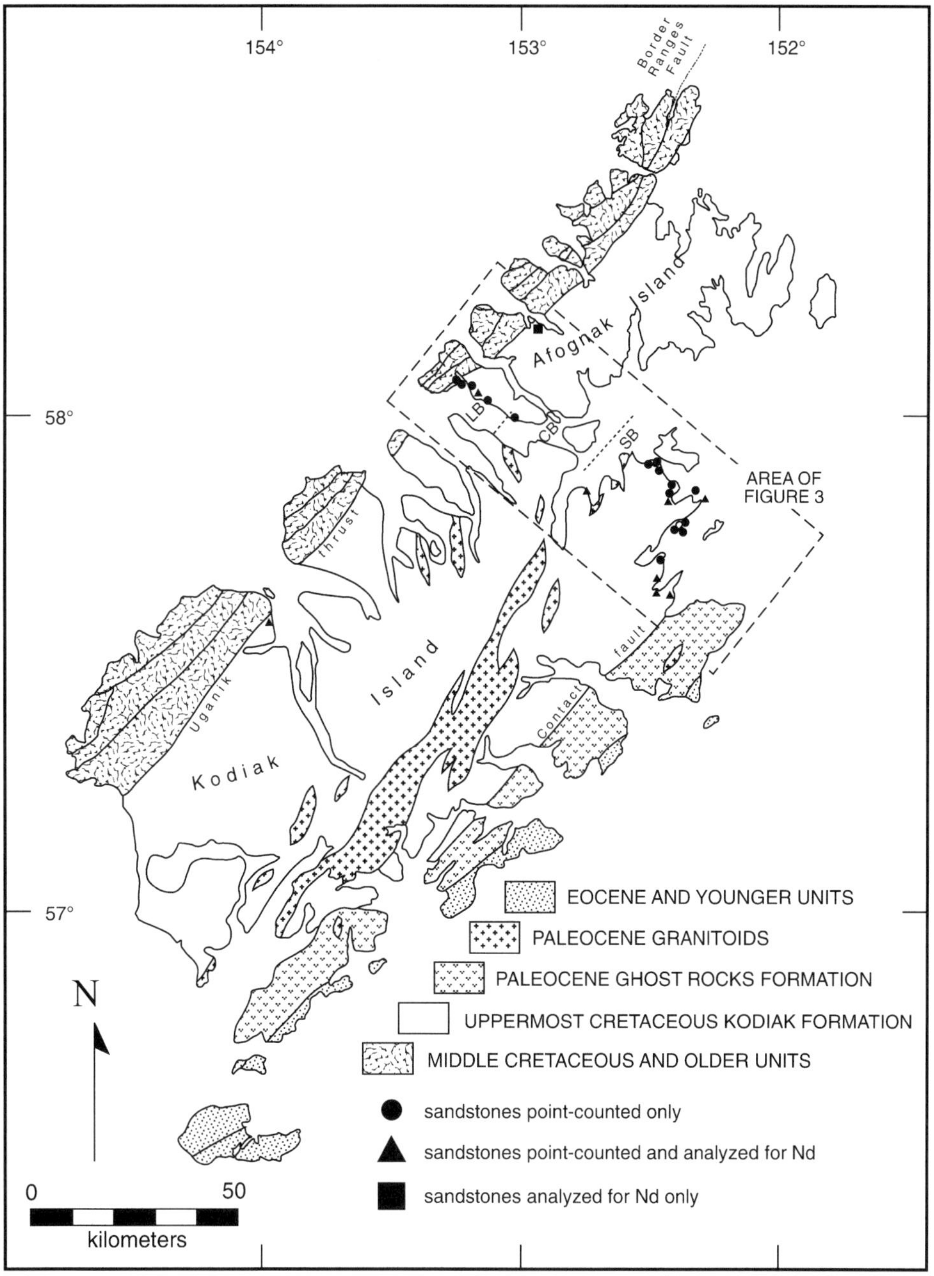

Figure 2. Geologic map of the Kodiak accretionary complex. The Chugach flysch here is the Kodiak Formation. The landward belt (LB), central belt (CB), and seaward belt (SB) are divided by dashed lines. Geology after Sample and Moore (1987) and references therein. Location of Figure 3 shown by box.

TECTONIC SETTING

The Kodiak Formation is part of an accretionary complex that ranges in age from Jurassic to Oligocene and decreases in metamorphism from blueschist to zeolite facies in a seaward direction (Fig. 2). All five units of the complex were assembled by underplating or offscraping in an accretionary wedge (Byrne, 1986; Connelly, 1978; Moore and Allwardt, 1980; Roeske, 1986; Sample and Moore, 1987). Granodioritic or tonalitic plutons and dikes of Paleocene age intrude all units of the complex, but regional metamorphism does not appear to be related to this event (Davies and Moore, 1984; Sample and Moore, 1987).

The Chugach terrane represents the older part of the accretionary complex and consists of a landward Jurassic to middle Cretaceous mélange belt (Raspberry-Seldovia-Iceberg schists and Uyak-McHugh complex) and a seaward Upper Cretaceous slate belt (Shumagin Formation, Kodiak Formation, Valdez Group, Sitka Graywacke) that were accreted to North America in middle(?) and Late Cretaceous time, respectively (Plafker et al., 1977; Connelly, 1978; Decker et al., 1979; Moore et al., 1983; Page et al., 1986; Sample and Moore, 1987; Plafker et al., 1994).

The Kodiak Formation consists of slates with southeast-verging folds and thrust faults related to accretion. Systematic variations in dips of these structures have been used to divide the formation into three structural belts (Fig. 3) (Sample and Fisher, 1986; Fisher and Byrne, 1987). Syn-deformation prehnite-pumpellyite facies metamorphism and fluid inclusions from syntectonic quartz veins suggest that the Kodiak Formation was deformed at >12 km depth as it was underplated to an accretionary wedge as duplexes (Sample and Fisher, 1986; Sample and Moore, 1987; Vrolijk et al., 1988). Because of the low grade of metamorphism and coherent style of underplating, sedimentary structures are well preserved in most of the Kodiak Formation.

ESTIMATED SIZE OF CHUGACH FLYSCH

The areal extent and presumed structural thickness of the Chugach flysch indicates that a large volume of sediment was

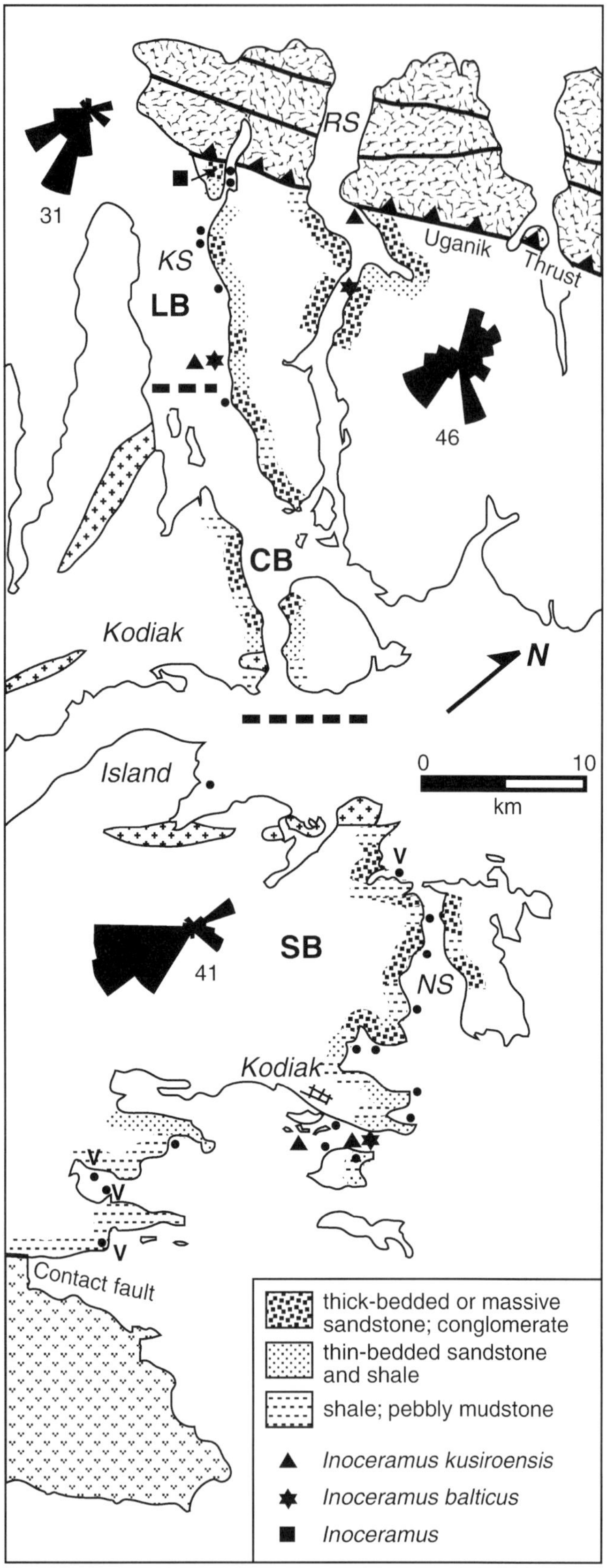

Figure 3. Geologic map of transect. Kodiak Formation is unpatterned except along mapped portions of coastline, where lithofacies are shown. Bold, dashed lines separate landward belt (LB), central belt (CB), and seaward belt (SB). Coquille pattern of northwestern units denotes all middle Cretaceous and older rocks. Paleocene Ghost Rocks Formation shown by "V" pattern. Paleocene granitic rocks shown with crosses. Paleocurrent rose diagrams are totals of absolute indicators for Kupreanof Strait (KS), Raspberry Strait (RS), and the seaward portion of Kodiak Island and Narrow Strait (NS). Total number of data is shown for each section. Paleocurrents corrected to horizontal by two-tilt method (Ramsay, 1961). Small filled circles are locations of point-counted sandstones; large Vs denote location of volcaniclastic sandstones in Kodiak Formation. All fossil locales are from this study except for large island east of Kodiak (Woody Island). Fossils identified by J. Miller and D. Jones of the U.S. Geological Survey.

delivered to the underthrusting oceanic plate in latest Cretaceous time. A direct calculation of sedimentation rate during deposition of the Chugach flysch cannot be made because of multiple repetition of strata by thrust faults and the lack of a distinctive stratigraphy. Nevertheless, comparison of the flysch to modern turbidite-filled basins with known sedimentation rates can be made by considering cross-sectional areas of the different turbidite fans and normalizing these areas to stratigraphic ages. Cross-sectional areas have been chosen rather than volumes because good three-dimensional control on the geometry of the Chugach flysch is not available, and geophysical surveys of modern turbidite fans considered here are two-dimensional.

Kodiak Formation Volume

The Kodiak Formation is the portion of the Chugach flysch with the best age control on the time interval of deposition. The age range of the Kodiak Formation is constrained by two species of mollusc: *Inoceramus kusiroensis* and *Inoceramus balticus* (Table 1; Fig. 3). Both appear to have been limited to early Maastrichtian time based on correlations between Japanese and European stages (Nagao and Matsumoto, 1940; Jones and Clark, 1973; J. Miller, 1984, written commun.). Published age ranges for early Maastrichtian time vary among different time scales, from ca. 4 Ma (Lanphere and Jones, 1978; Harland et al., 1982; Palmer, 1983; Kent and Gradstein, 1985) to 1.9 Ma (Gradstein et al., 1995). The latter is based on the current globally accepted time scale published by the International Union of Geological Sciences, International Commission on Stratigraphy. *Inoceramus* fossils of the same species occur in the Chugach flysch belt from the Shumagin Islands to the central Chugach Mountains (Jones and Clark, 1973) and in general have been found only in middle and upper latitudes around the north Pacific rim (Jones and Clark, 1973; Nagao and Matsumoto, 1940; Vereshchagin et al., 1965). Although the stratigraphic ages strictly apply to the ~50% of the flysch belt in which fossils have been found, we will include the unfossiliferous Chugach flysch in our discussions because of similarities in tectonic position and lithologic character along the length of the southern Alaska margin (Fig. 1).

With the exception of the lower boundary, the cross-sectional geometry of the Kodiak Formation is well constrained (Fig. 4). The landward boundary (the Uganik thrust) dips 60° northwest (Moore and Wheeler, 1978) and the seaward boundary (the Contact fault) is vertical (Moore, 1967). Even substantial errors in the estimated dips of the two bounding faults would not greatly affect the value of the cross-sectional area. The position of the lower boundary can be projected from deep seismic reflection data collected along strike from the Kodiak Formation in the Gulf of Alaska (Fisher et al., 1987; Fisher et al., 1983). These reflection data show coherent reflectors beginning at depths of about 12 km beneath the sea floor extending beneath the Chugach flysch (Fig. 1); the intervening material shows no coherent reflectivity. The coherent reflectors have been interpreted as the top of a thick packet of underplated Eocene-Oligocene sedimentary rocks (Byrne, 1986). We interpret the top of the coherent reflectors as the base of the Chugach flysch and extend this lower boundary at a constant 12 km depth beneath the Kodiak Formation on Kodiak Island (Fig. 4). For our calculation, the addition of mass by intrusion of Paleocene granitoids is assumed to approximately balance

TABLE 1. INOCERAMIDS FROM THE KODIAK FORMATION

Locality	USGS number	Latitude	Longitude	Fossil	Age	Source
Raspberry Island, Onion Bay	M7849	58°02'06"	−153°14'	*Inoceramus* sp.	Cretaceous (?)	this study
Raspberry Island, West of Ustia Point	M7850	58°01'06"	−153°02'14"	*Inoceramus balticus* var. *kunimiensis*	early Maastrichtian	this study
Raspberry Island, West of Ustia Point	M7850	58°01'06"	−153°02'14"	*Inoceramus kusiroensis*	early Maastrichtian	this study
East side of Raspberry Strait	M8029	58°05'17"	−153°03'47"	*Inoceramus balticus* var. *kunimiensis*	early Maastrichtian	this study
Southwest end of Crooked Island	M8028	57°46'25"	−152°23'50"	*Inoceramus kusiroensis*(?) (crushed specimen)	early Maastrichtian	this study
Woody Island	M5962			*Inoceramya concentrica(?)*		Jones and Clark, 1973
Woody Island	M5963			*Inoceramya concentrica(?)*		Jones and Clark, 1973
Woody Island	M5963			*Inoceramus kusiroensis*	early Maastrichtian	Jones and Clark, 1973
Woody Island	M5964			*Inoceramus kusiroensis(?)*	early Maastrichtian	Jones and Clark, 1973

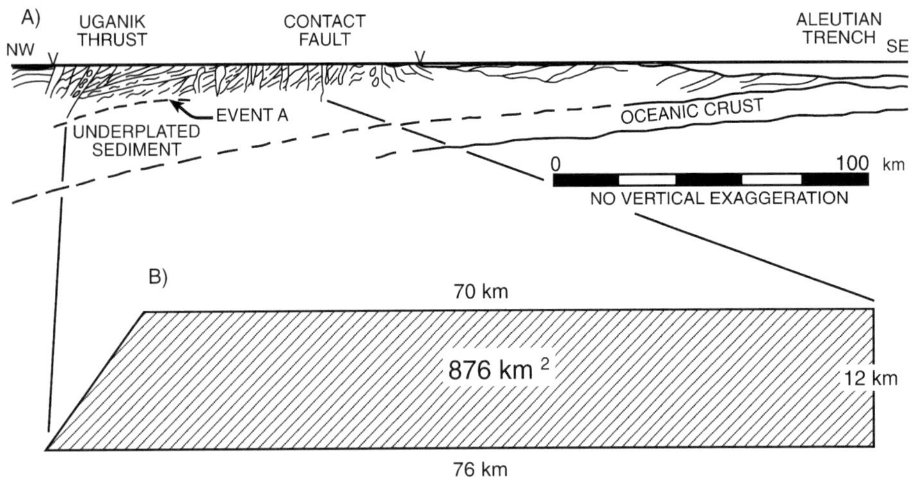

Figure 4. Estimate of cross-sectional area of Kodiak Formation. A: Cross section of Kodiak Shelf from Shelikof Strait in the northwest to the Aleutian Trench. Shorelines of Kodiak accretionary complex delimited by Vs. Kodiak Formation extends from Uganik thrust to Contact fault. Event A is lower boundary interpreted from deep seismic reflection profile of Fisher et al. (1983). Offshore geology from von Huene et al. (1985). B: Cross-sectional area determined by using Uganik thrust, Contact fault, and event A as boundaries.

the loss of structurally higher parts of the formation to erosion. This is a conservative part of the estimate, because over 10 km of overburden may have been eroded from the slate belt since accretion (Sample and Moore, 1987).

Based on the geometry shown in Figure 4, the total cross-sectional area of the Kodiak Formation is estimated as 876 km^2 (Table 2). Based on constraints from geophysical data in the Chugach Mountains (Page et al., 1986) and a constant outcrop width along strike, this area is approximately valid at least from Kodiak Island to Prince William Sound. There is evidence for major right-lateral shear overprinting the Border Ranges fault system in the Chugach Mountains during accretion of the Chugach flysch and continuing into Paleocene time (the "Hanagita fault system" of Roeske et al., this volume, Chapter 6). This may have modified the across-strike width of the flysch in this area, but most of the deformation appears to have occurred north of the Valdez Group (the local unit of the Chugach flysch in the Chugach Mountains), and the outcrop width does not appear to vary considerably.

Comparison to Large Modern Submarine Fans

To put the magnitude of the Kodiak Formation cross-sectional area into perspective, it may be compared with three large submarine fan bodies: the Bengal Fan, Amazon Cone, and Zodiac Fan. The Bengal Fan and Amazon Cone are two of the three largest continuous sediment bodies in the world (Bouma, 1985). The Zodiac Fan is chosen because it covers a large area (280,000 km^2) in the northeast Pacific, and its trajectory on the Pacific plate is sending it towards eventual accretion at the Aleutian Trench (Stevenson et al., 1983). As such, it is a potential modern analog for the Chugach flysch. To be conservative in our comparison of cross-sectional areas, cross sections of the three fans were chosen from published profiles in the thickest portion, where possible, to maximize our estimate of their cross-sectional areas (Damuth and Kumar, 1975; Stevenson et al., 1983). Uncorrected cross-sectional areas are shown in Table 2.

Depositional intervals for the sediments included in each cross section are constrained by nannofossils, seismic stratigra-

TABLE 2. CROSS-SECTIONAL AREAS (CSA) AND TIME SPAN OF DEPOSITION FOR THE KODIAK FORMATION AND MODERN SUBMARINE FANS

Sediment body	Uncorrected CSA (square km)	CSA dense rock equivalent (square km)	Lateral dimension (km)	Age range* (Ma)
Kodiak Formation	876	832	2000	1–4
Bengal Fan	7770†	5594	500	7–9
Amazon Cone	3415§	2783	600	16.5–22
Zodiac Fan	770#	287	1100	8–16

*Ranges are based on data and uncertainties cited in references.
†Only "w" and "y" sediments used for calculation.
§From cross section E–F in Damuth and Kumar (1975).
#References: Kodiak—this study; Bengal—Curray and Moore (1971); Amazon—Damuth and Kumar (1975), Kumar (1978), Damuth and Embley (1981); Zodiac—Stevenson et al. (1983).

phy, or inferred depositional rates and allow particularly wide possible ranges for the Amazon Cone and Zodiac Fan (Table 2) (Curray and Moore, 1971; Damuth and Embley, 1981). Nannofossils collected from Deep Sea Drilling Project (DSDP) hole 183 in the Zodiac Fan give a minimum age range of 8 m.y. (40–32 Ma) but the hole did not penetrate the thickest portion of the fan. The maximum age range allowable is probably 16 m.y. based on depositional patterns on the Alaska Abyssal Plain (Stevenson et al., 1983). Before comparisons can be made between highly porous sediments and the lithified and compacted turbidites of the Chugach flysch, both must be corrected to dense rock equivalents (DRE) that eliminate porosity from volume estimates. The submarine fans were corrected to DRE by integrating the cross-sectional area over a depth-porosity function (Table 2) (Sample and Karig, 1982). All three corrected cross-sectional areas decreased significantly from uncorrected values, particularly for the Zodiac Fan. However, porosity values calculated for the base of the Zodiac Fan from the porosity function are very close to porosities measured in a DSDP borehole (Kulm and von Huene, 1973). The Kodiak Formation was converted to a DRE by using an average compressional acoustic velocity of 5.9 km/s, taken from refraction velocities in the Chugach Mountains (Page et al., 1986), and empirical velocity-porosity curves. From this calculation, a porosity of 5% was removed from the Kodiak Formation.

When examined in light of large modern submarine fans, the Kodiak Formation represents a very large mass of sediment deposited rapidly. Simply on the basis of cross-sectional area, its relative depositional rate was probably greater than that of the Amazon Cone, the third largest modern fan body in the world (Table 2). Comparisons of cross-sectional areas inherently underestimate the size of the Kodiak Formation relative to modern submarine fans because its area was calculated in the narrowest outcrop dimension, whereas for all of the submarine fans the area was calculated in the widest outcrop dimension. The third dimension of the Chugach flysch is as much as 2000 km, whereas the third dimensions of the Bengal fan, Amazon Cone, and Zodiac Fan are approximately 500, 600, and 1100 km, respectively. In addition, the submarine fans thin radially, while the Chugach flysch only thins longitudinally. Although accurate volumes cannot presently be determined, the absolute size of the Chugach flysch is probably at least as large as that of the Amazon fan. Moreover, the depositional rate of the Chugach flysch was probably comparable to that of the Bengal Fan, the largest modern submarine fan complex, and one resulting from unroofing of a continental collision zone.

Possible Causes of Rapid Sedimentation

There are numerous possible causes of rapid sedimentation of the Chugach flysch related to northeast Pacific paleogeography. Two likely possibilities were input from an energetic volcanic arc or rapid and extensive uplift along the margin. Provenance types have been evaluated using sandstone petrography and Nd isotopic analyses of the Kodiak Formation in conjunction with some constraints from lithofacies analysis.

SEDIMENTOLOGY

Lithofacies

The depth of deposition of the Kodiak Formation is probably bathyal because it contains *Nereites* facies trace fossils (Nilsen and Moore, 1979), and the lack of preserved primary carbonate layers indicates deposition below the carbonate compensation depth. The degree of fault disruption makes detailed facies analysis problematical, particularly in the central belt where strain is highest, but generalizations can be made about lithologic changes across the formation. Along the mapped transect the northwestern three quarters of the Kodiak Formation consists of sections of thickly bedded coarse-grained sandstone and lesser shale interleaved with smaller amounts of thinly bedded fine-grained sandstone and shale. The southeastern quarter lacks thickly bedded lithologies and includes mostly thin-bedded sandstone and shale or massive shale.

Because of structural complexity the formation has been divided into three simple lithofacies mappable in the field (Fig. 3). Variations across strike are more detailed than can be shown at the map scale.

Thickly Bedded Facies

This facies consists of thickly bedded (0.2–3 m), medium- to coarse-grained sandstone with lesser shale, massive sandstone, and conglomerate (Fig. 5). This facies crops out mainly in the northwestern three-quarters of the formation. Sandstone:shale ratios in thickly bedded sandstone vary from 5:1 to >>10:1. Individual sandstone beds are generally plane-parallel, and in the few outcrops where cyclicity is observed, beds are organized into thickening-upward sequences over 5 m intervals. Locally, Bouma sequences include T_{a-d} to T_{c-d}. Beds are internally structureless or planar-laminated but locally contain cross-beds, convolute laminations, sandstone dikes, slump folds, or flame structures. Some of the thicker coarse-grained layers exhibit lenticular shapes with bases scoured into thinly laminated sandstone or shale.

Conglomerates are clast-supported and contain quartzite, black chert, green chert, granitic rock fragments, sandstone, greenstone, and volcanic rock fragments. Clasts have average diameters of 2–5 cm, but can range up to 20 cm; shale clasts can be larger.

Thinly Bedded Facies

This facies is composed of thinly bedded (≤20 cm), fine-grained sandstone and shale (Fig. 5) and crops out in all areas of the mapped transect (Fig. 3). Sandstone:shale ratios for the most part vary from 0.2:1 to 1:1, but can reach as high as 5:1. Individual beds are exclusively plane-parallel and thickening- or thinning-upwards trends were not observed. Cross lamination or graded bedding dominates internal structures in sandstone lay-

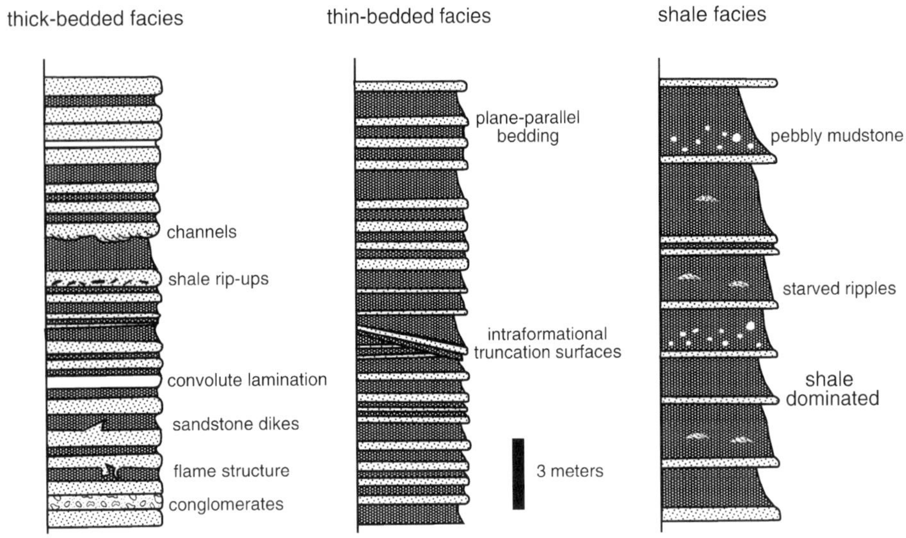

Figure 5. Schematic stratigraphic columns of three lithofacies mapped in the Kodiak Formation.

ers; Bouma sequences are lacking. Intraformational truncational surfaces resulting from slumping are common in this facies in the landward belt, particularly in Raspberry Strait (Fig. 3).

Shale Facies

This facies is dominantly shale, with minor thin, fine-grained sandstone and siltstone interbeds or pebbly mudstone. These lithologies are volumetrically least significant along the transect. They occur chiefly in the southeasternmost quarter (Fig. 3). Shale units can include thinly bedded, graded siltstone-shale couplets on the centimeter scale. Small-scale cross laminations and starved ripples are locally present. Intraformational truncation surfaces are present in the landward belt. Pebbly mudstones commonly contain intraformational clasts such as sandstone and siltstone but can include well-rounded quartzite and plutonic rock fragments. Outcrops of this lithofacies are commonly sheared into scaly argillite; criteria for distinguishing pebbly mudstone from zones of tectonically disrupted strata have been discussed in Sample and Moore (1987).

The general trend for lithofacies in the Kodiak Formation is from proximal submarine-fan facies in the northwest to distal facies in the southeast. Based upon the occurrence of coarse detritus and channel geometries throughout most of the formation, it appears that few turbidites were deposited more distally than outer fan. The plane-parallel beds of the thickly and thinly bedded facies that dominate the Kodiak Formation could signify deposition in a distal submarine fan or basin-plain environment (Mutti and Ricci-Luchi, 1972). This interpretation is supported by the local occurrence of thickening-upward sequences characteristic of distal-fan facies and probably holds true for much of the seaward belt. However, in the landward belt lenticular sandstones and conglomerates representing channel fill are more abundant, and channelized sandstones of the thickly bedded facies and accompanying siltstone-shale couplets probably represent mid to upper fan turbidites and overbank deposits, respectively. Similar stratigraphic relationships hold for part of the seaward belt along Narrow Strait, where lenticular massive sandstones channel into underlying thinly bedded sandstone and shale. Furthermore, the presence of some conglomerate in the southeastern quarter of the formation suggests proximity to the source region for sediments deposited in this seaward portion of the Kodiak Formation.

Paleocurrents

Sole markings are abundant in outcrops throughout the Kodiak Formation with the exception of the central belt where tectonic strains are higher. Paleocurrent features include flute casts, groove casts, bounce marks, and tool marks. Only absolute indicators of paleocurrents are reported here because they allow unambiguous interpretation of flow direction. Where fold axes or cleavage-bedding intersections were plunging, all current orientations were corrected by a standard two-tilt method (Ramsay, 1961). Caution must be used in the interpretation of paleocurrent data derived from thrust belts because linear features like sole markings might be rotated toward the extension direction of finite strain ellipse as a result of ductile strain (Ramsay, 1967). Finite strain data and fold axes orientations indicate that the Kodiak Formation underwent northwest-oriented extension during deformation (Fisher, 1984; Sample and Moore, 1987). No data have been used from rocks of the central belt where strain magnitudes are highest and ductile rotation of markings is most likely.

The results of 118 paleocurrent measurements show southerly dominated flow directions in Kupreanof Strait (KS) and in the seaward belt (SB), but indicators show a strong bimodal north-south orientation in Raspberry Strait (RS) (Fig. 3). Most of the paleocurrent data trend north-south, which is 45° from the direction expected, if paleocurrents were rotated towards maximum extension during tectonic deformation. Thus, paleocurrent features do not appear to have been rotated by penetrative deformation and can be interpreted reliably after correction for tilt.

SANDSTONE PETROGRAPHY

Methods

In order to investigate the lithologic character of possible source regions, 23 sandstones from across the Kodiak Formation were point-counted by standard petrographic techniques. Intensely deformed sandstones, as indicated by cleavage development, were avoided to reduce bias introduced by tectonic destruction of grains. For this reason few samples were used from the heart of the central belt. Samples were prepared as standard 25 × 45 mm thin sections and were stained for calcic plagioclase and potassium feldspar according to the methods of Houghton (1980) and Wilson and Sedeora (1979). Samples were counted on a petrographic microscope using a 0.5 × 0.5 mm grid. 400–500 framework grains were counted for each sample. To reduce operator bias, samples were counted without prior knowledge of location, and some samples were recounted months after the initial count to check data consistency.

Results

Most sandstones are lithic wackes (classification of Pettijohn, 1975) and have undergone diagenetic and metamorphic alteration typical of convergent margins. Plagioclase grains and volcanic rock fragments are albitized and sericitized. Matrix percentages range from <1% to 26% and average 14% (Table 3). Relic textures suggest that the matrix is composed largely of altered and crushed sedimentary rock fragments with lesser amounts of volcanic or metamorphic rock fragments. Evidence for the conversion of sedimentary rock fragments to epimatrix or pseudomatrix includes the stringing out of mixtures of silt-size quartz fragments and phyllosilicate particles between larger individual grains and along cleavage planes.

Point-count data plotted on standard ternary diagrams show that Kodiak Formation sandstones are split into two fields dominated by quartz-rich and quartz-poor sandstones (Fig. 6). The most abundant quartz fragments are monocrystalline quartz or distinctive cherty grains (Table 3). Dominant lithic fragments are felsitic and microlitic volcanic rock fragments. Albite, in part the product of albitization of more calcic plagioclase, is the most prevalent feldspar, and less numerous grain types include foliated quartzite, quartz-mica schist, chlorite-mica schist, siltstone, and argillite.

The cherty grains represent a problem in classification because of their degree of alteration and their microcrystalline nature. They most commonly exhibit homogeneous grain size, although locally they include patches of larger grains (Fig. 7). Most are grayish-green and contain opaques. Some have aligned micas or thin quartz veins. The large number of these grains in many of the samples makes understanding their origin important (Table 3). The most likely origins for these grains are meta-volcanic rock fragments or meta-chert fragments. If they are meta-volcanic rock fragments, they should contain some albitic component, but if they are meta-cherts, they should be pure quartz. In order to more accurately count cherty grain types, representative samples were selected from both the quartz-rich and quartz-poor categories for microprobe analysis. Grids to determine compositions were run across several grains using the energy-dispersive system and electron beam luminescence; plagioclase was easily distinguished from quartz in this manner. In the quartz-rich sample, 16 of 19 cherty grains were pure quartz (excluding opaques), while in the quartz-poor sample, 10 of 10 cherty grains contained small amounts of albite. This suggests that cherty grains in the quartz-rich samples have a predominantly meta-chert origin, and cherty grains in the quartz-poor samples are meta-volcanic rock fragments. It was impractical to point-count each sample with the microprobe, so for simplicity in the ternary diagrams, all cherty grains in quartz-rich samples are assigned to the category of polycrystalline quartz, and similar grains in the quartz-poor samples have been counted as volcanic rock fragments. A separate category for cherty grains has been maintained in the data table.

A significant result of the point-count data is the existence of separate quartzose and volcaniclastic sandstones (Fig. 6). In the QFL diagram, quartzose sandstones fall into the recycled orogen (RO) field, whereas volcaniclastic sandstones fall into the undissected arc/transitional arc (UA/TA) fields in the QFL and Q_mFL_t plots (abbreviations from Table 3; provenance fields from Dickinson et al., 1983, and Dickinson and Suczek, 1979). The Q_mFL_t plot differs from the QFL diagram only in assigning all polycrystalline quartz to total lithic fragments; the shift of the quartzose Kodiak sandstones towards the L_t apex signifies its high content of cherty grains. The plot of lithic fragments on a Q_pVS diagram is a useful means of discriminating between recycled orogen and magmatic arc sources. In the Kodiak Formation this differentiates among (meta-)chert, (meta-)volcanic, and (meta-)sedimentary contributions to the lithic component of sandstones. Lithic fragments are clearly dominated by volcanic and sedimentary rock fragments, the latter being characteristic of recycled orogens. It appears, however, compaction and diagenesis have greatly reduced the number of sedimentary rock fragments. If the matrix component were added to sedimentary rock fragments, the quartz-rich sandstones, being higher in matrix, would shift much closer to the S apex of the triangle than quartz-poor sandstones, which are lower in total matrix. A similar trend is observed in sediments from the Bengal/Nicobar fans and off-scraped upper Oligocene–lower Miocene trench deposits from Nias Island (Ingersoll and Suczek, 1979; Moore, 1979). The off-scraped deposits contain fewer sedimentary rock fragments, suggesting that greater compaction, deformation, and time of burial has destroyed many of the original sedimentary grains (Fig. 6). Thus, quartz-rich sandstones in the Kodiak Formation are probably a more deformed equivalent of recycled orogen sediments that were originally rich in sedimentary rock fragments.

TABLE 3. POINT COUNT DATA FOR KODIAK FORMATION GRAYWACKES

Sample Number	N (total counts)	Total grains	Percent matrix	1 Quartz (mono)	2 Quartz (poly)	3 Chert	4 Cherty	5 Plag	6 Kspar	7 MM	8 Sed	9 Volc	C(cherty)/Q	V/L	C/Q	V(cherty)/L	P/F
Quartz-rich																	
J3-006-OB	667	573	14	41	2	1	9	19	0	1	9	19	0.19	0.66	0.01	0.74	1.00
U13a	778	563	17	19	15	2	11	9	0	2	5	36	0.28	0.64	0.05	0.84	1.00
J3-058-0B	668	580	13	34	3	0	18	21	1	4	13	6	0.33	0.25	0.00	0.58	0.98
J3-108-NK	688	540	21	25	7	<1	18	25	0	3	14	7	0.37	0.29	0.01	0.60	1.00
J3-126-SK	756	646	15	36	8	0	3	14	0	1	20	19	0.07	0.47	0.00	0.51	1.00
J3-068-SK	710	602	15	47	3	1	5	9	0	2	13	19	0.12	0.57	0.02	0.63	1.00
J3-096-SK	621	459	26	41	2	<1	5	25	0	1	10	15	0.11	0.57	0.01	0.64	1.00
J4-061-SB	677	524	23	23	2	<1	20	14	0	0	25	16	0.44	0.39	0.01	0.58	1.00
J4-096-NS	693	580	16	27	2	<1	5	19	0	2	21	23	0.16	0.50	0.01	0.55	1.00
J4-103-NS	723	658	9	27	5	<1	12	8	1	4	20	23	0.28	0.50	0.01	0.60	0.93
J4-116-NS	686	594	13	27	4	0	13	20	0	1	24	11	0.30	0.30	0.00	0.49	1.00
J3-193-MoB	841	674	20	25	3	2	22	9	<1	1	21	15	0.46	0.40	0.05	0.62	0.95
J3-201-KO	685	581	15	27	5	1	19	18	0	1	7	22	0.39	0.74	0.02	0.84	1.00
J3-218-MP	662	577	13	24	1	<1	27	11	0	0	30	6	0.53	0.17	0.01	0.52	1.00
J3-195-KO	687	603	12	41	2	3	12	10	0	0	9	22	0.26	0.69	0.05	0.78	1.00
J4-131-WI	689	569	17	36	3	0	18	13	0	2	14	14	0.33	0.46	0.01	0.67	1.00
J4-136-WI	715	652	9	30	2	1	9	8	<1	2	25	24	0.24	0.47	0.03	0.55	0.98
J3-159-MB	722	599	17	27	3	0	11	10	0	1	26	20	0.27	0.43	0.00	0.53	1.00
Transitional																	
J3-169-KB	608	557	8	10	3	1	17	22	<1	4	14	29	0.56	0.62	0.03	0.72	0.99
Volcanic-rich																	
J4-088-NS	688	685	<1	6	0	0	4	15	0	0	2	73	0.44	0.97	0.00	0.97	1.00
J4-150-MoB	685	626	9	14	0	0	6	22	2	1	2	53	0.30	0.95	0.00	0.95	0.91
J3-153-MB	680	631	7	9	0	0	5	21	1	0	5	59	0.36	0.92	0.00	0.93	0.97
J3-180-MB	636	540	15	5	2	0	7	26	0	0	2	58	0.53	0.97	0.00	0.97	1.00

Note: Numbered categories in percent of total grains; mono—monocrystaline; poly—polycrystalline; MM—metamorphic rock fragments; sed—sedimentary rock fragments; volc—volcanic rock fragments; cherty—explained in text; C(cherty)/Q—ratio of chert plus cherty to total quartz fragments; C/Q—ratio of chert to quartz; V/L—ratio of volcanic to total lithic fragments; for V(cherty)/L, V includes "cherty"; P/F—ratio of plagioclase to total feldspar. In Figure 6, for QFL plot, Q = 1 + 2 + 3 + 4, F = 5 + 6, and L = 7 + 8 + 9 in quartzose rocks; 4 is assigned to L in volcaniclastic rocks. For Q_mFL_t, Q_m = 1, F = 5 + 6, L_t = 2 + 3 + 4 + 7 + 8 + 9; for Q_pVS, Q_p = 2 + 3 + 4, V = 9, S = 7 + 8 in quartzose rocks; in vocaniclastic rocks, 4 is assigned to V. For comparative purposes, dual categories are shown for C/Q and V/L ratios—C(cherty) places 4 with chert grains; V(cherty) places 4 with volcanic rock fragments.

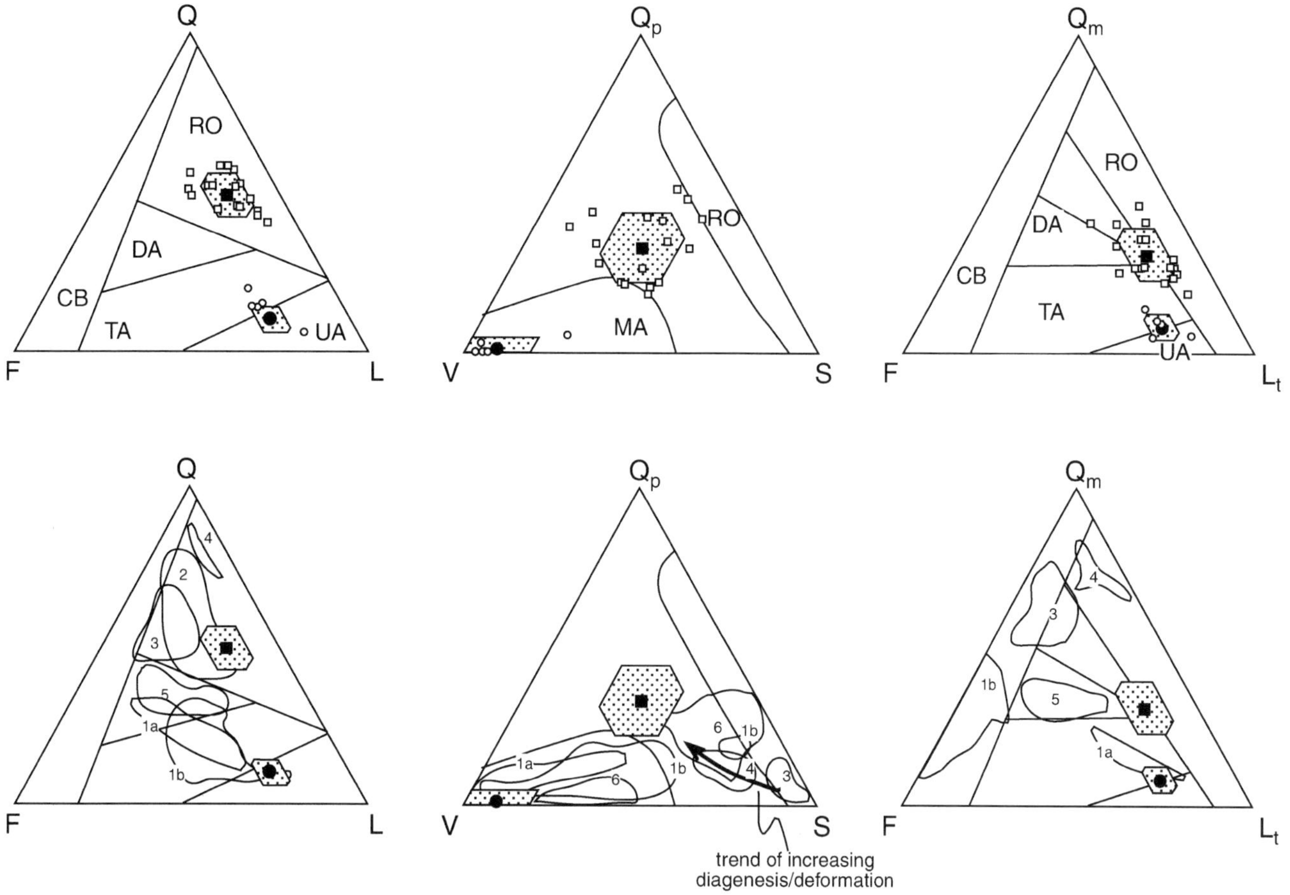

Figure 6. Ternary diagrams of sandstone compositions. Volcaniclastic sandstones shown with open circles; quartzose sandstones shown with open squares (filled symbols are means, standard deviations are shaded polygons). Apices defined in Table 3. Provenance fields are from Dickinson and Suczek (1979) and Dickinson et al. (1983). Abbreviations: RO—recycled orogen; CB—continental block; DA—dissected arc; TA—transected arc; UA—undissected arc; MA—magmatic arc. Separate fields of Kodiak data in Q_pVS and Q_mFL_t diagrams are based on two fields from QFL plot. Numbers designate fields for comparison: 1a—Chugach flysch (Dickinson, 1982; Zuffa et al., 1980); 1b—Chugach flysch (Dumoulin, 1987); 2—Alaska Peninsula (Mancini et al., 1978); 3—Bengal and Nicobar fans (Ingersoll and Suczek, 1979); 4—Nias Island offscraped rocks (Moore, 1979); 5—Cretaceous and Paleogene rocks of western North America (Dickinson et al., 1983); 6—Kodiak sandstones if matrix is included with sedimentary rock fragments; 7—strike-slip continental arc field (Marsaglia and Ingersoll, 1992). Arrow shows trend from modern Bengal and Nicobar fan sands toward offscraped and deformed ancient equivalents on Nias Island. The same trend would presumably apply to the Kodiak Formation as the graywackes underwent diagenesis and deformation related to underthrusting and underplating.

The other major component of detritus in Kodiak sedimentary rocks derives from a relatively undissected volcanic arc terrane, as indicated by the group of quartz-poor sandstones on the QFL and Q_mFL_t plots (Fig. 6). This component also appears in quartz-rich sandstones on the Q_pVS plot. Numerous granitic cobbles in conglomerates suggest deeper levels of exposure in the arc terrane as well.

Nd ISOTOPIC DATA

Nd isotope analyses have been performed on selected sandstones from the Kodiak Formation in order to further constrain their provenance. The utility of Nd isotope analyses of sediments arises from the largely quantitative transfer of rare earth elements through cycles of erosion, lithification, and low-grade metamorphism. This is primarily a consequence of the low solubility of the rare earth elements in aqueous solutions. To a first approximation the Nd isotope signature of a clastic sedimentary rock averages the Nd isotopic composition of its provenance, weighted by the Nd concentrations of the mineral grains. In addition, because of the relatively systematic fractionation of ^{147}Sm, the radiogenic parent isotope, from Nd during production of continental crust, crustal terranes of distinct isotopic compositions can be characterized. The Nd isotope signatures of sedimentary rocks act as isotopic and temporal fingerprints of their sources, whether they are direct sources of detritus or sources, which contributed as contaminants to a magmatic arc terrane.

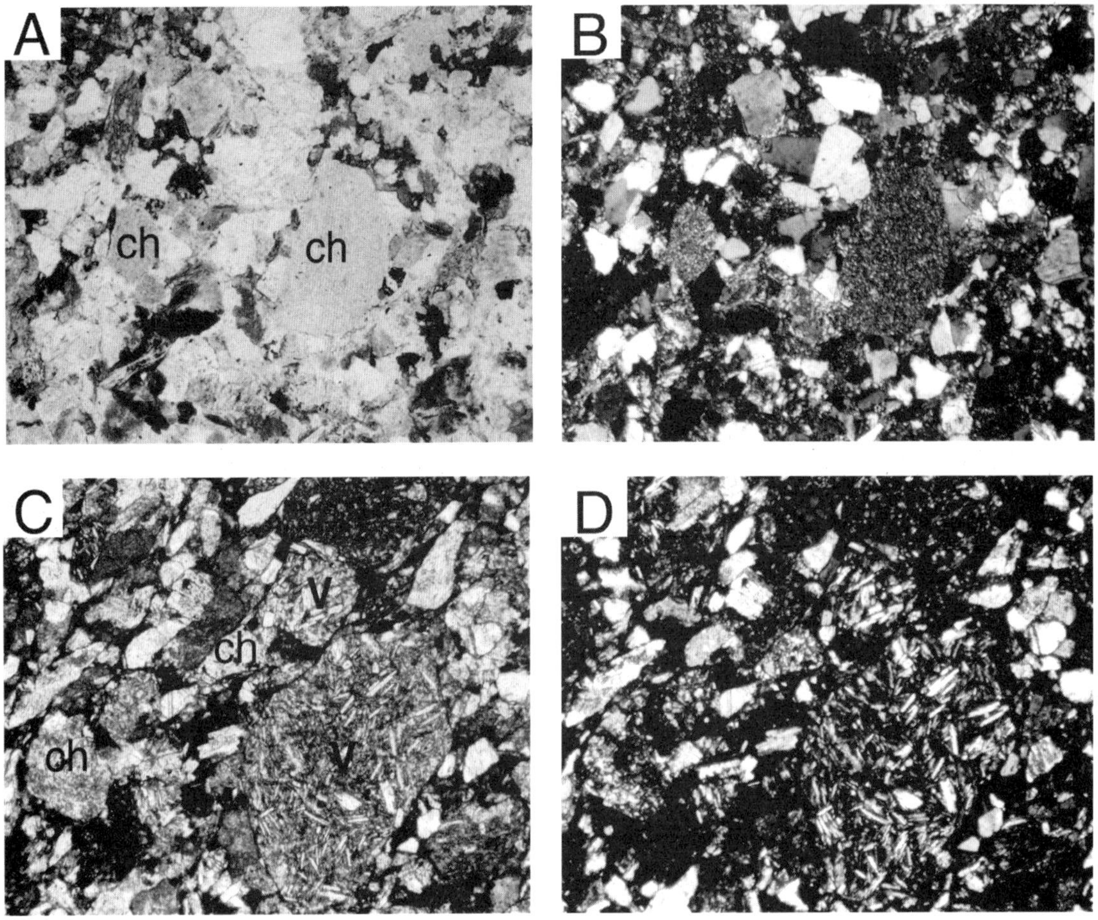

Figure 7. Photomicrographs of sandstones. Field of view for each photo approximately 2 mm. A: J3-058-OB (plane polarized light) shows typical quartz-rich lithology, including cherty grains ("ch"), counted as polycrystalline quartz in quartz-rich lithologies and volcanic rock fragments in volcaniclastic sandstones. B: J3-058-OB (crossed nicols). Note abundant monocrystalline quartz grains. C: J3-153-MB (plane polarized light). Volcanic rock fragments ("V") commonly contain plagioclase laths and are abundant in this lithology. D: J3-153-MB (crossed nicols).

TABLE 4. ND ISOTOPE ANALYSES, KODIAK FORMATION SANDSTONES

	Nd*	Sm*	$^{147}Sm/^{144}Nd$*	$^{143}Nd/^{144}$†	ε_{Nd} (0)§	ε_{Nd} (70)#
Quartzose sandstones						
J4–061–SB	17.44			0.512611	−0.5	1.2
J3–108–NK	14.25			0.512621	−0.3	1.4
J3–195–KO	14.72			0.512639	0.0	1.8
U13a	17.65	3.575	0.123	0.512562	−1.5	−0.8
B6a	15.88	3.371	0.128	0.512538	−2.0	−1.3
Transitional sandstone						
J4–169–KB	17.46			0.512736	1.9	3.7
Volcaniclastic sandstones						
J4–150–MoB	16.17			0.512799	3.1	4.9
J3–153–MB	10.89	2.628	0.145	0.512875	4.6	5.1
J3–180–MB	17.95	3.784	0.127	0.512763	2.4	3.1

*Measured by isotope dilution; see Reid (1995) for analytical procedure.

† $^{143}Nd/^{144}Nd$ corrected for fractionation using $^{146}Nd/^{144}Nd$ = 0.7219. Measured values for La Jolla averaged 0.511910; data normalized to $^{143}Nd/^{144}Nd$ (La Jolla) = 0.511860.

§at t = 0 Ma; estimated for some samples.

#at t = 70 Ma; estimated for some samples.

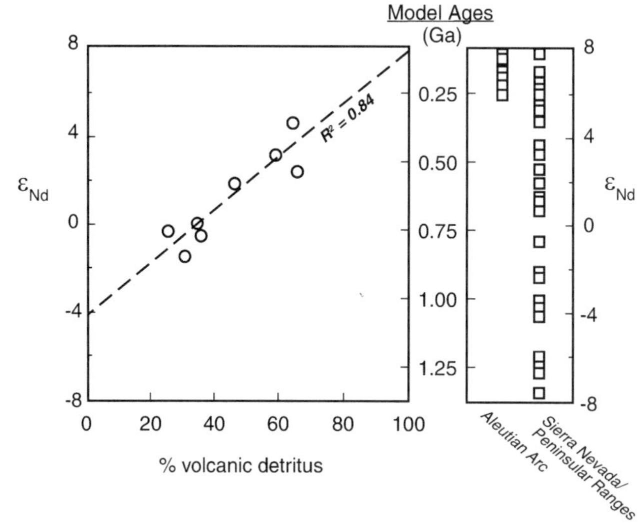

Figure 8. Present-day Nd isotope signatures of Kodiak Formation sandstones plotted against % volcanic component. Right-hand axis shows Nd model ages corresponding to $^{143}Nd/^{144}Nd$ signatures and assumes extraction from a depleted mantle source. Also shown for comparison are isotopic data for volcanic rocks of the Aleutian Arc (von Drach et al., 1986) and granitoids of the Sierra Nevada/Peninsular Ranges (DePaolo, 1981).

Nd isotope analyses of two sandstones from the Kodiak Formation were performed previously as part of a geochemical study of Kodiak Island (M.R. Reid, 1998, written commun.; summarized in Table 4). Both of these samples are from the quartz-rich portion of the formation, and a portion of sample U13a was still available for point-counting. Seven additional sandstones, three more from the quartz-rich portions, three from the volcaniclastic portions, and one from a sandstone transitional in composition of the formation, were analyzed for this study (Table 4). It is evident that the Nd isotopic compositions of the volcaniclastic and transitional sandstones are distinctly higher than those of the quartzose sandstones, corroborating the distinction between these two groups based on petrographic evidence. These observations are combined in Figure 8, where ε_{Nd} ratios are plotted against the percentage of volcanic rock fragments determined from point counts.

DISCUSSION

Our results place significant constraints on the tectonic origin of the Chugach flysch in latest Cretaceous time. The paleogeography and source region of the flysch must satisfy requirements of inferred sedimentation rate, paleocurrent data, sandstone compositions, and Nd isotopic data. Taken together, these suggest rapid deposition adjacent to an uplifting orogen along a continental margin. The following discussion addresses these requirements systematically, and considers the additional constraint of paleomagnetic data, to speculate on a type of source region for the Chugach flysch.

Significance of Sedimentological Data

Our estimates of the large volume of sediment that accumulated in a short time period indicate that sedimentation must have occurred close to the western margin of North America and that rates of sedimentation were high along this margin (Sample and Reid, 1986). Paleocurrents oriented transverse to the structural grain (and presumed orientation of the accretionary margin) of the Kodiak Formation indicate that sediment was supplied to the oceanic plate transverse to the subduction margin. This suggests that the source of sediment was close to the site of accretion of the Kodiak Formation. A previous paleocurrent study of smaller parts of the Kodiak Formation yielded data oriented exclusively southwest and southeast in current coordinates (Nilsen and Moore, 1979). These data, coupled with paleocurrent and petrographic results from the rest of the Chugach flysch, were used to suggest a model for deposition of the Kodiak Formation in which the dominant sediment source was turbidity currents traveling northeastward and then southwestward along a continuous trench, with the ultimate point source in central British Columbia (Nilsen and Zuffa, 1982). If the structural grain of the Chugach flysch represents the orientation of the "paleotrench," then our paleocurrent data suggest that the model of continuous, along-trench transport is incorrect for at least the Kodiak Formation. Its dominant southward paleocurrent direction is oriented approximately 45° obliquely to the contacts of the formation. Considering the large volume of sedimentary rocks in the Kodiak Formation, it is also unlikely that a "paleotrench" would have existed in latest Cretaceous time. A more likely interpretation is that the Chugach flysch represents a large submarine fan complex that filled the bathymetric trench and spilled far out onto the oceanic plate. Taken together, the dominance of coarse material and the paleocurrent data suggest that the Kodiak Formation must have been deposited close to its sediment supply along the continental margin.

Based on the sandstone provenance data, we favor a mixture of Kodiak source terranes that includes an uplifting magmatic arc with varying levels of exposure to explain the volcanic and plutonic component, built in and on a mixture of terranes with abundant cherty, sedimentary, and metamorphic lithologies to explain the variety of non-volcanic rock fragments. This interpretation is supported by sandstone compositions, which lie in the recycled orogen field, the magmatic arc field, or along a mixing curve between the two (Fig. 6). A previous provenance study of Chugach flysch in the Kodiak islands (Zuffa et al., 1980; Nilsen and Zuffa, 1982) also suggested that a mixture of recycled orogen and magmatic source terranes contributed to the formation of the Chugach flysch. These results implied that the main source of the Chugach flysch was a volcanic arc increasingly dissected from west to east, with minor contributions of sediment derived from a mixture of sedimentary and metamorphic source terranes. But the trend of their data is not simple. Their data show that some eastern portions of the Chugach flysch contain some of the sandstones richest in volcanic rock fragments. Our results differ from these previous studies in that we observe a higher percentage of quartzose rock fragments in the Kodiak Formation, derived from a recycled orogen (Fig. 6), which does not support a decreasing ratio of volcanic rock fragments to total lithic fragments along strike. Rather we observe an increase in volcanic lithics across strike from northwest to southeast in the Chugach flysch on Kodiak Island, presumably representing a relative increase in volcanic input over time. We have more confidence in the results of our provenance study of the Kodiak Formation than that of Zuffa et al. (1980) and Nilsen and Zuffa (1982) because (1) our sample size is considerably larger (n = 23 vs. n = 5); (2) we considered all grain types rather than just lithic fragments; (3) we distinguished dual origins of cherty grains with microprobe analysis, thus avoiding the potential for misidentification of cherty grains as volcanic rock fragments in quartzose sandstones (Nilsen and Zuffa, 1982, assign all "undetermined cherty grains" to the volcanic rock fragment category); and (4) the compositions of the two populations of sandstones we observe are consistent with the Nd isotopic data.

Sandstone compositions from the Chugach flysch (Valdez Group) in Prince William Sound (Dumoulin, 1987) are similar in some respects to the Kodiak results (Fig. 6). Valdez Group lithic fragments contain substantial percentages of sedimentary rock fragments and overlap on the Q_pVS with Kodiak flysch samples. Valdez Group sandstones are generally lower in total quartz and significantly higher in feldspar, which shifts their position on the QFL and Q_mFL_t ternary diagrams into the magmatic arc and continental block fields (Dumoulin, 1987). Dumoulin (1987) further demonstrated that the provenances of the Valdez Group and the Paleocene Orca Group to the south are indistinguishable, suggesting the mixed orogen-arc source continued into Tertiary time.

Although all studies of the provenance of the Chugach flysch agree upon a mixture of a recycled orogen and magmatic arc source, we conclude that the dominant source in the Kodiak islands was clearly a recycled orogen. Some of the more quartz-rich sandstones lie in an arc provenance field that has been related to strike slip–dominated arcs (Marsaglia and Ingersoll, 1992); this suggests that the arc source may have formed along an oblique convergent margin. A transpressive origin for uplift in the source region is consistent with observations of major dextral shear along the margin in latest Cretaceous time (Roeske et al., this volume, Chapter 6). While volcanic rock fragments occur in all of the sandstones examined here, volcanic detritus is overshadowed in most of the formation by more numerous quartz fragments from the recycled-orogen source. It is noteworthy that the increase in number of volcaniclastic sandstones coincides with an increase in fine-grained, thinly bedded lithologies in the seaward belt (Fig. 3). The presence of the volcaniclastic sandstones may have resulted from blocking of the quartz-rich source, possibly as a result of changes in the location of feeder canyons to fans on the underthrusting plate.

Significance of the Nd Isotopic Data

The disparity in Nd isotopes between the volcaniclastic and quartzose sandstones permits general observations about the likely provenance of the sediments. As inspection of Figure 8 shows, the isotopic composition of the volcaniclastic sediments is only somewhat lower than the isotopic range of volcanic rocks from the Aleutians and Alaska Peninsula, which are representative of an oceanic arc with some continental influence (Fig. 8). Inspection of the Nd model ages of both the arc lavas and the sandstones suggests that the volcaniclastic component of the Kodiak Formation may be derived from an analogous volcanic arc that is <~500 m.y. old. Alternatively, the volcaniclastic sandstones of the Kodiak Formation may be dominated by detritus derived from a relatively juvenile volcanic arc with a minor input of older continental material. This continental component to the sandstones may arise as either continental detritus or as a basement contaminant in the arc.

Although lower than the volcaniclastic sandstones, the ε_{Nd} of the quartzose sandstones are still within the range of continental margin volcanic arcs, as illustrated by ε_{Nd} data for the Sierra Nevada/Peninsular Ranges. However, given the petrographic evidence for metamorphic and sedimentary detritus in the quartzose sandstones, a probable origin for the quartzose sandstones is as a mixture of recycled orogen and volcanic components. For example, the correlation line between the two sedimentary groups shown in Figure 8 could be interpreted as mixing between the volcaniclastic sediments and sediments derived from a metamorphic/sedimentary source (i.e., % volcanic grains = 0) with $\varepsilon_{Nd} \sim -4$. The model age represented by this endmember is ca. 1.0 Ga. Since there is little difference in the Nd concentrations of the two groups of sediments, a linear mixing line such as that of Figure 8 may be justified. Thus, combined petrographic and isotopic evidence for the Kodiak Formation sandstones require at least two distinct sources, one of which is a relatively juvenile volcanic source and the other of which contains older metamorphic and sedimentary material.

The proportion of volcanics shown in Figure 8 is likely to significantly underestimate the true volcanic component of the volcaniclastic sediments since plagioclase and the matrix were not included in the estimate of a volcanic component. If most of the plagioclase and matrix of these sediments were considered to have the same source as the volcanic clasts (i.e., % volcanics > 80), then the mixing line would rotate toward a lower ε_{Nd} for the continental endmember. Further constraints on the sedimentary provenances for the Kodiak Formation require consideration of the mass balance of Nd in the sediments and are speculative since each end-member probably reflects more than one sedimentary provenance. In any case, it appears that a provenance with $\varepsilon_{Nd} \leq -1$ contributed a significant volume of sediments to the Kodiak Formation; likely crustal residence times for this component are 0.9–1.3 Ga. Taken together, the combined petrographic and isotopic data suggest a provenance that progressed from a recycled orogen composed in part of

Proterozoic materials to a continental margin volcanic arc that may lie on or adjacent to the uplifting recycled orogen.

Potential Source Terranes

Before identifying a possible source terrane that satisfies all of the geological evidence, we must consider paleomagnetic evidence bearing on the paleolatitude of the Kodiak Formation during its deposition and accretion. Paleomagnetic studies from southern Alaska have led to terrane models, which make the Chugach flysch, along with the rest of the accretionary complex, allochthonous with respect to the Cretaceous volcanic arc of southern Alaska (Coe et al., 1985; Moore et al., 1983; Plumley et al., 1983; Thrupp and Coe, 1986). The most reliable paleomagnetic data indicating an allochthonous origin of the Kodiak accretionary complex are from volcanic rocks in the Paleocene Ghost Rocks Formation, an accreted turbidite unit just seaward of the Kodiak Formation (Fig. 3). These data from pillow basalts are considered applicable to the Chugach flysch by virtue of a plutonic tie of Paleocene granitoids that intrude both units (Davies and Moore, 1984). The paleomagnetic data from the two regions in the Ghost Rocks Formation are of good quality, showing clean secondary and primary components, and magnetizations that were stable to relatively high temperatures. Each site passed either the fold test or reversal test (Plumley et al., 1983). The magnetic inclinations of the two sites differ by nearly 16° at a statistically significant level, suggesting a complicated terrane history for the Ghost Rocks Formation. The paleolatitudes of the two sites were averaged to yield a mean paleolatitude of 40.3° ± 6.2° N for the Ghost Rocks Formation in Paleocene time (Plumley et al., 1983). A later paleomagnetic study of possibly time-equivalent rocks along strike in the same terrane indicate 13° ± 6° of northward translation with respect to North America since 60 m.y. ago (Bol et al., 1992). In order to take into account the paleomagnetic data, even conservative interpretations must allow for the possibility that the Kodiak accretionary complex was 1500 to 2000 km south of its expected position during deposition of the Kodiak Formation.

The major Late Cretaceous orogenic terranes of western North America that can satisfy the constraints of sediment volume, sandstone compositions, Nd-isotope characteristics, paleocurrents, and lithofacies of the Chugach flysch are the Alaska Range–Talkeetna Mountains belt (hereafter referred to as the Alaska Range for convenience) of southern Alaska, terranes in the Coast Mountains belt of southeastern Alaska and western British Columbia, the Sierra Nevada batholith in eastern California, the Salinian block in west-central California, and the

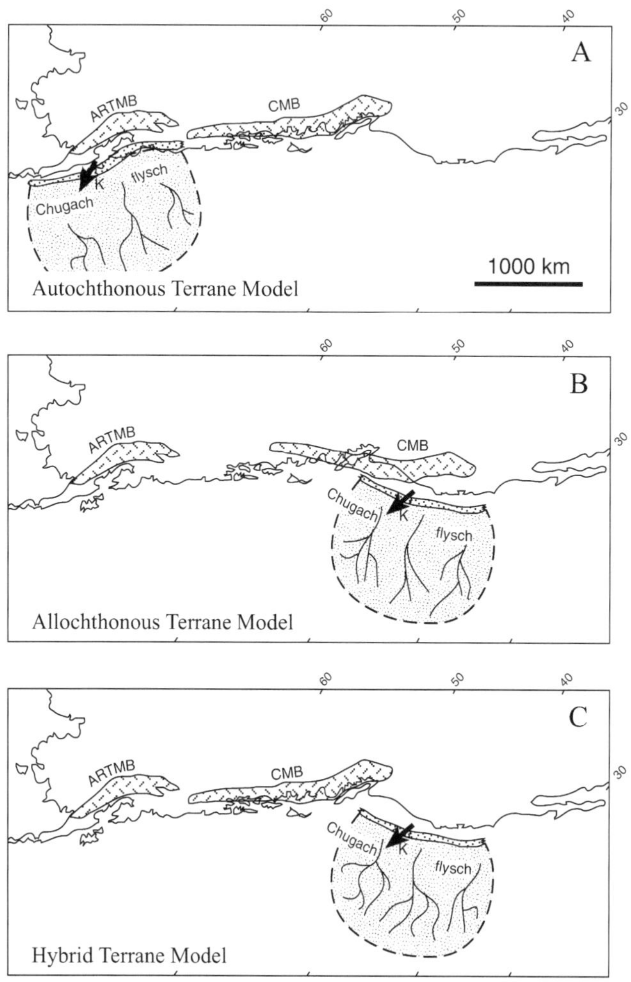

Figure 9. Tectonic models depicting combinations of allochthonous and autochthonous tectonic settings of Chugach flysch during sedimentation and subduction-accretion. Base map is Mercator projection; areas at lower latitudes appear smaller than those at higher latitudes. Location of North America from latest Cretaceous paleopole in Dickinson and Butler (1998), derived from data of Gunderson and Sheriff (1991) and Diehl (1991). K denotes Kodiak accretionary complex; arrows show mean direction of paleocurrents in orientation relative to interformational contacts. Tertiary oroclinal bending of southern Alaska (Coe et al., 1985) has been removed. ARTMB—Alaska Range–Talkeetna Mountains belt; CMB—outline of Coast Mountains belt (Coast Plutonic Complex) of western British Columbia. A: Autochthonous terrane model. Both Chugach flysch and CMB are shown in place relative to North America. Paleocurrent orientation suggests that flysch would have to be derived in large part from the ARTMB or interior Alaska, inconsistent with sandstone compositions and Nd isotopic data. B: Terranes in allochthonous positions at ca 70 Ma. There are probably large uncertainties in actual locations of allochthonous terranes. Chugach flysch located using 47° N paleolatitude of study region in Ghost Rocks Formation with least northward transport (see text for discussion). CMB is located by linearly extrapolating motion between southerly latitude at 90 Ma and in-place latitude by Middle Tertiary time (Cowan et al., 1997). Paleocurrent orientations and Nd isotopic data suggest Chugach flysch could have been derived from CMB. This is consistent with voluminous sediments of the Chugach flysch being shed from rapid latest Cretaceous uplift of the CMB. C: Hybrid model showing CMB in place and Chugach flysch at more southerly latitude. This model is inconsistent with paleocurrent data, and raises additional problem of no sufficiently uplifted terrane present in this region of the western United States.

Peninsular Ranges batholith of southern California and Baja California (Fig. 9). The three southernmost orogenic belts have low probabilities as source terranes for various tectonic reasons. The Sierra Nevada batholith and Salinian block were largely inactive by 80 Ma and are not appropriate as magmatic arcs during subduction-accretion of the Chugach flysch (Mattinson and James, 1985; Stern et al., 1981). Neither the Sierra Nevada batholith nor Peninsular Ranges batholith has experienced widespread uplift of the necessary magnitude, most of the uplift of the Sierra Nevada batholith has occurred in the last 10 m.y. (Chase and Wallace, 1986; Unruh, 1991), and much of the sediment shed off the Sierra Nevada batholith may have been trapped in the Great Valley forearc basin (Bateman, 1983; Gastil, 1983; Huber, 1981). Both the Alaska Range and the Coast Mountains belt have higher probabilities for being the source of the Chugach flysch because of their magmatic histories, favorable paleogeography, and, for the Coast Mountains belt, an appropriate uplift history. We will consider in detail the relevant characteristics of the Alaska Range and the Coast Mountains belt to evaluate them as source terranes. The former would require little or no northward movement of the Chugach flysch, whereas the latter is consistent with substantial northward translation since deposition of the Kodiak Formation.

Alaska Range Source

The Alaska Range consists of predominantly granitic and granodioritic rocks intruding terranes of Upper Paleozoic to Upper Mesozoic volcanic and volcaniclastic rocks along with shelf and flyschoid sedimentary rocks (e.g., Hudson, 1983). This magmatic arc records a pulse of magmatism from 75 to 55 Ma that is related to subduction during latest Cretaceous and Paleocene time (Wallace and Engebretson, 1984). This pulse of magmatism is consistent with plate motion models that show rapid convergence of the Kula plate at this part of North America throughout much of Cretaceous and Tertiary time (Engebretson et al., 1985; Stock and Molnar, 1988). Paleomagnetic data from the Alaska Range show it to be essentially autochthonous with respect to North America by Paleocene time (Hillhouse and Gromme, 1982; Coe et al., 1985; Thrupp and Coe, 1986; Stamatakos et al., 1989).

The most compelling justification of the Alaska Range as a source terrane is its tectonic position with respect to the present position of the Chugach flysch. The Chugach flysch and the Alaska Range would make an obvious paired subduction system similar to the current pairing of the Aleutian arc and trench. Rapid convergence in this region is consistent with the growth of the large Kodiak accretionary complex. Scant QFL data show that Upper Cretaceous sandstones on the Alaska Peninsula are similar in composition to those from the Kodiak Formation and allow a tie between the two (Fig. 6), but QFL data alone are not a unique indicator. A major problem associated with the Alaska Range as a source is that it shows no evidence for major uplift since its formation. Country rocks intruded by the granitoids are only mildly metamorphosed (Hudson, 1983), and there is no evidence for deep-level rocks exposed at the surface (Dusel-Bacon et al., 1996). The Alaska Range lacks quartzite to serve as a source of the abundant quartz fragments in the Kodiak flysch. Finally, there is no indication of Precambrian basement or evidence of sedimentary rocks derived from old basement in southern Alaska to act as the source of unradiogenic Nd signatures measured in the Chugach flysch. Sedimentary and volcanic rocks taken to be representative of older continental basement beneath the active Aleutian arc have ε_{Nd} values from 1.3 to 6.5, significantly more radiogenic than values measured in quartzose sandstones of the Kodiak Formation (von Drach et al., 1986).

Coast Mountains Belt Source

The Coast Mountains belt has been proposed as a source for the Chugach flysch previously (Sample and Reid, 1986; Farmer et al., 1993; Plafker et al., 1994). As described previously, the Chugach flysch may have been as far south as the current western margin of the United States during its formation. Seemingly contradictory paleomagnetic and geologic data have led to the development of two major hypotheses for the position of the Coast Mountains belt in Cretaceous time (summarized in Cowan et al., 1997). One hypothesis, based mainly on paleomagnetic data from igneous rocks of British Columbia and the Yukon Territory, suggests the Coast Mountains belt was south of present-day California at 90 Ma (Irving et al., 1985; "Baja B.C." hypothesis of Umhoefer, 1987). The other hypothesis based on geologic evidence proposes the Coast Mountains belt was north of California at 90 Ma and is autochthonous with respect to North America (Cowan et al., 1997; Mahoney et al., 1999; McClelland and Mattinson, 2000; Rusmore et al., 2000). A model for the history of terrane motions in the western Coast Mountains belt and outboard Insular Superterrane consistent with geologic and paleomagnetic data has not yet been developed. If the positions of the Kodiak accretionary complex and the Coast Mountains belt were reconstructed accepting the more southerly hypothesis, this would place the Chugach flysch and western Coast Mountains belt in juxtaposition in latest Cretaceous time (Fig. 9). Since the Coast Mountains belt magmatic arc currently has no outboard subduction complex, this tectonic reconstruction would restore the pairing of subduction complex and arc system. If the autochthonous model for the Coast Mountains belt is correct, then the Chugach flysch would lie southward along the margin from the Coast Mountains belt in Late Cretaceous time. The required southward transport of sediment is possible, but would not be consistent with paleocurrent indicators.

The Coast Mountains belt includes numerous terranes with a broad range of tectonic origins that are intruded by a large batholith belt of Jurassic to Eocene granitoids that include Late Paleozoic(?) to Early Mesozoic(?) high-grade paragneisses and metasediments (Monger and Price, 1972; Parrish, 1983; Roddick, 1983; Cui and Russell, 1995; Boghossian and Gehrels, 2000; Hollister and Andronicos, 2000; Andronicos et al., 1999). Crystallization and cooling ages of magmatic rocks range from about Middle Triassic to late Neogene in age. The major magmatic episodes occurred between 150 and 90 Ma and between 75 and 40 Ma (Armstrong, 1988; Cui and Russell, 1995; Gehrels and Boghossian, 2000; McClelland and Mattinson, 2000; Rubin and Saleeby, 2000; Saleeby, 2000). Spectacular uplift centered on the Coast Plutonic Complex appears to have occurred in Paleocene to Eocene time, based on cooling ages determined from a variety of techniques on different mineral types (Donelick and Miller, 1986; Parrish, 1983; Crawford et al., 1987). There is thermochronologic and geologic evidence that portions of the Coast Plutonic Complex were uplifted significantly in latest Cretaceous time (Donelick and Miller, 1986; Gehrels et al., 1989; Parrish, 1983). In particular a large area of dextral transpression and southwest-directed thrusting was active from 75 to 65 Ma (Hollister and Andronicos, 1997). Although much of the record of this deformation has been overprinted by Tertiary structures, the zone of Cretaceous dextral thrusting appears to have been widespread (Andronicos et al., 1999). This latest Cretaceous to early Tertiary uplift may have been responsible for rapid erosion at the time of deposition of the Kodiak Formation.

The heritage of the Coast Mountains belt as a magmatic arc and recycled orogen, the magnitude and history of uplift of the Coast Plutonic Complex, and the possibly more southerly position of the Coast Mountains belt during latest Cretaceous time make the Coast Mountains belt an attractive candidate as the source terrane for the Chugach flysch. Sediments eroded from the Coast Mountains belt would have sampled a magmatic terrane and a wide variety of metamorphic terranes to account for the range in Chugach sandstone compositions. Nd isotopic characteristics of terranes in the Coast Mountains belt span a large range in values, from rocks reflecting juvenile crustal signatures to those with clear Proterozoic signatures (Fig. 10). The ε_{Nd} of the quartzose sandstones overlap strongly with ε_{Nd} data from continental margin assemblages in the Coast Mountains belt (Fig. 10; Samson et al., 1991). The average volcaniclastic sandstones overlap better with "oceanic" terranes and middle Cretaceous plutons intruded them. We suggest that the source of the Kodiak Formation was a rapidly uplifting orogenic belt within the Coast Mountains belt, associated with a volcanic arc built on crust with a Proterozoic model age. This conclusion is similar to other reconstructions of the Chugach source terrane (Farmer et al., 1993), but our results differ in two significant ways. First, the paleocurrent data suggest that long-distance transport of sediment along a trench was not the dominant mode of transport, rather the Chugach flysch was immediately seaward of its source during deposition and accretion. Second, our combined Nd and petrographic data suggest a bimodal source terrane consisting of a juvenile arc and Proterozoic crustal rocks. We suggest that the Chugach flysch was a latest Cretaceous subduction complex paired with an arc orogen in the Coast Mountains belt that was also experiencing rapid uplift of more evolved crustal material. The arc was formed in an oblique convergent setting, recorded by dextral transpression in the Coast Mountains belt (Andronicos et al., 1999; Roeske et al., this volume, Chapter 6),

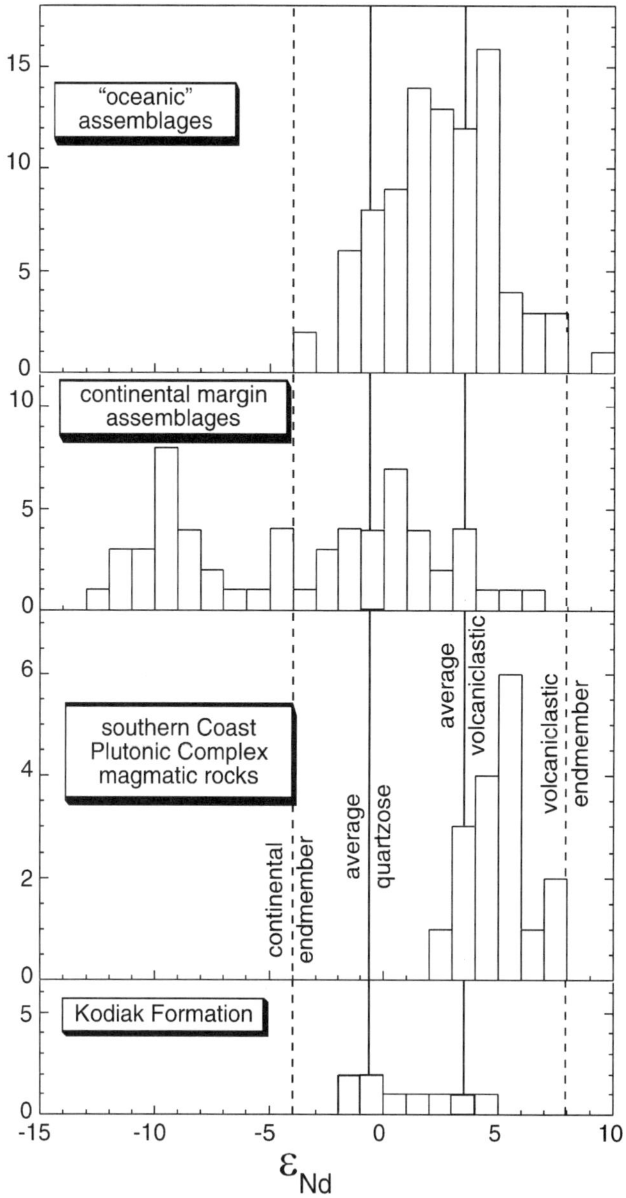

Figure 10. Comparison of Nd signatures of Kodiak Formation and various units in the Coast Mountains belt. ε_{Nd} values are given in present-day values for ease of comparison. Reference lines given for Kodiak Nd data, including endmembers calculated from Figure 8. Data sources are as follows: "oceanic" assemblages (Alexander terrane, Stikine terrane, Porcher Island assemblage, central belt of Burke Channel area, Wrangellia)—Boghossian and Gehrels, 2000; Samson et al., 1989; Samson et al., 1991; "continental margin assemblages" (Burke Channel area western and eastern belts, Ruth assemblage, Port Houghton assemblage, Tracy Arm assemblage, Taku terrane, Endicott Arm assemblage)—Boghossian and Gehrels, 2000; Samson et al., 1991; southern Coast Plutonic Complex magmatic rocks—Cui and Russell, 1995.

as well as by a wide zone of right-slip faults in the seaward belt of the Kodiak Formation.

SUMMARY

One of the most important results of our study is the high deposition rate of the Kodiak Formation, and by extrapolation, the entire Chugach flysch. Comparisons of the size of the Chugach flysch with large submarine fan systems indicates that the Chugach sediments were comparable to the Bengal fan system, implying that Chugach sediments were derived from a collision-driven uplift along the western margin of North America in latest Cretaceous time. Paleocurrent indicators oriented normal to the ancient subduction margin and proximal turbidite lithofacies suggest that the Kodiak Formation was deposited relatively near the North American margin and directly offshore from its source terrane. Two populations of sandstones, the dominant one quartzose and the minor one volcaniclastic, imply a mixture of recycled orogen and magmatic arc terranes. Nd isotope characteristics show the volcaniclastic sandstones to be derived from a continentally influenced magmatic arc and the quartzose sandstones to originate in a recycled orogen with a significant component of Proterozoic influence. Our data, coupled with terrane reconstructions based on paleomagnetic data and plate motion models, suggest that the Cordillera in western British Columbia was the most likely source of the Chugach flysch. In particular, the potentially large magnitude of uplift of the Coast Mountains belt in latest Cretaceous time makes this region an attractive source terrane. However, our data cannot rule out the possibility of other source terranes. The presence of the voluminous Chugach flysch strongly suggests that a large magnitude of uplift occurred somewhere along the western margin of North America in latest Cretaceous time; accretion of this flysch resulted in very rapid growth of the southern Alaska accretionary complex before Tertiary time. More information on Nd isotopic characteristics and uplift histories of southern Alaska and western British Columbia would be particularly useful in constraining the likely source of Chugach flysch sediments.

ACKNOWLEDGMENTS

JCS acknowledges the guidance of J. Casey Moore in the early stages of this research, and the able field assistance of S. Roeske, D. Davies and G. Myers. JCS is also thankful for support from the National Science Foundation (OCE-189396 to JCS and grants to J.C. Moore), Arco, Union, Sohio, and Mobil oil companies, the U.S. Geological Survey, the U.S. Coast Guard, the Geological Society of America, and numerous residents of the Kodiak islands, whose generous hospitality greatly assisted completion of the field work. MRR thanks the Presidents Fellowship Program of the University of California for research support. R. Garrison, T. Byrne, W. Dickinson, and R. Ingersoll reviewed earlier versions of this manuscript. Thorough reviews by K. Marsaglia and G. Gehrels greatly improved the final version.

REFERENCES CITED

Andronicos, C.L., Hollister, L.S., Davidson, C., and Chardon, D., 1999, Kinematics and tectonic significance of transpressive structures within the Coast Plutonic Complex, British Columbia: Journal of Structural Geology, v. 21, p. 229–243.

Armstrong, R.L., 1988, Mesozoic and early Cenozoic magmatic evolution of the Canadian Cordillera, *in* Clark, S.P., et al., eds., Processes in continental lithospheric deformation: Boulder, Colorado, Geological Society of America Special paper 218, p. 55–92.

Bateman, P.C., 1983, A summary of critical relations in the central part of the Sierra Nevada batholith, California, USA, *in* Roddick, J.A., ed., Circum-Pacific plutonic terranes: Boulder, Colorado, Geological Society of America Memoir 159, p. 241–254.

Boghossian, N.D., and Gehrels, G.E., 2000, Nd isotopic signature of metasedimentary pendants in the Coast Mountains between Prince Rupert and Bella Coola, British Columbia, *in* Stowell, H.H., and McClelland, W.C., eds., Tectonics of the Coast Mountains, southeastern Alaska and British Columbia: Boulder, Colorado, Geological Society of America Special Paper 343, p. 77–88.

Bol, A.J., Coe, R.S., Grommé, C.S., and Hillhouse, J.W., 1992, Paleomagnetism of the Resurrection Peninsula, Alaska: Implications for the tectonics of southern Alaska and the Kula-Farallon ridge: Journal of Geophysical Research, v. 97, p. 17,213–17,232.

Bouma, A.H., 1985, Introduction to submarine fans and related turbidite systems, *in* Bouma, A.H., et al., eds., Submarine fans and related turbidite systems: New York, Springer Verlag, p. 3–5.

Byrne, T., 1986, Eocene underplating along the Kodiak Shelf, Alaska: Implications and regional correlations: Tectonics, v. 5, p. 403–422.

Chase, C.G., and Wallace, T.C., 1986, Uplift of the Sierra Nevada of California: Geology, v. 14, p. 730–733.

Coe, R.S., Globerman, B.R., Plumley, P.W., and Thrupp, G.A., 1985, Paleomagnetic results from Alaska and their tectonic implications, *in* Howell, D.G., ed., Tectonostratigraphic terranes of the circum-Pacific region: Houston, Texas, Circum-Pacific Council for Energy and Mineral Resources, Earth Science Series, v. 1, p. 85–108.

Connelly, W., 1978, Uyak Complex, Kodiak Islands, Alaska: A Cretaceous subduction complex: Geological Society of America Bulletin, v. 89, p. 755–769.

Cowan, D.S., Brandon, M.T., and Garver, J.I., 1997, Geologic tests for hypotheses of large coastwise displacements—A critique illustrated by the Baja British Columbia controversy: American Journal of Science, v. 297, p. 117–173.

Crawford, M.L., Hollister, L.S., and Woodsworth, G.J., 1987, Crustal deformation and regional metamorphism across a terrane boundary, Coast Plutonic complex, British Columbia: Tectonics, v. 6, p. 343–362.

Cui, Y., and Russell, J.K., 1995, Nd-Sr-Pb isotopic studies of the southern Coast Plutonic Complex, southwestern British Columbia: Geological Society of America Bulletin, v. 107, p. 127–138.

Curray, J.R., and Moore, D.G., 1971, Growth of the Bengal deep-sea fan and denudation in the Himalayas: Geological Society of America Bulletin, v. 82, p. 563–572.

Damuth, J.E., and Embley, R.W., 1981, Mass transport processes on the Amazon Cone; western equatorial Atlantic: American Association of Petroleum Geologists Bulletin, v. 65, p. 629–643.

Damuth, J.E., and Kumar, N., 1975, Amazon Cone: Morphology, sediments, age, and growth pattern: Geological Society of America Bulletin, v. 86, p. 863–878.

Davies, D.L., and Moore, J.C., 1984, 60 m.y. intrusive rocks from the Kodiak Islands link the Peninsular, Chugach, and Prince William terranes: Geological Society of America Abstracts with Programs, v. 16, no. 5, p. 277.

Decker, J., Nilsen, T.H., and Karl, S., 1979, Turbidite facies of the Sitka Graywacke, southeastern Alaska, *in* Johnson, K.M., and Williams, J.R., eds., The U.S. Geological Survey in Alaska: Accomplishments during 1978: U.S. Geological Survey Circular 804B, p. B125–B129.

DePaolo, D.J., 1981, A neodymium and strontium isotopic study of the Mesozoic calc-alkaline granitic batholiths of the Sierra Nevada and Peninsular Ranges, California: Journal of Geophysical Research, v. 86, p. 10,470–10,488.

Dickinson, W.R., 1982, Compositions of sandstones in circum-Pacific subduction complexes and forearc basins: American Association of Petroleum Geologists Bulletin, v. 66, p. 121–137.

Dickinson, W.R., and Butler, R.F., 1998, Coastal and Baja California paleomagnetism reconsidered: Geological Society of America Bulletin, v. 110, p. 1268–1280.

Dickinson, W.R., and Suczek, C.A., 1979, Plate tectonics and sandstone compositions: American Association of Petroleum Geologists Bulletin, v. 63, p. 2164–2182.

Dickinson, W.R., Beard, L.S., Brakenridge, G.R., Erjavec, J.L., Ferguson, R.C., Inman, K.F., Knepp, R.A., Lindberg, F.A., and Ryberg, P.T., 1983, Provenance of North American Phanerozoic sandstones in relation to tectonic setting: Geological Society of America Bulletin, v. 94, p. 222–235.

Diehl, J.F., 1991, The Elkhorn Mountains revisited: New data for the Late Cretaceous paleomagnetic field of North America: Journal of Geophysical Research, v. 96, p. 9887–9894.

Donelick, R.A., and Miller, D.S., 1986, Low-temperature geothermometry of the Coast plutonic complex, Whitehorse Trough and Atlin terrane, from Atlin, BC, to Haines, AK: Geological Society of America Abstracts with Programs, v. 18, no. 2, p. 102.

Dumoulin, J.A., 1987, Sandstone composition of the Valdez and Orca Groups, Prince William Sound, Alaska: U.S. Geological Survey Bulletin 1774, 37 p.

Dusel-Bacon, C., Doyle, E.O., and Box, S.E., 1996, Distribution, facies, ages, and proposed tectonic associations of regionally metamorphosed rocks in southwestern Alaska and the Alaska Peninsula: U.S. Geological Survey Professional Paper 1497-B, p. B1–B30.

Engebretson, D.C., Cox, A., and Gordon, R.G., 1985, Relative motions between oceanic and continental plates in the Pacific basin: Boulder, Colorado, Geological Society of America Special Paper 206, 59 p.

Farmer, G.L., Ayuso, R., and Plafker, G., 1993, A Coast Mountains provenance for the Valdez and Orca groups, southern Alaska, based on Nd, Sr, and Pb isotopic evidence: Earth and Planetary Science Letters, v. 116, p. 9–21.

Fisher, D.M., 1984, Kinematics of duplex accretion as determined by incremental strain studies, Kodiak Island, Alaska: Geological Society of America Abstracts with Programs, v. 16, no. 6, p. 509.

Fisher, D.M., and Byrne, T., 1987, Structural evolution of underthrusted sediments, Kodiak islands, Alaska: Tectonics, v. 6, p. 775–794.

Fisher, M.A., von Huene, R., Smith, G.L., and Bruns, T.R., 1983, Possible seismic reflections from the downgoing Pacific plate, 275 kilometers arcward from the eastern Aleutian Trench: Journal of Geophysical Research, v. 88, p. 5835–5849.

Fisher, M.A., von Huene, R., and Smith, G.L., 1987, Reflections from midcrustal rocks within the Mesozoic subduction complex near the eastern Aleutian Trench: Journal of Geophysical Research, v. 92, p. 7907–7915.

Gastil, G., 1983, Mesozoic and Cenozoic granitic rocks of southern California and western Mexico, *in* Roddick, J.A., ed., Circum-Pacific plutonic terranes: Boulder, Colorado, Geological Society of America Memoir 159, p. 265–276.

Gehrels, G.E., and Boghossian, N.D., 2000, Reconnaissance geology and U-Pb geochronology of the west flank of the Coast Mountains between Bella Coola and Price Rupert, coastal British Columbia, *in* Stowell, H.H., and McClelland, W.C., eds., Tectonics of the Coast Mountains, southeastern Alaska and British Columbia: Boulder, Colorado, Geological Society of America Special Paper 343, p. 61–76.

Gehrels, G.E., McClelland, W.C., Samson, S.D., and Patchett, P.J., 1989, Structural evolution of the northern Coast Mountains batholith, southeastern Alaska: Eos (Transactions, American Geophysical Union), v. 70, p. 1308.

Gradstein, F.M., Agterberg, F.P., Ogg, J.G., Hardenbol, J., van Veen, P., Thierry, J., and Huang, Z., 1995, A Triassic, Jurassic and Cretaceous time scale, *in* Berggren, W.A., et al., eds., Geochronology, time scales, and global stratigraphic correlation: SEPM (Society for Sedimentary Geology) Special Publication No. 54, p. 95–126.

Gunderson, J.A., and Sheriff, S.D., 1991, A new Late Cretaceous paleomagnetic pole from the Adel Mountains, west-central Montana: Journal of Geophysical Research, v. 96, p. 317–326.

Harland, W.B., Cox, A.V., Llewellyn, P.G., Pickton, C.A.G., Smith, A.G., and Walters, R., 1982, A geologic time scale: Cambridge, Cambridge University Press, 128 p.

Hillhouse, J.W., and Grommé, C.S., 1982, Limits to northward drift of the Paleocene Cantwell Formation, central Alaska: Geology, v. 10, p. 552–556.

Hollister, L.S., and Andronicos, C.L., 1997, A candidate for the Baja British Columbia fault system in British Columbia: GSA Today, v. 7, p. 1–7.

Hollister, L.S., and Andronicos, C.L., 2000, The Central Gneiss Complex, Coast Mountains, British Columbia, *in* Stowell, H.H., and McClelland, W.C., eds., Tectonics of the Coast Mountains, southeastern Alaska and British Columbia: Boulder, Colorado, Geological Society of America Special Paper 343, p. 45–60.

Houghton, H.F., 1980, Refined techniques for staining plagioclase and alkali feldspars in thin section: Journal of Sedimentary Petrology, v. 50, p. 629–631.

Huber, N.K., 1981, Amount and timing of late Cenozoic uplift and tilt of the central Sierra Nevada, California—Evidence from the upper San Joaquin River basin: U.S. Geological Survey Professional Paper 1197, 28 p.

Hudson, T., 1983, Calc-alkaline plutonism along the Pacific rim of southern Alaska, *in* Roddick, J.A., ed., Circum-Pacific plutonic terranes: Boulder, Colorado, Geological Society of America Memoir 159, p. 159–169.

Hudson, T., Plafker, G., and Peterman, Z.E., 1979, Paleogene anatexis along the Gulf of Alaska margin: Geology, v. 7, p. 573–577.

Ingersoll, R.V., and Suczek, C.A., 1979, Petrology and provenance of Neogene sand from Nicobar and Bengal Fans, DSDP sites 211 and 218: Journal of Sedimentary Petrology, v. 49, p. 1217–1228.

Irving, E., Woodsworth, G.J., Wynne, P.J., and Morrison, A., 1985, Paleomagnetic evidence for displacement from the south of the Coast Plutonic Complex, British Columbia: Canadian Journal of Earth Science, v. 22, p. 584–598.

Jones, D.L., and Clark, S.H.B., 1973, Upper Cretaceous (Maestrichtian) fossils from the Kenai-Chugach Mountains, Kodiak and Shumagin Islands, southern Alaska: U.S. Geological Survey Journal of Research, v. 1, p. 125–136.

Kent, D.V., and Gradstein, F.M., 1985, A Cretaceous and Jurassic geochronology: Geological Society of America Bulletin, v. 96, p. 1419–1427.

Kulm, L.D., and von Huene, R., 1973, Initial reports of the Deep Sea Drilling Project, Leg 18: Washington, D.C., U.S. Government Printing Office, 1077 p.

Kumar, N., 1978, Sediment distribution in western Atlantic off northern Brazil—Structural controls and evolution: AAPG Bulletin, v. 62, p. 273–294.

Lanphere, M.A., and Jones, D.L., 1978, Cretaceous time scale from North America, *in* Cohee, G.V., and Glaessner, M.F., eds., The geologic time scale: Tulsa, Oklahoma, AAPG Studies in Geology, p. 259–268.

Mahoney, J.B., Mustard, P.S., Haggart, J.W., Friedman, B.M., Fanning, C.M., and McNicoll, V.J., 1999, Archean zircons in Cretaceous strata of the western Canadian Cordillera: The "Baja B.C." hypothesis fails a "crucial test": Geology, v. 27, p. 195–198.

Mancini, E.A., Deeter, T.M., and Wingate, F.H., 1978, Upper Cretaceous arc-trench gap sedimentation on the Alaska Peninsula: Geology, v. 6, p. 437–439.

Marsaglia, K.M., and Ingersoll, R.V., 1992, Compositional trends in arc-related, deep-marine sand and sandstone: A reassessment of magmatic-arc provenance: Geological Society of America Bulletin, v. 104, p. 1637–1649.

Mattinson, J.M., and James, E.W., 1985, Salinian block U/Pb age and isotopic variations: Implications for origin and emplacement of the Salinian terrane, *in* Howell, D.G., ed., Tectonostratigraphic terranes of the circum-Pacific region: Houston, Texas, Circum-Pacific Council for Energy and Mineral Resources, Earth Science Series, v. 1, p. 215–226.

McClelland, W.C., and Mattinson, J.M., 2000, Cretaceous-Tertiary evolution of the western Coast Mountains, central southeastern Alaska, *in* Stowell, H.H., and McClelland, W.C., eds., Tectonics of the Coast Mountains, southeastern Alaska and British Columbia: Boulder, Colorado, Geological Society of America Special Paper 343, p. 159–182.

Monger, J.W.H., and Price, R.A., 1972, Geodynamic evolution of Canadian Cordillera—Progress and problems: Canadian Journal of Earth Science, v. 16, p. 770–791.

Moore, G.F., 1979, Petrography of subduction zone sandstones from Nias Island, Indonesia: Journal of Sedimentary Petrology, v. 49, p. 71–84.

Moore, G.W., 1967, Preliminary geologic map of Kodiak Island and vicinity, Alaska: U.S. Geological Survey Open-File Report 271.

Moore, J.C., 1973, Cretaceous continental margin sedimentation, southwestern Alaska: Geological Society of America Bulletin, v. 84, p. 595–614.

Moore, J.C., and Allwardt, A., 1980, Progressive deformation of a Tertiary trench slope, Kodiak Islands, Alaska: Journal of Geophysical Research, v. 85, p. 4741–4756.

Moore, J.C., and Wheeler, R.W., 1978, Structural fabric of a melange, Kodiak Islands, Alaska: American Journal of Science, v. 278, p. 739–765.

Moore, J.C., Byrne, T., Plumley, P.W., Reid, M., Gibbons, H., and Coe, R.S., 1983, Paleogene evolution of the Kodiak Islands, Alaska: Consequences of ridge-trench interaction in a more southerly latitude: Tectonics, v. 2, p. 265–293.

Mutti, E., and Ricci-Luchi, F., 1972, Le torbiditi dell'Appennino settentrionale: Introduzione all'analisi di facies: Memorie della Societa Geologica di Italiana, v. 11, p. 161–199 (in Italian).

Nagao, T., and Matsumoto, T., 1940, A monograph of the Cretaceous Inoceramus of Japan, part 2: Hokkaido University Faculty of Science Journal, v. 6, p. 1–64.

Nilsen, T.H., and Moore, G.W., 1979, Reconnaissance study of Upper Cretaceous to Miocene stratigraphic units and sedimentary facies, Kodiak and adjacent islands: U.S. Geological Survey Professional Paper 1093, 34 p.

Nilsen, T.H., and Zuffa, G.G., 1982, The Chugach terrane, a Cretaceous trench-fill deposit, southern Alaska, *in* Leggett, J.K., ed., Trench-forearc geology: Oxford, Blackwell Scientific Publications, p. 213–227.

Page, R.A., Plafker, G., Fuis, G.S., Nokleberg, W.J., Ambos, E.L., Mooney, W.D., and Campbell, D.L., 1986, Accretion and subduction tectonics in the Chugach Mountains and Copper River Basin, Alaska: Initial results of the Trans-Alaska Crustal Transect: Geology, v. 14, p. 501–506.

Palmer, A.R., 1983, The decade of North American geology 1983 geologic time scale: Geology, v. 11, p. 503–504.

Parrish, R.L., 1983, Cenozoic thermal evolution and tectonics of the Coast Mountains of British Columbia, 1: Fission track dating, apparent uplift rates, and patterns of uplift: Tectonics, v. 2, p. 601–632.

Pettijohn, F.J., 1975, Sedimentary Rocks: New York, Harper & Row, 628 p.

Plafker, G., Jones, D.L., and Pessagno, E.A., Jr., 1977, A Cretaceous accretionary flysch and melange terrane along the Gulf of Alaska margin: U.S. Geological Survey Circular 751-B, p. B41–B43.

Plafker, G., Moore, J.C., and Winkler, G.E., 1994, Geology of the southern Alaska margin, *in* Pafker, G., and Berg, H.C., eds., The geology of Alaska: Boulder, Colorado, Geological Society of America, Geology of North America, v. G-1, p. 389–451.

Plumley, P.W., Coe, R.S., and Byrne, T., 1983, Paleomagnetism of the Paleocene Ghost Rocks Formation, Prince William terrane, Alaska: Tectonics, v. 2, p. 295–314.

Ramsay, J.G., 1961, The effects of folding on the orientation of sedimentation structures: Journal of Geology, v. 71, p. 84–100.

Ramsay, J.G., 1967, Folding and fracturing of rock: New York, McGraw-Hill, 568 p.

Reid, M.R., 1995, Processes of mantle enrichment and magmatic differentiation in the eastern Snake River Plain: Th isotope evidence: Earth and Planetary Science Letters, v. 131, p. 239–254.

Roddick, J.A., 1983, Geophysical review and composition of the Coast plutonic complex, south of latitude 55° N, *in* Roddick, J.A., ed., Circum-Pacific plutonic terranes: Boulder, Colorado, Geological Society of America Memoir 159, p. 195–212.

Roeske, S., 1986, Field relations and metamorphism of the Raspberry schist, Kodiak Islands, Alaska, *in* Evans, B.W., and Brown, E.H., eds., Blueschists and eclogites: Boulder, Colorado, Geological Society of America Memoir 164, p. 169–184.

Roeske, S.M., Snee, L.W., and Pavlis, T.L., 2003, Dextral-slip reactivation of an arc-forearc boundary during Late Cretaceous–Early Eocene oblique convergence in the northern Cordillera, *in* Sisson, V.B., Roeske, S.M., and Pavlis, T.L., eds., Geology of a transpressional orogen developed during ridge-trench interaction along the North Pacific margin: Boulder, Colorado, Geological Society of America Special Paper 371, p. 141–169 (this volume).

Rubin, C.M., and Saleeby, J.B., 2000, U-Pb geochronology of mid-Cretaceous and Tertiary plutons along the western edge of the Coast Mountains, Revillagigedo Island, and Portland Peninsula, southeast Alaska, *in* Stowell, H.H., and McClelland, W.C., eds., Tectonics of the Coast Mountains, southeastern Alaska and British Columbia: Boulder, Colorado, Geological Society of America Special Paper 343, p. 145–158.

Rusmore, M.E., Woodsworth, G.J., and Gehrels, G.E., 2000, Late Cretaceous evolution of the eastern Coast Mountains, Bella Coola, British Columbia, *in* Stowell, H.H., and McClelland, W.C., eds., Tectonics of the Coast Mountains, southeastern Alaska and British Columbia: Boulder, Colorado, Geological Society of America Special Paper 343, p. 89–106.

Saleeby, J., 2000, Geochronologic investigations along the Alexander-Taku terrane boundary, southern Revillagigedo Island to Cape Fox areas, southeast Alaska, *in* Stowell, H.H., and McClelland, W.C., eds., Tectonics of the Coast Mountains, southeastern Alaska and British Columbia: Boulder, Colorado, Geological Society of America Special Paper 343, p. 107–143.

Sample, J.C., and Fisher, D.M., 1986, Duplexes and underplating in an ancient accretionary complex, Kodiak islands, Alaska: Geology, v. 14, p. 160–163.

Sample, J.C., and Karig, D.E., 1982, A volcanic production rate for the Mariana island arc: Journal of Volcanology and Geothermal Research, v. 13, p. 73–82.

Sample, J.C., and Moore, J.C., 1987, Structural style and kinematics of an underplated stale belt, Kodiak and adjacent islands, Alaska: Geological Society of America Bulletin, v. 99, p. 7–20.

Sample, J.C., and Reid, M.R., 1986, Bearing of sediment volume, sandstone petrography, and Nd model ages on the origin of the Kodiak Formation, Alaska: Geological Society of America Abstracts with Programs, v. 18, no. 2, p. 180.

Samson, S.D., McClelland, W.C., Patchett, P.J., Gehrels, G.E., and Anderson, R.G., 1989, Evidence from neodymium isotopes for mantle contributions to Phanerozoic crustal genesis in the Canadian Cordillera: Nature, v. 337, p. 705–709.

Samson, S.D., Patchett, P.J., McClelland, W.C., and Gehrels, G.E., 1991, Nd isotopic characterization of metamorphic rocks in the Coast Mountains, Alaskan and Canadian Cordillera: Ancient crust bounded by juvenile terranes: Tectonics, v. 4, p. 770–780.

Stamatakos, J.A., Kodama, K.P., Vittorio, L.F., and Pavlis, T.L., 1989, Paleomagnetism of Cretaceous and Paleocene sedimentary rocks across the Castle Mountain fault, south-central Alaska: American Geophysical Union Geophysical Monograph Series, v. 50, p. 151–177.

Stern, T.W., Bateman, P.C., Morgan, B.A., Newell, M.F., and Peck, D.L., 1981, Isotopic U-Pb ages of zircon from granitoids of the central Sierra Nevada, California: U.S. Geological Survey Professional Paper 1185, 17 p.

Stevenson, A.J., Scholl, D.W., and Vallier, T.L., 1983, Tectonic and geologic implications of the Zodiac fan, Aleutian Abyssal Plain, northeast Pacific: Geological Society of America Bulletin, v. 93, p. 259–273.

Stock, J., and Molnar, P., 1988, Uncertainties and implications of the Late Cretaceous and Tertiary position of North America relative to the Farallon, Kula, and Pacific plates: Tectonics, v. 7, p. 1339–1384.

Thrupp, G.A., and Coe, R.S., 1986, Early Tertiary paleomagnetic evidence and the displacement of southern Alaska: Geology, v. 14, p. 213–217.

Umhoefer, P.J., 1987, Northward translation of "Baja British Columbia" along the Late Cretaceous to Paleocene margin of western North America: Tectonics, v. 6, p. 377–394.

Unruh, J.R., 1991, The uplift of the Sierra Nevada and implications for late Cenozoic epeirogeny in the western Cordillera: Geological Society of America Bulletin, v. 103, p. 1395–1404.

Vereshchagin, V.N., Kinasov, V.P., Paraketsov, K.V., and Terekhova, G.P., 1965, Polevoi Atlas: Melovoi fauni Severo-Vostoka S.S.S.R., Magadan, Gosudarst. Proizvodst. Geol. Dom. RSFSR, Servero-Vostoch. Geol. Upravlenie, 66 p. (in Russian).

von Drach, V., Marsh, B.D., and Wasserburg, G.J., 1986, Nd and Sr isotopes in the Aleutians: Multicomponent parenthood of island-arc magmas: Contributions to Mineralogy and Petrology, v. 92, p. 13–34.

von Huene, R., Box, S., Detterman, R., Moore, J.C., and Pulpan, H., compilers, 1985, Geodynamics Cross Section, A-2 Kodiak to Kuskokwim, Alaska: Boulder, Colorado, Geological Society of America, Centenial Continent/Ocean Transects, v. 6, 1 sheet, scale 1:500,000.

Vrolijk, P., Myers, G., and Moore, J.C., 1988, Warm fluid migration along tectonic melanges in the Kodiak accretionary complex, Alaska: Journal of Geophysical Research, v. 93, p. 10,313–10,324.

Wallace, W.K., and Engebretson, D.C., 1984, Relationships between plate motions and Late Cretaceous to Paleogene magmatism in southwestern Alaska: Tectonics, v. 3, p. 295–315.

Wilson, M.D., and Sedeora, S.S., 1979, An improved thin section stain for potash feldspar: Journal of Sedimentary Petrology, v. 49, p. 637–638.

Zuffa, G.G., Nilsen, T.H., and Winkler, G.R., 1980, Rock fragment petrography of the Upper Cretaceous Chugach terrane, southern Alaska: U.S. Geological Survey Open-file Report 80-713, 28 p.

Manuscript Accepted by the Society February 5, 2003

Printed in the USA

Geological Society of America
Special Paper 371
2003

Cooling and exhumation history of the Kodiak accretionary prism, southwest Alaska

W.S. Clendenen
ExxonMobil Upstream Research Center, Box 2189, Houston, Texas 77252-2189, USA

Donald Fisher
Department of Geosciences, Pennsylvania State University, University Park, Pennsylvania 16802, USA

Timothy Byrne
Department of Geology and Geophysics, University of Connecticut, Storrs, Connecticut 06269, USA

ABSTRACT

In this paper, we combine zircon and apatite fission track dating with $^{40}Ar/^{39}Ar$ mineral dating to establish the thermal history of accreted units from the Kodiak Accretionary Complex in southwest Alaska. The Kodiak archipelago exposes a series of northeast-trending subduction-related units that range in age from Jurassic on the northwest to Eocene in the southeast. Thus, this complex was assembled and has resided along a margin with a long-lived history of convergence. This analysis focuses on the thermal history of the Late Triassic through Late Cretaceous units: the Early Jurassic Afognak pluton, Jurassic Raspberry schist, the Late Cretaceous Uyak Complex and the early Maastrichtian Kodiak Formation. This tectonic assemblage straddles the arc-accretionary prism boundary (marked by the Border Ranges fault) and constitutes over 60% of the exposed accretionary complex.

$^{40}Ar/^{39}Ar$ data of feldspar and fission track data from the Afognak pluton suggest a relatively simple history of slow cooling through Ar closure from 150 to 120 Ma. Apatite and zircon fission track analyses for these units show that the pluton and schist underwent similar thermal histories through the closure of zircon and apatite fission tracks, suggesting that these two diverse units may have been in proximity to one another since at least the Late Jurassic. These results also suggest that the Border Ranges fault, which separates these units on Kodiak Island, has not experienced any vertical throw that has impacted the post-Jurassic cooling history.

Thermal modeling based on fission track lengths in apatite suggests rapid cooling of the pluton, schist, and the Uyak Complex in the Late Cretaceous–early Tertiary. In the highest structural levels of the Kodiak Formation (the landward and seaward belts), zircon fission track dates are consistently ca. 50 Ma. The lowest structural level (the central belt) records 44 Ma ages for both zircon and apatite, indicating rapid cooling in the Eocene within the core of a regional-scale antiform. Apatite fission track ages for the landward and seaward belts range from 44 to 25 Ma from northwest to southeast, suggesting a seaward wave of cooling within the higher structural levels. The antiformal structure and its associated cooling history could be a response to underplating, with seaward propagation of a forearc high leading to erosional unroofing.

The two rapid cooling events in the latest Cretaceous–early Tertiary and Eocene-Oligocene correspond with known events of rapid accretion. Thus, rapid accretion may lead to rapid cooling of the prism, probably due to uplift and exhumation of the accretionary complex. The prism can experience extremely slow cooling in the absence of fluxes of sediment into and off the top of the wedge during accretion/unroofing events.

Keywords: Alaska, accretion, geochronology, cooling, exhumation.

Clendenen, W.S., Fisher, D., and Byrne, T., 2003, Cooling and exhumation history of the Kodiak accretionary prism, southwest Alaska, *in* Sisson, V.B., Roeske, S.M., and Pavlis, T.L., eds., Geology of a transpressional orogen developed during ridge-trench interaction along the North Pacific margin: Boulder, Colorado, Geological Society of America Special Paper 371, p. 71–88.

INTRODUCTION

It is widely recognized that two common processes along convergent plate margins are accretion and subduction erosion, with rates that depend on factors such as volume of sediment input, subduction angle, basal friction, and seafloor roughness (e.g., Cloos and Shreve, 1988a, 1988b; Stern, 1991; von Huene and Scholl, 1991; Lallemand et al., 1992; Lallemand et al., 1994; Bangs and Cande, 1997; Gutscher et al., 1998a, 1998b). Based on these characteristics, margins can be classified into (1) type I margins where a thick sediment pile on the subducting plate is accreted into a prism that thickens and grows seaward by a combination of underplating and frontal accretion (i.e., offscraping) and (2) type II margins where there is very little sediment input and the outer forearc can be embayed and scarred by incoming seamounts (see von Huene and Scholl, 1991). The processes by which accretionary wedges grow are well documented through studies of exposed accretionary complexes and submarine accretionary prisms (e.g., Moore et al., 1990; Sample and Fisher, 1986; Brown and Westbrook, 1988; Byrne and Fisher, 1990; Byrne et al., 1993; and many others). The processes by which margins experience basal erosion are more enigmatic, however, as they must be inferred from the absence of material, the subsidence of the margin, or the truncation of structure along the plate boundary. Nevertheless, subduction erosion has been inferred along margins such as the Japan, Costa Rica, and Peru margins based on trench retreat as well as subsidence of the upper slope.

The thermal structure of the forearc may be influenced by the age of oceanic crust being subducted and by processes of accretion and subduction erosion, with transient thermal responses to variations in margin behavior. Along an accreting margin, rapid cooling will result if the prism can build above sea level, whereby subaerial erosion can lead to shorter residence times within the wedge. Local uplift due to the underplating of sediment packages can also lead to oversteepening of the surface and development of a supercritical wedge that reequilibrates through submarine erosion and/or extension. Along a margin that undergoes subduction erosion, the outer forearc experiences subsidence, burial, and heating while the inner forearc is uplifted, eroded, and cooled (e.g., Fisher et al., 1998). Subduction of a spreading center may result in an increase in thermal gradient followed by slow cooling as older crust is progressively consumed. Given the fact that thermal histories provide a constraint on mass fluxes within the forearc of a subduction zone, it is unfortunate that there are few accretionary prisms (e.g., the Shimanto belt; Ohmori et al., 1997) where there are well-documented spatial variations in cooling histories.

In this paper, we combine $^{40}Ar/^{39}Ar$ mineral dating with zircon and apatite fission track dating to establish the thermal history of accreted units from the Kodiak Accretionary Complex in southwest Alaska. The Kodiak archipelago exposes series of northeast-trending subduction-related units that range in age from Jurassic on the northwest to Eocene in the southeast. Thus, this complex was assembled and has resided along a margin with a long-lived history of convergence. By comparing the thermal history of accreted packages, we will address three questions concerning accretionary processes: (1) When did different units cool through the closure of different thermochronometers and how does this history relate to the documented history of episodic accretion (Byrne and Fisher, 1987) and ridge subduction (Moore et al., 1983) within the complex? (2) How do cooling histories vary across major tectonic boundaries within the complex (e.g., the Border Ranges fault) and what constraint does this result place on the movement history along the boundaries? (3) Are there local variations in thermal history within accreted units that relate to spatial variations in exhumation history?

GEOLOGIC SETTING

This analysis focuses on the thermal history of, from northwest to southeast, the Late Triassic to Early Jurassic Afognak pluton, the early Jurassic Raspberry schist, the Late Cretaceous Uyak Complex, and the early Maastrichtian Kodiak Formation (Fig. 1). This tectonic assemblage straddles the Border Ranges fault and Uganik thrust and constitutes over 60% of the exposed accretionary complex. The Border Ranges fault (Fig. 1) is a major terrane boundary that separates arc rocks of the Peninsular terrane on the northwest from accreted rocks of the Chugach and Prince William terranes on the southeast (Silberling et al., 1992). On the Kodiak Islands, this fault juxtaposes a magmatic arc assemblage and a blueschist-bearing schist belt of early Mesozoic age. The Afognak pluton (Fig. 2) is a calc-alkaline dioritic to tonalitic intrusion (Hill, 1979; Hill et al., 1981) of Late Triassic to Early Jurassic age (Roeske, 1986; Roeske et al., 1989) and is hosted by Late Triassic volcanics and volcaniclastics of the Shuyak Formation (Connelly, 1978). The Afognak pluton and correlative rocks to the northeast on Kenai Peninsula (Cowan and Boss, 1978) and in the Chugach Mountains are probably remnants of a primitive island arc (Burns, 1985; Hudson et al., 1985). In addition, six small tectonic slivers of mafic and ultramafic rocks are present within the earlier Mesozoic section exposed on the Kodiak Islands and in several other locations along strike to the northeast. These rocks probably represent the plutonic core of an intra-oceanic island arc that was emplaced at depths of 18–30 km (Burns, 1985). Together, these units represent the tectonically dismembered Late Triassic to Early Jurassic Kodiak/Chugach magmatic arc assemblage (Hudson et al., 1985), which has been subsequently juxtaposed with Mesozoic accreted rocks along the Border Ranges fault (Roeske et al., 1989).

Southeast of the Border Ranges fault lies the Early Jurassic Raspberry schist (Fig. 2). This schist is composed primarily of metabasite and quartzite (80%) derived from basalt and chert, respectively, and little detrital material (about 10% pelite) (Roeske, 1986). The presence of metabasite and hemipelagic sediments and the absence of any thick, sandstone-shale sequences indicates that this unit probably represents ocean

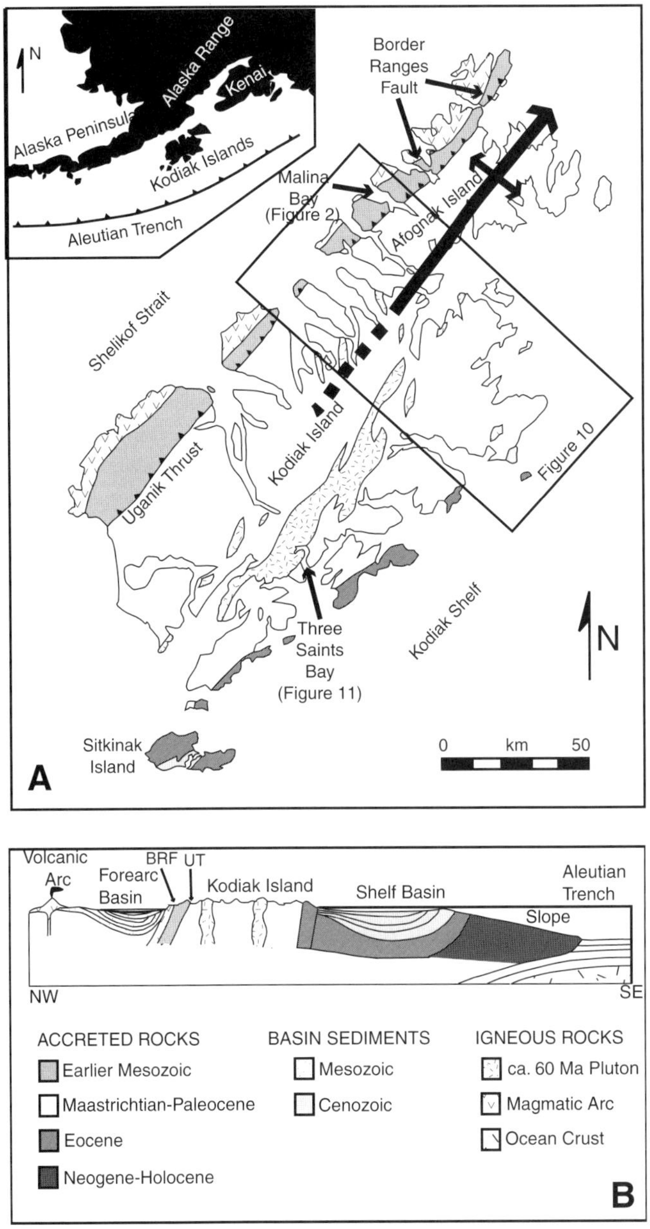

Figure 1. Regional geology of the Kodiak Islands. A: Generalized geologic map of the Kodiak Islands. Geology from Moore (1967), Connelly and Moore (1979), Moore and Allwardt (1980), Moore et at (1983), and Fisher and Byrne (1992). B: Schematic cross section through the southern Alaska convergent margin. Basin geometries from Fisher and von Huene (1980) and Fisher et al. (1987a). BRT—Border Ranges thrust; UT—Uganik thrust.

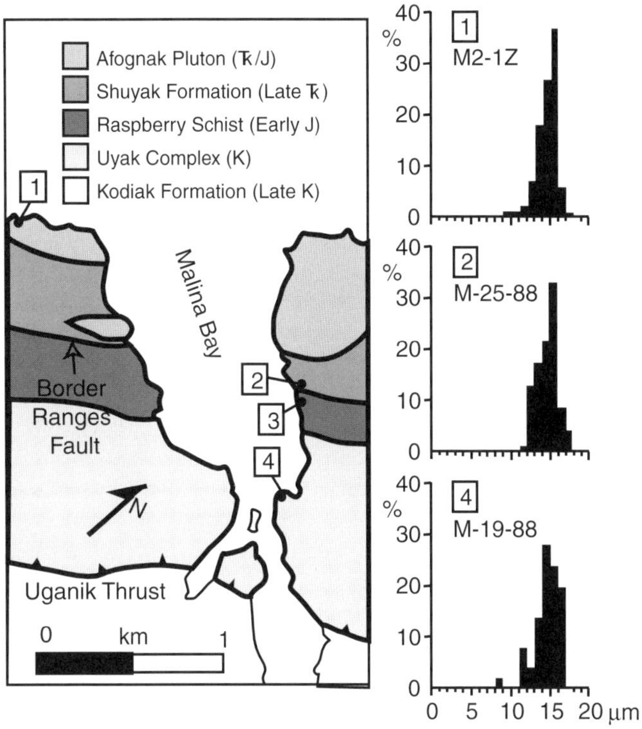

Figure 2. Geologic map of Malina Bay showing sample locations and apatite fission track length distributions. Statistical details of the track length distributions are given in Table 4. Geology from Connelly and Moore (1979) and Roeske (1986).

crust with very little clastic cover. The schist is glaucophane-bearing, recording transitional blueschist facies metamorphic conditions, and lacks a greenschist facies overprint (Carden et al., 1977; Roeske, 1986; Roeske et al., 1989). The Raspberry schist correlates with other blueschist-bearing units found sporadically for over 1000 km along strike to the northeast and record similar metamorphic conditions (Forbes and Lanphere, 1973; Carden et al., 1977; Cowan and Boss, 1978; Sisson and Onstott, 1986). This package was accreted and metamorphosed in the Late Triassic–Early Jurassic (Roeske, 1986).

The Uyak Complex (Fig. 2) lies oceanward of the Raspberry schist and is a tectonic melange containing argillite, graywacke, greenstone, gray and red radiolarian chert, gabbroic and ultramafic rocks, tuff, and limestone (Connelly, 1978). The complex is correlative with the Seldovia Bay Complex and McHugh Complex on Kenai Peninsula (Clark, 1973; Cowan and Boss, 1978; Kusky and Bradley, 1999) and with other similar units exposed for over 1000 km along strike (Plafker et al., 1977). Fossils within the Uyak Complex are scarce and range in age from Permian to mid–Early Cretaceous (Connelly, 1978). This and the fact that Maastrichtian rocks to the southeast were added to the margin after the Uyak Complex indicate that the melange was accreted sometime during the Cretaceous. The Uyak Complex has relatively little clastic detritus, containing only about 35% foliated shale and graywacke, with transitional composition of the graywacke between that found at continental convergent margins and oceanic island arcs (Connelly, 1978). The Uyak Complex most likely represents material accreted to the convergent margin while ocean crust with a relatively thin clastic veneer was subducted during the Cretaceous.

The accretionary prism experienced a significant transition in the latest Cretaceous and Paleocene when a large volume of clastic-rich sediments comprising the bulk of the Kodiak Islands outcrop was accreted in a relatively short period of time (Raymond, 1980; Byrne and Fisher, 1987). This event represents the first of the three episodes of clastic sediment accretion during the latest Cretaceous and Cenozoic. The latest Cretaceous transition from the accretion of clastic-poor to clastic-dominated material may have taken place as the plate boundary shifted from an intraoceanic to a continental margin setting, permitting continentally derived detritus to reach the trench (Clendenen and Byrne, 1989). Latest Cretaceous and Paleocene sediments (Fig. 1) are exposed for over 1000 km along strike (Moore, 1973; Plafker et al., 1977) and are represented on the Kodiak Islands by the Kodiak and Ghost Rocks formations.

The Kodiak Formation is a largely structurally coherent unit consisting almost entirely of interbedded graywacke and shale (slate) of Maastrichtian age with a slaty cleavage that formed prior to intrusion of 60 Ma plutons (Jones and Clark, 1973; Nilsen and Moore, 1979; Sample and Moore, 1987; Fisher and Byrne, 1992). The formation has been divided into three structural units on the basis of structural history and intensity of deformation associated with slaty cleavage development (Sample and Fisher, 1986; Sample and Moore, 1987; Fisher and Byrne, 1992). The differences between structural units reflect variations in structural level related to development of a regional antiform that postdates cleavage development (Fisher and Byrne, 1987; 1992). The central belt is a regional culmination that exposes the lowest structural level of the Kodiak Formation in the core of this antiform (Fisher and Byrne, 1992). Based on incremental strain histories from pressure shadows and other kinematic indicators such as rotated veins and asymmetric folds, this unit experienced a high degree of top-to-the southeast noncoaxial bulk shear, perhaps near the basal decollement (Fisher and Byrne, 1992). In contrast, the landward and seaward belts, which flank the central belt, preserve higher structural levels of the accretionary prism and did not experience the same degree of metamorphism and deformation.

Vitrinite reflectance and illite crystallinity data indicate a maximum regional metamorphic temperature of about 200 to 250 °C in the Kodiak Formation (Sample and Moore, 1987). Coexisting H_2O-rich and CH_4- or CO_2-rich fluid inclusions within cleavage-related quartz veins record trapping temperatures of 215–255 °C and trapping pressures of approximately 300 MPa (Vrolijk et al., 1988; Brantley et al., 1997). In the central belt, oxygen isotopic analyses of vein quartz and wall rock suggest temperatures of approximately 280 °C during cleavage formation (Brantley et al., 1997). These results are consistent with complete resetting of apatite fission tracks and partial to complete resetting of zircon fission tracks during metamorphism of the central belt. These data also indicate that ~10 km of material has been removed from above the Kodiak Formation at some time in the Tertiary.

During the early Tertiary and contemporaneous with deposition of the Ghost Rocks Formation, the accretionary prism and the Triassic arc were intruded by an extensive igneous suite (Moore et al., 1983; Bradley et al., 1993). These ca. 60 Ma intrusives (Fig. 1) have been variously interpreted to have resulted from anatectic melting of subducted sediments (Hudson et al., 1979), an outboard shift of the volcanic arc (Kienle and Turner, 1976), or interaction of the subduction zone with a spreading ridge (Hill et al., 1981; Moore et al., 1983; Kusky et al., 1997). Isotopic data from basalts and andesites of the Ghost Rocks Formation (Moore et al., 1983) and from slightly younger plutons intruding the Kodiak and Ghost Rocks Formations (Hill et al., 1981) indicate mixing of MORB and metasedimentary sources, strongly supporting the idea that these intrusives formed when a spreading ridge entered the subduction zone. In any case, the plutons signify a distinct thermal culmination during the early Tertiary, making at least the deepest parts of the accretionary prism hotter than expected for a normal subduction regime (Sample and Moore, 1987; Paterson and Sample, 1988). The plutons that mark this event stitch all the units used for thermochrometry in this study, indicating conclusively that all these units were present in the accretionary prism at 60 Ma.

Two additional sediment packages were accreted during the Cenozoic (Fig. 1). The rapid sediment accretion recorded by these events is relevant to the Tertiary cooling history of the preexisting accretionary prism. On Kodiak Island, the second episode of clastic-dominated accretion is represented by the Eocene Sitkalidak Formation (Moore and Allwardt, 1980). Eocene age rocks also extend beneath the offshore shelf basins, where they have been penetrated by three Deep Stratigraphic Test wells (Turner, 1987). In addition, a large package of domed seismic reflectors deep within the accretionary prism (Fisher et al., 1983, 1987b) have been attributed to an underplated equivalent of these Eocene rocks (Byrne, 1986; Moore et al., 1991). The third episode of clastic sediment accretion is represented by rocks underlying the slope (Kulm et al., 1973) that were accreted largely during the Plio-Pleistocene (Fig. 1).

THERMOCHRONOMETRY

A total of 19 rock samples (Table 1) were collected for thermochronometric analysis in order to investigate (1) the thermal history of rocks along the arc-prism boundary (Malina Bay, northwest Kodiak Island) and (2) the variations in thermal history across the exposed prism (along a NW-trending transect from Malina Bay to near the town of Kodiak). Of the 19 samples, 18 are from the Kodiak Islands and one is from the Kenai Peninsula. Samples of the Tertiary plutonic rocks from Three Saints Bay in south-central Kodiak Island and the metasedimentary rock from the Kenai Peninsula, northwest of Kodiak Island, have been projected into this across-strike transect. The samples yielded 2 mineral separates suitable for $^{40}Ar/^{39}Ar$ step-heating analysis, 12 zircon fission track ages, including 3 sets of individual grain ages and 15 apatite fission track ages. Track length data were obtained in 12 of the apatite samples. Data used to calculate apatite and zircon zeta factors based on samples from

TABLE 1. SAMPLES USED FOR FISSION TRACK AND $^{40}Ar/^{39}Ar$ ANALYSES

Sample no.	Sample identifier	Latitude	Longitude	Elevation (m)	Description
1	M2-IZ	N58°12'13"	W153°12'42"	0	Late Triassic Afognak Pluton
2	M-25-88	N58°14'04"	W153°2'54"	0	Dike cutting Triassic Shuyak Formation
3	M-26-88	N58°13'84"	W153°2'18"	0	Jurassic Raspberry Schist
4	M-19-88	N58°12'64"	W153°00'08"	0	Dike cutting Cretaceous Uyak Complex
5	P-1-87	N57°56'30"	W153°4'28"	0	Pluton in Kodiak Formation Central Belt
6	P-2-87	N57°55'09"	W152°49'25"	0	Pluton in Kodiak Formation
7	P-5-87	N57°52'17"	W152°38'11"	14	Pluton in Kodiak Formation Seaward Belt
8	TLP-95	N57°39'44"	W152°58'25"	400	Pluton at Terror Lake hydroelectric tunnel
9	TL-2-87	N57°39'29"	W152°58'33"	922	Pluton at peak above hydroelectric tunnel
10	TS-1-87	N57°11'56"	W153°29'92"	770	Pluton at Three Saints Bay peak
11	TS-4-87	N57°11'43"	W153°30'28"	463	Pluton at Three Saints Bay peak
12	TS-7-87	N57°11'25"	W153°29'82"	230	Pluton at Three Saints Bay peak
13	TS-9-87	N57°11'06"	W153°30'14"	0	Pluton at Three Saints Bay
14	H1-10Z	N57°31'34"	W153°29'88"	0	Pluton at Three Saints Bay
15	P-6-87	N57°31'34"	W152°27'10"	120	Pluton in Ghost Rocks Formation
16	K-2-88	N58°10'32"	W152°53'66"	0	Kodiak Formation graywacke
17	K-5-87	N57°56'95"	W152°56'23"	0	Kodiak Formation graywacke
18	K-16-87	N57°41'40"	W152°28'45"	30	Kodiak Formation graywacke
19	V-3-87	N60°59'03"	W149°33'04"	10	Valdez Group graywacke near Anchorage

the Fish Canyon Tuff, Durango apatite, and Mount Dromedary Igneous Complex are in Clendenen (1991).

The Arc-Prism Boundary, Northwest Kodiak Island

Four samples were collected along the coastal exposures in Malina Bay: one each from the Afognak pluton and the Raspberry schist, and two from the early Tertiary igneous dikes (Fig. 2). One of the dikes intrudes the Shuyak Formation whereas the other intrudes the Uyak Complex; both dikes cross-cut the dominant structural fabrics present in the wall rocks. All four of the samples yielded sufficient apatite for dating and, in three samples, track lengths could be measured. In addition, three of the samples yielded sufficient zircon for dating and two of the samples yielded minerals that could be analyzed with argon step heating.

Argon Step Heating

Two of the samples from the earlier Mesozoic section in Malina Bay yielded amphibole and feldspar that were sutable for analysis with $^{40}Ar/^{39}Ar$ incremental heating (see Tables 2 and 3 for analytical data). Amphibole from an approximately one meter wide igneous dike intruding the Uyak Complex (sample M-19-88; loc. 4, Fig. 2) was dated to attempt to constrain its age of emplacement. The age spectrum (Fig. 3A) shows anomalously old ages in the first 13% of gas released, probably due to excess Ar, but has a clean plateau over 75% of total gas released. Data from these four steps were used to calculate an isochron age of 59.3 ± 2.2 Ma (Fig. 3B). The $^{40}Ar/^{39}Ar$ ratio for the trapped argon component is apparently lower than atmospheric (Fig. 3B). This is most likely due to the clustering of the limited data and is not statistically significant. The isochron age is interpreted to be the age of dike emplacement because the ~500 °C closure temperature of amphibole (Harrison, 1981) is much higher than the maximum temperature experienced by the unit it intrudes (i.e., the Uyak Complex, 270–290 °C (Vrolijk et al., 1988).

The second mineral sample analyzed with $^{40}Ar/^{39}Ar$ step-heating is feldspar separated from the Afognak pluton (sample M2-1Z; loc. 1, Fig. 2). Low K/Ca ratios (Fig. 4; K_2O = 0.56%, Table 3) suggest that the sample contains albite. The spectrum is somewhat ambiguous (Fig. 4). Three small steps, comprising

TABLE 2. $^{40}Ar/^{39}Ar$ INCREMENTAL RELEASE DATA FROM SAMPLE M-19-18

Temp (˚C)	$^{40}Ar/^{39}Ar$	$^{37}Ar/^{39}Ar$	$^{36}Ar/^{39}Ar$ (10^{-3})	^{39}Ar (x10^{-13} moles)	Cumulative ^{39}Ar (%)	^{40}Ar* (%)	$^{40}Ar^*/^{39}Ar$	$^{39}Ar/^{40}Ar$	$^{36}Ar/^{40}Ar$ (10^{-4})	Age (Ma)
750	64.71	0.8520	183.8	0.555	6.65	16.1	10.43			83.3 ± 15.6
850	18.93	2.478	35.16	0.168	8.67	44.7	8.684			69.6 ± 10.1
900	12.76	6.692	11.70	0.100	9.87	71.5	9.744			77.9 ± 12.8
930	11.11	5.154	4.71	0.063	10.6	81.0	10.05			80.4 ± 13.1
960	11.51	7.385	37.84	0.072	11.5	6.07	0.7694			6.3 ± 26.7
990	11.71	12.28	16.15	0.147	13.3	62.7	7.753			62.3 ± 17.1
1020	10.64	18.16	20.28	0.659	21.2	53.5	5.834			47.1 ± 2.8
1040	9.475	20.30	12.58	1.67	41.2	73.5	7.104	0.1042 ± 0.0003	8.85 ± 0.29	57.2 ± 1.0
1070	8.761	20.32	9.965	2.81	75	80.3	7.165	0.1127 ± 0.0002	6.65 ± 0.17	57.6 ± 0.6
1110	10.02	20.11	15.46	0.765	84.2	66.0	6.786	0.0985 ± 0.0003	11.32 ± 0.75	54.6 ± 2.4
1140	8.781	19.64	11.05	1.03	96.5	75.8	6.814	0.1125 ± 0.0010	8.01 ± 0.77	54.9 ± 2.5
1200	7.728	10.46	24.33	0.053	97.1	12.7	1.175			9.6 ± 48.2
1450	9.506	17.17	4.559	0.241	100.0	93.6	9.323			74.6 ± 7.7

Note: Samples were irradiated in a fast neutron beam in the Ford reactor, University of Michigan. Flux monitors (Fe-Mica), used to determine J values, were placed above, below, and between samples. Incremental heating experiments were performed at M. Harrison's laboratory at State University of New York at Albany on an automated Nuclide 4.5-60-RSS mass spectrometer operated with a sensitivity of about 2 x 10^{-15} mol Ar mV^{-1}. The correction factors used were $(^{40}Ar/^{39}Ar)_K$ = 0.0218, $(^{36}Ar/^{37}Ar)_{Ca}$ = 0.00021, and $(^{39}Ar/^{37}Ar)_{Ca}$ = 0.00075. Other details regarding analysis and data reduction protocol are given in Harrison et al. (1986). All errors are one standard deviation. J—0.004531; Ample weight—302.20 mg; Total gas age—58.3 Ma; K_2O—0.23%. α data are given only for those increments used to calculate an isochron age.

*Denotes radiogenic Ar.

TABLE 3. $^{40}Ar/^{39}Ar$ INCREMENTAL RELEASE DATA FROM SAMPLE M2-1Z FELDSPAR

Temp (˚C)	$^{40}Ar/^{39}Ar$	$^{37}Ar/^{39}Ar$	$^{36}Ar/^{39}Ar$ (10^{-3})	^{39}Ar (x10^{-13} moles)	Cumulative ^{39}Ar (%)	^{40}Ar* (%)	$^{40}Ar^*/^{39}Ar$	Age (Ma)
500	59.76	0.4569	123.6	0.343	10.6	38.7	23.24	121.4 ± 3.7
570	29.39	0.5234	16.12	0.579	28.25	83.3	24.65	128.5 ± 3.9
630	31.00	0.7693	12.85	0.612	47.4	87.4	27.24	141.5 ± 6.2
700	30.33	1.939	27.96	0.136	51.7	71.2	22.17	115.9 ± 12.3
770	29.24	3.761	14.96	0.241	59.1	84.1	25.05	130.5 ± 3.5
850	27.42	4.137	1.237	0.123	62.9	96.4	27.31	141.8 ± 3.2
950	31.08	3.798	13.05	0.338	73.4	87.2	27.46	142.5 ± 3.6
1050	28.04	3.134	20.47	0.167	78.6	77.2	22.17	115.9 ± 14.3
1100	30.53	4.173	30.55	0.114	82.1	69.0	21.73	113.7 ± 9.2
1500	52.24	11.74	81.38	0.579	100.0	54.6	28.96	150.0 ± 3.5

Note: Samples were irradiated in a fast neutron beam in the Ford reactor, University of Michigan. Flux monitors (Fe-Mica), used to determine J values, were placed above, below, and between samples. Incremental heating experiments were performed on an automated Nuclide 4.5-60-RSS mass spectrometer operated with a sensitivity of about 2 × 10^{-15} mol Ar mV^{-1}. The correction factors used were $(^{40}Ar/^{39}Ar)_K$ = 0.0106, $(^{36}Ar/^{37}Ar)_{Ca}$ = 0.00013, and $(^{39}Ar/^{37}Ar)_{Ca}$ = 0.00095. Other details regarding analysis and data reduction protocol are given in Harrison et al (1986). All errors are one standard deviation. J—0.002994; Ample weight—72.37 mg; Total gas age—134.5 Ma; K_2O—0.56%.

*Denotes radiogenic Ar.

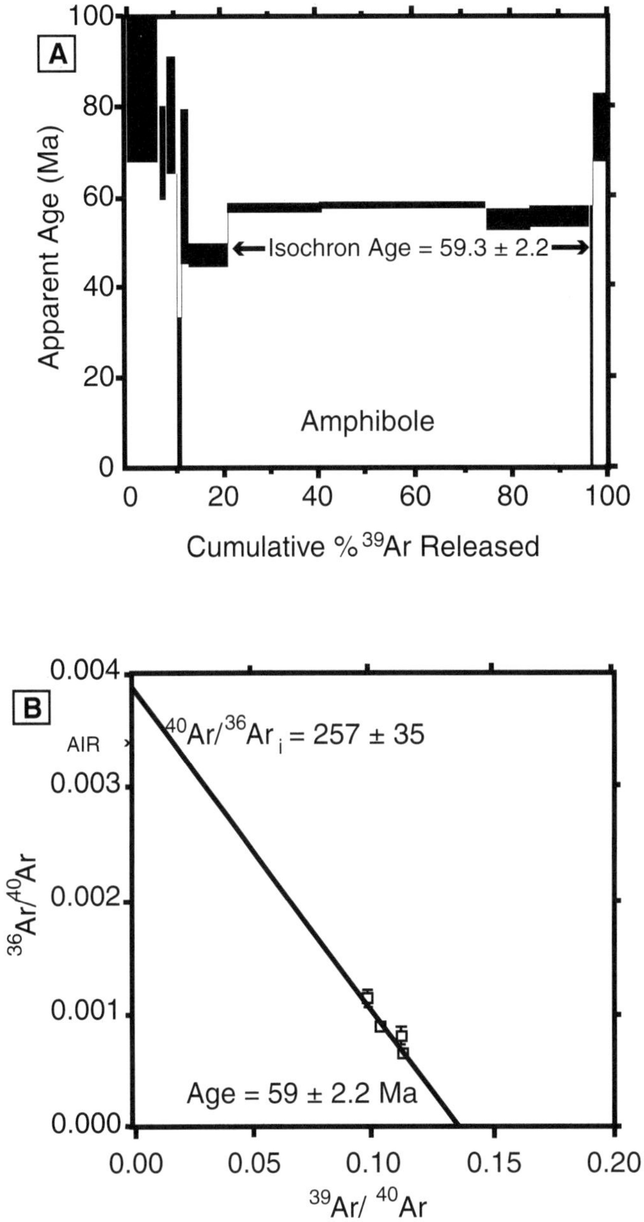

Figure 3. Amphibole $^{40}Ar/^{39}Ar$ data from sample M-19-88, an igneous dike intruding the Cretaceous Uyak Complex. A: Incremental release spectrum indicating the four steps used to calculate an isochron age, which comprise 7.5% of the total gas released. B: Isochron plot and linear fit of data giving an age of 59.3 ± 2.2 Ma (mean square weighted deviates = 1.96; York, 1969; McDougall and Harrison, 1988). Analytical data are provided in Table 2. Error bars and uncertainty in age are one standard deviation.

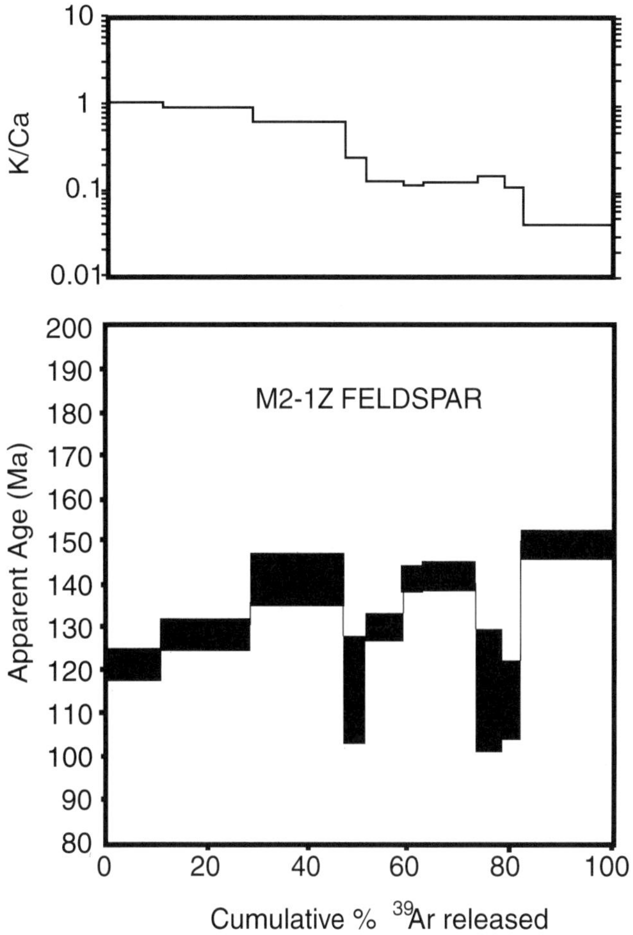

Figure 4. Feldspar $^{40}Ar/^{39}Ar$ incremental release spectrum from the Afognak pluton, sample M2-1Z. Analytical data are provided in Table 3. Uncertainty in age is one standard deviation.

13% of the total gas released, fall below the general trend in age and appear anomalous. These steps do not have anomalous K/Ca ratios (Fig. 4), suggesting that a distinct phase was not sampled. With the exception of these three steps, the spectrum shows a poorly defined age gradient that extends from ca. 150 Ma for the last increment of gas released to ca. 121 Ma for the first increment. This age gradient may have resulted from slow cooling through Ar closure from 150 to 120 Ma (e.g., Harrison and McDougall, 1982). In any case, there is no evidence in the age spectrum for any significant thermal event after ca. 120 Ma that might have caused episodic loss of radiogenic Ar, suggesting a relatively simple cooling history.

Apatite and Zircon Fission Track Ages

Zircon and apatite ages do not differ between the Afognak pluton and the Raspberry schists (see Tables 4 and 5 for analytical data); the former are ca. 150 Ma and the latter are ca. 60–70 Ma (Fig. 5). The pluton and schist, therefore, cooled through zircon and apatite closure temperatures at the same time, indicating that these units have not experienced significantly different cooling histories since the late Jurassic.

If all of the thermochronologic data available from the Afognak pluton and Raspberry schist are considered (Fig. 6),

TABLE 4. ANALYTICAL DATA FOR APATITE FISSION TRACK AGE AND TRACK LENGTH MEASUREMENTS

Sample ID	No. of grains	U (ppm)	ρ_s* (N_s)† ($\times10^6$cm^{-2})	ρ_i§ (N_i)† ($\times10^6$cm^{-2})	ρ_d# (N_d)† ($\times10^6$cm^{-2})	r**	P(χ^2) (%)	Age†† (Ma)	No. of tracks	Mean length (μm)§§	Standard deviation (μm)
M2-1Z	19	24	0.676 (373)	2.36 (1303)	4.25 (7841)	0.72	7	64 ± 4	100	14.53 ± 0.13	1.29
M-25-88	13	14	0.369 (115)	1.65 (515)	4.64 (7254)	0.61	63	55 ± 6	69	14.66 ± 0.17	1.37
M-26-88	15	5	0.164 (81)	0.518 (256)	4.60 (7254)	0.62	93	77 ± 10			
M-19-88	20	10	0.250 (180)	1.05 (754)	4.21 (7841)	0.63	60	53 ± 5	50	14.57 ± 0.23	1.63
P-1-87	20	27	0.497 (155)	2.46 (766)	4.17 (7841)	0.72	53	44 ± 4	100	13.68 ± 0.16	1.61
P-2-87	13	16	0.259 (84)	1.62 (524)	4.78 (7254)	0.46	43	40 ± 5			
P-5-87	20	39	0.546 (166)	4.51 (1370)	4.56 (7254)	0.80	4	29 ± 3	50	13.17 ± 0.28	2.00
P-5-87								36 ± 4##			
TLP-95	15	98	1.44 (260)	9.88 (1779)	4.47 (7254)	0.53	75	34 ± 2	106	13.92 ± 0.19	1.95
TL-2-87	15	138	1.85 (318)	12.6 (2170)	4.42 (7254)	0.63	57	34 ± 2	102	14.45 ± 0.14	1.36
TS-1-87	18	53	0.807 (410)	4.96 (2518)	4.06 (5922)	0.82	66	35 ± 2	200	14.51 ± 0.08	1.11
TS-4-87	19	65	0.923 (277)	6.34 (1902)	4.11 (5922)	0.73	13	32 ± 2	100	14.22 ± 0.11	1.12
TS-7-87	20	75	0.925 (322)	7.34 (2554)	4.19 (5922)	0.92	99	28 ± 2	100	14.20 ± 0.12	1.16
TS-9-87	20	73	0.966 (313)	7.09 (2297)	4.23 (5922)	0.84	64	30 ± 2	100	13.94 ± 0.14	1.35
H1-10Z	20	89	1.03 (263)	10.3 (2638)	4.73 (7254)	0.89	77	25 ± 2	100	14.13 ± 0.12	1.21
V-3-87	22	5	0.175 (122)	0.609 (424)	4.34 (7841)	0.85	95	66 ± 7			

*Spontaneous track density from internal mineral surface.
†Number of tracks counted.
§Induced track density from low-U muscovite external detector.
#Standard track density from low-U muscovite external detector.
**Correlation coefficient between individual crystal spontaneous and induced track counts.
†† Age (one standard deviation error) is calculated using pooled sample ρ_s/ρ_i and ζ-105.7 for dosimeter glass CN1 (see table 2)
§§One standard error.
##Mean of individual crystal ρ_s/ρ_i is used for age calculation when pooled data fail χ^2 test at 5% level (Galbraith, 1981).

the coincidence of the thermal histories of the schist and pluton from late Jurassic until the present is clearly evident. The simple temperature-time paths shown require slow cooling of both the pluton and schist at an average rate of 1.5 °C/m.y. over the last 150 m.y. This interpretation is based on the fission track ages alone, but slow cooling between zircon and apatite fission track closure is also supported by the feldspar $^{40}Ar/^{39}Ar$ data from the pluton (Fig. 3). These ages provide no constraint on the cooling path below 100 °C.

Fission Track Length Data

The nature of the temperature-time path below 100 °C can be inferred from consideration of the mean track length and age. For any arbitrary age, short tracks indicate prolonged residence in the temperature range 70–125 °C where tracks are more easily annealed, while longer tracks result from rapid cooling through this temperature interval. Apatite track length information is available from the Afognak pluton and the two igneous dikes (Fig. 7, samples 1, 2, and 4). No track lengths were measured in the Raspberry schist because no horizontal confined tracks were observed in these low-U apatites (Table 4).

Based on the track length data, the Afognak pluton cooled relatively rapidly through the temperature range 70–125 °C (Fig. 7). In addition, both igneous dikes cooled very rapidly after intrusion at 60 Ma to temperatures significantly below 100 °C (Figs. 5 and 7). The apatite ages from the pluton and schist (Fig. 5), however, require that the earlier Mesozoic section had just cooled through ~100 °C near the end of the Cretaceous. Together, these data require that the earlier Mesozoic section cooled rapidly through the apatite "annealing zone" (70–125 °C) (Gleadow et al., 1983, 1986) in the 5–15 m.y. period from the Maastrichtian to late Paleocene. These data further require that the earlier Mesozoic section was not reheated above ~100 °C during the 60 Ma igneous event or any time thereafter.

TABLE 5. ANALYTICAL DATA FOR ZIRCON FISSON TRACK AGE MEASUREMENTS

Sample ID	No. of grains	U (ppm)	ρ_s* (N_s)† ($\times10^6$cm^{-2})	ρ_i§ (N_i)† ($\times10^6$cm^{-2})	ρ_d# (N_d)† ($\times10^6$cm^{-2})	r**	P(χ^2) (%)	Age†† (Ma)
M2-IZ	12	151	9.79 (1215)	3.92 (486)	0.340 (2178)	0.72	15	153 ± 10
M-25-88	8	571	12.9 (447)	14.7 (510)	0.346 (2178)	0.52	2	55 ± 4
M-25-88								56 ± 7##
M-26-88	5	190	11.9 (413)	4.86 (168)	0.333 (2178)	0.92	78	148 ± 14
P-1-87	15	284	4.91 (1228)	7.20 (1802)	0.349 (2178)	0.81	6	43 ± 2
P-2-87	9	338	7.44 (696)	8.46 (792)	0.330 (2178)	0.83	16	53 ± 3
P-5-87	2	468	8.35 (68)	10.9 (89)	0.343 (2178)	1.00	86	48 ± 8
TS-1-87	8	306	6.11 (510)	8.15 (680)	0.358 (2164)	0.99	95	49 ± 3
TS-9-87	8	368	7.47 (669)	9.62 (861)	0.354 (2164)	0.92	30	50 ± 3
P-6-87	8	536	10.7 (305)	13.1 (372)	0.336 (2178)	0.92	98	50 ± 4
K-2-87	6	213	5.85 (226)	4.60 (178)	0.336 (2178)	0.77	55	72 ± 8
K-5-87	7	273	3.05 (373)	3.20 (391)	0.317 (2178)	0.99	87	55 ± 4
K-16-87	6	392	9.21 (300)	8.94 (291)	0.314 (2178)	0.91	48	59 ± 5

*Spontaneous track density from internal mineral surface.
†Number of tracks counted.
§Induced track density from low-U muscovite external detector.
#Standard track density from low-U muscovite external detector.
**Correlation coefficient between individual crystal spontaneous and induced track counts.
††Age (one standard deviation error) is calculated using pooled sample ρ_s/ρ_i and ζ-105.7 for dosimeter glass CN1 (see table 2).
§§One standard error.
##Mean of individual crystal ρ_s/ρ_i is used for age calculation when pooled data fail χ^2 test at 5% level (Galbraith, 1981).

Using the forward model of Green et al. (1989), a temperature-time history consistent with the measured age and track length distribution of the sample from the Afognak pluton (sample M2-1Z) can be determined. This sample is particularly well suited for modeling because its composition (Cl/[F + Cl]) is nearly the same as Durango apatite (Table 6). By trial-and-error fitting, a *T-t* path was found that produced an age and track length distribution that closely fit the data (Fig. 8). The measured and best-fit (rapid cooling) distributions are the same, with a χ^2 probability of 0.85, and have similar means and standard deviations (Table 7). In contrast, the linear cooling distribution has a mean that falls outside the two standard error uncertainty of the measured mean, is significantly broader than the measured distribution, and has a probability of only 0.11 of matching the measured distribution (Table 7). From this analysis, it is clear that the linear cooling path does not satisfy the apatite track length information.

The resulting best-fit temperature-time path (dashed line, Fig. 9) has relatively rapid cooling rates of 5 °C/m.y. from 75 to 60 Ma and 2 °C/m.y. until ca. 30 Ma. In addition, the path shows a constant, near surface temperature throughout most of the Neogene, with a final, more rapid cooling to surface

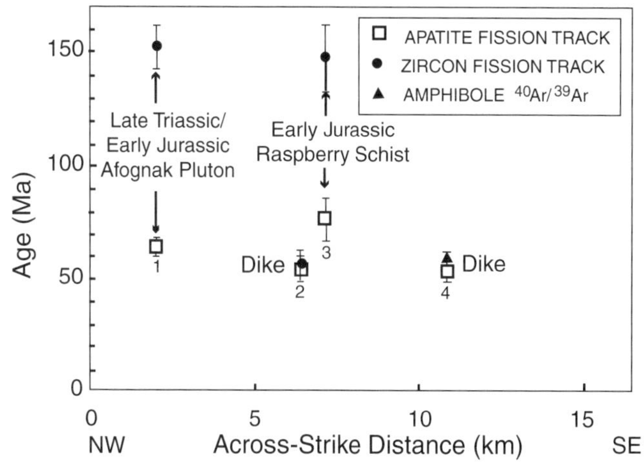

Figure 5. Fission track and $^{40}Ar/^{39}Ar$ ages from rocks collected in Malina Bay. Age is plotted versus across-strike distance, which increases from 0 km at the forearc basin on the northwest (Figure 1). The two dikes are part of the ca. 60 Ma igneous suite that intrudes the accretionary prism. The numbers below the data pairs refer to the sample numbers listed in Table 1. Error bars are one standard deviation.

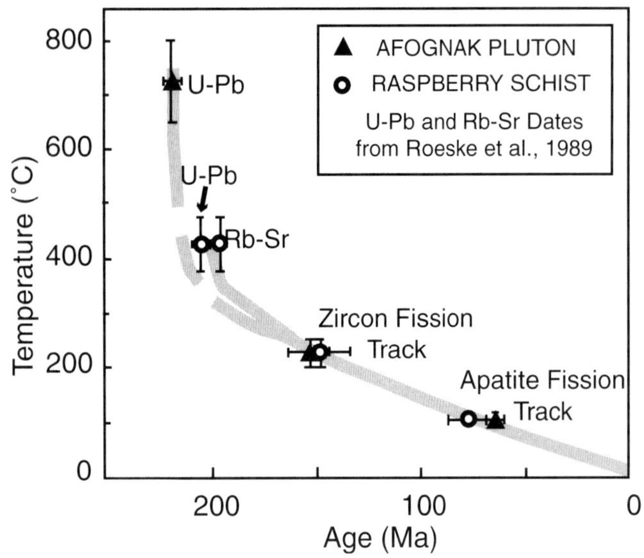

Figure 6. Plot of all ages available from the Afognak pluton and Raspberry schist exposed in Malina Bay versus the appropriate closure/annealing temperature (Roeske et al., 1989; and this study). The time of intrusion of the pluton at a temperature of 725 ± 75 °C (e.g., Harrison et al., 1979) is dated using U-Pb (Roeske et al., 1989), the time of maximum metamorphism of the schist at a temperature of approximately 425 ± 50 °C is given by both U-Pb (sphene-albite-amphibole and sphene-phengite-chlorite isochrons) and Rb-Sr (whole rock–white mica isochron) dates (Roeske, 1986; Roeske et al., 1989), and fission track ages provide the time of cooling through their closure temperatures. In addition, K-Ar ages of the schist indicate that it cooled from its peak metamorphic temperature to below 350 °C within 10 m.y. (Roeske et al., 1989). The shaded temperature-time paths, based on the ages alone, indicate constant, slow cooling at a rate of 1.5 °C/m.y. since at least 150 Ma. Uncertainty in age is one standard deviation.

temperatures sometime during the last few m.y., although this portion of the *T-t* path is poorly constrained. The apatite track length data indicate that cooling was relatively rapid between 75 and 60 Ma, however, they do not constrain the thermal history before 75 Ma. It is possible that rapid cooling was underway before 75 Ma, which would indicate that cooling during much of the Cretaceous was even slower than suggested by the fission track ages (Fig. 6).

Thermal History Variations across the Exhumed Prism, Central Kodiak Island

A total of fifteen samples of plutonic (11) and metasedimentary rocks (4) were used to study the thermal history across the Late Cretaceous-Paleocene portion of the prism southeast of the Uganik thrust (Figs. 10 and 11). These samples yielded a total of 11 apatite fission track ages (of which 8 also yielded track length data) and 9 zircon fission track ages (of which 3 also included individual grain ages).

Zircon Fission Track Ages

Zircon fission track ages from the plutons are generally constant across strike at ca. 50 Ma (dark shading, Fig. 12), with the exception of sample 5, which is from the structurally lower central belt of the Kodiak Formation. This sample has a zircon age younger than the general trend in plutonic zircon ages but identical to the apatite age, suggesting that this sample experienced rapid cooling at ca. 44 Ma. The average cooling rate of this sample between the apatite and zircon fission track closure temperatures, calculated using two standard deviation uncertainty in both ages, was at least 11 °C/m.y. In contrast, the apatite age and track length data from this same sample (see below) suggest an average linear cooling of 5 °C/m.y. for lower temperatures (i.e., below ~100 °C).

Both pooled and individual crystal zircon fission track ages were obtained from the three metasedimentary samples collected from the Kodiak Formation (Fig. 13). Individual zircon crystal ages were obtained using the external detector method (Hurford and Green, 1981). All of the individual crystal ages from the central belt (sample 17, Fig. 13B) are younger than the age of deposition (Maastrichtian) and are clustered in the early Tertiary, indicating that this sample was reset, probably totally, during the early Tertiary regional metamorphism and ridge-trench interaction. This inference is consistent with the temperature of ~280 °C estimated for regional metamorphism in the central belt (Brantley et al., 1997). The 55 Ma zircon fission track age for this sample, therefore, dates the time of cooling after the early Tertiary thermal peak. In contrast, individual ages from the structurally higher landward belt (sample 16, Fig. 13A) straddle the age of deposition, suggesting that the zircon fission tracks in this sample have been only partially annealed and that the sample age of 72.8 Ma is a "mixed age" that does not date a specific event. Sample 18 (Fig. 13C), from the seaward belt, has been at least partially reset; it may retain an older, pre-depositional component, although the data are insufficient to fully resolve this.

The zircon ages from the sediments provide insights into the relative peak thermal conditions experienced by the three structural belts of the Kodiak Formation. The zircon fission track ages from the central belt appear to be completely annealed, indicating that this belt experienced a higher peak temperature than either the seaward or landward belts. In contrast, Sample and Moore (1987) found no across-strike difference in peak metamorphic temperature on the basis of vitrinite reflectance and illite crystallinity. It is probable that the central belt was at a peak temperature near the top of the temperature range documented by Sample and Moore (1987) (~250 °C) whereas the landward and seaward belts were at temperatures near the lower end of their range (~200 °C).

Apatite Ages

The apatite ages from the across-strike transect (including samples from Three Saints Bay) decrease monotonically from 44 Ma on the northwest (landward) to 25 Ma on the southeast (seaward) (Fig. 12). In addition, all but one of the samples have track lengths that require rapid cooling through fission track closure at 35 to 25 Ma (Fig. 7).

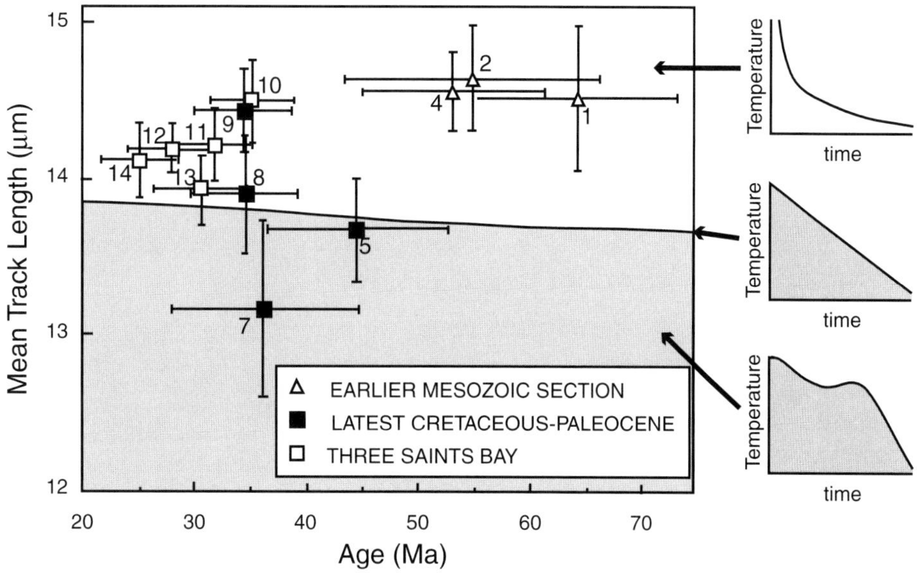

Figure 7. Scatter plot of apatite fission track mean length versus age, showing the nature of the thermal history inferred from this information (generalized examples shown to the right of the graph). The semihorizontal line, which divides the graph into two fields, was constructed using the model of Green et al. (1989) and marks the locus of length-age pairs that would result from linear cooling from high temperature to the present-day surface temperature (5 °C for Kodiak [Turner, 1987]). The upper field, shown in light shading, delineates length-age pairs generated by thermal histories with rapid initial cooling through the apatite fission track "annealing zone" (70–125 °C [Gleadow et al., 1983]), while the lower field, shown in dark shading, contains length-age pairs resulting from more prolonged thermal exposure within this temperature range, such as a reheating episode (Gleadow et al., 1986). The numbers next to the data points refer to sample numbers listed in Table 1. Two standard deviation uncertainty is shown to emphasize the fact that, based on the age and mean track length information, most of the data from the Mesozoic section and Three Saints Bay are consistent with rapid cooling through the apatite fission track annealing zone.

TABLE 6. ELECTRON MICROPROBE ANALYSES OF APATITE

Sample ID	No. of analyses	CaO (weight %)	P_2O_5 (weight %)	F (weight %)	Cl (weight %)	Cl/(F+Cl)
Durango*		54.02	40.78	3.53	0.41	0.104
Durango[†]	7	54.43 ± 0.53	40.61 ± .33	3.90 ± 0.40	0.43 ± 0.05	0.100 ± 0.011
M2-1Z	4	56.81 ± 0.28	42.80 ± 0.12	2.67 ± 0.39	0.27 ± 0.11	0.094 ± 0.040
P-1-87	4	55.84 ± 0.27	42.58 ± 0.54	2.64 ± 0.12	0.29 ± 0.02	0.098 ± 0.005
TS-1-87	4	54.90 ± 0.20	42.28 ± 0.21	2.47 ± 0.09	0.50 ± 0.06	0.167 ± 0.022
TS-7-87	4	55.57 ± 0.14	42.15 ± 0.72	2.89 ± 0.24	0.40 ± 0.05	0.121 ± 0.022
TS-9-87	5	55.09 ± 0.70	41.95 ± 0.81	2.37 ± 0.06	0.55 ± 0.04	0.187 ± 0.011

Note: Wilberforce apatite used as standard. All errors are one standard deviation.
*Previous analysis for comparison (Young et al., 1969).
[†]This study.

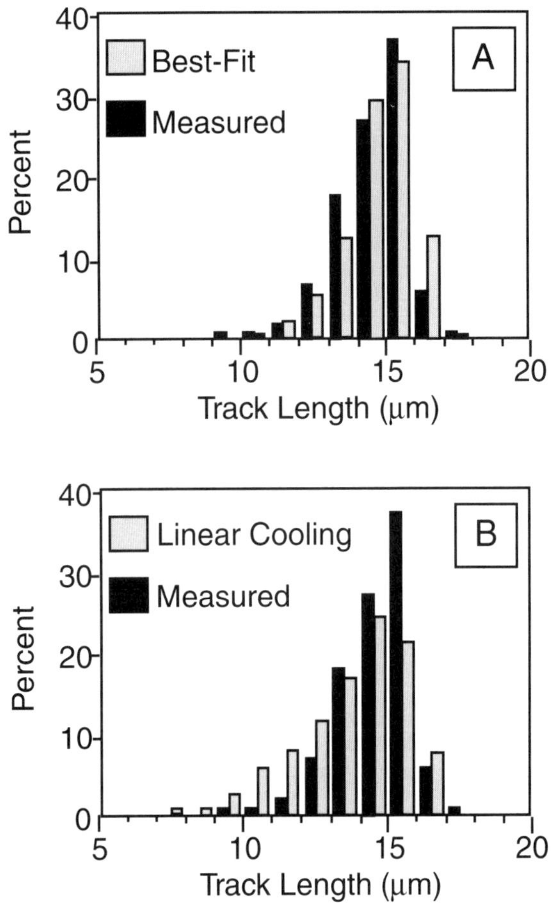

Figure 8. Results of apatite fission track modeling (Green et al., 1989) of sample M2-1Z from the Afognak pluton. A: Comparison of the measured apatite fission track-length distribution with the best-fit model distribution. The temperature-time path producing this model distribution is shown in Figure 9 (dashed line). B: Comparison of the same measured distribution with that which would result from linear cooling. Statistical details of these three distributions and results of the chi-square test are provided in Table 7.

DISCUSSION

The post-Jurassic cooling history of the Kodiak accretionary complex in southwest Alaska depicts a pattern of cooling ages and cooling rates that show a remarkable correspondence with variations in sediment input along the margin as inferred from the volume and character of accreted packages. For example, zircon and apatite fission track ages and $^{40}Ar/^{39}Ar$ feldspar cooling ages indicate slow cooling through much of the Cretaceous. During this time, there were no voluminous turbidite packages accreted; in contrast, the rocks thought to have been accreted at this time (the Uyak Complex) consist of greenstones and pelagic sedimentary sequences with only relatively thin intervals of sandstone and foliated shale. Track length distributions in apatite from these units indicate rapid cooling between the latest Cretaceous and the intrusion of plutons at 60 Ma. Contemporaneous with this rapid cooling was the accretion of a massive turbidite unit that accounts for over 60% of the exposed accretionary complex (the Kodiak Formation).

Apatite and zircon in the Kodiak Formation and associated intrusions were also reset in the early Tertiary (Central Belt in Fig. 14A) and subsequently cooled through the partial annealing zone in the Eocene through Oligocene (Fig. 14B). This cooling was contemporaneous with seaward growth through the addition of a thick turbidite sequence (the Eocene Sitkalidak Formation) and may have been coincident with underplating of a large sediment package at the base of the prism (Byrne, 1986) (Figs. 14A and 14B). Thus, the cooling history of the prism involved slow cooling followed by two events of rapid cooling in the Tertiary, and these two rapid cooling events were coincident with accretion of thick sediment packages. Collectively, these results from the Kodiak Accretionary Complex indicate that the accretionary flux dominates the thermal history of the wedge.

Cooling in the middle Tertiary could be in part due to the increase in age of subducting plate following the ridge-trench encounter at 60 Ma. However, the cooling in the aftermath of ridge subuction would not alone be capable of reducing the temperature

TABLE 7: RESULTS OF APATITE FISSION TRACK MODELING OF SAMPLE M2-1Z

	Age (Ma)	Mean track length (μm)	Standard deviation (μm)	χ^{2*}	$P(\chi^2)$
Measured	64.0 ± 4.0†	14.53 ± 0.13§	1.29		
Best-Fit	64.2	14.59	1.68	4.04 (8)	0.85
Linear Cooling	64.0	13.69	2.07	15.66 (10)	0.11

*Chi-square test to determine whether the modeled best-fit or linear cooling distributions are different than the measured apatite fission track length distribution using the CHSTWO routine (Press et al., 1986, p. 472). Value in parentheses is the number of degrees of freedom.

†One standard deviation.

§One standard error.

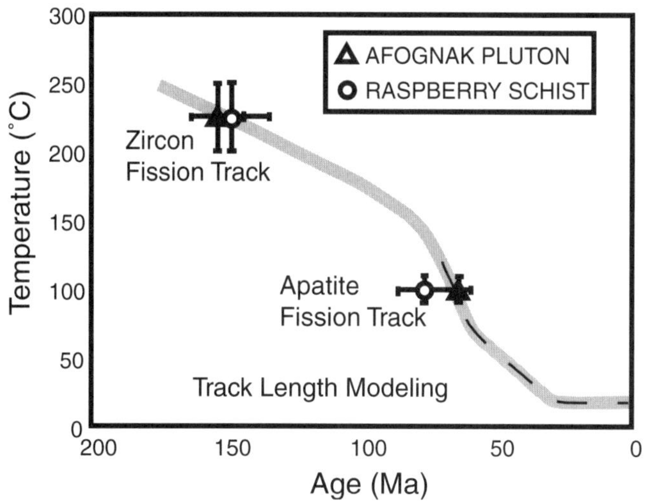

Figure 9. Plot of the model temperature-time path for sample M2-1Z from the Afognak pluton and the apatite fission track ages from the Raspberry schist. The dashed line shows thermal history obtained from forward modeling (Green et al., 1989) of the fission track age and length data. The shaded cooling path is drawn to be consistent with both the age and track length data, which require relatively rapid cooling in the latest Cretaceous and early Tertiary. Error bars are one standard deviation.

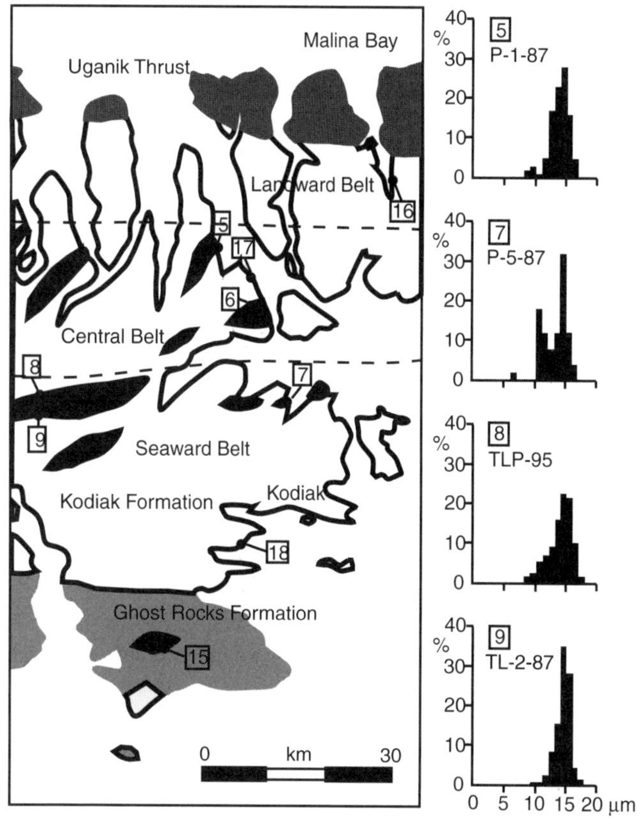

Figure 10. Geologic map of an across-strike portion of the Kodiak Islands showing sample locations and apatite fission track length distributions from the Maastrichtian and early Tertiary rocks (see Figure 1 for location and legend). Dashed lines divide the Kodiak Formation into the landward, central, and seaward belts (Sample and Moore, 1987; Fisher, 1988). The significance of these structural subdivisions is discussed in the text. Statistical details of the track length distributions are given in Table 4. Geology from Moore (1967), Connelly and Moore (1979) and Moore et al. (1983).

in the central belt of the Kodiak Formation from 280 °C to less than 70 °C at a depth of 10 km (resulting in closure of both zircon and apatite fission tracks). Given the initial ~10 km burial depth of the complex in the latest Cretaceous–early Tertiary, rapid cooling in the middle Tertiary most likely involved exhumation. Episodic events of rapid exhumation are also a probable response to pulses of incoming sediment along the base and at the toe of the wedge. Exhumation can occur by submarine or subaerial erosion or by extensional exhumation. Forearc and slope basin deposits along the Kodiak margin record subaerial erosion during the two major Tertiary cooling events. For example, in the neighboring forearc basin, a regional angular unconformity is present where steeply dipping Mesozoic beds are overlain by only slightly deformed Tertiary basin fill (Fig. lB) (Magoon, 1986; Fisher et al., 1987a). Evidence from a Chevron exploratory well (OCS-Y-0248 No. 1) drilled into the forearc basin near Kodiak Island in Shelikof Strait shows that the steeply northwest dipping late Campanian to early Maastrichtian units immediately below the angular unconformity were deposited in shelf or deeper water depths. In marked contrast, the shallowly dipping late Eocene and younger units above the unconformity were deposited under inner neritic to nonmarine conditions (R. Turner, Minerals Management Service, 1990, written commun.). This angular uncomformity indicates a major early Tertiary episode of shortening, uplift, and erosion. We correlate this event with the accretion of the Kodiak Formation.

The middle Tertiary erosion event is recorded by slope deposits of the Sitkinak Formation, a sequence of conglomerate and sandstone intercalated with fine-grained sandstone, siltstone, coal, and carbonaceous shale. These rocks are primarily terrestrial and were deposited as channelized, braided-stream gravels interbedded with interchannel, lagoonal, and interdistributary bay deposits (Nilsen and Moore, 1979), with a well-preserved deciduous and coniferous flora of probable Oligocene age (Wolfe, 1977). Abundant granite cobbles attest to erosion of the Kodiak batholith or related intrusions at this time as these represent the only potential proximal sources. These deposits overlie the Sitkalidak Formation along an angular unconformity. The Sitkinak Formation is constrained to be no younger than late Oligocene, based on the latest Oligocene age of the overlying siltstone of Trinity Islands (Allison, 1978; Allison and Marincovich, 1981). Thus, there is evidence for uplift and subaerial erosion at the time of Middle Tertiary cooling and sediment accretion.

Fission track dating of apatite and zircon from plutons that intrude the Kodiak Formation suggests that middle Tertiary

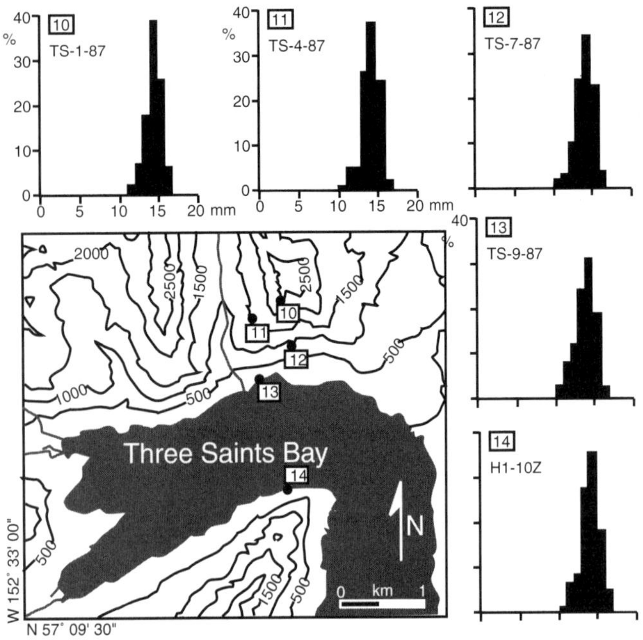

Figure 11. Sample locations and apatite fission track length distributions from Three Saints Bay (see Fig. 1). The location map, digitized from U.S. Geological Survey quadrangle Kodiak A-5, shows topographic contours in feet. All of the track length histograms have the same axis labels as those shown for sample number 10. Statistical details of the track length distributions are given in Table 4.

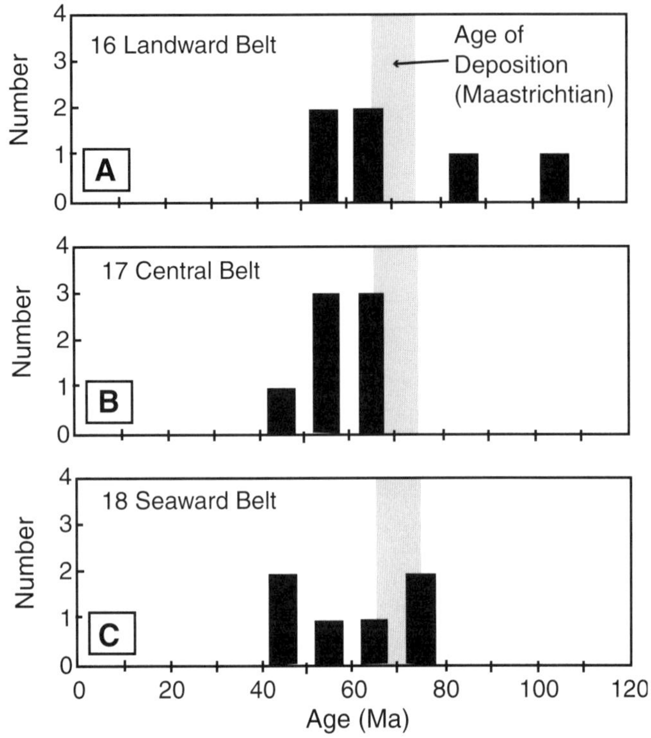

Figure 13. Histograms of zircon fission track ages from individual crystals separated from three graywacke samples of the Kodiak Formation. The Maastrichtian age of deposition of the Kodiak Formation (Jones and Clark, 1973) is shaded for reference. Samples are from the (A) landward, (B) central, and (C) seaward structural belts of the Kodiak Formation (Sample and Fisher, 1986; Sample and Moore, 1987; Fisher and Byrne, 1992).

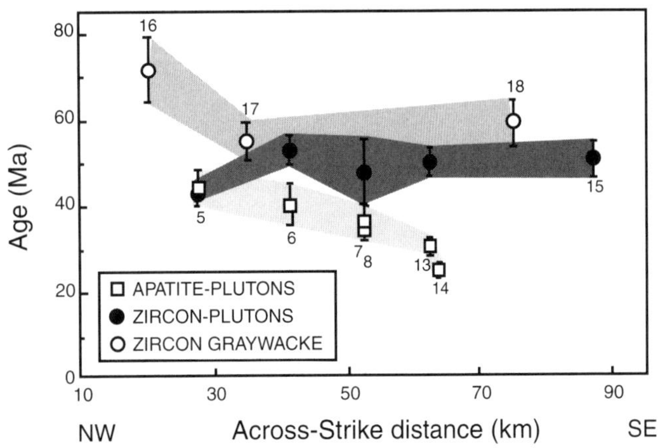

Figure 12. Plots of fission track ages from the Maastrichtian and early Tertiary rocks, which comprise most of the Kodiak Islands outcrop (Fig. 1), versus across-strike distance (defined in Fig. 5). All of the apatite ages and most of the zircon ages are from ca. 60 Ma plutons, which intrude the accreted sediments. Three of the zircon ages are from graywacke of the Maastrichtian Kodiak Formation. Multiple samples were collected from Terror Lake (locations 8 and 9, Fig. 10) and Three Saints Bay (Figure 11), but only data from the lowest elevation are plotted here. The numbers next to the data points refer to sample numbers listed in Table 1. Uncertainty in age is one standard deviation.

cooling of the accretionary wedge migrated seaward through time. A probable explanation for this cooling pattern is that it records a thermal response to uplift and unroofing of a forearc high that migrates trenchward (Fig. 14B and 14C). This kinematic wave affects rocks that were a large distance landward of the toe of the wedge, so it is not clear that offscraping of sediment at the toe of the prism could produce such a pattern. An alternative possibility is that underplating is localized beneath the forearc high—a likely possibility if duplex accretion is the underplating mechanism (e.g., Sample and Fisher, 1986). In such a case, material from the underthrusting plate is transferred to the overriding plate due to footwall collapse along a frontal ramp and seaward migration of the sole thrust into the underthrusting pile. Alternatively, uplift may have been related to progressive migration of active out-of-sequence thrust faults (e.g., Morley, 1988; Lucas, 1989).

The close similarity in thermal histories of the Afognak pluton and Raspberry schist, which lie on either side of the Border Ranges fault, suggests that these rocks were juxtaposed no later than the late Jurassic. Roeske et al. (1989) favored the idea that the Raspberry schist was formed near a magmatic arc now preserved on Vancouver Island, and that it was subsequently moved northward to its present position relative to

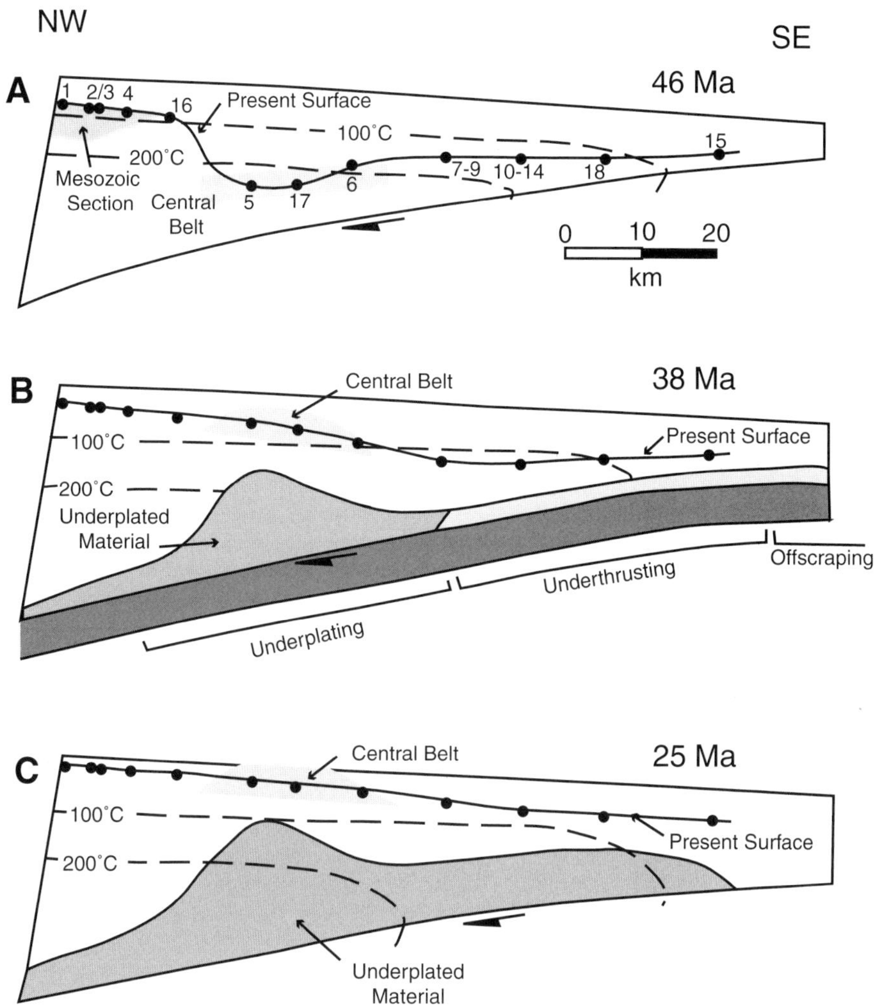

Figure 14. Schematic cross sections showing the thermal evolution of the accretionary prism during the middle Tertiary. Any horizontal shortening that took place during accretion is neglected for simplicity. Vertical scale is unknown because the geothermal gradient and position of the subducting plate (basal decollement) are not constrained. Sample locations are shown using the sample numbers from Table 1. Mesozoic Section refers to the high temperature Shuyak Formation and the high pressure Raspberry schist, which are juxtaposed along the Border Ranges fault. Central Belt refers to the central belt of the Late Cretaceous Kodiak Formation and Underplated Material refers to rocks of probable Eocene age. Cooling, uplift and exhumation of the central belt rocks in the Eocene is driven by underplating of Eocene sedimentary rocks (B). Underthrusting of progressively younger rocks beneath the forearc causes uplift and cooling of more seaward units in the Oligocene (C). See text for details.

the Kodiak/Chugach magmatic arc assemblage by strike-slip faulting during the Cretaceous. However, the fact that crystallization of the Afognak pluton and metamorphism of the Raspberry schist occurred so closely in time (U-Pb and Rb-Sr ages; Fig. 6) (Roeske et al., 1989) also makes it more likely that the magmatic arc assemblage represented by the Afognak pluton was formed along the same convergent margin as the Raspberry schist. Dextral overprinting of the early history along the Border Ranges fault has been documented to the northeast (Roeske et al., this volume, Chapter 6) and, although such an overprint has not been clearly identified on Kodiak Island, it is not clear what impact this deformation would have on the cooling history in the absence of significant dipslip deformation.

The mechanism(s) responsible for the juxtaposition and exhumation of the high-pressure/low-temperature Raspberry schist and the high-temperature magmatic arc assemblage remains obscure. One possibility is that subduction along the southwest Alaska margin was initiated in the middle Cretaceous and relatively cool oceanic crust and sediments were thrust beneath the arc crust and mantle rocks (Pavlis et al., 1988). As subduction continued, the subduction channel may have become constricted, causing a return flow of previously accreted materials (see Cowan and Silling, 1978; Cloos, 1982; Cloos and Shreve, 1988a, 1988b; Escher and Beaumont, 1987), including the Raspberry schist and the Afognak arc rocks. Although cooling of this returning material was relatively slow initially, it increased substantially in the Late Cretaceous when the Kodiak Formation was accreted. The present juxtaposition of these contrasting rocks, their similar ages and their nearly parallel cooling histories strongly suggests that the units were juxtaposed early in their histories and that they remained in close proximity throughout much of their history. In addition, the thermal histories show only slow cooling and no sign of reheating, indicating that blueschists can be preserved during relatively slow uplift, in accord with other studies (Ernst, 1988; Baldwin, 1988; Baldwin and Harrison, 1989).

CONCLUSIONS

The Cenozoic thermal history of the sediment-dominated accretionary prism has been marked by episodes of rapid cooling, which contrasts with the earlier Mesozoic history of slow cooling. Two rapid cooling events in the Cenozoic are contemporaneous with major influxes of sediment into the accretionary prism, with evidence for erosional exhumation of a subaerial accretionary wedge. These results illustrate the importance of variations in the incoming sediment flux for the thermal history of the forearc.

The prism experienced a thermal peak coincident with the emplacement of the ca. 60 Ma igneous complex soon after a large package of Maastrichtian and Paleocene sediments was accreted. These deformed and metamorphosed sedimentary rocks then cooled heterogeneously in the middle Tertiary, with fastest cooling rates in the core of a regional scale antiform. Decreasing cooling ages toward the trench are consistent with a seaward wave of exhumation brought on by underplating and footwall collapse of the underthrust sediment pile beneath the forearc high.

ACKNOWLEDGMENTS

This work was supported by NSF grant EAR-8707993 to Byrne and by grants from the Geological Society of America and Sigma Xi to Clendenen; it was completed while Clendenen was a graduate student at Brown University. Don Miller and Mark Harrison are respectively acknowledged for graciously allowing Clendenen to use the fission track laboratory at Rensselaer Polytechnic Institute and the $^{40}Ar/^{39}Ar$ laboratory at SUNY, Albany. Don Miller, Mary Roden, and Ray Donelick are thanked for educating Clendenen in the fine art of fission track analysis and Matt Heizler is acknowledged for also guiding Clendenen through all phases of the $^{40}Ar/^{39}Ar$ analyses. We thank Andrew Gleadow who allowed us to use the SUPER-TRACK modeling program, Ronald Tumer and John Larson of the Minerals Management Service for kindly providing us with data from the Shelikof Strait well, Chuck Naeser for doing a few preliminary fission track analyses, and Malcolm Hill for providing rock samples. Finally, we want to thank the U.S. Coast Guard, Bruce Heine and the good people of Kodiak Island who assisted us a great deal in the field. An earlier version of this paper benefited greatly through reviews by Peter Gromet, Jan Tullis, Don Miller, Kip Hodges, and members of the Minerals Management Service, Anchorage, Alaska. A penultimate version was reviewed by Jim Ryan and Harold Stowell and their comments are greatly appreciated.

REFERENCES CITED

Allison, R.C., 1978, Late Oligocene through Pleistocene molluscan faunas in the Gulf of Alaska region: Veliger, v. 21, p. 171–188.

Allison, R.C., and Marincovich, L., Jr., 1981, A late Oligocene or earliest Miocene molluscan fauna from Sitkinak Island, Alaska: U.S. Geological Survey Professional Paper 1233, 11 p.

Baldwin, S.L., 1988, Thermochronology of a subduction complex in western Baja Califomia [Ph.D. thesis]: Albany, State Univeristy of New York, 247 p.

Baldwin, S.L., and Harrison, T.M., 1989, Geochronology of blueschists from west-central Baja Califomia and the timing of uplift in subduction zones: Journal of Geology, v. 21, p. 149–163.

Bangs, N.L., and Cande, S.C., 1997, Episodic development of a convergent margin inferred from structures and processes along the southern Chile margin: Tectonics, v. 16, p. 489–503.

Bradley, D.C., Haeussler, P.J., and Kusky, T.M., 1993, Timing of early Tertiary ridge subduction in southern Alaska, *in* Till, A., and Dusel-Bacon, C., eds, Geologic studies in Alaska by the U.S. Geological Survey, 1992: U.S. Geological Survey Bulletin 2068, p. 163–177.

Brantley, S.L., Fisher, D.M., Deines, P., Clark, M.B., and Myers, G., 1997, Segregation veins: Evidence for the deformation and dewatering of a low-grade metapelite, *in* Holness, M., ed., Deformation-enhanced fluid transport in the Earth's crust and mantle: London, Chapman and Hall, p. 267–288.

Brown, K.M., and Westbrook, G.K., 1988, Mud diapirism, and subcretion in the Barbados Ridge complex: Tectonics, v. 7, p. 613–640.

Burns, L.E., 1985, The Border Ranges ultramafic and mafic complex, south-central Alaska: Cumulate fractionates of island-arc volcanics: Canadian Journal of Earth Science, v. 22, p. 1020–1038.

Byrne, T., 1986, Eocene underplating along the Kodiak shelf, Alaska: Implications and regional correlations: Tectonics, v. 5, p. 403–421.

Byrne, T., and Fisher, D., 1987, Episodic growth of the Kodiak convergent margin: Nature, v. 325, p. 338–341.

Byrne, T., and Fisher, D., 1990, Evidence for a weak and overpressured decollement beneath sediment-dominated accretionary prisms: Journal of Geophysical Research, v. 95, p. 9081–9097.

Byrne, T., Maltman, A., Stephenson, E., and Knipe, R., 1993, Deformation structures and fluid flow in the toe region of the Nankai accretionary prism: Proceedings of the Ocean Drilling Program, Scientific results, Volume 131: College Station, Texas, Ocean Drilling Program, p. 83–92.

Carden, J.R., Connelly, W., Forbes, R.B., and Tumer, D.L., 1977, Blueschists of the Kodiak Islands, Alaska: An extension of the Seldovia schist terrane: Geology, v. 5, p. 529–533.

Clark, S.H.B., 1973, The McHugh Complex of south-central Alaska: U.S. Geological Survey Bulletin 1372, p. Dl–Dll.

Clendenen, W.S., 1991, Thermal history and vertical tectonics of the southern Alaska convergent margin [Ph.D. thesis]: Providence, Rhode Island, Brown University, 177 p.

Clendenen, W.S., and Byrne, T., 1989, Late Cretaceous increase in uplift and cooling rates of the Kodiak accretionary prism and evidence for a change from an intra-oceanic to a continental margin setting: Geological Society of America Abstracts with Programs, v. 21, p. A314.

Cloos, M., 1982, Flow melanges: Numerical modeling and geologic constraints on their origin in the Franciscan subduction complex: Bulletin of the Seismological Society of America, v. 93, p. 330–345.

Cloos, M., and Shreve, R.L., 1988a, Subduction-channel model of prism accretion, melange formation, sediment subduction, and subduction erosion at convergent margins, 1: Background and description: Pure and Applied Geophysics, v. 128, p. 455–500.

Cloos, M., and Shreve, R.L., 1988b, Subduction-channel model of prism accretion, melange formation, sediment subduction, and subduction erosion at convergent margins, 2: Implications and discussion: Pure and Applied Geophysics, v. 128, p. 501–545.

Connelly, W., 1978, Uyak Complex, Kodiak Islands, Alaska: A Cretaceous subduction complex: Geological Society of America Bulletin, v. 89, p. 755–769.

Connelly, W., and Moore, J.C., 1979, Geologic map of the northwest side of the Kodiak and adjacent islands, Alaska: U.S. Geological Survey Miscellaneous Field Studies Map MF-1057, scale 1:250,000, 2 sheets.

Cowan, D.S., and Boss, R.F., 1978, Tectonic framework of the southwestern Kenai Peninsula, Alaska: Geological Society of America Bulletin, v. 89, p. 155–158.

Cowan, D.S., and Silling, R.M., 1978, A dynamic, scaled model of accretion at trenches and its implications for the tectonic evolution of subduction complexes: Journal of Geophysical Research, v. 83, p. 5389–5396.

Ernst, W.G., 1988, Tectonic history of subduction zones inferred from retrograde blueschist *P-T* paths: Geology, v. 16, p. 1081–1084.

Escher, A., and Beaumont, C., 1987, Formation, burial and exhumation of basement nappes at crustal scale: Geometric model based on the western Swiss-Italian Alps: Journal of Structural Geology, v. 19, p. 955–974.

Fisher, D.M., 1988, Structural evolution of a thickly sedimented convergent margin: Evidence from macroscopic structures, micro-structures, and

incremental strain studies [Ph.D. thesis]: Providence, Rhode Island, Brown University, 120 p.
Fisher, D.M., and Byrne, T., 1987, Structural evolution of underthrusted sediments: Evidence from Kodiak Island, Alaska: Tectonics, v. 6, p. 775–793.
Fisher, D.M., and Byrne, T., 1992, Strain variations in an ancient accretionary wedge: Implications for forearc evolution: Tectonics, v. 11, p. 330–347.
Fisher, M.A., and von Huene, R., 1980, Structure of upper Cenozoic strata beneath Kodiak Shelf, Alaska: American Association of Petroleum Geologists Bulletin, v. 64, p. 1014–1033.
Fisher, M.A., von Huene, R., Smith, G.L., and Bruns, T.R., 1983, Possible seismic reflections from the downgoing Pacific plate, 275 kilometers arcward from the eastern Aleutian Trench: Journal of Geophysical Research, v. 88, p. 5835–5849.
Fisher, M.A., Detterman, R.L., and Magoon, L.B., 1987a, Tectonics and petroleum geology of the Cook-Shelikof Basin, southern Alaska, *in* Scholl, et al., eds., Geology and resource potential of the continental margin of western North America and adjacent ocean basins—Beaufort to Baja California: Houston, Texas, Circum-Pacific Council for Energy and Mineral Resources, Earth Science Series, v. 6, p. 213–228.
Fisher, M.A., von Huene, R., and Smith, G.L., 1987b, Reflection from midcrustal rocks within the Mesozoic subduction complex near the eastern Aleutian Trench: Journal of Geophysical Research, v. 92, p. 7907–7915.
Fisher, D.M., Gardner, T.W., Marshall, J.S., Sak, P., and Protti, M., 1998, Effect of subducting seafloor roughness on forearc kinematics, Pacific Coast, Costa Rica: Geology, v. 26, p. 467–470.
Forbes, R.B., and Lanphere, M.A., 1973, Tectonic significance of mineral ages of blueschists near Seldovia, Alaska: Journal of Geophysical Research, v. 78, p. 1383–1386.
Galbraith, R.F., 1981, On statistical models for fission track counts: Mathematical Geology, v. 13, p. 471–478.
Gleadow, A.J.W., Duddy, I.R., and Lovering, J.F., 1983, Fission track analysis: A new tool for the evaluation of thermal histories and hydrocarbon potential: Australian Petroleum, Exploration Association Journal, v. 23, p. 93–102.
Gleadow, A.J.W., Duddy, I.R., Green, P.F., and Lovering, J.F., 1986, Confined fission track lengths in apatite: A diagnostic tool for thermal history analysis: Contributions to Mineralogy and Petrology, v. 94, p. 405–415.
Green, P.F., Duddy, I.R., Laslett, G.M., Hegarty, K.A., Gleadow, A.J.W., and Lovering, J.F., 1989, Thermal annealing of fission tracks in apatite 4: Quantitative modelling techniques and extension to geological time scales: Chemical Geology, v. 79, p. 155–182.
Gutscher, M.-A., Kukowski, N., Malavieille, J., and Lallemand, S., 1998a, Episodic imbricate thrusting and underthrusting: Analog experiments and mechanical analysis applied to the Alaskan accretionary wedge: Journal of Geophysical Research, v. 103, p. 10,161–10,176.
Gutscher, M.-A., Kukowski, N., Malavieille, J., and Lallemand, S., 1998b, Material transfer in accretionary wedges from analysis of a systematic series of analog experiments: Journal of Structural Geology, v. 20, p. 407–416.
Harrison, T.M., 1981, Excess ^{40}Ar in metamorphic rocks from Broken Hill, New South Wales: Implications for $^{40}Ar/^{39}Ar$ age spectra and the thermal history of the region: Earth and Planetary Science Letters, v. 55, p. 123–149.
Harrison, T.M., and McDougall, I., 1982, The thermal significance of potassium feldspar K-Ar ages inferred from $^{40}Ar/^{39}Ar$ age spectrum results: Geochimica et Cosmochimica Acta, v. 46, p. 1811–1820.
Harrison, T.M., Armstrong, R.L., Naeser, C.W., and Harakal, J.E., 1979, Geochronology and thermal history of the Coast Plutonic Complex, near Prince Rupert, British Columbia: Canadian Journal of Earth Science, v. 16, p. 400–410.
Harrison, T.M., Morgan, P., and Blackwell, D.D., 1986, Contstraints on the age of heating at the Finton Hill site, valles Caldera, New Mexico: Journal of Geophysical Research, v. 91, p. 1899–1908.
Hill, M.D., 1979, Volcanic and plutonle rocks of the Kodiak-Shumagin shelf, Alaska: Subduction deposits and near-trench magmatism [Ph.D. thesis]: Santa Cruz, University of California, 274 p.
Hill, M.D., Morris, J., and Whelan, J., 1981, Hybrid granodiorites intruding the accretionary prism, Kodiak, Shumagin, and Sanak Islands, southwest Alaska: Journal of Geophysical Research, v. 86, p. 10,569–10,590.
Hudson, T.L., Plafker, G., and Peterman, Z.E., 1979, Paleogene anatexis along the Gulf of Alaska margin: Geology, v. 7, p. 573–577.
Hudson, T.L., Arth, J.G., Winkler, G.W., and Stern, T.W., 1985, Jurassic plutonism along the Gulf of Alaska: Geological Society of America Abstracts with Programs, v. 17, no. 6, p. 362.
Hurford, A.J., and Green, P.F., 1981, A reappraisal of neutron dosimetry and Uranium-238 lf values in fission-track dating: Nuclear Tracks, v. 5, p. 53–61.
Jones, D.L., and Clark, S.H.B., 1973, Upper Cretaceous (Maastrichtian) fossils from the Kenai-Chugach Mountains, Kodiak and Shumagin Islands, southern Alaska: U.S. Geological Survey Journal of Research, v. 1, p. 125–136.
Kienle, J., and Tumer, D.L., 1976, The Shumagin-Kodiak batholith—A Paleocene magmatic arc?: Juneau, Alaska Department of Natural Resources, Geological Report 51, p. 9–11.
Kulm, L.D., von Huene, R., Duncan, J.R., Ingle, J.C., Kling, S.A., Musich, L.F., Piper, D.J.W., Pratt, R.M., Schroder, H.J., Weser, O., and Wise, S.W., 1973, Initial reports of the Deep Sea Drilling Project: Washington, D.C., U.S. Govemment Printing Office, v. 18, 1077 p.
Kusky, T.M., and Bradley, D.C., 1999, Kinematic analysis of melange fabrics; examples and applications from the McHugh Complex, Kenai Peninsula, Alaska: Journal of Structural Geology, v. 21, p. 1773–1796.
Kusky, T.M., Bradley, D.C., and Haeussler, P.J., 1997, Progressive deformation of the Chugach accretionary complex, Alaska, during a Paleogene ridge-trench encounter: Journal of Structural Geology, v. 19, p. 139–157.
Lallemand, S.E., Schnurle, P., and Manoussis, S., 1992, Reconstruction of subduction zone paleogeometries and quantification of upper plate material losses caused by tectonic erosion: Journal of Geophysical Research, v. 97, p. 217–239.
Lallemand, S.E., Schnurle, P., and Malavieille, J., 1994, Coulomb theory applied to accretionary and nonaccretionary wedges: Possible causes for tectonic erosion and/or frontal accretion: Journal of Geophysical Research, v. 99, p. 12,033–12,055.
Lucas, S.B., 1989, Structural evolution of the Cape Smith thrust belt and the role of out-of-sequence faulting in the thickening of mountain belts: Tectonics, v. 8, p. 655–676.
Magoon, L.B., editor, 1986, Geologic studies of the Lower Cook Inlet COST no. 1 Well, Alaska outer continental shelf: U.S. Geological Survey Bulletin 1596, 99 p.
McDougall, I., and Harrison, T.M., 1988, Geochronoloigy and thermochronology by the $^{40}Ar/^{39}Ar$ method: New York, Oxford University Press, 212 p.
Moore, G.F., Shipley, T.H., Stoffa, P.L., Karig, D.E., Taira, A., Kuramoto, S., Tokuyama, H., and Suyehiro, K., 1990, Structure of the Nankai trough accretionary zone from multichannel seismic reflection data: Journal of Geophysical Research, v. 95, p. 8753–8765.
Moore, G.W., 1967, Preliminary geologic map of Kodiak Island and vicinity, Alaska: U.S. Geological Survey Open-File Report 271.
Moore, J.C., 1973, Cretaceous continental margin sedimentation, southwestern Alaska: Geological Society of America Bulletin, v. 84, p. 595–614.
Moore, J.C., and Allwardt, A., 1980, Progressive deformation of a Tertiary trench slope, Kodiak Islands, Alaska: Journal of Geophysical Research, v. 85, p. 4741–4756.
Moore, J.C., Byrne, T., Plumley, P.W., Reid, M., Gibbons, H., and Coe, R.S., 1983, Paleogene evolution of the Kodiak Islands, Alaska: Consequences of ridge-trench interaction in a more southerly latitude: Tectonics, v. 2, p. 265–293.
Moore, J.C., Diebold, J., Fisher, M.A., Sample, J., Brocher, T., Talwani, M., Ewing, J., von Huene, R., Rowe, C., Stone, D., Stevens, C., and Sawyer, D., 1991, EDGE deep seismic reflection transect of the eastern Aleutian arc-trench; layered lower crust reveals underplating and continental growth: Geology, v. 19, p. 420–424.
Morley, C.K., 1988, Out-of-sequence thrusts: Tectonics, v. 7, p. 539–561.
Nilsen, T.H., and Moore, G.W., 1979, Reconnaissance study of Upper Cretaceous to Miocene stratigraphic units and sedimentary facies, Kodiak and adjacent islands, Alaska: U.S. Geological Survey Professional Paper 1093, 34 p.
Ohmori, K., Taira, A., Sakaguchi, A., Okamura, M., and Aihara, A., 1997, Paleothermal structural of the Shimanto accretionary prism, Shikoku, Japan: role of an out-of-sequence thrust: Geology, v. 25, p. 327–330.
Paterson, S.R., and Sample, J.C., 1988, The development of folds and cleavages in slate belts by underplating in accretionary complexes: A comparison of the Kodiak Formafion, Alaska and the Calaveras Complex, Califomia: Tectonics, v. 7, p. 859–874.
Pavlis, T.L., Monteverde, D.H., Bowman, J.R., Rubenstone, J.L., and Reason, M.D., 1988, Early Cretaceous near-trench plutonism in southern Alaska; a tonalite trondhjemite intrusive complex injected during ductile thrusting along the Border Ranges fault system: Tectonics, v. 7, p. 1179–1199.
Plafker, G., Jones, D.L., and Pessagno, E.A., Jr., 1977, A Cretaceous accretionary flysch and melange terrane along the Gulf of Alaska margin: U.S. Geological Survey Circular 751B, p. 1341–1342.

Press, W.H., Flannery, B.P., Teukolsky, S.A., and Vetterling, W.T., 1986, Numerical recipes, the art of scientific computing: New York, Cambridge University Press, 818 p.

Raymond, L., 1980, Episodic accretion and plutonism in southwestern Alaska: Earth and Planetary Science Letters, v. 49, p. 29–33.

Roeske, S.M., 1986, Field relations and metamorphism of the Raspberry schist, Kodiak Islands, Alaska: Boulder, Colorado, Geological Society of America Memoir 164, p. 169–184.

Roeske, S.M., Mattison, J.M., and Armstrong, R.L., 1989, Isotopic ages of glaucophane schists on the Kodiak Islands, southern Alaska, and their implications for the Mesozoic tectonic history of the Border Ranges fault system: Geological Society of America Bulletin, v. 101, p. 1021–1037.

Roeske, S.M., Snee, L.W., and Pavlis, T.L., 2003, Dextral slip reactivation of an arc-forearc boundary during Late Cretaceous–Early Eocene oblique convergence in the northern Cordillera, *in* Sisson, V.B., Roeske, S.M., and Pavlis, T.L., eds., Geology of a transpressional orogen developed during ridge-trench interaction along the North Pacific margin: Boulder, Colorado, Geological Society of America Special Paper 371, p. 141–169 (this volume).

Sample, J.C., and Fisher, D., 1986, Duplex accretion and underplating in an ancient accretionary complex, Kodiak Islands, Alaska: Geology, v. 14, p. 160–163.

Sample, J.C., and Moore, J.C., 1987, Structural style and kinematics of an underplated slate belt, Kodiak and adjacent islands, Alaska: Geological Society of America Bulletin, v. 99, p. 7–20.

Silberling, N.J., Jones, D.L., Monger, J.W.H., and Coney, P.J., 1992, Lithotectonic terrane map of the North American Cordillera: U.S. Geological Survey Miscellaneous Investigations Map I-2176, scale 1:5,000,000.

Sisson, V.B., and Onstott, T.C., 1986, Dating blueschist metamorphism: A combined $^{40}Ar/^{39}Ar$ and electron microprobe approach: Geochimica et Cosmochimica Acta, v. 50, p. 2111–2117.

Stern, C.R., 1991, Role of subduction erosion in the generation of Andean magmas: Geology, v. 19, p. 78–81.

Turner, R.F., editor, 1987, Geological and operational summary, Kodiak Shelf stratigraphic test wells, Western Gulf of Alaska: Anchorage, Alaska, U.S. Department of the Interior, Minerals Management Service, 341 p.

von Huene, R., and Scholl, D.W., 1991, Observations at convergent margins concerning sediment subduction, subduction erosion, and the growth of continental crust: Reviews of Geophysics, v. 29, p. 279–316.

Vrolijk, P., Myers, G., and Moore, J.C., 1988, Warm fluid migration along tectonic melanges in the Kodiak Accretionary Complex, Alaska: Journal of Geophysical Research, v. 22, p. 10,313–10,324.

Wolfe, J.A., 1977, Paleogene floras from the gulf of Alaska region: U.S. Geological Survey Professional Paper 997, 108 p.

York, D., 1969, Least squares fitting of a straight line: Earth and Planetary Science Letters, v. 5, p. 320–324.

Young, E.J., Myhers, A.T., Munson, E.L., and Conklin, N.M., 1969, Mineralogy and geochemistry of fluorapatite from Cerro de mercado, Durango, Mexico: U.S. Geological Survey Professional Paper 650-D, p. D84–D93.

Manuscript Accepted by the Society February 5, 2003

Printed in the USA

Geological Society of America
Special Paper 371
2003

Sedimentary record of transpressional tectonics and ridge subduction in the Tertiary Matanuska Valley–Talkeetna Mountains forearc basin, southern Alaska

Jeffrey M. Trop*
Department of Geology, Bucknell University, Lewisburg, Pennsylvania 17837, USA

Kenneth D. Ridgway
Department of Earth and Atmospheric Sciences, Purdue University, West Lafayette, Indiana 47907, USA

Terry L. Spell
Department of Geosciences, University of Nevada, Las Vegas, Nevada 89154, USA

ABSTRACT

The Chickaloon, Arkose Ridge, Wishbone, and Tsadaka Formations consist of more than 2800 m of Paleocene-Oligocene sedimentary and volcanic strata that are the products of sedimentation, volcanism, and faulting in the Matanuska Valley–Talkeetna Mountains forearc basin. These deposits provide a record of early Tertiary tectonic processes that formed the southern Alaska convergent margin. The northern margin of the forearc basin is characterized by nonmarine sandstone, conglomerate, and minor mudstone that interfinger with volcanic strata. On the basis of lithofacies, paleocurrent, and compositional data, the northern basin margin deposits are interpreted to represent southward prograding alluvial-fluvial systems. New $^{40}Ar/^{39}Ar$ ages from detrital feldspars in volcaniclastic sandstone and from igneous clasts in conglomerate suggest that these deposits were derived from Middle Jurassic and Paleocene-Eocene volcanic arc-related rocks. Stratigraphic and structural data from the northern basin margin adjacent to the Castle Mountain fault document syndepositional faulting that produced footwall growth synclines in the forearc basin. Paleocene-Oligocene strata exposed along the southern margin of the forearc basin are characterized by nonmarine sedimentary deposits that lack volcanic strata. Lithofacies, paleocurrent, and compositional data from these deposits are interpreted as recording northward prograding alluvial-fluvial systems that were derived from metavolcanic and metasedimentary source terranes of the accretionary prism. Both northern and southern basin-margin deposits merge into basin-axis deposits characterized by thick sections of carbonaceous mudstone and coal, and minor channelized sandstone. Basin-axis strata are interpreted as products of high-sinuosity fluvial and lacustrine environments that drained southwestward into the ancestral Cook Inlet basin.

Unlike most previously studied ancient forearc basins, the Matanuska Valley–Talkeetna Mountains basin contains a fairly complete stratigraphic record of nonmarine sedimentation and volcanism. These deposits record multiple episodes of transpressional deformation that may be related to northward translation of the forearc basin along the continental margin, oroclinal bending of Alaska, and/or subduction of a spreading ridge. To evaluate the record of ridge subduction in Paleocene-Oligocene forearc basin deposits, two reconstructions are presented. In one reconstruction, the forearc basin and accretionary prism

*Corresponding author: jtrop@bucknell.edu

Trop, J.M., Ridgway, K.D., and Spell, T.L., 2003, Sedimentary record of transpressional tectonics and ridge subduction in the Tertiary Matanuska Valley–Talkeetna Mountains forearc basin, southern Alaska, *in* Sisson, V.B., Roeske, S.M., and Pavlis, T.L., eds., Geology of a transpressional orogen developed during ridge-trench interaction along the North Pacific margin: Boulder, Colorado, Geological Society of America Special Paper 371, p. 89–118.

were translated northward as a single block with most displacement accommodated along inboard dextral strike-slip faults such as the Castle Mountain and Denali fault systems. In this reconstruction, ridge subduction beneath the forearc basin would have occurred at ca. 54–50 Ma, coeval with basinward progradation of coarse-grained deposystems, and with syndepositional displacement and growth-syncline development along the Castle Mountain fault. In the second reconstruction, in addition to displacement on inboard strike-slip faults, significant northward displacement was accommodated along the Border Ranges and Hanagita faults. These fault systems separated the forearc basin from the accretionary prism. In this reconstruction, the forearc basin and accretionary prism were translated separately and have different displacement histories; ridge subduction beneath the forearc basin would have occurred at ca. 61–58 Ma. The sedimentary record of ridge subduction in this reconstruction is represented by a basinwide unconformity and/or deposition of relatively fine-grained deposits in the forearc basin. We prefer the first reconstruction, but until additional high-resolution geochronological data are available, and the displacement histories of major fault systems are better known, both reconstructions are feasible.

Keywords: ridge subduction, Chugach terrane, Peninsular terrane, Wrangellia composite terrane, geochronology, sedimentology, forearc basin, tectonics, transpression.

INTRODUCTION

Forearc basin deposits provide a long-term record of tectonic processes responsible for the growth of convergent plate margins. Due to their proximity to both magmatic arcs and accretionary prisms, forearc basin deposits may record both inboard arc-related volcanism and outboard deformation within the subduction complex (e.g., Dickinson and Seely, 1979). Most of the conceptual framework for forearc basins is based on studies of the Great Valley Group in California (e.g., Dickinson and Rich, 1972; Ingersoll, 1978a, 1978b, 1979, 1983; Dickinson and Seely, 1979; Moxon and Graham, 1987; Linn et al., 1991, 1992; Williams, 1997) and the Peninsular Ranges in Baja Mexico (e.g., Busby-Spera, 1986; Morris and Busby-Spera, 1988; Busby et al., 1998). The general sedimentologic model for forearc basins predicts a progressive evolution from deep-marine through shallow-marine to nonmarine deposystems, with progressive filling of the basin (Ingersoll, 1979; Dickinson, 1995). Previous research focused mainly on the sedimentology and petrofacies of marine and deltaic deposits of ancient forearc basins; notably lacking from the literature are detailed studies of ancient nonmarine deposits within the context of forearc basin development (e.g., Fulford and Busby, 1993). The Matanuska Valley–Talkeetna Mountains forearc basin in south-central Alaska contains 2800 m of well-exposed and relatively undeformed Paleocene-Oligocene nonmarine strata (Fig. 1; Winkler, 1992), and thus provides an excellent opportunity to evaluate the stratigraphic and structural development of an ancient nonmarine forearc basin. In this study, we present new $^{40}Ar/^{39}Ar$ geochronologic ages, measured stratigraphic sections, lithofacies analyses, and compositional data from Paleocene-Oligocene deposits of the Matanuska Valley–Talkeetna Mountains forearc basin. New stratigraphic and structural cross sections are presented to better define relationships between sedimentation and deformation in the forearc basin. Integration of these new data sets also allows us to develop tectonic reconstructions that demonstrate the early Tertiary tectonic development of the Matanuska Valley–Talkeetna Mountains forearc basin. The stratigraphy of this forearc basin records regional transpression related to terrane

Figure 1. A: Index map showing location of general study area in southern Alaska (shaded rectangle). B: Map showing the present distribution of tectonostratigraphic terranes, sedimentary basins, and magmatic arcs exposed in southern Alaska. The Yukon composite terrane represents the late Mesozoic continental margin of southern Alaska and contiguous areas in Canada. The allochthonous Wrangellia composite terrane consists of Late Proterozoic(?) and younger magmatic arc and oceanic plateau assemblages that accreted to the Yukon composite terrane during the Mesozoic. The Southern Margin composite terrane (SMCT) represents an accretionary prism that formed in response to Mesozoic and Cenozoic northeastward to northwestward subduction beneath the Wrangellia composite terrane. Uplifted deposits of the Matanuska Valley–Talkeetna Mountains forearc basin (MT) unconformably overlie the Peninsular terrane segment (P) of the Wrangellia composite terrane mainly between the Border Ranges (BRF) and Castle Mountain (CMF) fault systems. During the Tertiary this forearc basin was located between the coeval accretionary prism (SMCT) and magmatic arc (dashed line labeled AT). The Cook Inlet basin (CIB) represents part of an actively subsiding forearc basin located between the modern Aleutian magmatic arc to the northwest (dashed line labeled AL) and the accretionary prism to the southeast (SMCT). F-QCF—Fairweather-Queen Charlotte fault; HF—Hanagita fault. Modern Pacific–North America plate motion of 5.4 cm/yr (black arrow) from DeMets et al. (1990, 1994). Modern Yakatat–North America plate motion of 4.6 cm/yr from Fletcher and Freymueller (1997). Map adapted from Nokleberg et al. (1994) and Plafker et al. (1994). X–X′, Y–Y′, and Z–Z′ represent lines of cross sections shown on Figure 2. Rectangle denotes area of detailed geologic map shown in Figure 3. C: Digital elevation map showing the uplifted Matanuska Valley–Talkeetna Mountains basin (area within black polygon). MV—Matanuska Valley; A—Anchorage; V—Valdez. See Figure 1B for explanation of other abbreviations. Adapted from Riehle et al. (1997).

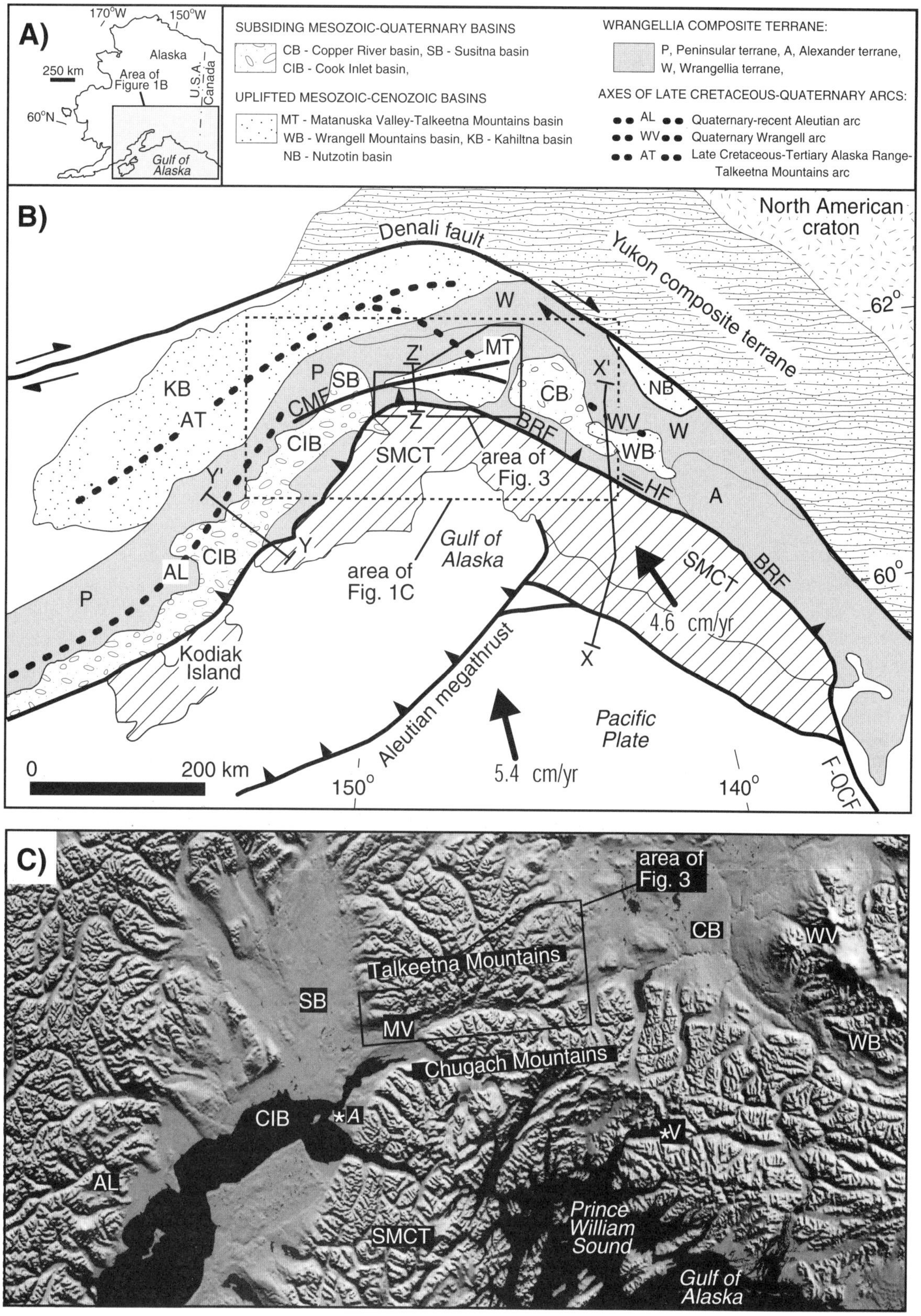
A)
170°W
150°W
Alaska
250 km
Area of Figure 1B
U.S.A.
Canada
60°N
Gulf of Alaska
SUBSIDING MESOZOIC-QUATERNARY BASINS
CB - Copper River basin, SB - Susitna basin
CIB - Cook Inlet basin,
UPLIFTED MESOZOIC-CENOZOIC BASINS
MT - Matanuska Valley-Talkeetna Mountains basin
WB - Wrangell Mountains basin, KB - Kahiltna basin
NB - Nutzotin basin
WRANGELLIA COMPOSITE TERRANE:
P, Peninsular terrane, A, Alexander terrane,
W, Wrangellia terrane,
AXES OF LATE CRETACEOUS-QUATERNARY ARCS:
AL Quaternary-recent Aleutian arc
WV Quaternary Wrangell arc
AT Late Cretaceous-Tertiary Alaska Range-Talkeetna Mountains arc
B)
North American craton
Denali fault
Yukon composite terrane
62°
W
MT
Z'
X'
P
SB
CB
NB
KB
AT
CMF
Z
BRF
WV
W
CIB
SMCT
area of Fig. 3
WB
Y'
HF
A
Gulf of Alaska
Y
CIB
AL
area of Fig. 1C
SMCT
BRF
60°
P
4.6 cm/yr
Kodiak Island
Aleutian megathrust
X
Pacific Plate
0
200 km
150°
5.4 cm/yr
140°
F-QCF
C)
area of Fig. 3
CB
WV
Talkeetna Mountains
SB
MV
WB
Chugach Mountains
CIB
*A
*V
AL
Prince William Sound
SMCT
Gulf of Alaska

translation, subduction of oceanic crust, ridge subduction, and/or oroclinal bending along the early Tertiary accretionary convergent margin of southern Alaska.

GEOLOGIC SETTING

Forearc Basin, Accretionary Prism, and Volcanic Arc

Over 7600 m of Upper Cretaceous–Lower Tertiary sedimentary strata are exposed in a ~90 km long and 20 to 70 km wide outcrop belt in the Matanuska Valley, southern Talkeetna Mountains, and northern Chugach Mountains (Fig. 1). We refer to these deposits as the Matanuska Valley–Talkeetna Mountains forearc basin. This basin developed between a coeval accretionary prism to the south (present coordinates) and a volcanic arc to the north (present coordinates) (Little, 1988; Little and Naeser, 1989; Winkler, 1992).

The southern margin of the forearc basin deposits is bounded by metamorphosed Triassic-Paleogene marine sedimentary deposits, oceanic basalt, and mélange of the Southern Margin composite terrane (Figs. 1, 2, 3; Plafker et al., 1994). These deposits represent one of the largest accretionary prisms on Earth (Plafker et al., 1994). The age of strata, degree of structural deformation, and grade of metamorphism within the accretionary prism progressively decreases southward. The Border Ranges/Hanagita fault system marks the boundary between the accretionary prism and the forearc basin (Figs. 1, 2; MacKevett and Plafker, 1977; Roeske et al., this volume). This fault system accommodated mainly Mesozoic dip-slip and Cenozoic dextral strike-slip displacement (e.g., Pavlis, 1982; Little and Naeser, 1989; Little, 1990; Pavlis and Crouse, 1989; Roeske et al., 1993, this volume; Smart et al., 1996). The amount of Cenozoic strike-slip displacement is presently unclear, with inferred displacements ranging from a few tens of km (e.g., Little and Naeser, 1989) to more than 700 km (e.g., Smart et al., 1996; Roeske et al., this volume). Along the northern margin of the forearc basin, strata depositionally onlap Upper Cretaceous-Quaternary arc-related volcanic and plutonic rocks (TKg and Tv on Figs. 2B, 3; Nokleberg et al., 1994). The northeast-trending Castle Mountain fault system bisects the arcward margin of the forearc basin (Figs. 1, 3). Cenozoic dextral offset of several tens of kilometers has been proposed along this fault system (Grantz, 1966; Clardy, 1974; Detterman et al., 1976; Fuchs, 1980).

Forearc basin, accretionary prism, and volcanic arc construction are the products of northeastward to northwestward subduction of the Farallon, Kula, and Pacific oceanic plates beneath the northwestern Cordillera margin throughout the Late Cretaceous and Tertiary (Plafker and Berg, 1994; Plafker et al., 1994). Matanuska Valley–Talkeetna Mountains forearc basin sedimentation, volcanism, and plutonism apparently ceased during the Neogene. The southwestern extension of the Matanuska Valley–Talkeetna Mountains basin is marked by the modern Cook Inlet, which includes an active volcanic arc, forearc basin, and accretionary prism located above the northwestward-subducting Pacific plate (Fig. 2C; Magoon et al., 1976; Haeussler et al., 2000).

Accreted Terranes and Translation History

The Matanuska Valley–Talkeetna Mountains forearc basin and associated volcanic arc were constructed upon the allochthonous Peninsular terrane (Fig. 1B), a ~7 km thick sequence of Lower to Upper Jurassic volcanic rocks, granitic plutons, and marine sedimentary strata (Nokleberg et al., 1994). The Peninsular terrane represents the westernmost segment of the Wrangellia composite terrane, a subcontinent-scale terrane consisting of three tectonostratigraphic terranes (P, A, and W on Fig. 1B) that were juxtaposed by the late Paleozoic. Paleomagnetic, paleontologic, and lithologic evidence indicate that the composite terrane was located ~30° south of its present latitude with respect to cratonic North America during the Late Triassic (Hillhouse and Coe, 1994) and was subsequently translated northward throughout the late Mesozoic and early Cenozoic (Plafker and Berg, 1994). Paleomagnetic studies indicate that Late Cretaceous sedimentary strata of the Matanuska Valley–Talkeetna Mountains forearc basin (Km on Figs. 3, 4) were deposited ~2800 ± 2000 km south of their present latitude relative to cratonic North America (Stamatakos et al., 1989). Paleomagnetic data from Paleocene-Eocene strata (Tar and Tc on Figs. 3, 4) indicate that the forearc basin was positioned ~1600 ± 1200 km south of its present latitude during deposition (Stamatakos et al., 1989). Similarly, paleomagnetic data from Late Cretaceous-Paleocene ophiolites of the accretionary prism imply ~1400 ± 1000 km of northward latitudinal displacement from 57 to 45 Ma (SMCT on Fig. 1; Coe et al., 1985; Bol et al., 1992). Paleomagnetic data indicate that the basin and accretionary prism were at their present latitude by ca. 50–45 Ma (Stamatakos et al., 1988; Hillhouse and Coe, 1994). Thus, the forearc basin and accretionary prism probably originated south of their present latitude and were translated northward during the Late Cretaceous–early Tertiary. Northward displacements were accommodated by orogen-parallel dextral strike-slip fault systems, including the Denali (Fig. 1; Nokleberg et al., 1994) and Border Ranges/Hanagita fault systems (Roeske et al., 1993, this volume; Smart et al., 1996).

STRATIGRAPHIC AND GEOCHRONOLOGIC DATA

Sedimentary strata of the Matanuska Valley–Talkeetna Mountains forearc basin consist of a lower marine package and an upper nonmarine package. The marine package consists of Cretaceous mudstone, sandstone, limestone and minor conglomerate, and has a maximum preserved thickness of ~4800 m (Kn and Km on Figs. 3, 4; Grantz and Jones, 1960; Csejtey et al., 1978; Winkler, 1992). The nonmarine package, the focus of this study, consists of Paleocene-Oligocene conglomerate, sandstone, mudstone, coal, and minor interbedded lava flows and tuffs. The maximum preserved thickness of the Paleocene-Oligocene non-

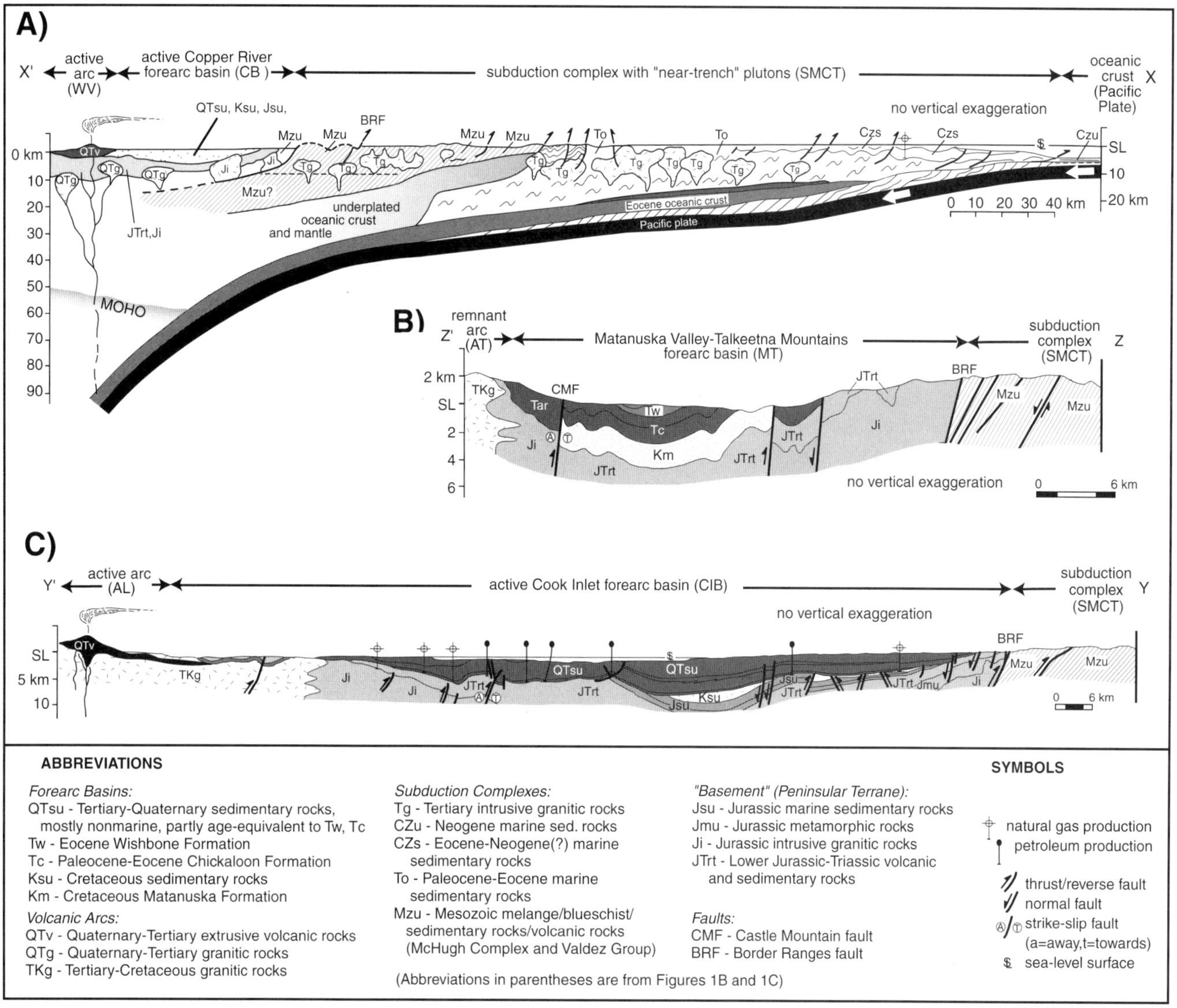

Figure 2. Cross sections through the southern Alaska convergent margin showing magmatic arcs, forearc basins, and accretionary complexes. See Figure 1B for lines of cross sections. Figures 2A and 2C are sections through regions located southeast and southwest, respectively, of the Matanuska Valley–Talkeetna Mountains basin. In these areas, arc volcanism and forearc basin subsidence are active. A: Cross section through the Copper River forearc basin (CB) in south-central Alaska. The Copper River forearc basin (CB) is an actively subsiding depocenter located between an active magmatic arc (WV) and accretionary prism (SMCT). Active volcanoplutonism along the arc and uplift along the accretionary prism are controlled by northwestward subduction of the Pacific plate. Section adapted from Plafker et al., 1994 (their Figure 3A). B: Cross section through the uplifted Matanuska Valley–Talkeetna Mountains forearc basin (MT) in south-central Alaska. This remnant forearc basin formed between a Late Cretaceous-Eocene volcanoplutonic arc (AT) and a late Mesozoic-Cenozoic accretionary prism (SMCT). The uplifted forearc basin deposits are northeastern extensions of age-equivalent subsurface strata in the modern Cook Inlet forearc basin (Figs. 1B, 2C). Section adapted from Barnes (1962) and Little (1988). C: Cross section through the Upper Cook Inlet in south-central Alaska. The Cook Inlet forearc basin (CIB) is an actively subsiding depocenter located between an active magmatic arc (AL) and accretionary prism (SMCT). Active volcanoplutonism along the arc and uplift along the accretionary prism are controlled by northwestward subduction of the Pacific plate (not shown on cross section, see Figs. 1A, 2A). Section adapted from Plafker et al. (1982).

marine deposits is ~2800 m (Little and Naeser, 1989; Winkler, 1992; this study). The Paleocene-Oligocene stratigraphy consists of four formations—the Chickaloon, Arkose Ridge, Wishbone and Tsadaka Formations (Tc, Tar, Tw, and Tsd on Figs. 3, 4; Csejtey et al., 1978; Winkler, 1992). The age, lithofacies, thickness, and stratigraphic equivalents of each formation is briefly described below based on our new measured stratigraphic sections and $^{40}Ar/^{39}Ar$ age determinations (Table 1) in combination with data reported from previous studies. Procedures for the $^{40}Ar/^{39}Ar$ analyses are presented in Appendix 1.

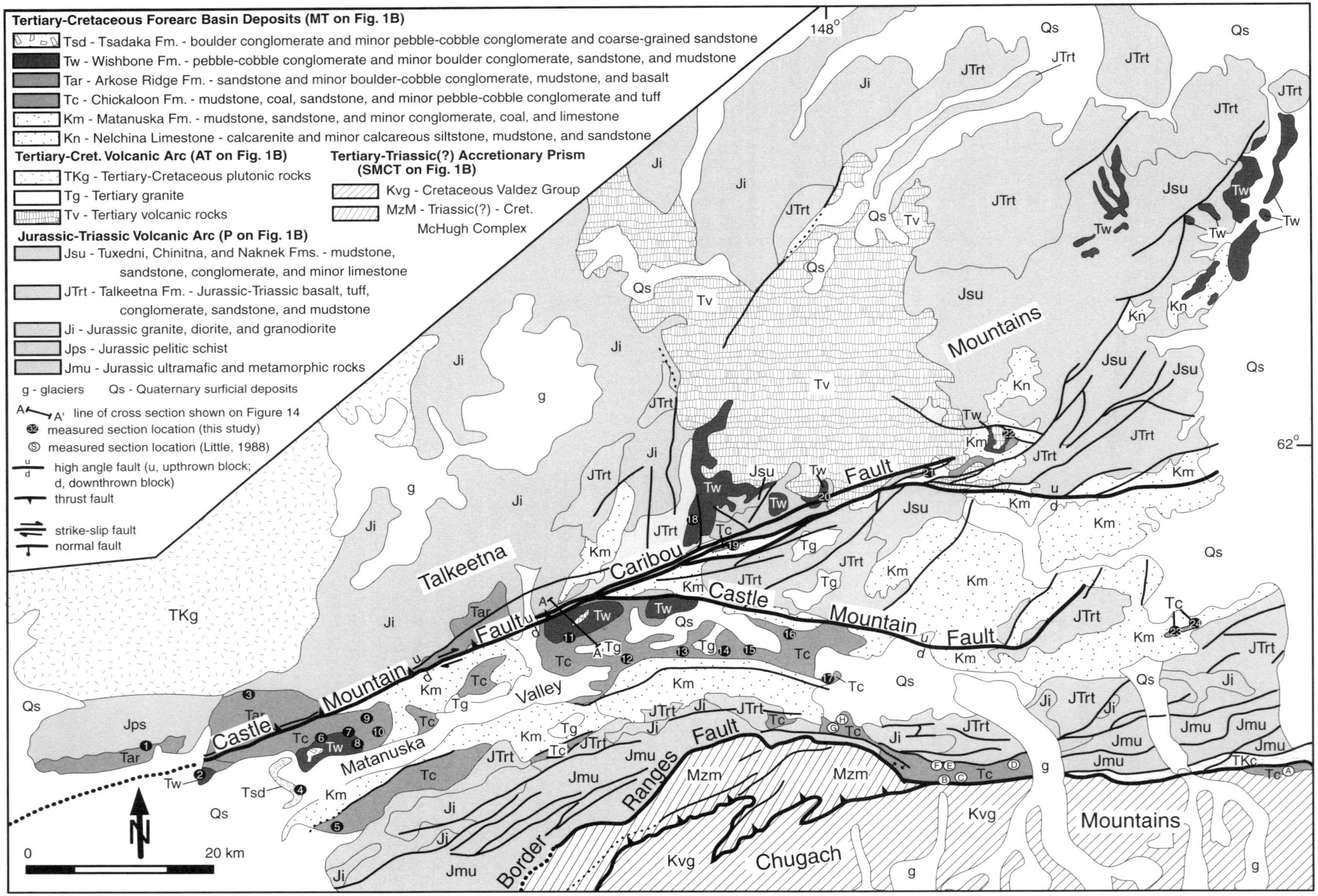

Figure 3. Generalized geologic map of the Matanuska Valley and southern Talkeetna Mountains in the Anchorage and Talkeetna Mountains quadrangles. Dark gray areas (Tar, Tc, and Tw) represent mostly nonmarine Tertiary sedimentary strata that are the focus of this study. Numbered black circles represent the locations of new measured stratigraphic sections discussed in text. Lettered open circles represent locations of measured sections from Little (1988). See Figures 1B and 1C for map location. Geology modified from Winkler (1992), Csejtey et al. (1978), and Wilson et al. (1998).

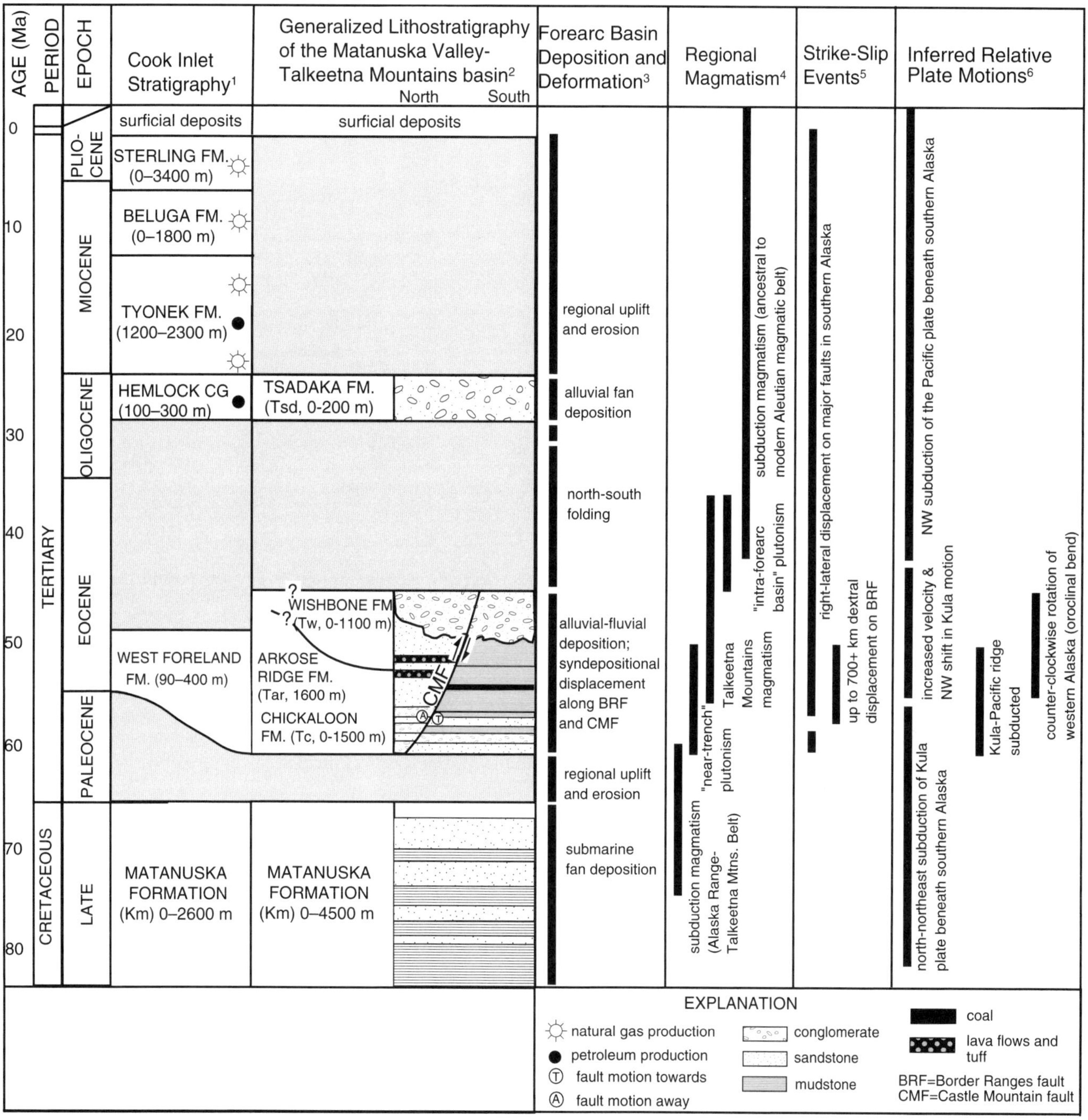

Figure 4. Age-event diagram for late Mesozoic and Cenozoic stratigraphy of the Matanuska Valley, Talkeetna Mountains, and Cook Inlet. 1 and 2—rock units as described by Winkler (1992), Trop and Ridgway (1999), and this study; 3—depositional and deformational events based on results of this study, Clardy (1974), Fuchs (1980), and Little and Naeser (1989); 4—after Hudson (1983), Wallace and Engebretson (1984), Bradley et al. (2000), and Moll-Stalcup (1994); 5—Nokleberg et al. (1994), Plafker and Berg (1994), includes Denali fault (Lanphere, 1978; Nokleberg et al., 1994; Ridgway et al., 1995, 1999), Tintina fault (Gabrielse, 1985; Dover, 1994), Castle Mountain fault (Fuchs, 1980; Silberman and Grantz, 1984), and Border Ranges fault (Little, 1990; Roeske et al., 1993, this volume; Smart et al., 1996); 6—Engebretson et al. (1985), Hillhouse and Coe (1994), Plafker et al. (1994), Plafker and Berg (1994), Pavlis and Sisson (1995), and Bradley et al. (2000).

TABLE 1. $^{40}Ar/^{39}Ar$ AGES FOR TERTIARY VOLCANIC ROCKS, MATANUSKA VALLEY–TALKEETNA MOUNTAINS

Sample number	Geographic location*	Map unit† (map abbreviation)	Stratigraphic position	Rock type	Mineral	Age (Ma)§	Notes#
BM1-54	Billy Mountain Section 22	Eocene (unnamed, Tv)	8 m above top of Wishbone Formation	Tuff	Plagioclase	37.3 ± 2.1	Single-crystal laser fusion, 11 single crystals
WC1-CC1-BR (basalt clast)	Wishbone Hill Section 8	Paleocene-Eocene Wishbone Fm. (Tw)	12 m above base of Wishbone Formation	Basalt	Whole rock	56.8 ± 0.3	Furnace step-heat
BOX2-CC2-2 (granite clast)	Box Canyon Section 18	Paleocene-Eocene Wishbone Fm. (Tar)**	252 m above base of Wishbone Formation	Granite	Biotite	59.7 ± 0.3	Furnace step-heat
CA1-TV-1300	Castle Mountain Section 11	Eocene (unnamed, Tv)	4 m above top of Wishbone Formation	Basalt	Whole rock	<61.0 Ma	Furnace step-heat
SUN1-3.6	Boulder Creek Section 19	Paleocene-Eocene Chickaloon Fm. (Tc)	231 m above faulted base of Chickaloon Formation	Volcaniclastic sandstone	Sanidine	177.0 ± 1.2	Single-crystal laser fusion, 4 single crystals

*See Figure 3 for measured section locations.
†Stratigraphic units and unit abbreviations from Anchorage 1:250 000 quadrangle of Winkler (1992) and Figure 3 (this study).
§Preferred ages reported in text. Ages for BM1-54, WC1-CC1-BR, and SUN1-3.6 represent isochron analyses. Age for BOX2-CC2-2 represents plateau age from step-heating analysis. Age of CA1-TV-1300 represents maximum age value from step-heating analysis.
#Explanation of furnace step-heat and laser fusion analytical methods presented in Appendix 1. Analytical data and age spectra available from the authors upon request.
**Originally interpreted as Arkose Ridge Formation (Tar) by Winkler (1992). Interpreted as Wishbone Formation (Tw, this study) on the basis of new stratigraphic, sedimentologic, and compositional data.

Chickaloon Formation

The Chickaloon Formation consists of ~1500 m of mudstone, sandstone, and coal with minor conglomerate and tuff (Fig. 4; Little, 1988; Flores and Stricker, 1993a, 1993b). This formation is exposed between the Castle Mountain and Border Ranges fault systems (Tc on Figs. 2B, 3) and overlies Upper Cretaceous to Lower Jurassic strata along an angular unconformity (Figs. 3, 4; Winkler, 1992). The Chickaloon Formation is gradationally overlain by the Eocene Wishbone Formation along a conformable contact at most locations, but locally this contact is an angular unconformity (Figs. 3, 4; Clardy, 1974; Fuchs, 1980). A late Paleocene–early Eocene age for the Chickaloon Formation is defined by plant megafossils (Wolfe et al., 1966; Little, 1988), palynomorphs (Little, 1988), isotopic ages of interbedded airfall tuffs from the upper part of the formation (55.8 to 52.2 Ma; n = 5; Triplehorn et al., 1984), and fission track ages of crosscutting dikes (47.8–41.3 Ma, n = 2; Little and Naeser, 1989). Whereas the age of the uppermost Chickaloon Formation is well constrained by dated airfall tuffs, the lower Chickaloon Formation may represent all or part of the Late Paleocene prior to 55.8 Ma (oldest dated tuff).

Arkose Ridge Formation

The Arkose Ridge Formation consists of sandstone, conglomerate, and minor mudstone, coal, tuff, and lava flows, and has a maximum preserved thickness of 1600 m (Silberman and Grantz, 1984; Trop and Ridgway, 1999). Outcrops of the Arkose Ridge Formation are confined to the north side of the Castle Mountain fault system (Tar on Fig. 3; Winkler, 1992). The base of the Arkose Ridge Formation unconformably overlies Jurassic igneous and metamorphic rocks (Ji and Jps on Fig. 3); the top of the formation is unconformably overlain by Quaternary surficial deposits (Qs on Fig. 3) or unnamed Tertiary volcanic rocks (Tv on Fig. 3). K-Ar ages from interbedded volcanic rocks (56–46 Ma; n = 6; Silberman and Grantz, 1984) indicate that the Arkose Ridge Formation is Late Paleocene to Middle Eocene (Fig. 4). Accordingly, the Arkose Ridge Formation is partly coeval with the Upper Paleocene–Lower Eocene Chickaloon Formation (Winkler, 1992), which is exposed south of the Castle Mountain fault system (Tc on Fig. 3). Five of the six K-Ar ages reported from the Arkose Ridge Formation, however, are younger than the fission track and K-Ar ages reported from tuffs of the uppermost Chickaloon Formation. Thus, the Arkose Ridge Formation may be correlative with the uppermost Chickaloon Formation and/or the Wishbone Formation, which gradationally overlies the Chickaloon Formation (see description below).

Wishbone Formation

The Wishbone Formation consists mainly of conglomerate and sandstone (Clardy, 1974; Fuchs, 1980; this study). The maximum preserved thickness of the Wishbone Formation ranges from <150 m north of the Castle Mountain fault system to >1100 m south of the fault system (Tw on Fig. 5). South of the fault system the Wishbone Formation gradationally overlies early Eocene strata of the Chickaloon Formation, but north of the fault system the formation unconformably overlies folded Lower Jurassic to Tertiary strata along an angular unconformity (Figs. 3, 5; Csejtey et al., 1978; Fuchs, 1980; Winkler, 1992). The Wishbone Formation is overlain by unnamed Tertiary volcanic rocks, the Oligocene Tsadaka Formation, or Quaternary surficial deposits along a disconformity or angular unconformity (Figs. 3, 5).

The age of the Wishbone Formation was previously interpreted as Eocene (ca. 55–34 Ma) on the basis of (1) sparse age-diagnostic palynomorphs (K. Piel, personal commun., cited in Clardy, 1974), and (2) a gradational lower contact with early Eocene strata of the Chickaloon Formation (Fuchs, 1980). Due to the lack of datable material in the Wishbone Formation, we analyzed plagioclase feldspar from an airfall tuff exposed ~7 m above uppermost strata of the Wishbone Formation (unnamed Tertiary volcanic rocks, Tv on Fig. 3) to provide a minimum age for the Wishbone Formation. Eleven single-crystal $^{40}Ar/^{39}Ar$ analyses define a late Middle Eocene isochron age of 37.3 ± 2.1 Ma (BM1-54 on Table 1 and Figs. 5, 6). A basalt flow located ~4 m above the uppermost Wishbone Formation was also analyzed to constrain the minimum age of the Wishbone Formation (CA1-TV-1300 on Table 1 and Figs. 5, 6). $^{40}Ar/^{39}Ar$ whole rock step-heating ages are very discordant and no valid isochrons were derived. Attempted isochrons have $^{40}Ar/^{39}Ar$ values greater than atmospheric values, suggesting that excess argon is present. Thus, the age of the basalt is interpreted as being younger than the minimum value in the age spectrum (ca. 61 Ma; Fig. 6). $^{40}Ar/^{39}Ar$ analysis of two igneous clasts contained in conglomerates of the Wishbone Formation provides crude maximum ages of deposition. Step heating analysis of biotite from a granite clast yielded a moderately concordant age spectrum with a plateau age of 59.7 ± 0.3 Ma (sample BOX-CC2-2 on Table 1 and Figs. 5, 6). A basalt clast yielded a discordant age spectrum with ages from ca. 36 to 55 Ma. An isochron age of 56.8 ± 0.3 Ma is defined by steps 3–7 (sample WC1-CC1-BR on Table 1 and Figs. 5, 6). These age interpretations are consistent with the previously reported Eocene palynomorphs from the Wishbone Formation and with isotopic and fission track ages of airfall tuffs from the upper part of the Chickaloon Formation (55.8 to 52.8 Ma; n = 5; Triplehorn et al., 1984; Figs. 4, 5). In summary, no direct age data are available for the formation; the possible age of the Wishbone Formation ranges from 52.2 ± 1.6 Ma (youngest dated tuff in uppermost Chickaloon Formation) to 37.3 ± 2.1 Ma (tuff overlying uppermost Wishbone Formation).

Tsadaka Formation

The Tsadaka Formation is limited in outcrop distribution to the lower Matanuska Valley (Tsd on Fig. 3) and has a maximum preserved thickness of ~200 m (Fig. 4). This formation consists of poorly sorted, boulder to cobble conglomerate with minor mudstone and sandstone. Palynomorphs recovered from the Tsadaka Formation indicate a middle to late(?) Oligocene age (ca. 29–24 Ma; Clardy, 1974; Fig. 4). The Tsadaka Formation overlies folded Paleocene-Eocene strata of the Chickaloon and Wishbone Formations along an angular unconformity (Clardy, 1974). The top of the Tsadaka Formation is eroded and overlain by Quaternary surficial deposits.

DEPOSITIONAL SYSTEMS

New measured stratigraphic sections, lithofacies analyses, paleocurrent data, and maximum particle size data from 22 study sites (sections 1–22 on Fig. 3) provide a database for reconstructions of the depositional systems of the early Tertiary Matanuska Valley–Talkeetna Mountains forearc basin. These new data are integrated with previous geologic mapping studies (Fuchs, 1980) and sedimentologic analyses that were limited to more localized parts of the basin (Clardy, 1974; Little, 1988; Flores and Stricker, 1993a, 1993b).

Proximal Lithofacies: Stream-Dominated Alluvial-Fan Deposits

The northern and southern margins of the forearc basin are characterized by interbedded conglomerate and sandstone deposits that range in thickness from 50 to 700 m (Figs. 5, 7, 8A). These deposits are most common adjacent to the Castle Mountain fault along the northern basin margin (northern outcrops of Tw and Tc on Figs. 5, 7) and adjacent to the Border Ranges fault along the southern basin margin (Tc outcrops adjacent to BRF on Figs. 3, 5, 7; Little, 1988). Three distinct lithofacies characterize these deposits.

Lithofacies 1: Unorganized, Boulder-Cobble Conglomerate

Description. This lithofacies is characterized by unorganized, clast-supported conglomerate with angular to subrounded clasts. Conglomerate beds are broadly lenticular (10–20 m) and are arranged into 5 to 15 m thick upward-fining packages. Subordinate sandstone and mudstone beds cap the upward-fining sequences. Interbedded with the poorly organized, clast-supported conglomerates are minor matrix-supported cobble-pebble conglomerates that often contain outsized boulders.

Interpretation. The framework support, lenticular bed geometries, and crude upward-fining trends of Lithofacies 1 indicate deposition by high-energy streamflow processes. The poorly organized, clast-supported conglomerates of this lithofacies were most likely deposited by stream-driven, high-density, gravelly traction carpets (Todd, 1989) and/or hyperconcentrated flood flows (Pierson, 1980; Smith, 1986). These types of flows are reported from modern humid alluvial-fans that are influenced by seasonal floods (Wells and Harvey, 1986). The matrix-supported conglomerates are interpreted as being deposited by

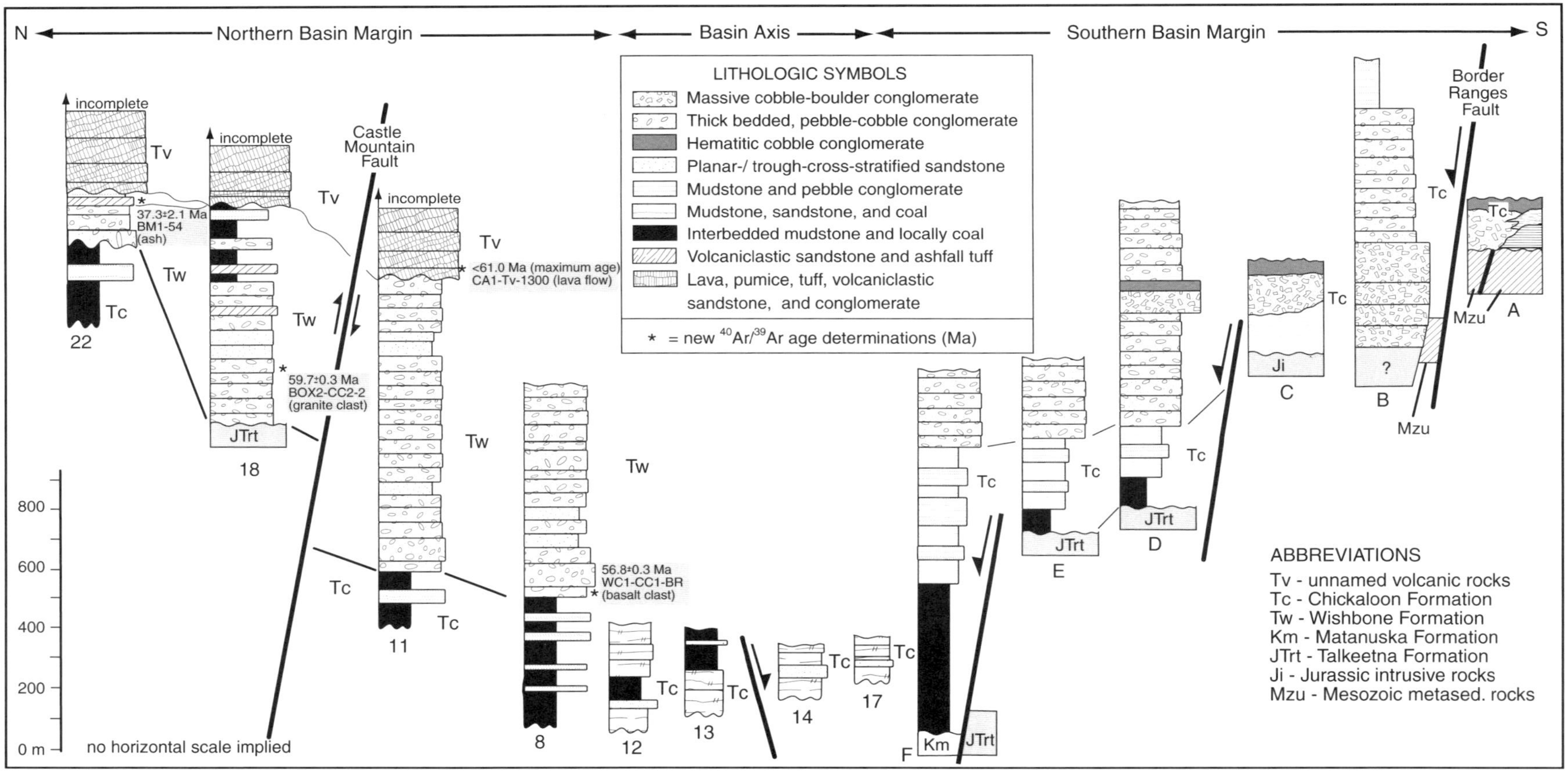

Figure 5. Generalized measured stratigraphic sections, stratigraphic positions of new $^{40}Ar/^{39}Ar$ age determinations, and inferred correlations of Tertiary strata within the Matanuska Valley–Talkeetna Mountains forearc basin. Stratigraphic sections are from bed-by-bed measurements using a Jacob staff. Numbered sections (22, 18, 11, 8, 12, 13, 14) are from this study; lettered sections (A–F) are from Little (1988). See Figure 3 for section locations and Figure 6 for $^{40}Ar/^{39}Ar$ data.

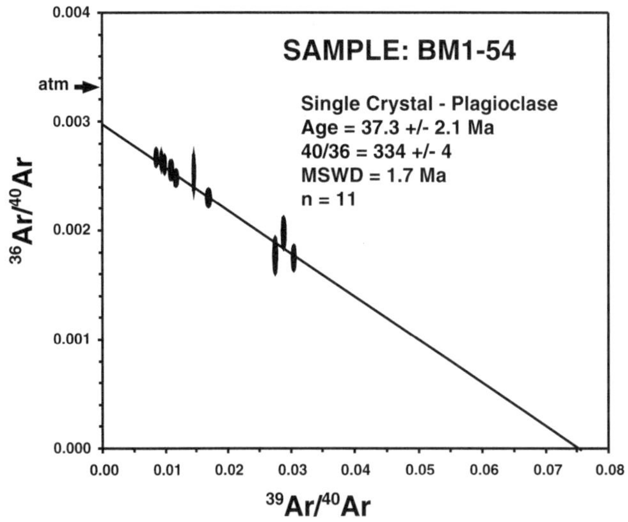

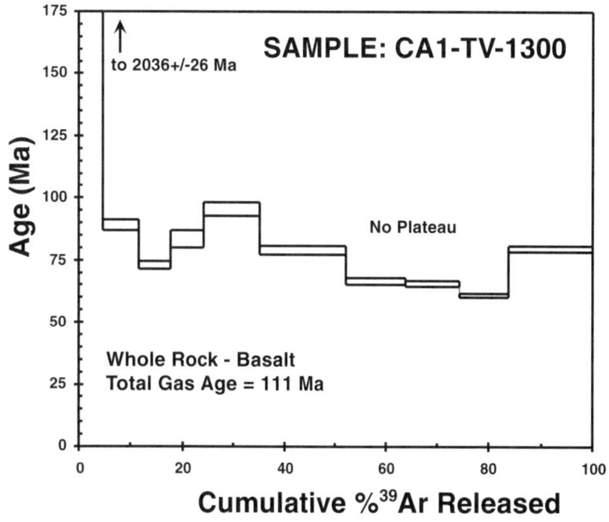

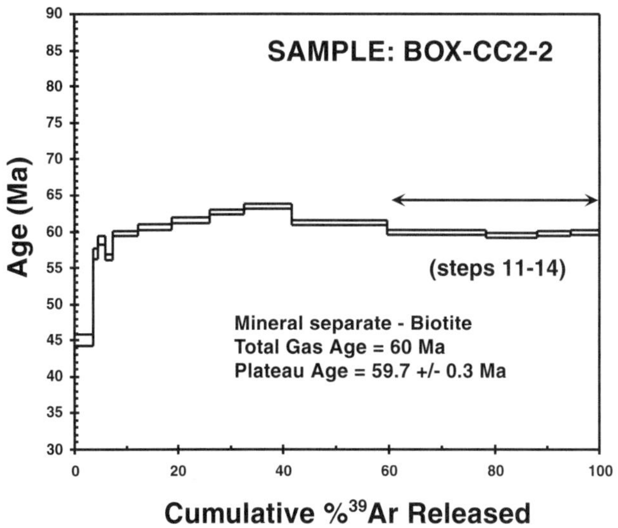

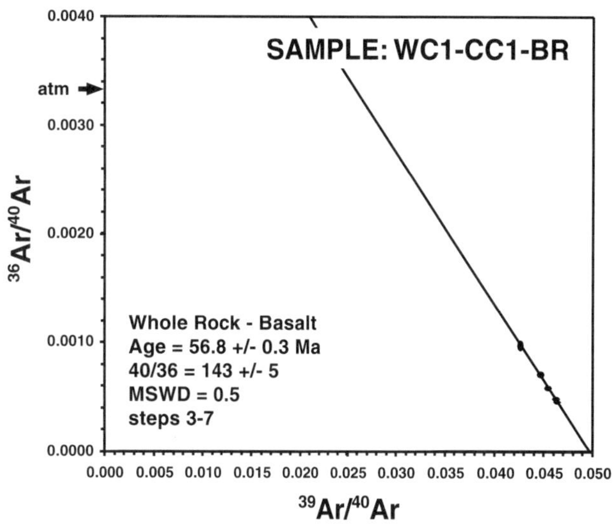

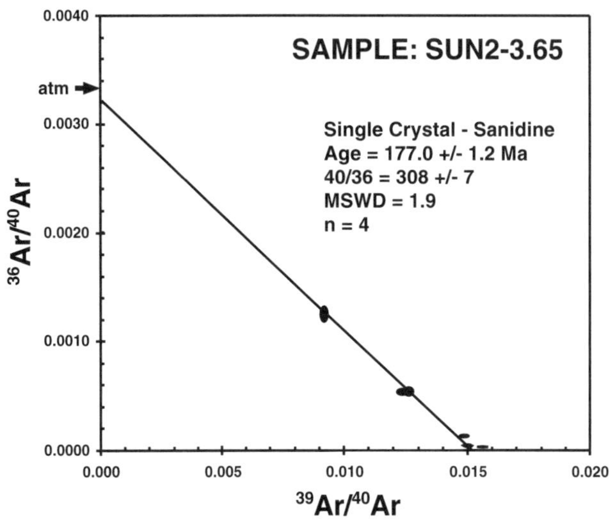

Figure 6. ^{40}Ar/^{39}Ar age data for feldspar phenocrysts from an airfall tuff (BM1-54), whole rock basalt (CA1-TV-1300), biotite and whole rock from igneous clasts (BOX-CC2-2, WC1-CC1-BR) from conglomerates, and detrital feldspars from a volcaniclastic sandstone (SUN2-3.65). See Table 1 and Figure 5 for geographic location and stratigraphic position of each sample. See Appendix 1 for ^{40}Ar/^{39}Ar analytical procedures and data treatment.

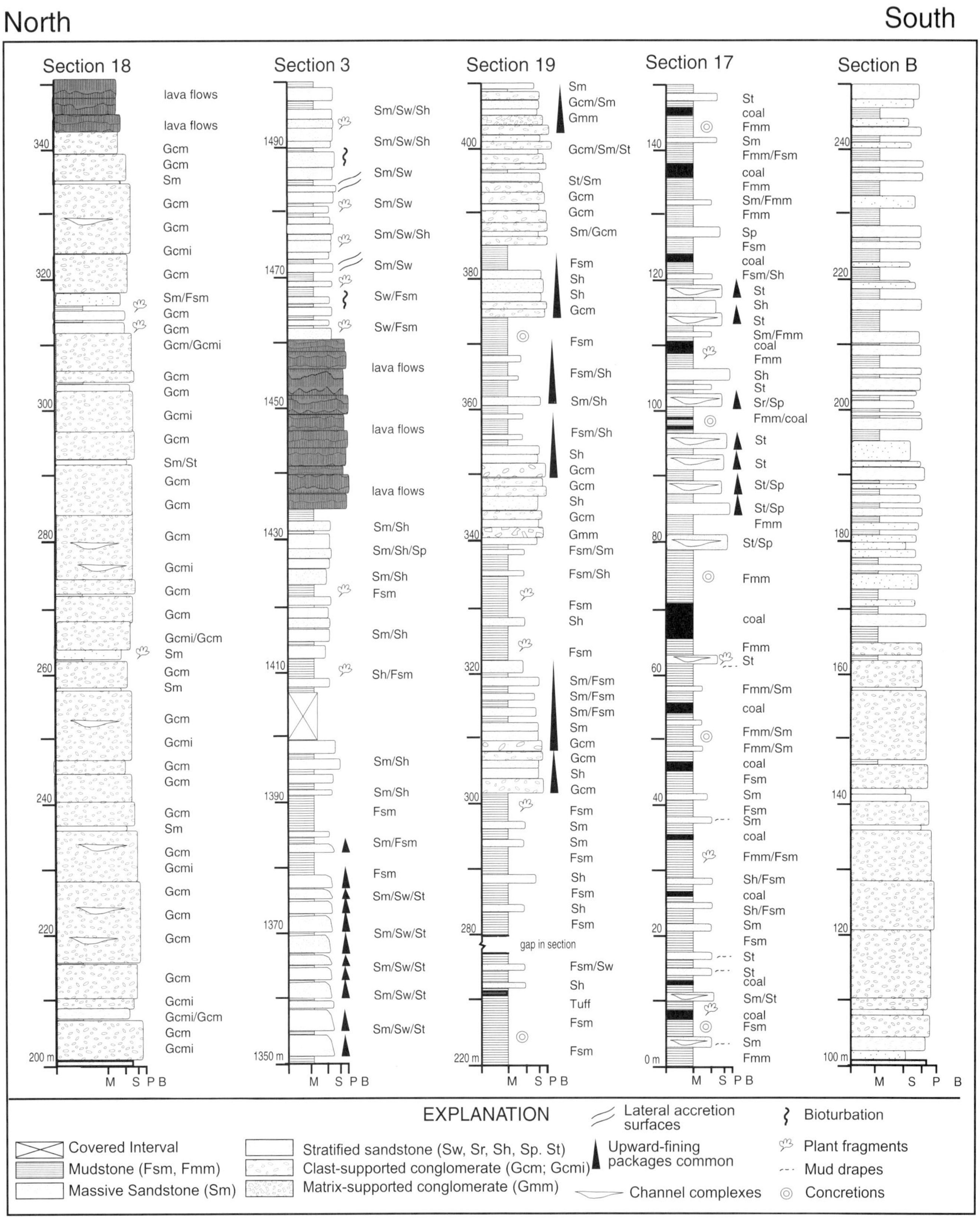

Figure 7. Representative measured stratigraphic sections of the Matanuska Valley–Talkeetna Mountains basin showing changes in Paleocene-Oligocene lithofacies from the northern basin margin to the southern basin margin. See Figure 3 for section locations. See Figure 5 for explanation of lithofacies symbols. Due to space limitations, only a 150 m representative part of each measured section is shown.

debris flows on the basis of the poor sorting and coarseness, in combination with a lack of tractive transport indicators or basal scour (Johnson, 1970).

Lithofacies 2: Organized, Imbricated Conglomerate

Description. Well-organized, clast-supported conglomerate is the most common lithofacies along the northern and southern basin margins. Pebble- and cobble-size clasts dominate this lithofacies. Upward-fining packages are well developed in this lithofacies (Figs. 8B, 8C). Individual packages grade upward from imbricated or massive cobble conglomerate to massive or planar cross-stratified pebbly sandstone and/or carbonaceous mudstone. Clast imbrication and low-angle cross-stratification are common sedimentary structures (Fig. 8D). Individual beds are 0.5 to 2.5 m thick and extend tens of meters laterally.

Interpretation. The upward-fining trends, clast imbrication, lenticular bed geometries, and presence of current-generated sedimentary structures all indicate that deposition of this lithofacies was mainly by streamflow processes. The organized conglomerate lithofacies is closely comparable to gravel and coarse-grained sand deposited in braided streams under waning flow conditions by accretion of progressively smaller clasts in channels and on longitudinal bars (Collinson, 1986). Interbedded mudstones probably record stagnation and fluvial channel abandonment.

Lithofacies 3: Cross-Stratified Sandstone

Description. A third lithofacies common to basin margin deposits is characterized by massive to planar- and trough-cross-stratified sandstone. These sandstones are medium- to coarse-grained, moderately sorted, and have tabular geometries. Individual sandstone units are amalgamated into broad sheet-like complexes with internal discontinuities such as scour and reactivation surfaces. Individual foresets in the stratified sandstones have average thicknesses of 40–70 cm. Plant fossils are common in this lithofacies.

Figure 8. Photographs of alluvial-fan and braided-stream lithofacies along the northern margin of the Matanuska Valley–Talkeetna Mountains forearc basin. See Figure 3 for measured section locations. A: Boulder and cobble conglomerate and minor sandstone of the uppermost Wishbone Formation (Tw) at section 11. The conglomerates are overlain by unnamed Tertiary volcanic rocks (Tv). Exposure is approximately 200 m thick. B and C: Broadly lenticular upward-fining conglomeratic packages at sections 19 and 18, respectively. Typical packages consist of 2–7 m thick imbricated pebble-cobble conglomerate (Gcmi) that grades upward into massive conglomerate, massive and trough-cross-stratified sandstone (Sm/St), and mudstone (Fsm). Person (right center, arrow) for scale in Figure 8C. Exposure in Figure 8B is approximately 200 m thick. D: Organized, clast-supported boulder and cobble conglomerate common along the northern basin margin (section 22). Bedding dips ~14° to the northwest (right in photograph). Hammer (lower center) for scale.

Interpretation. Lithofacies 3 is interpreted as having been formed by streamflow processes. Evidence for streamflow processes includes the trough and planar cross-stratification. The abundant stratification, with foresets in the 40–70 cm thick range, indicates that deposition was primarily by subaqueous three-dimensional ripples (dunes) and two-dimensional ripples (sandwaves) in shallow channels (Harms et al., 1982). The sheetlike geometries and average foreset thicknesses documented for Lithofacies 3 are indicative of deposition in unconfined fluvial systems with multiple shallow channels.

Summary

We interpret the three lithofacies common along the forearc basin margins to represent deposits of stream-dominated alluvial-fan systems (e.g., Boothroyd and Nummedal, 1978; Ridgway and DeCelles, 1993a). Stream-dominated alluvial fans are fans whose surface processes are characterized by perennial stream flow (Collinson, 1986). These types of fans have also been referred to as wet fans (Schumm, 1977; Schumm et al., 1987) and humid fans (Fraser and Suttner, 1986). In our depositional interpretation, the boulder-cobble conglomerate of Lithofacies 1 is interpreted as the product of localized hyperconcentrated flows in the proximal parts of the fan system associated with flood-flow conditions. The organized, imbricated conglomerate of Lithofacies 2 represents normal-flow deposition in proximal and medial parts of the alluvial-fan system characterized by streamflow processes. The cross-stratified sandstones of Lithofacies 3 are interpreted as representing distal sand-rich parts of stream-dominated alluvial fans. Distal parts of these types of fans are commonly dominated by sandy linguoid bar deposition (Boothroyd and Nummedal, 1978; Boothroyd and Ashley, 1975). Stream-dominated alluvial fans adjacent to the Castle Mountain fault along the northern basin margin transported sediment southward based on paleocurrent data (Fig. 9). Similar alluvial-fan deposystems adjacent to the Border Ranges fault along the southern basin margin transported sediment northwestward (Fig. 9; Little, 1988).

Medial Lithofacies: Sandy Braided Stream Deposits

Amalgamated 700–1000 m thick sections of channelized sandstone are present north of the Castle Mountain fault in the northwestern part of the basin (Tar outcrops on Fig. 3; section 3 on Fig. 7; Trop and Ridgway, 1999) and adjacent to the Border Ranges fault system along the southern basin margin (Tc outcrops adjacent to BRF on Fig. 3; section 3 on Fig. 5; Little, 1988).

Lithofacies 1: Channelized Planar Cross-Stratified Sandstone

Description. This lithofacies consists of lenticular, fine- to medium-grained sandstone beds that are 0.6–2.0 m thick and

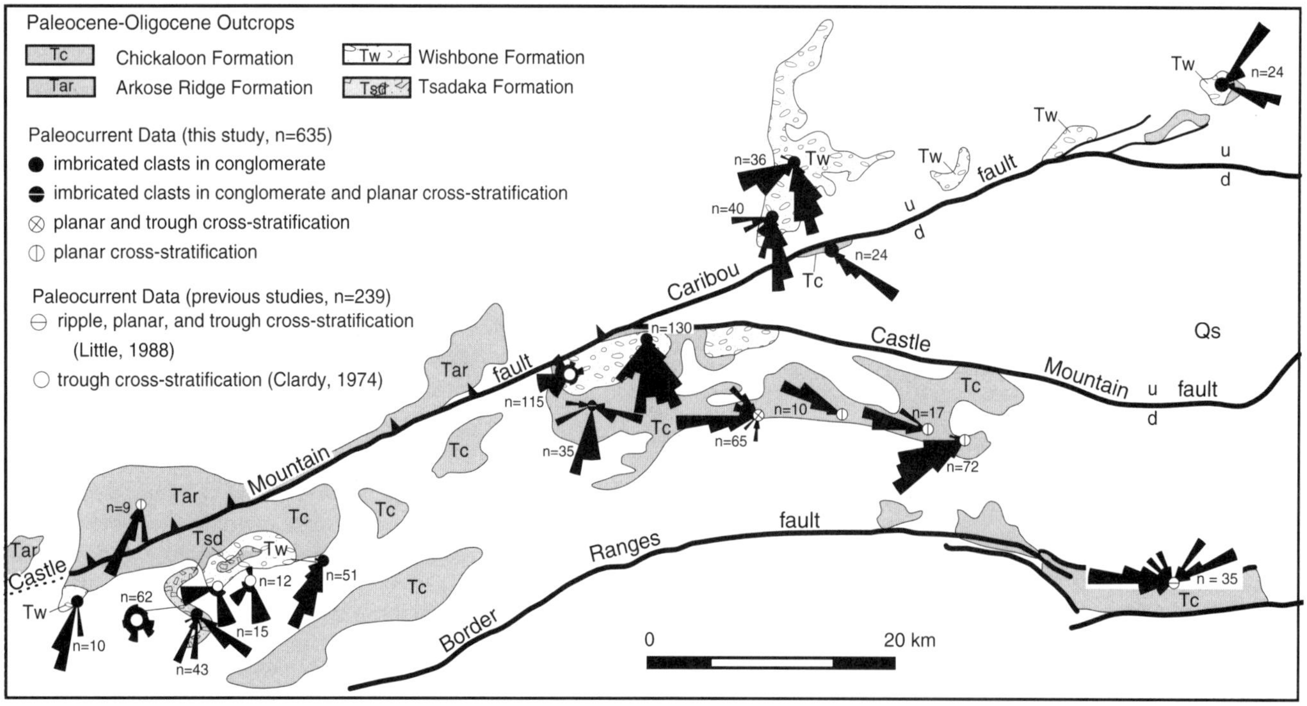

Figure 9. Map showing sediment transport directions measured in Paleocene-Oligocene strata. n = number of paleocurrent measurements for individual rose diagrams. All measurements were corrected for structural tilt. Note southeastward- to southwestward-directed paleoflow along the northern basin margin, westward-directed paleoflow along the axis of the basin, and northeastward- to northwestward-directed paleoflow along the southeastern basin margin.

over 10–20 m in length. Amalgamated lenticular sandstone beds form broad ~20 m thick sandstone sheets that fine upward (Fig. 10A). These sandstone sheets consist of massive to planar cross-stratified, medium-grained sandstone that grades upward into ripple-laminated, fine-grained sandstone capped by structureless to laminated mudstones. Trough cross-stratification and horizontal stratification are also common sedimentary structures in this lithofacies. Basal scours have relief of ~0.25–0.80 m and commonly include fossilized tree branches and leaves. Subordinate mudstone beds are 0.2–10 m thick, structureless to laminated, and are laterally continuous at the outcrop scale (10–50 m). Thin interbeds of carbonaceous shale and coal are also present in this lithofacies.

Interpretation. The amalgamated sandstones are interpreted as being formed by streamflow processes in low-sinuosity, fluvial systems. Evidence for fluvial deposition includes the presence of cross-stratification, upward-fining packages, and lenticular bed geometries. The amalgamated sheetlike architecture of sandstones and the scale of individual sedimentary structures (25–50 cm thick planar foresets) suggest that deposition occurred by migration of transverse and longitudinal sandbars in low-sinuosity streams (Miall, 1978; Rust, 1978; Harms et al., 1982). The amalgamation of sandstone beds is attributed to high-frequency switching of channels across an unconfined depositional surface whereas upward-fining packages likely resulted from eventual abandonment of channel complexes. The

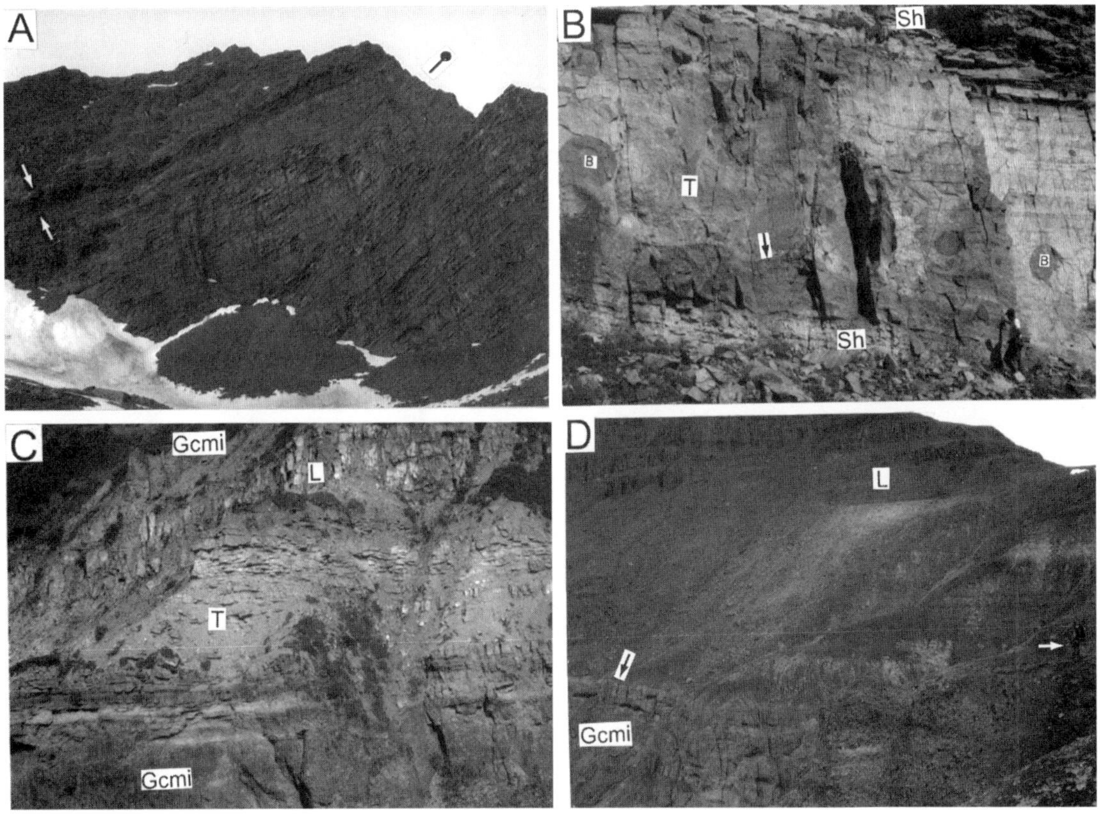

Figure 10. Photographs of interbedded volcanic and sedimentary strata along the northern margin of the Matanuska Valley–Talkeetna Mountains forearc basin. A: Sandstone and lava flows characteristic of sandy braided stream deposits of the Arkose Ridge Formation at section 3. Exposure includes massive and trough cross-stratified sandstone and 2–5 m thick basaltic lava flows. White arrows (upper left) denote upper and lower contact of a 25 m thick section of basalt flows. Sandstones consist of highly amalgamated, broad shallow channel deposits. Exposure is approximately 250 m thick. Tadpole symbol (upper right) shows dip of bedding. B: Interbedded fluvial volcaniclastic sandstone (Sh) and lithic-pumice tuff breccia (T) with outsized lithic blocks (B) of tuff and pumice. These volcanic deposits are interbedded with nonvolcanic fluvial conglomerates (not shown). Outcrop is from the upper part of the Wishbone Formation at section 18. Person (lower right) for scale. C: Nonvolcanic pebble-cobble conglomerate (Gcmi) with interbedded lithic tuff and pumice (T), and felsic lava flows (L). Outcrop is from the Wishbone Formation at section 18. Exposure is approximately 35 m thick. D: Nonvolcanic pebble-cobble conglomerate (Gcmi) overlain by nonvolcanic massive mudstone, lithic tuff and pumice, and airfall tuff. Section is overlain by lava flows (L) of the unnamed Tertiary volcanic rocks (Tv on Figure 3). Black arrow (lower left) marks top of channelized nonvolcanic conglomerate. Outcrop is from the Wishbone Formation at section 18. Exposure is approximately 240 m thick. Person (white arrow, lower right) for scale.

thick sections of this lithofacies (700–1000 m thick) suggest that extensive braidplains were well developed along the northern and southern basin margins. Most mudstones were probably deposited in overbank areas by suspension fallout of silt and clay during waning flow following floods. Limited paleocurrent data (n = 9) from planar cross-stratified sandstones along the northwestern basin margin (section 3 on Figs. 3, 7, 9) indicate southward sediment transport.

Basin Axis Lithofacies: Sinuous Stream, Swamp, and Lacustrine Deposits

The axis of the forearc basin is characterized by abundant fine-grained siliciclastic and coal deposits (Figs. 5, 7, 11A). Two distinct lithofacies make up the basin-axis deposits.

Lithofacies 1: Fine-Grained Sandstone, Mudstone, and Coal

Description. This lithofacies includes channelized, fine-grained sandstone, laminated sandstone and shale, and carbonaceous shale and coal (Fig. 11A). Large-scale, inclined heterolithic strata consisting of laminated siltstone and sandstone are common in this lithofacies. Clay-draped trough and ripple cross-stratification, flaser bedding, wavy bedding, climbing-ripple stratification, and bioturbation are also characteristic of this lithofacies. Plant megafossils are extremely abundant with fossilized leaves being exquisitely preserved (Wolfe et al., 1966).

Interpretation. We interpret the intricate mixture of channelized sandstone and mudstone, in addition to the type, scale, and arrangement of sedimentary structures as indicative of deposition in tidally influenced fluvial and estuary systems (Clifton, 1982; Eisma, 1998). Channelized, fine-grained sandstone deposits probably represent deposition in upper estuary channels, whereas tabular carbonaceous mudstones and coals were deposited in nearby fluvial marshes and tidal flats. Flaser to wavy bedding and clay-draped ripples suggest repeated fluctuations in current velocity, a depositional feature characteristic of tidally influenced environments (Eisma, 1998). Inclined heterolithic strata common in this lithofacies are characteristic of

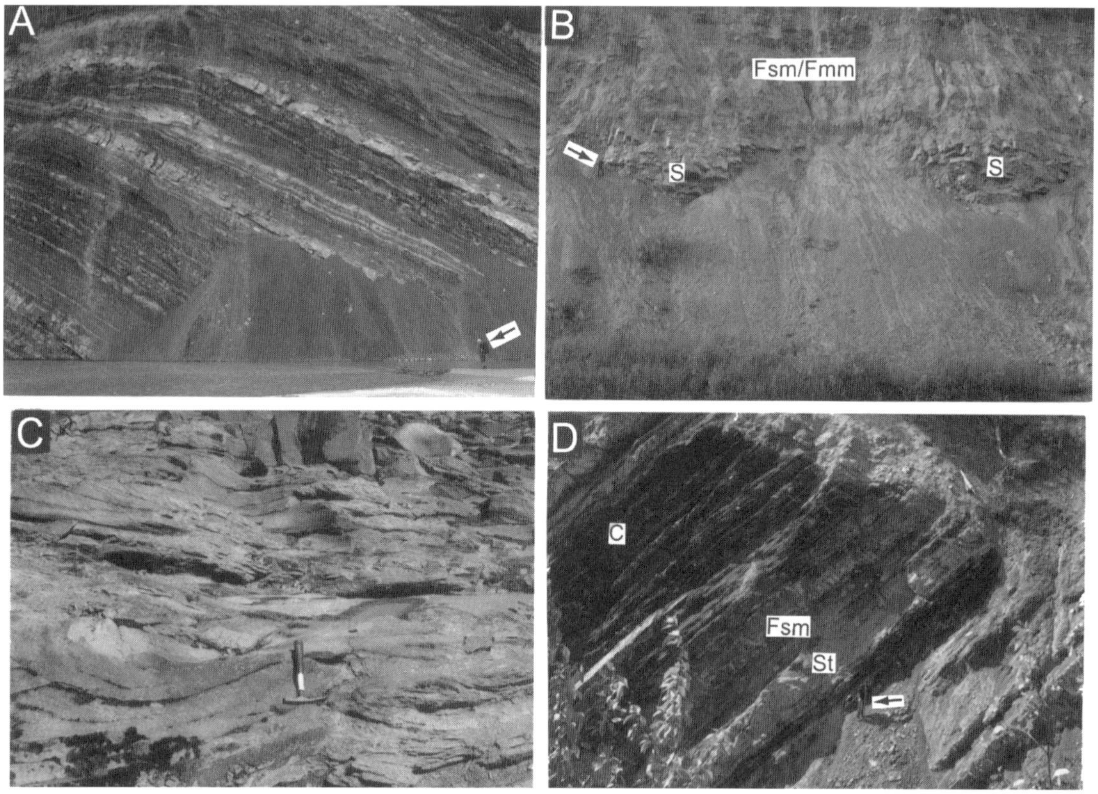

Figure 11. Photographs of sedimentary strata deposited along the axis of the Matanuska Valley–Talkeetna Mountains forearc basin. All photos are of the Chickaloon Formation. See Figure 3 for measured section locations. A: Carbonaceous mudstone- and coal-rich section (dark beds) containing broadly channelized, medium-grained sandstone (white beds) with abundant plant fragments. Person (arrow, lower right) for scale. Photo from section 17. B: Trough cross-stratified and massive sandstone (Sm/St) in 10 m wide channels encased in carbonaceous mudstone (Fsm/Fmm). Person (arrow, upper left) for scale. Photo from section 7. C: Close-up of trough cross-stratified, medium-grained sandstone of the basin-axis deposits. Hammer (lower center) for scale. Photo from section 12. D: Upward-fining packages common along the basin axis. Typical package consists of medium-grained trough cross-stratified sandstone (St) overlain by carbonaceous mudstone (Fsm) with abundant well-preserved plant fossils and coal (C). Person (arrow, lower center) for scale. Photo from section 6.

deposits formed by sandy fluvial point bars in micro- to mesotidally-influenced reaches of river systems (Smith, 1985; Thomas et al., 1987). This tidally influenced lithofacies is restricted to the Chickaloon and Arkose Ridge Formations and is only found in the western part of the basin axis (sections 3, 7, and 8 on Fig. 3).

Lithofacies 2: Trough Cross-Stratified Sandstone and Ripple-Laminated Siltstone

Description. This lithofacies consists of medium- to fine-grained sandstone and ripple laminated siltstone. The sandstones form upward-fining, lenticular units that are 1–2 m thick, tens of meters wide, and encased in carbonaceous ripple-laminated siltstone (Fig. 11B). Typical upward-fining packages consist of basal medium-grained, trough cross-stratified sandstone (Fig. 11C) that grade into fine-grained, ripple-laminated sandstone. Individual sandstones are separated by thick packages of micaceous, ripple-laminated siltstones. Coal seams up to 3.5 m thick are present (Fig. 11D). Paleocurrent data (n = 164) from trough cross-stratification in this lithofacies indicate westward sediment transport (Fig. 9).

Interpretation. Basin-axis strata are interpreted to have been deposited by longitudinal, westward-flowing, high-sinuosity fluvial systems. The upward-fining, trough cross-stratified sandstones of the basin-axis are typical of point bar and levee deposits of high-sinuosity stream systems (Allen, 1970a, 1970b; Jordan and Pryor, 1992). The thick ripple-laminated siltstone deposits represent deposition in overbank fluvial environments.

VOLCANIC DEPOSITS

Interbedded volcanic and sedimentary strata are well exposed along the northern margin of the Matanuska Valley–Talkeetna Mountains forearc basin. Lava flows and volcaniclastic deposits are the two most common volcanic units documented in our measured stratigraphic sections.

Lava Flows

Description

Basaltic and felsic lava flows are locally interbedded with sedimentary deposits along the northern basin margin (section 3 on Fig. 7; Figs. 10A, 10B, 10C). Lava flows are particularly abundant in the Arkose Ridge Formation north of the Castle Mountain fault (Silberman and Grantz, 1984; this study; Tar on Fig. 3). Most lava beds are massive, 2 to 5 m thick, and traceable for hundreds of meters.

Interpretation

Most of the lava flows were probably deposited in subaerial settings based on the absence of pillow structures or interbedded lacustrine facies, and the association with interbedded alluvial-fluvial deposits (see descriptions above). Several pillowed lava flows, however, are present in the upper part of the Arkose Ridge Formation. These deposits are interpreted to represent subaqueous deposition in local lacustrine environments (Trop and Ridgway, 1999). Fluvial sandstones that depositionally overlie basalt flows in the Arkose Ridge Formation contain abundant basaltic lithic fragments, indicating that basalt flows were subaerially exposed and were a sediment source for younger sedimentary deposits of the Arkose Ridge Formation (Trop and Ridgway, 1999).

Volcaniclastic Deposits

Volcaniclastic deposits are common north of the Castle Mountain fault system in the Arkose Ridge and Wishbone Formations (Silberman and Grantz, 1984; this study; sections 18, 20, and 22 on Figs. 3, 5). Minor volcaniclastic strata are present south of the fault system in the Chickaloon Formation (Triplehorn et al., 1984; Little, 1988). Three distinct volcaniclastic lithofacies were documented in our measured stratigraphic section data.

Lithofacies 1: Pumice-Rich Sandstone

Description. Pumice-rich sandstones are the most common volcaniclastic deposits along the northern basin margin. These deposits are characterized by horizontal stratification (Sh on Fig. 10B), trough- and planar-cross-stratification, and lenticular geometries with scoured bases. Volcaniclastic sandstone beds are 0.20 to 0.80 m thick, are medium- to coarse-grained, and consist of pumice fragments, epiclastic volcanic lithic fragments, and plagioclase and quartz crystals. Framework grains are moderately sorted with grain shapes ranging from subangular to subrounded.

Interpretation. The high proportion of stratified, juvenile volcanic detritus indicates that this lithofacies formed by fluvial reworking of unconsolidated pyroclastic deposits. These deposits formed contemporaneously with, or relatively shortly after pyroclastic eruptions.

Lithofacies 2: Pumice-Rich Tuff

Description. Massive, tabular, pumice-rich tuffs are exposed in the Arkose Ridge and Wishbone Formations along the northern basin margin (Figs. 10B, 10C, 10D). This lithofacies is poorly sorted with grain sizes ranging from 0.5 mm to 250 cm (Fig. 10B). Framework grains include abundant pumice with subordinate amounts of basaltic lithic fragments, euhedral quartz and plagioclase crystals, and vitric shards. Individual beds can be traced for several hundred meters and bed thicknesses range from 10 cm to >5 m. Beds tend to thicken into underlying fluvial channels.

Interpretation. The lateral persistent of individual beds, lack of evidence for tractive transport, and textural features suggests that the tuffs were deposited by pyroclastic flows (Fisher and Schminke, 1984; Cas and Wright, 1987).

Lithofacies 3: Volcanic Airfall Tuff

Description. Airfall tuffs are common in the Chickaloon Formation and the uppermost Wishbone Formation and range in thickness from >25 cm north of the Castle Mountain fault (section 22 on Figs. 5, 10D) to a few millimeters south of the Castle Mountain fault (sections 6 and 7 on Fig. 3; Triplehorn et al., 1984). Framework grains consist mainly of plagioclase with minor volcanic and sedimentary rock fragments in a kaolinitic matrix (Triplehorn et al., 1984; this study). The average estimated diameter of framework grains range from 0.5 mm north of the Castle Mountain fault (section 9 on Fig. 3; this study) to <0.25 mm south of the fault (section 9 on Fig. 3; Triplehorn et al., 1984).

Interpretation. On the basis of textural features described by Triplehorn et al. (1984), we interpret this lithofacies as being deposited by airfall tuffs associated with suspension fallout from pyroclastic eruptions along the northern basin margin.

CONGLOMERATE COMPOSITIONAL DATA

New compositional data from Paleocene-Oligocene conglomerates of the Matanuska Valley–Talkeetna Mountains forearc basin, in combination with our geochronologic data and previously published compositional data (Clardy, 1974; Little, 1988; Trop and Ridgway, 1999), help identify source terranes that contributed sediment to the basin. Clast composition data from conglomerates were collected by counting all pebble- and cobble-size clasts in a delineated rectangle on an outcrop face, usually yielding a population of 100 or more clasts. A total of ~5900 clasts were counted from 59 different conglomerate beds. Thin sections were made of representative clast types to check field identifications. Recalculated conglomerate compositional data are presented on Figures 12 and 13.

Conglomerate Clast Types

Paleocene-Oligocene conglomerates are generally polymictic with a dominance of igneous clast types, mainly andesite, tuff, basalt, pumice, and granite (Figs. 12, 13). Andesite clasts are brown or tan with plagioclase feldspar phenocrysts in a fine-grained glassy matrix. Basalt clasts are dark gray or black. Metabasalt (greenstone) clasts exhibit low-grade metamorphic features, including foliated textures and chlorite alteration. The metabasalt clasts are grouped with the basalt clasts due to their petrologic similarities. We identified three main types of tuff clasts: laminated green and white crystalline tuffs, gray siliceous tuffs, and gray crystal-vitric tuffs. Pumice clasts are white or gray, and altered. Plutonic clasts include mostly granite and granodiorite and subordinate tonalite and diorite; some of the plutonic clasts have a weakly foliated texture. Sedimentary and metamorphic clast types include chert, argillite, quartz-mica schist, amphibolite, lithic sandstone, siltstone, and limestone.

Clast composition of conglomerates from the Matanuska Valley–Talkeetna Mountains forearc basin deposits can be divided into three distinct petrofacies: volcanic, plutonic, and mixed

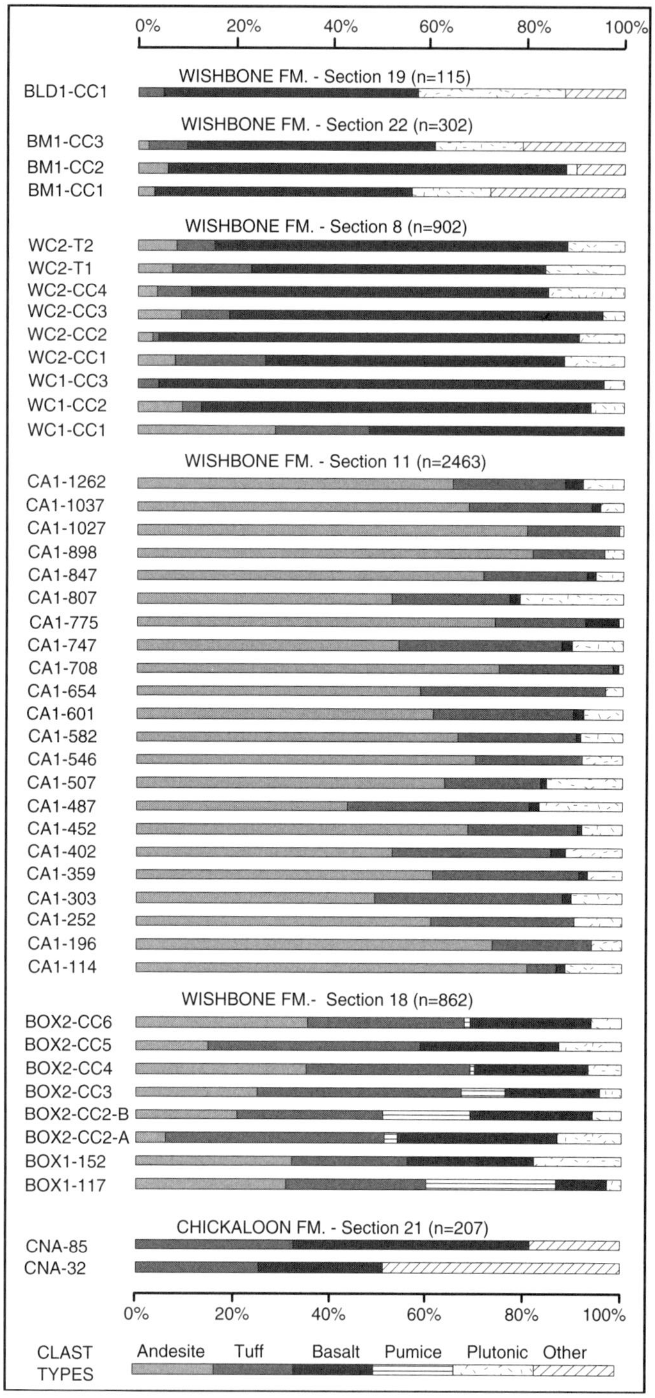

Figure 12. Histograms showing composition of clasts from Paleocene-Eocene conglomerates characteristic of the volcanic petrofacies common along the northern basin margin. Note the relative abundance of volcanic clast types (andesite, tuff, basalt, and pumice) compared to plutonic and other (mainly sedimentary and metamorphic) clast types. Histograms are positioned in relative stratigraphic order within individual measured sections. n = total number of clasts counted in measured section. See Figure 3 for measured section locations.

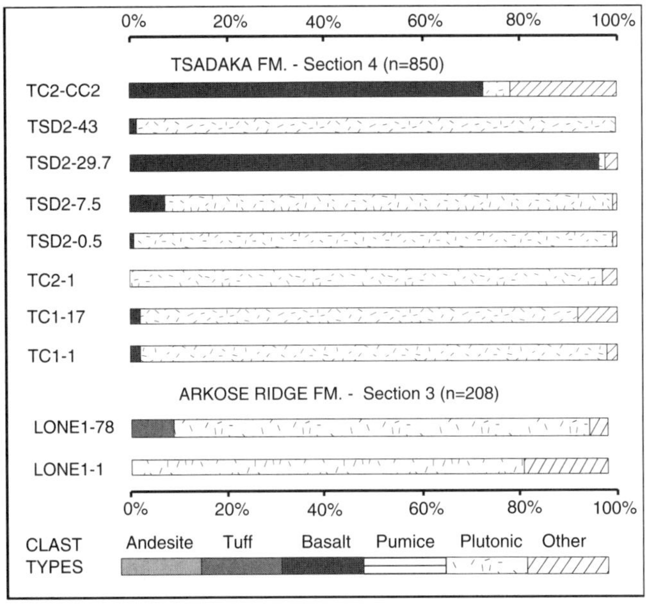

Figure 13. Histograms showing composition of clasts from Paleocene-Oligocene conglomerates characteristic of plutonic petrofacies along the northern basin margin. Histograms are positioned in relative stratigraphic order within individual measured sections. n = total number of clasts counted within measured section. See Figure 3 for measured section locations.

volcanic-metamorphic. Conglomerates of the volcanic and plutonic petrofacies crop out along the northern basin margin. Conglomerates of the mixed volcanic-metamorphic petrofacies are exposed along the southern basin margin.

Volcanic Petrofacies

The volcanic petrofacies includes the Wishbone Formation and the Chickaloon Formation along the northern basin margin (Tw and Tc on Fig. 3). Conglomerates in this petrofacies are dominated by volcanic clast types (84%) with minor plutonic (12%), sedimentary (2%), and metamorphic (2%) clasts (Fig. 12). Volcanic clasts consist of basalt/metabasalt, andesite, tuff, and pumice. Diagnostic clast types can be matched with volcanic source terranes exposed along the northern basin margin. For example, highly altered metabasalt (greenstone) and red jasper clasts are petrographically similar to the Early Jurassic Talkeetna Formation (JTrt on Fig. 3). Unaltered to moderately altered black basalt, andesite, pumice, and crystal-vitric tuff clasts are petrographically similar to Paleocene-Eocene volcanic rocks exposed north of the Castle Mountain fault (southernmost exposures of Tv on Fig. 3).

New paleocurrent and $^{40}Ar/^{39}Ar$ age determinations provide additional evidence that the volcanic petrofacies was derived from northern source terranes. Paleocurrent indicators (n = 429) show consistent southeastward- to southwestward-directed sediment transport along the northern margin of the basin (Fig. 9). $^{40}Ar/^{39}Ar$ ages of biotite from a granite clast (59.7 ± 0.3 Ma, BOX-CC2-2 on Table 1 and Fig. 6) and a basalt clast (56.8 ± 0.3 Ma, WC1-CC1-BR on Table 1 and Fig. 6) are compatible with previously published K-Ar ages of Tertiary igneous rocks exposed along the northern basin margin (TKg, Tar, Tv on Fig. 3). $^{40}Ar/^{39}Ar$ step heating and laser fusion analyses of detrital K-feldspars from a Paleocene-Eocene volcaniclastic sandstone (site 19 on Fig. 3) yield an isochron age of 177.0 ± 1.2 Ma (SUN2-3.65 on Table 1 and Fig. 6). The ages for the detrital feldspars (170–180 Ma) are consistent with the age of Jurassic granitic plutons exposed north of the Castle Mountain fault system. Locally, these plutons have K-Ar ages of 169–173 Ma (e.g., sites 51 and 55 of Winkler, 1992) and are located ~7 km north of the location of the dated feldspars in the forearc basin. Thus, the $^{40}Ar/^{39}Ar$ age determinations and diagnostic conglomerate clasts indicate that volcanic detritus of the forearc basin was derived from a coeval volcanoplutonic arc (Tv and TKg on Figs. 2B and 3) as well as from an older Jurassic volcanoplutonic arc (Ji and JTrt on Figs. 2B and 3). Our new data and interpretations from the volcanic petrofacies are consistent with previously reported petrofacies data from the Wishbone Formation (Clardy, 1974).

Plutonic Petrofacies

The plutonic petrofacies includes the Tsadaka Formation (Tsd on Fig. 3) and the Arkose Ridge Formation (Tar on Fig. 3). The modal clast composition of the plutonic petrofacies is 76% plutonic, 18% volcanic, 3% metamorphic, and 3% sedimentary (Fig. 13). The most abundant clast types are granite, granodiorite, diorite, amphibolite, quartz-mica schist, and quartz. Most of the clast types are petrographically similar to Lower to Upper Jurassic and Upper Cretaceous–Tertiary plutonic and metamorphic rocks that are presently exposed along the northern basin margin (Ji, Jps, and TKg on Fig. 3; Clardy, 1974; Trop and Ridgway, 1999). Derivation from these northern source terranes is supported by consistent southeastward- to southwestward-directed paleocurrent indicators along the northern margin of the forearc basin (Fig. 9). Our compositional data and provenance interpretations for the plutonic petrofacies are consistent with previous studies by Clardy (1974) and Winkler (1978).

Mixed Volcanic-Metamorphic Petrofacies

The mixed volcanic-metamorphic petrofacies is best developed along the southern basin margin. Conglomerate clast counts obtained during this study (site 23 on Fig. 3) and from a previous study (sites A–H on Fig. 3; Little, 1988) document abundant metabasalt/basalt, metasiltstone, quartz-mica schist, chert, and granitic clasts in conglomerates along the southern basin margin. These clast types are similar to lithologies exposed in the metamorphosed accretionary prism deposits (Mzm and Kvg on Fig. 3) and the southernmost outcrops of the Talkeetna Formation (JTrt on Fig. 3). Upper Triassic and possibly Jurassic-Cretaceous radiolaria extracted from chert clasts in these conglomerates are consistent with derivation from the adjacent McHugh Complex of the accretionary prism (MzM on Fig. 3; C.D. Blome, 1988, personal commun., cited in Little and Naeser, 1989). Northeastward- to northwestward-directed paleocurrent indicators (n = 35; Fig. 9) and the arrangement of alluvial-fluvial lithofacies transitions along the southern basin margin indicate derivation from southern source terranes (Fig. 5; Little, 1988; Little and Naeser, 1989).

STRUCTURAL CONTROLS ON BASIN DEVELOPMENT

Syndepositional Paleocene-Eocene Displacement on the Castle Mountain Fault

Cross sections shown on Figures 2B and 14A show the general structural characteristics of the northern forearc basin margin. Note that Paleocene-Eocene strata exposed in the footwall of the north-dipping Castle Mountain fault system form asymmetric synclines that have steep northern limbs (Fig. 14A). A particularly well exposed footwall syncline is located at Castle Mountain, near measured section 11 on Figure 3. At this location, Eocene strata exposed immediately south of the Castle Mountain fault (Tw on Fig. 14A) dip 85° to 65° SW whereas stratigraphically younger strata exposed further south have progressively shallower dips ranging from 65° to 5° SW (Tw and Tv on Fig. 14A). The stratigraphically youngest Paleocene-Eocene sedimentary strata (uppermost Tw on Fig. 14A) are separated from overlying unnamed Tertiary volcanic rocks (lowermost Tv on Fig. 14A) along an angular unconformity with ~8° discordance. The angular unconformity is easily identified by abrupt contrasts in color and lithology (Figs. 8A, 14B, 14C). A footwall syncline with similar stratigraphic and structural relationships is also exposed south of the Castle Mountain fault at Puddingstone Hill (Tw outcrop north of section 13 on Fig. 3). The thickest accumulations of Paleocene-Eocene alluvial-fan deposits are concentrated in footwall synclines along the north-dipping Castle Mountain fault system (e.g., sections 8 and 11 on Figs. 3, 5).

We interpret the progressively tilted beds documented in Paleocene-Eocene strata as forming when these deposits, located in the footwall of the Castle Mountain fault system, were progressively tilted basinward (southward) during north-

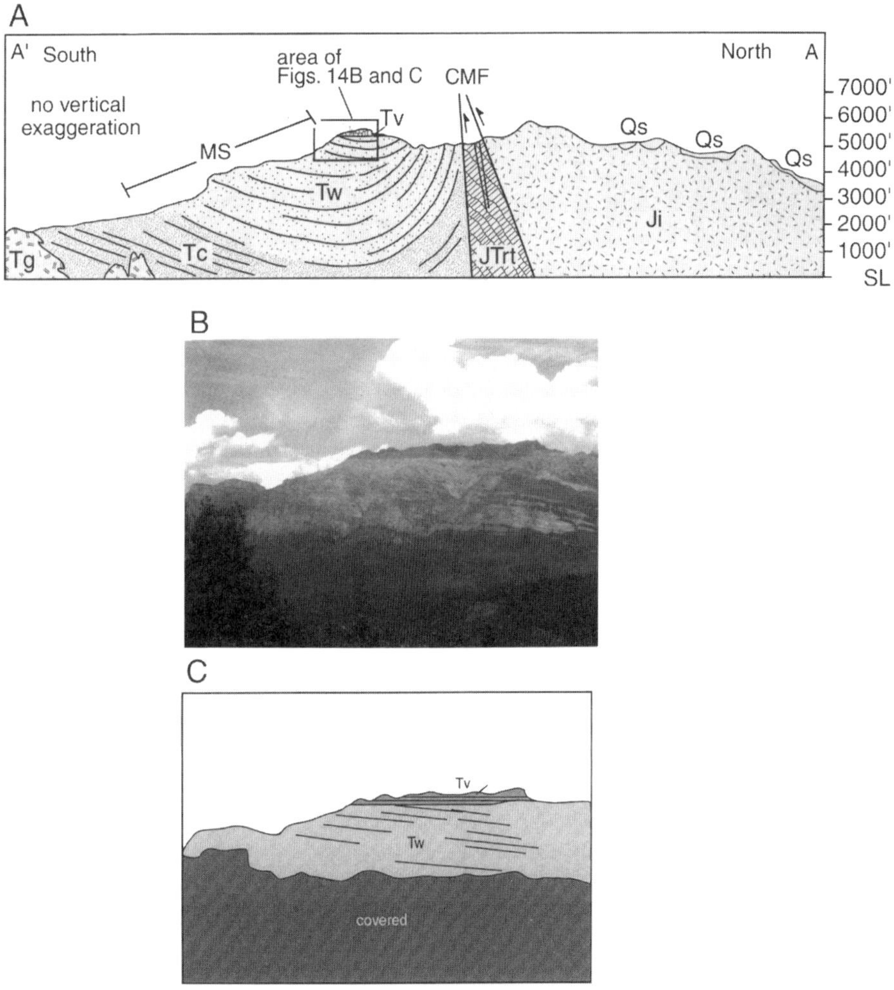

Figure 14. A: Cross section through the Castle Mountain fault at Castle Mountain. See Figure 3 (lower center) for line of cross section and explanation of abbreviations. Structural and geological data used to construct cross section are from Detterman et al. (1976), Fuchs (1980), and this study. Note that immediately south of the Castle Mountain fault, attitudes of beds decrease progressively upsection (from ~80° to 10°), indicating active footwall rotation during deposition of the Eocene Wishbone Formation (Tw). Progressive tilting of Eocene deposits is attributed to syndepositional displacement along the south-vergent Castle Mountain fault. Along this fault, Jurassic-Triassic volcanic (JTrt) and plutonic (Ji) rocks are juxtaposed against Tertiary sedimentary (Tc and Tw) and volcanic (Tv) deposits in a footwall syncline. A distinct angular unconformity is located between the Eocene Wishbone Formation (Tw) and overlying unnamed Tertiary volcanic strata (Tv). B and C: Close-up photograph and interpretive line sketch of the angular unconformity between the Eocene Wishbone Formation (Tw) and overlying unnamed Tertiary volcanic rocks (Tv). Area of photograph and line sketch is shown on Figure 14A. Conglomerates of the uppermost Wishbone Formation (light units, Tw) dip ~10° to the right (north). Unnamed Tertiary volcanic rocks (dark units, Tv) dip <5° to the right (north). Exposure is approximately 400 m thick.

side-up fault displacement and incorporated into a footwall growth syncline. Footwall growth synclines are footwall folds in which deposition occurs during structural growth of the syncline (e.g., Hoy and Ridgway, 1997, 2002). During each episode of north-side-up displacement on the Castle Mountain fault, sediments in the proximal footwall area were tilted and younger sediments were deposited on an angular unconformity. Packages of progressively tilted strata are common elements of deposits related to syndepositional shortening (i.e., Riba, 1976; DeCelles et al., 1991; Ridgway et al., 1997).

DISCUSSION

Nonmarine Forearc Basin Development

The ~2800 m of Paleocene-Oligocene deposits in the Matanuska Valley and Talkeetna Mountains contain a long term record of nonmarine sedimentation in a forearc basin. Most previous sedimentological studies of ancient forearc basins have focused on marine deposits (e.g., Ingersoll, 1979; Heller and Dickinson, 1985; Busby-Spera, 1986; Morris and Busby-Spera, 1988; among others) with only a few studies addressing nonmarine deposits (e.g., Kuenzi et al., 1979; Vessel and Davies, 1981; Fulford and Busby, 1993). In addition, preservation of arc-proximal deposits in most ancient forearc basin is rare (Smith and Landis, 1995; Dickinson, 1995). Our analysis of the Matanuska Valley–Talkeetna Mountains forearc basin deposits, in contrast, documents well-preserved sedimentary and volcanic strata along the basin margin that was adjacent to the coeval volcanic arc.

Paleocene-Oligocene coarse-grained sedimentary strata exposed along the southern and northern forearc basin margins are interpreted as mainly stream-dominated alluvial fan and braided stream deposits. Northern basin margin sedimentary deposits are interbedded with lava flows, tuff, and volcaniclastic sandstone. These volcanic deposits are interpreted as the products of episodic eruptions of coeval volcanoes that bordered the northern basin margin. Southern basin margin deposits, in contrast, lack interbedded volcanic strata. Basin margin coarse-grained deposits interfinger with and locally overlie carbonaceous mudstone, coal, and fine- to medium-grained sandstone along the basin axis (Fig. 5). We interpret the finer-grained strata as representing deposits of sinuous fluvial systems and overbank swamps and lakes. Unlike lithofacies exposed along the northern basin margin, basin axis strata include few volcanic deposits; only thin airfall tuffs are reported from Paleocene-Eocene formations exposed south of the Castle Mountain fault system (Fig. 5; Triplehorn et al., 1984).

Along the northern and southern basin margins, as well as the basin axis, the nonmarine sedimentary deposits are characterized by an overall upward-coarsening megasequence. This megasequence is clearly seen in the measured sections shown on Figure 5. Note that the lower part of most of the measured sections consists of mudstones, overlain by sandstones, which are in turn overlain by conglomerates. We interpret this megasequence as the product of basinward progradation of alluvial-fan and fluvial deposystems in response to syndepositional uplift on the Castle Mountain and Border Ranges fault systems. Growth synclines preserved in the footwall of the Castle Mountain fault (Fig. 14) are interpreted as the result of north-side-up displacement during deposition of the Wishbone Formation. Fault-related sedimentation associated with the Border Ranges fault was a product of uplift of the accretionary prism and associated syndepositional displacement along northward-dipping oblique-slip faults (Little, 1988; Little and Naeser, 1989). Our analysis indicates that in the Matanuska Valley–Talkeetna Mountains forearc basin, nonmarine deposition was closely linked to displacement on major basin-bounding faults. There is also a distinctive upsection change in volcanic lithofacies in the megasequence along the northern, arc-proximal basin margin. In the lower part of the megasequence, sedimentary strata are interbedded with mainly airfall tuffs, in the middle part of the megasequence volcaniclastic sandstones become more common, and in the upper part of the megasequence sedimentary strata are interbedded with lava flows. In addition, overlying the megasequence along the northern basin margin are thick sections of primary volcanic deposits (Tv on sections 22, 18, and 11 on Fig. 5). We interpret this upsection change in volcanic lithofacies as representing trenchward migration of the volcanic arc and/or progradation of volcanic deposits during Paleocene-Oligocene development of the Matanuska Valley–Talkeetna Mountains forearc basin.

Our petrofacies, geochronologic, and paleocurrent data provide insight on the provenance of the Matanuska Valley–Talkeetna Mountains forearc basin deposits. The volcanic and plutonic petrofacies (Figs. 12, 13), common along the northern basin margin, indicate derivation from both remnant and coeval volcanic arcs. $^{40}Ar/^{39}Ar$ ages of 59.7 Ma and 56.8 Ma from igneous clasts in conglomerates are interpreted as having been derived from the coeval volcanic arc, whereas detrital feldspars with ages of 170–180 Ma are interpreted as being derived from an older Jurassic arc exposed along the northern basin margin (Table 1). Paleocene-Oligocene deposits along the southern margin of the forearc basin were derived mainly from metabasalts, metasiltstones, and granitic plutons that characterize the uplifted accretionary prism deposits. Our data suggest that in this nonmarine forearc basin, both arc-margin and trench-margin source terranes contributed large amounts of sediment to the basin.

Forearc Basin Development and Regional Tectonics

This section presents a combined synthesis and tectonic interpretation of sedimentological, compositional, and structural data from the Tertiary Matanuska Valley–Talkeetna Mountains forearc basin. Plate motions and velocities are from the reconstructions of Engebretson et al. (1985); these plate reconstructions are not universally accepted (e.g., Bradley et al., 1993). The time scale and absolute ages are from Gradstein and Ogg (1996).

Latest Maastrichtian–Early Paleocene (ca. 67–62 Ma)

A major basinwide angular unconformity marks the contact between deep-marine deposits of the Upper Cretaceous Matanuska Formation and Upper Paleocene–Lower Eocene fluvial-estuarine deposits of the Chickaloon Formation in the Matanuska Valley–Talkeetna Mountains forearc basin (Fig. 4; Csejtey et al., 1978; Winkler, 1992). Near the contact with overlying fluvial deposits of the Chickaloon Formation, the Matanuska Formation contains marine fossils suggestive of deposition in bathyl water depths (>500 m below sea level; Jones, 1967; Bergquist, 1961). Eustatic sea-level changes during the latest Cretaceous–earliest Paleocene were less than 100 m (Haq et al., 1988). Thus, the magnitude of base level change between deposition of the uppermost Matanuska Formation (>500 m below sea level) and the lowermost Chickaloon Formation (sea level or above sea level) requires tectonic uplift of the forearc basin (Fig. 15A). Latest Cretaceous–Early Paleocene uplift is also indicated along the trenchward margin of the forearc basin, where locally the Chickaloon Formation depositionally overlies metamorphosed Late Cretaceous accretionary prism deposits (section A on Fig. 3; Little, 1988; Winkler, 1992). Early Maastrichtian (ca. 75 Ma) accretionary prism deposits (Kvg on Fig. 3) were underthrust, metamorphosed, and rapidly uplifted (~0.55 mm/yr) prior to unconformable onlap of the nonmarine Upper Paleocene (ca. 61–52 Ma) Chickaloon Formation (Little and Naeser, 1989).

From a regional perspective, the latest Cretaceous (Maastrichtian) to Early Paleocene unconformity and associated

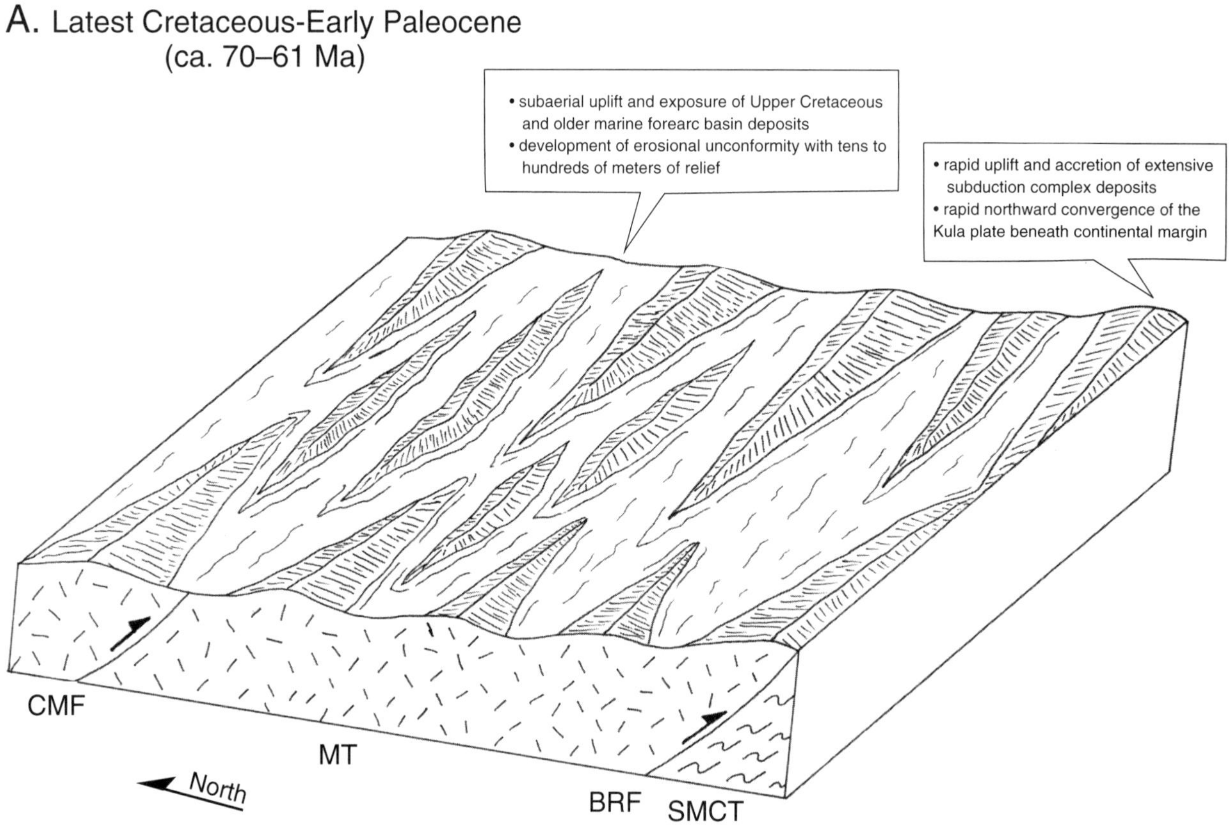

Figure 15 (*on this and following page*). Simplified block diagrams showing the latest Cretaceous to Eocene development of the Matanuska Valley–Talkeetna Mountains forearc basin. A: Latest Cretaceous to Early Paleocene tectonic uplift and erosion of Late Cretaceous deep marine sedimentary strata (Matanuska Formation) and older forearc basin deposits. B: Late Paleocene to Early Eocene renewed forearc basin subsidence and nonmarine/estuarine sedimentation. Alluvial fans and braided streams deposited coarse-grained strata (northern outcrops of Arkose Ridge and Chickaloon Formations) along the northern basin margin coeval with arc magmatism (AT). Alluvial fans and braided streams also deposited coarse-grained strata along the southern basin margin (southern outcrops of Chickaloon Formation) coeval with uplift in the accretionary prism (SMCT). Alluvial fans and braided streams merged with fine-grained sinuous streams, swamps, and lakes along the basin axis (Chickaloon Formation). C: Early to Middle(?) Eocene alluvial fans along the northern basin margin (Wishbone Formation) prograded southward into the basin axis in response to syndepositional displacement on the Castle Mountain fault (CMF). Along the southern basin margin, alluvial fans and braided streams (southern outcrops of upper Chickaloon Formation) prograded northward into the basin axis coeval with uplift along the Border Ranges fault (BRF). The basin was deformed and intruded by plutons and sills during the Middle to Late Eocene. With the exception of a brief, localized episode of alluvial fan deposition during the Late Oligocene (Tsadaka Formation), the forearc basin remained topographically inverted throughout the Early Oligocene to recent.

B. Late Paleocene-Early Eocene
(ca. 60–50 Ma)

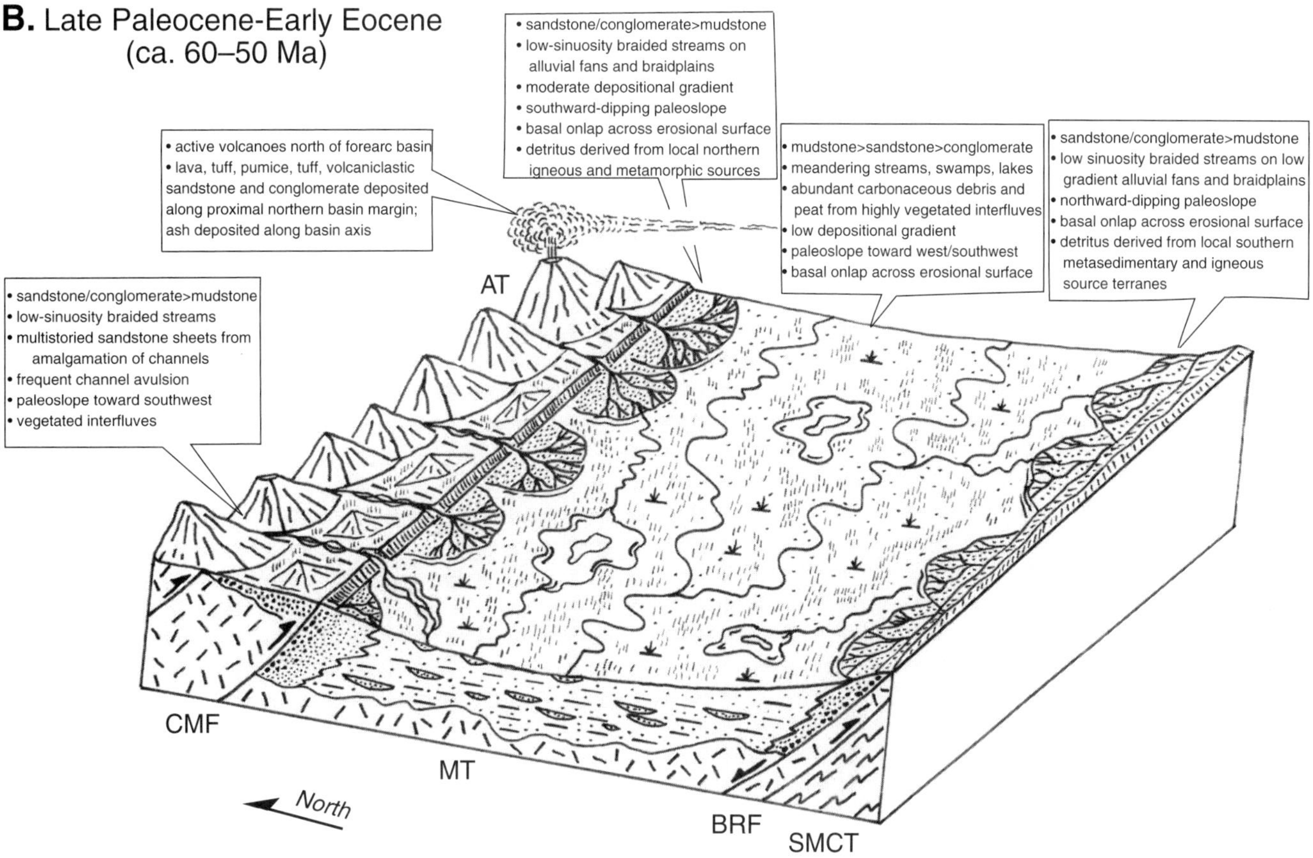

C. Early to Middle Eocene
(ca. 49–37? Ma)

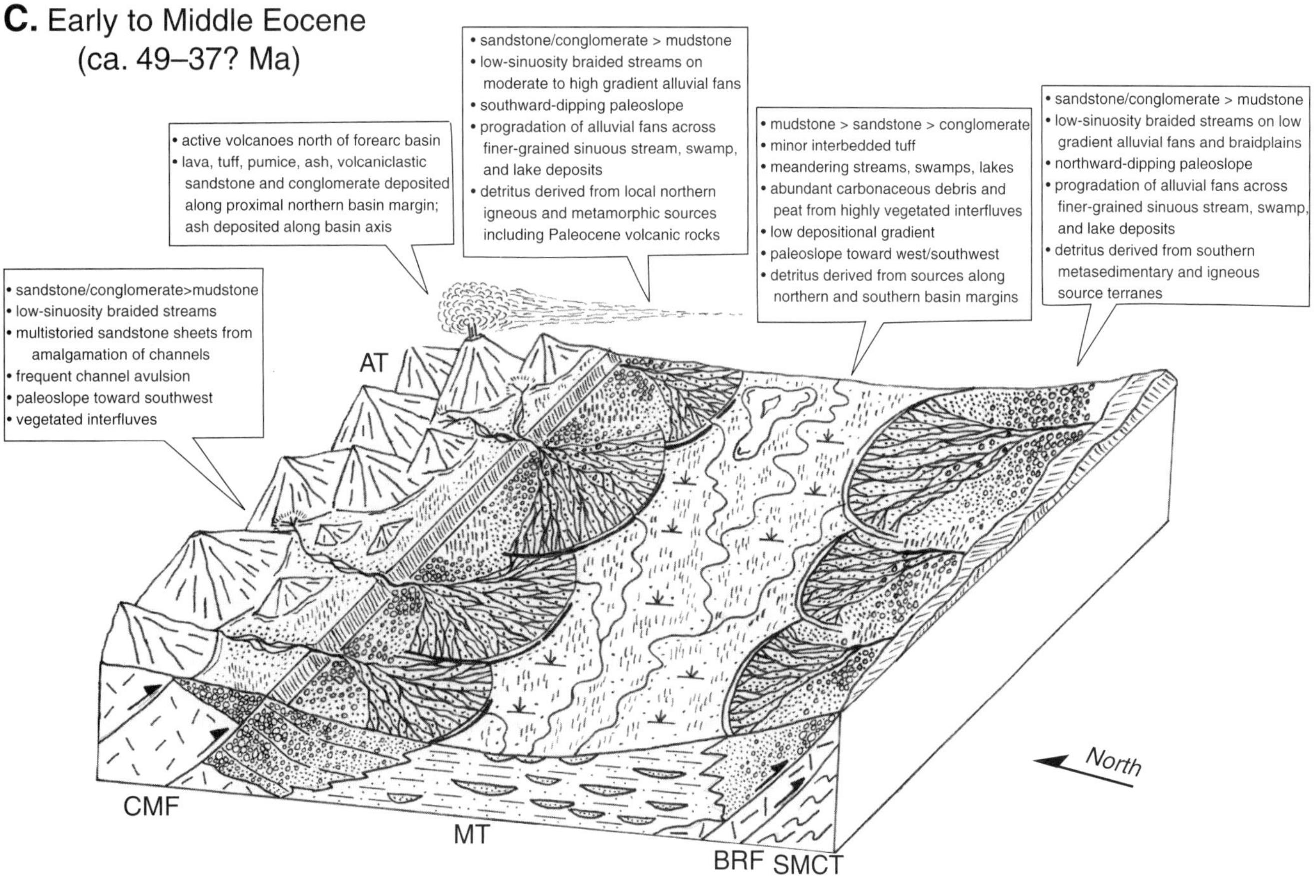

deformation in the forearc basin was contemporaneous with crustal shortening throughout the northwestern Cordillera (see discussion in McClelland et al., 1992, and Trop et al., 1999). This regional-scale deformation may have been related to Late Cretaceous accretion of the allochthonous Wrangellia composite terrane (e.g., Nokleberg et al., 1994; Plafker and Berg, 1994), rapid subduction of the Kula plate (Wallace and Engebretson, 1984; Engebretson et al., 1985), and/or accretion of an enormous wedge of metasedimentary strata in the accretionary prism (Mzu on Figs. 2, 3; Plafker et al., 1994).

Deformation in the forearc basin may have also been partly coeval with northward latitudinal translation of the forearc basin and underlying Wrangellia composite terrane. Paleomagnetic studies by Stamatakos et al. (1988, 1989) from the Matanuska, Chickaloon, and Arkose Ridge Formations, as well as unnamed Middle to Late Eocene plutons and sills (Tg on Fig. 3), indicate that the forearc basin was positioned ~2800 ± 2000 km south of its present latitude relative to cratonic North America during the latest Cretaceous (ca. 80–70 Ma), ~1600 ± 1200 km south of its present latitude during the Upper Paleocene–Lower Eocene (ca. 61–50 Ma), and at or near its present position by the Middle to Late Eocene (45.5 to 36.8 Ma).

Late Paleocene–Middle Eocene (ca. 61–50 Ma)

Following late Maastrictian–early Paleocene uplift and subaerial exposure of the forearc basin, the mainly nonmarine Chickaloon, Arkose Ridge, and Wishbone Formations were deposited over deformed Upper Cretaceous marine strata (Fig. 15B). These formations represent progradation of proximal, gravel-rich alluvial-fluvial deposystems over more distal, mud-rich sinuous fluvial environments (discussed above). We attribute progradation of proximal deposystems to syndepositional displacement on the Castle Mountain and Border Ranges fault systems (discussed above).

The timing of Late Paleocene–Early Eocene (ca. 56–50 Ma) fault-related, coarse-grained sedimentation in the Matanuska Valley–Talkeetna Mountains forearc basin was broadly coeval with multiple tectonic events important in the development of the northwestern Cordillera continental margin. These include (1) a northwestward shift in Kula plate motion; (2) subduction of the Kula-Farallon ridge or another unidentified spreading center (discussed in a later section); (3) northward translation of the forearc basin and accretionary prism; (4) dextral displacement on regional faults; and (5) regional counterclockwise oroclinal bending of Alaska. These tectonic events are briefly discussed below within the context of forearc basin development. The Upper Paleocene–Lower Eocene Chickaloon and Arkose Ridge Formations were deposited during a northwestward shift in Kula plate motion starting at ca. 54 Ma (56 Ma event of Wallace and Engebretson [1984] and Lonsdale [1988] recalibrated to magnetic time scale of Cande and Kent [1992]). This plate shift prompted dextral oblique plate convergence along the northwestern Cordilleran continental margin (Engebretson et al., 1985; Lonsdale, 1988). Dextral oblique convergence may have resulted in northward displacement of the forearc basin and accretionary prism along orogen-parallel dextral faults (Plafker and Berg, 1994). The timing of possible northward translation of the forearc basin and accretionary prism based on paleomagnetic data (Coe et al., 1985; Stamatakos et al., 1988, 1989; Bol et al., 1992) was broadly coeval with large dextral displacements along regional fault systems. For example, estimates of up to 400 km of dextral displacement have been proposed along the eastern part of the Denali fault system during the interval from ca. 57 to 38 Ma (Fig. 1; Eisbacher, 1976; Nokleberg et al., 1985). Similarly, Roeske et al. (1993, this volume) and Smart et al. (1996) proposed 700 to >1000 km of Early Eocene (ca. 58 to 50 Ma) dextral displacement along the Border Ranges/Hanagita fault system (Fig. 1).

Late Paleocene–Early Eocene coarse-grained deposition in the Matanuska Valley–Talkeetna Mountains forearc basin may have also been coeval with oroclinal bending of Alaska. Geologic and paleomagnetic data indicate that from ca. 65 to 50 Ma parts of southwestern and south-central Alaska were rotated 45° to 60° counterclockwise about a vertical axis (see Plafker and Berg, 1994, for summary). The Matanuska Valley–Talkeetna Mountains forearc basin is presently located in the axial region of the orocline, a position of expected localized compression during orocline formation (Grantz, 1966).

Middle Eocene–Early Oligocene (49–30 Ma)

The forearc basin deposits were deformed, uplifted, and partly eroded during the Early Eocene to Early Oligocene as recorded by an angular unconformity separating the Lower Eocene Wishbone Formation from the Upper Oligocene Tsadaka Formation (Fig. 3; Barnes, 1962; Clardy, 1974). Along the northern basin margin, oblique slip along the Castle Mountain fault system may have been coeval with unconformity development (Fuchs, 1980). During this stage of forearc basin development, silicic dikes and plutons (Tg on Fig. 3) intruded Tertiary sedimentary strata throughout the forearc basin (36.8–45.5 Ma, n = 10; fission track and K-Ar, Little and Naeser, 1989; Silbermann and Grantz, 1984), accretionary prism (42.9–49.6 Ma; n = 4; fission track; Little and Naeser, 1989), and volcanic arc (39.9–43.6 Ma, n = 2; K-Ar, Adams et al., 1985). Coeval with silicic intrusions was basinward (southward) progradation of volcanic deposits along the northern basin margin and/or basinward migration of the volcanic arc. These processes are documented in the upsection change in volcanic lithofacies (discussed earlier) and in the hundreds of meters of unnamed Tertiary volcanic rocks (Tv on Figs. 3, 5) that directly overlie Upper Paleocene-Eocene sedimentary strata (Chickaloon and Wishbone Formations) along the northern basin margin (sections 11, 18, and 22 on Fig. 5; Csejtey et al., 1978; Winkler, 1992).

Middle Eocene–Early Oligocene deformation and magmatism in the forearc basin was coeval with plate reorganization in the Pacific basin (47 Ma event of Engebretson et al., 1985; Lonsdale, 1988). Following subduction of the Kula plate, there was a change to north to northwest motion of the Pacific plate beginning at ca. 50 Ma (Engebretson et al., 1985; Plafker and Berg, 1994). After this change, convergence with the continental margin was relatively slower and more orthogonal than during the Late Paleocene–Early Eocene (ca. 61–50 Ma). Middle Eocene–Early Oligocene changes in forearc magmatism may record an increase in the dip of the subducting oceanic slab caused by a decrease in the rate of relative plate convergence, an increase in the age of the subducting crust following plate reorganization (47 Ma event of Engebretson et al., 1985), and/or subduction of the Kula-Farallon spreading ridge, or another unidentified spreading center (Bradley et al., 2000, this volume). Paleomagnetic studies of 45.5 to 36.8 Ma plutons indicate that the forearc basin was positioned at or near its present position relative to North America during the Middle to Late Eocene (Stamatakos et al., 1988). Some strike-slip displacement continued on the Denali fault system, located north of the forearc basin, during the Late Eocene–Early Oligocene (Fig. 1B; Ridgway and DeCelles, 1993b; Cole et al., 1999; Ridgway et al., 2002).

Late Oligocene–Recent (ca. 29–0 Ma)

Sedimentation resumed in the forearc basin during deposition of the Upper Oligocene Tsadaka Formation. Conglomerates of the Tsadaka Formation were deposited across an angular unconformity above folded Paleocene-Eocene strata (Chickaloon and Wishbone Formations). Our petrofacies and paleocurrent data (Figs. 9, 13) show that coarse-grained granitic detritus was derived from the dissected Paleocene-Eocene volcanic arc and older plutonic source terranes that are presently exposed along the northern basin margin. Outcrops of the Tsadaka Formation are only found in the lower part of the Matanuska Valley (Tsd on Fig. 3) and thus may reflect localized alluvial-fan deposition.

Neogene deformation in the forearc basin included dextral strike-slip displacement on the east-west–trending Castle Mountain fault system, dextral motion on west-northwest–trending strike-slip faults, and sinistral movement on northeast-trending strike-slip faults (Bruhn and Pavlis, 1981; Detterman et al., 1974; Fuchs, 1980; Lahr et al., 1986). Recent earthquakes (1984, M5.7; 1996, M4.6) indicate that the Castle Mountain fault remains an active transpressional structure (Haeussler et al., 2000). Late Oligocene to Recent (ca. 30–0 Ma) dextral transpression in the forearc basin was probably driven by northwestward underthrusting of the Pacific plate (Wallace and Engebretson, 1984; Engebretson et al., 1985) and collision of the allochthonous Yakutat terrane along the outboard margin of southern Alaska (Plafker et al., 1994).

Record of Ridge Subduction in Forearc Basin Deposits

The Kula-Farallon spreading ridge, or another unidentified spreading center, was subducted beneath accretionary prism deposits of the Southern Margin composite terrane (Fig. 1B) between 61 and 50 Ma (Pavlis and Sisson, 1995; Kusky et al., 1997; Kusky and Young, 1999; Bradley et al., 2000, this volume). Geologic evidence for ridge subduction in the accretionary prism deposits includes diachronous "near-trench" magmatism (Bradley et al., 2000), gold mineralization (Haeussler et al., 1995), and regional high-grade metamorphism and anatectic melting (Sisson et al., 1989; Barker et al., 1992; Pavlis and Sisson, 1995). The timing of ridge subduction was apparently diachronous along the continental margin in response to subduction of a ridge that was oriented subparallel to the trench (Bradley et al., 1993). For example, "near-trench" plutons show a distinct progression from ca. 61 Ma $^{40}Ar/^{39}Ar$ and K-Ar ages in southwestern Alaska to ca. 50 Ma $^{40}Ar/^{39}Ar$ and K-Ar ages in southeastern Alaska (Bradley et al., 2000). This age progression suggests that a triple junction, possibly the Kula–Farallon–North America junction, migrated southeastward during the early Tertiary (modern coordinates in Fig. 1; Sisson and Pavlis, 1993; Bradley et al., 2000). Migration of a trench-ridge-trench triple junction past a given point would likely be accompanied by subduction of progressively more buoyant, topographically higher lithosphere (the spreading ridge) followed by less buoyant, topographically lower lithosphere (e.g., Thorkelson and Taylor, 1989). Thus, subduction of a sufficiently large spreading ridge could prompt adjustment of the topographic profile of the accretionary prism and possibly the forearc basin through vertical uplift. In this section, we evaluate the possible record of early Tertiary ridge subduction in the Matanuska Valley–Talkeetna Mountain forearc basin deposits. Two reconstructions are presented to address contrasting interpretations concerning the amount of Tertiary strike-slip displacement along the Border Ranges/Hanagita fault system (Fig. 1). We prefer the minimum strike-slip displacement reconstruction because the timing of ridge subduction in this reconstruction would have been coeval with progradation of coarse-grained deposystems and intrusion of primitive igneous melts into the Matanuska Valley–Talkeetna Mountains forearc basin (discussed below). However, until additional high-resolution geochronologic data are available for the ages of the forearc basin deposits, and the displacement histories of major fault systems are better known, both reconstructions are feasible. It is also important to note that early Tertiary ridge subduction was broadly coeval with several tectonic events that may have also influenced forearc basin deposition (discussed above).

Reconstruction 1—Minimum Tertiary Strike-Slip Displacement on Border Ranges Fault

Several observations from previous studies indicate a maximum of a few tens of kilometers of Late Paleocene to Middle Eocene dextral displacement along the Border Ranges fault system in south-central Alaska. For example, geologic mapping along the southern margin of the forearc basin reports Upper Paleocene–Lower Eocene forearc basin deposits (the Chickaloon Formation) on both sides of the Border Ranges fault (sections A, B, C on Fig. 3; Little, 1988; Little and Naeser, 1989). In addition, middle Eocene plutons reportedly intrude the Border Ranges fault along the southern margin of the forearc basin (Little and Naeser, 1989) as well as east of our study area in the Cordova and Middleton Island quadrangles (Winkler and

Plafker, 1981). In this minimum strike-slip reconstruction, the forearc basin and accretionary prism were translated northward as a single block with most displacement accommodated along inboard strike-slip faults such as the Denali fault system (Fig. 1). In this reconstruction, the age of "near-trench" plutons presently exposed directly south of the forearc basin across the Border Ranges fault (between longitudes 150°–145° W) records the time during which the spreading ridge was subducted outboard of the forearc basin. $^{40}Ar/^{39}Ar$ ages (53.4–52.2 Ma; Bradley et al., 2000) and K-Ar ages (54.5 to 50 Ma; Bradley et al., 1993) of these near-trench plutons indicate Early Eocene ridge subduction in this area. Ridge subduction in this reconstruction would have been coeval with deposition of coarser-grained deposits of the uppermost Chickaloon Formation and Wishbone Formation in the forearc basin, and with syndepositional displacement along the Castle Mountain and Border Ranges/Hanagita fault systems. Previous studies of ridge subduction processes have predicted vertical uplift of the accretionary prism and forearc basin during ridge subduction (e.g., Thorkelson and Taylor, 1989). In the minimum strike-slip reconstruction, uplift of the forearc region may be recorded in the thick coarse clastic wedges that prograded basinward from both the northern and southern forearc basin margins during the Early Eocene. In this reconstruction, upflow of hot mantle through the slab window may have increased the crustal elevation of the forearc basin and reactivated major basin-bounding fault systems (i.e., the Castle Mountain and Border Ranges faults). The footwall growth synclines along the Castle Mountain fault may be a product of this vertical uplift. This deformation may also have resulted in basinward progradation of coarse-grained deposystems producing the overall upward-coarsening megasequence that characterizes Paleocene-Oligocene deposits of the Matanuska Valley–Talkeetna Mountains forearc basin (Fig. 5).

In the minimum strike-slip reconstruction, ridge subduction was also coeval with igneous activity along both the arcward and trench margins of the forearc basin. Along the southern margin of the forearc basin, dike swarms intruded the accretionary prism along the Border Ranges fault between 57 and 52 Ma (Little, 1988). Age-equivalent lava flows (56–46 Ma) of the Arkose Ridge Formation along the northern basin margin have primitive initial $^{86}Sr/^{87}Sr$ isotopic ratios (0.7042–0.7047; Silberman and Grantz, 1984), indicating little crustal contamination and possible development of a "slab-free" window during ridge subduction beneath the forearc basin (e.g., Plafker et al., 1989; Winkler, 1992). This interpretation would be consistent with studies of Tertiary ridge subduction in California that have suggested that the slab window provides a pathway for the upward flux of mantle into the overlying continental lithosphere (Dickinson and Snyder, 1979; Cole and Basu, 1992, 1995; Dickinson, 1997).

Reconstruction 2—Maximum Tertiary Strike-Slip Displacement on Border Ranges Fault

In contrast to estimates of minimum Tertiary strike-slip displacement (tens of km) on the central part of the Border Ranges fault near the Matanuska Valley (Little and Naeser, 1989), large amounts (700 to >1000 km) of dextral displacement have been proposed for the southeastern part of the fault system (Roeske et al., 1993, this volume; Smart et al., 1996). In the maximum strike-slip reconstruction, in addition to displacement on inboard strike-slip fault systems, significant northward displacement was accommodated along the Border Ranges/Hanagita fault system. This fault system separated the forearc basin from the adjacent accretionary prism (Figs. 1B, 3). In the maximum strike-slip displacement reconstruction, the forearc basin and accretionary prism were not translated northward as a single block (i.e., they have different displacement histories). Using the estimated 700 km of dextral translation, the Matanuska Valley–Talkeetna Mountains forearc basin would have a restored Paleocene-Eocene position adjacent to accretionary prism deposits that are presently exposed near Kodiak Island (Fig. 1B). In this reconstruction, near-trench plutons that would have been proximal to the Matanuska Valley–Talkeetna Mountains forearc basin have $^{40}Ar/^{39}Ar$ and U-Pb ages of 61.1 to 57.8 Ma (ages from Kodiak Island and the Shumagin Islands in Bradley et al., 2000). Ridge subduction in this reconstruction would be coeval with unconformity development, and/or with initial subsidence and fine-grained sedimentation of the lower Chickaloon Formation in the forearc basin (Fig. 5). In this reconstruction, the predicted increase in crustal elevation associated with subduction of progressively more buoyant, topographically higher lithosphere (i.e., the spreading ridge) would be represented by the regional unconformity between the Upper Cretaceous Matanuska Formation and Upper Paleocene–Lower Eocene Chickaloon Formation. Subduction of buoyant lithosphere would have been followed by subduction of less buoyant, topographically lower lithosphere, which may be represented in the forearc basin record by reinitiation of subsidence during the beginning of Chickaloon Formation deposition. This interpretation is presently limited by the poor resolution of the maximum age of the Chickaloon Formation (see earlier discussion). Igneous rocks in the forearc basin with ages that are consistent with ridge subduction in the maximum strike-slip reconstruction have not been reported.

ACKNOWLEDGMENTS

This research was supported by the National Science Foundation (EAR-9725587; EAR-9406078) and the Donors of the Petroleum Research Fund administered by the American Chemical Society. The Geological Society of America, the American Association of Petroleum Geologists, and the Purdue University Research Foundation also provided support. Field work was facilitated by discussions with Arthur Grantz and assistance from Jeff Manuszak, Ryan Nicols, and Jonathon Epps. Kathleen Zanetti assisted with $^{40}Ar/^{39}Ar$ geochronology at the University of Nevada—Las Vegas. The Nevada Isotope Geochronology Laboratory was funded by National Science Foundation grant EPS-9720162 to T. Spell. We thank Ric Wilson for providing digital geologic maps, Dwight Bradley for discussions on Tertiary ridge subduction in southern Alaska, and Ron Cole for discussion on Tertiary volcanism and regional geology of the Talkeetna Mountains. The manuscript was improved through careful reviews by Ray Ingersoll, John Stamatakos, and Terry Pavlis.

APPENDIX 1. ^{40}AR/^{39}AR ANALYTICAL PROCEDURES

^{40}Ar/^{39}Ar Analytical Procedures

Samples analyzed by the ^{40}Ar/^{39}Ar method at the University of Nevada, Las Vegas, were wrapped in Al foil and stacked in 6 mm inside diameter pyrex tube. Individual packets averaged 3 mm thick and neutron fluence monitors (ANU 92-176, Fish Canyon Tuff sanidine) were placed every 5–10 mm along the tube. Synthetic K-glass and optical grade CaF_2 were included to monitor neutron induced argon interferences from K and Ca. Loaded tubes were packed in an Al container and irradiated for 14 hours in the D3 position on the core edge of the 1 MW TRIGA type reactor at the Nuclear Science Center at Texas A&M University. Samples are shielded against thermal neutrons by a 5 mm thick jacket of B_4C powder, which rotates about its axis at a rate of 0.7 revolutions per minute to mitigate horizontal flux gradients. Correction factors for interfering neutron reactions on K and Ca were: $(^{40}Ar/^{39}Ar)_K = (2.54 \pm 3.18) \times 10^{-2}$, $(^{36}Ar/^{37}Ar)_{Ca} = (2.89 \pm 0.06) \times 10^{-4}$, and $(^{39}Ar/^{37}Ar)_{Ca} = (7.06 \pm 0.01) \times 10^{-4}$. J factors were determined by fusion of 3–4 individual crystals of ANU 92-176, which gave reproducibilities of 0.1 to 0.2%. No significant neutron fluence gradients were present within individual packets of crystals as indicated by the excellent reproducability of the single-crystal fluence monitor fusions. Variation in neutron fluence along the 100 mm length of the irradiation tubes was <2%. An error in J of 0.5% was used in all age calculations.

Irradiated crystals together with CaF_2 and K-glass fragments were placed in a Cu sample tray in a high-vacuum extraction line and fused using a 20 W CO_2 laser. Sample viewing during laser fusion was by a video camera system and positioning was via a motorized sample stage. Samples analyzed by the furnace step heating method utilized a double vacuum resistance furnace similar to the Staudacher et al. (1978) design. Reactive gases were removed by a single MAP and two GP-50 SAES getters prior to being admitted to a MAP 215-50 mass spectrometer by expansion. The relative volumes of the extraction line and mass spectrometer allow 80% of the gas to be admitted to the mass spectrometer for laser fusion analyses and 76% for furnace heating analyses. Peak intensities were measured using a Balzers electron multiplier by peak hopping through seven cycles; initial peak heights were determined by linear regression to the time of gas admission. Mass spectrometer discrimination and sensitivity was monitored by repeated analysis of atmospheric argon aliquots from an on-line pipette system. Measured ^{40}Ar/^{36}Ar ratios were 289.55 ± 0.17 and 290.14 ± 0.49 during this work, thus discrimination corrections of 1.02055–1.01848 (4 AMU) were applied to measured isotope ratios. The sensitivity of the mass spectrometer was $\sim 6 \times 10^{-17}$ mol mV^{-1} with the multiplier operated at a gain of 30 over the Faraday cup. Line blanks averaged ~5 mV for mass 40 and 0.02 mV for mass 36 during laser fusion analyses and 40 mV for mass 40 and 0.15 mV for mass 36 for furnace heating analyses. Computer automated operation of the sample stage, laser, extraction line and mass spectrometer as well as final data reduction and age calculations were done using LabSpec software written by B. Idleman (Lehigh University). An age of 27.9 Ma (Steven et al., 1967; Cebula et al., 1986) was used for the Fish Canyon Tuff sanidine fluence monitor in calculating ages for samples. All analytical errors are reported at the 1σ level.

^{40}Ar/^{39}Ar Data Treatment

For step heating apparent age spectra total gas ages are calculated based on weighting by the amount of ^{39}Ar released in each step. A plateau is defined if a significant proportion of the gas released (>40–50%) over three or more contiguous steps are indistinguishable at the 2σ internal precision level. Plateau ages are calculated by utilizing a weighted mean (weighting by inverse variance) with internal (analytical) errors; a 0.5% error in J is then added quadratically to yield a total error estimate. Inverse isochron analysis (York, 1969) for step heating experiments utilizes similar criteria to that for defining a plateau, with the exception that the mean square of weighted deviates (MSWD) be below the limit defined by the number of data points on the isochron (Wendt and Carl, 1991) and at least four data points must define an isochron. For single-crystal data analyses outside of 2σ of the mean were excluded and a weighted mean calculated from the remaining analyses as for step heating plateaus. For single-crystal isochrons initial regressions utilized all data from a population; if MSWD was unacceptably high (Wendt and Carl, 1991) the sample contributing the most to the MSWD was omitted and the isochron regressed again using the trimmed population. This process was repeated until an acceptable MSWD was obtained.

REFERENCES CITED

Adams, D.D., Burns, L.E., Pessel, G.H., Little, T.A., Newberry, R.J., and Flynn, L.R., 1985, Preliminary geologic map of the central Talkeetna Mountains, Alaska, State of Alaska Division of Geological and Geophysical Surveys, Public Data File 85-20, scale 1:25,000.

Allen, J.R.L., 1970a, Studies in fluviatile sedimentation: A comparison of fining upwards cyclothems with special reference to coarse-member composition and interpretation: Journal of Sedimentary Petrology, v. 40, p. 298–323.

Allen, J.R.L., 1970b, A quantitative model of grain size and sedimentary structures in lateral deposits: Journal of Geology, v. 7, p. 129–146.

Barker, F., Farmer, G.L., Ayuso, R.A., Plafker, G., and Lull, J.S., 1992, The 50 Ma granodiorite of the eastern Gulf of Alaska: Melting of the accretionary prism in the forearc: Journal of Geophysical Research, v. 97, p. 6757–6778.

Barnes, F.F., 1962, Geologic map of lower Matanuska Valley, Alaska: U.S. Geological Survey Miscellaneous Investigations Map I-359, scale 1:63,360.

Bergquist, H.R., 1961, Foraminiferal zonation in Matanuska Formation, Squaw Creek–Nelchina River area, south-central Alaska: American Association of Petroleum Geologists Bulletin, v. 45, p. 1944–2011.

Bol, A.J., Coe, R.S., Gromme, C.S., and Hillhouse, J.W., 1992, Paleomagnetism of the Resurrection Peninsula: Implications for the tectonics of southern Alaska and the Kula-Farallon Ridge: Journal of Geophysical Research, v. 97, p. 17,213–17,232.

Boothroyd, J.C., and Ashley, G.M., 1975, Processes, bar morphology, and sedimentary structures on braided outwash fans, northeastern Gulf of Alaska, *in* Jopling, A.V., and McDonald, B.C., eds., Glaciofluvial and glaciolacustrine sedimentation: Society of Economic Paleontologists and Mineralogists Special Publication 23, p. 193–222.

Boothroyd, J.C., and Nummedal, D., 1978, Proglacial braided outwash: A model for humid alluvial fan deposits, *in* Miall, A.D., ed., Fluvial sedimentology: Canadian Society of Petroleum Geologists Memoir 5, p. 641–668.

Bradley, D.C., Haeussler, P.J., and Kusky, T.M., 1993, Timing of early Tertiary ridge subduction in southern Alaska: U.S. Geological Survey Bulletin 2068, p. 163–177.

Bradley, D.C., Parrish, R., Clendenen, W., Lux, D., Layer, P.W., Heizler, M., and Donley, D.T., 2000, New geochronological evidence for the timing of early Tertiary ridge subduction in southern Alaska: U.S. Geological Survey Professional Paper 1615, p. 5–21.

Bradley, D., Kusky, T., Haeussler, P., Goldfarb, R., Miller, M., Dumoulin, J., Nelson, S.W., and Karl, S., 2003, Geologic signature of early Tertiary ridge subduction in Alaska *in* Sisson, V.B., Roeske, S.M., and Pavlis, T.L., eds, Geology of a transpressional orogen developed during ridge-trench

interaction along the North Pacific margin: Boulder, Colorado, Geological Society of America Special Paper 371, p. 19–49 (this volume).
Bruhn, R.L., and Pavlis, T.L., 1981, Late Cenozoic deformation in the Matanuska Valley, Alaska: Three-dimensional strain in a forearc region: Geological Society of America Bulletin, v. 92, p. 282–293.
Busby, C., Smith, D., Morris, W., and Fackler-Adams, B., 1998, Evolutionary model for convergent margins facing large ocean basins: Mesozoic Baja California, Mexico: Geology, v. 26, p. 227–230.
Busby-Spera, C.J., 1986, Depositional features of rhyolitic and andesitic volcaniclastic rocks of the Mineral King submarine caldera complex, Sierra Nevada, California: Journal of Volcanology and Geothermal Research, v. 27, p. 43–76.
Cande, S.C., and Kent, D.V., 1992, A new geomagnetic polarity time scale for the Late Cretaceous and Cenozoic: Journal of Geophysical Research, v. 97, p. 13,917–13,951.
Cas, R.A.F., and Wright, J.V., 1987, Volcanic successions—Modern and ancient: London, Allen and Unwin, 528 p.
Cebula, G.T., Kunk, M.J., Mehnert, H.H., Naeser, C.W., Obradovich, J.D., and Sutter, J.F., 1986, The Fish Canyon Tuff, a potential standard for the ^{40}Ar-^{39}Ar and fission-track dating methods [abs.]: Terra Cognita, 6th International Conference on Geochronology, Cosmochronology and Isotope Geology, v. 6, p. 139.
Clardy, B.I., 1974, Origin of the lower and middle Tertiary Wishbone and Tsadaka Formations, Matanuska Valley, Alaska [M.S. thesis]: Fairbanks, Alaska, University of Alaska, 50 p.
Clifton, H.E., 1982, Estuarine deposits, *in* Scholl, P.A., and Spearing, D., eds., Sandstone depositional environments: American Association of Petroleum Geologists Memoir 31, p. 179–189.
Coe, R.S., Globerman, B.R., Plumley, P.R., and Thrupp, G.A., 1985, Paleomagnetic results from Alaska and their tectonic implications, *in* Howell, D.G., ed., Tectonostratigraphic terranes of the circum-Pacific region: Houston, Texas, Circum-Pacific Council for Energy and Mineral Resources, Earth Science Series, v. 1, p. 85–108.
Cole, R.B., and Basu, A.R., 1992, Middle Tertiary volcanism during ridge-trench interactions in western California: Science, v. 258, p. 793–796.
Cole, R.B., and Basu, A.R., 1995, Nd-Sr geochemistry and tectonics of ridge subduction and middle Cenozoic volcanism in western California: Geological Society of America Bulletin, v. 107, p. 167–179.
Cole, R.B., Ridgway, K.D., Layer, P.W., and Drake, J., 1999, Kinematics of basin development during the transition from terrane accretion to strike-slip tectonics, Late Cretaceous early Tertiary Cantwell Formation, south-central Alaska: Tectonics, v. 18, p. 1224–1244.
Collinson, J.D., 1986, Alluvial sediments, *in* Reading, H.G., ed., Sedimentary environments and facies: Oxford, Blackwell Scientific, p. 20–62.
Csejtey, B., Nelson, W.H., Jones, D.L., Silberling, N.J., Dean, R.M., Morris, M.S., Lanphere, M.A., Smith, J.G., and Silberman, M.L., 1978, Reconnaissance geologic map and geochronology, Talkeetna Mountains quadrangle, north part of Anchorage quadrangle, and southwestern corner of Healy quadrangle, Alaska: U.S. Geological Survey Open-File Report 78-558-A, 60 p., 1 sheet, scale 1:250,000.
DeCelles, P.G., Gray, M.B., Ridgway, K.D., Cole, R.B., Srivastava, P., Pequera, N., and Pivnik, D.A., 1991, Kinematic history of a foreland uplift from Paleocene synorogenic conglomerate, Beartooth Range, Wyoming and Montana: Geological Society of America Bulletin, v. 103, p. 1458–1475.
DeMets, C., Gordon, R.G., Argus, D.F., and Stein, S., 1990, Current plate motions: Geophysical Journal International, v. 101, p. 425–478.
DeMets, C., Gordon, R.G., Argus, D.F., and Stein, S., 1994, Effect of recent revisions to the geomagnetic reversal time scale on estimates of current plate motions: Geophysical Research Letters, v. 21, p. 2191–2194.
Detterman, R.L., Plafker, G., Hudson, T., Tysdal, R.G., and Pavoni, N., 1974, Surface geology and Holocene breaks along the Susitna segment of the Castle Mountain fault, Alaska: U.S. Geological Survey Miscellaneous Field Studies Map MF-618, scale 1:24,000.
Detterman, R.L., Plafker, G., Russell, G.T., and Hudson, T., 1976, Features along part of the Talkeetna segment of the Castle Mountain-Caribou fault system, Alaska: U.S. Geological Survey Miscellaneous Field Studies Map MF-738, scale 1:63,360.
Dickinson, W.R., 1995, Forearc basins, *in* Busby, C.J., and Ingersoll, R.V., eds., Tectonics of sedimentary basins: Cambridge, Blackwell Science, p. 221–262.
Dickinson, W.R., 1997, Tectonic implications of Cenozoic volcanism in coastal California: Geological Society of America Bulletin, v. 109, p. 936–954.
Dickinson, W.R., and Rich, E.I., 1972, Petrologic intervals and petrofacies in the Great Valley Sequence, Sacramento Valley, California: Geological Society of America Bulletin, v. 83, p. 3007–3024.
Dickinson, W.R., and Seely, D.R., 1979, Structure and stratigraphy of forearc regions: American Association of Petroleum Geologists Bulletin, v. 63, p. 2164–2182.
Dickinson, W.R., and Snyder, W.S., 1979, Geometry of triple junctions related to San Andreas transform: Journal of Geophysical Research, v. 84, p. 561–572.
Dover, J.H., 1994, Geology of part of east-central Alaska, *in* Plafker, G., and Berg, H.C., eds., The geology of Alaska: Boulder, Colorado, Geological Society of America, Geology of North America, v. G-1, p. 153–204.
Eisbacher, G.H., 1976, Sedimentology of the Dezadeash flysch and its implications for strike-slip faulting along the Denali fault, Yukon Territory and Alaska: Canadian Journal of Earth Sciences, v. 13, p. 1495–1513.
Eisma, D., 1998, Intertidal deposits: River mouths, tidal flats, and coastal lagoons, with P.L. Boer, G.C. Cadee, K. Dikema, H. Ridderinkhof, and C. Philippart: Boca Raton, CRC Press, 525 p.
Engebretson, D.C., Cox, A., and Gordon, R.G., 1985, Relative motions between oceanic and continental plates in the Pacific basin: Boulder, Colorado, Geological Society of America Special Paper 206, 59 p.
Fisher, R.V., and Schminke, H.U., 1984, Pyroclastic rocks: Berlin, Springer Verlag, 472 p.
Fletcher, H.J., and Freymueller, J.T., 1997, New GPS constraints on the tectonics of the Yakutat block: Eos (Transactions, American Geophysical Union), v. 78, p. F628.
Flores, R.M., and Stricker, G.D., 1993a, Early Cenozoic depositional systems, Wishbone Hill District, Matanuska coal field, Alaska: U.S. Geological Survey Bulletin 1992, p. 101–117.
Flores, R.M., and Stricker, G.D., 1993b, Responses of coal-splitting and associated drainage pattern to syntectonism in the Paleocene and Eocene Chickaloon Formation, Matanuska Coal Field, Alaska, *in* Rao, P.D., and Walsh, D.E., eds., Focus on Alaska's coal in 1993: Proceedings of the conference held at Anchorage, Alaska, on May 5–7, 1993, p. 113–131.
Fraser, G.S., and Suttner, L., 1986, Alluvial fans and fan deltas—A guide to exploration for oil and gas: Boston, International Human Resources Development Corporation, 199 p.
Fuchs, W.A., 1980, Tertiary tectonic history of the Castle Mountain fault system in the Talkeetna Mountains [Ph.D. thesis]: Salt Lake City, University of Utah, 152 p.
Fulford, M.M., and Busby, C.J., 1993, Tectonic controls on non-marine sedimentation in a Cretaceous forearc basin, Baja California, Mexico, *in* Frostick, L.E., and Steel, R.J., eds., Tectonic controls and signatures in sedimentary successions: International Association of Sedimentologists Special Publication 20, p. 301–333.
Gabrielse, H., 1985, Major dextral transcurrent displacements along the Northern Rocky Mountain Trench and related lineaments in north-central British Columbia: Geological Society of America Bulletin, v. 96, p. 1–14.
Gradstein, F.M., and Ogg, J.G., 1996, A Phanerozoic time scale: Episodes, v. 19, p. 1–2.
Grantz, A., 1966, Strike-slip faults in Alaska: U.S. Geological Survey Open-File Report 66-53, 82 p.
Grantz, A., and Jones, D.L., 1960, Stratigraphy and age of the Matanuska Formation, south-central Alaska: U.S. Geological Survey Professional Paper 400-B, p. 347–350.
Haeussler, P.J., Bradley, D., Goldfarb, R., Snee, L., and Taylor, C., 1995, Link between ridge subduction and gold mineralization in southern Alaska: Geology, v. 23, p. 995–998.
Haeussler, P.J., Bruhn, R.L., and Pratt, T.L., 2000, Potential seismic hazards and tectonics of the upper Cook Inlet basin, Alaska, based on analysis of Pliocene and younger deformation: Geological Society of America Bulletin, v. 112, p. 1414–1429.
Haq, B.U., Hardenbol, J., and Vail, P.R., 1988, Mesozoic and Cenozoic chronostratigraphy and cycles of sea-level change, *in* Wilgus, C.K., et al., eds., Sea-level changes: An integrated approach: Society of Economic Paleontologists and Mineralogists Special Publication 42, p. 78–108.
Harms, J.C., Southard, J.B., and Walker, R.G., 1982, Structure and sequences in clastic rocks: Society of Economic Paleontologists and Mineralogists, Short Course No. 9, p. 3-1–3-51.
Heller, P.L., and Dickinson, W.R., 1985, Submarine ramp facies model for delta-fed, sand-rich turbidite systems: American Association of Petroleum Geologists Bulletin, v. 69, p. 960–976.

Hillhouse, J.W., and Coe, R.S., 1994, Paleomagnetic data from Alaska, *in* Plafker, G., and Berg, H.C., eds., The geology of Alaska: Boulder, Colorado, Geological Society of America, Geology of North America, v. G-1, p. 797–812.

Hoy, R.G., and Ridgway, K.D., 1997, Structural and sedimentological development of footwall growth synclines along an intraforeland uplift, east-central Bighorn Mountains, Wyoming: Geological Society of America Bulletin, v. 109, p. 915–935.

Hoy, R.G., and Ridgway, K.D., 2002, Syndepositional thrust-related deformation and sedimentation in an Ancestral Rocky Mountains basin, Central Colorado trough, Colorado, USA: Geological Society of America Bulletin, v. 114, no. 7, p. 804–828.

Hudson, T., 1983, Calc-alkaline plutonism along the Pacific rim of southern Alaska, *in* Roddick, J.A., ed., Circum-Pacific plutonic terranes: Boulder, Colorado, Geological Society of America Memoir 159, p. 159–169.

Ingersoll, R.V., 1978a, Paleogeography and paleotectonics of the late Mesozoic forearc basin of northern and central California, *in* Howell, D.G., and McDougall, K.A., eds., Mesozoic paleogeography of the western United States (Pacific Coast Paleogeography Symposium 2): Pacific Section, Society of Economic Paleontologists and Mineralogists, Los Angeles, p. 335–352.

Ingersoll, R.V., 1978b, Submarine fan facies of the Upper Cretaceous Great Valley Sequence, northern and central California: Sedimentary Geology, v. 21, p. 205–230.

Ingersoll, R.V., 1979, Evolution of the Late Cretaceous forearc basin, northern and central California: Geological Society of America Bulletin, v. 90, p. 813–826.

Ingersoll, R.V., 1983, Petrofacies and provenance of a late Mesozoic forearc basin, northern and central California: American Association of Petroleum Geologists Bulletin, v. 67, p. 1125–1142.

Johnson, A.M., 1970, Physical processes in geology: San Francisco, Freeman, Cooper, and Co., 577 p.

Jones, D.L., 1967, Cretaceous ammonites from the lower part of the Matanuska Formation, southern Alaska, with a stratigraphic summary by Arthur Grantz: U.S. Geological Survey Professional Paper 547, 49 p.

Jordan, D.W., and Pryor, W.A., 1992, Hierarchical levels of heterogeneity in a Mississippi River meander belt and application to reservoir systems: American Association of Petroleum Geologists Bulletin, v. 76, p. 1601–1624.

Kuenzi, W.D., Horst, O.H., and McGehee, R.V., 1979, Effect of volcanic activity on fluvial-deltaic sedimentation in a modern arc-trench gap, southwestern Guatemala: Geological Society of America Bulletin, v. 90, p. 827–838.

Kusky, T.M., and Young, C.P., 1999, Emplacement of the Resurrection Peninsula ophiolite in the southern Alaska forearc during a ridge-trench encounter: Journal of Geophysical Research, v. 104, p. 29,025–29,054.

Kusky, T.M., Bradley, D.C., and Haeussler, P.J., 1997, Progressive deformation of the Chugach accretionary complex, Alaska, during a Paleogene ridge encounter: Journal of Structural Geology, v. 19, p. 139–157.

Lahr, J.C., Page, R.A., Stephens, C.D., and Fogleman, K.A., 1986, Sutton, Alaska, earthquake of 1984: Evidence for activity on the Talkeetna segment of the Castle Mountain fault system: Bulletin of the Seismological Society of America, v. 76, p. 967–983.

Lanphere, M.A., 1978, Displacement history of the Denali fault system, Alaska and Canada: Canadian Journal of Earth Sciences, v. 15, p. 815–822.

Linn, A.M., DePaolo, D.J., and Ingersoll, R.V., 1991, Nd-Sr isotopic provenance analysis of Upper Cretaceous Great Valley forearc sandstones: Geology, v. 19, p. 803–806.

Linn, A.M., DePaolo, D.J., and Ingersoll, R.V., 1992, Nd-Sr isotopic, geochemical, and petrographic stratigraphy and paleotectonic analysis: Mesozoic Great Valley forearc sedimentary rocks of California: Geological Society of America Bulletin, v. 104, p. 1264–1279.

Little, T.A., 1988, Tertiary tectonics of the Border Ranges fault system, north-central Chugach Mountains, Alaska—Sedimentation, deformation, and uplift along the inboard edge of a subduction complex [Ph.D. thesis]: Stanford, California, Stanford University, 343 p.

Little, T.A., 1990, Kinematics of wrench and divergent-wrench deformation along a central part of the Border Ranges fault system, northern Chugach Mountains, Alaska: Tectonics, v. 9, p. 585–611.

Little, T.A., and Naeser, C.W., 1989, Tertiary tectonics of the Border Ranges fault system, southern Alaska, USA: Journal of Structural Geology, v. 14, p. 343–359.

Lonsdale, P., 1988, Paleogene history of the Kula plate: Offshore evidence and onshore implications: Geological Society of America Bulletin, v. 100, p. 733–754.

MacKevett, E.M., and Plafker, G., 1977, The Border Ranges fault in south-central Alaska: U.S. Geological Survey Journal of Research, v. 2, p. 323–329.

Magoon, L.B., Adkison, W.L., and Egbert, R.M., 1976, Map showing geology, wildcat wells, Tertiary plant fossil localities, K-Ar age dates, and petroleum operations, Cook Inlet area, Alaska: U.S. Geological Survey Miscellaneous Investigations Series Map I-1019, scale 1:250,000.

McClelland, W.C., Gehrels, G.E., and Saleeby, J.B., 1992, Upper Jurassic–Lower Cretaceous basinal strata along the Cordilleran margin: Implications for the accretionary history of the Alexander-Wrangellia-Peninsular terrane: Tectonics, v. 11, p. 823–835.

Miall, A.D., 1978, Lithofacies types and vertical profile models in braided river deposits: A summary, *in* Miall, A.D., ed., Fluvial sedimentology: Canadian Society of Petroleum Geology Memoir 5, p. 597–604.

Moll-Stalcup, E.J., 1994, Latest Cretaceous and Cenozoic magmatism in mainland Alaska, *in* Plafker, G., and Berg, H.C., eds., The geology of Alaska: Boulder, Colorado, Geological Society of America, Geology of North America, v. G-1, p. 589–619.

Morris, W.R., and Busby-Spera, C.J., 1988, Sedimentologic evolution of a submarine canyon in a forearc basin, Upper Cretaceous Rosario Formation, San Carlos, Mexico: American Association of Petroleum Geologists Bulletin, v. 72, p. 717–737.

Moxon, I.W., and Graham, S.A., 1987, History and controls of subsidence in the Late Cretaceous–Tertiary Great Valley forearc basin, California: Geology, v. 15, p. 626–629.

Nokleberg, W.J., Jones, D.L., and Silberling, N.J., 1985, Origin and tectonic evolution of the Maclaren and Wrangellia terranes, eastern Alaska Range, Alaska: Geological Society of America Bulletin, v. 96, p. 1251–1270.

Nokleberg, W.J., Plafker, G., and Wilson, F.H., 1994, Geology of south-central Alaska, *in* Plafker, G., and Berg, H.C., eds., The geology of Alaska: Boulder, Colorado, Geological Society of America, Geology of North America, v. G-1, p. 311–365.

Pavlis, T.L., 1982, Origin and age of the Border Ranges fault of southern Alaska and its bearing on the late Mesozoic tectonic evolution of Alaska: Tectonics, v. 1, p. 343–368.

Pavlis, T.L., and Crouse, G.W., 1989, Late Mesozoic strike-slip movement on the Border Ranges fault system in the eastern Chugach Mountains, southern Alaska: Journal of Geophysical Research, v. 94, p. 4321–4332.

Pavlis, T.L., and Sisson, V.B., 1995, Structural history of the Chugach metamorphic complex in the Tana River region, eastern Alaska: A record of Eocene ridge subduction: Geological Society of America Bulletin, v. 107, p. 1333–1355.

Pierson, T.C., 1980, Erosion and deposition by debris flows at Mt. Thomas, North Canterbury, New Zealand: Earth Surficial Processes, v. 5, p. 227–247.

Plafker, G., and Berg, H.C., 1994, Overview of the geology and tectonic evolution of Alaska, *in* Plafker, G., and Berg, H.G., eds., The geology of Alaska: Boulder, Colorado, Geological Society of America, Geology of North America, v. G-1, p. 989–1021.

Plafker, G., Bruns, T.R., Winkler, and Tysdal, R.G., 1982, Cross section of the eastern Aleutian Arc, from Mount Spurr to the Aleutian Trench near Middleton Island, Alaska: Geological Society of America Map and Chart Series MC-28P, 1 sheet, scale 1:1,000,000.

Plafker, G., Nokleberg, W.J., and Lull, J.S., 1989, Bedrock geology and tectonic evolution of the Wrangellia, Peninsular, and Chugach terranes along the Trans-Alaska Crustal Transect in the northern Chugach Mountains and southern Copper River basin: Journal of Geophysical Research, v. 94, p. 4255–4295.

Plafker, G., Casey Moore, J., and Winkler, G.R., 1994, Geology of the southern Alaska margin, *in* Plafker, G., and Berg, H.C., eds., The geology of Alaska: Boulder, Colorado, Geological Society of America, Geology of North America, v. G-1, p. 389–450.

Riba, O., 1976, Syntectonic unconformities of the Alto Cardener, Spanish Pyrenees: A genetic interpretation: Sedimentary Geology, v. 15, p. 213–233.

Ridgway, K.D., and DeCelles, P.G., 1993a, Stream-dominated alluvial fan and lacustrine depositional systems in Cenozoic strike-slip basins, Denali fault system, Yukon Territory, Canada: Sedimentology, v. 40, p. 645–666.

Ridgway, K.D., and DeCelles, P.G., 1993b, Petrology of Mid-Cenozoic strike-slip basins in an accretionary orogen, St. Elias Mountains, Yukon Territory, Canada, *in* Johnsson, M., and Basu, A., eds., Processes controlling the composition of siliciclastic sediments: Boulder, Colorado, Geological Society of America Special Paper 284, p. 67–89.

Ridgway, K.D., Sweet, A.R., and Cameron, A.R., 1995, Climatically induced floristic changes across the Eocene-Oligocene transition in the northern high latitudes, Yukon Territory, Canada: Geological Society of America Bulletin, v. 107, p. 676–696.

Ridgway, K.D., Trop, J.M., and Sweet, A.R., 1997, Thrust-top basin development along a suture zone, Cantwell basin, Alaska Range: Implications for development of the Denali fault system: Geological Society of America Bulletin, v. 109, p. 505–523.

Ridgway, K.D., Trop, J.M., and Sweet, A.R., 1999, Stratigraphy, depositional systems, and age of the Tertiary White Mountain basin, Denali fault system, southwestern Alaska: Short Notes on Alaskan Geology, Alaska Division of Geological and Geophysical Surveys, Professional Report 119, p. 77–84.

Ridgway, K.D., Trop, J.M., Nokleberg, W.J., Davidson, C.M., and Eastham, K.R., 2002, Mesozoic and Cenozoic tectonics of the eastern and central Alaska range: Progressive basin development and deformation in a suture zone: Geological Society of America Bulletin, v. 114, no. 12, p. 1480–1504.

Riehle, J.R., Fleming, M.D., Molnia, B.F., Dover, J.H., Kelley, J.S., Miller, M.L., Nokleberg, W.J., Till, A.B., and Plafker, G., 1997, A digital shaded relief map of Alaska: U.S. Geological Survey Miscellaneous Investigation Map I-2585, scale 1:2,500,000.

Roeske, S.M., Snee, L.W., and Bunds, M.P., 1993, $^{40}Ar/^{39}Ar$ dates from the Border Ranges fault system, a hydrothermally altered brittle-ductile transition strike-slip shear zone, southern Alaska: Geological Society of America Abstracts with Programs, v. 25, no. 6, p. A419.

Roeske, S.M., Snee, L.W., Pavlis, T.L., 2003, Dextral-slip reactivation of an arc-forearc boundary during Late Cretaceous–Early Eocene oblique convergence in the northern Cordillera, *in* Sisson, V.B., Roeske, S.M., and Pavlis, T.L., eds, Geology of a transpressional orogen developed during ridge-trench interaction along the North Pacific margin: Boulder, Colorado, Geological Society of America Special Paper 371, p. 141–169 (this volume).

Rust, B.R., 1978, Depositional models for braided alluvium, *in* Miall, A.D., ed., Fluvial sedimentology: Canadian Society of Petroleum Geologists Memoir 5, p. 605–625.

Schumm, S.A., 1977, The fluvial system: New York, Wiley & Sons, 338 p.

Schumm, S.A., Mosley, M.P., and Weaver, W.E., 1987, Experimental fluvial geomorphology: New York, Wiley, 413 p.

Silberman, M.L., and Grantz, A., 1984, Paleogene volcanic rocks of the Matanuska Valley area and the displacement history of the Castle Mountain fault, *in* Coonrad, W.L., and Elliot, R.L., eds., The U.S. Geological Survey in Alaska—Accomplishments during 1981: U.S. Geological Survey Circular 868, p. 82–86.

Sisson, V.B., and Pavlis, T.L., 1993, Geologic consequences of plate reorganization: An example from the Eocene southern Alaska forearc: Geology, v. 21, p. 913–916.

Sisson, V.B., Hollister, L.S., and Onstott, T.C., 1989, Petrologic and age constraints on the origin of a low-pressure/high-temperature metamorphic complex, southern Alaska: Journal of Geophysical Research, v. 94, p. 4392–4410.

Smart, K.J., Pavlis, T.L., Sisson, V.B., Roeske, S.M., and Snee, L.W., 1996, The Border Ranges fault system in Glacier Bay National Park, Alaska: Evidence for major early Cenozoic dextral strike-slip motion: Canadian Journal of Earth Sciences, v. 33, p. 1268–1282.

Smith, D.G., 1985, Modern analogs of the McMurray Formation channel deposits, sedimentology of mesotidal-influenced meandering point bars with inclined beds of alternating mud and sand: Alberta Oil Sands Technology and Research Authority, Final Report for Research Project No. 391, 78 p.

Smith, G.A., 1986, Coarse-grained nonmarine volcaniclastic sediment: Terminology and depositional processes: Geological Society of America Bulletin, v. 90, p. 1–10.

Smith, G.A., and Landis, C.A., 1995, Intra-arc basins, *in* Busby, C.J., and Ingersoll, R.V., eds., Tectonics of sedimentary basins: Cambridge, Blackwell Science, p. 263–298.

Stamatakos, J.A., Kodama, K.P., and Pavlis, T.L., 1988, Paleomagnetism of Eocene plutonic rocks, Matanuska Valley, Alaska: Geology, v. 16, p. 618–622.

Stamatakos, J.A., Kodama, K.P., Vittorio, L.F., and Pavlis, T.L., 1989, Paleomagnetism of Cretaceous and Paleocene sedimentary rocks across the Castle Mountain fault, south-central Alaska, *in* Hillhouse, J.W., ed., Deep structure and past kinematics of accreted terranes: American Geophysical Union Geophysical Monograph Series, v. 50, p. 151–177.

Staudacher, T.H., Jessberger, E.K., Dorflinger, D., and Kiko, J., 1978, A refined ultrahigh-vacuum furnace for rare gas analysis: Journal of Physics and Earth Science Instrumentation, v. 11, p. 781–784.

Steven, T.A., Mehnert, H.H., and Obradovich, J.D., 1967, Age of volcanic activity in the San Juan Mountains, Colorado: U.S. Geological Survey Professional Paper 575-D, p. 47–55.

Thomas, R.G., Smith, D.G., Wood, J.M., Visser, J., Calverly-Range, E.A., and Kostar, E.H., 1987, Inclined heterolithic stratification—Terminology, description, interpretation, and significance: Sedimentary Geology, v. 53, p. 123–179.

Thorkelson, D.J., and Taylor, R.P., 1989, Cordilleran slab windows: Geology, v. 17, p. 833–836.

Todd, S.P., 1989, Stream-driven, high-density gravelly traction carpets: Possible deposits in the Trabeg Conglomerate Formation, SW Ireland and some theoretical considerations of their origin: Sedimentology, v. 36, p. 513–530.

Triplehorn, D.M., Turner, D.L., and Naeser, C.W., 1984, Radiometric age of the Chickaloon Formation of south-central Alaska—Location of the Paleocene-Eocene boundary: Geological Society of America Bulletin, v. 95, p. 740–742.

Trop, J.M., and Ridgway, K.D., 1999, Sedimentology and provenance of the Paleocene-Eocene Arkose Ridge Formation, Cook Inlet–Matanuska Valley forearc basin, southern Alaska: Short Notes on Alaskan Geology, Alaska Division of Geological and Geophysical Surveys, Professional Report 119, p. 129–144.

Trop, J.M., Ridgway, K.D., Sweet, A.R., and Layer, P.W., 1999, Submarine fan deposystems and tectonics of a Late Cretaceous forearc basin along an accretionary convergent plate boundary, MacColl Ridge Formation, Wrangell Mountains, Alaska: Canadian Journal of Earth Sciences, v. 36, p. 433–458.

Vessel, R.K., and Davies, D.K., 1981, Nonmarine sedimentation in an active forearc basin, *in* Ethridge, F.G., and Flores, R.M., eds., Recent and ancient nonmarine depositional environments: Models for exploration: Society of Economic Paleontologists and Mineralogists Special Publications 31, p. 31–45.

Wallace, W.K., and Engebretson, D.C., 1984, Relationships between plate motions and Late Cretaceous to Paleogene magmatism in southwestern Alaska: Tectonics, v. 3, p. 295–315.

Wells, S.G., and Harvey, H.A., 1986, Sedimentologic and geomorphic variations in storm-generated alluvial-fans, Howgill Fells, Northwest England, *in* Sediments down under, 12th International Sedimentological Congress, p. 327–328.

Wendt, I., and Carl, C., 1991, The statistical distribution of the mean squared weighted deviation: Chemical Geology, v. 86, p. 275–285.

Williams, T.A., 1997, Basin-fill architecture and forearc tectonics, Cretaceous Great Valley Group, Sacramento basin, northern California [Ph.D. thesis]: Stanford, California, Stanford University, 412 p.

Wilson, F.H., Dover, J.H., Bradley, D.C., Weber, F.R., Bundtzen, T.K., and Haeussler, P.J., 1998, Geologic map of central (interior) Alaska: U.S. Geological Survey Open-File Report 98-133, scale 1:500,000.

Winkler, G.R., 1978, Framework grain mineralogy and provenance of sandstones from the Arkose Ridge and Chickaloon Formations, Matanuska Valley, *in* Johnson, K.M., ed., The United States Geological Survey in Alaska—Accomplishments during 1977: U.S. Geological Survey Circular 772-B, p. B70–B73.

Winkler, G.R., 1992, Geologic map and summary geochronology of the Anchorage 1° × 3° quadrangle, southern Alaska: U.S. Geological Survey Miscellaneous Investigations Series Map I-2283, scale: 1:250,000.

Winkler, G.R., and Plafker, G., 1981, Geologic map and cross sections of the Cordova and Middleton Island quadrangles, southern Alaska: U.S. Geological Survey Open-File Report 81-1164, scale 1:250,000.

Wolfe, J.A., Hopkins, D.M., and Leopold, E.B., 1966, Tertiary stratigraphy and paleobotany of the Cook Inlet region, Alaska: U.S. Geological Survey Professional Paper 398-A, p. A1–A29.

York, D., 1969, Least squares fitting of a straight line with correlated errors: Earth and Planetary Science Letters, v. 5, p. 320–324.

Manuscript Accepted by the Society February 5, 2003

Printed in the USA

Geological Society of America
Special Paper 371
2003

Brittle deformation along the Gulf of Alaska margin in response to Paleocene-Eocene triple junction migration

Peter J. Haeussler
Dwight C. Bradley
U.S. Geological Survey, 4200 University Drive, Anchorage, Alaska 99508, USA

Richard J. Goldfarb
U.S. Geological Survey, Box 25046, MS 973, Denver Federal Center, Denver, Colorado 90225, USA

ABSTRACT

A spreading center was subducted diachronously along a 2200 km segment of what is now the Gulf of Alaska margin between 61 and 50 Ma, and left in its wake near-trench intrusions and high-*T*, low-*P* metamorphic rocks. Gold-quartz veins and dikes, linked to ridge subduction by geochronological and relative timing evidence, provide a record of brittle deformation during and after passage of the ridge. The gold-quartz veins are typically hosted by faults, and their regional extent indicates there was widespread deformation of the forearc above the slab window at the time of ridge subduction. Considerable variability in the strain pattern was associated with the slab window and the trailing plate. A diffuse network of dextral, sinistral, and normal faults hosted small lode-gold deposits (<50,000 oz) in south-central Alaska, whereas crustal-scale dextral faults in southeastern Alaska are spatially associated with large gold deposits (up to 800,000 oz).

We interpret the gold-quartz veins as having formed above an eastward-migrating slab window, where the forearc crust responded to the diminishing influence of the forward subducting plate, the increasing influence of the trailing plate, and the thermal pulse and decreased basal friction from the slab window. In addition, extensional deformation of the forearc resulted from the diverging motions of the two oceanic plates at the margins of the slab window. Factors that complicate interpretations of fault kinematics and near-trench dike orientations include a change in plate motions at ca. 52 Ma, northward translation of the accretionary complex, oroclinal bending of the south-central Alaska margin, and subduction of transform segments. We find the pattern of syn-ridge subduction faulting in southern Alaska is remarkably similar to brittle faults near the Chile triple junction and to earthquake focal mechanisms in the Woodlark basin—the two modern sites of ridge subduction. Therefore, extensional and strike-slip deformation above slab windows may be a common occurrence.

Keywords: Alaska, Gulf of Alaska, triple junction, Paleocene, Eocene, lode gold, mesothermal gold, ridge subduction, slab window

INTRODUCTION

Subduction of an oceanic spreading center, commonly known as ridge subduction, perturbs the overriding plate both thermally and mechanically (e.g., Delong and Fox, 1977). Two diverging oceanic plates will continue to separate even after they are subducted, but new oceanic crust no longer forms in the gap between them (e.g., Thorkelson, 1996). The overriding forearc thus comes into direct contact with hot, upwelling asthenosphere at a slab window, resulting in anomalous igneous

Haeussler, P.J., Bradley, D.C., and Goldfarb, R.J., 2003, Brittle deformation along the Gulf of Alaska margin in response to Paleocene-Eocene triple junction migration, *in* Sisson, V.B., Roeske, S.M., and Pavlis, T.L., eds., Geology of a transpressional orogen developed during ridge-trench interaction along the North Pacific margin: Boulder, Colorado, Geological Society of America Special Paper 371, p. 119–140.

and hydrothermal activity (Dickinson and Snyder, 1979). In the general case where a subducting ridge is neither parallel nor perpendicular to the subduction zone, ridge subduction will be diachronous along strike. During oblique ridge subduction, a trench-ridge-trench triple junction migrates along the convergent margin, accompanied by a change in the plate convergence vector, and presumably, by a change in deformation regime in the upper plate. Moreover, the subduction of any topographically high object—whether seismically active or aseismic—would be expected to disrupt the friction at the base of the accretionary complex and cause it to deform (see Davis et al., 1983).

In this paper, we examine the structural effects of an inferred early Tertiary ridge subduction event along the Pacific margin of Alaska. Four suites of geologic features, each of them linked by precise geochronology to subduction of an oceanic spreading center, might reveal something about the mechanics of ridge subduction: (1) plutons, (2) high-temperature, low-pressure metamorphic terranes, (3) dikes, and (4) gold-quartz mineralized faults. In this paper, we focus on the regional structural trends of the dikes and the gold-mineralized faults. Kusky et al. (this volume, Chapter 12), Sisson et al. (this volume, Chapter 13), and Groome et al. (this volume, Chapter 14) focus on the plutons along this belt. Large plutons complicate the regional structures as their intrusion probably led to local mechanical instability in the accretionary wedge, perhaps masking the regional stress and strain regime. The high-grade metamorphic rocks have been studied locally in considerable detail (Hudson and Plafker, 1982; Goldfarb et al., 1986; Pavlis and Sisson, 1995; Sisson et al., 1989; and Zumsteg et al., this volume, Chapter 11). Our data include the attitudes and kinematics of 681 dikes and 104 gold-quartz mineralized faults and joints, representing regions along a strike-length of 1400 km. Although the structures show broad similarities from one location to the next, they also reveal considerable along-strike variation in the structural response of the accretionary complex to ridge subduction.

Geologic and Tectonic Setting

The Mesozoic and early Cenozoic accretionary complex along Alaska's Pacific margin, referred to as the Chugach terrane or Chugach accretionary complex, stretches from Sanak Island in the southwest to Baranof Island in southeastern Alaska (Fig. 1). Three accreted rock units recognized on the basis of age, rock type, and structural style are germane to the present paper. From landward to seaward and from west to east: (1) The Uyak Complex (Connelly, 1978), McHugh Complex (Clark, 1973), and Kelp Bay Group (Decker, 1980) are mélanges consisting of an argillaceous matrix and more competent blocks of graywacke, greenstone, tuff, Triassic to middle Cretaceous chert, gabbro, plus rare ultramafic rocks and Permian limestone. (2) The Campanian-Maastrichtian Shumagin Formation (Moore, 1973), Kodiak Formation (Nilsen and Moore, 1979), Valdez Group (Nilsen and Zuffa, 1982), and Sitka Graywacke (Decker, 1980) consist of siliciclastic turbidites thought to have originated as trench-fill deposits derived from uplift of the Coast Plutonic Complex

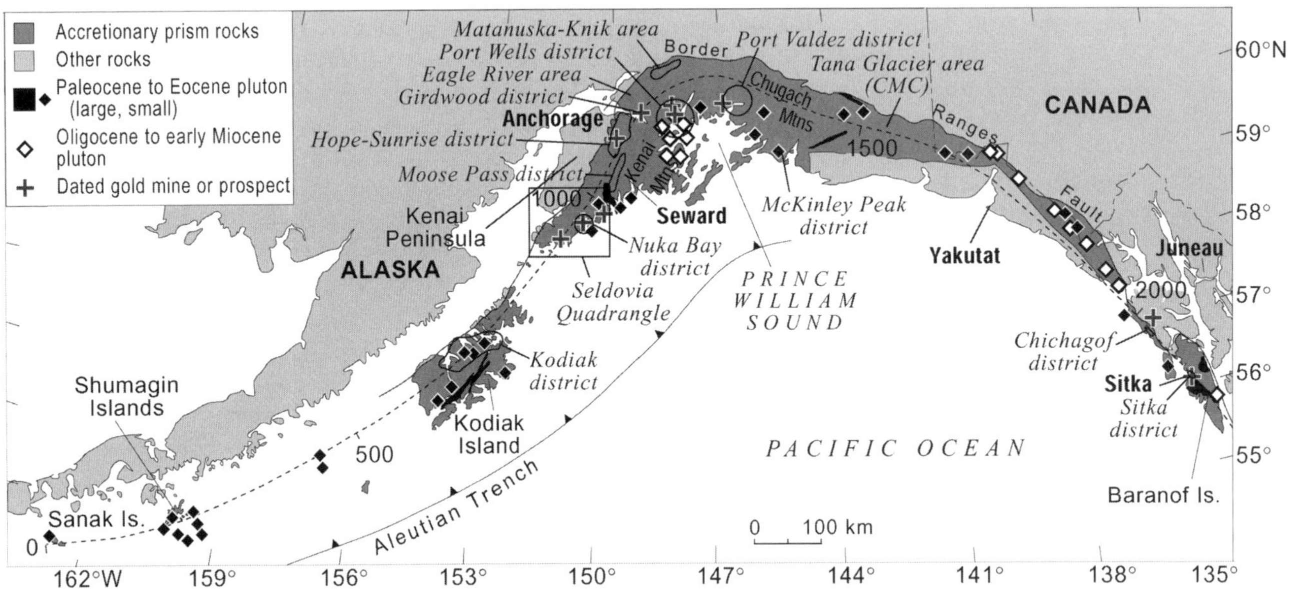

Figure 1. Overview of the southern Alaska Mesozoic–early Cenozoic accretionary complex and areas discussed in the text. The distance (in kilometers) along strike from Sanak to Baranof Island is also shown. Note that all the units in the accretionary complex are lumped together (see Plafker et al., 1994, for more details). CMC—Chugach metamorphic complex, which is an anomalous tract of high-*T*, low-*P* metamorphism (Sisson et al., 1989).

in southeastern Alaska and western British Columbia (Nilsen and Zuffa, 1982; Plafker et al., 1994). Tholeiitic metabasalts are also found in the Valdez Group and its correlatives (Plafker et al., 1994). (3) The Paleocene–middle Eocene Ghost Rocks Formation (Moore et al., 1983) and Orca Group (Winkler, 1976; Helwig and Emmet, 1981) also consist of siliciclastic turbidites along with pillow lavas. These rock units were progressively accreted to the southern Alaskan margin, and all of the units have been deformed and underwent low-grade metamorphism. We use the term "south-central Alaska" to refer to the region of coastal southern Alaska—from Kodiak Island to the Yakutat area. We refer to the panhandle of Alaska as "southeastern Alaska."

The regional structural grain—defined by foliation in the mélange units and by bedding and cleavage in the more coherent units—roughly parallels the coastline of Alaska and sweeps around a ~90° bend in the Prince William Sound area. This bend defines the south-central Alaska orocline (see Fig. 1). Oroclinal bending has been proposed on the basis of structural and paleomagnetic evidence (Carey, 1955; Grantz, 1966; Coe et al., 1985, 1989; Bol and Gibbons, 1992; Bol and Roeske, 1993; Haeussler and Nelson, 1993), although it is not universally accepted (e.g., Csejtey, 1992). Bending is thought to have involved a ~45° to 90° counterclockwise rotation of southwestern Alaska about a vertical axis, some time between ca. 65 and ca. 35 Ma (Coe et al., 1985, 1989). Bol and Gibbons (1992) proposed that a 45° initial bend was further tightened during accretion of the Orca Group by dextral strike-slip faulting in eastern Prince William Sound, which was accompanied by northwest-directed thrust faulting in western Prince William Sound.

Paleomagnetic data from Kodiak Island and the Kenai Peninsula have been interpreted to show that, in the early Tertiary, the accretionary complex was located 13° to 25° to the south, off southern British Columbia or Oregon and Washington (Plumley et al., 1983; Coe et al., 1985; Bol et al., 1992). The rock types, isotopic composition, and age of the Orca and Valdez groups are also consistent with a provenance in the Coast Mountains of British Columbia or southeastern Alaska (Dumoulin, 1987; Plafker et al., 1994; Farmer et al., 1993). Thus, although the presence of the Late Cretaceous and early Tertiary parts of the accretionary complex has been linked with uplift of the Coast Mountains of southeastern Alaska and Canada, neither the exact location of the accretionary complex, nor its orientation with respect to the present margin of southern Alaska, are precisely known for early Tertiary time.

Post-cleavage, "late" faults are locally abundant in the accretionary complex (Byrne, 1984; Sample and Moore, 1987; Bradley and Kusky, 1990, 1992; Kusky et al., 1997). These faults cut mélange fabrics in the McHugh Complex and folds and cleavage in the Valdez Group. Near Anchorage, three fault sets have been identified (Bradley and Kusky, 1990; Kusky et al., 1997): (1) a conjugate set of reverse faults with strikes subparallel to the present-day Aleutian trench, (2) a conjugate set of vertical strike-slip faults, whose acute bisector is perpendicular to the Aleutian trench, and (3) a conjugate set of normal faults that strike perpendicular to the Aleutian trench. Kusky et al. (1997) suggested that fault sets 2 and 3 comprise an orthorhombic fault system that was active at the time of ridge subduction, based on similarities in structural style to dated gold-mineralized faults elsewhere in the accretionary complex rocks.

Evidence for Ridge Subduction beneath Southern Alaska

Six independent lines of evidence suggest a spreading ridge was subducted beneath Alaska's Pacific margin during early Tertiary time. (1) Global plate reconstructions require that a Kula-Farallon spreading center was subducted somewhere along the western margin of North America in Paleocene and Eocene time (Engebretson et al., 1985; Atwater, 1989). Unfortunately, not enough Kula plate marine magnetic anomalies have been preserved to reveal where the ridge intersected the continental margin. (2) A belt of near-trench intrusions along Alaska's Pacific margin has been interpreted as having formed during ridge subduction, and thus they mark the migrating site of the elusive trench-ridge-trench triple junction (Marshak and Karig, 1977; Hill et al., 1981; Helwig and Emmet, 1981; Moore et al., 1983; Bradley et al., 1993; Pavlis and Sisson, 1995; Harris et al., 1996). These intrusions, collectively referred to as the Sanak-Baranof belt (Hudson et al., 1979), extend for 2200 km from Sanak Island on the west to Baranof Island on the east (Fig. 1). They are located between 75 and 250 km seaward of the coeval early Tertiary magmatic arc. The largest intrusions are elongate granodiorite plutons, such as the 100 km long Kodiak batholith, and there are many smaller stocks and dike swarms. Petrogenetic modeling of Paleocene granitoids on Kodiak Island suggests interaction between a parent magma similar to MORB and anatectically melted turbidites (Hill et al., 1981). Similarly, Barker et al. (1992) explained the geochemistry and isotopic composition of granitoids in eastern Prince William Sound by invoking heat from a subducted spreading center to melt the base of the accretionary complex. Harris et al. (1996) concluded from geochemistry that there were two magma sources for felsic intrusions in the eastern Chugach Mountains: metasedimentary rocks from the accretionary complex and a mafic source, possibly mantle derived, which may have been underplated oceanic material. (3) Near-trench magmatism progressed from ca. 61 Ma in the west to ca. 50 Ma in the east (Fig. 2; see Bradley et al., 1993 and this volume, Chapter 1; Poole, 1996. (4) Ophiolites and mafic rocks within the accretionary complex have unusual characteristics consistent with a ridge subduction origin. Two ophiolites have pillow lavas interbedded with turbidites, suggesting proximity of a spreading center to a sediment source (Bol et al., 1992). One of these has been dated at 57 ± 1 Ma (U/Pb zircon; Nelson et al., 1989)—slightly older than nearby near-trench plutons. This date suggests a spreading center was close to the location of near-trench magmatism (see Fig. 3 in Bradley et al., 1993). Basalts from two ophiolites have the geochemical signature of sediment-contaminated MORB (Nelson and Nelson, 1993; Hill et al., 1981). (5) High-*T* and low-*P* metamorphism—opposite of what

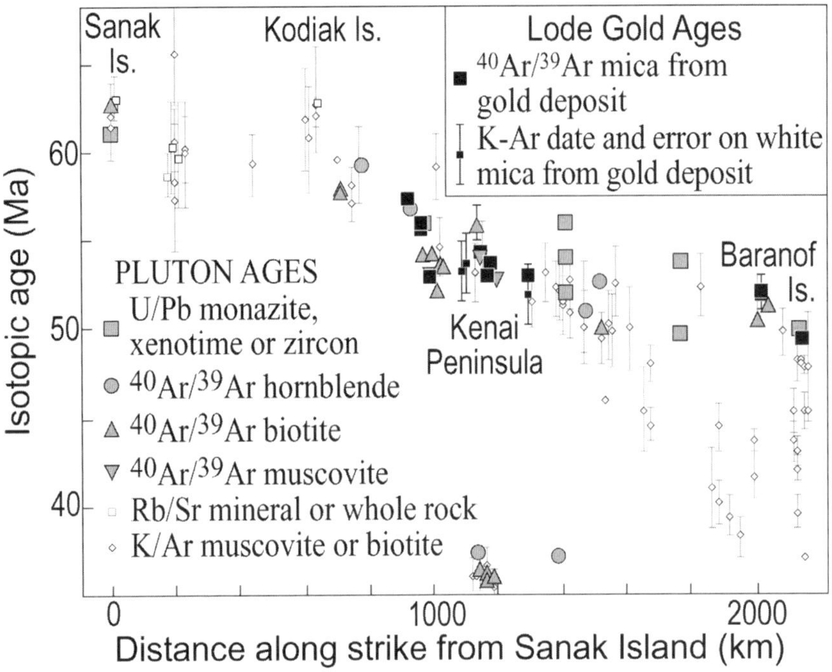

Figure 2. Age versus distance from Sanak Island of near-trench intrusions in the Sanak-Baranof belt. See Figure 1 for distance from Sanak Island. Note the large-scale diachronous trend in ages of older to the west and younger to the east, indicating a migrating trench-ridge-trench triple junction (Bradley et al., 1993). From Bradley et al. (2000).

is normally expected in an accretionary complex—occurred in south-central Alaska at the same time as nearby near-trench magmatism (see compilation of Bradley et al., 1993; Onstott et al., 1989). In the Chugach metamorphic complex, rocks in the accretionary complex were locally heated to temperatures of 650 °C at pressures of 2.5 kbar (Sisson et al., 1989; Pavlis and Sisson, 1995; Loney and Brew, 1987). An unusual heat source is required for these metamorphic conditions in an accretionary wedge setting. (6) The presence of abundant quartz veins, many containing gold, in fault zones and joints throughout the accretionary complex also supports a ridge-subduction interpretation. Temperatures of at least 300 to 400 °C are required for significant gold transport and large fluid volumes, both of which are compatible with a large heat pulse beneath the prism. This heat drove prograde metamorphic reactions with associated vein-forming fluid release at isograd reaction boundaries. The age of the veins decreases from west to east along the continental margin in a pattern similar to that shown by the near-trench intrusions (Haeussler et al., 1995; see Fig. 2). The mineralogy of the gold-bearing quartz veins and the temperatures and pressures of ore formation are fairly consistent over the entire area indicating a consistent regional process of gold mineralization (Goldfarb et al., 1986; Borden et al., 1992; Taylor et al., 1994).

We view the evidence for ridge subduction, slab window magmatism, and lode-gold deposition as compelling.

DISTRICT-BY-DISTRICT DIKE AND FAULT DATA

In this section, we focus on regional structural data for the dikes and the gold-mineralized faults (Fig. 3). Some of the dike orientation data are from our own measurements (principally in the lower Kenai Peninsula), some are from our compilation of structural data at gold occurrences, and the rest are from published geologic maps (see Fig. 3 caption for data sources). The dikes are commonly subvertical, and thus we present the data as rose diagrams of dike strike, which allows us to include dike orientation data from published maps (Fig. 3E).

Fault zones have been studied in greatest detail in areas of lode-gold mines and prospects found throughout much of the south-central Alaska accretionary complex (Fig. 1). We compiled published structural data from gold mines and prospects (Haeussler and Bradley, 1993), and synthesized this with our own data collected both from gold districts and elsewhere along the prism. The strength of the published data is that it covers a very broad region, from Kodiak to Baranof Island, and there are some structural measurements available from most mines and prospects, many of which are now inaccessible or unlocatable. We used reports from the early part of this century principally to determine if a gold-quartz vein occupies a fault zone, its orientation and dimensions, and any crosscutting relations (Haeussler and Bradley, 1993). Unfortunately, most of the published data are vintage 1910–1935, and they lack information on the sense of offset of faults.

In order to address this lack of kinematic data, we collected such information from 22 mines, prospects, or mineral occurrences. We then discarded old data that were vague or inconsistent and compiled all the new and old structural and kinematic data giving information on a total of 361 mines, prospects, and mineral occurrences in nine regions (Fig. 1; stereonets on Fig. 3). In the following sections we abstract the most important structural

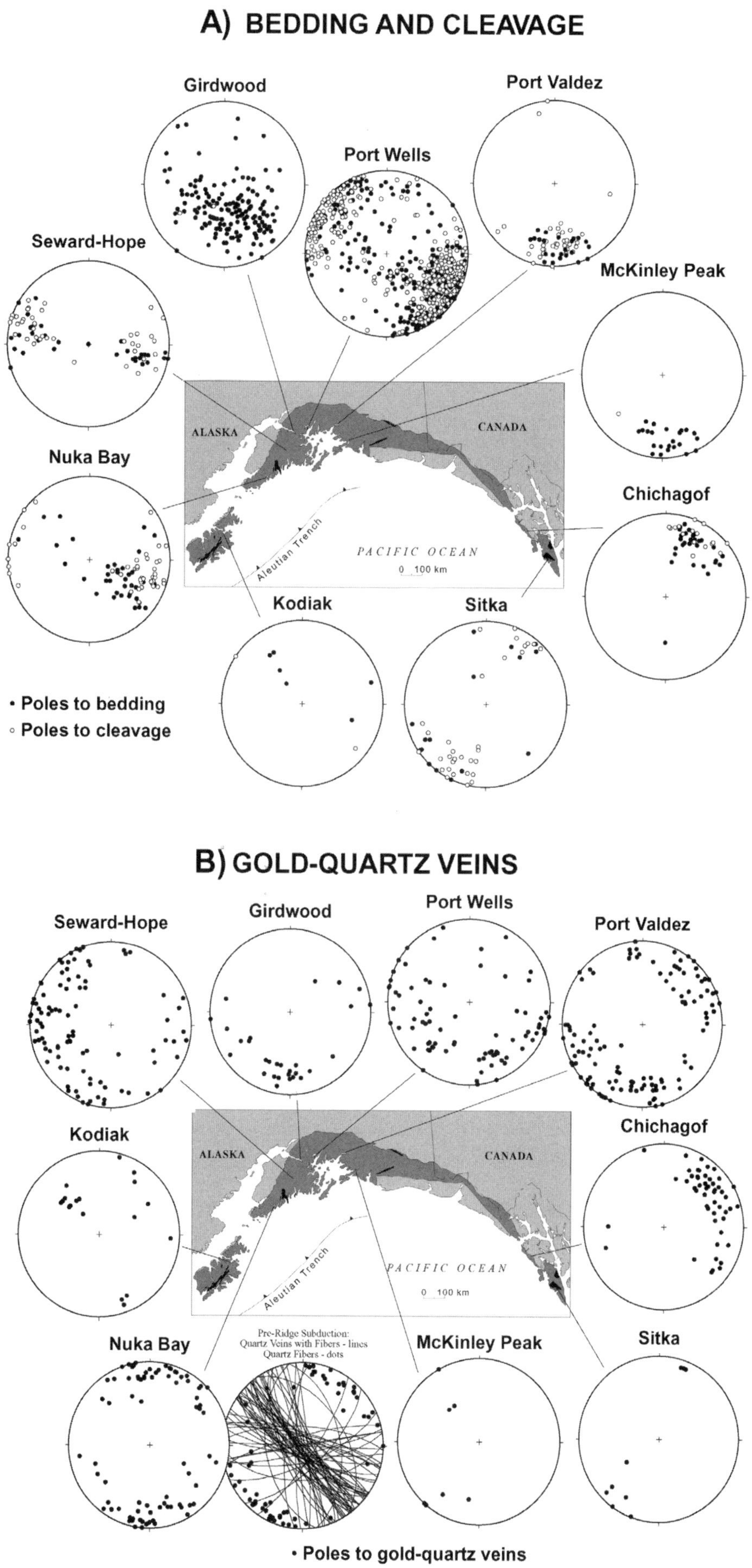

Figure 3 (*on this and following two pages*). Equal-area stereonet plots of structural data from regions of gold mineralization in the southern and southeastern Alaska accretionary complex. The index map has the same legend as Fig. 1. Data for the gold-bearing quartz veins and host faults come from our fieldwork and the 31 sources listed in the compilation of Haeussler and Bradley (1993). Data from southeastern Alaska come from Becker (1898), Wright and Wright (1905), Knopf (1912), and Reed and Coats (1941). Data on dike orientations come from the previously listed sources and Park (1933), Winkler (1992), and Plafker et al. (1992). A: Poles to bedding and cleavage. B: Poles to gold-quartz veins with additional plot of Nuka Bay area pre-Sanak-Baranof belt gold-quartz veins plotted as great circles and quartz fibers as dots. C: Gold-quartz mineralized strike-slip faults and slickenlines plotted on solid great circles, and nonmineralized faults and slickenlines plotted on dashed circles. D: Gold-quartz mineralized dip-slip faults and slickenlines plotted on solid great circles, and nonmineralized faults and slickenlines plotted on dashed circles. E: Rose diagrams of Sanak-Baranof belt dike strikes. Each petal has a 10° width, n is the number of samples plotted, and "longest" is the number of samples in the longest petal. F: Same data as in Figure 3E but rotated into a possible pre-oroclinal bending configuration. See text for discussion. Amount of clockwise rotation for each area is listed below each rose diagram.

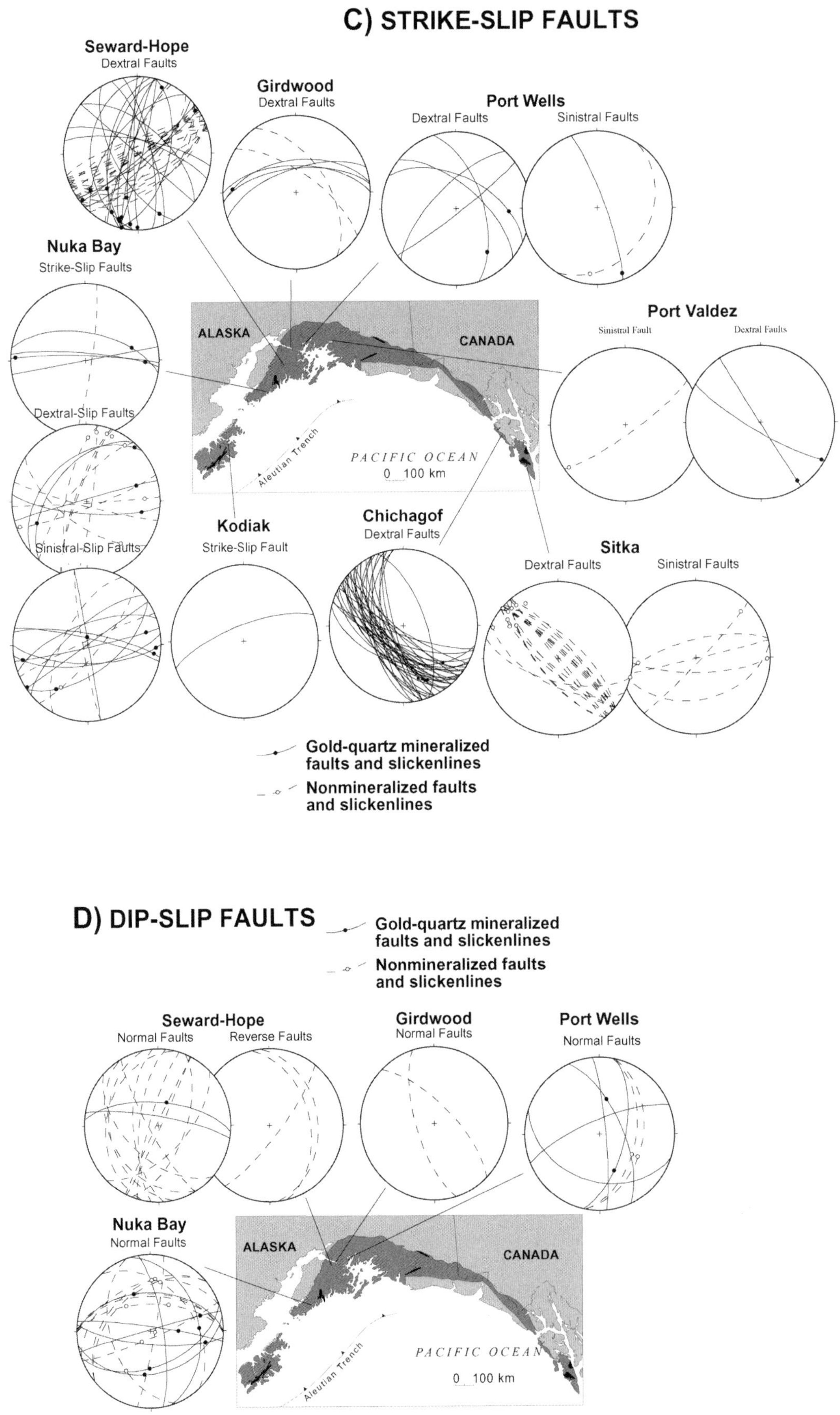

Figure 3 (*continued*).

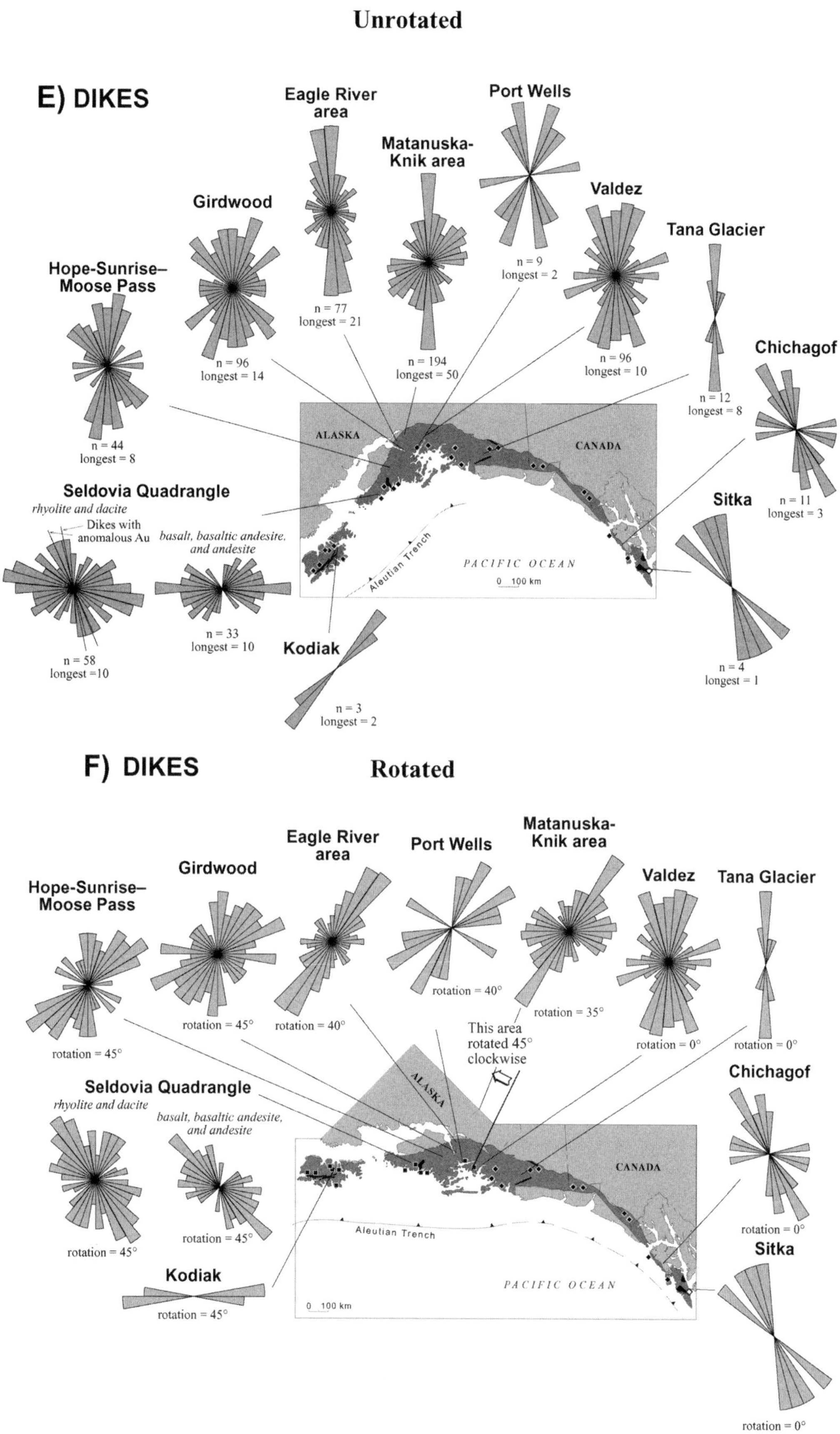

Figure 3 (*continued*).

information. A more complete description of the structural characteristics, data sources, geologic setting and crosscutting relationships, and age constraints can be found in the GSA Data Repository[1]. The repository differs significantly from the compilation of Haeussler and Bradley (1993) in that it summarizes information from each of the regions in more detail and from a number of areas not discussed below. It also includes data from the Sitka and Chichagof districts, which were not in the compilation of Haeussler and Bradley (1993).

Sitka District

This area encompasses lode-gold mines and prospects near Sitka of Baranof Island in southeastern Alaska (Fig. 1). It is underlain by the prehnite-pumpellyite facies Sitka Graywacke and Kelp Bay Group. Bedding and cleavage in the Sitka Graywacke, and compositional layering in the Kelp Bay Group, strike northwest and dip steeply (Fig. 3A). Four Sanak-Baranof belt plutons lie within 15 km of the Sitka district on Baranof Island, including the large (~760 km^2) Crawfish Inlet pluton. Dikes cut both the Sitka Graywacke and the Kelp Bay Group; these are not dated but are compositionally the same as isotopically dated plutons. On a broad scale, the Paleogene dikes (felsite and diorite) appear to strike parallel to the structural grain (Fig. 3E); in detail, however, dikes are at a 15°–30° clockwise angle to the strike of bedding and cleavage.

A $^{40}Ar/^{39}Ar$ age of 49.4 ± 0.4 Ma on sericite from vein quartz at the Lucky Chance Mine (Haeussler et al., 1995) is similar to a U/Pb zircon date of 50 Ma on the Crawfish Inlet pluton. Therefore, this mine and presumably seven other mines and prospects are part of the Sanak-Baranof belt. The veins, which typically occupy faults, have remarkably consistent attitudes: They strike northwest or west-northwest, and dip steeply to the north (Fig. 3B). The most compelling evidence of the fault kinematics is from the Golden Eagle (or Liberty) mine, where a gold-mineralized northwest-striking fault displays dextral slickenside steps. Numerous other gold-quartz veins in the area have the same general attitude and are likely along dextral faults but have not yielded kinematic data. In summary, the mineralized faults suggest a north-trending maximum shortening axis. The relative age of dikes and gold mineralization on Baranof cannot be proven, because isotopic ages of the two overlap (Haeussler et al., 1995). However, faults similar in structural style and orientation to mineralized faults do locally cut some Tertiary dikes, which suggests the gold-quartz mineralized faults are younger than the intrusions—consistent, as will be shown, with crosscutting relationships in other districts.

[1]GSA Data Repository item 2003124, Summary of structural data from gold mines, prospects, and mineral occurrences in the accretionary complex of southern and southeastern Alaska, is available on request from Documents Secretary, GSA, P.O. Box 9140, Boulder, CO 80301-9140, USA, editing@geosociety.org, at www.geosociety.org/pubs/ft2003.htm, or on the CD-ROM accompanying this volume.

Chichagof District

Western Chichagof Island is underlain by the prehnite-pumpellyite facies Sitka Graywacke and Kelp Bay Group. Bedding and cleavage in the Sitka Graywacke, and compositional layering in the Kelp Bay Group, strike northwest and dip steeply (Johnson and Karl, 1985; Fig. 3A). Plutons lie 10 km to the south and southwest of the locus of mining activity, as well as 15–20 km to the northwest (Loney et al., 1975). $^{40}Ar/^{39}Ar$ ages on biotite from these intrusions range between 50.3 and 51.4 ± 0.1 Ma (Taylor et al., 1994; see also Fig. 2). A few aplite dikes cut the Sitka Graywacke in mines and prospects in the Chichagof District. They appear to be compositionally similar to the Sanak-Baranof intrusions and different from basaltic dikes of late Tertiary age (Johnson and Karl, 1985). Similar to the Sitka District, the dikes mostly strike north-northwest, nearly parallel, but commonly at a 15°–30° clockwise angle, to the structural grain (Fig. 3E).

The Chichagof district had by far the largest gold production of all the ridge subduction-related gold deposits, with more than 24,900 kg (800,000 oz) produced primarily from the Chichagof and Hirst-Chichagof mines. A total of 60 mines and prospects are within this district. There are no isotopic ages on any of the gold-quartz veins cutting the sedimentary rocks. Taylor et al. (1994) obtained $^{40}Ar/^{39}Ar$ ages on muscovite and fuchsite of 51.9 and 52.1 ± 1.1 Ma from the Apex and El Nido mines, which are part of the same gold district but lie 5 km east of the Chugach accretionary complex. (The gold-bearing quartz veins in these mines have a similar structural style to those at the larger Chichagof and Hirst-Chichagof mines.)

All but a few of the gold-bearing quartz veins in this district are within faults that strike northwest and dip steeply southwest (Fig. 4; Reed and Coats, 1941), and many of these lie near the Border Ranges fault (Roeske et al., 1992; Smart et al., 1996). Further evidence for focused fluid flow associated with the Border Ranges fault are the Golden Hand and New Chichagof Syndicate groups of mines that lie virtually on top of the Border Ranges fault, and the other mines define a trend parallel to the fault. At the Chichagof and Hirst-Chichagof mines, both of which are on northwest-striking faults, slickenlines consistently plunging 30° NNW indicate reverse-dextral motion (Figs. 3C, 4). In all but one instance, faults with gold-quartz veins cut Sanak-Baranof belt intrusions. The one exception is on the Hirst-Chichagof fault, where an aplite dike intruded between layers of banded quartz (see Reed and Coats, 1941, plate 23) and was later cut by the fault and gold-quartz veins. This is the only gold occurrence where a dike intruded along a fault, and the relationship indicates that faulting, gold-mineralization, and magmatism were broadly coeval. Fault zones containing thick gouge zones with no quartz veins cut all gold-quartz mineralized features and are subparallel to the main mineralized faults.

Tana Glacier Area

In this region, the earliest deformation is a regional WNW-striking layer-parallel slaty and phyllitic cleavage that

Figure 4. Photograph looking northwest along the Chichagof fault at the Chichagof Mine in southeastern Alaska. The gold-quartz mineralized fault dips steeply to the southwest. Near the scale bar (long bar is 10 cm), slickenlines plunging 30° to the southwest show mineralization was along a reverse-dextral fault. The main gold-quartz vein is visible in the top center of the photograph and is highly ribboned. Gold and sulfides are also found in veins throughout the fault zone. Rock fragments within the fault zone to the right (northeast) of the main vein and above the scale bar are cut by quartz veins, and these are offset on later faults. To the left of the main vein is a gouge zone, with no quartz veining. Similar gouge zones offset all gold-bearing quartz veins, both within the Chichagof Mine and at all mines in the southern Alaska accretionary complex.

developed during accretion and subsequent shortening (D_1 of Pavlis and Sisson, 1995). High-T (~650 °C), low-P (~300 MPa) amphibolite-facies metamorphism of the Chugach metamorphic complex followed (Sisson et al., 1989), and it was accompanied first by orogen-parallel extension with vertical contraction (D_2 of Pavlis and Sisson, 1995) and then by subhorizontal contraction perpendicular to the margin (D_3). D_2 and D_3 were ductile events, and both were accompanied by injection of melts. Large plutons are elongate parallel to regional structural trends (Winkler and Plafker, 1993; Pavlis and Sisson, 1995). Notable among these is the Mt. Tom White pluton, which outcrops over an 80 km length but is rarely wider than 5 km (Hudson and Plafker, 1982). Within the Chugach metamorphic complex, the second-youngest structures are a set of tonalitic dikes, which are inferred to be the same age as the large intrusions (Ti_3 of Pavlis and Sisson, 1995; Fig. 3E). Isolated dikes restore to a north-striking orientation, but larger bodies strike east-west and dip steeply to the north, parallel to regional structural trends. These are the only dikes in the Chugach metamorphic complex that can be confidently related to other structures in the present compilation. Pavlis and Sisson (1995) concluded they were intruded during dextral transpression. The youngest dikes in the Chugach metamorphic complex (Ti_4 of Pavlis and Sisson, 1995) are undated. These dikes locally have chilled margins, which were interpreted as indicating significant cooling of the metamorphic complex prior to their cooling. These dikes also have north-south strikes (Fig. 3E).

Gold mineralization is unknown in the amphibolite-facies rocks in this area, and moreover, none would be expected because the fluids involved in gold mineralization are thought to have originated from greenschist-to-amphibolite facies dehydration reactions (Goldfarb et al., 1986). Brabb and Miller (1962) reported lode gold in the lower-grade greenschist facies rocks in this region. Thus, the rocks in the Chugach metamorphic complex may be typical of the deeper-seated source region for the fluids involved in overlying gold mineralization.

Port Valdez District

This area includes road-accessible areas northeast of the town of Valdez and the Port Valdez gold district. Bedding and cleavage in the Valdez Group, and compositional layering in the McHugh Complex, strike west-southwest and dip steeply (Fig. 3A). Although most mines and prospects lie within the Valdez Group, a few small prospects in the southwestern part of this area lie in the Orca Group, both of which consist of turbidites at middle greenschist facies (Goldfarb et al., 1986). Orogen-parallel stretching lineations have also been measured in these rocks and are inferred to postdate development of phyllitic cleavage (Marty, 1994; Pavlis et al., this volume, Chapter 7). Marty (1994) and Pavlis et al. (this volume, Chapter 7) correlated these with orogen-parallel strains in the Tana Glacier area, which occurred at the time of peak metamorphism and thus at the time of ridge subduction. There are no large intrusions in the area, but there are at least a dozen andesite and dacite dikes that have been correlated with Sanak-Baranof belt intrusions (Plafker et al., 1992). The dikes most commonly strike north-south (Fig. 3E), which is roughly perpendicular to the strike of bedding and cleavage.

Structural data are available from 57 gold mines and prospects in the Port Valdez area. Gold-quartz veins generally strike northwest and dip steeply between 50° and 90° (Fig. 3B). At least 24 of the 57 gold occurrences with some structural data are apparently hosted by faults; shear sense is known for a few faults. Two dextral faults dip steeply and strike northwest, like the gold-quartz veins. Pickthorn (1982, p. 24 and 27) implied that these northwest-striking dextral faults are common and stated they are cut by east-striking sinistral faults. If these fault sets are considered together, the sense of offset is opposite what would be expected with a conjugate set. This suggests the fault sets are not related to the same stress regime. Only two of the 57 gold-quartz mineral occurrences are reported to be near dikes, and in both cases the veins cut the dikes; thus, gold mineralization is younger.

Port Wells District

This large area includes lode-gold occurrences around Port Wells and adjacent areas of western Prince William Sound.

Bedding in the middle-greenschist facies turbidites of the Valdez and Orca Groups generally strikes northeast (Fig. 3A). This area includes two age groups of plutons. The older set (ca. 54 Ma), represented by granitic plugs at the Granite and Homestake mines, belongs to the Sanak-Baranof intrusive suite (Haeussler et al., 1995); the younger set is 34–36 Ma (Tysdal and Case, 1979; Nelson et al., 1985; Bradley et al., 1993; Nelson et al., 1999). Dikes, which are undated and could belong to either set, generally strike north-northeast and dip steeply at a high angle to bedding.

The orientation and sense-of-offset of gold-quartz mineralized faults is variable. Most subeconomic quartz veins in the district strike northwest and dip steeply (Fig. 3B). Many of the mined gold veins, however, strike northeast. About half of the gold-quartz veins are demonstrably within a fault. We measured three dextral faults with northwest strikes, and Hoekzema et al. (1987) reported a northeast-striking dextral fault. We found one north-northwest–striking gold-quartz mineralized sinistral fault. We measured five north-south–striking normal faults with steep easterly dips, three of which were gold mineralized.

Gold-quartz veins lie within a few hundred meters of intrusions at 17 of 49 localities. At 11 occurrences the veins cut intrusions, and thus the gold mineralization postdates magmatism. It is remarkable there are this many gold-quartz veins that cut dikes, because of the abundance of ca. 35 Ma intrusions in the area, which postdate the Sanak-Baranof belt. Therefore, these dikes must be related to the Sanak-Baranof belt.

Stüwe (1986) reported three types, or generations, of quartz veins in the Port Wells district. Type 1, the oldest, are boudinage-related veinlets cutting competent beds and veinlets paralleling the structural fabric. Goldfarb et al. (1986) referred to these early, bedding-parallel veins as metamorphic quartz segregation features, and we find they are common throughout the Kenai and Chugach Mountains. Fisher and Byrne (1990) documented similar veins on Kodiak Island (their D_2 veins). Stüwe's (1986) Type 2 veins are the gold-quartz veins, and are younger than the intrusions. Stüwe's (1986) Type 3 veins occur as networks, stockworks, and healed fracture fillings with no consistent orientations. The only reported crosscutting relationship among the gold-quartz veins is in Harriman Fiord, where Stüwe (1984) found an older sulfide-mineralized vein (N41°E/65°N) cut by a younger, barren, drusy white vein (N64°E/86°N). The fact that both veins are similar in orientation to veins elsewhere, such as in the Hope-Sunrise–Moose Pass area (see below), and that the younger veins are clockwise of the older veins, suggests similar structural histories for both areas.

Chugach Mountains near Anchorage (Matanuska-Knik and Eagle River Areas)

Bedding and cleavage in the Valdez Group, and compositional layering in the McHugh Complex, strike east-west in the drainages of the Matanuska and Knik Glaciers (Winkler, 1992). There are no large intrusions, but dacitic dikes that cut across the regional lower greenschist facies fabric in the Valdez Group and McHugh Complex (Burns et al., 1991; Winkler, 1992) are common. Locally, these intrusions are sills (Little and Naeser, 1989). Winkler's (1992) map of the area shows a similar dike swarm in the Eagle River drainage (Fig. 3E), which also cuts across the regional structural fabric in the McHugh Complex and Valdez Group rocks. Both dike swarms show a strong north-south preferred orientation (Fig. 3E). Two gold occurrences are present within this region (Capps, 1916), and their post-cleavage, high-angle, brittle faulting structural style is similar to the veins found elsewhere.

Girdwood District

The Girdwood district lies within laumontite to lower greenschist facies (Goldfarb et al., 1986) Valdez Group rocks and has a structural style typical of the Valdez Group, but it has anomalous orientations. Bedding and cleavage strike east-northeast and dip steeply (Fig. 3A), in contrast to the more typical north-northeasterly strikes in this region. Lode-gold mines and prospects in the Girdwood district are concentrated near small granitic intrusions, including the Crow Pass pluton, which has the map pattern of a tadpole-like dike. Most dikes have north-northeasterly strikes (Park, 1933). Seven of nine gold occurrences are usually within a few tens of meters of one or more dikes. Two of the nine documented gold occurrences crosscut a dike or associated hornfels; therefore, gold mineralization is younger than near-trench magmatism. Most of the gold-quartz veins, including the most productive ones (Hoekzema et al., 1987), strike roughly east-west; some of these veins are dextral faults. A few gold-quartz veins with unknown offset strike approximately north-south. Two non-mineralized, probably younger, gouge-filled dextral faults strike northwest-southeast.

Upper Kenai Peninsula (Hope-Sunrise and Moose Pass Districts)

These areas lie entirely within lower greenschist facies Valdez Group rocks (Mitchell, 1979). Bedding and cleavage strike north-northeast and dip steeply. Large plutons are absent, but felsic dikes are abundant. Mitchell (1979, p. 35) found that most dikes have northwest strikes, but the two longest and economically most significant dikes trend north-northeast (015°) (Hoekzema et al., 1987, p. 38). These are the Gilpatrick and Palmer Creek dikes, which are mapped as 18 and 12 km long, right-stepping intrusions.

Gold-bearing quartz veins have a wide range of orientations, and most are steeply dipping, although veins with a north-south strike are slightly more common. Twenty-two of 63 gold occurrences in our database are clearly along faults, and most are along north-northeast–striking dextral faults that are subparallel to, and at seven localities, cut the 015°-striking dikes. The mineralized faults are cut by a later set of dextral faults that strike northeast to east-northeast. No sinistral- or reverse-motion mineralized faults were recognized in this area, but there are two east-west–striking normal faults.

The approximate parallelism between gold-quartz mineralized dextral faults and bedding in the Hope-Sunrise–Moose Pass area is similar to the Girdwood area. In the Girdwood area, however, most dikes are perpendicular to bedding, whereas in the Hope-Sunrise–Moose Pass area, the dikes are subparallel to bedding. Another similarity of the two areas is the presence of barren faults oriented about 30° clockwise of the gold-bearing faults. Curiously, the orientations of bedding, mineralized dextral faults, and barren dextral faults in the Girdwood area are all oriented approximately 70° clockwise of the same structures in the Hope-Sunrise–Moose Pass areas and 70° clockwise with respect to regional orientations of bedding (Winkler, 1992). We therefore suggest a clockwise vertical-axis rotation of the Girdwood area after gold-quartz mineralization and after movement on the barren faults. Although dikes are not obviously rotated, they have a much wider range of orientations in the Girdwood area than in the Hope-Sunrise–Moose Pass area.

Figure 5. A: These veins with subhorizontal quartz fibers in the Nuka Bay area are oriented perpendicular to bedding and indicate orogen-parallel extension. The veins cut the graywacke but not the slate. B: Hornfelsed vein with subhorizontal quartz fibers parallel to the pencil from the margin of a pluton in the McCarty Fiord area. This relationship demonstrates that orogen-parallel extension predated near-trench magmatism. C: Photomicrograph of quartz vein in Fig. 5B. Note the abundant small equant quartz grains that comprise the crystalloblastic fabric. The few larger and elongate quartz grains are probably relict fibrous quartz crystals.

Lower Kenai Peninsula (Nuka Bay District and Seldovia Quadrangle)

The lower Kenai Peninsula is underlain by the lower greenschist facies Valdez Group and the McHugh Complex. The regional orientation of bedding and cleavage is north-northeast–striking and moderately to steeply dipping to the northwest (Bradley et al., 1999).

The oldest structures that postdate accretion record orogen-parallel extension. In the Nuka Bay mining district (Richter, 1970; Fig. 1), barren, northwest-southeast–striking quartz veins strike perpendicular to thick sandstone beds (Fig. 5A). These veins are unrelated to faults and are steeply dipping with subhorizontal quartz mineral fibers. At one locality, stretched pebbles are parallel to the quartz fibers. At another, a quartz vein perpendicular to bedding was hornfelsed in the margins of Sanak-Baranof belt intrusions. The quartz vein in outcrop has relict fibrous forms (Fig. 5B). A thin section shows evidence of hornfels with abundant small equant quartz grains in a crystalloblastic fabric, and a few relict larger elongate quartz grains (Fig. 5C). Both the geometry and microscopy of the quartz vein indicate that margin-parallel extension occurred prior to intrusion of the pluton.

The Seldovia quadrangle includes three large biotite-granodiorite plutons: the Nuka, Harris Bay, and Tustumena, which are likely connected at depth because of their proximity and similar orientation (see also Kusky et al., this volume, Chapter 12). The three largest plutons are elongate along a north-south trend, and swarms of mafic, intermediate, and felsic dikes are common. The mafic and felsic dikes have westerly strikes, with a subordinate northwest-striking group of felsic dikes (Fig. 3E). Geochronology indicates dikes have the same ages as plutons (Bradley et al., this volume, Chapter 1). At one locality north of Harris Lagoon, however, sills (dipping ~35° to the WNW) subparallel to the larger intrusions are cut by steeply dipping dikes (Fig. 6). This suggests a change in the strain regime between intrusion of the sills and plutons and the dikes.

The greatest concentration of gold occurrences is in the Nuka Bay area. Most gold-bearing quartz veins are within steeply dipping sinistral- faults that strike west—approximately perpendicular

Figure 6. Photograph is of a ~200 m high cliff face just north of Harris Lagoon in the eastern part of the Seldovia quadrangle (Fig. 1). The low-angle sills were intruded into Valdez Group graywacke before the steeply dipping dikes. This suggests there was a significant change in stress directions during intrusion of the sills versus intrusion of the dikes. We consider the dikes as related to extension during the later stages of ridge subduction.

to regional structural trends (Fig. 7). Some gold-quartz veins occur along dextral-lateral faults that strike northeast. The dextral and sinistral faults form a conjugate set and appear to be the same age. The acute bisector of the conjugate fault sets strikes northeast and indicates contraction subparallel to the regional strike of bedding. Most of the fractured dikes along which there was gold also strike northeast. This relationship suggests these dikes were intruded into the same stress regime that produced the strike-slip faults. Five of 15 mineral occurrences in the Nuka Bay area lie within, or at the margins of, dikes. There are also mineralized normal faults but no mineralized thrust faults. Mineralization in Thunder Bay, 25 km northeast of Nuka Bay, was on dextral-normal or normal-dextral faults. Most mineral occurrences in the Nuka Bay area were on faults with the same orientation but opposite sense of offset. The differences in fault offset between these two areas demonstrate significant differences in fault kinematics between areas not very far apart.

Sinistral, dextral, normal faults, and one thrust fault were observed to cut gold-bearing quartz veins. At least some brittle faulting postdated mineralization at all locations. Some of the late faulting is manifested by clay-rich gouge zones at the margins of the gold-bearing quartz veins/faults (Fig. 7). These gouge zones are typically less than 30 cm wide and consist of brecciated and altered wall rock fragments. Some gouge zones also contain a few thin quartz veins. It is uncertain whether these gouge zones represent movement on the faults soon after cessation of hydrothermal activity or whether they are substantially younger faults that reactivated older structures.

The youngest phase of brittle faulting occurred on faults filled with clay-rich gouge zones up to 10 m wide and with no quartz veins. These faults form prominent topographic lineaments that cut across all structures. In our examination of several of these faults in shoreline exposures, there was no indication of significant offset. In summary, there was orogen-parallel extension before Sanak-Baranof magmatism and orogen-parallel contraction during gold mineralization and dike intrusion.

Kodiak District

The prospects on Kodiak and adjacent islands are either in the Kodiak Formation or in diorite or granodiorite intrusions that cut the Kodiak Formation, with the exception of one prospect that cuts the Uyak Complex. All of the intrusions appear related to the northeast-striking Kodiak batholith because of similarities in composition, geometry, and age. Four of twelve prospects were described as being near numerous and persistent dikes, three are in structures that cut an intrusion, and one is at the contact of an intrusion. Extensive work has been conducted on early quartz veins associated with underthrusting of sediments beneath the accretionary prism (Fisher and Byrne, 1987, 1990; Vrolijk, 1987; Meyers and Vrolijk, 1986; Fisher and Brantley, 1992). These veins formed during or before zeolite to prehnite-pumpellyite facies (Sample and Moore, 1987) cleavage formation (Fisher and Byrne, 1990) and are thus easily distinguished from the gold-bearing quartz veins and faults that crosscut the cleavage.

Bedding and cleavage strike northeast (Moore, 1969). Dikes strike northeast, dip steeply, and are subparallel to the massive and elongate Kodiak batholith (Fig. 1). There are two main orientations of gold-quartz veins: northeast-striking and northwest-striking. The northwest-striking veins are approximately perpendicular to cleavage, and two of the gold-quartz veins are demonstrably in a fault zone. Two faults with an

Figure 7. This east-west–striking vein in the Nuka Bay area (locality 12 of Richter, 1970) is typical of many gold-bearing quartz veins. It dips steeply, is 30 cm wide, was deposited in a strike-slip fault (in this case sinistral), and has a 10 cm wide clay-rich gouge zone at the right (northern) margin. The wall rocks on the right side of the fault are graywackes, and the wall rock on the left margin is an aphanitic intermediate dike.

unknown sense of offset are at a 60° angle to bedding and are steeply dipping, and thus the late-fault orientations are not controlled by a pre-existing fabric.

DISCUSSION

Structural History of the Accretionary Complex

The Sanak-Baranof belt of intrusions along Alaska's Pacific margin provide us with the special opportunity to discriminate between pre-, syn-, and post-ridge subduction structures (Fig. 8). By examination of crosscutting relationships, we find a common sequence of events along the length of the accretionary complex.

1. Underplating, offscraping, regional metamorphism, and contractional deformation. Offscraping and underplating of sediments on the subducting plate was the fundamental step in the construction of the accretionary complex. Numerous studies have focused on this aspect of the development of the southern Alaska accretionary complex (Sample and Moore, 1987; Byrne, 1984, 1986; Moore and Allwardt, 1980; Bradley and Kusky, 1992; Kusky et al., 1993; Kveton, 1989; Clendenen, 1991). On Kodiak Island, Fisher and Byrne (1990) found there were two major episodes of fracturing and quartz mineralization related to underthrusting and underplating. Subsequently, the Valdez Group turbidites and its correlatives were then metamorphosed up to middle greenschist facies (M. Miller, 1993, personal commun.; Goldfarb et al., 1986; Sample and Moore, 1987). This low-grade metamorphism and deformation is exhibited in a common slaty and phyllitic cleavage and associated metamorphic segregation quartz veins (Goldfarb et al., 1986; Stüwe, 1986). All of these events predated passage of a trench-ridge-trench triple junction. If this triple junction is related to the Kula–Farallon–North America triple junction, then the Valdez Group and its correlatives were offscraped or underplated from the Farallon plate (Bradley et al., 1993), not from the Kula plate (Plafker et al., 1994; Wallace and Engebretson, 1984), because the Kula–Farallon–North America triple junction was farther north than previously assumed.

2. Orogen-parallel extension. We find evidence of orogen-parallel extension in the Nuka Bay and Sitka areas, and other workers find evidence for it in the Valdez and Tana Glacier areas (Marty, 1994; Davis et al., 1998; Pavlis et al., this volume, Chapter 7). Orogen-parallel extension may have occurred either prior to ridge subduction, as in the Nuka Bay area, or it may have occurred at the onset of ridge subduction, as in the Tana Glacier area. Orogen-parallel extension is a common occurrence in modern and ancient forearc settings (e.g., Avé Lallemant and Guth, 1990; McCaffrey, 1996), and it need not be related to ridge subduction processes.

3. Near-trench magmatism and high-T, low-P metamorphism. The locus of near-trench magmatism migrated along the Gulf of Alaska margin, from ca. 61 Ma near Sanak Island to ca. 50 Ma on Baranof Island (Bradley et al., 1993 and this volume, Chapter 1). Near-trench magmatism postdated margin-parallel extension in the Nuka Bay and Sitka areas and was synchronous with margin-parallel dextral shear in the Tana Glacier area. Plutons were emplaced in some areas, but steeply dipping dikes are most common. Sills are uncommon.

4. Gold mineralization in strike-slip or extensional setting. In all cases but one, gold-quartz veins cut intrusions. The one exception is at the Hirst-Chichagof mine in southeastern Alaska where some veining and faulting occurred prior to intrusion of a dike along a fault. This exception indicates that faulting, veining, and near-trench magmatism were broadly coeval. Moreover, the $^{40}Ar/^{39}Ar$ ages of gold-bearing quartz veins and nearby near-trench intrusions are indistinguishable and therefore link mineralization to magmatism (Haeussler et al., 1995). There was widespread brittle deformation of the accretionary complex when there was a slab window beneath the forearc. All of the faults hosting the gold-quartz veins are either strike-slip or extensional (we discuss this more completely in a following section). Contractional deformation played no role in the emplacement of these veins. Upper greenschist to amphibolite facies dehydration reactions appear necessary to produce the appropriate vein-forming fluids (Goldfarb et al., 1986). In all cases, we consider it likely that the sources of the vein forming fluids are rocks below the present level of exposure. Rocks like the high-*T*, low-*P* Chugach metamorphic complex (Sisson et al., 1989) were probably in the source region for the vein-forming fluids.

5. Right-lateral brittle faulting without fluid flow on margin-parallel faults. Unmineralized right-lateral brittle faults with gouge zones were active subsequent to veining and crosscut mineralized faults. In southeastern Alaska, these late faults are subparallel to the earlier mineralized structures. In south-

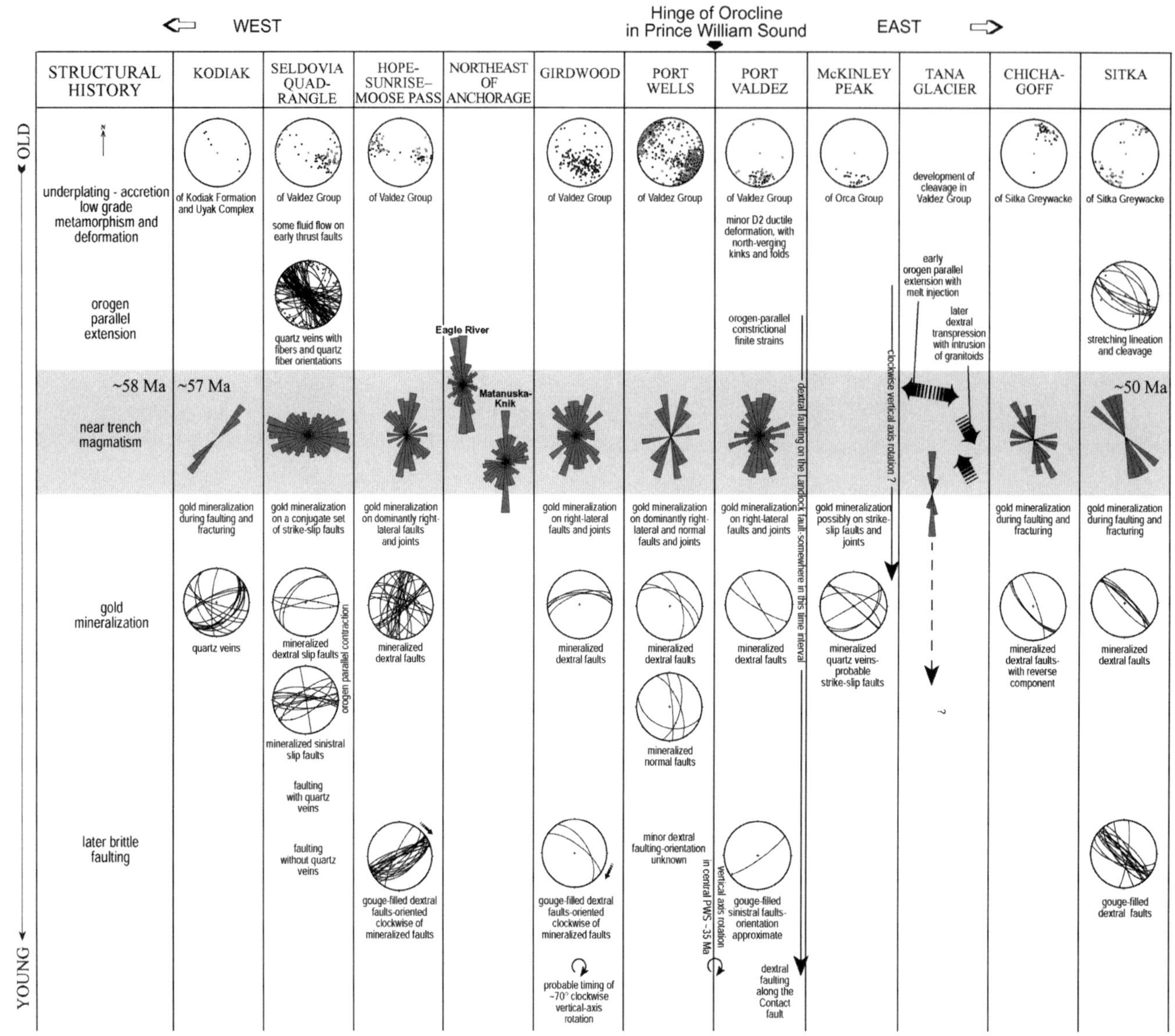

Figure 8. Graphical structural history of the southern and southeastern Alaska accretionary complex. Stereonets and rose diagrams (from Fig. 3) show orientations of structures. Note near-trench magmatism is time transgressive from ca. 57 to 58 Ma near Kodiak Island to ca. 50 Ma on Baranof Island. Gold mineralization is also time transgressive (Haeussler et al., 1995), and other events may be time transgressive. Events younger than ca. 30 Ma are not shown. Blank boxes do not imply an event did not occur, but rather an event has not been observed in a particular area.

central Alaska, these dextral faults are parallel to the Aleutian trench, strike northeast-southwest, and are typically oriented 15°–30° clockwise of the largest mineralized structures.

It is puzzling why these faults in south-central Alaska are oriented clockwise of the older mineralized faults. Continued dextral shear along the margin and vertical-axis block rotations do not explain the relationships between the fault sets. The faults might be synthetic Riedel shears that emanated from vein-parallel dextral faults. This explanation is consistent with the present data, and it may provide the simplest interpretation.

6. Vertical axis rotations. Sometime between deposition of the gold-quartz veins and ca. 30 Ma, vertical axis rotations occurred in central Prince William Sound near the hinge of the orocline (Haeussler and Nelson, 1993; Nelson et al., 1999). These rotations were probably caused by oroclinal bending in the area and are not documented elsewhere. As previously indicated, ~70° clockwise vertical-axis rotation may have occurred in the Girdwood area. Another area within the accretionary complex where vertical axis rotations have been proposed is in eastern Prince William Sound. Bol and Roeske (1993) argued for clockwise rotations during accretion of Orca Group sedimentary rocks and during tightening of the hinge of the orocline. Finally, Stamatakos et al. (1988) found paleomagnetic evidence for clockwise block rotations of about 50° along the Border Ranges fault in the Matanuska Valley. Rotations may also have affected rocks in the adjacent accretionary complex to the south.

Interpretation of Fault Kinematics and Structures Hosting Gold-Quartz Veins

We conclude there was widespread deformation of the forearc above the slab window at the time of ridge subduction because the gold-quartz veins are linked to ridge subduction by geochronology (Haeussler et al., 1995) and because the veins typically lie in faults distributed over broad regions of the accretionary complex.

All gold-quartz mineralized faults are strike-slip or normal motion. We found no mineralized thrust faults. There is a remarkable consistency of right-lateral strike-slip faults in southeastern Alaska and no dominant orientation of gold-quartz mineralized faults in south-central Alaska. Nonetheless, right-lateral faults at a low, but variable, angle to the regional strike of bedding and cleavage are perhaps most common in south-central Alaska. Despite the general consistency in sense of offset in some areas, considerable variation exists in the kinematics of mineralized faults over distances of a few tens to a hundred kilometers. In Nuka Bay, for example, east-west–striking sinistral faults are most common, in contrast to at Thunder Bay, 25 km to the northeast, where east-west–striking normal-dextral faults are most common. A different orientation and sense of offset is most common in the upper Kenai Peninsula area, 125 km to the northeast, where north-northeast–striking dextral faults are most common. However, these faults could be considered as conjugate to the east-west–striking sinistral faults in Nuka Bay. In Nuka Bay, the kinematics of the gold-quartz mineralized conjugate fault set imply margin-parallel contraction during gold mineralization. There are not enough measured normal-fault orientations with which to draw firm conclusions (Fig. 3D), but most are oriented at a high angle to the regional strike of bedding and cleavage, perhaps allowing strike- or margin-parallel extension.

Bradley and Kusky (1990) and Kusky et al. (1997) described three sets of quartz-mineralized brittle faults along Turnagain Arm, southeast of Anchorage, which together accommodated northwest-southeast–directed contraction. These faults are probably age-equivalent to the gold-quartz mineralized faults because of relative timing relationships and because their structural style is comparable. The kinematics of these fault sets differ from the adjacent upper Kenai Peninsula areas, because along Turnagain Arm the dextral faults are E-W, whereas in the upper Kenai Peninsula areas the dextral faults strike NNE. Thus, although local structural and kinematic patterns can be found, the pattern of orthorhombic faulting in Turnagain Arm cannot be readily extended to other areas. Perhaps the principle of orthorhombic faulting may apply to other areas if enough faults are measured.

The gold-quartz mineralized faults in southeastern Alaska lie along late faults near a major crustal-scale strike-slip fault system—the Border Ranges fault (e.g., Smart et al., 1996), whereas those in south-central Alaska do not. Presumably, ore-forming fluids were channeled more effectively near this large structure, which led to larger deposits than in south-central Alaska. The Chichagof district in southeastern Alaska produced more than 800,000 oz of gold, in contrast to smaller but more widespread veining in south-central Alaska, where none of the mines produced more than 50,000 oz of gold. Gold mineralization is not associated with the Border Ranges fault in south-central Alaska (Fig. 1). This might be due to a smaller component of margin-parallel shear in south-central Alaska during the time of ridge subduction, which resulted in the diffuse network of strike-slip and normal faults.

It is likely most of the gold-quartz veins were deposited within active faults because many faults have crosscutting relationships showing evidence of movement during veining. Thus, the movement on the faults allowed multiple pulses of fluid to migrate upward and be deposited in the faults and nearby joints. Joints are probably subsidiary to the faults because, for example at the Chichagof Mine, there are ribbon quartz veins within the main fault zone and within joints oriented subperpendicular to the main fault zone. Thus, it appears ore-forming fluids entered the joint systems in multiple pulses, in the same way as in the main fault zone. After gold-quartz deposition, faulting without fluid flow formed gouge zones along the margins of most vein systems. These faults are ubiquitously margin-parallel gouge-filled dextral faults, which suggests (1) there was a significant component of simple shear on this margin, and (2) the faults formed at a time the trailing oceanic plate had a significant (≥30°) obliquity of subduction (Jarrard, 1986).

During west-to-east passage of the slab window during ridge subduction, we consider the stress and strain regime in the accretionary complex to be related to at least several factors: (1) the relative motions of the forward and trailing oceanic plates, (2) the slab window (perhaps an absence of relative motion with upwelling asthenospheric mantle beneath the forearc), and (3) the geometry of the plate margin. The fact that the gold-quartz veins cut near-trench intrusions formed during passage of the slab window indicates the gold-quartz veins and their host faults and joints formed either above the slab window or above the trailing plate. The relative motion of the trailing oceanic plate would probably affect some region above the slab window as well.

It is not possible to provide a single kinematic framework for all the gold-quartz mineralized faults, but this may be due, in part, to the geometry of the early Tertiary plate margin and the relative motions of the subducting plates. In southeastern Alaska, dextral strike-slip faults developed at a low angle to the regional structural fabric at the time of gold-quartz mineralization. Jarrard (1986) found that when convergent plate obliquities exceed 30°, trench-parallel strike-slip faults develop. The fact that all gold-quartz mineralization in southeastern Alaska was on dextral faults indicates a strong dextral component of the trailing plate relative motion. However, in south-central Alaska, the gold-quartz mineralized faults are not as tightly clustered in their orientations (see Figs. 3C, 3D). We interpret these characteristics as indicating (1) there was a higher degree of simple shear in southeastern Alaska versus southern Alaska, and (2) the obliquity between the trailing plate and the forearc was not as high as in southeastern Alaska. This may be explained by either a change in the relative motion of the trailing, possibly Kula, plate with respect to the forearc, or there was some pre-existing bend to the south-central Alaska margin, or both. There was a significant change in the Farallon–North America plate motion during anomaly 23R (Atwater and Severinghaus, 1989), which corresponds to ages of 51.75–52.35 Ma (Berggren et al., 1995). At this time, the relative motion of subduction became more dextrally oblique by about 15°–20° with respect to the Alaskan margin. Isotopic ages of near-trench intrusions and the gold-quartz veins indicate the subducted ridge lay between south-central and southeastern Alaska at that time (Bradley et al., 1993 and this volume, Chapter 1; Haeussler et al., 1995). Therefore, it is likely that at least some of the change in the structural style of the gold-quartz mineralized faults is due to this plate motion change. As discussed earlier, the timing of ridge subduction with respect to development of the orocline is uncertain based on paleomagnetic and geochronologic constraints, but structural considerations may favor development ca. 35 Ma (Haeussler and Nelson, 1993; Nelson et al., 1999). Nonetheless, the most recent model for development of the orocline suggests a pre-existing ~45° bend in the Alaskan margin that was tightened later in Tertiary time (Bol and Gibbons, 1992). Our data from the gold-quartz mineralized faults are also consistent with a pre-existing bend in the south-central Alaska margin. We consider the diffuse set of strike-slip and normal faults in south-central Alaska as more likely to have formed where trailing plate convergence vectors are not highly oblique (>30°), whereas in southeastern Alaska there apparently was a high degree of obliquity. Therefore, both the geometry of the ancient Alaskan margin as well as a change in relative plate motions may be responsible for the kinematics and orientation of gold-quartz mineralized faults in different areas.

Interpretation of Dike Orientations

Dikes can also provide information on the direction of extension above the slab window. If a dike is emplaced into a homogenous isotropic medium undergoing extension, the dike will be oriented normal to the least compressive stress direction (σ_3). For a dike to intrude along pre-existing planes of weakness in Earth's crust, magmatic pressure must exceed the normal component of the regional stresses on the fracture plane. We conclude that the regional strike of bedding and cleavage did not strongly influence the orientation of dikes for the following reasons. (1) Away from large plutons, dike swarms show broad regional trends (Fig. 3E). Dike orientations do not appear randomly oriented nor serially correlated with a particular regional strike of bedding or cleavage. (2) Although there are locations where intrusions are parallel to bedding and cleavage, such as in the northern Chugach Mountains (Little and Naeser, 1989), in detail dikes most commonly cut across the pre-existing fabric, commonly at a low angle. The strike of bedding and cleavage is usually consistent within a region and it is remarkable how few dikes are parallel to it. (3) Delaney et al. (1986) found that when dikes are emplaced along pre-existing joints in regions of active tectonism, joints suitably oriented for dike intrusion have a narrow range of orientations, and dikes are likely to be good indicators of principal stress directions at the time of emplacement even when the difference in magnitude between the principal and intermediate stress directions is not large. Finally, we reiterate that the one paleostress orientation that we can be confident of is σ_3, which must be roughly perpendicular to the dike orientation. σ_1 and σ_2 should lie within the plane of the dike, but the orientation of the axes is uncertain.

Although it is simplest to consider dike orientations in their present orientation, it is also useful to evaluate their orientation prior to oroclinal bending. We use the oroclinal bending model of Bol and Gibbons (1992), and we choose to "un"rotate the dike orientations in the Kodiak to Port Wells region 45° clockwise, the Matanuska-Knik area 35° clockwise—because it is closer to the hinge of the orocline—and consider the remaining regions in their in situ orientation (see Fig. 3F).

In southeastern Alaska, most dikes in both the Sitka and Chichagof districts strike northwest, indicating a southwest-northeast component of extension. Because a dike intruded a northwest-striking dextral fault in the Hirst-Chichagof mine, it appears that dextral shear, not pure shear, occurred at the time of dike emplacement. In south-central Alaska, the areas between Hope-Sunrise–Moose Pass and Matanuska-Knik all have northeasterly restored dike strikes, and the Valdez and Tana Glacier area have northerly dike strikes (Fig. 3F). If we consider these areas together, the dike strikes indicate northwest-southeast extension during emplacement of the dikes. The two regionally extensive right-stepping en echelon dikes in the upper Kenai Peninsula area are indicative of a component of margin-parallel dextral shear during intrusion of the dikes. Pavlis and Sisson (1995) concluded there was right-lateral shear soon after intrusion of the Ti_3 generation of dikes in the Tana Glacier area. Therefore, although the inferred timing of deformation in these two areas is slightly different, they both appear related to margin-parallel dextral shear. Nonetheless, in the lower Kenai Peninsula, the dikes are approximately perpendicular to the margin, and thus extension was subparallel to the margin. This extension direction is similar to that inferred prior to ridge subduction based on the subhorizontal quartz mineral fibers. This extension direction is perpendicular to the σ_3 orientation in the upper Kenai Peninsula to Tana Glacier region and could be due to either a flip in σ_1 or σ_2, in addition to the σ_3 orientation.

Overall, differences in dike strikes between different areas cannot be explained by a single kinematic regime, but indicate considerable variability in the strain pattern associated with the slab window and the trailing plate.

A Model for Complex Brittle Deformation at a Migrating Trench-Ridge-Trench Triple Junction

At any convergent margin with an accretionary complex where ridge subduction occurs at a migrating trench-ridge-trench (TRT) triple junction, we envision three configurations in which the forearc can respond (see Fig. 9). Before ridge subduction, Plate A (Fig. 9) influences the shape of the accretionary complex and the processes in the forearc by controlling the basal traction by its relative motion with respect to the forearc and by a characteristic basal friction. After ridge subduction, Plate B (Fig. 9) influences the shape of the accretionary complex for the same reasons. The forearc will have different relative motions to Plate B than Plate A, and this difference will be accommodated in the vicinity of the triple junction.

The effects of the passing slab window also need to be considered. At the base of the forearc, there is no longer an oceanic plate beneath the accretionary complex, and there will be less basal friction than during "normal" subduction. Moreover, because both the forward and trailing oceanic plates are moving apart relative to each other away to the sides of the slab window, the region above the slab window would inevitably experience extension (unless the velocity of the spreading center toward the trench is significantly greater than the spreading rate). In addition, the region above the slab window receives a large thermal pulse. Therefore, in the region above the slab window, the forearc crust will respond to extension caused by the relative motion of Plate A to B, the decreasing influence of the pull of Plate A, the increasing influence of the pull of Plate B, and the thermal pulse and decreased(?) basal friction from the slab window.

This model provides insights into the structural geology of the gold-quartz occurrences and the near-trench dikes in southern Alaska by predicting that the zone above the slab window will have a complex but dominantly extensional or strike-slip

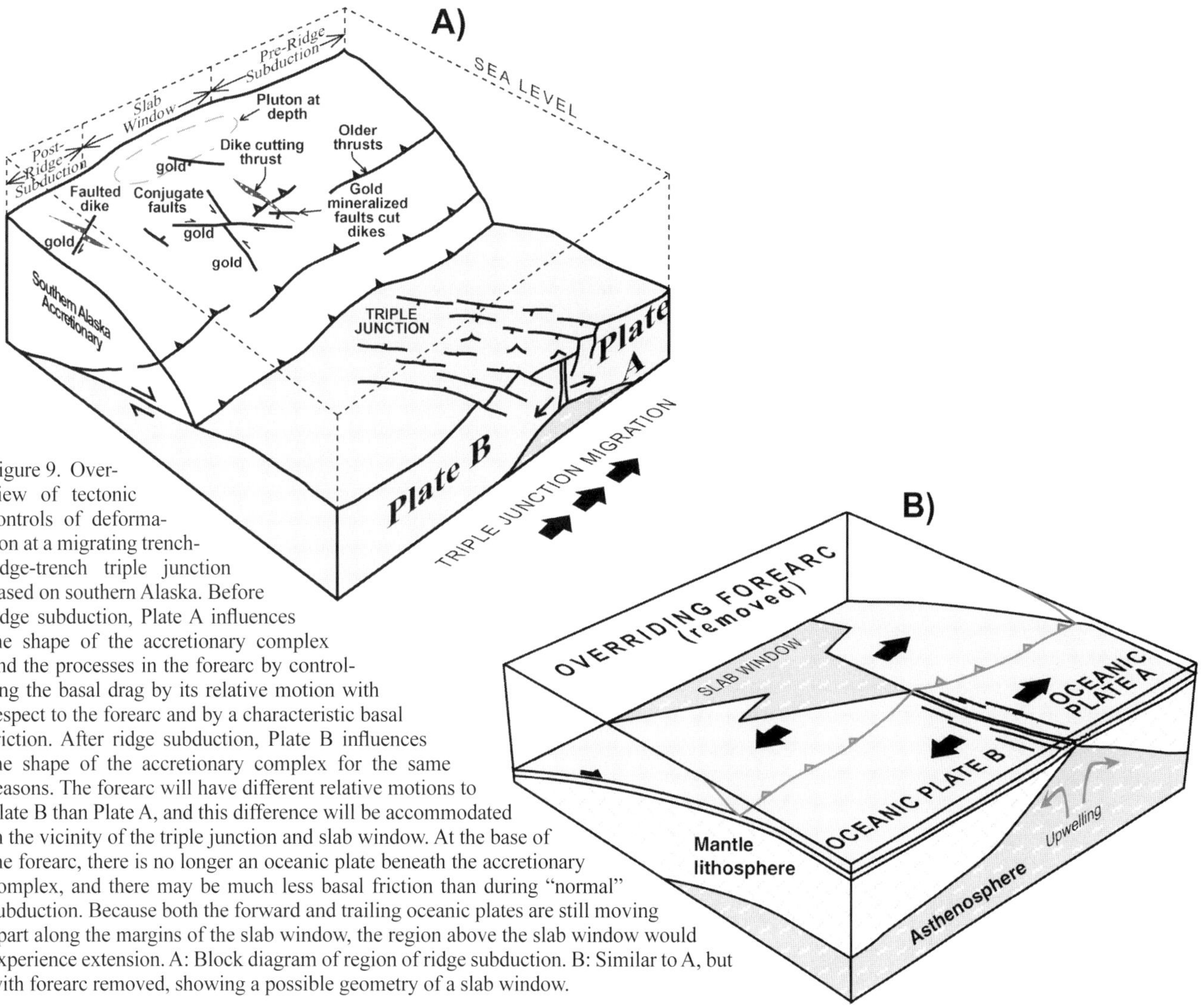

Figure 9. Overview of tectonic controls of deformation at a migrating trench-ridge-trench triple junction based on southern Alaska. Before ridge subduction, Plate A influences the shape of the accretionary complex and the processes in the forearc by controlling the basal drag by its relative motion with respect to the forearc and by a characteristic basal friction. After ridge subduction, Plate B influences the shape of the accretionary complex for the same reasons. The forearc will have different relative motions to Plate B than Plate A, and this difference will be accommodated in the vicinity of the triple junction and slab window. At the base of the forearc, there is no longer an oceanic plate beneath the accretionary complex, and there may be much less basal friction than during "normal" subduction. Because both the forward and trailing oceanic plates are still moving apart along the margins of the slab window, the region above the slab window would experience extension. A: Block diagram of region of ridge subduction. B: Similar to A, but with forearc removed, showing a possible geometry of a slab window.

structural history. First, we suggest that prior to passage of the triple junction, there was a significant margin-parallel component to the relative plate motions, as indicated by the margin-parallel extension in the Nuka Bay, Valdez, and Sitka areas. The margin-parallel extension in the Tana Glacier area at the time of near-trench magmatism indicates this extension continued until the triple junction migrated beneath the region.

The diffuse network of strike-slip and normal faults in south-central Alaska along which there was gold-quartz mineralization would be expected in the region above the slab window. Many of the mineralized faults in the Chichagof, Sitka, Port Valdez, Moose Pass, and Hope-Sunrise areas are dextral faults with orientations at a low angle to the regional strike of bedding and cleavage. They presumably were also at a low angle to the ancient continental margin, which suggests that despite the variability in kinematics between some areas (e.g., lower versus upper Kenai Peninsula), a number of areas of gold-quartz mineralization were affected by the same deformational regime, which had a component of dextral shear.

The younger, gouge-filled dextral faults also record a significant dextral component of the relative motion, consistent with inferred Kula–North America relative plate motions. We infer there must have been a greater dextral component of plate motion after triple junction passage. Pavlis and Sisson (1995), in their study of the Tana Glacier area, also found evidence for dextral-transpressional deformation in the region that would have been above the slab window. Both studies are consistent with inferred Kula–North America relative plate motions and rates, in which the Kula–North America relative velocity is north-northwest oriented and approximately double the Farallon–North America relative motion, which was north-northeasterly directed (Engebretson et al., 1985; Lonsdale, 1988). If it had been the Kula-Farallon triple junction that migrated from west to east beneath southern Alaska, the relative plate motions would have been much higher after passage of the triple junction, with an associated greater component of dextral shear—particularly along the southeastern Alaska margin.

Deformation and magmatism near transform segments would be markedly different from areas where ridge segments were consumed. We can assume that some transform segments were subducted beneath southern Alaska, which would have caused a pause and a jump in the locus of near-trench magmatism, a change in velocity and possibly a change in sense of margin parallel-shear, in addition to deformation caused by topography of the transform segment (Fig. 10). In the Woodlark Basin, where ridge subduction is occurring today, there is nearly 4 km of relief across transform offsets (Taylor and Exon, 1987).

If it was the Kula-Farallon ridge that was being subducted, plate reconstructions indicate most transform segments were right-stepping (e.g., Engebretson et al., 1985). We cannot link any particular sequence of structural events to this process, but along-strike changes in the sense of offset of mineralized faults with the same orientation (such as between Nuka Bay and Thunder Bay) could perhaps be due to this mechanism. Moreover, the effects of subducting a right-stepping transform segment would be markedly different from a left-stepping segment. Poole (1996) proposed subduction of a left-stepping transform segment and a resultant jump in the TRT triple junction to the west to account for slight variations in the trend of intrusive ages in the Sanak-Baranof belt.

Differences in the level of exhumation of the accretionary complex may have caused the variations in the observed structural response to ridge subduction. However, the *P-T* estimates from gold-quartz vein fluid inclusions (1.5 to 3.0 kbar; Goldfarb et al., 1986; Borden et al., 1992; Pickthorn, 1982; Dadoly, 1987) and from the Chugach metamorphic complex (~2.5 kbar; Sisson et al., 1989) are within the same broad range of crustal depths between about 5 and 10 km. Therefore, although they appear to have formed in the upper crust, it is beyond the resolution of the data to determine if there are real differences in amount of uplift between different regions. In the northern Kenai Peninsula, gold-quartz mineralized faults cut gouge-filled fault zones, and, as in all areas, the gold-quartz veins are cut by gouge-filled fault zones. Thus, there were approximately the same deformational conditions (*P* and *T*) both before and after ridge-subduction related gold mineralization.

Comparison with Other Ridge Subduction Events

We see a remarkable similarity in the pattern of seismicity and focal mechanisms between the Woodlark Basin, where ridge subduction is occurring today, and southern Alaska. In the Woodlark basin, away from the region where the ridge is being subducted, focal mechanisms indicate thrust faults parallel to the trench, as would be expected (Cooper and Taylor, 1987). However, in the region of ridge subduction, there are no thrust faults, only strike-slip faults and nearly vertical dip-slip faults. These focal mechanisms are almost identical to our finding of strike-slip and normal, but no reverse, mineralized faults that formed during ridge subduction in southern Alaska.

The Chile triple junction, the other location where ridge subduction is occurring today, may be more analogous to southern Alaska in early Tertiary time because the triple junction is migrating along the continental margin. Forsythe and Nelson (1985) found only strike-slip and normal faults in the region near the triple junction. These faults, like those in southern Alaska, also had different orientations in different regions. Despite the wide range of fault orientations between and within regions, they concluded the bulk of the data indicated regional E-W contraction—perpendicular to the margin. Kusky et al. (1997) also inferred margin perpendicular contraction for syn-ridge subduction faults in the Turnagain Arm area.

Hibbard and Karig (1990) studied probable effects of Miocene subduction of a backarc spreading center in the Shimanto accretionary complex of southwestern Japan. Direct evidence for ridge subduction is limited to a small area (~20 km^2) where they found a pervasive set of brittle faults that cut intrusives related to ridge subduction. They found thrust, strike-slip, and normal faults sets that are relatively homogenous and taken together

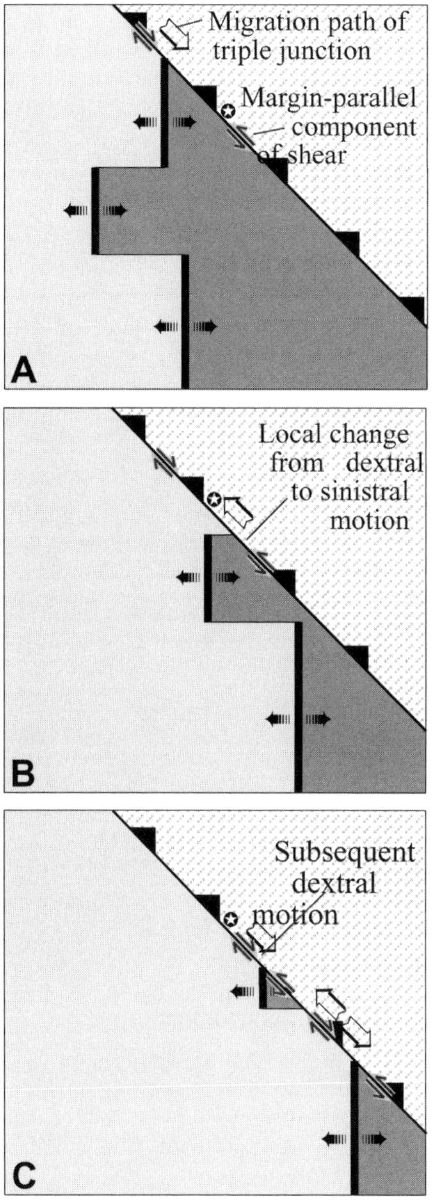

Figure 10. Hypothetical tectonic map showing how subduction of transform segments can cause jumps in the location of the trench-ridge-trench (TRT) triple junction. A: During subduction of a ridge segment, TRT triple junction is migrating to the southwest, and margin-parallel component of shear is opposite on either side of the ridge. B: During subduction of a transform segment, the triple junction becomes a trench-transform-trench triple junction and migrates to the northwest. C: Continued subduction of the transform leads to another ridge segment and and return to a southeast migrating TRT triple junction. Note how at the star (✪) the sense of shear along the continental margin (the overriding plate) changed from right-lateral, to left-lateral, and back to right lateral. This mechanism may explain some of the complexity in the record of brittle faulting in southern Alaska.

indicate contraction roughly perpendicular to the paleotrench. Again, this is similar to the relationships Kusky et al. (1997) found in the Turnagain Arm area of Alaska, with the exception of the presence of thrust faults. In Turnagain Arm, Kusky et al. (1997) concluded that thrust faults were early and were followed by normal and strike-slip faults. Perhaps the same relationship exists in Japan but was not noted.

Therefore, strike-slip and normal faulting above a slab window at the time of ridge subduction may be a common occurrence. If so, these faults may be related to deformation in the region above the slab window. Because both the forward and trailing oceanic plates are still moving apart at the margins of the slab window, we consider it likely the region above the slab window would experience extension. Thus, the brittle faults reflect deformation in the transfer zone above and between the subducting oceanic plates.

CONCLUSIONS

- There was widespread brittle deformation of the southern Alaska accretionary complex associated with Paleocene-Eocene triple junction migration.
- In southeastern Alaska, gold-quartz mineralization during ridge subduction occurred on margin-parallel dextral faults that effectively focussed fluid flow forming the largest deposits, whereas in south-central Alaska, mineralization occurred on a diffuse set of dextral and normal faults, not on reverse faults.
- The structural history of the southern Alaska accretionary complex is remarkably consistent from Kodiak Island to southeastern Alaska, and includes (1) underplating, offscraping, regional metamorphism, and shortening, (2) orogen-parallel extension that occurred prior to, and at the early stages of ridge subduction, (3) near-trench magmatism and locally regional high-T, low-P metamorphism, (4) lode-gold mineralization along strike-slip and/or normal faults, (5) right-lateral brittle faulting without fluid flow on margin-parallel faults, and (6) local vertical-axis rotations.
- Brittle deformation during ridge subduction and triple junction migration is related to the increasing influence of the forward subducting plate, the decreasing influence of the trailing plate, and the thermal pulse and decreased basal friction from the slab window. In addition, extensional deformation of the forearc will be caused by divergence of the subducting oceanic plates at the margins of the slab window.
- Additional factors make a detailed comparison of structures between regions in Alaska difficult. These include the timing and extent of oroclinal bending of southern Alaska, changes in plate motions, and subduction of transform segments with associated jumps in the locus of near-trench magmatism and changes in velocity and possibly sense-of-offset of margin parallel faults. The fundamental difference in the character of the fault sets between south-central and southeastern Alaska indicates there was some bend in the proto-Alaskan margin at the time of ridge subduction.

• Extensional and strike-slip faulting during ridge subduction in southern Alaska appears similar to brittle fault patterns near the Chile triple junction, and to the pattern of shallow earthquake focal mechanisms in the Woodlark basin, where ridge subduction is occurring today.

• Extensional and strike-slip faulting above a slab window may be a common occurrence during ridge subduction, reflecting deformation in a thermally weakened transfer zone above and between the subducting and still diverging oceanic plates.

ACKNOWLEDGMENTS

We thank Tim Kusky, Sue Karl, Alison Till, Larry Snee, Clifford Taylor, and Steve Nelson for stimulating discussions and assistance in field work. Reviews of an early version of the manuscript by Ron Bruhn and Tim Kusky were greatly appreciated. Later reviews by Jim Hibbard, Terry Pavlis, Sarah Roeske, Jinny Sisson, Mike Underwood, and an anonymous reviewer were very constructive. Field work was supported by the Alaska Mineral Resource Assessment Program of the U.S. Geological Survey within the Seldovia and Sitka quadrangles. Support for Haeussler was provided in part by a post-doctoral appointment at the U.S. Geological Survey.

REFERENCES CITED

Atwater, T., 1989, Plate tectonic history of the northeast Pacific and western North America, *in* Winterer, E.L., et al., eds., The eastern Pacific Ocean and Hawaii: Boulder, Colorado, Geological Society of America, Geology of North America, v. N, p. 21–72.

Atwater, T., and Severinghaus, J., 1989, Tectonic map of the Northeast Pacific, *in* Winterer, E.L., et al., eds., The eastern Pacific Ocean and Hawaii: Boulder, Colorado, Geological Society of America, Geology of North America, v. N, 3 sheets, scale 1:5,000,000.

Avé Lallemant, H.G., and Guth, L.R., 1990, Role of extensional tectonics in exhumation of eclogites and blueschists in an oblique subduction setting, northwest Venezuela: Geology, v. 18, p. 950–953.

Barker, F., Farmer, G.L., Ayuso, R.A., Plafker, G., and Lull, J.S., 1992, The 50 Ma granodiorite of the eastern Gulf of Alaska: Melting of the accretionary prism in the forearc: Journal of Geophysical Research, v. 97, p. 6757–6778.

Becker, G.F., 1898, Reconnaissance of the gold fields of southern Alaska: Geological Survey 18th Annual Report, part 3, p. 80–81.

Berggren, W.A., Kent, D.V., Swisher, C.C., III, and Aubry, M.-P., 1995, A revised Cenozoic geochronology and chronostratigraphy, *in* Berggren, W.A., et al., eds., Geochronology, time scales, and global stratigraphic correlation: SEPM (Society for Sedimentary Geology) Special Publication 54, p. 129–212.

Bol, A.J., and Gibbons, H., 1992, Tectonic implications of out-of-sequence faults in an accretionary prism, Prince William Sound, Alaska: Tectonics, v. 11, p. 1288–1300.

Bol, A.J., Coe, R.S., Gromme, C.S., and Hillhouse, J.W., 1992. Paleomagnetism of the Resurrection Peninsula, Alaska: Implications for the tectonics of southern Alaska and the Kula- Farallon ridge: Journal of Geophysical Research, v 97, p. 17,213–17,232.

Bol, A.J., and Roeske, S.M., 1993, Strike-slip faulting and block rotation along the Contact fault system, eastern Prince William Sound, Alaska: Tectonics, v. 12, p. 49–62.

Borden, J.C., Goldfarb, R.J., Gent, C.A., Burruss, R.C., and Roushey, B.H., 1992, Geochemistry of lode-gold deposits, Nuka Bay district, southern Kenai Peninsula: U.S. Geological Survey Bulletin 2041, p. 13–22.

Brabb, E.E., and Miller, D.J., 1962, Reconnaissance traverse across the eastern Chugach Mountains, Alaska, U.S. Geological Survey Miscellaneous Investigations Map I-341, scale 1:96,000.

Bradley, D.C., and Kusky, T.M., 1990, Kinematics of late faults along Turnagain Arm, Mesozoic accretionary complex, south-central Alaska, *in* Dover, J.H., and Galloway, J.P., eds., Geologic Studies in Alaska by the U.S. Geological Survey, 1989: U.S. Geological Survey Bulletin1946, p. 3–10.

Bradley, D.C., and Kusky, T.M., 1992, Deformation history of the McHugh Complex, Seldovia quadrangle, south-central Alaska, *in* Bradley, D.C., and Ford, A.B., eds., Geologic studies in Alaska by the U.S. Geological Survey, 1990: U.S. Geological Survey Bulletin 1999, p. 17–32.

Bradley, D.C., Haeussler, P.J., and Kusky, T.M., 1993, Timing of early Tertiary ridge subduction in southern Alaska, *in* Till, A., and Dusel-Bacon, C., eds., Geologic studies in Alaska by the U.S. Geological Survey, 1992: U.S. Geological Survey Bulletin 2068, p. 163–177.

Bradley, D.C., Kusky, T.M., Haeussler, P.J., Karl, S.M., and Donley, T., 1999, Geologic map of the Seldovia Quardrangle, south-central Alaska: U.S. Geological Survey Open-File Report 99-18, 1 sheet, scale 1:250,000.

Bradley, D.C., Parrish, R., Clendenen, W., Lux, D., Layer, P., Heizler, M., and Donley, T.D., 2000, New geochronological evidence for the timing of early Tertiary ridge subduction in southern Alaska, *in* Kelley, K., and Gough, L., eds., Geologic studies in Alaska by the U.S. Geological Survey, 1998: U.S. Geological Survey Professional Paper 1615, p. 5–21.

Bradley, D., Kusky, T., Haeussler, P., Goldfarb, R., Miller, M., Dumoulin, J., Nelson, S.W., and Karl, S., 2003, Geologic signature of early Tertiary ridge subduction in Alaska *in* Sisson, V.B., Roeske, S.M., and Pavlis, T.L., eds., Geology of a transpressional orogen developed during ridge-trench interaction along the North Pacific margin: Boulder, Colorado, Geological Society of America Special Paper 371, p. 19–49 (this volume).

Brew, D.A., Hammarstrom, J.M., Himmelberg, G.R., Wooden, J.L., Loney, R.A., and Karl, S.M., 1991, Crawfish Inlet pluton, Baranof Island, southeastern Alaska—A north-tilted Eocene body or an untilted enigma?: Geological Society of America Abstracts with Programs, v. 23, no. 2, p. 8.

Burns, L.E., Pessel, G.H., Little, T.A., Pavlis, T.L., Newberry, R.J., Winkler, G.R., and Decker, J., 1991, Geology of the northern Chugach Mountains, south-central Alaska: Alaska Division of Geological and Geophysical Surveys Professional Report 94, 63 p.

Byrne, T., 1984, Structural geology of melange terranes in the Ghost Rocks Formation, *in* Raymond, L., ed., Melanges: Their origin and significance: Boulder, Colorado, Geological Society of America Special Paper 198, p. 21–51.

Byrne, T., 1986, Eocene underplating along the Kodiak Shelf, Alaska: Implications and regional correlations: Tectonics, v. 5, p. 403–421.

Capps, S.R., 1916, The Turnagain-Knik region: U.S. Geological Survey Bulletin 642, p. 147–194.

Carey, S.W., 1955, The orocline hypothesis in geotectonics: Proceedings of the Royal Society of Tasmania, v. 89, p. 255–288.

Clark, S.H.B., 1973, The McHugh Complex of south-central Alaska: U.S. Geological Survey Bulletin 1372-D, p. 1–11.

Clendenen, W.S., 1991, Thermal history and vertical tectonics of the southern Alaska convergent margin [Ph.D. thesis]: Providence, Rhode Island, Brown University, 177 p.

Coe, R.S., Globerman, B.R., Plumley, P.W., and Thrupp, G., 1985, Paleomagnetic results from Alaska and their tectonic implications, *in* Howell, D.G., ed., Tectonostratigraphic terranes of the circum-Pacific region: Houston, Texas, Circum-Pacific Council for Energy and Mineral Resources, Earth Science Series, v. 1, p. 85–108.

Coe, R.S., Globerman, B.R., Plumley, P.R., and Thrupp, G.A., 1989, Rotation of central and southern Alaska in the Early Tertiary: Oroclinal bending by megakinking?, *in* Kissel, C., and Laj, C., eds., Paleomagnetic rotations and continental deformation: Boston, Kluwer Academic Publishers, NATO-ASI series, p. 327–339.

Connelly, W., 1978, Uyak Complex, Kodiak Islands, Alaska: A Cretaceous subduction complex: Geological Society of America Bulletin, v. 89, p. 755–769.

Cooper, P., and Taylor, B., 1987, The spatial distribution of earthquakes, focal mechanisms, and subducted lithosphere in the Solomon Islands, *in* Taylor, B., and Exon, N.F., eds., Marine geology, geophysics, and geochemistry of the Woodlark Basin–Solomon Islands: Houston, Texas, Circum-Pacific Council for Energy and Mineral Resources, Earth Science Series, v. 7, p. 67–88.

Csejtey, B., Jr., 1992, Discrepancies between geologic evidence and rotational models—Talkeetna Mountains and adjacent areas of south-central Alaska, *in* Bradley, D.C., and Ford, A.B., eds., Geologic studies in Alaska by the U.S. Geological Survey, 1990: U.S. Geological Survey Bulletin 1999, p. 71–80.

Dadoly, J.P., 1987, Gold mineralization in a regional metamorphic terrane: A wall-rock alteration study of the Chichagof and Hirst-Chichagof gold mines, southeast Alaska [M.S. thesis]: Rapid City, South Dakota, South Dakota School of Mines and Technology, 100 p.

Davis, D., Suppe, J., and Dahlen, F.A., 1983, Mechanics of fold-and-thrust belts and accretionary wedges: Journal of Geophysical Research, v. 88, p. 1153–1172.

Davis, S.J., Roeske, S.M., and Karl, S.M., 1998, Late Cretaceous to Early Tertiary transtension and strain partitioning in the Chugach accretionary complex, SE Alaska: Journal of Structural Geology, v. 20, p. 639–654.

Decker, J.E., 1980, Geology of a Cretaceous subduction complex, western Chichagof Island, southeastern Alaska [Ph.D. thesis]: Stanford, California, Stanford University, 134 p.

Delaney, P.T., Pollard, D.D., Ziony, J.I., and McKee, E.H., 1986, Field relations between dikes and joints; emplacement processes and paleostress analysis: Journal of Geophysical Research, v. 91, p. 4920–4938.

Delong, S.E., and Fox, P.J., 1977, Geological consequences of ridge subduction, *in* Talwani, M., and Pitman, W.C., III, eds., Island arcs, deep sea trenches, and backarc basins: American Geophysical Union, Maurice Ewing Series, v. 1, p. 221–228.

Dickinson, W.R., and Snyder, W.S., 1979, Geometry of subducted slabs related to San Andreas transform: Journal of Geology, v. 87, p. 609–627.

Dumoulin, J.A., 1987, Sandstone composition of the Valdez and Orca Groups, Prince William Sound, Alaska: U.S. Geological Survey Bulletin 1774, 37 p.

Engebretson, D.C., Cox, A.V., and Gordon, R.G., 1985, Relative motions between oceanic and continental plates in the Pacific Basin: Boulder, Colorado, Geological Society of America Special Paper 206, 59 p.

Farmer, G.L., Ayuso, R., and Plafker, G., 1993, A Coast Mountains provenance for the Valdez and Orca groups, southern Alaska, based on Nd, Sr, and Pb isotopic evidence: Earth and Planetary Science Letters, v. 116, p. 9–21.

Fisher, D.M., and Brantley, S.L., 1992, Models of quartz overgrowth and vein formation; deformation and episodic fluid flow in an ancient subduction zone: Journal of Geophysical Research, v. 97, p. 20,043–20,061.

Fisher, D.M., and Byrne, T., 1987, Structural evolution of underthrusted sediments, Kodiak Islands, Alaska: Tectonics, v. 6, p. 775–793.

Fisher, D.M., and Byrne, T., 1990, The character and distribution of mineralized fractures in the Kodiak Formation, Alaska: Implications for fluid flow in an underthrust sequence: Journal of Geophysical Research, v. 95, p. 9069–9080.

Forsythe, R., and Nelson, E., 1985, Geological manifestations of ridge collision: Evidence from the Golfo de Penas-Taitao basin, southern Chile: Tectonics, v. 4., p. 477–495.

Goldfarb, R.J., Leach, D.L., Miller, M.L., and Pickthorn, W.I., 1986, Geology, metamorphic setting, and genetic constraints of epigenetic lode-gold mineralization within the Cretaceous Valdez Group, south-central Alaska, *in* Keppie, J.D., et al., eds., Turbidite-hosted gold deposits: Geological Association of Canada Special Paper 32, p. 87–105.

Grantz, A., 1966, Strike-slip faults in Alaska: U.S. Geological Survey Open-File Report 66-53, 82 p.

Groome, W.G., Thorkelson, D.J., Friedman, R.M., Mortensen, J.K., Massey, N.W.D., Marshall, D.D., and Layer, P.W., 2003, Magmatic and tectonic history of the Leech River Complex, Vancouver Island, British Columbia: Evidence for ridge-trench intersection and accretion of the Crescent Terrane, *in* Sisson, V.B., Roeske, S.M., and Pavlis, T.L., eds., Geology of a transpressional orogen developed during ridge-trench interaction along the North Pacific margin: Boulder, Colorado, Geological Society of America Special Paper 371, p. 327–353 (this volume).

Haeussler, P.J., and Bradley, D.C., 1993, Map and compilation of structural data from lode-gold mineral occurrences in the Chugach–Prince William terrane of southern Alaska: U.S. Geological Survey Open-File Report 93-325, 53 p. plus 1 sheet, scale 1:500,000.

Haeussler, P.J., and Nelson, S.R., 1993, Structural evolution of the Chugach–Prince William terrane at the hinge of the orocline in Prince William Sound, and implications for ore deposits, *in* Till, A., and Dusel-Bacon, C., eds., Geologic studies in Alaska by the U.S. Geological Survey, 1992: U.S. Geological Survey Bulletin 2068, p. 143–162.

Haeussler, P.J., Bradley, D.C., Goldfarb, R.J., Snee, L.W., and Taylor, C., Jr., 1995, Link between ridge subduction and gold mineralization in southern Alaska: Geology, v. 23, p. 995–998.

Harris, N.R., Sisson, V.B., Wright, J.E., and Pavlis, T.L., 1996, Evidence for Eocene mafic underplating during forearc intrusive activity, eastern Chugach Mountains, Alaska: Geology, v. 24, p. 263–266.

Helwig, J., and Emmet, P., 1981, Structure of the early Tertiary Orca Group in Prince William Sound and some implications for the plate tectonic history of southern Alaska: Journal of the Alaska Geological Society, v. 1, p. 12–35.

Hibbard, J.P., and Karig, D.E., 1990, Structural and magmatic responses to spreading ridge subduction: An example from southwest Japan: Tectonics, v. 9, p. 201–230.

Hill, M., Morris, J., and Whelan, J., 1981, Hybrid granodiorites intruding the accretionary prism, Kodiak, Shumagin, and Sanak Islands, southwest Alaska: Journal of Geophysical Research, v. 86, p. 10,569–10,590.

Hoekzema, R.B., Fechner, S.A., and Kurtak, J.M., 1987, Evaluation of selected lode gold deposits in the Chugach National Forest, Alaska: U.S. Bureau of Mines Information Circular 9113, 62 p.

Hudson, T., and Plafker, G., 1982, Paleogene metamorphism of an accretionary flysch terrane, eastern Gulf of Alaska: Geological Society of America Bulletin, v. 93, p. 1280–1290.

Hudson, T., Plafker, G., and Peterman, Z.E., 1979, Paleogene anatexis along the Gulf of Alaska margin: Geology, v. 7, p. 573–577.

Jarrard, R.D., 1986, Relations among subduction parameters: Reviews of Geophysics, v. 24, p. 217–284.

Johnson, B.R., and Karl, S.M., 1985, Geologic map of eastern Chichagof and Yakobi Islands, southeastern Alaska: U.S. Geological Survey Miscellaneous Investigations Map I-1506, scale 1:125,000, 1 sheet, 15 p.

Knopf, A., 1912, The Sitka mining district, Alaska: U.S. Geological Survey Bulletin, Report: B504, 32 p.

Kusky, T.M., Bradley, D.C., Haeussler, P.J., Karl, S.M., and Donley, D.T., 1993, The Chugach Bay thrust, a major tectonic boundary in the southern Alaska accretionary prism: Geological Society of America Abstracts with Programs, v. 25, no. 6, p. A-282.

Kusky, T.M., Bradley, D.C., and Haeussler, P.J., 1997, Progressive deformation of the Chugach accretionary complex, Alaska, during a Paleogene ridge-trench encounter: Journal of Structural Geology, v. 19, p. 139–157.

Kusky, T.M., Bradley, D., Donley, D.T., Rowley, D., and Haeussler, P., 2003, Controls on intrusion of near-trench magmas of the Sanak-Baranof Belt, Alaska, during Paleogene ridge subduction, and consequences for forearc evolution, *in* Sisson, V.B., Roeske, S.M., and Pavlis, T.L., eds., Geology of a transpressional orogen developed during ridge-trench interaction along the North Pacific margin: Boulder, Colorado, Geological Society of America Special Paper 371, p. 269–292 (this volume).

Kveton, K.J., 1989, Structure, thermochronology, provenance, and tectonic history of the Orca group in southwestern Prince William Sound, Alaska [Ph.D. thesis]: Seattle, University of Washington, 173 p.

Little, T.A., and Naeser, C.W., 1989, Tertiary tectonics of the Border Ranges fault system, Chugach Mountains, Alaska: Deformation and uplift in a forearc setting: Journal of Geophysical Research, v. 94, p. 4333–4359.

Loney, R.A., and Brew, D.A., 1987, Regional thermal metamorphism and deformation of the Sitka Graywacke, southern Baranof Island, southeastern Alaska: U.S. Geological Survey Bulletin 1779, 17 p.

Loney, R.A., Brew, D.A., Muffler, L.P.J., and Pomeroy, J.S., 1975, Reconnaissance geology of Chichagof, Baranof, and Kruzof Islands, southeastern Alaska: U.S. Geological Survey Professional Paper 792, 105 p.

Lonsdale, P., 1988, Paleogene history of the Kula plate: Offshore evidence and onshore implications: Geological Society of America Bulletin, v. 100, p. 733–754.

Marshak, R.S., and Karig, D.E., 1977, Triple junctions as a cause for anomalously near-trench igneous activity between the trench and volcanic arc: Geology, v. 5, p. 233–236.

Marty, K., 1994, Finite strain analysis of the Chugach accretionary complex: Implications for tectonic processes, Richardson Highway transect, Alaska [M.S. thesis]: New Orleans, Louisiana, University of New Orleans, 88 p.

McCaffrey, R., 1996, Estimates of modern arc-parallel strain rates in forearcs: Geology, v. 24, p. 27–30.

Meyers, G., and Vrolijk, P., 1986, Fluid evolution associated with the accretion of the Kodiak Formation, Kodiak Islands, Alaska: Eos (Transactions, American Geophysical Union), v. 67, p. 1219.

Mitchell, P.A., 1979, Geology of the Hope-Sunrise (gold) mining district, north-central Kenai Peninsula, Alaska [M.S. thesis]: Palo Alto, California, Stanford University, 122 p.

Moore, G.W., 1969, New formations on Kodiak and adjacent islands, Alaska, *in* Cohee, G.V., et al., compilers, Changes in stratigraphic nomenclature by the U.S. Geological Survey, 1967: U.S. Geological Survey Bulletin 1274A, p. A27–A35.

Moore, J.C., 1973, Cretaceous continental margin sedimentation, southwestern Alaska: Geological Society of America Bulletin, v. 84, p. 595–614.

Moore, J.C., Byrne, T., Plumley, P.W., Reid, M., Gibbons, H., and Coe, R.S., 1983, Paleogene evolution of the Kodiak Islands, Alaska—Consequences of ridge-trench interaction in a more southerly latitude: Tectonics, v. 2, p. 265–293.

Nelson, S.W., and Nelson, M.S., 1993, Geochemistry of ophiolitic rocks from Knight Island, Prince William Sound, Alaska, *in* Till, A., and Dusel-Bacon, C., eds., Geologic studies in Alaska by the U.S. Geological Survey, 1992: U.S. Geological Survey Bulletin 2068, p. 130–142.

Nelson, S.W., Dumoulin, J.A., and Miller, M.L., 1985, Geologic map of the Chugach National Forest, Alaska: U.S. Geological Survey Miscellaneous Field Studies Map MF-1645-B, 16 p., 1 plate, scale 1:250,000.

Nelson, S.W., Miller, M.L., and Dumoulin, J.A., 1989, Resurrection Peninsula ophiolite, *in* Nelson, S.W., and Hamilton, T.W., eds., Guide to the geology of Resurrection Bay, eastern Kenai Fjords area, Alaska: Anchorage, Geological Society of Alaska, p. 10–20.

Nelson, S.W., Miller, M.L., Haeussler, P.J., Snee, L.W., Phillips, P.J., and Huber, C., 1999, Preliminary geologic map of the Chugach National Forest Special Study Area, Alaska: U.S. Geological Survey Open-File Report 99-362, scale 1:63,360, http://wrgis.wr.usgs.gov/open-file/of99-362/ (March 2003).

Nilsen, T.H., and Moore, G.W., 1979, Reconnaissance study of Upper Cretaceous to Miocene stratigraphic units and sedimentary facies, Kodiak and adjacent islands, Alaska: U.S. Geological Survey Professional Paper 1093, 34 p.

Nilsen, T.H., and Zuffa, G.G., 1982, The Chugach terrane, a Cretaceous trench-fill deposit, southern Alaska, p. 213–227, *in* Legget, J.K., ed., Trench-forearc geology—Sedimentation and tectonics on modern and ancient active plate margins: London, Blackwell, 582 p.

Onstott, T.C., Sisson, V.B., and Turner, D.L., 1989, Initial argon in amphiboles from the Chugach Mountains, southern Alaska: Journal of Geophysical Research, v. 94, p. 4361–4372.

Park, C.F., Jr., 1933, The Girdwood District, Alaska: U.S. Geological Survey Bulletin 849-G, p. 381–424.

Pavlis, T.L., and Sisson, V.B., 1995, Structural history of the Chugach metamorphic complex in the Tana River region, eastern Alaska: A record of Eocene ridge subduction: Geological Society of America Bulletin, v. 107, p. 1333–1355.

Pavlis, T.L., Marty, K., and Sisson, V.B., 2003, Constrictional flow within the Eocene forearc of southern Alaska: An effect of dextral shear during ridge subduction, *in* Sisson, V.B., Roeske, S.M., and Pavlis, T.L., eds., Geology of a transpressional orogen developed during ridge-trench interaction along the North Pacific margin: Boulder, Colorado, Geological Society of America Special Paper 371, p. 171–190 (this volume).

Pickthorn, W.J., 1982, Stable isotope and fluid inclusion study of the Port Valdez gold district, southern Alaska [M.S. thesis]: Los Angeles, California, University of California, 66 p.

Plafker, G., Lull, J.S., Nokleberg, W.J., Pessel, G.A., Wallace, W.K., and Winkler, G.R., 1992, Geologic map of the Valdez A-4, B-3, B-4, C-3, C-4 and D-4 quadrangles, northern Chugach Mountains and southern Copper River basin, Alaska: U.S. Geological Survey Miscellaneous Investigations Series Map I-2164, 1 sheet, scale 1:125,000.

Plafker, G., Moore, J.C., and Winkler, G.R., 1994, Geology of the southern Alaska margin, *in* Plafker, G., and Berg, H.C., eds., The geology of Alaska: Boulder, Colorado, Geological Society of America, Geology of North America, v. G-1, p. 389–449.

Plumley, P.W., Coe, R.S., and Byrne, T., 1983, Paleomagnetism of the Paleocene Ghost Rocks Formation, Prince William Terrane, Alaska: Tectonics, v. 2, p. 295–314.

Poole, A., 1996, Age and geochemical constraints on ridge subduction for igneous rocks of the eastern Chugach Mountains, Alaska [M.A. thesis]: Houston, Texas, Rice University, 83 p.

Reed, J.C., and Coats, R.R., 1941, Geology and ore deposits of the Chichagof mining district Alaska: U.S. Geological Survey Bulletin 929, 148 p.

Richter, D.H., 1970, Geology and lode-gold deposits of the Nuka Bay area, Kenai Peninsula, Alaska: U.S. Geological Survey Professional Paper 625-B, 16 p.

Roeske, S.M., Pavlis, T.L., Snee, L.W., and Sisson, V.B., 1992, $^{40}Ar/^{39}Ar$ isotopic ages from the combined Wrangellia-Alexander terrane along the Border Ranges fault system in the eastern Chugach Mountains and Glacier Bay, Alaska, *in* Bradley, D.C., and Ford, A.B., eds., Geologic studies in Alaska by the U.S. Geological Survey, 1990: U.S. Geological Survey Bulletin 1999, p. 180–195.

Sample, J.C., and Moore, J.C., 1987, Structural style and kinematics of an underplated slate belt, Kodiak and adjacent islands, Alaska: Geological Society of America Bulletin, v. 99, p. 7–20.

Sisson, V.B., Hollister, L.S., and Onstott, T.C., 1989, Petrologic and age constraints on the origin of a low-pressure/high-temperature metamorphic complex, southern Alaska: Journal of Geophysical Research, v. 94, p. 4392–4410.

Sisson, V.B., Poole, A.R., Harris, N.R., Cooper Burner, H., Pavlis, T.L., Copeland, P., Donelick, R.A., and McLelland, W.C., 2003, Geochemical and geochronologic constraints for genesis of a tonalite-trondhjemite suite and associated mafic intrusive rocks in the eastern Chugach Mountains, Alaska: A record of ridge-transform subduction, *in* Sisson, V.B., Roeske, S.M., and Pavlis, T.L., eds., Geology of a transpressional orogen developed during ridge-trench interaction along the North Pacific margin: Boulder, Colorado, Geological Society of America Special Paper 371, p. 293–326 (this volume).

Smart, K.J., Pavlis, T.L., Sisson, V.B., Roeske, S.M., and Snee, L.W., 1996, The Border Ranges fault system in Glacier Bay National Park, Alaska; evidence for major early Cenozoic dextral strike-slip motion: Canadian Journal of Earth Sciences, v. 33, p. 1268–1282.

Stamatakos, J.A., Kodama, K.P., and Pavlis, T.L., 1988, Paleomagnetism of Eocene plutonic rocks, Matanuska Valley, Alaska: Geology, v. 16, p. 618–622.

Stüwe, K., 1984, Granitoid intrusions and gold-mineralization in the western Port Wells mining district, Prince William Sound Alaska [M.S. thesis]: Leoben, Austria, Mining University of Leoben, 188 p.

Stüwe, K., 1986, Structural evolution of the Port Wells gold mining district, Prince William Sound, south-central Alaska: Implications for the origin of the gold lodes: Mineralium Deposita, v. 21, p. 288–295.

Taylor, B., and Exon, N.F., 1987, An investigation of ridge subduction in the Woodlark-Solomons region: Introduction and overview, *in* Taylor, B., and Exon, N.F., eds., Marine geology, geophysics, and geochemistry of the Woodlark Basin-Solomon Islands: Houston, Texas, Circum-Pacific Council for Energy and Mineral Resources, Earth Science Series, v. 7, p. 1–24.

Taylor, C.D., Goldfarb, R.J., Snee, L.W., Gent, C.A., Karl, S.M., and Haeussler, P.J., 1994, New age data for gold deposits and granites, Chichagof mining district, SE Alaska: Evidence for a common origin: Geological Society of America Abstracts with Programs, v. 26, no. 7, p. A-140.

Thorkelson, D.J., 1996, Subduction of diverging plates and the principles of slab window formation: Tectonophysics, v. 255, p. 47–63.

Tysdal, R.G., and Case, J.E., 1979, Geologic map of the Seward and Blying Sound quadrangles, Alaska: U.S. Geological Survey Miscellaneous Geologic Investigation Series Map I-1150, 12 p., 1 plate, scale 1:250,000.

Vrolijk, P., 1987, Tectonically driven fluid flow in the Kodiak accretionary complex, Alaska: Geology, v. 15, p. 466–469.

Wallace, W.K., and Engebretson, D.C., 1984, Relationships between plate motions and Late Cretaceous to Paleogene magmatism in southwestern Alaska: Tectonics, v. 3, p. 293–315.

Winkler, G.R., 1976, Deep-sea fan deposition of the lower Tertiary Orca Group, eastern Prince William Sound, Alaska, *in* Miller, T.P., ed., Recent and ancient sedimentary environments in Alaska: Anchorage, Alaska Geological Society Symposium Proceedings, p. R1–R20.

Winkler, G.R., 1992, Geologic map and summary geochronology of the Anchorage 1° × 3° quadrangle, southern Alaska: U.S. Geological Survey Miscellaneous Investigations Map I-2283, scale 1:250,000, 1 sheet.

Winkler, G.R., and Plafker, G., 1993, Geologic map of the Cordova and Middleton Island quadrangles, southern Alaska: U.S. Geological Survey Miscellaneous Investigations Map I-1984, scale 1:250,000, 1 sheet.

Wright, F.E., and Wright, C.W., 1905, Economic developments in southeastern Alaska: U.S. Geological Survey Bulletin, Report B259, p. 47–68.

Zumsteg, C.L., Himmelberg, G.R., Karl, S.M., and Haeussler, P.J., 2003, Metamorphism within the Chugach accretionary complex on southern Baranof Island, southeastern Alaska, *in* Sisson, V.B., Roeske, S.M., and Pavlis, T.L., eds., Geology of a transpressional orogen developed during ridge-trench interaction along the North Pacific margin: Boulder, Colorado, Geological Society of America Special Paper 371, p. 253–267 (this volume).

Manuscript Accepted by the Society February 5, 2003

Printed in the USA

Geological Society of America
Special Paper 371
2003

Dextral-slip reactivation of an arc-forearc boundary during Late Cretaceous–Early Eocene oblique convergence in the northern Cordillera

Sarah M. Roeske*
Geology Department, University of California, One Shields Avenue, Davis, California 95616, USA

Lawrence W. Snee
U.S. Geological Survey, MS 974, Denver Federal Center, Denver, Colorado 80225, USA

Terry L. Pavlis
Department of Geology and Geophysics, University of New Orleans, New Orleans, Louisiana 70148, USA

ABSTRACT

The Border Ranges fault system forms the most fundamental crustal boundary in southern Alaska, separating crystalline basement with Paleozoic and Mesozoic arc sequences, the Wrangellia terrane, from a long-lived accretionary complex, the Chugach terrane. Although the Border Ranges fault system originated as a subduction zone megathrust, few strands of it preserve that original history. This study shows that a dextral fault zone, herein named the Hanagita fault system, has overprinted the subduction zone megathrust. This dextral fault zone is 5–10 km wide and has been traced continuously for ~250 km. The Hanagita fault system is probably continuous with dextral slip systems described in Glacier Bay and Baranof Island in southeast Alaska, which would extend its length to ~700 km. Within this fault zone, elements of Wrangellia and the Chugach terrane are complexly intermixed.

Geologic mapping and $^{40}Ar/^{39}Ar$ geochronology of white mica crystallized during dextral slip indicate the strike-slip faulting may have begun by ~85 Ma and was continuous from ~70 Ma until ~51 Ma. Accretion of a thick flysch sequence along this margin began in the Maastrichtian (latest Cretaceous) and continued uninterrupted through middle Eocene. Migration of a ridge-trench-trench triple junction along this margin during the late Paleocene–early Eocene introduced hot fluids to the fault zone and plutons on the southern edge of the fault system. Dextral slip occurred before, during, and after the triple junction migration. Thus, the Hanagita fault system records long-lived oblique convergence between two separate oceanic plates and North America and is a relatively well-preserved example of displacement partitioning along the backstop of an accretionary prism.

The Hanagita fault system includes several through-going dextral high-angle faults that are continuous on the scale of 10s to 100+ km. These strike WNW, subparallel to the structural grain in the accretionary prism to the south. Between these strands are numerous shear zones and faults that are continuous on the scale of several 100 m to several km. The majority of these also show subhorizontal to oblique slip, but the dip of the fault planes is more variable than the through-going faults. The western part of the fault system has north-dipping thrust and oblique slip faults that formed approximately synchronously with south-dipping oblique slip faults 100 km to the east. The faults formed at

*roeske@geology.ucdavis.edu

Roeske, S.M., Snee, L.W., and Pavlis, T.L., 2003, Dextral-slip reactivation of an arc-forearc boundary during Late Cretaceous–Early Eocene oblique convergence in the northern Cordillera, *in* Sisson, V.B., Roeske, S.M., and Pavlis, T.L., eds., Geology of a transpressional orogen developed during ridge-trench interaction along the North Pacific margin: Boulder, Colorado, Geological Society of America Special Paper 371, p. 141–169.

progressively lower temperatures through time based on cross-cutting relations and $^{40}Ar/^{39}Ar$ white mica dates of fault fabrics. The oldest shear zones formed at low greenschist facies, and the youngest structures are discrete brittle faults and gouge zones.

Rocks within the fault system include crystalline rocks from the inboard side of the fault system and mélange of the McHugh Complex, the oldest part of the accretionary prism in this area. The crystalline rocks include abundant Late Jurassic diorite (U/Pb zircon date) with $^{40}Ar/^{39}Ar$ hornblende cooling ages of 142–146 Ma, which correlates with the Chitina batholith of the Wrangell Mountains and the St. Elias Plutonic Suite of the St. Elias range. Other crystalline rocks include an ~170 Ma diorite (hornblende $^{40}Ar/^{39}Ar$) associated with abundant metabasite and leucocratic plutonic rock. No known correlative to these units occurs for a minimum of 600 km to the southeast along the Border Ranges fault system. Thus, the total amount of dextral displacement along the Hanagita fault system from the Late Cretaceous to Middle Eocene is conservatively estimated at 600 km but could be >1000 km, based on correlation of the 170 Ma pluton and related rocks to a similar suite on the west side of Vancouver Island. This displacement is also consistent with regional geology elsewhere in the accretionary prism. The timing of displacement on the fault system coincides with dextral slip on several other major faults, parallel but more inboard from the margin, including the Denali and Tintina fault systems and the Coast Shear zone. Movement on these faults is commonly attributed to highly oblique convergence between the Kula and North American plates, which would require that the Kula-Farallon triple junction be located south of these faults. Further study of the timing and magnitude of displacement on all of these faults will help resolve questions of northern Cordillera plate reconstructions and will improve the ability to recognize accommodation of major translation along ancient plate margins.

Keywords: oblique convergence, dextral strike-slip, Chugach accretionary complex, Late Cretaceous, southern Alaska.

INTRODUCTION

Studies of modern oblique convergent margins show that a majority of the translational component of the plate motion is taken up by strike-slip faults within the overriding plate, a process often referred to as strain or displacement partitioning. The location of the main zones of strike slip appear to be controlled by sites with major contrasts in strength, most typically in the arc and at the backstop of the accretionary prism. The classic example of Sumatra (Fitch, 1972) shows the strike-slip component concentrated within the arc (Sieh et al., 1994; Prawirodirdjo et al., 1997) but also splaying into the forearc, continuing at least as far as the outer-arc high (Karig et al., 1978). The Hikurangi margin of New Zealand has strike-slip fault motion occurring over two major fault strands, one along the backstop and one along the outer part of a forearc basin (Cashman et al., 1992). The location of the dominant strike-slip faults along the Aleutian margin is not well constrained but 30%–50% of the translation component of the relative plate motion is occurring outboard of the main axis of the arc (Avé Lallemant and Oldow, 2000). Numerical solutions for oblique convergent margins indicate the strike-slip displacement will be greatly concentrated along the backstop, in contrast to the rest of the prism, with the exact distribution of the displacement controlled by the bulk rheology of the accretionary prism (Platt, 2000). Thus, if one were to look for significant strike-slip displacement along an ancient convergent margin, the arc and the backstop are the most likely sites.

Late Cretaceous–Early Eocene plate models for the Cordillera margin from ~40° N latitude to Alaska show consistently northeast-directed oblique convergence between the oceanic plates and the North American plate. The magnitude of obliquity varies greatly with time and nature of the down-going plate, however. The Kula plate had a significantly higher velocity and a higher degree of obliquity than the more southerly Farallon plate, especially in the time period between 71 and 57 Ma (Engebretson et al., 1985; Lonsdale, 1988; Stock and Molnar, 1988). Many authors have invoked this episode of oblique convergence of the Kula plate with respect to North America to explain northward displacement of terranes along the North American margin (Beck, 1983; Umhoefer, 1987; Cowan, 1994; Hollister and Andronicos, 1997; and many others) but the location of the Kula–Farallon–North America triple junction has been poorly constrained. Smart et al. (1996) identified the Border Ranges fault system (Fig. 1A), the backstop of the accretionary prism, as a potential major dextral-slip fault active in the Late Cretaceous to early Tertiary. This study provides further evidence for a long-lived, complex dextral fault system along the trace of the Border Ranges fault system and shows that the displacement occurred both before and after passage of a ridge-trench-trench triple junction.

REGIONAL GEOLOGY

The Border Ranges fault system separates crystalline rocks of the Wrangellia composite terrane from a Mesozoic to recent

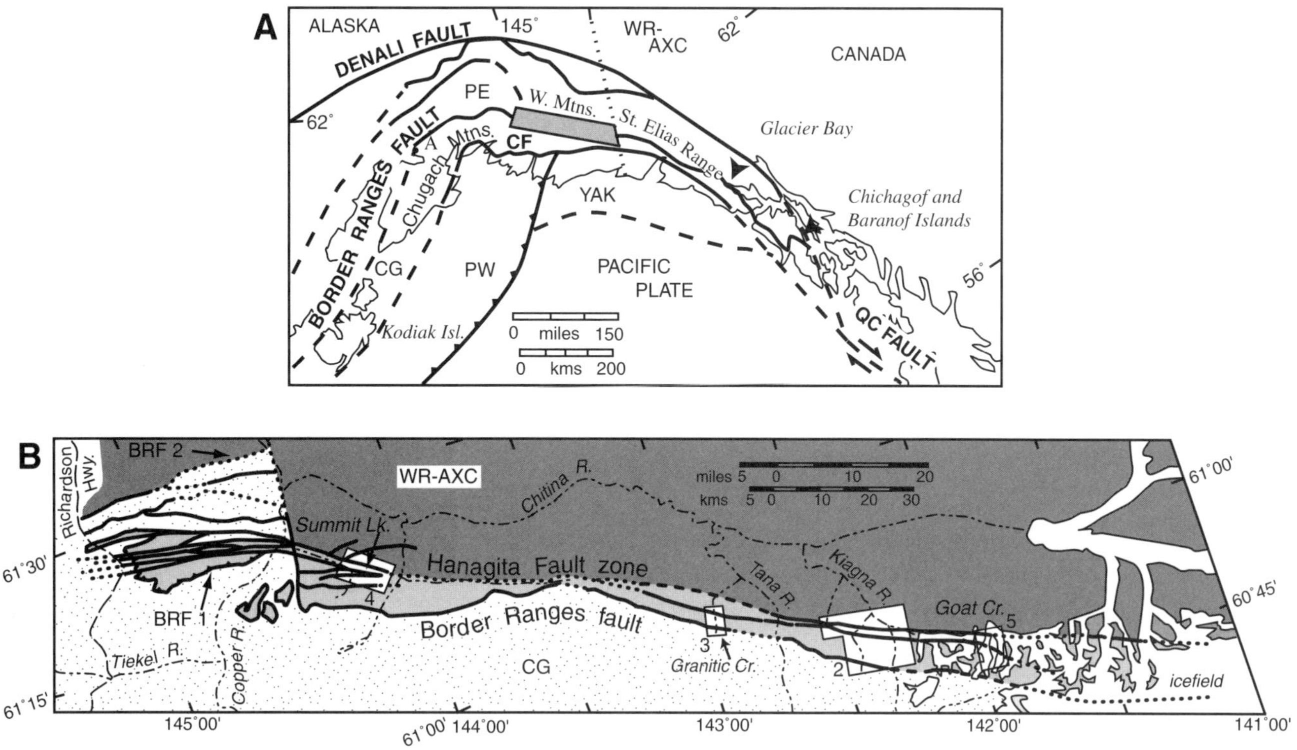

Figure 1. Location maps. A: Border Ranges fault system separates PE (Peninsular) and WR-AXC (Wrangellia-Alexander) terranes from the Chugach accretionary complex (CG—Chugach terrane; PW—Prince William terrane). Sense of motion on faults reflects current plate boundary between North America and Pacific plates. Abbreviations: A—Anchorage, CF—Contact fault, QC Fault—Queen Charlotte fault, W. Mtns.—Wrangell Mountains, YAK—Yakutat terrane. Shaded block indicates region covered by Figure 1B. B: Generalized geologic map of the eastern Chugach Mountains. Geology based on MacKevett (1978), Plafker et al. (1989), Winkler et al. (1981), this study, and unpublished mapping of S. Roeske and T. Pavlis. Medium gray—WR-AXC and PE terranes; light gray—crystalline rocks, slate, and mélange of the Hanagita fault zone; fine dots—west of the Copper River, undifferentiated Chugach accretionary complex, east of the Copper River, flysch and graywacke of the Valdez Group. BRF 1, 2—Border Ranges Fault 1 and 2 respectively (see text). White boxes with numbers outline map areas shown in those figure numbers. The two strands of the Hanagita fault are merged locally at this scale.

accretionary complex (Fig. 1B). The Border Ranges fault originated as a subduction zone megathrust (MacKevett and Plafker, 1974; Pavlis, 1982) but was extensively overprinted and modified by subsequent tectonic events along most of its strike length of 2000+ km (Roeske et al., 1989; Plafker et al., 1989; Little, 1990; Clendenen, 1991; Haeussler et al., 1994; Smart et al., 1996). The Wrangellia composite terrane is an amalgamation of three ocean-island arc complexes, the Peninsular, Wrangellia, and Alexander terranes (Fig. 1A). By the Late Cretaceous, these terranes were contiguous and had collided with North America; thus the structural history of the Border Ranges fault system and the accretionary prism to the south reflect the relative motion between North America and either the Kula, Farallon (Engebretson et al., 1985), or recently postulated Resurrection plate (Bradley et al., this volume), depending on where the ridge-trench-trench triple junctions were.

This study examined the section of the Border Ranges fault system in the eastern Chugach Mountains, where rocks of Wrangellia are juxtaposed against the accretionary complex, the Chugach terrane (Fig. 1B). North of the fault, the southern Wrangellia terrane is composed primarily of late Paleozoic sedimentary and volcanic rocks, Late Triassic basalt and limestone, Pennsylvanian and Late Jurassic plutonic rocks, and Late Jurassic-Cretaceous clastic rocks (MacKevett, 1978; Hudson, 1983; Nokleberg et al., 1994). Immediately north of the fault are greenschist and amphibolite facies calcareous schists, metabasites, marble, and quartz-mica schist of the Strelna metamorphics. A conodont age of early Pennsylvanian (Plafker et al., 1985) provides the oldest date for rocks from north of the fault zone. Because no clear stratigraphic link exists between the Strelna metamorphics and the Triassic stratigraphic section in Wrangellia to the north, some authors prefer to keep the Strelna metamorphics and intrusive rocks that intrude it as a separate unit (Grantz et al., 1991; Nokleberg et al., 1994). Others prefer to interpret the Strelna metamorphics as a deeper and/or more metamorphosed section of the Wrangellia rocks to the north (Plafker et al., 1989). For the purposes of this study we will include the Strelna metamorphics as part of Wrangellia, as they were clearly in their present relative positions based on overlapping sediments of Late Cretaceous age (MacKevett, 1978).

Within the Chugach terrane are two distinct accretionary packages, which occur in variable widths but are continuous along the whole length of the Border Ranges fault system. The older mélange facies, named the McHugh Complex in the eastern Chugach Mountains (Plafker et al., 1977), has fossil ages ranging from Triassic to Albian just west of the study area (Winkler et al., 1981). Rocks with late Albian to late Campanian fossils are very rare in the prism, but whether that gap reflects erosion of the prism, lack of accretion, or absence of subduction, is unknown. The gap coincides with a dramatic change in offscraped material, from the mélange facies, which is lithologically very heterogeneous, to a homogeneous slate and graywacke, known as the flysch facies of the Chugach terrane (Plafker et al., 1977). The Valdez Group is part of this facies (Fig. 1B) and although unfossiliferous in the study area it can be correlated along strike with flysch containing late Campanian to Maastrichtian fauna (Winkler et al., 1981; Plafker et al., 1994; Sample and Reid, this volume).

NATURE OF THE HANAGITA FAULT SYSTEM

The Hanagita fault system is a zone of structural reworking along the ancient backstop of the Chugach accretionary complex. The fault zone includes Strelna metamorphics and overlying sediments, McHugh Complex mélange, and rocks of unknown terrane affinity all tectonically interleaved at map scale (Figs. 2, 3, 4, and 5). Within the fault system are two throughgoing faults, named here the north and south strands of the Hanagita fault. The two strands of the Hanagita fault produce a profound topographic lineament, from the Copper River east until the faults disappear under the ice fields of the St. Elias Range (Fig. 1B). The lineament is caused by the preferential weathering of sedimentary units, which are the dominant lithology between the two fault strands, in contrast to the dominantly crystalline units north and south of the main fault strands (Figs. 2, 4, and 5). The two strands have a parallel trend of 293°–294° for over 100 km to the east of Summit Lake, then merge and become one dominant strand, oriented 287°–290°, for another 75 km to the Canadian border (longitude 141°W) (Fig. 1B). No rocks of the McHugh Complex occur inboard of the north strand of the Hanagita fault. By former definitions (MacKevett and Plafker, 1974), this would be the Border Ranges fault. However, because Wrangellia terrane rocks are also found outboard, or south, of this fault, interleaved with the mélange, we introduce the name north strand of the Hanagita fault for this structure. This fault is roughly correlative with the Border Ranges Fault 2 of Plafker et al. (1994), west of the Copper River (Fig. 1B).

South of the south strand of the Hanagita fault an anastamosing zone of strike-slip, oblique-slip, thrust, and normal faults occurs until another major fault is crossed, which separates the collage of mélange and crystalline rocks from slate and graywacke of the Valdez Group. This fault, shown as the Border Ranges fault in this study, correlates with Border Ranges Fault 1 of the TACT transect (Plafker et al., 1994) (Fig. 1B), the Spirit Mountain fault in the Summit Lake region (Winkler et al., 1981), and the Border Ranges fault of MacKevett (1978). The geometry and structural history of this fault are complex, with low-angle, south-verging thrust motion documented (Nokleberg et al., 1989; Winkler et al., 1981) as well as subvertical dextral slip (this study; Fig. 6A). Because McHugh Complex mélange is abundant inboard of this fault, it does not fit the original definition of the Border Ranges fault (MacKevett and Plafker, 1974), but for continuity with existing literature we retain the name. Thus, structures mapped as the Border Ranges fault differ in age and structural history along strike. East of longitude 145°W it is primarily a high-angle dextral-slip structure overprinting an earlier contractional history; west of 145°W, in the western Chugach range and on the Kodiak islands, it has a variable history of reverse, dextral, and normal slip that has reactivated, but still occurs along, the innermost boundary of the Chugach accretionary prism (Pavlis, 1982; Roeske et al., 1989; Little, 1990; Winkler, 1992; Clendenen et al., Chapter 3, this volume).

Rock Units within the Hanagita Fault System

A wide range of protoliths exists within the Hanagita fault system. Many of the rocks are hydrothermally altered at very low grades and strongly deformed, locally making recognition of the original composition difficult to impossible (Fig. 6B). The protoliths can be subdivided into mélange, clastic rocks, and crystalline rocks (schist, gneiss and intrusive rocks) (Figs. 2–5). The crystalline rocks have been further divided into four map units: Strelna metamorphics (Pzs) (Figs. 2, 4, and 5), hornblende diorite, quartz diorite and tonalite (Jd) (Figs. 2–5), crystalline rocks of the Hanagita fault zone (MzPzs) (Figs. 2–5), and Tertiary biotite granodiorite (Tg) (Fig. 5).

We designated the metamorphic rocks within the fault system (MzPzs) as a separate unit from the Strelna metamorphics to make a clear distinction between schists that clearly meet the definition of the latter (see above) and are along strike with the type locale (Plafker et al., 1989; Nokleberg et al., 1994) and those schists that may or may not include Strelna metamorphic rocks. Rocks within the MzPzs unit that appear similar to the Strelna metamorphics include amphibolite facies pelite and carbonate; these are most abundant in the northern part of the MzPzs unit. These schists are usually faulted against all other units but locally still in intrusive contact with the diorite and quartz diorite. Hornblende cooling ages from most of the schist and plutonic rock within the fault zone are earliest Cretaceous (below), identical to those north of the fault system (Roeske et al., 1992).

Other crystalline rocks within the Hanagita fault system do not appear to have a correlative unit immediately to the north (Winkler et al., 1981; this study). These include massive amphibolite, intruded extensively by muscovite granite and trondhjemite (Fig. 6C), and Middle Jurassic diorite (below). Because of the difficulty in separating the schists and smaller intrusive bodies at the 1:24,000 scale at which the mapping was done, they have

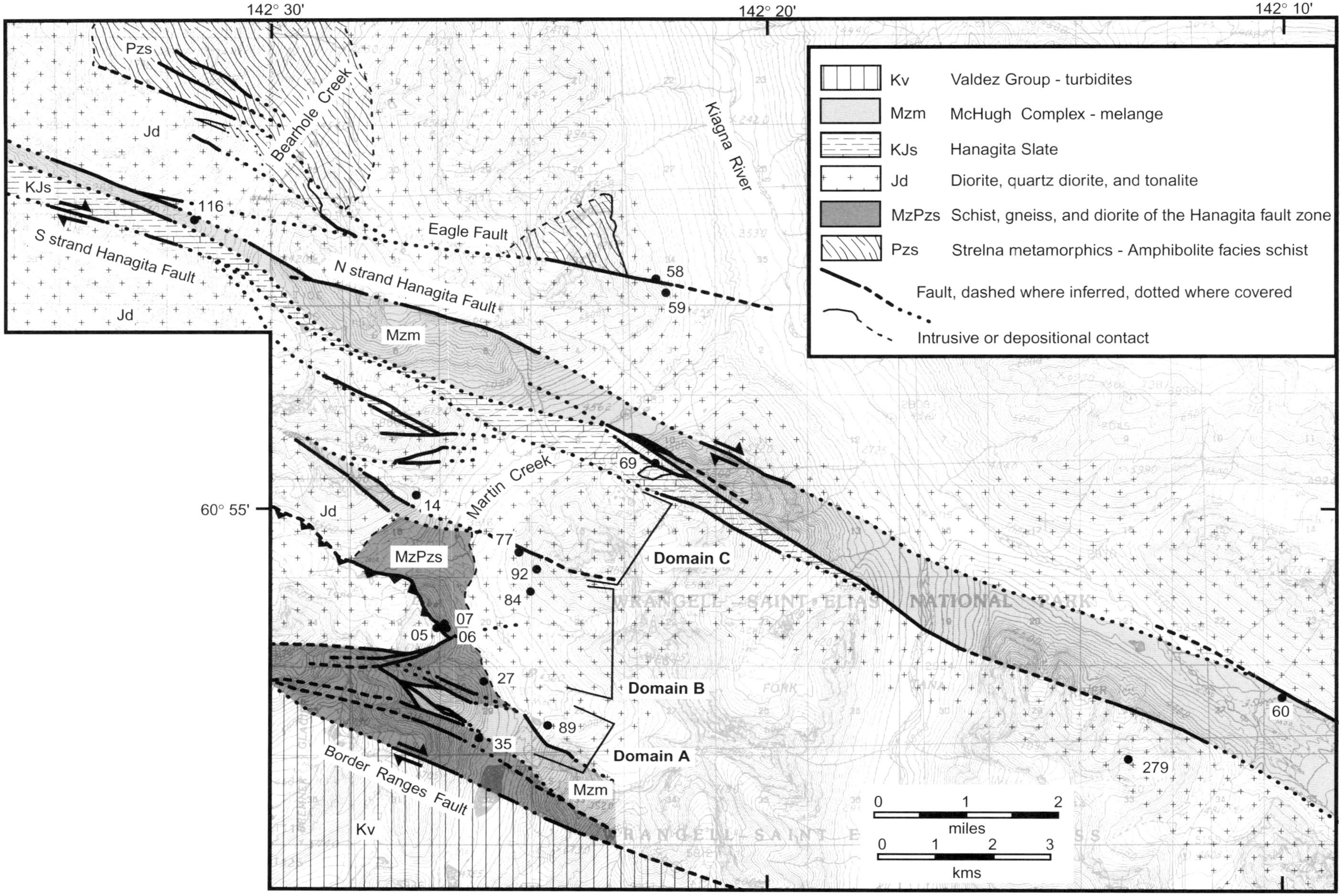

Figure 2. Map of Kiagna River area. Numbers are sample locales of isotopic ages or fossils referred to in text, tables, and figures. (A color version of this figure is on the CD-ROM accompanying this volume.)

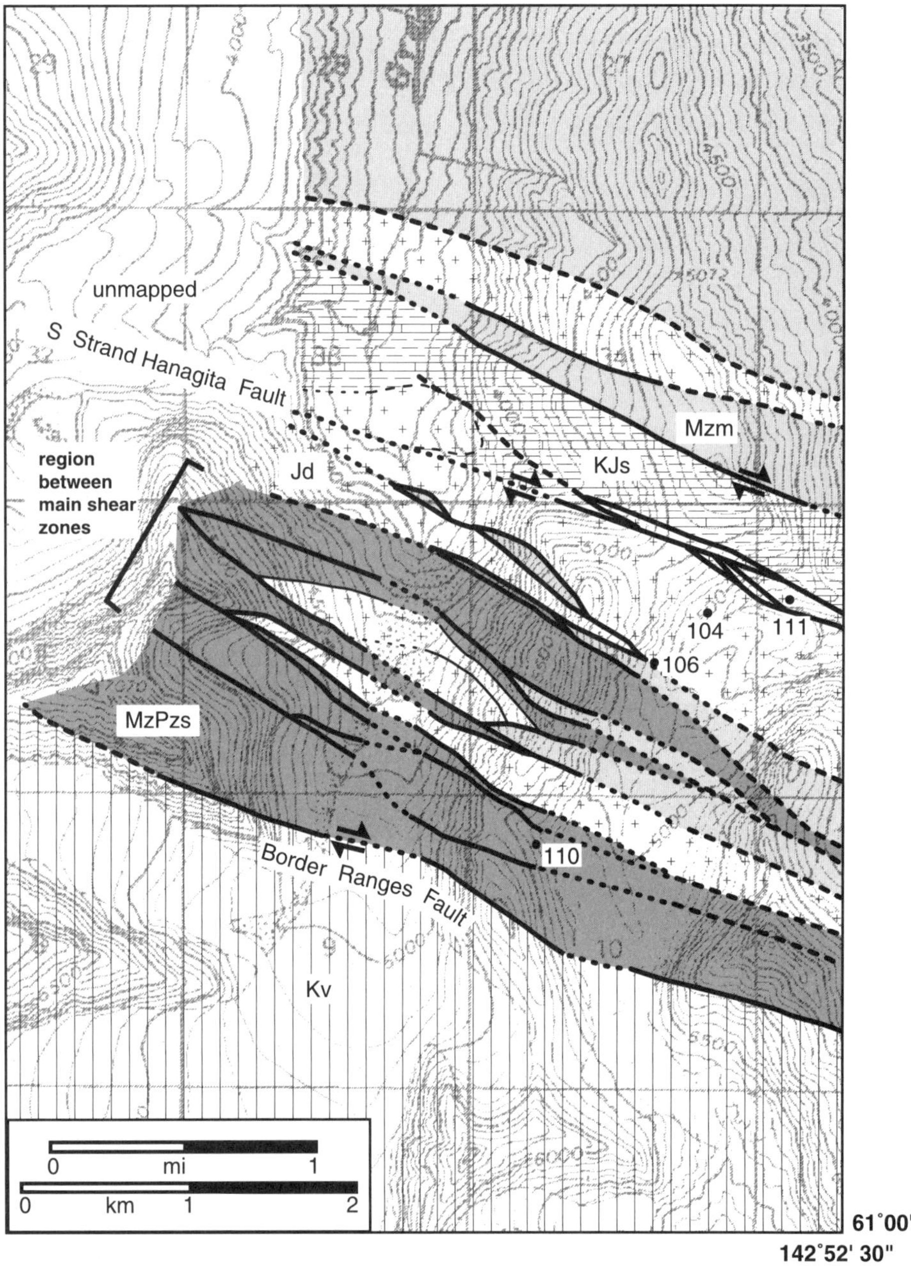

Figure 3. Map of Granitic Creek area. Key same as Figure 2. (A color version of this figure is on the CD-ROM accompanying this volume.)

been grouped as Paleozoic and Mesozoic schist, gneiss and diorite (Fig. 2). Our MzPzs unit is equivalent to plutonic rocks and schists of the Haley Creek terrane in the Summit Lake region and west of the Copper River (Wallace, 1981; Winkler et al., 1981) (Fig. 1B). Although most previous authors interpret all of the crystalline rocks to be from the overriding plate of a middle to late Mesozoic subduction zone, it is possible that the amphibolite and leucocratic plutonic rock may have formed in a setting similar to rocks within the Border Ranges fault exposed in the Anchorage area, the Knik River terrane. In this area, amphibolite has been interpreted to be McHugh Complex, metamorphosed at a higher grade during early stages of subduction (Pavlis et al., 1988).

A few undeformed plutonic rocks crosscut the fabrics of the Hanagita fault system. Most of these occur as dikes of fine-grained, leucocratic, muscovite granite to biotite tonalite, less than 20 m thick. The only large pluton of this generation occurs in the Goat Creek region (Figs. 5 and 6D). This pluton is similar in age (see below) and composition to Early Eocene plutonic rocks in the accretionary prism immediately to the south and record passage of a slab window along the convergent margin (Pavlis and Sisson, 1995).

Two metasedimentary units, a mélange (Mzm) and a clastic unit (KJs), occur consistently within the Hanagita fault system. Neither of these units has been mapped in this area by previ-

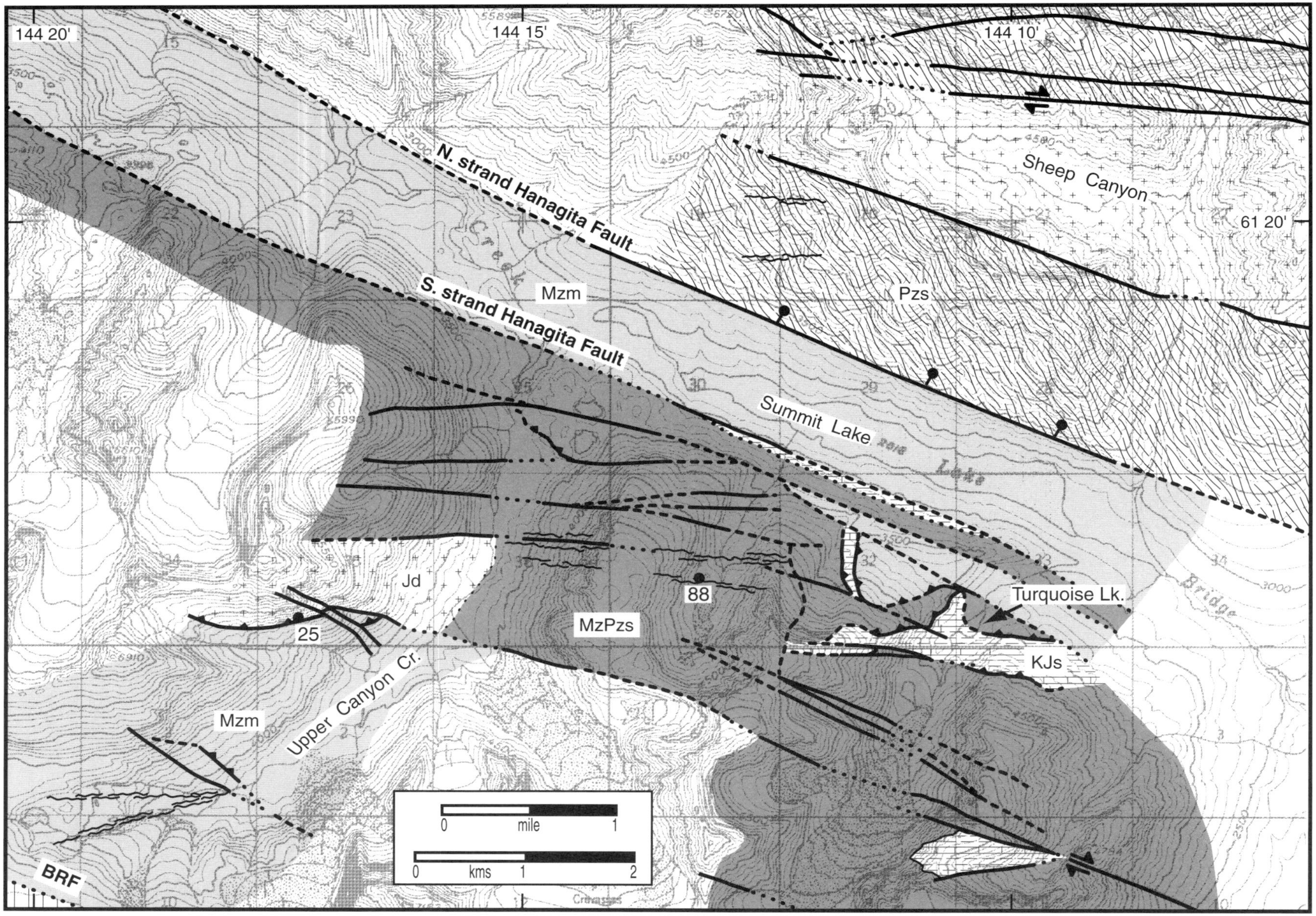

Figure 4. Map of Summit Lake area. Key same as Figure 2, with addition of wavy lines—ductile shear zones. (A color version of this figure is on the CD-ROM accompanying this volume.)

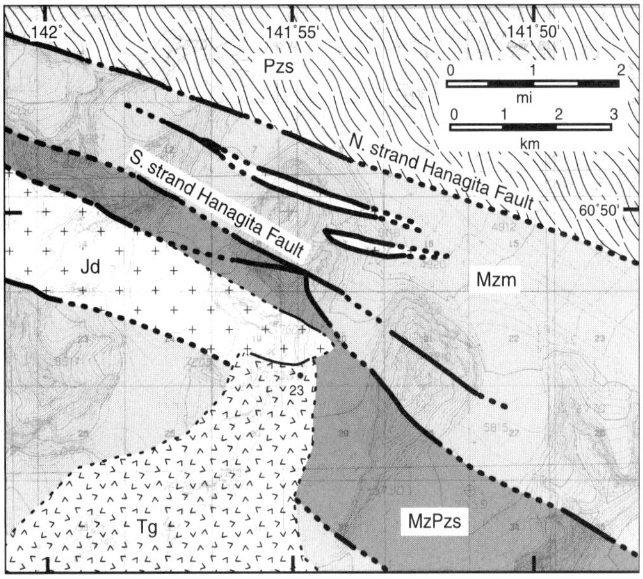

Figure 5. Map of Goat Creek area. Key same as Figure 2, with addition of Tg—Tertiary granodiorite. (A color version of this figure is on the CD-ROM accompanying this volume.)

ous workers. The mélange is composed of red to gray chert, black and green mudstone, metabasalt, and minor graywacke (Figs. 6E and 6F). Poorly preserved radiolarian from two chert samples in the Kiagna River drainage (nos. 116 and 60, Fig. 2) give an Early Cretaceous age (C. Blome, 1992 and 1994, written commun.). We interpret the mélange to be part of the McHugh Complex, based on this unit's similarity in lithology and age to type locales of the McHugh Complex along strike to the west. Previous regional mapping had not identified McHugh Complex east of the Copper River and had included these rocks as part of the Valdez Group (MacKevett, 1978) or as part of the Haley Creek terrane (Winkler et al., 1981).

The second metasedimentary unit is dominantly a black slate, informally named here the Hanagita slate (KJs) (Fig. 2). It always occurs south of the northernmost McHugh Complex and is in fault contact with it. We interpreted it originally to be part of the Valdez Group, because of its location outboard of the furthest north McHugh Complex and the overall homogeneous character, which is similar to portions of the Valdez Group. However, field relations described below indicate that the Hanagita slate must have been deposited in the hanging wall of the subduction zone and thus is part of the Wrangellia composite terrane.

Although the slate is in fault contact with the crystalline rocks in most places, several key locales around the southeast side of Summit Lake (Fig. 4) and in the Granitic Creek drainage (Fig. 3) show that it is depositional on plutonic and carbonate rocks. At Summit Lake the slate is homogeneous dark mudstone until within a few tens of meters of its southern contact with schists and carbonates, where thinly layered sandy beds become visible. The sandy layers consist of carbonate-rich clasts and the slate is carbonate-cemented. Although the contact between the slate and carbonate is covered, the composition of the sandy unit requires derivation from a carbonate-rich unit and we infer it to be depositional on the adjacent calcareous schist. At Granitic Creek the slate is dominantly a fine-grained argillaceous slate with zones of more chloritic-rich slate and metasiltstone, interlayered on a scale of tens of cm to m. Within approximately 10 m of plutonic rocks the slate becomes a coarse cobble conglomerate, with clasts of diorite (Fig. 6G). Although highly deformed, one can see that the clasts are similar in composition to the adjacent quartz diorite. Thus, we interpret the boundary in Granitic Creek as a faulted depositional contact and the age of the slate is only constrained to be Cretaceous, because it contains clasts of latest Jurassic plutonic rocks and is involved in Paleocene deformation (see below).

Exposures of the Hanagita slate in the Kiagna region show that a Cretaceous age cannot apply to this entire unit because the slate locally is intruded by Latest Jurassic plutonic rock. This relationship is seen at several locales in the Martin Creek area, including sites 69 and 92 (Figs. 2 and 6H). The plutonic composition is variable, but generally more leucocratic and biotite-rich than the typical quartz diorite of the area. Igneous rocks that intrude this part of the slate are approximately the same age as those that the slate is depositional on elsewhere. We obtained an U/Pb zircon date of ca. 150 Ma (Fig. 7 and Appendix, Table A-1[1]) (J. Wright, 1992, written commun.) from site 92 and a muscovite cooling age of ca. 141 Ma (Table 1) from site 69. Thus, the Hanagita slate was deposited over some time span and perhaps should not be considered one unit.

The Hanagita slate is almost certainly correlative to the "enigmatic phyllite" of the Haley Creek terrane (Pavlis and Crouse, 1989) (Wallace, 1981), exposed west of the Copper River on strike with the Hanagita Slate in Summit Lake (Figs. 1B and 4). The phyllite of the Haley Creek region includes marble, which prompted previous workers to question correlation with the Valdez Group, which lacks carbonate-bearing rocks. Shear zones in the Haley Creek terrane that correlate with the brittle faults of the Summit Lake region (see below) are greenschist facies, consistent with a deeper structural level being preserved west of the Copper River.

No specific correlation of the Hanagita slate with any units inboard of the Hanagita fault system is possible because the slate's age and depositional setting are poorly known and because unknown amounts of displacement have occurred along the fault system. However, the Hanagita slate usually occurs along inner strands of the fault system, where displacement is probably least. Thus, several different clastic units north

[1]GSA Data Repository item 2003125, Appendix Figures A-1 and A-2, and Appendix Tables A-1, A-2, and A-3, is available on request from Documents Secretary, GSA, P.O. Box 9140, Boulder, CO 80301-9140, USA, editing@geosociety.org, at www.geosociety.org/pubs/ft2003.htm, or on the CD-ROM accompanying this volume.

Figure 6 (*on this and following page*). Field photographs. A: Arrow indicates a subvertical section of the Border Ranges fault, looking east from uppermost Granitic Creek (Fig. 3). Dark rocks to south (right) are Valdez Group; mixed units to north include highly altered and sheared trondhjemite, schist, and amphibolite. B: Retrograde shear zone, low greenschist facies, completely replacing mesoscopic crystals in diorite. S-C relation between shear planes indicates dextral slip. Site 91SR90, near site 91SR89, upper Martin Creek (Fig. 2). C: Trondhjemite intruding amphibolite. Near site of a Middle Jurassic diorite (90SR35), ~1 km north of Border Ranges fault, Slender Lake (Fig. 2). D: Intrusive contact between undeformed Tertiary biotite granite (top) (94SR23) and Jurassic–Early Cretaceous quartz diorite with retrograde shear zone. Goat Creek (Fig. 5). E: Ductile shear zone in McHugh Complex, upper Canyon Creek, Summit Lake (Fig. 4). Siliceous clast orientation relative to the through-going foliation indicates dextral shear. F: Greenstone layer within McHugh Complex elongate in high-strain fabric, crosscut by high-angle brittle faults with apparent dextral slip. Upper Canyon Creek, Summit Lake (Fig. 4). G: Cobble conglomerate at south strand of Hanagita fault, Granitic Creek. The fault follows the depositional contact between the Hanagita slate and Late Jurassic diorite. Cobbles pulled apart, rotated, in a dextral shear plane. H: Faulted intrusive contact between Hanagita slate and Late Jurassic–Early Cretaceous diorite (91SR92). U/Pb zircon date from diorite is 150 ± 8 Ma (Fig. 7; Table A-1, Appendix [see footnote 1]). I: Ductile shear zone (S-C) fabric in diorite. Float, from upper Martin Creek (Fig. 2). J: Anastamosing very low grade shear zones in quartz diorite, south strand of Hanagita fault, Bearhole Creek (Fig. 2). K: Penetrative, subvertical, very low grade fabric formed during dextral slip in quartz diorite. South strand Hanagita fault, Granitic Creek region (Fig. 3). L: Brittle fault in Hanagita slate, showing dextral slip from steps in slickenlines. East side of Martin Creek, near south strand Hanagita fault (Fig. 2). M: Folds in MzPzs metasediment overturned to the north, tens of m below south-dipping thrust fault exposed on west side of upper Martin Creek (Fig. 2). N: South-directed thrust placing schist over Hanagita slate. View looking west, from cliff east of Turquoise Lake, Summit Lake region (Fig. 4).

Figure 6 (*continued*).

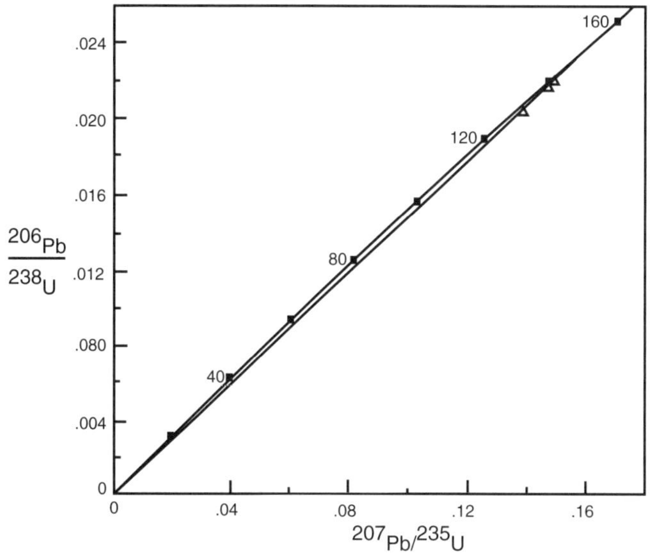

Figure 7. U/Pb zircon data from sample 91SR92, plutonic rock that intrudes Hanagita slate (Fig. 6H). Age of 150 ± 8 Ma based on the upper intercept with concordia of a line through three fractions (triangles) and a force fit through 0 Ma. Upper Martin Creek (Fig. 2). U/Pb data in Table A-1, Appendix (see footnote 1).

of the Hanagita fault may correlate with the younger part of the Hanagita slate. The closest locale is the Berg Creek Formation, 8 km west-northwest from the Hanagita slate exposure at Summit Lake, on the east side of the confluence between the Chitina and Copper Rivers (Fig. 1B) (Winkler et al., 1981). The Neocomian Berg Creek Formation is a locally preserved clastic unit; in this site it is interpreted to rest unconformably on the schist, but the contact in many places appears to be a fault (Grantz et al., 1966). The section is dominantly siltstone, with thin interbeds of sandstone, overlain by a calcareous conglomerate containing clasts of local origin (Grantz et al., 1966). Another possible correlative unit is the Kennicott Formation, an early Albian clastic unit in the Wrangell mountains (Jones and Berg, 1964; Jones and MacKevett, 1969). The closest exposures to the Hanagita slate occur approximately 20 km due north of the Kiagna river region, just north of the Chitina River (Fig. 1B). The Kennicott Formation contains a basal unit of conglomerate and sandstone overlain by a thicker section of fine siltstone, shale and sandstone (Jones and MacKevett, 1969; Trop et al., Chapter 4, this volume). Paleocurrent data indicate transport to the west and southwest (Trop et al., this volume), which would allow the Hanagita slate to be a distal facies of this formation.

STRUCTURE

General Description

All of the rock units described above have at least one tectonic fabric, with the exception of the Eocene intrusive rocks. The schist, gneiss, and mélange had a tectonic fabric prior to the development of the Hanagita fault system. The schist and gneiss have a penetrative foliation that generally strikes WNW to NW and dips moderately to steeply NE and SW. This foliation formed under amphibolite facies conditions, as seen by sparsely preserved garnet and biotite in the pelitic sections and hornblende and biotite in the amphibolite. Mesoscopic folds are common within the schist in the Summit Lake region, with fold axial surfaces forming subparallel to the schistosity. Evidence for a pre-existing fabric in the McHugh Complex comes primarily from outcrops outside of this study area. The typical fabric of the McHugh and related mélange units within the Chugach terrane is a classic mélange texture formed by very low-grade, anastamosing cleavage. Few sites of McHugh Complex within the Hanagita fault system have this fabric preserved and instead the characteristic fabric of the McHugh is a subvertical plane striking WNW (see below).

The Hanagita slate has one foliation, a slaty to phyllitic cleavage, in the Granitic Creek and Kiagna River regions. Bedding is well preserved locally and cleavage-bedding intersections produce a pencil-fabric in the slate. The slate exposed south of the Summit Lake region shows both the penetrative slaty cleavage and a secondary spaced cleavage, with the latter locally developed near faults (see below).

The diorite and quartz diorite are the most important rock types for identifying the deformation history associated with the Hanagita fault system. Based on our mapping in diorites north of the Hanagita fault system (Figs. 2 and 5), the only fabric these rocks had prior to faulting is a locally developed very high-temperature foliation. This foliation is defined by the preferred orientation of hornblende and is parallel to the contacts of the plutons. The foliation has no asymmetric fabrics within it or lineation associated with it. It grades into a gneissic fabric at the contact with the schists. Hornblende cooling ages from the gneiss are earliest Cretaceous, the same as those from the undeformed plutons (Roeske et al., 1992). This high-temperature fabric appears to be primarily igneous and/or a flow fabric associated with emplacement of the plutons.

The diorites' homogeneous igneous mineralogy and texture makes it ideal for recording the deformation associated with the Hanagita fault system. Thus, most of the geochronology is from plutonic rocks and wherever possible the orientations of shear zones were measured within plutonic rocks. Crosscutting relationships and mineral assemblages also provide us with excellent evidence that the deformation mechanisms associated with the Hanagita fault system changed through time. Fabrics within the plutonic rocks range from ductile to brittle, with a complete continuum between them. The highest-temperature fabrics formed under greenschist facies conditions, as seen by the formation of biotite, muscovite, epidote, chlorite, and actinolite in the shear zone fabric (Figs. 6B, 6I, and 8A). Igneous quartz grains are highly modified by dislocation creep and subgrain growth (Fig. 8B). These fabrics are described as ductile shear zones (below) and grade into those described as retrograde

TABLE 1. SUMMARY OF AR/AR DATA FROM THE HANAGITA FAULT ZONE*

Mineral	Region	Sample no.	Latitude	Longitude	Plateau date or weighted mean date (wm)	High-*T* isochron date	Low-*T* isochron date
Hornblende							
#Primary	Summit Lake	89SR88a	61°18'13"	144°13'15"	no	148.7 ± 0.9	85.2 ± 0.3
		†91SR25c	61°17'59"	144°17'24"	146.1± 0.2	145.9 ± 0.2	no
		†91SR25d	61°17'59"	144°17'24"	143.2 ± 0.4 (wm)	143.6 ± 0.4	no
	Kiagna River	90SR05	60°54'05"	142°26'40"	144.6 ± 0.6	144.2 ± 0.3	no
		**90SR06	60°54'05"	142°26'42"	no	142.2 ± 0.5	67.0 ± 0.5
		**90SR07	60°54'07"	142°26'44"	141.9 ± 0.6	144.2 ± 0.3	69.1 ± 0.5
		**90SR14b	60°55'17"	142°27'13"	143.3 ± 0.6	143.9 ± 0.2	63.0 ± 0.9
		90SR35	60°52'55"	142°26'00"	170.6 ± 0.6 (wm)	171.5 ± 1.1	no
		†91SR58h	60°57'22"	142°23'03"	143.6 ± 0.4	144.1 ± 0.2	no
		**91SR59a	60°57'16"	142°22'48"	no	140.5 ± 0.3	64.4 ± 1.5
		**91SR77d	60°54'53"	142°25'14"	no	142 ± 4	60 ± 2
		†91SR77g	60°54'53"	142°25'14"	143.0 ± 1.9	144.5 ± 0.4	no
		**91SR84b	60°54'25"	142°25'04"	no	140.5 ± 0.3	64.4 ± 0.5
		91SR89a	60°53'05"	142°24'36"	no	141.9 ± 0.4	64.1 ± 1.4
		**73AH279a	60°52'37"	142°13'02"	142.2 ± 0.5	142.4 ± 0.9	70.4 ± 1.5
	Granitic Creek	**91SR106d	61°01'48"	142°53'53"	no	139.2 ± 1.2	56.6 ± 0.7
Muscovite							
Primary	Kiagna River	90SR27 (coarse and fine)	60°53'30"	142°25'55"	no		
		91SR58a	60°57'22"	142°23'03"	54.4 ± 0.5		
		91SR69b	60°55'35"	142°22'36"	141.4 ± 0.3		
	Granitic Creek	91SR110c	61°01'15"	142°54'37"	no		
#Secondary	Kiagna River	90SR14b	60°55'17"	142°27'13"	~58		
	Granitic Creek	91SR104h	61°01'57"	142°53'24"	54.8 ± 0.2		
		91SR111d	61°02'00"	142°52'49"	51.4 ± 0.1		
Biotite							
Primary	Goat Creek	94SR23a	60°48'10"	141°55'08"	51.48 ± .51 (wm)	52.2 ± 0.1 (steps 3-12)	
K-feldspar							
Primary	Kiagna River	91SR69b	60°55'35"	142°22'36"	no	46.0 ± 0.1 (steps 11-16)	

*Complete Ar data presented in Appendix, Table A-2 (see footnote 1).
#Primary refers to mineral that crystallized from magma or grew during amphibolite facies metamorphism. Secondary refers to mineral that grew during very low grade metamorphism associated with shear zones.
†Spectrum in Appendix, Figure A-1 (see footnote 1).
**Spectrum in Appendix, Figure A-2 (see footnote 1).

shear zones. The latter is typically associated with abundant hydrothermal alteration and the original mineralogy is more obscure than in the ductile shear zones. The shear zones may contain calcite and quartz as well as chlorite, epidote, muscovite, and wairikite (Bunds, 1994). The latter mineral is a high-*T*, low-*P* phase typically seen in hot spring environments and its occurrence in the Hanagita fault zone reflects the high heat flow associated with passage of the triple junction. The retrograde shear zones can form as discrete, anastamosing faults (Fig. 6J) and as an intense, planar fabric (Fig. 6K). Deformation mechanisms in the retrograde shear zones include cataclasis and pressure solution (Fig. 8C; Bunds, 1994). Brittle faults are quite common, occurring as discrete fault planes with slickenlines (Figs. 6H and 6L) and as zones of pervasive cataclastic deformation. Alteration along the brittle faults is especially severe in the Summit Lake region, where orange-stained lineaments mark the faults with the youngest displacement.

Orientation and Kinematics of Shear Zones and Faults

Kiagna and Granitic Creek Regions—Shear Zones in the Vicinity of Major Fault Strands

The main strands of the Hanagita fault zone and the Border Ranges fault consist of ductile and retrograde shear zones and brittle faults, all of which have dominantly a west-northwest strike and subvertical dip (Figs. 6A, 9A, 9B, 10A, 10B, 11A, and

Figure 8. Photomicrographs of plutonic rocks deformed in Hanagita fault zone. A: Ductile shear zone in quartz diorite (91SR89a), Slender Lake, Kiagna region (Fig. 2). Fabric defined by muscovite, chlorite, and fine-grained epidote. Igneous hornblende fractured along cleavage planes. Image is 1.2 mm high. See Figure 17A for low- and high-*T* isochron dates from this sample. B: Quartz-trondhjemite (89SR86) from ductile shear zone, south of Summit Lake. Ribbon quartz and fine muscovite define the quartz. Image is 1.2 mm high. C: Retrograde shear zone: cataclastic and pressure solution fabric in felsic plutonic rock (91SR104h), south strand Hanagita fault, Granitic Creek (Fig. 3). Image is 3.9 mm high. Plateau date from white mica separate of this sample is 54.8 ± 0.2 Ma (Table 1; Fig. 21). D: Retrograde shear zone in diorite. Hornblende pulled apart and rotated, indicating dextral shear.

12A). Rare shear zones and faults strike ~north-south to northeast, orthogonal to the main set. Mineral lineations from plutonic rocks deformed by the ductile and retrograde shear zones primarily plunge gently to the east-southeast (Figs. 9A, 10A, and 11B). The sense of slip was determined in the field by the geometric relation (S-C) between the foliation (S) and the shear zone (C) planes steps in slickenlines or grooves on brittle fault planes (Figs. 6I and 6L). The sense of slip for WNW-striking faults is overwhelmingly dextral to dextral oblique. Kinematics on the NNW- to NE-striking faults are dominantly sinistral, consistent with their formation as antithetic faults to the main WNW-striking fault system. Oriented thin sections were also examined and rotated and fractured hornblende (Fig. 8D), garnet, and feldspar grains consistently show dextral-slip displacement.

Recognition of the shear zones is difficult in the Hanagita slate and McHugh Complex, but field data indicate the penetrative cleavage in the sedimentary units formed during the development of the shear zones. This is particularly well exposed in the Granitic Creek region, where the slaty cleavage in the Hanagita slate intensifies into a shear zone fabric within 100–300 meters of the south strand of the Hanagita fault. Conglomerate cobbles deformed in this fabric show a clockwise rotation, demonstrating they were deformed during simple or general shear in a dextral shear zone (Fig. 6G). The orientation of the cleavage in the Hanagita slate and McHugh Complex is also consistent with forming during dextral slip on the shear zones. These units' cleavage (Figs. 9C and 10C) consistently strikes ~12–15° counterclockwise from the average strike of the shear

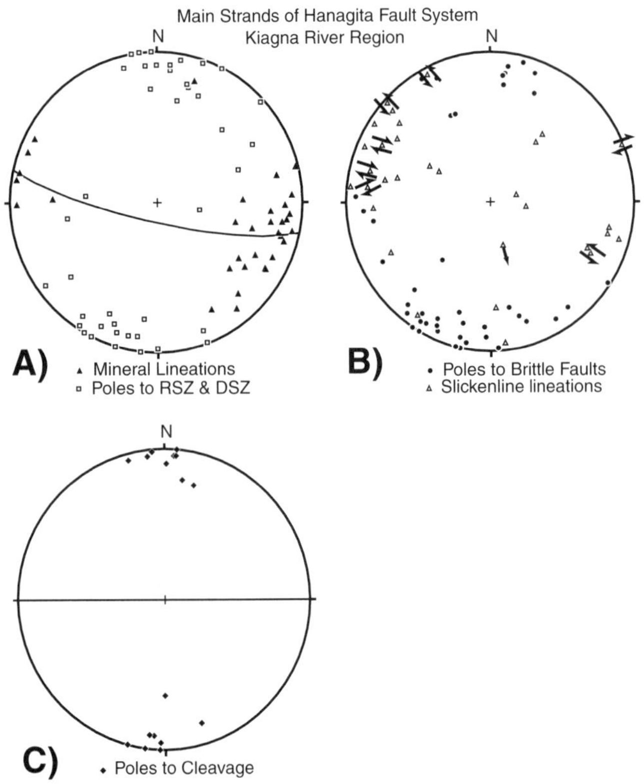

Figure 9. Equal-area lower-hemisphere projection of structures in the Kiagna River region (Fig. 2). A: Ductile shear zone (DSZ) and retrograde shear zone (RSZ) data from sites along the Border Ranges fault and between the main strands of the Hanagita fault system and 1–2 km on either side. The average of the shear zones, shown as a line, is oriented 282°, 79°S. Kamb contour plot of mineral lineations (not shown) has maximum at 109°, 19°. B: Brittle fault data from sites between the north and south strands of the Hanagita fault and 1–2 km on either side. Average brittle fault plane is ~297°, 87°N. Arrows indicate sense of slip, determined by slickenlines or grooves on fault planes. Pairs of arrows show strike-slip sense for lineations with rake ≤30°. Single arrows show sense of slip of hanging wall relative to footwall. C: Cleavage planes from the Hanagita slate and McHugh mélange between the north and south strands of the Hanagita fault. Average cleavage plane, shown as a line, is ~270°, 90°.

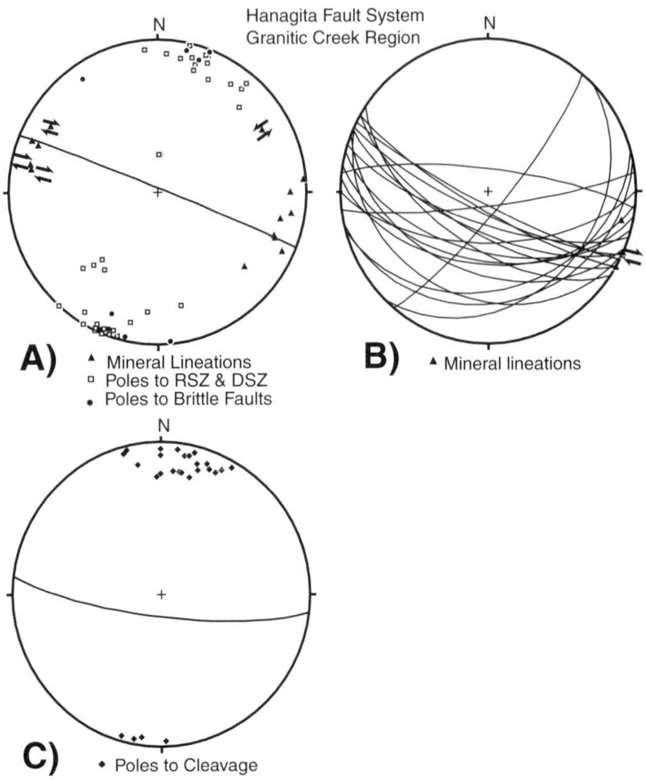

Figure 10. Equal-area lower-hemisphere projection of structures in the Granitic Creek region (Fig. 3). A: Brittle fault and shear zone data from sites 1.5 km north of the south strand of the Hanagita fault and ~750 m south of it, and from the Border Ranges fault and ~750 m to north. The average of the shear zones, shown as a line, is oriented 292°, 88°N. Arrows—same as Figure 9B. B: Shear zone planes (great circles) from sites between the main strands of the Hanagita fault and Border Ranges fault (see Fig. 3 for outline of structural domain). All mineral lineations (total = 4) plotted with their respective shear zones. C: Cleavage planes from the Hanagita slate and McHugh mélange near the south strand of the Hanagita fault. Average cleavage plane, shown as a line, is ~277°, 78°S.

zones (Figs. 9A and 10A), as would be expected if the cleavage is a foliation forming in a dextral shear system.

Kiagna and Granitic Creek Regions—Shear Zones and Brittle Faults between the Main Strands of the Hanagita Fault System

The frequency, width, and continuity of the shear zones diminish 500–750 m south of the south strand of the Hanagita fault and north of the Border Ranges fault. Between these through-going fault strands is a region 2.0 km wide that contains clusters of complex shear zones and brittle faults, none of which can be followed more than a kilometer along strike. The shear zones vary more in strike and dip compared to those near the main strands of the Hanagita fault system. The dominant difference in orientation from the main strands of the Hanagita fault system is that, in addition to the high-angle WNW-striking set of faults and shear zones, a number of shallow to moderately SE- to SW-dipping shear zones exist (compare Figs. 9A, 10A, and 12A with 10B, 11A, and 12B). Field relations on the west side of Martin Creek clearly show several 100 m- to km-scale low-angle, southwest-dipping faults (Fig. 2). Petrography of samples from within these fault zones show they formed under similar conditions to high-angle ductile and retrograde shear zones in the immediate vicinity. Pre-existing foliation in the schist beneath the SW-dipping faults have tight asymmetric folds overturned to the NE (Fig. 6M). Other smaller, outcrop-scale south-dipping shear zones show oblique dextral slip (Figs. 10B and 11B). The data from the west side of Martin

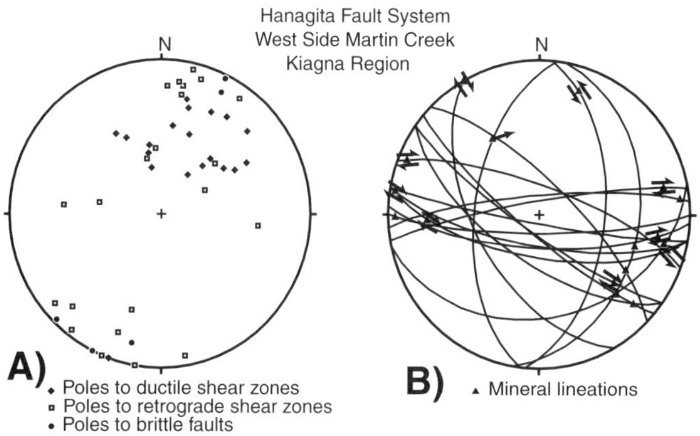

Figure 11. Equal-area lower-hemisphere projection of structures on the west side of Martin Creek, in the Hanagita fault zone from south of the south strand to the Border Ranges fault (see Fig. 2 for location). A: Poles to all fault and shear zone fabrics. B: All mineral lineations, with their respective shear planes (plotted as great circles). Arrows same as Figure 9B.

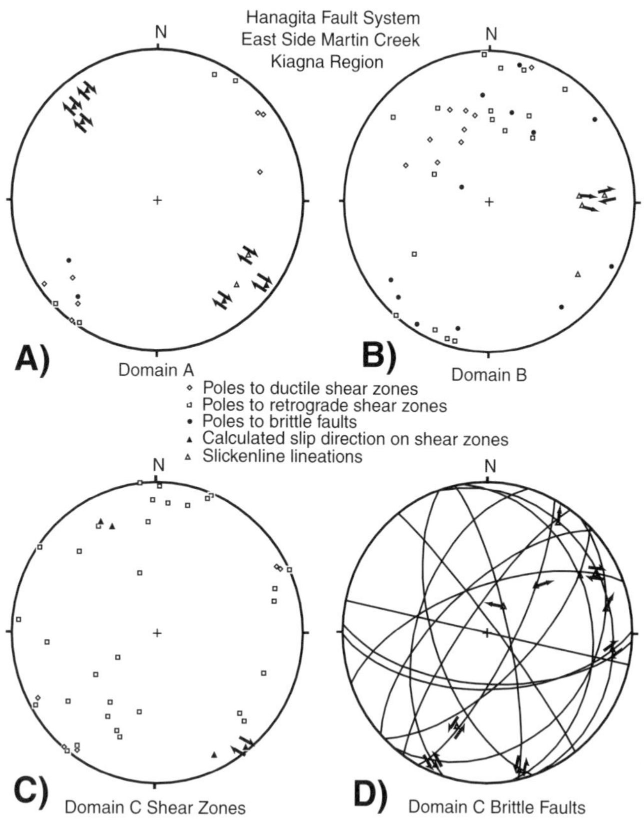

Figure 12. Equal-area lower-hemisphere projection of structures on the east side of Martin Creek, in the Hanagita fault zone, from south of the south strands of the Hanagita fault to the Border Ranges fault (see Fig. 2 for location of domains). Arrows same as Figure 9B. Calculated slip directions (in Figs. 12A and 12C) are determined for shear zones where S and C fabrics are well determined. The calculated slip direction is derived by taking the lineation located in the C plane that is 90° from the intersection of the S and C planes.

Creek and Domain B on the east side indicate that the shear zone geometry appears to evolve with time, because the ductile shear zones are dominantly moderately S-dipping, the brittle faults are dominantly subvertical, parallel to the regional WNW strike, and the retrograde shear zones range between these two orientations (Figs. 11A and 12B).

Shear zones and faults on the east side of Martin Creek have a wide range in orientations compared to the other sites. By separating the region into domains, a pattern begins to emerge, becoming more similar to the regional patterns as one goes from north to south (Figs. 2 and 12). The shear zone and fault geometries in the southern part (domains A and B, Figs. 12A and 12B) are fairly similar to regions along strike in the west side of Martin Creek and Granitic Creek (see above), with the addition of a number of NE-striking, moderately S-dipping shear zones in Domain B. Shear zones in Domain C (Fig. 12C) have a very broad distribution of orientations, dominantly high-angle, but also including some moderately N-dipping planes. The lack of kinematic indicators from most of these precludes a clear interpretation. In contrast, the brittle faults from this region have well-developed slickenlines that are consistent with overall dextral WNW-striking slip. The E-W faults are dextral and those with N to NE strikes are sinistral or commonly sinistral-normal (Fig. 12D).

Hanagita Fault System, Summit Lake Region

The overall orientations of the shear zones and their mineral lineations in the Summit Lake region are very similar to those of the Kiagna and Granitic Creek regions (Figs. 4, 13A, and 13B), with WNW strikes, generally steep N dips, and gently WNW-ESE–plunging lineations. The kinematic history of the shear zones is not as easily discerned as in the Kiagna region because of the preexisting fabrics in the schist and mélange. Where WNW-striking shear zones occur in plutonic rocks they consistently show dextral slip (Fig. 13A). Microstructure from shear zones in the schists also show right-lateral slip, based on offset porphyroblasts and S-C orientations of ribbon quartz (Fig. 8B). These more ductile shear zones, although not as abundant as in the Hanagita fault to the east, occur over one km N of the north strand of the Hanagita fault and over 2 km S of the south strand. The strong subvertical fabric in the McHugh complex between the main strands of the Hanagita fault may also be a shear zone fabric, but the rocks are too uniform in composition and too low grade to find evidence for kinematic indicators. Many surfaces contain lineations, but it is not clear if these are intersection lineations, between the high-angle and a subsequent lower-angle fabric in the McHugh, or whether they are stretching lineations related to slip. Their orientation is consistent with either interpretation (Fig. 14A). Zones of intense, high-angle fabric also occur in the McHugh complex up to 5 km south of the Hanagita fault (Fig. 4), and within 1–2 km of the projection of the Border

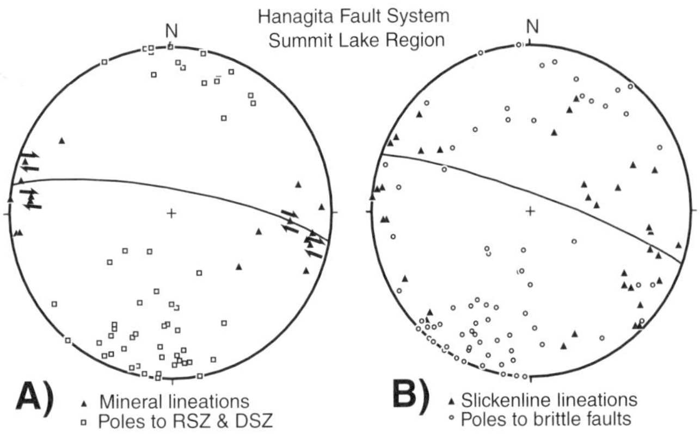

Figure 13. Equal-area lower-hemisphere projection of structures in the Hanagita fault zone, from Summit Lake south to end of map area (see Figs. 1 and 4 for location). A: Mineral lineations in shear zones and poles to shear zones. Average of the shear zones, shown as a line, is oriented 280°, 78°N. Arrows—same as Figure 9B. B: Slickenline lineations and poles to brittle faults, same region as Figure 13A. Average orientation of brittle faults, shown as a line, is 290°, 82°N.

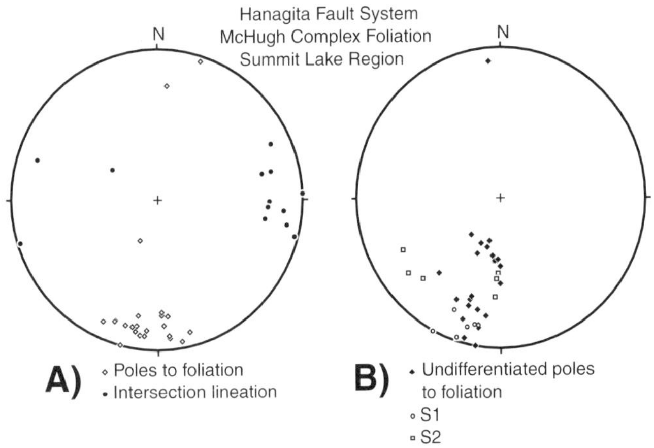

Figure 14. Equal-area lower-hemisphere projection of poles to foliations in the McHugh Complex, Summit Lake region (see Fig. 4 for location). A: Intersection lineation on the foliation plane and poles to dominant foliation in McHugh within Summit Lake Valley. B: McHugh foliation south of the Hanagita fault, in the Turquoise Lake drainage and in the headwaters of upper Canyon Creek. S1 and S2 are from sites with two crosscutting foliations, undifferentiated foliations are from sites with one foliation.

Ranges fault through the ice fields. These ductile shear zones are quite distinct from the mélange fabric and show dextral slip on them (Fig. 6E).

The main difference between this area and the eastern Hanagita fault zone is that the Summit Lake region has a number of moderately N-dipping shear zones and no moderately S-dipping shear zones. These moderately N-dipping shear zones are of regional extent and locally play a significant role in determining the outcrop pattern, particularly around Turquoise Lake and upper Canyon Creek (Fig. 4). Kinematic indicators from the regional N-dipping shear zones show both dip slip and oblique slip. Rocks in the hanging walls of these oblique thrusts include all of the units, with some of the thrusts being within the McHugh Complex (uppermost Canyon Creek) and others placing mélange over schists, and schists over the Hanagita slate (Fig. 6N). The field relations indicate these S-directed thrusts occurred prior to the final stages of brittle, high-angle faulting that cut the thrusts (Fig. 4). Structural data from the McHugh Complex and Hanagita slate also indicate that the south-directed thrusts postdate the high-angle ductile shear zone fabric, if one interprets the dominant sub-vertical foliation in the McHugh Complex as forming during the development of dextral shear zones seen in the adjacent metaplutonic rocks. Two distinct orientations of penetrative, planar foliation occur (Figs. 14A and 14B), a dominant subvertical fabric and a less abundant, moderately N-NE-dipping fabric. The subvertical fabric occurs throughout the region but is the only fabric in McHugh complex between the main strands of the Hanagita fault (Fig. 14A). South of the south strand of the Hanagita fault usually only one fabric is developed, but it may have either a steep or moderate dip. In areas near the N-dipping thrusts a strong second cleavage is developed, clearly overprinting the subvertical cleavage. The lower-angle cleavage becomes the dominant fabric within ~100–200 m below the thrusts seen in upper Canyon Creek.

The brittle fault history in the Summit Lake region is more extensive and complex than sites along the eastern Hanagita fault zone. Although many brittle faults have a similar orientation to the shear zones (Fig. 13B), a significant number show a wide range of high- and low-angle orientations. Through-going subvertical, NW-striking faults with subhorizontal slickenlines are the dominant set, occurring both along the south strand of the Hanagita fault as well as 3 km away both north and south of the Hanagita fault strands (Fig. 4). They range in scale from 100 m wide fault zones with unknown amounts of offset to closely spaced fault planes with a few cm of dextral slip (Fig. 6F). Mesoscopic brittle faults N of Summit Lake, in Sheep canyon, are in a more westerly striking orientation compared to the dominant regional strike but have dominantly gently plunging mineral lineations and dextral slip, consistent with their formation as P-shears relative to the through-going brittle faults (Fig. 15A). South of the south strand of the Hanagita the faults are more closely similar to the through-going strands (Fig. 15B), striking NW and with dextral slip seen on subhorizontal slicks. Of those faults that strike NE, the two that have kinematic indicators show sinistral-slip and sinistral-oblique normal with increasing plunge, suggesting they formed as anti-Riedel faults in the dextral-slip system.

The latest phase of slip on the west-northwest–striking faults in the Summit Lake region is normal, with N side down. This displacement is seen at all scales and appears to be concentrated within 1–2 km of either side of the main strands of the Hanagita fault. The northwest-striking fault with normal displacement plotted on Fig. 15B is a few cms thick and has

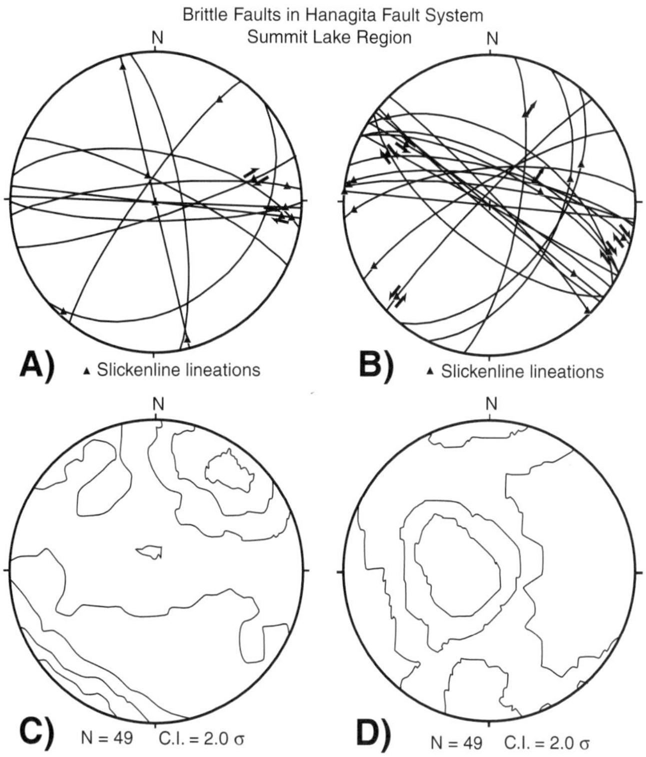

Figure 15. Equal-area lower-hemisphere projection, brittle fault slip data, Summit Lake region. A: Brittle faults with lineations, Sheep Canyon region (north of Hanagita fault) (Fig. 4). B: Subset of brittle faults in Figure 13B—brittle faults with lineations, Turquoise creek and drainage to the southeast. C: Kamb contour plot of poles to faults in the north strand of the main Hanagita fault. D: Kamb contour plot of slickenline lineations on faults of Figure 14A.

less than 2 m of apparent offset. On the scale of 500 m high cliffs, apparent offset of marble layers on a scale of tens of meters is clearly visible, with sub-vertical, N-side-down displacement occurring on orange-stained faults. Normal fault motion also reactivated the north strand of the Hanagita fault, forming a broad, ~150 m wide brittle fault zone with abundant slickenline-covered fault planes. The faults strike parallel to the regional trend of the Hanagita fault (Fig. 15C) but have almost pure dip-slip motion on them (Fig. 15D).

ARGON THERMOCHRONOLOGY

We used argon thermochronology to determine the cooling and crystallization history of rocks within shear zones and rocks adjacent to the shear zones in order to both constrain and directly date the timing of displacement within the Hanagita fault system. Because igneous hornblende is ubiquitous in the older crystalline rocks and clearly in places has been deformed during later shearing, much effort was focused on defining argon systematics in undeformed and deformed hornblende. The temperatures of deformation in the shear zones were not sufficient to crystallize new hornblende, but many of the hornblende separates from the ductile shear zones have two distinct trapped argon components, distinguishable on isotopic correlation diagrams. These diagrams, in conjunction with electron images, microprobe analysis of the phases, and K/Ca plots, allow us to make meaningful interpretations of complex spectra. White mica is also plentiful in sheared rocks and because the Mesozoic dioritic rocks are generally void of white mica, white mica in foliations and shear zones likely formed during younger periods of fluid and/or heat flow related to metamorphism and deformation. Igneous biotite is generally replaced by chlorite in the deformed rocks, and with a few exceptions (see samples 59a and 77d, Appendix, Table A-3 [see footnote 1]) did not occur as a secondary phase. K-feldspar is generally absent from the Mesozoic igneous rocks as an igneous phase but in one case provided constraints on the low-*T* portion of the regional cooling history.

Argon data for samples of this study are summarized in Table 1 with detailed data presented for each sample in Table A-2 (Appendix). Argon geochronologic interpretive and diagrammatic methods are described in Snee (2002), McDougall and Harrison (1999), and York (1969) and laboratory methods are summarized in the notes at the end of Table A-2, Appendix, and in Snee (2002). Age spectra and isochrons for selected samples are shown in Figures 16, 17, 19–21 and in Figures A-1 and A-2 in the Appendix.

Hornblende Separates

Hornblende separates are all from variably deformed and retrograded metaplutonic rocks, and the $^{40}Ar/^{39}Ar$ data from the majority of the separates yield dates on two distinct events: a latest Jurassic–earliest Cretaceous cooling from igneous temperatures, and crystallization of new phases during low-grade metamorphism associated with deformation of the plutonic rocks. Nine of 16 hornblende separates give plateau or weighted mean plateau dates and all samples yield high-temperature isochron dates. These dates, which give the cooling ages from intrusion of the plutons, are remarkably consistent with the plateau dates falling between 142 and 146 Ma, with one notable exception (Table 1; Figs. 16 and 17; Figs. A-1 and A-2 and Table A-2, Appendix). A diorite ~800 m north of the Border Ranges fault (90SR35) (Figs. 2 and 16), and thus at the southern edge of the Hanagita fault zone, has a cooling age of 171.5 ± 1.1 Ma. This plutonic rock is petrographically identical to those that have Late Jurassic–Early Cretaceous cooling ages, and without extensive further dating of rocks in the fault zone, the distribution and extent of the older diorites is unknown.

We collected the metaplutonic samples from shear zones with the intent of determining the timing of deformation, because they appear to have a ductile fabric in hand sample and thus amphibole may have recrystallized. Although new amphibole did form in fractures within the hornblende, it is K-poor actinolite (Table A-3, Appendix). Back-scattered electron images and microprobe analyses of fractions from the mineral separates

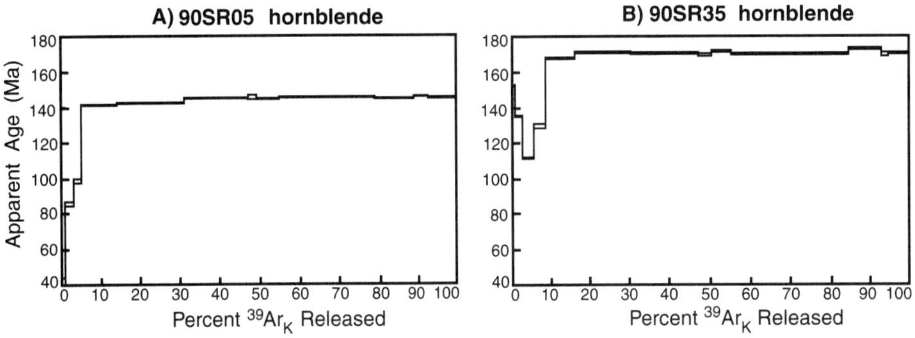

Figure 16. Argon release spectra from hornblende separates from weakly deformed diorites, Hanagita fault zone, Kiagna River region (Fig. 2). A: 90SR05—plateau date of 144.6 ± 0.6 Ma. B: 90SR35—weighted mean date of 170.6 ± 0.6 Ma.

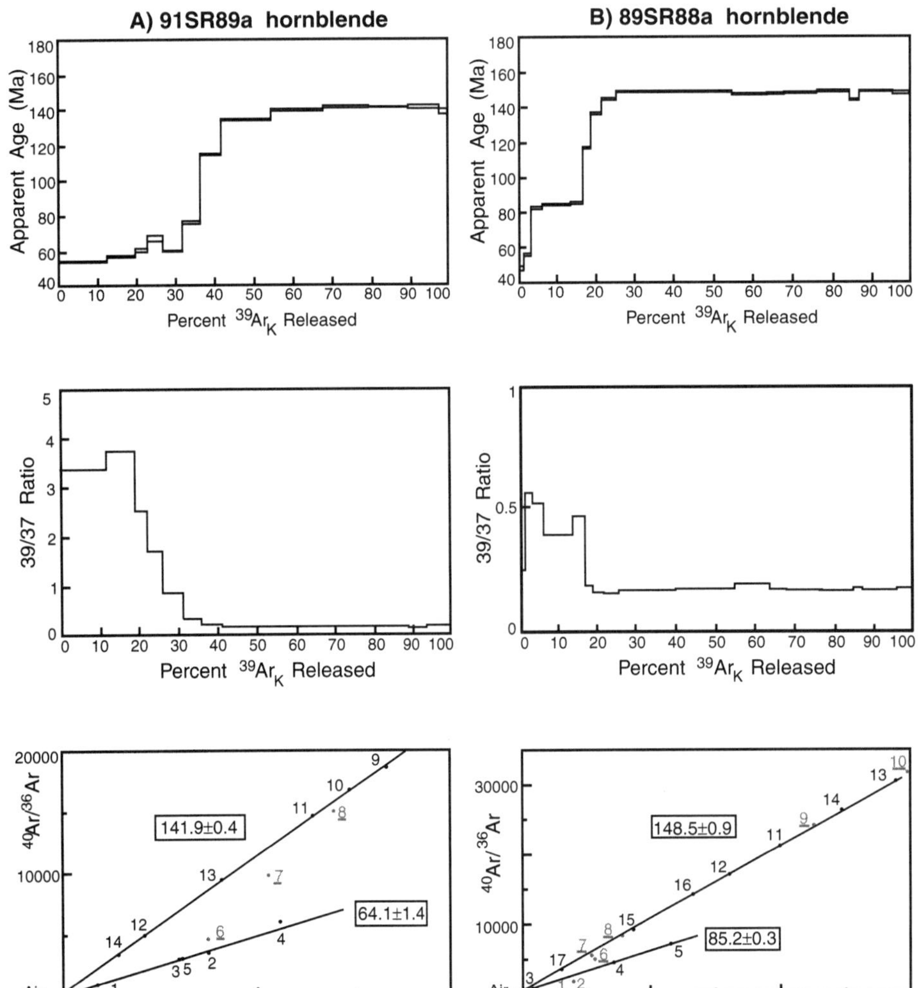

Figure 17. Ar data from two representative hornblende separates from deformed plutonic rocks. Top figure for each is Ar release spectrum. Middle figure is $^{39}Ar_K/^{37}Ar_{Ca}$ value for each heating step (equal to 2 × the K/Ca ratio) (see notes, Table A-2, Appendix [see footnote 1]). Bottom figure is isochron diagram. Date in box derived from low- and high-*T* isochrons. Black numbers refer to steps included in the isochrons; gray underlined numbers were not included. See Table 1 for summary of dates and text for discussion and Table A-2, Appendix, for complete data. A: 91SR89a from quartz diorite with greenschist facies shear zone fabric (Fig. 8A), Kiagna River region. Sample location shown on Figure 2. B: 89SR88a from plastically deformed diorite, Summit Lake region.

show that many of these samples also have K-rich phases growing in the fractures, including muscovite (Fig. 18), biotite, and locally K-feldspar (Table A-3, Appendix). This mixture resulted in almost all of the hornblende samples showing a higher $^{39}Ar_K/^{37}Ar_{Ca}$ value at the lower temperature steps (Fig. 17; Fig. A-2 and Table A-2, Appendix). Because the phases are fine-grained and closely intergrown, pure mineral separates were not obtained and total gas ages from these samples are mixed ages with no geologic significance. The muscovite and biotite have about ten times the K_2O value of the hornblende (Table A-3, Appendix), however, and during step-heating muscovite releases the great majority of its radiogenic Ar at lower temperature intervals than hornblende does (700–950 °C compared to 1000–1250 °C; see Table A-2, Appendix). This allows the mixed phases to be separated on isochron diagrams and two distinct events can be distinguished in these metaplutonic rocks. Ten out of sixteen of the hornblende separates have sufficient "contamination" by K-rich phases in their fractures that a low-temperature isochron can be determined (Table 1; Fig. 17; Fig. A-2 and Table A-2, Appendix). The selection of which steps to include in the low-*T* isochrons was based on the fit to a line of the release steps' isotopic ratios. The slope of the line yields an age. Thus, each isotope correlation diagram with two lines is from separates that contain two Ar components with distinct ages. Ages from these low-*T* isochrons range from 56 to 85 Ma, with 80% of them between 60 and 70 Ma. We interpret the low-*T* isochrons ages as maximum ages for the deformation event that pulled apart the hornblende and the accompanying metamorphism that crystallized secondary phases in these fractures. Some small but unknown amount of gas from the older hornblende may be mixed in with the lower-*T* fractions of the isochrons.

Sample 91SR89a is one example of a deformed quartz diorite that has phengite in fractures within the hornblende (Fig. 18) (Table A-3, Appendix). Two Ar components can be distinguished on the isotope correlation diagram. The muscovite within the hornblende separate released most of its Ar between 700 °C and 950 °C, yielding an isochron line over these temperature values that gives a date of 64.1 ± 1.4 Ma (Fig. 17A). The next 2–3 steps appear to represent mixtures between the hornblende and muscovite Ar reservoirs, based on the data plotting off the two isochron lines. The steps for $T \geq 1050$ (step 8) have a constant $^{39}Ar_K/^{37}Ar_{Ca}$ and the isochron age is 141.9 ± 0.4 Ma, similar to the plateau dates from hornblende separates from less deformed samples that do not have multiple Ar reservoirs (Fig. 16A). Sample 89SR88a, from a ductilely deformed diorite in a dextral shear zone in the Summit Lake region, yields the oldest low-*T* isochron age (85 Ma). The mineral separate from this sample contains K-feldspar, muscovite, biotite, and minor chlorite and albite in fractures within hornblende. The K-rich phases are not as abundant as in the separate from 91SR89a, and there are fewer steps in the low-*T* isochron (Fig. 17B). The low-*T* isochron age appears to be geologically meaningful, however, because the higher-K-content phase(s) clearly yielded a separate population of radiogenic Ar (Fig. 17B). Because the

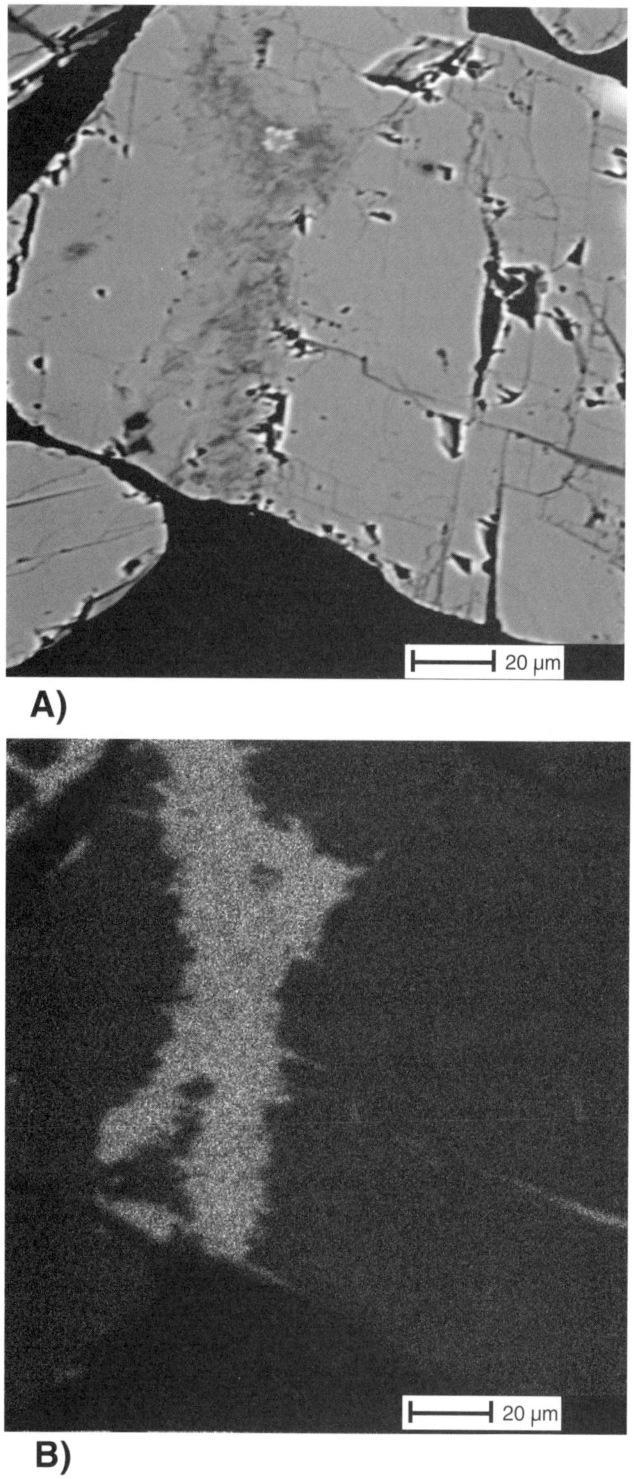

Figure 18. Hornblende separate from 91SR89a. A: Back-scattered electron image of hornblende crystal with intergrown white mica and albite in fracture. B: K X-ray image of same view as Figure 18A. Dots represent X-ray counts from K k-alpha wavelength. Image shows the distribution of relatively high potassium phase (muscovite; see Table A-3, Appendix [see footnote 1]), in relatively low potassium hornblende.

age has only three gas releases defining the isochron the age is considered a maximum for the beginning of ductile dextral slip in the Summit Lake region.

Muscovite, Biotite, and K-Feldspar Separates

The muscovite cooling ages from metamorphic rocks and igneous muscovite from small leucocratic dikes within the Hanagita fault zone yield information about the regional thermal history following the Late Jurassic plutonic event. Similarly to the hornblende spectra, they show cooling from an Early Cretaceous high-temperature event, with loss occurring in the early Cenozoic (Fig. 19). The amount of loss on the Ar spectra is directly proportional to the deformation in the sample. A muscovite from a weakly deformed muscovite-leucogranite dike within diorite (91SR69b) has a plateau date only 1 Ma younger than hornblende plateau dates from the region. Primary (igneous) K-feldspar from the same sample is surprisingly young, with an isochron age of 46 Ma (Fig. 20C). Sample 91SR27 is from amphibolite facies pelite within a south-dipping retrograde shear zone. Although the coarse muscovite is deformed, it did not lose all of its radiogenic gas from the higher-T metamorphism. A finer-grained separate from the same sample was analyzed, to see if a lower-grade muscovite phase had formed. The spectrum suggests not, based on its similarity to the coarse muscovite. These muscovite ages indicate that the Kiagna and Granitic Creek regions cooled fairly quickly from hornblende through muscovite closure temperatures (~550–400 °C) following a Late Jurassic plutonic and metamorphic event, and no regionally significant high-temperature event affected the rocks after this time. However, the region stayed at temperatures elevated enough to keep the Ar from being retained in K-feldspar until after 46 Ma.

Two mica separates from igneous rocks provide useful constraints on the minimum time of penetrative deformation in their areas. A muscovite granite dike (sample 91SR58a) that intrudes deformed diorite trends subparallel to the Hanagita fault (Fig. 2) and has brittle deformation along its margins, but internally is undeformed. The muscovite age is 54.4 ± 0.5 Ma with no loss (Fig. 20A), which we interpret as a crystallization age for the 15 m wide dike. Brittle deformation may have continued for some time afterward, but shear zone formation, such as recorded in hornblende from 91SR59a (Table 1; Fig. A-2A and Table A-2, Appendix), had ceased. Farther east and south of the main trace of the Hanagita fault, an undeformed biotite granite pluton intrudes diorite that contains retrograde shear zones (Figs. 5 and 6D) and brittle faults. The biotite has a weighted mean age of 51.5 Ma (Table 1), indicating shear zones and faults in the southern Hanagita fault zone predate this time.

Direct dating of the dextral slip with white mica separates has proven difficult because of the low temperatures associated with the deformation and the intermixing of the white mica with chlorite and epidote. Very few samples contain sufficient white mica that is coarse enough to date by $^{40}Ar/^{39}Ar$ method. Two samples of K-white mica (91SR104h and 91SR111d) from very high-strain metaplutonic rocks within the main strand of the Hanagita fault (Figs. 3 and 6K) produced plateau ages (Figs. 21B and 21C). In both these samples the white mica is clearly defining the high-strain pressure solution fabric (Fig. 8C). Thus, the ages of 54.8 and 51.4 Ma date the formation of the dextral-slip

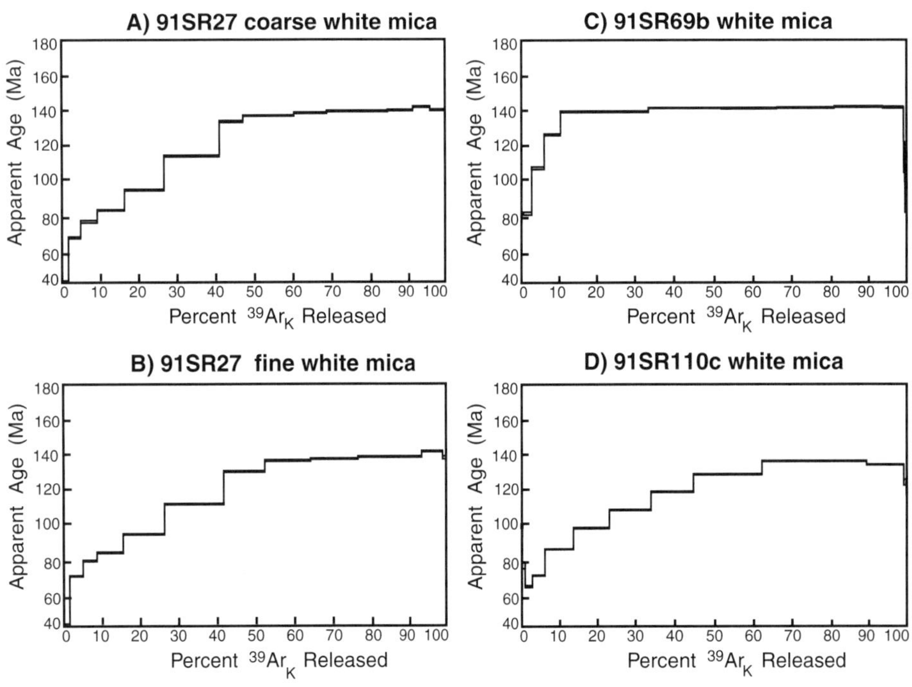

Figure 19. Ar release spectra from high-temperature ("primary") muscovite separates. A and B: An amphibolite facies pelitic schist within Hanagita fault zone. C: From muscovite leuco-granite. Sample locations shown on Figure 2. D: From muscovite trondhjemite, Granitic Creek region (Fig. 3).

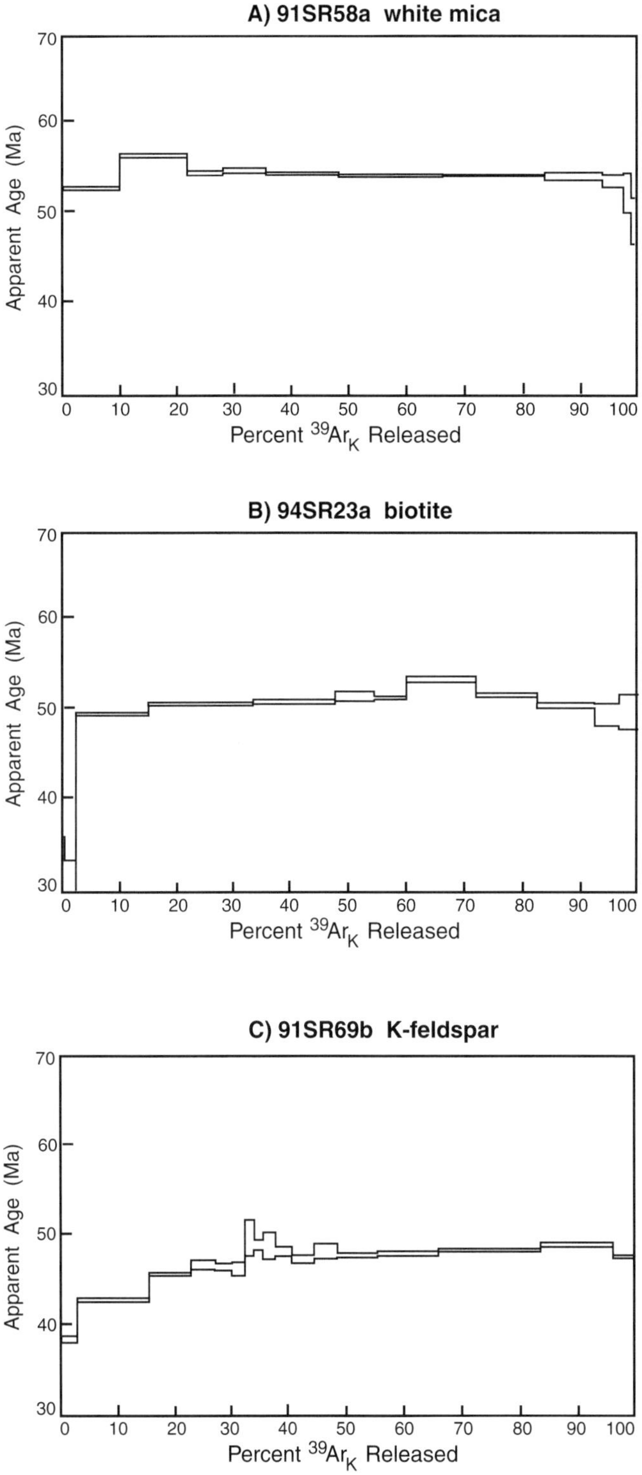

Figure 20. Ar release spectra from mica and K-feldspar within slightly deformed to undeformed igneous rocks in the Hanagita fault zone. See text and Table 1 for dates. A: 91SR58a—muscovite from Tertiary dike, north of north strand of Hanagita fault zone (Fig. 2). B: 94SR23a—biotite from undeformed granite (Fig. 5, 6D). C: K-feldspar from Late Jurassic–Early Cretaceous muscovite leuco-granite (Fig. 2).

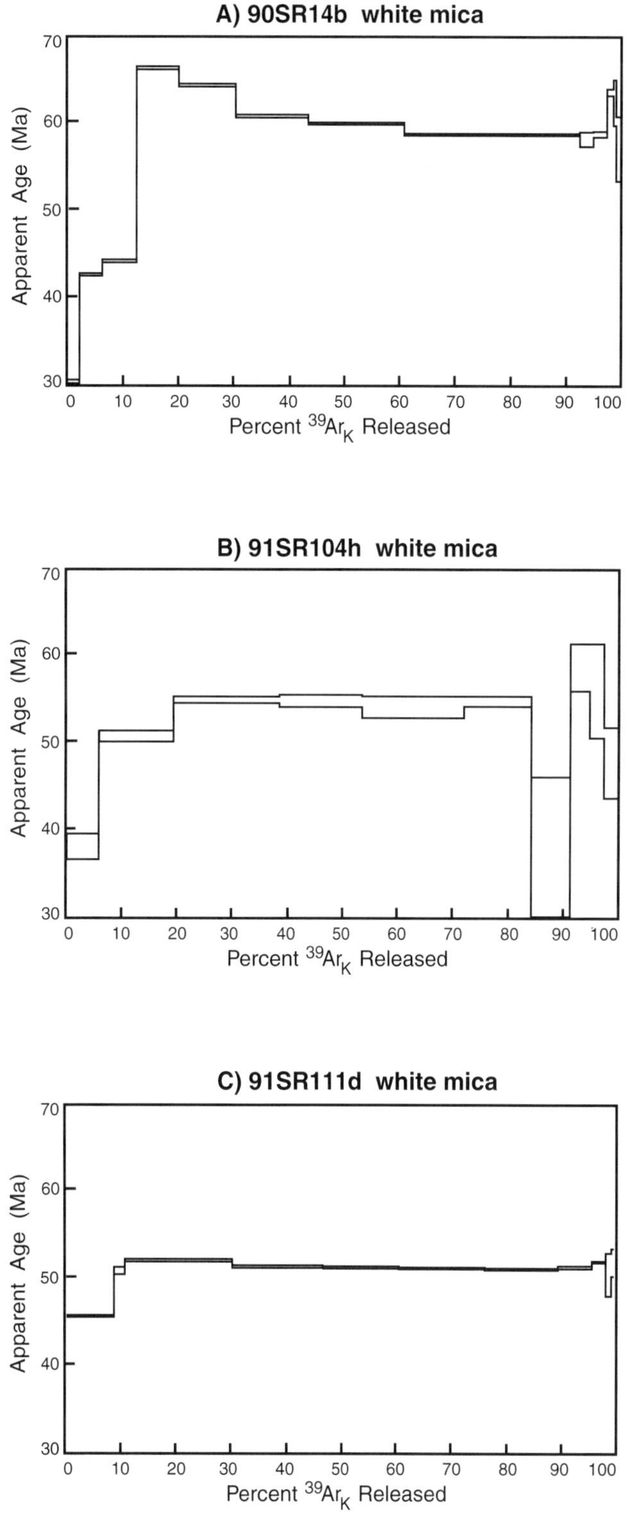

Figure 21. Ar release spectra from fine-grained muscovite within metaplutonic rocks. A: 90SR14b—weighted mean date 58.4 ± 0.2 Ma. Kiagna river region (Fig. 2). B and C: 91SR104h and 91SR111d—plateau dates 54.8 ± 0.2 and 51.4 ± 0.1 Ma respectively. Granitic Creek region (Fig. 3).

fabric on the south strand of the Hanagita fault in the Granitic Creek region. Sample 90SR14b (Fig. 21A) is from several km S of the Hanagita fault, along a brittle fault contact between red chert of the McHugh complex and a deformed diorite. The sample is from the diorite, which is pervasively altered but only locally penetratively deformed. The white mica separate from this sample is from veins that crosscut the fabric. The hump-shape of the spectra probably reflects contamination from chlorite in the mineral separate and we interpret the white mica to be ca. 58 Ma. Because the brittle fault shows evidence for right-slip fabrics, and the fluids that formed the veins most likely accompanied this deformation, this dates the formation of the brittle fault. This date is slightly younger than the low-T isochron age of 63 Ma from the hornblende separate of this sample (Table 1; Fig. A-2D and Table A-2, Appendix). The difference between these two dates may reflect the age of an overall transition to brittle behavior of the diorite at this site, but the data are not high enough quality to be certain of the significance.

CORRELATION OF THE HALEY CREEK TECTONIC ZONE WITH THE HANAGITA FAULT SYSTEM AND SUMMARY OF STRUCTURE AND GEOCHRONOLOGY

The Hanagita fault system continues at least 50 km to the west, in the region between the Copper River and the Richardson Highway (Fig. 1B). Several groups have studied the structural history in this region, but the lack of absolute ages of deformation events and the problems of separating out several generations of faults have resulted in varying interpretations. Wallace (1981) identified a fault zone, the Haley Creek tectonic zone, in the region along strike from the Hanagita fault system exposed in Summit Lake. The Haley Creek tectonic zone is 1–2.5 km wide zone of intense mylonitic deformation with a subvertical foliation. Wallace (1981) and Nokleberg et al. (1989) interpreted the subvertical fabric with subhorizontal lineations as primarily an intersection of two generations of convergent deformation. Pavlis and Crouse (1989) determined that the subhorizontal lineations indicated stretching direction and thus the mylonitic fabric formed in a strike-slip fault system. They did not find convincing evidence for the type or timing of slip, however. Extrapolating our results of the structural history of the Hanagita fault system to the west accounts for many of the ambiguous features previous authors described and helps explain the relative timing of some of the events in the Summit Lake region.

$^{40}Ar/^{39}Ar$ ages that record mineral growth during the low-T metamorphism along the Hanagita fault system are distributed evenly from 51.5 to 70 Ma, with one date as old as 85 Ma (Table 1). Because of the break in time between the 85 and 70 Ma dates and the possibility that the former is a mixed age, it is not clear if dextral slip occurred continuously during that time or not. Dates between 70 Ma and ca. 58–56 Ma are from samples that have a higher-T deformation fabric and are from shear zones south of the through-going strands. The dates of 51.4 and 54.8 Ma are from the south strand of the Hanagita fault and record dextral slip during cataclasis and extensive pressure solution. Thus, the dates and deformation mechanisms indicate that the dextral-slip system became narrower and developed at lower temperatures with time, becoming focused by the Early Eocene on the through-going strands of the Hanagita fault. The timing of this transition from a broad to narrow fault system correlates approximately with passage of the ridge-trench-trench triple junction, based on extrapolation of peak metamorphic age from the cooling ages of igneous rocks in the adjacent accretionary prism (Sisson et al., 1989). The narrowing of the fault system may be related to the high volume of fluids focused on the main strands of the Hanagita fault system during the triple junction passage. A minimum age of deformation has not been determined by this study because no undeformed plutonic rock crosscuts the whole fault system. Regional correlations, however, (see below) suggest major displacement did not occur post–early Eocene.

Despite the wide range of fault and shear zone orientations within the system, the geometry and absolute ages are consistent with a long-lived dextral fault system during accretion of the flysch-facies part of the prism (Fig. 22). This oblique convergent history probably accounts for the complex relationship between thrust and high-angle slip seen in the Hanagita fault zone. Absolute ages from northeast-directed shear zones in the Kiagna area, for example (samples 06 and 07, Figs. A-2E and A-2H, Appendix; Table 1), overlap ages from dextral-slip shear zones in the region, indicating that the thrust and strike-slip faults were broadly concurrent. The south-directed thrusts in the Summit Lake area are not dated, but relative timing and regional geology indicate they formed after the higher-T dextral deformation but prior to the brittle dextral slip and normal slip. The studies west of the Copper River identified a major phase of south-directed thrusting that placed crystalline rocks of the Haley Creek terrane over the Valdez Group (Wallace, 1981; Nokleberg et al., 1989; Plafker et al., 1989). This event is bracketed between the depositional age of the Valdez Group, Late Campanian-Maastrichtian, and the age of non-penetratively deformed dikes in the region, ca. 52 Ma (Plafker et al., 1989). Similar to the Summit Lake area, this thrust is offset by high-angle faults (Fig. 1B), which strike into the brittle faults of the Hanagita fault system. Thus, the south-directed thrust faults in the Summit Lake area are part of a regional event that affected the western Hanagita fault system during the Paleocene (Fig. 22).

The absence of Valdez Group flysch within the Hanagita fault zone east of the Copper River presents an intriguing problem. As discussed above, the Hanagita slate could be mistaken for the Valdez Group, but the field relations unequivocally show the former was deposited in a forearc basin. The absolute ages indicate dextral deformation was occurring during and after accretion of the Valdez Group, yet none of the anastamosing shear zones within the fault zone interleave flysch in with the schists and diorites. If one assumes that the structural history of the Hanagita fault zone is similar along either side of the

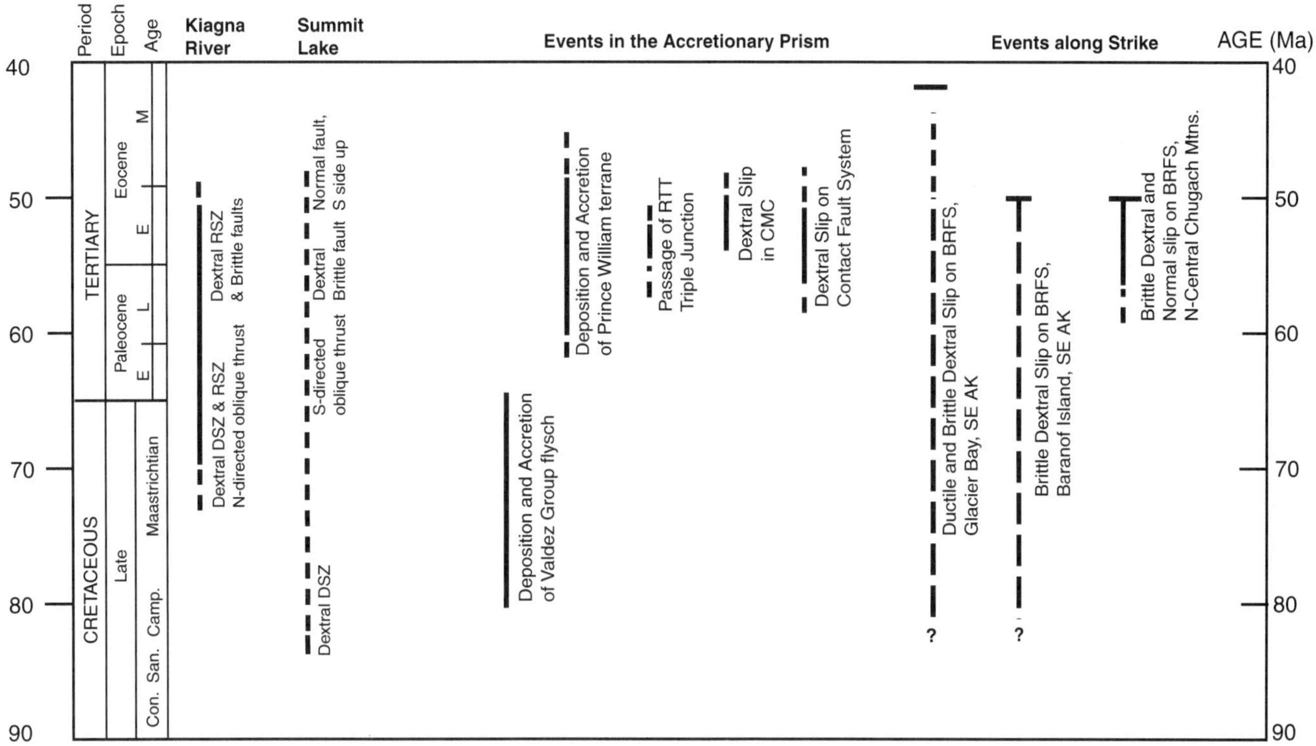

Figure 22. Timing of deformation along the Hanagita fault zone compared to events recorded in the Chugach accretionary complex and along strike of the Hanagita fault. See text for discussion. BRFS—Border Ranges fault system; CMC—Chugach metamorphic complex; DSZ—ductile shear zone; RSZ—retrograde shear zone, RTT—ridge-trench-trench.

Copper River, then Valdez Group flysch should be structurally below the Hanagita fault zone (Fig. 1B). The main difference between the two sites reflects the amount of vertical movement on later faults, with the Haley Creek region representing the most deeply exhumed portion of the fault zone.

DISCUSSION

Evidence for Hundreds of Kilometers of Dextral Slip

Length of Fault System

Major strike-slip fault zones that have tens of millions of years of displacement might be expected to have displacement on the same order or more of the length of the fault zone. Our mapping identifies a continuous fault zone 225 km long, from 145°W to 141°W (the Alaska/Canada border). The eastern end disappears under ice fields, but extrapolation of the dextral shear zone along strike to the southeast ties it into a very similar shear zone in Glacier Bay (Fig. 1A) (Smart et al., 1996). The dextral-slip history in Glacier Bay is more loosely constrained than in the eastern Chugach to between 123 Ma (Smart et al., 1996) and 42 Ma (Sisson et al., Chapter 13, this volume) (Fig. 22). Brittle dextral slip also has been identified on the southernmost extent of the Border Ranges fault system, on Baranof Island southwest of Chichagof Island (Fig. 1A). This strand of the fault is cut by a 50 Ma pluton (Haeussler et al., 1995) and is on strike with the shear zone in Glacier Bay (Fig. 22). Combining the dextral fault strands of the Border Ranges fault system, from Baranof Island to the Richardson Highway, gives an along strike length of ~750 km.

A possible western continuation of the younger phase of deformation on the Hanagita fault exists ~250 km to the west of the Summit Lake region. Little (1990) describes a dextral and oblique normal slip history along a complex series of brittle faults in the vicinity of the Border Ranges fault system. The particular deformation Little studied occurred in the Early Eocene, during deposition of a coarse clastic sequence, the Chickaloon Formation, immediately north of the Border Ranges fault (Fig. 22). This deformation predates Middle Eocene dikes (Little and Naeser, 1989) and is thought to have resulted in tens of km to no more than 100 km of dextral displacement within an eight million year time span (Little, 1990).

Offset Geologic Features

Identifying direct evidence for the magnitude of slip along large strike-slip fault systems is always difficult due to a common lack of definitive piercing points. This problem is exacerbated along the Hanagita fault system because the strike-slip fault runs parallel to regional geologic trends, and the protoliths in both the inboard and outboard terranes are non-unique.

Paleozoic schists and Mesozoic diorites are common along most of the Wrangellia superterrane margin, and the McHugh Complex is a uniformly heterogeneous mélange along its whole strike length of 2000+ km.

The best direct evidence for major displacement is from the 170 Ma diorite in the outer part of the Hanagita fault system in the Kiagna River region. This pluton is associated with schist and intrusive rock dissimilar from those inboard of the fault system (see above). The 170 Ma age is distinct from dates of diorite and quartz diorite immediately to the north in Wrangellia, the Chitina batholith and its along-strike equivalent in Canada, the Saint Elias Plutonic Suite. Hornblende K-Ar dates from the Chitina batholith and related rocks range from 130 to 165 Ma, with most between 140 and 160 Ma (Hudson, 1983; Dodds and Campbell, 1988; Plafker et al., 1989). The closest locale to the southeast with any plutons reported to have a ca. 170 Ma age system is on Chichagof Island, ~600 km on strike from the Kiagna river region (Figs. 1A and 23). A diorite with a biotite $^{40}Ar/^{39}Ar$ date of 169 Ma intrudes the Alexander terrane, immediately northeast of a Cenozoic fault separating it from the Chugach terrane (Karl, Layer, and Haeussler, 2000, written commun.). Most of the plutonic rocks in this region are substantially older than this date, however. This potential correlation of plutonic rocks, however, provides the best estimate for the minimum amount of total offset on the Hanagita fault system.

Another site, much farther south, provides a more convincing lithologic match with schists and plutonic rocks in the Hanagita fault system. The outermost part of Wrangellia on Vancouver Island contains a thin sliver of deformed rocks, the Westcoast Crystalline Complex (Fig. 23). Arc-related plutonic rocks have U-Pb zircon ages from 190 to 168 Ma and intrude amphibolite, calc-silicate schist, and metapsammite (Isachsen, 1987; DeBari et al., 1999). The plutonic rocks include diorite, hornblende gabbro, pyroxenite, and trondhjemite (Isachsen, 1987; DeBari et al., 1999), an assemblage identical to that found in the southern parts of the Hanagita fault system. The Westcoast Crystalline Complex is bounded on the west side by a brittle fault, the Westcoast fault (Fig. 23). To the west is the Pacific Rim Complex, a Cretaceous olistostrome and mélange complex (Brandon, 1989). Structural studies from Westcoast Crystalline Complex rocks adjacent to the Westcoast fault indicate it formed as a brittle dextral strike-slip fault (Isachsen, 1987). Based on paleogeographic reconstructions of the Pacific Rim Complex, Brandon (1989) interpreted the Westcoast fault to have been active from 85 to 50 Ma with ~220 km of dextral slip occurring during this time span. Thus, the Westcoast fault is a possible southern continuation of the Hanagita fault system, based on the similarity of rocks in the fault system, the location on the western (outboard) side of Wrangellia, and the displacement history. This correlation would help resolve a long-standing problem of when and how the Wrangellia margin was truncated in southernmost British Columbia and northern Washington State. Cowan (1982) proposed removal of the Wrangellian accretionary complex post-40 Ma, but Johnson (1984) suggested significant transcurrent faulting on the Westcoast fault occurred between Late Cretaceous and middle Eocene, based on the tectonic setting of basins in the proposed strike-slip system.

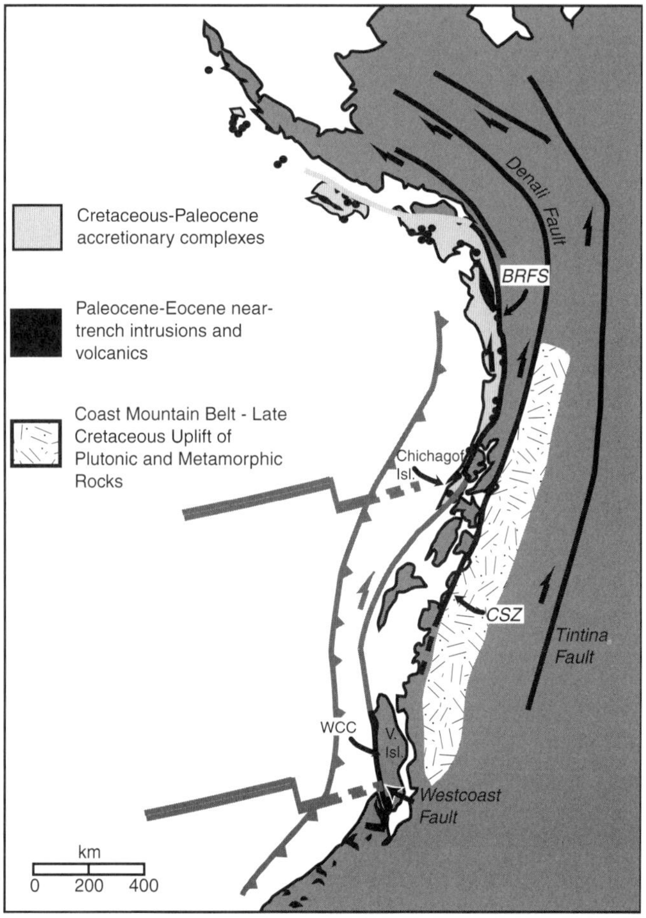

Figure 23. Two options for the location of the Kula-Farallon ridge-trench-trench triple junction at ca. 60 Ma. The northern option is permissible with known paleomagnetic and plate motion studies and allows for a minimum (~600 km) of displacement on the Hanagita fault system between 70 and 50 Ma. The southern option is also permissible with paleomagnetic data and explains anomalous volcanic and plutonic complexes of Vancouver Island and Washington State but requires unreasonably large velocities on the Hanagita fault system between 58 and 50 Ma. See text for discussion. Figure adapted from Haeussler et al., 2003. Abbreviations: CSZ—Coast shear zone, BRFS—Border Ranges fault system (includes Hanagita fault system), WCC—Westcoast Crystalline Complex, V Isl—Vancouver Island. Faults in dark gray are hypothetical locations of faults active in the Paleocene. Faults in black are Paleocene dextral-slip faults. BRFS shown in light gray (western extent) shows no evidence of strike-slip displacement in the Paleocene. Black bodies indicate anomalous magmatism interpreted to be caused by ridge-trench interaction (Bradley et al., 1993; Haeussler et al., 1995; Groome et al., Chapter 14, this volume; Haeussler et al., 2003).

Middle Jurassic plutons are common ~200–300 km to the west of the Summit Lake region, in the Peninsular terrane, but displacement from this site would require sinistral slip. Previous authors have argued for Late Jurassic–Early Cretaceous sinistral slip along this margin to explain offset between similar parts of Wrangellia on Vancouver Island and in the Wrangell Mountains and to match basins of similar age (Plafker et al., 1994; Monger et al., 1994). However, no structural evidence for through-going sinistral slip exists on the Border Ranges fault system. One might argue that Early Cretaceous sinistral slip was overprinted by Late Cretaceous dextral slip on the Hanagita fault system, but it seems unlikely that a diorite fault block from an earlier sinistral phase would end up in the outermost part of the dextral system, well outboard of numerous faults that show only dextral slip. Thus, we use the direct evidence of dextral slip from Chichagof Island to indicate a minimum displacement of 600 km.

Several geologic aspects of the Chugach accretionary complex provide supporting evidence for major dextral slip between it and the inboard crystalline terranes. A limestone clast in a cobble conglomerate in the McHugh Complex near Anchorage (Fig. 1A) contains a distinctive Early Pennsylvanian conodont, similar in age and color alteration index to conodonts from the type locale of the Strelna metamorphics (Plafker et al., 1985; Nelson et al., 1986). Rocks of this age are unknown from the Peninsular terrane and the closest occurrence of Strelna metamorphics is ~300 km to the east, just north of Summit Lake. Assuming the conglomerate did not travel a great distance from the source, the data suggest that at least some of the dextral slip on the Hanagita fault system may continue westward within the McHugh Complex, where it would be very difficult to identify.

Other indirect evidence for major dextral displacement comes from the provenance of the flysch member of the Chugach accretionary complex. Several authors have suggested that the Valdez Group and correlative Kodiak Formation were derived from the Coast Mountains belt in British Columbia, which underwent rapid exhumation during the latest Cretaceous (Fig. 23) (Nilsen and Moore, 1979; Farmer et al., 1993; Sample and Reid, Chapter 2, this volume). Although some authors have suggested the turbidites derived from this uplift flowed northward and were accreted approximately in situ (Farmer et al., 1993; Plafker et al., 1994), Sample and Reid (this volume) propose that the Coast Mountains belt and flysch of the Chugach accretionary complex formed as a paired arc/subduction zone complex and the latter was translated 1500–2000 km northward following its accretion to the margin.

Late Cretaceous to Middle Eocene Tectonic History of the Eastern Chugach Mountains, and Correlation to Events across Strike

Despite the complexity of the structural history of the Hanagita fault system, the first-order tectonic history of this late Mesozoic–early Cenozoic plate margin is very clear. Integration of the structure and geochronologic data presented above with the regional geology requires that the Chugach accretionary complex formed at a long-lived oblique convergent margin. The latest Cretaceous–Early Eocene dextral slip coincides with accretion of a massive volume of turbidites, the Valdez Group (Chugach terrane) and the Orca Group (Prince William terrane), against and progressively outward from the Hanagita fault system (Figs. 1A and 22). Accretion appears to have been continuous from Late Maastrichtian time through at least the Early Eocene (Plafker et al., 1994). The fabrics associated with accretion are also trench-parallel, which has led to previous interpretations that subduction was orthogonal (Wallace and Engebretson, 1984; Plafker et al., 1989, 1994). However, structural (Fitch, 1972; Beck, 1983; Cashman et al., 1992), geodetic (Avé Lallemant and Oldow, 2000) and seismic slip vector analysis (McCaffrey, 1992) of modern oblique convergent margins show that much of the translational component of oblique convergence may occur on a discrete strike-slip fault zone, separate from structures typically associated with accretion. The angle of convergence at which strike-slip faults develop in the overriding plate varies greatly along modern plate margins, from as little as 10° from orthogonal (Avé Lallemant and Oldow, 2000) to more typically 30°–50° from orthogonal (Jarrard, 1986; McCaffrey, 1992). Both the absolute ages from shear zones in the Hanagita fault system and the structural history are consistent with continuous dextral slip from 70 to 50 Ma. Thus, the Chugach accretionary complex formed along an oblique convergent margin both before and after passage of the ridge-trench-trench triple junction. The 85 Ma date on a dextral ductile shear zone may reflect an older phase of oblique convergence, with more orthogonal motion following, or the initiation of continuous oblique convergence.

The Hanagita fault system is not the only dextral-slip fault in the accretionary prism. Evidence for strike-slip displacement exists on shear zones within the Valdez Group (Pavlis and Sisson, Chapter 8, this volume; Pavlis et al., Chapter 7, this volume) and along the Contact fault system, between the Orca and Valdez Groups (Bol and Roeske, 1993) (Fig. 22). The timing of strike-slip within the prism is not well constrained but appears to have been more short-lived. In the Chugach metamorphic complex, a high-grade metamorphic welt in the Valdez Group, initiation of dextral shear coincides with passage of a ridge-trench-trench (RTT) triple junction at ca. 54 Ma and the transfer of the accretionary prism to the Kula plate (Pavlis and Sisson, 1995; Pavlis and Sisson, this volume) (Fig. 23). Dextral slip continued in the Chugach metamorphic complex until ca. 48 Ma (Pavlis and Sisson, this volume). Strike slip on the Contact fault and related splays occurred between late Paleocene to early Middle Eocene (Bol and Roeske, 1993).

Integration of Strike-Slip History with Plate Tectonic Models for the Northern Cordillera

One of the critical questions in northern Cordillera plate tectonics is the location of the Kula-Farallon triple junction between 85 and 50 Ma. Plate tectonic reconstructions project a reasonable extension of the Kula-Farallon ridge from the

Kula-Farallon-Pacific triple junction, which places the intersection along the west coast of North America between Mexico and Vancouver Island, migrating from south to north with time (Engebretson et al., 1985; Lonsdale, 1988; Stock and Molnar, 1988). Neither the southern (Mexico) or northern (Vancouver Island) options of this oft-cited plate model are compatible with a model of Kula-Farallon ridge migration sweeping from west to east in southern Alaska (Fig. 23). Thus, the results from this study, which indicate that the record of ridge interaction with the trench was moving with respect to North America, help resolve the contradictions between the plate models and geologic data sets. Paleomagnetic data from igneous rocks within the Chugach accretionary complex provide the only direct estimate of the paleolatitude of ridge-trench interaction, and they also locate the triple junction substantially south of the paleolatitude expected if the accretionary complex had not moved with respect to North America (Plumley et al., 1983; Bol et al., 1992). Despite large error bars on the data sets, they show northward displacement of 25° since 62 Ma (Plumley et al., 1983) and 13° since 59 Ma (Bol et al., 1992). Paleomagnetic data from Paleocene volcanics immediately north of the Border Ranges fault system indicate no northward displacement (Hillhouse and Grommé, 1982). Thus, the results from this study may help solve the dilemma of contradictory paleomagnetic data sets, because the dextral slip identified by this study occurs along the backstop to the prism, between the two sites.

One problem with using the Hanagita fault system to resolve the contradictory paleomagnetic data comes from our current knowledge of plate vectors and timing of slip on the Hanagita fault with respect to other significant faults in the northern Cordillera. The minimum age for intersection of the ridge with the trench at the location of the Kiagna river section is ca. 54–52 Ma, based on the cooling ages of a dike north of the fault zone and a granodiorite pluton within the southern part of the fault zone. Formation of the igneous rocks may lag the passage of the ridge subduction by several million years (DeLong et al., 1979), so kinematically the transfer of the prism to the Kula–North America boundary may have occurred several million years before this. Even if the forearc sliver is transferred to Kula–North America motion at 58 Ma, it leaves only ~8 m.y. to accomplish moving the Chugach terrane along the Hanagita fault system. Pavlis and Sisson (this volume) show that if all of the translational component of the Kula–North America motion is on the Hanagita fault, then the triple junction will appear to migrate south (east, in today's reference frame) while moving north with respect to North America. Complete slip partitioning of the Kula–North America motion would result in the forearc moving at 13.2 cm/yr northwest, resulting in ~1000 km of displacement in 8 m.y. The forearc sliver may have continued moving after 50 Ma, with the displacement stepping further out into the accretionary prism (Roeske et al., 1998).

The results from this study and related ones finding extensive evidence for dextral slip in the eastern Chugach accretionary complex are bringing the geologic and paleomagnetic data into closer agreement, but contradictory data sets still hinder a resolution of the location of the Kula-Farallon triple junction through time. Two options for the location of the triple junction are shown in Figure 23. The southern option, requiring complete slip partitioning of the Farallon–North America and Kula–North America motion onto the Hanagita fault, seems unreasonable, given it is rare along modern oblique convergent margins (McCaffrey, 1994). In addition, Late Cretaceous-Paleocene time coincides with dextral slip being identified on a number of faults inboard of the Hanagita fault system in southern and southeast Alaska and Canada (Fig. 23) (Gabrielse, 1985; Struik, 1993; Umhoefer and Kleinspehn, 1995; Hollister and Andronicos, 1997; Andronicos et al., 1999; Cole et al., 1999; McClelland and Mattinson, 2000; Stamatakos et al., 2001; and many others). The amount of slip on each of these faults is debated, but a conservative estimate of total slip on the Coast shear zone–Denali system and Tintina fault system is 1000 km. If one evokes oblique convergence to drive all of these faults, clearly it is important to establish whether they were all active at the same time or whether the motion was transferred across strike with time.

The northern option shown in Figure 23 still requires significant post-60 Ma displacement of the accretionary complex, which may not be compatible with the geology of Alaska inboard of the Border Ranges fault system. As summarized by Bradley et al. (Chapter 1, this volume), a number of magmatic, tectonic, and sedimentary basin features in the arc and backarc region of southern Alaska are consistent with subduction of a ridge under the southern Alaska margin in the Paleocene–Early Eocene. Thus, removing the record of the ridge-trench encounter in the prism from the record of anomalous features immediately inboard in the arc and backarc requires some explanation for the more interior events. Bradley et al. (this volume) have proposed that another plate, the Resurrection, existed between the Kula and Farallon. This may resolve some of the geologic contradictions but remains untestable as no evidence remains for the existence or relative motion of such a plate.

In conclusion, the geologic and geochronologic record of dextral slip on the backstop of the accretionary prism indicates that from the Late Cretaceous through the early Eocene the Chugach accretionary complex formed at a margin that had significantly oblique convergence. We favor a total displacement during this time span of >1000 km. A significant proportion of the slip may have happened before the passage of a ridge-trench-trench triple junction at ca. 56–58 Ma, but no direct method of estimating the rate of slip through time exists. The location of the Kula-Farallon ridge from 85 to 50 Ma with respect to North America, and the plates' relative velocities, need to be better established before undertaking serious paleogeographic reconstructions for the Canadian and Alaskan Cordillera.

ACKNOWLEDGMENTS

Invaluable field assistance for this project was provided by Mike Bunds, Katherine Long, and Jinny Sisson. Logistics in Wrangell–St. Elias National Park were smoothed by Russell Galipeau of the National Park Service. Expert flying by Ken Bunch of Glenallen provided access to remote lakes. Richard Goldfarb of the U.S. Geological Survey provided significant helicopter and field support. Funding for the project came from Louisiana Educational Quality Trust Fund grant #RD-A-25 (Pavlis), ARCO (Roeske), and NSF (Roeske; EAR-9105499). Ross Yeoman (U.S. Geological Survey, Denver) assisted with $^{40}Ar/^{39}Ar$ data acquisition. Charles Blome (U.S. Geological Survey, Denver) provided radiolaria age determinations. Jim Wright (University of Georgia) provided the U/Pb date. Structure data plots were made with Richard Allmendinger's Stereonet program. We thank George Plafker for sharing unpublished mapping and for providing the hornblende separate for sample 73AH279. Mike Gardner, then with ARCO, generously shared his discoveries of the geology of the Granitic Creek region, which prompted the initiation of this study. Discussions with Gary Winkler and George Plafker aided in our regional interpretation.

REFERENCES CITED

Andronicos, C.L., Hollister, L.S., Davidson, C., and Chardon, D., 1999, Kinematics and tectonic significance of transpressive structures within the Coast Plutonic Complex, British Columbia: Journal of Structural Geology, v. 21, p. 229–243.

Avé Lallemant, H.G., and Oldow, J.S., 2000, Active displacement partitioning and arc-parallel extension of the Aleutian volcanic arc based on Global Positioning System geodesy and kinematic analysis: Geology, v. 28, p. 739–742.

Beck, M.E., Jr., 1983, On the mechanism of tectonic transport in zones of oblique subduction: Tectonophysics, v. 93, p. 1–11.

Bol, A.J., and Roeske, S.M., 1993, Strike-slip faulting and block rotations along the Contact fault system, eastern Prince William Sound, Alaska: Tectonics, v. 12, p. 49–62.

Bol, A.J., Coe, R.S., Gromme, C.S., and Hillhouse, J.W., 1992, Paleomagnetism of the Resurrection Peninsula, Alaska: Implications for the tectonics of southern Alaska and the Kula-Farallon Ridge: Journal of Geophysical Research, v. 97, p. 17,213–17,232.

Bradley, D.C., Haeussler, P.J., and Kusky, T.M., 1993, Timing of early Tertiary ridge subduction in southern Alaska: U.S. Geological Survey Bulletin 2068, p. 63–77.

Bradley, D., Kusky, T., Haeussler, P., Goldfarb, R., Miller, M., Dumoulin, J., Nelson, S.W., and Karl, S., 2003, Geologic signature of early Tertiary ridge subduction in Alaska *in* Sisson, V.B., Roeske, S.M., and Pavlis, T.L., eds., Geology of a transpressional orogen developed during ridge-trench interaction along the North Pacific margin: Boulder, Colorado, Geological Society of America Special Paper 371, p. 19–49 (this volume).

Brandon, M.T., 1989, Deformational styles in a sequence of olistostromal mélanges, Pacific Rim Complex, western Vancouver Island, Canada: Geological Society of America Bulletin, v. 101, p. 1520–1542.

Bunds, M.P., 1994, Fluid infiltration, alteration, and deformation mechanisms in the Border Ranges fault system, eastern Chugach mountains, Alaska [M.S. thesis]: Davis, California, University of California, 88 p.

Cashman, S., Kelsey, H.M., Erdman, C.F., Cutten, H.N.C., and Berryman, K.R., 1992, Strain partitioning between structural domains in the forearc of the Hikurangi subduction zone, New Zealand: Tectonics, v. 11, p. 242–257.

Clendenen, W.S., 1991, Thermal history and vertical tectonics of the southern Alaska convergent margin [Ph.D. thesis]: Providence, Rhode Island, Brown University, 177 p.

Clendenen, W.S., Fisher, D., and Byrne, T., 2003, Cooling and exhumation history of the Kodiak Accretionary Prism, Southwest Alaska, *in* Sisson, V.B., Roeske, S.M., and Pavlis, T.L., eds., Geology of a transpressional orogen developed during ridge-trench interaction along the North Pacific margin: Boulder, Colorado, Geological Society of America Special Paper 371, p. 71–88 (this volume).

Cole, R.B., Ridgway, K.D., Layer, P.W., and Drake, J., 1999, Kinematics of basin development during the transition from terrane accretion to strike-slip tectonics, Late Cretaceous–early Tertiary Cantwell Formation, south-central Alaska: Tectonics, v. 18, p. 1224–1244.

Cowan, D.S., 1982, Geological evidence for post–40 m.y. B.P. large-scale northwestward displacement of part of southeastern Alaska: Geology, v. 10, p. 309–313.

Cowan, D.S., 1994, Alternative models for the mid-Cretaceous paleogeography of the western Cordillera: GSA Today, v. 4, p. 181–186.

Dalrymple, G.B., Alexander, E.C., Lanphere, M.A., and Kraker, G.P., 1981, Irradiation of samples for $^{40}Ar/^{39}Ar$ dating using the Geological Survey TRIGA reactor: U.S. Geological Survey Professional Paper 1176, 55 p.

DeBari, S.M., Anderson, R.G., and Mortensen, J.K., 1999, Correlation among lower to upper crustal components in an island arc: The Jurassic Bonanza arc, Vancouver Island, Canada: Canadian Journal of Earth Sciences, v. 36, p. 1371–1413.

DeLong, S.E., Schwarz, W.M., and Anderson, R.E., 1979, Thermal effects of ridge subduction: Earth and Planetary Science Letters, v. 44, p. 239–246.

Dodds, C.J., and Campbell, R.B., 1988, Potassium-argon ages of mainly intrusive rocks in the Saint Elias Mountains, Yukon and British Columbia: Geological Survey of Canada Paper 87-16, 43 p.

Engebretson, D.C., Cox, A., and Gordon, R.G., 1985, Relative motions between oceanic and continental plates in the Pacific basin: Boulder, Colorado, Geological Society of America Special Paper 206, 59 p.

Farmer, G.L., Ayuso, R., and Plafker, G., 1993, A Coast Mountains provenance for the Valdez and Orca groups, southern Alaska, based on Nd, Sr, and Pb isotopic evidence: Earth and Planetary Science Letters, v. 116, p. 9–21.

Fitch, T.J., 1972, Plate convergence, transcurrent faults and internal deformation adjacent to southeast Asia and the western Pacific: Journal of Geophysical Research, v. 77, p. 4432–4460.

Fleck, R.J., Sutter, J.F., and Elliot, D.H., 1977, Interpretation of discordant $^{40}Ar/^{39}Ar$ age spectra of Mesozoic tholeiites from Antarctica: Geochimica et Cosmochimica Acta, v. 41, p. 15–32.

Gabrielse, H., 1985, Major dextral transcurrent displacements along the Northern Rocky Mountain Trench and related lineaments in north-central British Columbia: Geological Society of America Bulletin, v. 96, p. 1–14.

Grantz, A., Jones, D.L., and Lanphere, M.A., 1966, Stratigraphy, paleontology, and isotopic ages of upper Mesozoic rocks in the southwestern Wrangell Mountains, Alaska: U.S. Geological Survey Professional Paper 550-C, p. C39–C47.

Grantz, A., Moore, T.E., and Roeske, S.M., 1991, A-3 Gulf of Alaska to Arctic Ocean: Boulder, Colorado, Geological Society of America, Centennial Continental/Ocean Transect no. 15, 3 sheets with text, scale 1:500,000.

Groome, W.G., Thorkelson, D.J., Friedman, R.M., Mortensen, J.K., Massey, N.W.D., Marshall, D.D., and Layer, P.W., 2003, Magmatic and tectonic history of the Leech River Complex, Vancouver Island, British Columbia: Evidence for ridge-trench intersection and accretion of the Crescent Terrane, *in* Sisson, V.B., Roeske, S.M., and Pavlis, T.L., eds., Geology of a transpressional orogen developed during ridge-trench interaction along the North Pacific margin: Boulder, Colorado, Geological Society of America Special Paper 371, p. 327–353 (this volume).

Haeussler, P.J., Davis, J.S., Roeske, S.M., and Karl, S.M., 1994, Late Mesozoic and Cenozoic faulting at the leading edge of North America, Chichagof and Baranof Islands, southeastern Alaska: Geological Society of America Abstracts with Programs, v. 26, no. 7, p. 317.

Haeussler, P.J., Bradley, D.C., and Goldfarb, R.J., Snee, L.W., and Taylor, C.D., 1995, Link between ridge subduction and gold mineralization in southern Alaska: Geology, v. 23, p. 995–998.

Haeussler, P.J., Bradley, D.C., Wells, R.E., and Miller, M.L., 2003, Life and death of the Resurrection plate: Evidence for an additional plate in the NE Pacific in Paleocene-Eocene time: Geological Society of America Bulletin, v. 115, p. 867–880.

Hillhouse, J.W., and Grommé, C.S., 1982, Limits to northward drift of the Paleocene Cantwell Formation, central Alaska: Geology, v. 10, p. 552–556.

Hollister, L.S., and Andronicos, C.L., 1997, A candidate for the Baja British Columbia fault system in the Coast Plutonic Complex: GSA Today, v. 7, no. 11, p. 1–7.

Hudson, T., 1983, Calc-alkaline plutonism along the Pacific rim of southern Alaska, *in* Roddick, J.A., ed., Circum-Pacific plutonic terranes: Boulder, Colorado, Geological Society of America Memoir 159, p. 159–169.

Isachsen, C.E., 1987, Geology, geochemistry, and cooling history of the Westcoast Crystalline complex and related rocks, Meares Island and vicinity,

Vancouver Island, British Columbia: Canadian Journal of Earth Science, v. 24, p. 2047–2064.

Jarrard, R.D., 1986, Relations among subduction parameters: Review of Geophysics, v. 24, p. 217–284.

Johnson, S.Y., 1984, Evidence for a margin-truncating transcurrent fault (pre–late Eocene) in western Washington: Geology, v. 12, p. 538–541.

Jones, D.L., and Berg, J.C., 1964, Cretaceous stratigraphy of the McCarthy A-4 quadrangle, southern Alaska: U.S. Geological Survey Bulletin 1180-A, 18 p.

Jones, D.L., and MacKevett, E.M., Jr., 1969, Summary of Cretaceous stratigraphy in part of the McCarthy quadrangle, Alaska: U.S. Geological Survey Bulletin 1274-K, 19 p.

Karig, D.E., Lawrence, M.B., Moore, G.F., and Curray, J.R., 1978, Structural framework of the fore-arc basin, NW Sumatra: Journal of the Geological Society of London, v. 137, part 1, p. 77–91.

Little, T.A., 1990, Kinematics of wrench and divergent-wrench deformation along a central part of the Border Ranges fault system, northern Chugach mountains, Alaska: Tectonics, v. 9, p. 585–611.

Little, T.A., and Naeser, C.W., 1989, Tertiary tectonics of the Border Ranges fault system, Chugach Mountains, Alaska: Deformation and uplift in a forearc setting: Journal of Geophysical Research, v. 94, p. 4333–4359.

Lonsdale, P., 1988, Paleocene history of the Kula plate: Offshore evidence and onshore implications: Geological Society of America Bulletin, v. 100, p. 733–745.

MacKevett, E.M., Jr., 1978, Geologic map of the McCarthy quadrangle, Alaska: U.S. Geological Survey Miscellaneous Geologic Investigations Map I-1032, 1 sheet, scale 1:250,000.

MacKevett, E.M., Jr., and Plafker, G., 1974, The Border Ranges fault in south-central Alaska: U.S. Geological Survey Journal of Research, v. 2, p. 323–329.

McCaffrey, R., 1992, Oblique plate convergence, slip vectors, and forearc deformation: Journal of Geophysical Research, v. 97, p. 8905–8915.

McCaffrey, R., 1994, Global variability in subduction thrust zone-forearc systems: Pageoph, v. 142, p. 173–224.

McClelland, W.C., and Mattinson, J.M., 2000, Cretaceous-Tertiary evolution of the western Coast Mountains, central southeastern Alaska, *in* Stowell, J.J., and McClelland, W.C., eds., Tectonics of the Coast Mountains, southeastern Alaska and British Columbia: Boulder, Colorado, Geological Society of America Special Paper 343, p. 159–182.

McDougall, I., and Harrison, T.M., 1999, Geochronology and thermochronology by the $^{40}Ar/^{39}Ar$ method (2nd edition): New York, Oxford, Oxford University Press, 261 p.

Monger, J.W.H., van der Heyden, P., Journeay, J.M., Evenchick, C.A., and Mahoney, J.B., 1994, Jurassic-Cretaceous basins along the Canadian Coast Belt: Their bearing on pre-mid-Cretaceous sinistral displacements: Geology, v. 22, p. 175–178.

Nelson, S.W., Blome, C.D., Harris, A.G., Reed, K.M., and Wilson, F.H., 1986, Late Paleozoic and Early Jurassic fossil ages from the McHugh Complex, *in* Bartsch-Winkler, S., and Reed, K.M., eds., Geological studies in Alaska by the U.S. Geological Survey during 1985: U.S. Geological Survey Circular 978, p. 60–64.

Nilsen, T.H., and Moore, G.W., 1979, Reconnaissance study of Upper Cretaceous to Miocene stratigraphic units and sedimentary facies, Kodiak and adjacent islands, Alaska: U.S. Geological Survey Professional Paper 1093, 34 p.

Nokleberg, W.J., Plafker, G., Lull, J.S., Wallace, W.K., and Winkler, G.R., 1989, Structural analysis of the southern Peninsular, southern Wrangellia, and northern Chugach terranes along the Trans-Alaska Crustal Transect, Northern Chugach mountains, Alaska: Journal of Geophysical Research, v. 94, p. 4297–4320.

Nokleberg, W.J., Plafker, G., and Wilson, F.H., 1994, Geology of south-central Alaska, *in* Plafker, G., and Berg, H.C., eds., The geology of Alaska: Boulder, Colorado, Geological Society of America, Geology of North America, v. G-1, p. 311–366.

Pavlis, T.L., 1982, Cretaceous deformation along the Border Ranges fault, southern Alaska: Brittle deformation during formation of a convergent margin: Tectonics, v. 1, p. 343–368.

Pavlis, T.L., and Crouse, G.W., 1989, Late Mesozoic strike-slip movement on the Border Ranges fault system in the eastern Chugach Mountains, southern Alaska: Journal of Geophysical Research, v. 94, p. 4321–4332.

Pavlis, T.L., and Sisson, V.B., 1995, Tectonic processes during ridge subduction: Chugach metamorphic complex, southern Alaska: Geological Society of America Bulletin, v. 107, p. 1333–1355.

Pavlis, T.L., and Sisson, V.B., 2003, Development of a subhorizontal decoupling horizon in a transpressional system, Chugach metamorphic complex, Alaska: Evidence for rheological stratification of the crust, *in* Sisson, V.B., Roeske, S.M., and Pavlis, T.L., eds., Geology of a transpressional orogen developed during ridge-trench interaction along the North Pacific margin: Boulder, Colorado, Geological Society of America Special Paper 371, p. 191–216 (this volume).

Pavlis, T.L., Monteverde, D.H., Bowman, J.R., Rubenstone, J.L., and Reason, M.D., 1988, Early Cretaceous near-trench plutonism in southern Alaska: A tonalite-trondhjemite intrusive complex injected during ductile thrusting along the Border Ranges fault system: Tectonics, v. 7, p. 1179–1199.

Pavlis, T.L., Marty, K., and Sisson, V.B., 2003, Constrictional flow within the Eocene forearc of southern Alaska: An effect of dextral shear during ridge subduction, *in* Sisson, V.B., Roeske, S.M., and Pavlis, T.L., eds., Geology of a transpressional orogen developed during ridge-trench interaction along the North Pacific margin: Boulder, Colorado, Geological Society of America Special Paper 371, p. 171–190 (this volume).

Plafker, G., Jones, D., and Pessagno, E.A., Jr., 1977, A Cretaceous accretionary flysch and mélange terrane along the Gulf of Alaska margin, *in* Blean, K.M., ed., The United States Geological Survey in Alaska: Accomplishments during 1976: U.S. Geological Survey Circular 751-B, p. B41–B43.

Plafker, G., Harris, A.G., and Reed, K.M., 1985, Early Pennsylvanian conodonts from the Strelna Formation, Chitina Valley, *in* Bartsch-Winkler, S., ed., The U.S. Geological Survey in Alaska: Accomplishments during 1984: U.S. Geological Survey Circular 967, p. 71–74.

Plafker, G., Nokleberg, W.J., and Lull, J.S., 1989, Bedrock geology and tectonic evolution of the Wrangellia, Peninsular, and Chugach terranes along the Trans-Alaska Crustal Transect in the Chugach Mountains and southern Copper River basin, Alaska: Journal of Geophysical Research, v. 94, p. 4255–4295.

Plafker, G., Moore, J.C., and Winkler, G.R., 1994, Geology of the southern Alaska margin, *in* Plafker, G., and Berg, H.C., eds., The geology of Alaska: Boulder, Colorado, Geological Society of America, Geology of North America, v. G-1, p. 389–449.

Platt, J.P., 2000, Calibrating the bulk rheology of active obliquely convergent thrust belts and forearc wedges from surface profiles and velocity distributions: Tectonics, v. 19, p. 529–548.

Plumley, P.W., Coe, R.S., and Byrne, T., 1983, Paleomagnetism of the Paleocene Ghost Rocks Formation, Prince William terrane, Alaska: Tectonics, v. 2, p. 295–314.

Prawirodirdjo, L., Bock, Y., McCaffrey, R., Genrich, J., Calais, E., Stevens, C., Puntodewo, S.S.O., Subarya, C., Rais, J., Zwick, P., and Fauzi, 1997, Geodetic observations of interseismic strain segmentation at the Sumatra subduction zone: Geophysical Research Letters, v. 24, p. 2601–2604.

Roddick, J.C., 1983, Generalized numerical error analysis with applications to geochronology and thermodynamics: Geochimica et Cosmochimica Acta, v. 51, p. 2129–2135.

Roeske, S.M., Mattinson, J.M., and Armstrong, R.L., 1989, Isotopic ages of glaucophane schists on the Kodiak Islands, southern Alaska, and their implications for the Mesozoic tectonic history of the Border Ranges fault system: Geological Society of America Bulletin, v. 101, p. 1021–1037.

Roeske, S.M., Pavlis, T.L., Snee, L.W., and Sisson, V.B., 1992, $^{40}Ar/^{39}Ar$ isotopic ages from the combined Wrangellia-Alexander terrane along the Border Ranges fault system in the eastern Chugach Mountains and Glacier Bay, Alaska, *in* Bradley, D.C., and Ford, A.B., eds., Geologic studies in Alaska by the U.S. Geological Survey, 1990: U.S. Geological Survey Bulletin 1999, p. 180–195.

Roeske, S.M., Pavlis, T.L., and Sisson, V.B., 1998, Trenchward migration of a strike-slip system through the forearc during Early Eocene evolution of the Chugach accretionary complex: Eos (Transactions, American Geophysical Union), v. 79, p. 913.

Sample, J.C. and Reid, M.R., 2003, Large-scale, latest Cretaceous uplift along the northeast Pacific Rim: Evidence from sediment volume, sandstone petrography, and Nd isotope signatures of the Kodiak Formation, Kodiak islands, Alaska, *in* Sisson, V.B., Roeske, S.M., and Pavlis, T.L., eds., Geology of a transpressional orogen developed during ridge-trench interaction along the North Pacific margin: Boulder, Colorado, Geological Society of America Special Paper 371, p. 51–70 (this volume).

Samson, S.D., and Alexander, E.C., 1987, Calibration of interlaboratory $^{40}Ar/^{39}Ar$ dating standard, MMhb-1: Isotope Geoscience, v. 66, p. 27–34.

Sieh, K., Zachariasen, J., Bock, Y., Edwards, L., Taylor, F., and Gans, P., 1994, Active tectonics of Sumatra: Geological Society of America Abstracts with Programs, v. 26, no. 7, p. 382.

Sisson, V.B., Hollister, L.S., and Onstott, T.C., 1989, Petrologic and age constraints on the origin of a low-pressure/high-temperature metamorphic complex, southern Alaska: Journal of Geophysical Research, v. 94, p. 4392–4410.

Sisson, V.B., Poole, A.R., Harris, N.R., Cooper Burner, H., Pavlis, T.L., Copeland, P., Donelick, R.A., and McLelland, W.C., 2003, Geochemical and geochronologic constraints for genesis of a tonalite-trondhjemite suite and associated mafic intrusive rocks in the eastern Chugach Mountains, Alaska: A record of ridge-transform subduction, *in* Sisson, V.B., Roeske, S.M., and Pavlis, T.L., eds., Geology of a transpressional orogen developed during ridge-trench interaction along the North Pacific margin: Boulder, Colorado, Geological Society of America Special Paper 371, p. 293–326 (this volume).

Smart, K.J., Pavlis, T.L., Sisson, V.B., Roeske, S.M., and Snee, L.W., 1996, The Border Ranges fault system in Glacier Bay National Park: evidence for major early Cenozoic dextral strike-slip motion: Canadian Journal of Earth Sciences, v. 33, p. 1268–1282.

Snee, L.W., 2002, Argon thermochronology of mineral deposits: A review of analytical methods, formulations, and selected applications of the U.S. Geological Survey, Denver, Colorado: U.S. Geological Survey Bulletin 2194, 39 p.

Stamatakos, J.A., Trop, J.M., and Ridgway, K.D., 2001, Late Cretaceous paleogeography of Wrangellia: Paleomagnetism of the MacColl Ridge Formation, southern Alaska, revisited: Geology, v. 29, p. 947–950.

Steiger, R.H., and Jäger, E., 1977, Subcomission on geochronology: convention on the use of decay constants in geo- and cosmochronology: Earth and Planetary Science Letters, v. 36, p. 359–362.

Stock, J., and Molnar, P., 1988, Uncertainties and implications of the Late Cretaceous and Tertiary position of North America relative to the Farallon, Kula, and Pacific plates: Tectonics, v. 7, p. 1339–1384.

Struik, L.C., 1993, Intersecting intracontinental Tertiary transform fault systems in the North American Cordillera: Canadian Journal of Earth Sciences, v. 30, p. 1262–1274.

Trop, J.M., Ridgway, K.D., and Spell, T.L., 2003, Sedimentary record of transpressional tectonics and ridge subduction in the Tertiary Matanuska Valley–Talkeetna Mountains forearc basin, southern Alaska, *in* Sisson, V.B., Roeske, S.M., and Pavlis, T.L., eds., Geology of a transpressional orogen developed during ridge-trench interaction along the North Pacific margin: Boulder, Colorado, Geological Society of America Special Paper 371, p. 89–118 (this volume).

Umhoefer, P.J., 1987, Northward translation of "Baja British Columbia" along the Late Cretaceous to Paleocene margin of western North America: Tectonics, v. 6, p. 377–394.

Umhoefer, P.J., and Kleinspehn, K.L., 1995, Mesoscale and regional kinematics of the northwestern Yalakom fault system: Major Paleogene dextral faulting in British Columbia, Canada: Tectonics, v. 14, p. 78–94.

Wallace, W.K., 1981, Structure and petrology of a portion of a regional thrust zone in the central Chugach Mountains, Alaska [Ph.D. thesis]: Seattle, University of Washington, 253 p.

Wallace, W.K., and Engebretsen, D.C., 1984, Relationships between plate motions and Late Cretaceous to Paleocene magmatism in southwestern Alaska: Tectonics, v. 3, p. 295–315.

Winkler, G.R., 1992, Geologic map and summary geochronology of the Anchorage 1° × 3° quadrangle, southern Alaska: U.S. Geological Survey Map I-2283, 1 sheet, scale 1:250,000.

Winkler, G.R., Silberman, M.L., Grantz, A., Case, J.E., and Pickthorn, W.J., 1981, Geologic map and summary geochronology of the Valdez quadrangle, southern Alaska: U.S. Geological Survey Open-File Report 80-892-A, 2 sheets, scale 1:250,000.

York, D., 1969, Least squares fitting of a straight line with correlated errors: Earth and Planetary Science Letters, v. 5, p. 320–344.

Manuscript Accepted by the Society February 5, 2003

Printed in the USA

Geological Society of America
Special Paper 371
2003

Constrictional flow within the Eocene forearc of southern Alaska: An effect of dextral shear during ridge subduction

Terry L. Pavlis
Kevin Marty
Department of Geology and Geophysics, University of New Orleans, New Orleans, Louisiana 70148, USA

Virginia B. Sisson
Department of Earth Science, MS-126, Rice University, Houston, Texas 77005-1892, USA

ABSTRACT

Eocene deformation in metasedimentary rocks of the Valdez Group was examined along the Richardson Highway between Valdez and the Copper River basin. A topographic lineament along Stuart Creek and the lower Tiekel River separates two distinct structural domains: a southern domain (Domain 1) with two fabrics, and a northern domain (Domain 2) with three. Both domains possess an early, layer-parallel phyllitic cleavage (S_1) associated with south-directed thrust systems. In Domain 1, S_1 is steeply dipping and overprinted by a second, steeply south-dipping cleavage (S_2) with a subhorizontal elongation lineation, and F_2 folds are rare to absent. To the north in Domain 2, however, S_1 was first tightly folded during D_2 into a series of gently inclined folds (F_2) and then refolded by tight to close, steeply inclined F_3 folds that are asymmetric to the north. Metamorphic mineral assemblages indicate greenschist facies conditions accompanied the deformation: Biotite grew during D_2 in Domain 1 and during D_3 in Domain 2; and metatuffs in Domain 1 contain actinolite-chlorite-epidote-feldspar-quartz. Fluid inclusion analyses of multiple vein sets indicate a complex fluid history but indicate a metamorphic temperature of 350–400 °C, consistent with the observed mineral assemblages. Uniform metamorphic grade across the belt together with spatial variations in fabrics suggests that D_2 in Domain 2 is time-equivalent with D_3 in Domain 2.

We evaluate the kinematics of linear and planar fabric development through a combination of finite strain analysis and geometry of syn-tectonic growth-fibers developed on sand grains. Finite strains are estimated by different methods including object shapes, object centers, and stretches estimated from growth fibers. The strain data show a predominance of constrictional strains, particularly in Domain 1, with subhorizontal stretching parallel to the mesoscopic lineation. Strain superposition models indicate that the main deformation (D_2) in Domain 1 was a constrictional flow indicating that D_2 was marked by subhorizontal orogen-parallel elongation and simultaneous vertical and horizontal shortening. Shear sense indicators generally indicate dextral shear on steeply dipping surfaces, but are not always well developed because the constrictional flow deviates strongly from simple shear. These characteristics are typical of the strain facies referred to as lengthening shear to lengthening–narrowing shear flow. We interpret this large-scale kinematic pattern as a combined effect of distributed dextral shear and vertical shortening related to a coeval ridge-subduction event. The vertical shortening could be associated with lateral flow driven by

Pavlis, T.L., Marty, K., and Sisson, V.B., 2003, Constrictional flow within the Eocene forearc of southern Alaska: An effect of dextral shear during ridge subduction, *in* Sisson, V.B., Roeske, S.M., and Pavlis, T.L., eds., Geology of a transpressional orogen developed during ridge-trench interaction along the North Pacific margin: Boulder, Colorado, Geological Society of America Special Paper 371, p. 171–190.

the migrating ridge, vertical displacement gradients related to a mid-crustal decoupling zone recognized in the Chugach metamorphic complex, or both. We infer that dextral shear is distributed through a wide zone but is concentrated along the major topographic lineament that is coincident with the structural domain boundary. We informally refer to this dextral structure as the Stuart Creek shear zone and conclude that the two structural domains were juxtaposed along this structure. D_3 structures recognized north of the Stuart Creek shear zone probably represent a back-thrusting event synchronous with the dextral shear system, which may record mid-crustal strain partitioning. Locating the westward continuation of the dextral shear system is difficult due to lack of detailed structural studies and the difficulty of recognizing strike slip within the homogeneous flysch of the Chugach terrane. The dextral shear system may transfer into complex out-of sequence thrust systems if the strike-slip systems were developed during or after the Alaskan orocline.

Keywords: finite-strain, strain facies, transpression, fluid-inclusions, southern Alaska.

INTRODUCTION

The Chugach terrane of southern Alaska has long been recognized as an ancient accretionary complex, and structural studies within the complex serve as type examples for structural processes within these systems (e.g., Moore et al., 1983; Fisher and Byrne, 1987, 1992). The North America Cordillera is also famous, however, for large-magnitude strike-slip systems that systematically displaced outboard terranes northward relative to the continental interior (e.g., Grantz, 1966; Oldow et al., 1989; Plafker et al., 1994; Hillhouse and Coe, 1994). Some of these strike-slip systems are well known regional dextral fault systems (e.g., Denali and Tintina faults). Nonetheless, the net slip on these structures estimated from displaced geologic markers continues to be far lower than estimates based on paleomagnetic discordance. This discrepancy is particularly acute for the outermost part of the accretionary complex (Prince William terrane) where several paleomagnetic studies suggest northward displacements of >1500 km since the Paleocene (e.g., Plumley et al., 1983; Bol et al., 1992). This problem is related to the Baja British Columbia problem of the central Cordillera. As summarized by Cowan et al. (1997), these observations imply either a problem with the paleomagnetic database or a problem in the recognition of all of the ancient strike-slip systems within the northern Cordillera. Recent work along the Border Ranges fault system (e.g., Smart et al., 1996; Roeske et al., this volume, Chapter 6) implies that this structure experienced a major period of early Cenozoic strike slip that could account for part of this discrepancy. In addition there is also evidence for Eocene dextral shear within the Chugach metamorphic complex, the high-grade metamorphic belt developed within the Chugach terrane east of the Copper River (Pavlis and Sisson, this volume, Chapter 8). The dextral zones emanating from the Chugach metamorphic complex had to transfer that slip westward toward the central Chugach Mountains, yet previous studies (e.g., Nokleberg et al., 1989) did not recognize significant strike-slip motion within the Chugach terrane.

In this paper, we reanalyze structural relationships in the Chugach terrane along the Richardson Highway between Valdez and Tonsina (Figs. 1 and 2). We begin by presentation of new evidence from fabric analyses indicating that the main deformation in this region was dominated by EW subhorizontal elongation on a steeply dipping foliation surface. We infer a regional structure, the Stuart Creek shear zone, developed along what is now a topographic lineament, and present evidence that this structure is a dextral strike-slip zone. We show that this deformation occurred during the Eocene, with the *P-T-t* history constrained by geochronology, mineral assemblages and fluid-inclusion data from syntectonic veins. We then evaluate this kinematic interpretation through microstructural observations and finite strain analyses. We conclude that the strain was dominantly constrictional throughout most of the region. Using simple strain superposition models together with mesoscopic fabric observations, we establish that the constrictional strains are not the product of superposition of plane strain or flattening strain events. Instead, they are best interpreted as the result of a progressive deformation involving constrictional flow with dextral shear concurrent with vertical shortening. Finally, we discuss the origin of the observed strain and consider the regional significance of dextral shear within the Chugach accretionary event during the Eocene ridge-trench interaction.

TECTONIC SETTING

The Chugach terrane is the oldest part of a Mesozoic to Cenozoic accretionary complex developed from northward underflow of oceanic lithosphere beneath the northern Cordilleran subduction zone. Presently this subduction zone is the Aleutian trench, but geologic observations in southern Alaska demonstrate that during most of the Mesozoic and Cenozoic this subduction zone extended throughout the northern Cordillera (e.g., Plafker et al., 1994). The present curved margin—the Alaskan orocline—is apparently a relatively young feature developed during the early Cenozoic (Hillhouse and Coe, 1994). Thus, during the accretion of the Chugach terrane in the late Mesozoic the margin was probably more linear than it is today.

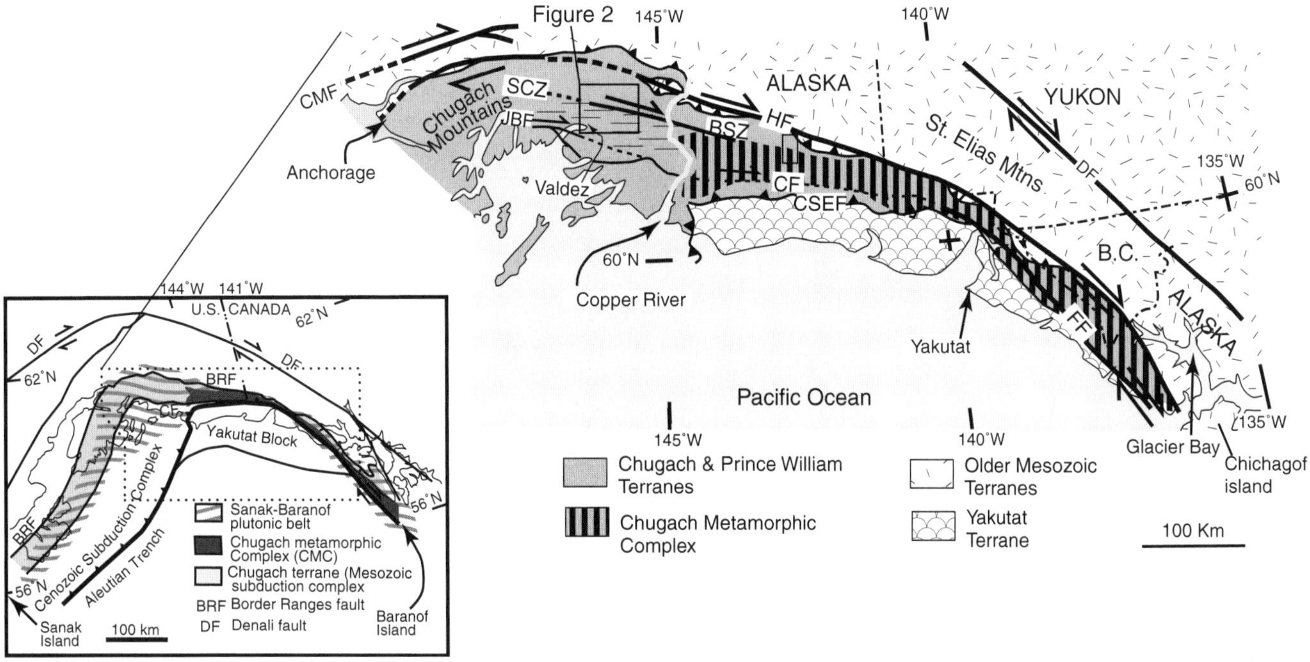

Figure 1. Tectonic map of southern Alaska showing position of major tectonic elements relevant to this study. Inset (lower left) shows the regional distribution of the Sanak-Baranof plutonic belt (after Hudson, 1979) relative to the Chugach metamorphic complex. More detailed map (after Plafker et al., 1994) shows the location of the study area and the inferred connection between the Stuart Creek shear zone and the Bremner shear zone within the Chugach metamorphic complex. BSZ—Bremner shear zone, CF—contact fault, CMF—Castle Mountain Fault, CSEF—Chugach St. Elias Fault, FF—Fairweather Fault, HF—Hanagita Fault, SCZ—Stuart Creek Zone.

The oldest parts of this subduction complex are now well exposed in a broad forearc ridge that extends northeastward from Sanak Island to the Chugach Mountains and continues eastward into the St. Elias Mountains and southward to Baranof Island in southeast Alaska (Fig. 1). Late Cenozoic uplift has exposed not only the accretionary complex but also the Mesozoic crystalline basement beneath which these rocks were accreted. The structural contact between older basement and the accretionary complex is referred to regionally as the Border Ranges fault (Fig. 1). The term Chugach terrane is generally used to describe this belt of Mesozoic accretionary complex rocks outboard of the Border Ranges fault (e.g., Plafker et al., 1994) and inboard of Tertiary accretionary assemblages. It is a composite terrane containing two distinctive rock assemblages: (1) a mélange subterrane exposed intermittently along, and complexly deformed with, crystalline rocks along the Border Ranges fault (e.g., Roeske et al., this volume, Chapter 6), and (2) a coherent interbedded graywacke and argillite assemblage referred to as the Valdez Group in the Chugach Mountains (Plafker et al., 1977, 1994). The mélange is an older assemblage that may include both early Mesozoic and Early Cretaceous accretionary rocks (e.g., Pavlis, 1982; Plafker et al., 1977, 1994). The coherent subterrane appears to be entirely latest Cretaceous in age representing a major submarine fan complex that was rapidly accreted in the latest Cretaceous-Paleocene (Tysdal and Plafker, 1978; Nilsen and Zuffa, 1982). It is this coherent "flysch subterrane" that is the primary focus of this study.

The Richardson Highway (Fig. 2) provides the only easily accessible crossing of the Chugach terrane between Anchorage and southeast Alaska (Fig. 1) and as a result has been the subject of some of the most detailed studies in Alaska (e.g., Nokleberg et al., 1989; Plafker et al., 1989). This area was the focus of the Trans-Alaska Crustal Transect (TACT) studies by the U.S. Geological Survey in the early 1980s, and contains one of the few published geologic maps at a scale larger than 1:250,000 in southern Alaska (Plafker et al., 1992). Nonetheless, these studies constitute reconnaissance studies by continental U.S. standards, and although previous work outlined the basic geologic framework, these studies did not specifically address details of fabric development that are key to kinematic interpretation.

In the last ten years, work east of the Richardson Highway has shown that Eocene dextral strike slip accompanied a major thermal event. The thermal event produced the high-grade forearc metamorphic belt that Hudson and Plafker (1982) described as the Chugach metamorphic complex (e.g., Pavlis and Sisson, 1995, this volume, Chapter 8). The most conspicuous zone of strike slip–related deformation occurs along the Border Ranges fault (Fig. 1) where the older mélange subterrane and Jurassic crystalline basement are deformed together in a 5–10 km wide band of ductile to brittle strike-slip faults and

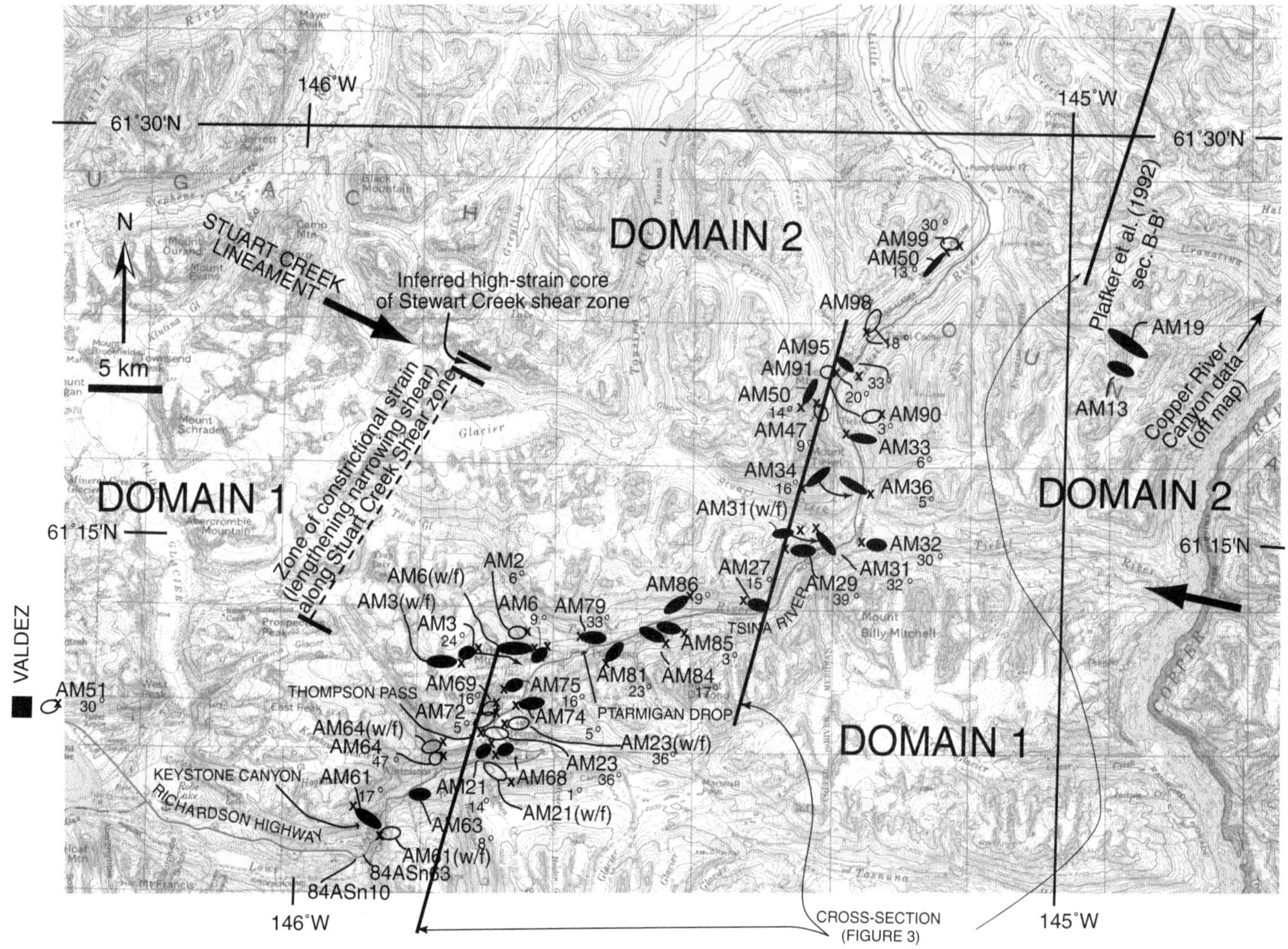

Figure 2. Location map for samples examined in this study. Base map is the 1:250,000 U.S. Geological Survey Valdez quadrangle. Ellipses plotted on map are estimated finite strains in the plane of the map as determined by Fry center-center methods as well as R_f/Φ + growth fiber methods (indicated by "w/f"). Open ellipses indicate sites where apparent flattening strains are indicated, whereas closed ellipses are for sites with plane strain to constrictional ellipsoids. The X mark at the end of each ellipse indicates the direction of plunge of the ellipsoid long axis, and the numbers adjacent to each sample are the plunge of the long axis (numbers are not repeated for samples with growth fibers). Heavy black lines show position of the cross section shown in Figure 3, but do not include the entire length of the northern end of the section, which is redrawn from Plafker et al.'s (1992) section B–B′. Bold arrows show positions of the Stuart Creek–Tiekel lineament.

shear zones (Roeske et al., this volume, Chapter 6). This strike-slip system can be traced from south of Glacier Bay on Chichagof Island and Baranof Islands, through the St. Elias Mountains and into the Chugach Mountains where it forms a prominent topographic trough along the Hanagita River (Fig. 1). Hence, Roeske et al. (this volume, Chapter 6) refer to this strike-slip rejuvenation of the Border Ranges fault as the Hanagita fault.

To the south of the Hanagita fault, the Chugach metamorphic complex shows a ductile deformational history that provides further evidence of extensive early Cenozoic dextral slip across the margin. In this region, Pavlis and Sisson (1995, this volume, Chapter 8) recognize a three-phase ductile deformation history that includes at least one period of dextral slip. Throughout the western half of the complex, early layer-parallel fabrics that are presumably related to subduction-related accretion (D_1) were overprinted by two kinematically distinct events. A second deformational event (D_2) is associated with vertical shortening and orogen-parallel (EW) elongation. A third deformational event (D_3) is related to a dextral shear system. It is this D_2-D_3 succession that is most significant to this study because mapping east of the Copper River (Pavlis and Sisson, this volume, Chapter 8) demonstrates that a major D_3 shear zone emanates from the western termination of the Chugach metamorphic complex and projects westward toward the Richardson Highway region (Fig. 1). The region between the Copper River and the Richardson Highway is known only from aerial reconnaissance and a few spot observations by the TACT group (Plafker et al., 1992). Nonetheless, there are no known crosscutting structures, and this shear zone should project along strike to the Richardson Highway. Thus, results from the Chugach metamorphic complex studies prompted a re-analysis of the structures exposed along the highway.

STRUCTURAL HISTORY OF THE RICHARDSON HIGHWAY AREA

Field Relationships

South of the Hanagita strike-slip system the Chugach and Prince William terranes show evidence of multiple phases of ductile deformation, but the kinematics of this fabric superposition were not well resolved in previous studies. Nokleberg et al. (1989) described field relationships along the Richardson Highway and recognized five phases of deformation. However, because one of these (G_1) occurs exclusively in crystalline assemblages north of the Border Ranges fault, we discard this terminology to clarify relationships within the Chugach terrane (Table 1). Four deformational phases are present within the Valdez Group, but these phases also vary across the Richardson Highway transect (Table 1). The youngest phase (D_4; equivalent to G_5 of Nokleberg et al., 1989) is limited to local open kink folds that lack a penetrative fabric and is not discussed further here.

We differentiate two structural domains along the Richardson Highway (Fig. 2) from variable overprinting of D_1 by D_2 and D_3 structures (Table 1; Figs. 3 and 4). Domain 2 corresponds to the area described by Nokleberg et al. (1989) as the "southern Valdez Group" and is distinguished by three distinct fabric-forming phases versus Domain 1 where only two phases are present (Table 1). Based on fabric geometry and kinematics described

TABLE 1. FABRIC SUMMARY

Fabric	Domain 1	Domain 2	Kinematics
D_4 (G_5)	Sporatic open NNE trending upright folds. No fabric.	Sporatic open NNE trending upright folds. No fabric.	local late shortening
D_3 (G_4)			
Folds (F_2 and F_3)	F_2—tight to close, upright folds with subhorizontal EW axes; developed only in areas of low-dipping layers.	F_3—Moderate to steeply inclined, horizontal to gently plunging, ~EW trending close to subisoclinal folds in older layering. Generally asymmetric to N.	D_2 structures in Domain 1 inferred as age equivalent of D_3 structures in Domain 2. Dextral shear concurrent with vert. Shortening in Domain 1, backthrusting in Domain 2?
Planar fabrics (S_2 and S_3)	S_2—Main phyllitic cleavage is subvertical with ~EW strike; intersects steeply N-dipping layers at a low-angle Crenulation cleavage overprint S_1 in argillite; rough cleavage in metagreywacke.	S_3—Moderately to steeply S-dipping foliation axial planar to F_3 folds. Crenulation cleavage in meta-argillite and rough cleavage in metagreywacke.	
Linear fabrics (L_2 and L_3)	S_1–S_2 intersection lineation in meta-argillites but finite strain shape fabric (extension lineation) in metagreywackes. Subhorizontall, EW plunge.	S_2-S_3 intersection lineation, subhorizontal EW plunge parallel to F_3 axes.	
D_2 (G_3)			
Folds (F_2)	Absent or indistinguishable from D_2 in Domain 1.	Close to subisoclinal folds coaxial with F_3 strong overprints by D_3 obscure geometry.	Uncertain due to intense D_3 overprints. May be partially age equivalent to D_2 in Domain 1 or distinct.
Planar fabrics (S_2)	Absent or indistinguishable from D_2 in Domain 1.	Crenulation cleavage in meta-argillites, obscure in metagreywackes. Axial planar to F_2 folds.	
Linear fabrics (L_2)	Absent or indistinguishable from D_2 in Domain 1.	S_1-S_2 intersection lineation in meta-argilites. Subhorizontal, shallow EW plunge parallel to F_2.	
D_1 (G_2)			
Folds (F_1)	Steeply inclined, subhorizontal to gently plunging, close to subisoclinal, south-vergent folds with associated thrust faults	Subisoclinal folds. Initial geometry uncertain due to complex oveprints of D_2 and D_3.	South-directed thrusting in Domain 1; presumed similar in Domain 2 but unconstrained.
Planar fabrics (S_1)	Spaced phyllitic cleavage axial planar to F_1 folds. Generally subparallel to layering (S_0)	Spaced phyllitic cleavage axial planar to F_1 folds. Generally subparallel to layering (S_0).	
Linear fabrics (L_1)	Bedding (S_0)–Cleavage (S_1) intersection.	Bedding (S_0)—Cleavage (S_1) intersection.	

Note: (G_1–G_5): Terminology of Nokleberg et al. (1989) for comparison.

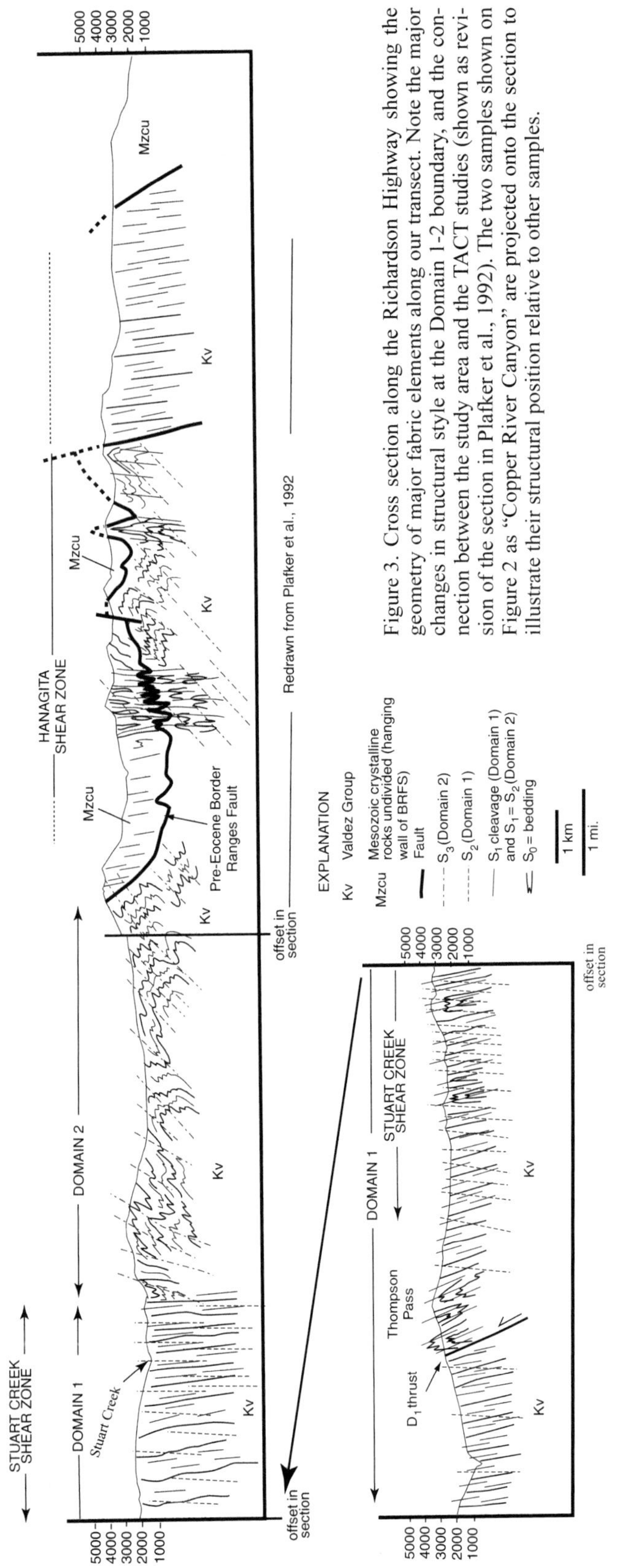

Figure 3. Cross section along the Richardson Highway showing the geometry of major fabric elements along our transect. Note the major changes in structural style at the Domain 1-2 boundary, and the connection between the study area and the TACT studies (shown as revision of the section in Plafker et al., 1992). The two samples shown on Figure 2 as "Copper River Canyon" are projected onto the section to illustrate their structural position relative to other samples.

below, we infer that D_2 in Domain 1 is equivalent to D_3 in Domain 2 (Table 1). We have not examined rocks south of Valdez, but based on TACT studies (Plafker et al., 1992) and work by Bol and Roeske (1993) it appears the general structure of Domain 1 continues at least as far south as the Contact fault (Fig. 1).

In both domains, the earliest fabric is a layer-parallel cleavage (S_1) overprinted by a one to two younger cleavages (S_2 and S_3), but the nature of the overprint varies with lithology. In argillaceous rocks D_2 and D_3 produced prominent crenulation cleavages and associated intersection lineations. In metagraywackes, however, the separate cleavages are generally difficult to distinguish in both hand specimen and in thin section. We interpret this observation as a consequence of finite strain superposition. That is, the metagraywackes do not record the overprinting as distinct fabrics. Instead they contain a shape fabric produced by superposition of finite strains from all of the ductile deformational events. This distinction is important because the overprints recorded in the argillaceous layers constrain the finite strain path and this constraint is critical to understanding the observed finite strains in the metagraywackes.

Fabric Development in Domain 1

Domain 1 is characterized by remarkably homogeneous structure over a significant width of the belt. Throughout Domain 1, bedding (S_0) is steeply dipping and two planar structural fabrics (S_1 and S_2) intersect bedding at a low angle (Figs. 3 and 4). The earliest planar fabric (S_1) is a layer-parallel, phyllitic cleavage in argillaceous units with a generally steep north dip. S_1 is overprinted by a second, crenulation cleavage (S_2) that intersects S_1 at a low angle but typically dips vertical to steeply south (Fig. 3). A few F_1 folds are apparent near Thompson Pass (Fig. 3) where they are recognized by bedding intersecting S_1 at an angle as beds roll through a fold closure with S_1 axial planar to the fold. However, these folds appear to be predominantly related to anticlinal closures along D_1 thrust faults because where facing is apparent the rocks generally face north, an observation suggesting repetition of strata by D_1 thrust faults rather than extensive D_1 folding. This D_1 structure is typical of the Chugach terrane from Kodiak Island (Sample and Fisher, 1986) to the eastern Chugach Mountains (Pavlis and Sisson, 1995) and has long been recognized as a product of subduction accretion along the Mesozoic subduction zone (Sample and Fisher, 1986; Sample and Moore, 1987).

Although D_1 produced both mesoscopic and macroscopic folds, F_2 folds are rare. Indeed, the only mesoscopic F_2 structures observed were in areas where layering dips moderately northward, and in those dip domains mesoscopic F_2 folds warp layering into open to tight F_2 folds and S_2 is axial planar to these warps. Elsewhere, the homoclinal $S_1 = S_0$ layering intersects S_2 at a low angle, and the only obvious manifestation of the two cleavages is a prominent S_1-S_2 intersection lineation in argillaceous units that is parallel to an elongation lineation in adjacent metagraywackes.

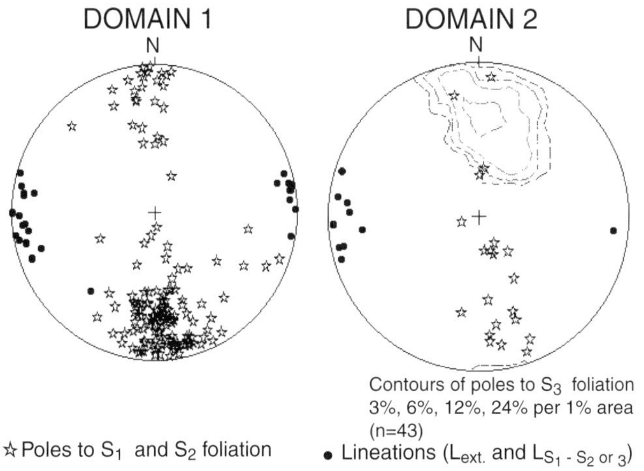

Figure 4. Equal-area stereographic projections of major fabric elements in Domains 1 and 2 along the Richardson Highway. See Nokleberg et al. (1989) for more detailed data from Domain 2.

We interpret these field relationships as evidence that D_2 was produced by subhorizontal shortening at a high angle to layers ($S_0 = S_1$) that had been rotated to moderate to steep north dips during D_1. As a result, D_1 left layering unfavorably oriented for F_2 folding and D_2 produced a second steeply dipping fabric without folding. This interpretation is supported by the scarcity of F_2 folds and the observed low-angle intersection between S_1 and S_2. The low-angle cleavage intersections imply either isoclinal folding during fabric superposition or superposition of two events with similarly oriented shortening axes. Given that no F_2 isoclinal folds were observed, the former is unlikely. In addition, the observation that F_2 folds are restricted to areas where S_1 has a relatively low dip is also consistent with this hypothesis.

Linear fabric elements are also a conspicuous feature of the D_1-D_2 succession. Throughout most of Domain 1, metagraywackes are LS or L tectonites with a prominent subhorizontal EW-trending lineation developed on the steeply dipping composite S_1-S_2 foliation (Figs. 3 and 4). Although this lineation is parallel to $L_{S1\text{-}S2}$ in adjacent argillaceous rocks, it is also a stretching lineation in metagraywackes. This relationship is demonstrated by both a dimensional preferred orientation of clasts as well as quartz-mica overgrowths that accentuate the lineation (Fig. 5). Elongation parallel to L is recognizable both in sections cut parallel to foliation (Figs. 5B and 5E) and perpendicular to foliation and parallel to L (Fig. 5C).

Fabric Development in Domain 2

The main structural feature that characterizes Domain 2 is a third phase of deformation (D_3) that overprints a composite D_1-D_2 fabric. Nokleberg et al. (1989) described the field relationships of the D_1-D_2-D_3 superposition in detail and our studies support these general observations. Specifically, the most conspicuous features in outcrop are D_3 structures that include subhorizontal, moderately to steeply inclined folds (F_3) with an axial-planar, south-dipping crenulation cleavage (S_3) (Fig. 4). F_3 is typically asymmetric to the north; i.e., Z-shaped profiles viewed toward the west (Nokleberg et al., 1989).

D_3 overprints two older generations of structures that are difficult to distinguish because of the intensity of D_3 overprints. D_1 fabrics are defined primarily by a layer-parallel phyllitic cleavage (S_1) and rare isoclinal folds (F_1); a general structure similar to Domain 1 and other parts of the Chugach terrane. The S_1 cleavage is, in turn, overprinted by close to subisoclinal F_2 folds and an axial planar crenulation cleavage S_2. As noted by Nokleberg et al. (1989), we observed several road cuts along the Richardson Highway where the F_2-F_3 superposition produces conspicuous type 3 fold interference patterns and in a few places all three generations of folds are present.

Linear elements are also present in the rocks of Domain 2. As in Domain 1, a prominent subhorizontal intersection lineation is parallel to F_3 fold axes in argillaceous rocks, but the intersection is a composite intersection of S_1~S_2 and S_3. Linear elements are less pronounced in metagraywackes, and this reflects a more complex strain history (see details below).

Kinematic Interpretation

Stuart Creek Shear Zone

The discrepancy in structural style and overprinting history between Domains 1 and 2 suggests that a significant regional structure separates these two domains. The domain boundary is coincident with a regional WNW trending topographic lineament that extends for at least 175 km from ~30 km west of the Richardson Highway transect in the upper Tonsina River to at least the Tana River ~140 km to the east (Figs. 1 and 2). We infer that the lineament is developed along a steeply dipping shear zone and the lineament is formed by differential erosion along the high-strain interior of the shear zone, a fault, or both. Along the Richardson Highway there are no rock outcrops more than about 30 m across in a ~1 km band parallel to the lineament and >99% of this band is covered by Quaternary deposits, a classic feature of a major fault or shear zone. We refer to this structure here as the Stuart Creek shear zone after the major drainage that parallels the lineament along the Richardson Highway (Figs. 1 and 2).

Fabric relationships between Domains 1 and 2. The geometry and spatial distribution of fabric overprints in the study area strongly suggest that the Stuart Creek shear zone is a dextral strike-slip system. This interpretation is supported most clearly in Domain 1 where D_2 fabric axes indicate subhorizontal, ~NS shortening and subhorizontal, ~EW elongation during D_2. This conclusion is based on the spatial geometry of linear fabric development, fibrous mineral growths, and foliation that together show a finite strain pattern consistent with D_2 strike slip–related shear. In all of our Domain 1 samples, subhorizontal EW extension is shown by growth fibers adjacent to sand grains in foliation

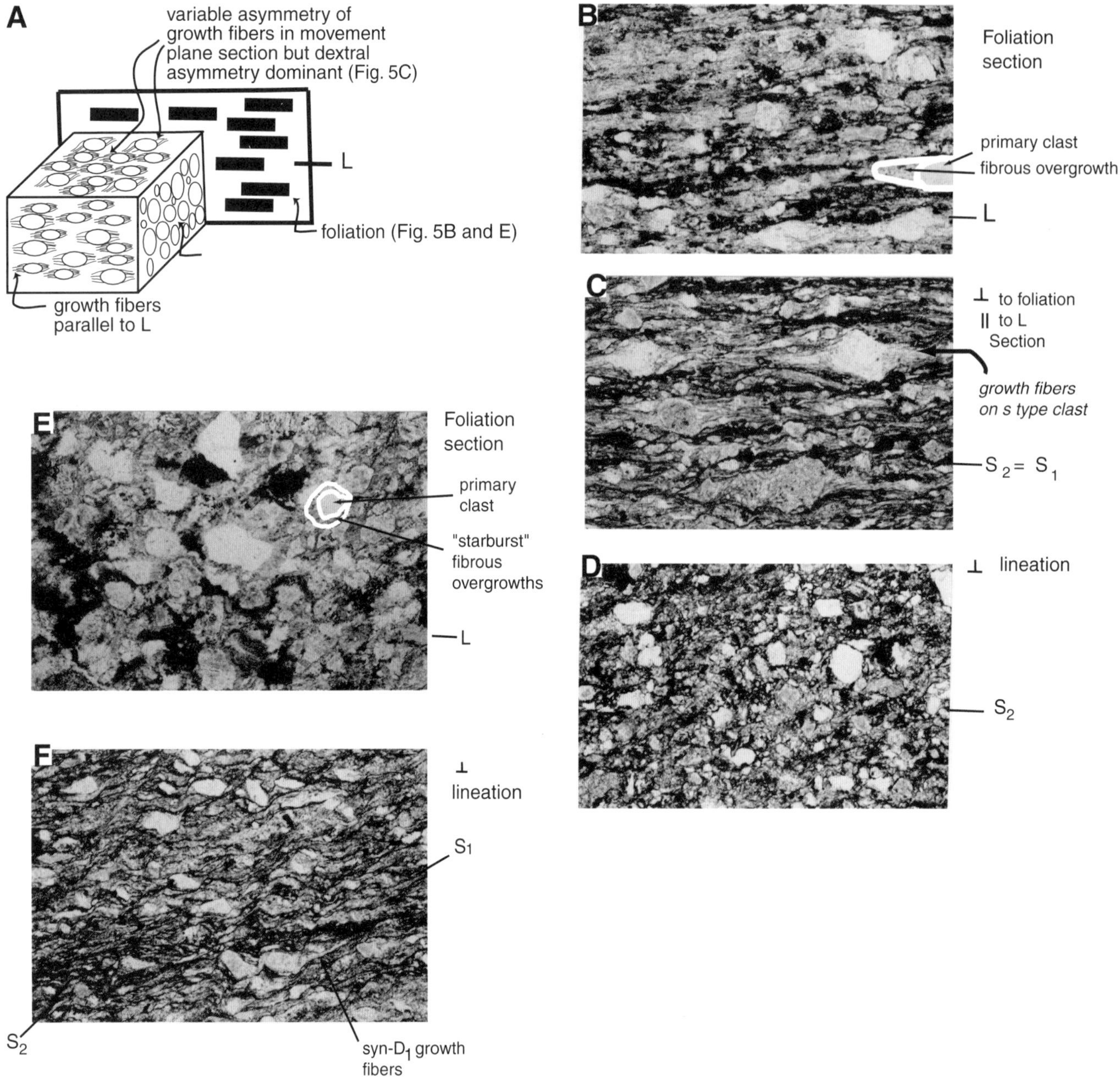

Figure 5. Typical microstructures observed in metagraywackes from the study area. A: Schematically illustrates the orientation of thin sections examined with three mutually perpendicular sections cut relative to the principal axes of the D_2 fabric. The figure also schematically shows the typical growth fiber patterns relative to the D_2 fabric. This also shows the section orientations relative to foliation and lineation. B–F: Low magnification photomicrographs of metagraywackes showing typical fabric development of the rocks along the Richardson Highway. B, C, and D: Different cuts from one sample (AM3). B and C: Mica growth fibers on sand grains demonstrating that the lineation is a stretching lineation; also note the fabric asymmetries in C suggestive of dextral shear. D: An almost nonexistent foliation consistent with constrictional strains. E: Section cut parallel to foliation (sample AM23) showing the "starburst" pattern of growth fibers around sand grains. Finite strain determined from this sample yielded apparent constrictional strain based on Fry and Rf/Φ methods but yielded a flattening strain when growth fibers were included. F: Section cut perpendicular to lineation (sample AM85) collected near the closure of an F_2 fold. This shows the intersection of S_1 and S_2 as well as fibrous overgrowths that were originally down the dip of S_1 and folded by F_2, a relationship suggesting D_1 may have been associated with downdip elongation, flattening, or both.

sections where the growth fibers are generally relatively straight and parallel to L (e.g., Fig. 5B). A few rocks show a "star-burst" pattern (Fig. 5E) on foliation sections suggesting local flattening. Nonetheless, these apparent flattening fabrics could also record early (D_1) downdip extension (e.g., Fig. 5F) followed by D_2 subhorizontal EW extension. Indeed, in some sections cut perpendicular to lineation, early mica overgrowths oriented down the dip of S_1 foliation are crenulated by S_2; a relationship that indicates D_1 was either a flattening event or was associated with down-dip elongation. Thus, because the principal elongation direction recorded by the metagraywackes is subhorizontal, we conclude this elongation was primarily the product of D_2, and therefore is consistent with D_2 strike-slip shear. Finally, the scarcity of F_2 folds is also consistent with D_2 strike-slip shear as D_2 strike slip would have re-used the steeply dipping S_1 foliation to produce penetrative strain without significant folding.

For Domain 2 we suggest that it is highly unlikely that D_2 is the same event as D_2 in Domain 1. Instead, based on geometry it is more likely that D_3 in Domain 2 is equivalent to D_2 in Domain 1 (Table 1), and that rocks with a different structural history have been juxtaposed along the Stuart Creek shear zone. The main evidence for this conclusion arises from two geometric relationships:

1. In Domain 1, S_2 is steeply south dipping and is the youngest foliation, yet the youngest foliation in Domain 2, S_3, is also moderately to steeply south-dipping. S_3 in Domain 2 also becomes successively steeper southward, and becomes geometrically indistinguishable from S_2 of Domain 1 just north of Stuart Creek (Fig. 3).

2. In cross section (Fig. 3), the macroscopic structure in Domain 2 is distinctly different from Domain 1. In Domain 1 the scarcity of F_2 folds strongly suggests that layering was steeply dipping prior to D_2 (see above). However, if the same geometric relationship seen in Domain 1 today—steeply dipping S_2 and S_1—were present in Domain 2 prior to D_3, there should be no prominent F_3 folds in Domain 2 for the same reason that F_2 folds are uncommon in Domain 1. Thus, we suggest that the presence of conspicuous F_3 folds requires that layering, and therefore also S_1 and S_2, were relatively shallowly dipping in Domain 2 prior to D_3.

Together these relationships suggest that Domain 2 contained two distinct fabrics prior to the development of the Stuart Creek shear zone whereas Domain 1 contained one. As the shear zone developed, rocks of Domain 1 were juxtaposed against those of Domain 2, producing D_3 structures in Domain 2 and D_2 structures in Domain 1.

Evidence for Dextral Shear

Although the orientation of D_2 kinematic axes is clear in Domain 1, and is consistent with strike slip–related shear, evidence for shear sense is less distinct. This observation is not necessarily surprising given the finite strain in these rocks (see below), but two general observations are consistent with dextral shear across the Stuart Creek shear zone:

1. We systematically examined movement plane sections for shear sense indicators, primarily asymmetric pressure shadows around original sand grains (Fig. 5C). Results indicate 11 samples show dextral shear, 2 show sinistral shear, and 14 are indeterminate. These observations are consistent with dextral shear but all samples contain significant contradictory evidence. Thus, these data alone are not diagnostic of shear sense.

2. The Stuart Creek shear zone strikes WNW, parallel to regional dextral strike-slip shear zones that emanate from the western end of the Chugach metamorphic complex (Pavlis and Sisson, this volume, Chapter 8). D_2 foliation in Domain 1, however, strikes essentially EW, which is oblique to these shear zones and the Stuart Creek topographic lineament (Figs. 1 and 4). This obliquity is indicative of distributed dextral shear if, as we infer here, the shear zones and foliation formed synchronously and finite strains decrease away from the shear zone core.

Together these observations are highly suggestive of dextral shear across the Stuart Creek zone. Late brittle faulting could account for the fabric obliquity, but this conclusion is ad hoc in the absence of other evidence. Thus, the simplest interpretation is that the Stuart Creek zone is a dextral shear zone.

Significance of D_3 in Domain 2

Our correlation of the fabric overprints between Domains 1 and 2, and our interpretation that the Stuart Creek shear zone is a dextral structure also has important implications for the significance of the prominent F_3 folding in Domain 2. Nokleberg et al. (1989) first recognized the regional significance of F_3 when they recognized the northward-asymmetry of the folds and inferred a major back-folding event late in the deformation. Our observations are consistent with this interpretation, but the connection to the D_2 event in Domain 1 implies a fundamentally different mechanism for F_3. Specifically, if D_3 occurred during strike-slip–related shear along the Stuart Creek shear zone, F_3 in Domain 2 could represent either wrench folds along the strike-slip structure or thrusting during strain-partitioned transpression. A clear resolution of D_3 kinematic axes would be required to distinguish these alternatives with hinge-parallel stretching in wrench folds versus down-dip elongate for thrusting. Unfortunately, available data do not yet constrain this problem (see below).

PRESSURE-TEMPERATURE CONDITIONS DURING DEFORMATION

Within the study area, there are up to three generations of quartz-calcite veins. The abundance of veins is dependent on the host lithology; in some metagraywackes, up to 15%–20% of the rock are veins compared to meta-argillite (e.g., near Keystone Canyon, Fig. 2) with up to 30% of the rock composed of veins. These veins, however, are variably developed throughout both Domains 1 and 2.

The earliest veins (V_1) are thin, parallel to S_0 or S_1, and in some locations are crenulated by D_2. The second generation are thin (mm to cm size) linear veins (V_2) that are parallel to the S_2.

In both of these veins, there are deformed quartz and calcite intergrowths. These are high-temperature veins (see below) that lack growth-fibers and thus, it is difficult to ascertain the direction of vein growth. In terms of the structural sequence in Domain 1, we interpret V_1 as late D_1 or early D_2 veins whereas V_2 represent syn- to late-D_2 veins. The third generation (V_3) of veins consists of euhedral quartz and crosscut all cleavages. These (V_3) have a lobate appearance, and thus, they are post-D_3.

Samples were collected from five locations along the Richardson Highway and where possible, multiple generations of veins were collected (Table 2). The metagraywacke host rocks for these veins typically contain chlorite and biotite indicating upper greenschist facies conditions. The peak metamorphic conditions are further constrained by the assemblage actinolite-chlorite-epidote-albite-feldspar-quartz in interlayered mafic metatuffs. This assemblage combined with lack of lawsonite + quartz or glaucophane implies metamorphic conditions between 300 and 420 °C above ~100 MPa and below 400 MPa (Bucher and Frey, 1994; Schiffman and Day, 1999). The pressure estimate is within the range predicted for the Chugach metamorphic complex to the east (Sisson et al., 1989).

Analysis of Fluid Inclusions

Microthermometric measurements were performed using standard techniques on a U.S. Geological Survey–style heating/freezing stage made by FLUID Inc. (e.g., Roedder, 1984). Fluid inclusions occur in both quartz and calcite and generally ranged in size from 1 to 20 µm, with most falling between 2 and 5 µm. Microthermometric measurements were only made on fluid inclusions in quartz to avoid uncertainties with stretching of fluid inclusions in calcite and the high birefringence of calcite. Those inclusions that appeared to have necked down, or re-equilibrated to lower temperatures and pressures after entrapment were avoided. The stage was calibrated using synthetic CO_2 inclusions (–56.6 °C) as well as synthetic H_2O inclusions (0 °C and 374.1 °C). Due to the small size and low salinity of many fluid inclusions, the first ice melting (eutectic temperature) was rarely

TABLE 2: FLUID INCLUSION RESULTS

Sample	Location and relative chronology*	Sample location (Fig. 2)	Structural domain	Composition	Homogenization temperature (°C) and number of measurements	T (°C) at 300 MPa[†]
84ASn6b	1.2 mi south of Tiekel Cache; V_2	33	2	early 2.5 wt% NaCl secondary 1.5 wt NaCl	125–143 (n = 7) 114–118 (n = 6)	280
84ASn6c	1.2 mi south of Tiekel Cache; V_1	33	2	3.5 wt% NaCl	170–220 (n = 52)	360–410[§]
84ASn8a	South of where Tsina turns parallel to highway	79	1	0.5–0 wt% NaCl	165–172 (n = 6)	360
84ASn8b	Metatuff south of where Tsina turns parallel to highway	79	1	3.5 wt% NaCl	173–180 (n = 6)	360
85ASn64b	South side of Thompson Pass near Blueberry Lake: V_1	67	1	3.5 wt% NaCl	192–260 (n = 7)	390–450[§]
85ASn64c	South side of Thompson Pass near Blueberry Lake: V_2	67	1	3.7–2 wt% NaCl	144–212 (n = 17)	330–380[§]
85ASn64d	South side of Thompson Pass near Blueberry Lake: V_3	67	1	1–2 wt% NaCl	216–313 (n = 13)	
84ASn10	Near VABM288 where highway is EW south of Keystone Canyon	59	1	Type 1: 3.5 wt% NaCl Type 2: H_2O–CO_2–8 wt% NaCl	201–255 (n = 7) 266–285 (n = 10, most decrepitated before T_H)	400
85ASn63	South end of Keystone Canyon	60	1	Type 1: early 3.5 wt% NaCl; secondary 6 wt% NaCl Type 2: H_2O–CO_2–3.5 wt% NaCl	Type 1: early 185–235 (n = 27); secondary 116–131 (n = 5) Type 2: 195–215 (n = 3)	440

* For outcrops with multiple vein sets: V_1—oldest vein set, V_2—second vein set, V_3—youngest vein set, n—number of measurements.
[†] The temperature represents the intersection of an aqueous isochore at 300 MPa. The isochores were calculated using equations of state in MacFlinCor (Brown and Hageman, 1995). Temperature was not calculated for V_3 veins as these probably were not trapped at 300 MPa.
[§] These veins have systematically higher T_H values in their core. Thus, the vein may have trapped hotter fluids during initial precipitation of quartz followed by cooler solutions.

reliably seen, but was recorded when possible. The ice melting temperature (T_M) measurements were recorded using cycling techniques with heating rates of 0.1 to 0.5 °C/min. Homogenization temperature (T_H) was measured at faster rates of 1–5 °C/min. The precision of T_M varied from ±0.1 to ±0.3 °C and the precision of T_H was generally around ±2 °C; both determined by repeat analysis of individual fluid inclusions.

Textural relations varied both between and within each sample. Two major textural types of fluid inclusions are (1) those found along healed micro-fractures and (2) those found as separate inclusions either distinct from all other inclusions, or in clusters. The latter inclusions do not cross grain boundaries; therefore, they are thought to have formed early and are designated as primary inclusions. Fluid inclusions found along healed microfractures cross grain boundaries and are therefore assumed to be secondary inclusions. Both generations of inclusions have relatively homogeneous compositions. There are two types of fluid compositions:

Type 1 Inclusions

Two-phase aqueous inclusions are the most common as they are present in all samples. They range in size from 2 to 15 µm. These aqueous inclusions are found as both primary and secondary inclusions. Shapes range from round to oblong with early inclusions tending to be round and secondary inclusions tending to be elongate.

Type 2 Inclusions

Two-phase aqueous inclusions with CO_2 present only as a clathrate hydrate also occur. They are much less common than type 1 inclusions. These inclusions are found only as early inclusions and are not found along healed fractures. They range in size from 4 to 10 µm and are oblong in shape.

T_{MELT} *Measurements*

The range in aqueous melting temperatures ($T_{M,\ ICE}$) is small. The initial melting temperature range between –22 and –20.1 °C indicating these are close to pure NaCl brines. Equilibrium melting temperatures range from –2.2 to 0 °C for type 1 inclusions corresponding to 3.5 to 0 wt% NaCl equivalent calculated using equations of Zhang and Frantz (1987). The small variation in salinity suggests that for most veins a single fluid was responsible for vein formation. The $T_{M,\ ICE}$ for secondary inclusions was lower at –4.0 to –5.0 °C corresponding to 6 to 8 wt% NaCl equivalents. These salinities overlap with those for gold-bearing quartz veins near Valdez (Goldfarb et al., 1986). No pure CO_2 melting behavior was observed as CO_2 is only present as a clathrate. The clathrate melting temperature was 9.0 °C corresponding to ~3.5 wt% NaCl equivalent.

T_H *Measurements*

A summary of homogenization results with melting temperatures, vein types, and inclusion types is presented in Table 2. In all cases, homogenization was to a liquid phase. T_H for both fluid inclusion types ranged from 165 to 350 °C in type 1 aqueous inclusions and ~120 °C for secondary aqueous inclusions. In most instances, the type 2 inclusions decrepitated before homogenization due to presence of minor CO_2. The range of T_H both between and within veins are important as T_H is used to calculate isochores (lines of constant volume, thus, density) for the various samples. Table 2 shows the entrapment temperature calculated using the intersection of aqueous isochores with an assumed peak metamorphic pressure of 300 MPa. This simplifying assumption does not account for possible variations in lithostatic and hydrostatic pressure as documented elsewhere for the Valdez Group (Vrolijk, 1987; Brantley et al., 1997; Weinberger and Sisson, this volume, Chapter 9). Nevertheless, the fluid inclusion data support the field observations about the timing of deformation relative to thermal history of the region as shown in Figure 6. The temperatures recorded by the second generation quartz veins synchronous with D_2 reflect metamorphic temperatures of 350–400 °C consistent with the mineral assemblages. The earlier veins may have been trapped at either higher temperature or lower pressure, and may record early D_2 high-T fluids related to regional heating (Sisson and Hollister, 1988). Finally, the post-metamorphic veins record exhumation and cooling of the region.

Age of Metamorphism and Deformation

Mineral assemblages, vein chronologies, and available geochronology together indicate that the deformation is entirely Eocene in age and developed at the same time as the high-grade metamorphism of the Chugach metamorphic complex to the east. The principal evidence for this absolute age is that the growth fibers associated with D_2 fabrics contain either biotite or biotite + chlorite that record peak metamorphic conditions

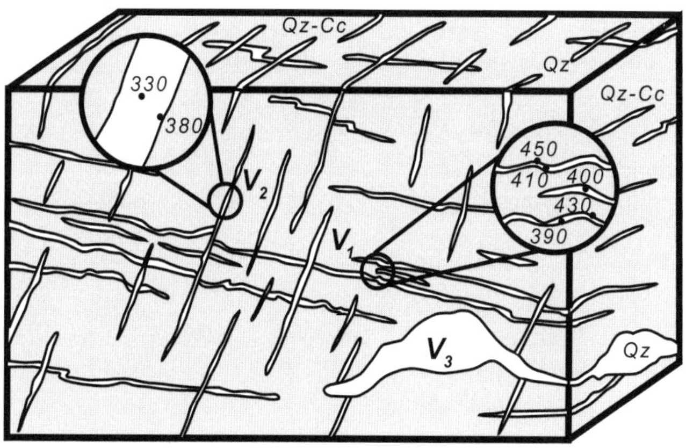

Figure 6. Schematic block diagram from structural Domain 2 (sample 84ASn6b) showing geometry of the different vein systems examined as part of the fluid-inclusion study. Qz—Quartz; CC—Calcite; V_1, V_2, V_3—vein generations from the oldest to the youngest.

with no evidence of recrystallization from an earlier phase. Moreover, the fluid-inclusion data indicate syn-tectonic veining consistent with these temperature conditions. In the Chugach metamorphic complex, immediately to the east, peak metamorphism is Eocene in age (Sisson et al., 1989, and this volume, Chapter 13), and thus, is presumably also Eocene along the Richardson Highway. Two minimum ages from K-Ar dating also support this conclusion: (1) a whole rock K-Ar date of 53.5 Ma (Winkler et al., 1981) from a metatuff near Thompson Pass; and (2) a muscovite K-Ar date of 51.6 Ma from a cross-cutting intrusive north of Valdez (Winkler et al., 1981).

FINITE STRAINS

Methods and Assumptions

To further constrain our kinematic interpretation, we collected a suite of metagraywacke samples for finite strain analysis. We conducted this analysis primarily to test our field-based hypothesis (described above) that the maximum elongation direction was horizontal. Testing this hypothesis is important because previous studies (Nokleberg et al., 1989) ascribed the overprints to superposed thrusting events, which would predict down-dip, nearly vertical, extension axes. Resolving this question would be straightforward in simply deformed rocks, but overprints complicate this picture.

Finite strains in multiply deformed rocks record only the final strain state, and some independent inference of strain history is needed to provide a record of the path. In our case, this independent evidence comes from fabric overprinting in argillaceous layers; yet, these overprints also contain ambiguities. In particular, overprinting foliations only constrain finite shortening axes under the assumption that planar fabric elements form perpendicular to the shortening axis for each event. The other kinematic axes and rotational components require additional assumptions or constraining observations. In this study, we resolved these issues through a combination of microstructural observations and strain superposition models (see below).

Finite strain measurement requires some knowledge of undeformed object shapes. Conglomerates would have been ideal, but are rare in this part of the Chugach terrane. Thus, our analysis was limited to clast shapes in coarse metasandstones. Several complicating factors handicapped our analysis, including:

1. Many of the original objects in the metagraywackes were feldspar clasts. During metamorphism, most of these feldspar clasts were completely recrystallized to a fine-grained "matrix" of albite and mica. Hence, many primary object shapes were destroyed or are poorly defined ghosts of the original grains.

2. The microstructure (Fig. 5) indicates that the primary deformation mechanism was pressure solution because micaceous cleavage domains are associated with dissolution surfaces on original quartz grains. This process did not involve a pure volume loss, however, because primary quartz grains are typically surrounded by quartz-mica fibrous overgrowths.

3. Many of our samples also displayed evidence of crystal plasticity (e.g., subgrain development in clasts, quartz-ribbon development, etc.). Crystal plastic fabrics are particularly conspicuous in quartz-rich metagraywackes that are clast-supported—i.e., rocks containing little albite-mica matrix—suggesting crystal plastic processes were activated where pressure solution was more difficult. Thus, interpretation of object shape fabrics is difficult when rocks contain both volume loss effects and crystal plastic effects.

These complications, particularly uncertainties in the magnitude of volume loss, introduce significant error to the actual strain magnitude. Nonetheless, object shapes are well defined (e.g., Fig. 5), and the kinematic axes as well as ellipsoid shapes—the two quantities of specific interest to our kinematic interpretation—should be well resolved.

Our strain analysis involved several steps and requires a set of simplifying assumptions. Samples were cut into three mutually perpendicular sections for two-dimensional strain analysis. Where possible we oriented slabs relative to foliation and lineation, but in most samples the foliation was weak and samples were oriented relative to arbitrary fracture surfaces. This approach has one advantage in that if the measured strains from these samples reproduce the fabric axes, the data provide an internal test of the consistency of the strain analysis. Indeed, in virtually all cases other than a few samples from Domain 2, the samples passed this test. The disadvantage of this sampling, however, is that growth fiber measurements (see below) were not possible in all samples because many samples did not contain a section parallel to lineation.

We began by estimating two-dimensional shape fabrics from quartz grains in the metagraywackes (Marty, 1994). In this analysis, we estimated the shape fabric using the R_f/Φ method by digitizing grain outlines and evaluating R_s using a program developed by A. Yonkee that uses a Chi Square test for best-fit R_s. For ~80% of our sections, it was possible to digitize a minimum of 100 grain outlines. From the grain boundary outlines, grain centers were also estimated and plotted as normalized Fry plots via the method of Erslev (1988). Both the R_f/Φ shapes and Fry shapes were then used to estimate a strain ellipsoid via the method of Owen (1984).

Ideally, the Fry method should have provided an estimate of the shape fabric, including the strain accommodated by fibrous overgrowths. However, because only quartz grains and a few rock fragments retained clear grain outlines due to destruction of feldspar during metamorphism, many of the sandstones were not true packed aggregates. Instead, most quartz grains or rock fragments are surrounded by "matrix," and only a small percentage of the deformed clasts remain in grain-grain contact. Thus, in the center-center analysis, grain center distances are often overestimated. Because of this problem about 1/3 of our samples showed at least one section with an uninterpretable Fry plot. Nonetheless, more than 80% of the sections yielded results broadly consistent with the R_f/Φ analysis. We rationalize this apparent success to the fact that there were enough quartz grains in grain-grain contact prior to deformation that the grain centers approximated a subset of the packed aggregate such that the deformed grain centers yielded a recognizable strain ellipse pattern on the Fry plot. Similarly, the R_f/Φ shapes alone are also suspect because this approach will generally underestimate the ellipticity in a given section. That is, because this method only records grain shapes, it fails to record finite elongations indicated by overgrowths (e.g., Figs. 5A, 5B, 5D).

To overcome these limitations, we selected eight samples from Domain 1 that showed easily recognized overgrowths as well as grain boundaries. In addition, these samples had been oriented relative to lineation and foliation rather than an arbitrary fracture surface. These orientations insured that the surfaces we analyzed are approximately principal sections of the strain ellipsoid, which allows an assessment of the finite strain using measurements of object shapes in combination with finite stretch from growth fibers. Figure 7 summarizes this method where the R_f/Φ strain is used to estimate the grain shape ellipticity and the fiber-based stretch is used to adjust the strain estimate for the fiber-related strain. The fiber stretch Lf/Li ratios were measured in 15–50 grains with the total number depending on how many grains had fibrous overgrowths in a given section. Standard deviation is <20% of the total stretch for all but two of 15 sections. In addition, two different sections contain the same stretching axis because the samples were cut from principal planes. Thus, in three samples, we were able to compare the stretch ratios between these two sections and the results were within one standard deviation of the data, indicating internally consistent results.

This approach (Fig. 7) works only for sections that display fibrous overgrowths (Figs. 5B, 5C, and 5E) with sections cut along principal planes of the strain ellipsoid. Fortunately, for samples cut along principal planes only ellipticities from two sections are needed to determine the ellipsoid (e.g., Ramsay and Huber, 1983), which allows calculation of ellipticities in the other principal section where growth fibers may be absent. Thus, to further test this method, we analyzed the fiber strain data in two ways: (1) we used two sections to calculate a third principal section to compare with estimates from the Fry and R_f/Φ methods; and (2) we used two sections using the fiber strain estimates together with a third section (the section perpendicular to lineation) and used the method of Owen (1984) to determine a best bit ellipsoid. In the first case the calculated ellipticities were consistent with the observed shapes and magnitudes estimated by the other strain methods. In the second case, the calculated ellipsoid axes were within 5 degrees of the observed fabric axes; a clear test that the data are robust at resolving finite strain axes.

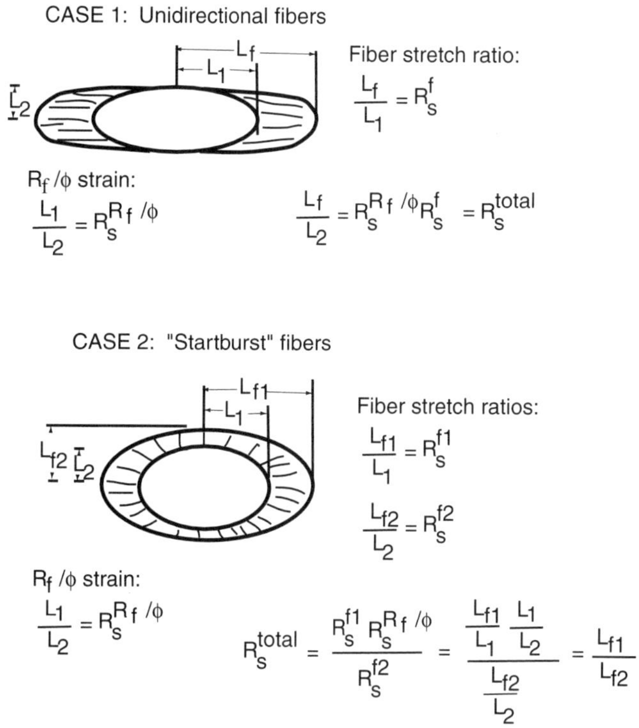

Figure 7. Illustration of the method used to estimate finite strain from the combination of R_f/Φ shapes and growth fibers.

RESULTS

The orientation of ellipsoid axes (Fig. 8) as well as apparent strain magnitudes and ellipsoid shapes (Fig. 9) determined from our strain analyses show several features that support our field interpretation of the finite strain. First, it is important to consider the orientation of calculated ellipsoid axes (Fig. 8) relative to field measurements of fabrics (Fig. 4) as well as the spatial variations in strain (Figs. 2 and 9). Two observations are particularly relevant: (1) a scatter of apparent ellipsoid shapes from constrictional to flattening types in both domains (Fig. 9), and (2) a general subhorizontal, EW orientation for ellipsoid long axes with ellipsoid short axes dispersed in an approximately vertical, NS great circle, regardless of strain analysis method.

Although interpretation of these results is complicated by both finite-strain overprinting and scatter due to measurement error, a number of conclusions can be made:

1. The finite strain data, regardless of technique used, are consistent with the thin-section evidence that the ~subhorizontal EW lineation is a stretching lineation.

2. Although the ellipsoid shapes are scattered on a Flinn diagram, this scatter is consistent with qualitative observations from the field and thin sections. The bulk of the calculated ellipsoid shapes are plane strain to constrictional ellipsoids, which is consistent with thin section observations in foliation-parallel sections where in most samples the growth fibers developed only along lines parallel to lineation (Fig. 5B). In contrast, most of the rocks in which apparent flattening ellipsoids were determined are from samples showing "starburst" fiber overgrowths in sections parallel to foliation (Fig. 5E). This starburst overgrowth

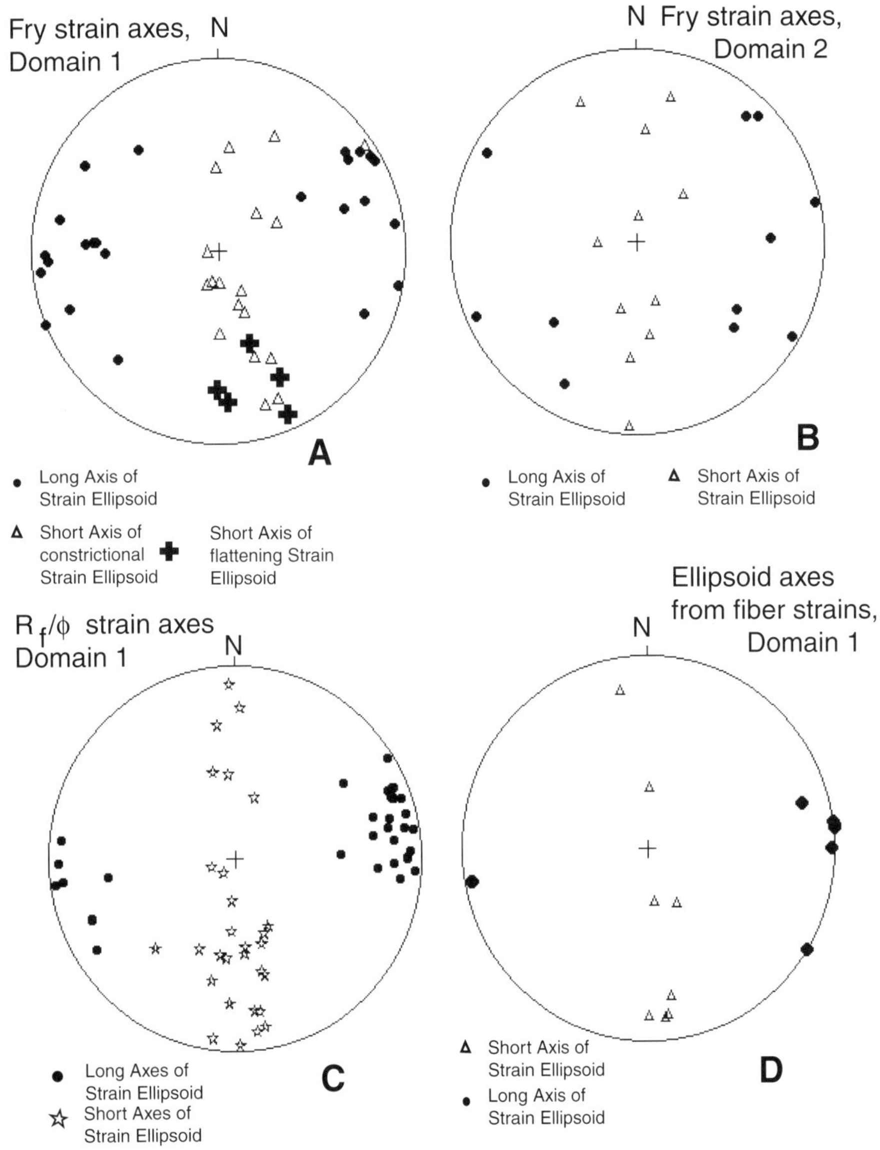

Figure 8. Equal-area stereographic projections showing calculated orientations for long axes and short axes of the strain ellipsoid using different methods. Note the consistent pattern in Domain 1 (A, C, and D) of shallow EW-plunging elongation axes indistinguishable from the lineations measured in the field (Fig. 4), but significantly greater scatter in lineations within Domain 2 (B). Also note that for all finite strain methods in both domains there is a scattering of shortening axes around an approximately NS-striking, steeply dipping great circle with a pole coincident with the concentration of elongation axes. This scatter is primarily due to low resolution of the orientation of the shortening axes in rocks with constrictional strains. This conclusion is supported in the case (A) where shortening axes for flattening ellipsoids are plotted separately from constrictional ellipsoids and show shortening axes plunging uniformly to the south at a low angle, consistent with observed steeply dipping foliations. See text for a more detailed discussion.

pattern clearly indicates two principal stretches >1 within the foliation plane and thus, must record an associated flattening finite strain, which is recorded by the measured strains.

3. The occurrence of flattening ellipsoids is spatially limited and broadly consistent with field observations. If only ellipsoid shapes determined by the Fry method and fiber strain methods are considered, all of the apparent flattening ellipsoids occur in either Domain 2 (Fig. 9) or in the southern half of Domain 1, south of Thompson Pass (Fig. 2). The latter is particularly significant because the mesoscopic lineation becomes less prominent to non-existent south of Thompson Pass; an observation consistent with the data showing a predominance of flattening strain. Marty (1994) separated this area as a distinct subdomain to clarify this relationship. Similarly, the occurrence of flattening ellipsoids in Domain 2 is not surprising given the complexity of the three-phase overprinting, particularly if, as suggested by Nokleberg et al. (1989), the D_3 event involved north-directed back thrusting. That is, superposition of north-directed thrust-related shear on the earlier fabrics could produce an apparent flattening.

4. The observed scatter of short and intermediate axes of the ellipsoid along a steeply dipping great circle is primarily due to poor resolution of these axes on constrictional ellipsoids. The significance of the constrictional strain requires further consideration through strain modeling (see below), but the scatter can be easily ascribed to random measurement error on three-dimensional ellipsoid data. Specifically, object ellipticities are reasonably large in sections parallel to lineation, but ellipticities in sections perpendicular to lineation are typically low (<2). As a result, ellipsoid axes are generally not well resolved in sections at a high angle to L. Thus, because this section is the main

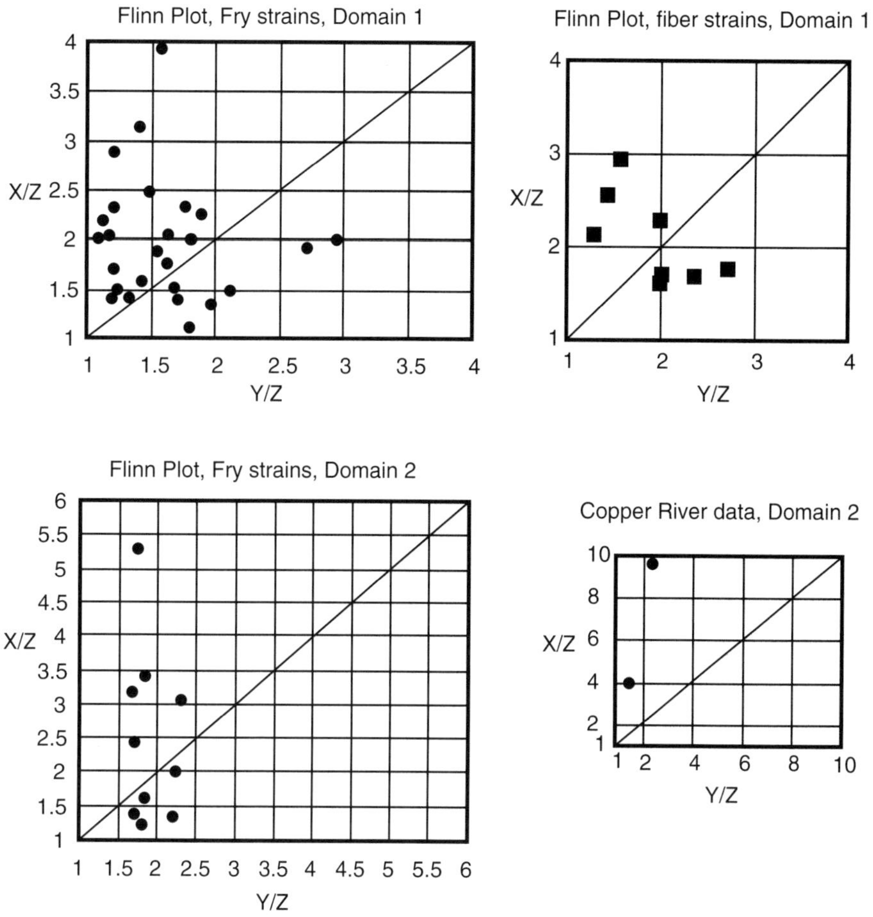

Figure 9. Flinn plots of ellipsoid shapes determined by Fry method and fiber strain methods in this study. Note the dominance of constrictional ellipsoids in Domain 1. See text for details.

surface defining the orientation of the short and intermediate axes of the strain ellipsoid, any errors in determining the orientation of the strain ellipse in this section is translated into scatter in the orientation of ellipsoid short and intermediate axes. That is, scatter tends to be radially distributed about a well-resolved ellipsoid long axis, as seen in our data.

Interpretation of Strain History

Domain 2

The three-phase strain superposition in Domain 2 makes it difficult to fully evaluate D_3 kinematics without examining D_3 fabric development over a larger region. The general subhorizontal EW trend of inferred ellipsoid long axes is broadly consistent with our interpretation that D_3 in Domain 2 is equivalent to D_2 in Domain 1. Scatter in the data, however, is sufficient to suspect this conclusion. For example, the observation of apparent flattening strain could be interpreted in the context of Nokleberg et al.'s (1989) conclusion that D_3 represents a back-thrusting event. That is, if D_1 to D_2 finite strains were similar to those in Domain 1, then the superimposed finite strain from a D_3 thrusting event would tend toward a flattening strain. Thus, we suggest that more detailed studies are needed before alternative kinematic interpretations can be fully tested.

Domain 1

Our field observations, together with strain data from Domain 1, can resolve the kinematic history of Domain 1 when the data are analyzed in the context of theoretical strain superposition models constrained by the field data. Even without the finite strain data, the observation of syn-D_2 growth fibers parallel to the subhorizontal EW-trending lineation in the rocks provides evidence that D_2 was an orogen-parallel elongation event. Thus, Nokleberg et al.'s (1989) inference that the S_1-S_2 intersection lineation was a null strain axis associated with a D_2 south-directed thrusting event is inconsistent with this observation. The strain analysis provides further evidence of the kinematics of D_2, although the full significance of these data can only be realized through modeling of the D_1-D_2 superposition.

Analysis of strain superposition is a difficult problem without information about strain path. Fortunately, interbedded meta-argillites record cleavage superposition to constrain the strain history. Throughout Domain 1, S_1 and S_2 show a low-angle intersection between 5 and 20 degrees with a general pattern of steeply north-dipping S_1 foliation and steeply south-dipping S_2 foliation (Figs. 3 and 4 and Nokleberg et al., 1989). Since F_2 folds are scarce, we infer that D_2 was dominated by shortening at a high angle to layering that was already steeply dipping prior to D_2. This conclusion is significant because it implies that little, if any, fold-related rotations accompanied D_2 and the effects of D_1-D_2 superposition can be modeled as a straightforward superposition of two homogeneous deformations. That is, the deformation can be modeled as a superposition of two coaxial (pure-shear) strain events with negligible fold-related spin. Note that D_2 was probably not a simple coaxial strain overprint because it appears to have a significant simple shear component (e.g., Fig. 5C). Nonetheless, simple shear is equivalent to a pure shear plus a rotation and since we are primarily dealing with ellipsoid shape, this distinction has little effect on the model results.

We calculated theoretical strain paths (Fig. 10) for different strain superposition cases under these geometric constraints. In these models, we assumed the horizontal EW lineation was a principal strain direction during both D_1 and D_2 and that D_1-D_2 superposition involved two shortening axes that were radially symmetric about this axis (Fig. 10). This assumption is justified by the parallelism of the observed stretching lineations in metagraywackes and S_1-S_2 intersection lineations in argillaceous rocks. This geometry simplifies calculation of the strain superposition essentially to two dimensions. D_1 and D_2 share a common principal axis (lineation) and angular variations are two-dimensional dispersions about the common axis (lineation). This problem is an elementary two-dimensional matrix multiplication of the deformation matrices (e.g., Ramsay and Huber, 1983, p. 291–292) with a modification that the strain is actually three dimensional. Thus, D_2 elongations occur along the common axis and must be calculated to determine the three-dimensional ellipsoid shape.

We constructed four models with the angle (θ) between the D_1 and D_2 finite shortening axes varying from 10 to 50 degrees (Fig. 10). We did not consider cases where this angle exceeded 50 degrees because these models produced unrealistic results in two ways: (1) at these angles layer-parallel compression would generally produce folds, yet F_2 folds are rare indicating this case is unrealistic; and (2) in the absence of folding, homogeneous D_2 strain would have to be much larger than observed to produce the low-angle S_1-S_2 intersection.

Model 1 is considered either two superimposed plane strain events with down dip elongation (not shown) or two superimposed flattening events. This model is a simulation of the kinematic interpretation of Nokleberg et al. (1989). Model 2 is a case of D_1 plane strain with down dip elongation on the foliation plane followed by D_2 plane strain with horizontal stretching on a vertical foliation plane. This case approximates D_1 thrusting followed by D_2 strike slip superimposed on S_1 foliation planes that were steeply dipping prior to D_2. Model 3 is D_1 pure flattening strain producing a vertical flattening foliation followed by

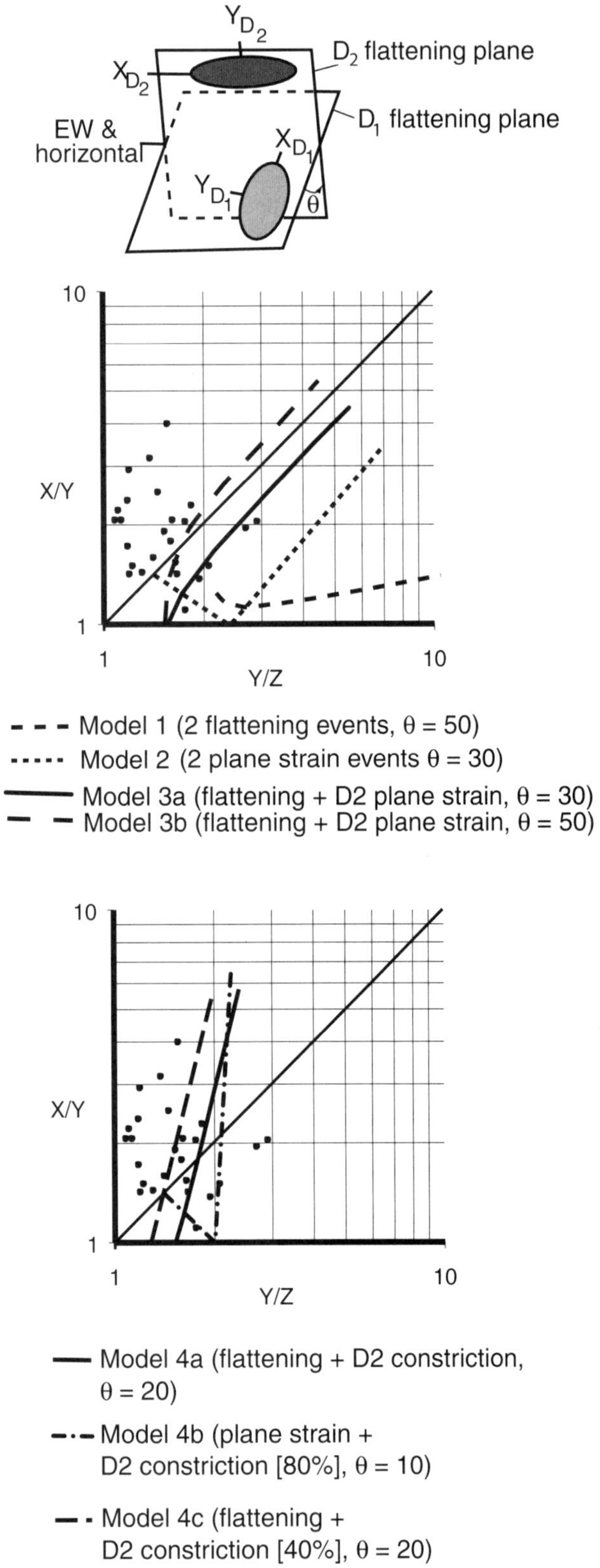

Figure 10. Theoretical strain path models for D_1-D_2 strain superposition together with Domain 1 strain data for reference. Note that variations on model 4 (D_2 constriction) provide the best match to the data. See text for discussion.

D_2 plane strain with horizontal stretching and shortening axes. This case is analogous to model 2 except the initial strain was a pure flattening to simulate a D_1 volume-loss slaty cleavage. Finally, model 4 is for D_1 pure flattening strain or plane strain with down dip elongation followed by D_2 constrictional strain with a horizontal maximum shortening axis and a vertical intermediate strain axis that is also a shortening axis with 40–80% of the horizontal shortening.

In line with other kinematic information, model 1 results do not match the data for any reasonable choice of model parameters. This model fails to match both ellipsoid shapes and orientation of principal kinematic axes. Thus, these models further substantiate that a two-phase thrusting history cannot explain the fabric development. A surprising result of the strain models, however, is that models 2 or 3, which seem most likely from field relationships and thin section textures, do not predict the observed ellipsoid shapes. Models 2 and 3 predict some ellipsoid shapes and finite strain axes consistent with the data, but fails to predict the characteristic constrictional strains. In contrast, model 4, the case of D_2 constrictional strain, provides a reasonable prediction for both ellipsoid shape and strain axes.

DISCUSSION

Significance of Apparent Constriction during D_2

Our modeling of the finite strain data suggests that D_2 was a constrictional strain event in Domain 1, with a subhorizontal, ~NS maximum shortening axis and an EW elongation axis. Shear sense indicators at thin section scale are generally consistent with dextral shear. In conjunction with regional relationships, the fabric geometries and asymmetries provide reasonable evidence that dextral shear was the driving process for D_2.

Constrictional strain fields are generally thought to be rare in geologic systems because they imply a convergent flow. The constrictional strains recognized here imply a horizontal orogen-parallel elongation as well as simultaneous vertical and horizontal shortening. In a dextral shear-dominated system, the simplest explanation for constrictional strain is distributed transtension (e.g., Teyssier et al., 1995). Indeed, Davis et al. (1998) called on transtension to explain constrictional strains in Chugach terrane metagraywackes from Baranof Island. We note here, however, that other scenarios are possible. Specifically, any flow that involves both vertical shortening and strike-slip shear will generate a constrictional finite strain because the strike-slip shear produces a horizontal incremental shortening axis synchronous with the vertical shortening. In transtension, vertical shortening is produced when the shear zone widens and material balance forces the crustal column to thin vertically. If the flow is unconfined laterally, however, this material balance condition is not required and it is possible to have extrusional flow in which the strike-slip shear zone narrows (transpression) at the same time as the crustal column is shortened vertically. Tikoff and Fossen (1999) described the possible range of kinematic variations due to flow geometry and referred to this type of flow as lengthening shear to lengthening–narrowing shear flow.

Lengthening shear or lengthening–narrowing shear provides a possible explanation for the observed constrictional strains along the Richardson Highway, and also provides a possible explanation of why shear sense indicators are not more diagnostic in these rocks. Shear sense indicators assume plane, simple-shear strain and divergence from this assumption produces greater ambiguity in practical applications (e.g., Simpson and De Paor, 1993). Tikoff and Fossen (1999) noted that in strain environments like this one, common shear sense indicators like shear bands are unlikely to develop and only rotated porphyroclasts can indicate the shear sense; a conclusion consistent with the observations in this study.

We further suggest that lengthening shear or lengthening–narrowing shear is a reasonable kinematic scenario based on correlation to the deformational history ~50 km to the east in the Chugach metamorphic complex (Pavlis and Sisson, this volume, Chapter 8). In the Chugach metamorphic complex (Fig. 1), D_1 fabrics analogous to those seen along the Richardson Highway, were subjected to two distinct overprints (Pavlis and Sisson, this volume, Chapter 8): an initial ~vertical shortening with EW elongation (D_2), followed by dextral shear and/or transpression (D_3). Locally within the Chugach metamorphic complex (e.g., Fig. 3 in Pavlis and Sisson, 1995) the deformation did not produce significant folds and steeply dipping layers were homogeneously strained to produce a structural geometry similar to the Richardson Highway area. In these homogeneously deformed domains, the observed finite strains are constrictional and were modeled as the result of strain superposition (Pavlis and Sisson, 1995). A similar strain superposition at lower strain magnitudes provides an attractive explanation for the observations along the Richardson Highway. Specifically, D_2 and D_3 were broadly contemporaneous and associated with peak metamorphism in the Chugach metamorphic complex (Sisson and Pavlis, 1993) indicating those events are coeval with D_2 and D_3 along the Richardson Highway. At the western termination of the Chugach metamorphic complex, the magnitude of D_2 finite strains clearly decreases westward into higher structural levels (Pavlis and Sisson, this volume, Chapter 8) whereas D_3 strains are concentrated in domainal shear zones such as the Bremner shear zone. The observed D_2 strain gradient presumably continues westward toward the Richardson Highway, and D_2 strains probably ultimately become sufficiently low that D_2 fabrics are cryptic. Assuming this relatively low amount of D_2 vertical shortening combined with distributed D_3 dextral shear, the net result would be a constrictional finite strain like that observed along the Richardson Highway.

Finally, regional tectonic relationships also are consistent with the interpretation of lengthening shear or lengthening–narrowing shear. The fundamental driving process for the deformation described here and in the Chugach metamorphic complex was closely linked to a ridge-subduction event. Large-magnitude strike slip was an important manifestation of this event including dextral shear along the Hanagita fault system (Roeske et al., this volume, Chapter 6) as well as strike-slip shear zones within the Chugach terrane (Fig. 1). Ridge subduction produces a transient three-dimensional disturbance in the taper of an accretionary wedge (e.g., Kusky et al., 1997) and the subduction of ridge topography would generally produce some vertical shortening within the accretionary wedge. In addition, triple junction migration would propagate this disturbance parallel to the margin and as a result material may be extruded laterally in the vicinity of the triple junction (e.g., Pavlis and Sisson, 1995). Thus, triple junction migration may account for the vertical shortening component in the constrictional strain, both as D_2 in the Chugach metamorphic complex and as the vertical shortening component of the constrictional strain along the Richardson Highway. Alternatively, vertical shortening in the Chugach metamorphic complex is at least partially due to subhorizontal shear along a mid-crustal decoupling horizon that separated supracrustal strike-slip shear zones from a homogeneously deformed gneissic core (Pavlis and Sisson, this volume, Chapter 8). Thus, this process may also have contributed a component of vertical shortening to the rocks along the Richardson Highway.

More information is needed before these alternative hypotheses can be distinguished. Although transtension is a possible mechanism to produce the observed constrictional strain (e.g., Davis et al., 1998), it is probably less likely than alternatives that are more closely linked to ridge-subduction processes. Ultimately, the distinction is important because it relates to the fundamental response of the accretionary complex to the ridge subduction event. The question of whether the dextral shear zones widened (true transtension) or remained constant (lengthening shear) or narrowed (lengthening-narrowing shear) has important implications for the large-scale response of the system.

Finally, D_2 fabric development in Domain 1 bears directly on the nature of the fabric superpositions in Domain 2. In homogeneous rocks such as the Valdez Group, structural geologists since Sander (1930) have assumed that fabrics of similar style and geometry are the same generation. However, we show above that D_2 in Domain 1 is probably equivalent to D_3 in Domain 2 and rocks with a different history have been juxtaposed along the Stuart Creek shear zone. This illustrates a clear problem with correlation of structural generation across structures that may possess significant transcurrent motion; displacement, partitioning of strain, or both may complicate local overprinting relations.

The correlation of structural histories across the Stuart Creek shear zone also has implications for processes in other transcurrent systems. The problem of slip partitioning along transpressional or transtensional systems is an important tectonic problem (e.g., Teyssier et al., 1995), but one of the most fundamental questions is how slip partitioning is accommodated at depth. In this system, the interpretation that D_3 and D_2 are age equivalents implies that the Richardson Highway transect records the mid-crustal, ductile effects of a slip-partitioned transpressional system. The strike-slip component was accommodated by D_2 flow in Domain 1 with strain concentrated in the Stuart Creek shear zone while the dip-slip component may have been accommodated as D_3 backthrusting and flow in Domain 2.

Implications for Regional Tectonics

Early Cenozoic dextral shear systems are prominent features throughout the northern Cordillera and include well-known regional strike-slip faults like the Denali and Tintina faults. Recent studies have recognized other prominent strike-slip systems that were originally overlooked because of complex ductile structures. These include the Border Ranges fault (Pavlis and Crouse, 1989; Roeske et al., 1993, and this volume, Chapter 6; Little and Naeser, 1989; Little, 1992; Smart et al., 1996; and Davis et al., 1998) and the Coast shear zone (e.g., Chardon et al., 1999). In light of these studies, the recognition of a strike-slip system within the Chugach terrane is not surprising. However, the development of the shear zone entirely within the Chugach terrane is disturbing because without a focused study of the type reported here, the structure would never have been recognized by conventional geologic mapping because lithologically indistinguishable rocks lie on both sides of the structure. Moreover, even in this case the structure would never have been recognized without a detailed analysis of the ductile fabrics. This raises the specter that if the structure were entirely brittle it might never have been discovered unless unusual exposures allowed recognition of fault rocks. Thus, we believe this study raises the question of how many similar structures may have been overlooked in the northern Cordillera, including the westward extension of the Stuart Creek shear zone.

The region to the west of the Richardson Highway is a vast heavily glaciated wilderness that has been rarely visited except by mountaineers. Thus, it is unknown where the Stuart Creek shear zone continues. The zone may link up with complex early Tertiary strike-slip faulting in the western Chugach Mountains and the Matanuska Valley (Pavlis et al., 1988; Little and Naeser, 1989; Little, 1992), but available reconnaissance maps (e.g., Winkler et al., 1981) provide no clues.

Evaluation of the regional geometry of these structures is ultimately critical to tectonic models for the region. To the southwest on Kodiak Island and the Kenai Peninsula the structures within the Chugach terrane are classic examples of syn-accretionary structures (e.g., Fisher and Byrne, 1987; Sample and Fisher, 1986; Sample and Moore, 1987; Kusky and Bradley, 1999). Thus, somewhere between these two areas there is a transition from the complex transpressional overprints of the eastern Chugach Mountains to these accretionary dominated systems.

One possible scenario is that development of the Alaskan orocline may have played a role in the large-scale evolution. Paleomagnetic studies generally suggest that the Alaskan orocline originated in the early Cenozoic and was complete by the time of final accretion of the Prince William terrane in the Eocene (Hillhouse and Coe, 1994). Thus, the development of the orocline was broadly synchronous with, or slightly prior to, the development of the Chugach metamorphic complex and the Stuart Creek shear zone. If this bend in the margin were already present, then oblique convergence on the eastern limb of the orocline would have driven forearc slivers along dextral shear systems. As these slivers reached the bend of the orocline, the motion would have been converted to thrusting; a prediction that is consistent with regional observations.

CONCLUSIONS

Eocene deformation in the Chugach accretionary complex along a NS transect near Valdez, Alaska, was dominated by a dextral shear system that was localized in a zone between Stuart Creek and the Worthington Glacier. This "Stuart Creek shear zone" was superimposed on rocks that had been previously deformed in the latest Cretaceous-Paleocene by thrust-related accretion that produced a regional steeply dipping layering. Deformation occurred during greenschist facies metamorphism at 350–400 °C at ~300 MPa. Vertical shortening broadly concurrent with dextral shear produced constrictional strains throughout the ductile deformation zone. The strain history along the Richardson Highway is interpreted as a higher structural level equivalent of a two-phase deformation recognized to the east in the Chugach metamorphic complex, but the two distinct fabrics of the Chugach metamorphic complex are manifest as a single constrictional strain event. A high-strain shear zone core is probably present along Stuart Creek, but poor exposure makes this conclusion uncertain. Thus, work along strike in areas of better exposure is required to address this question.

The structural history to the north of the Stuart Creek shear zone remains poorly resolved but the latest deformation was probably a back-thrusting event, an interpretation similar to that of Nokleberg et al. (1989). It seems likely that this backthrusting was closely associated with the dextral shear to the south in the Stuart Creek shear zone and probably was produced by transpression along the zone. Projection of the dextral shear system to the west is difficult. The dextral shear system may transfer into complex out-of sequence thrust systems to the west if the strike-slip systems were active after development of the Alaskan orocline.

ACKNOWLEDGMENTS

The field studies and finite strain studies were supported by NSF grants NSF EAR-9307169 and EAR-9018717 to Pavlis and EAR-9020391 and EAR-9304062 to Sisson. We thank Adolph Yonkee for use of the computer programs used in the strain analyses, and Rob Holcombe for the program Georient used in stereonet plotting. The fluid inclusion study was supported by the U.S. Geological Survey TACT project under the guidance of George Plafker. Lew Ashwal kindly provided access to the fluid inclusion stage at NASA-Johnson Space Center for some of the analyses. We thank Steve Davis, Jonathan Lewis, and Sarah Roeske for insightful reviews of the manuscript.

REFERENCES CITED

Bol, A.J., Coe, R.S., Gromme, C.S., and Hillhouse, J.W., 1992, Paleomagnetism of the Resurrection Peninsula, Alaska—Implications for the tectonics of southern Alaska and the Kula-Farallon ridge: Journal of Geophysical Research, v. 97, p. 17,213–17,232.

Bol, A.J., and Roeske, S.M., 1993, Strike-slip faulting and block rotation along the Contact fault system, eastern Prince William Sound, Alaska: Tectonics, v. 12, p. 49–62.

Brantley, S.L., Fisher, D.M., Deines, P., Clark, M.B., and Myers, G., 1997, Segregation veins: evidence for the deformation and dewatering of a low-grade metapelite, *in* Holness, M.B., ed., Deformation-enhanced fluid transport in the Earth's crust and mantle: Mineralogical Society Series 8, p. 267–288.

Brown, P.E., and Hagemann, S.G., 1995, MacFlinCor and its application to fluids in Archean lode-gold deposits: Geochimica et Cosmochimica Acta, v. 59, p. 3943–3952.

Bucher, K., and Frey, M., 1994, Petrogenesis of metamorphic rocks: Berlin, Springer Verlag, 318 p.

Chardon, D., Andronicos, C.L., and Hollister, L.S., 1999, Large-scale transpressive shear zone patterns and displacements within magmatic arcs; the Coast plutonic complex, British Columbia: Tectonics, v. 18, p. 278–292.

Cowan, D.S., Brandon, M.T., and Garver, J.I., 1997, Geological tests of hypotheses for large coastwise displacements; a critique illustrated by the Baja British Columbia controversy: American Journal of Science, v. 297, p. 117–173.

Davis, J.S., Roeske, S.M., and Karl, S.M., 1998, Late Cretaceous to Early Tertiary transtension and strain partitioning in the Chugach accretionary complex, SE Alaska: Journal of Structural Geology, v. 20, p. 639–654.

Erslev, E., 1988, Normalized center-to-center strain analysis of packed aggregates: Journal of Structural Geology, v. 10, p. 201–210.

Fisher, D.M., and Byrne, T., 1987, Structural evolution of underthrust sediments, Kodiak Islands, Alaska: Tectonics, v. 6, p. 775–793.

Fisher, D.M., and Byrne, T., 1992, Strain variations in an ancient accretionary complex: Implications for forearc evolution: Tectonics, v. 11, p. 330–347.

Goldfarb, R.J., Leach, D.L., Miller, M.L., and Pickthorn, W.J., 1986, Geology, metamorphic setting, and genetic constraints of epigenetic lode-gold mineralization within the Cretaceous Valdez Group, south-central Alaska, *in* Keppie, J.D., et al., eds., Turbidite hosted gold deposits: Geological Association of Canada Special Paper 32, p. 87–105.

Grantz, A., 1966, Strike-slip faults in Alaska: U.S. Geological Survey Open-File Report, 92 p.

Hillhouse, J.W., and Coe, R.S., 1994, Paleomagnetic data from Alaska, *in* Plafker, G., and Berg, H.C., eds., The geology of Alaska: Boulder, Colorado, Geological Society of America, Geology of North America, v. G-1, p. 797–812.

Hudson, T., 1979, Mesozoic plutonic belts of southern Alaska: Geology, v. 7, p. 230–234.

Hudson, T.L., and Plafker, G., 1982, Paleogene metamorphism of an accretionary flysch terrane, eastern Gulf of Alaska: Geological Society of America Bulletin, v. 93, p. 1280–1290.

Kusky, T.M., and Bradley, D.C., 1999, Kinematic analysis of mélange fabrics: Examples and applications from the McHugh Complex, Kenai Peninsula, Alaska: Journal of Structural Geology, v. 21, p. 1773–1796.

Kusky, T.M., Bradley, D.C., and Haeussler, P., 1997, Progressive deformation of the Chugach accretionary complex, Alaska, during a Paleogene ridge-trench encounter: Journal of Structural Geology, v. 19, p. 139–157.

Little, T.A., 1992, Development of wrench folds along the Border Ranges fault system, southern Alaska, USA: Journal of Structural Geology, v. 14, p. 343–359.

Little, T.A., and Naesar, C.W., 1989, Tertiary tectonics of the Border Ranges fault system, Chugach Mountains, Alaska: Deformation and uplift in a forearc setting: Journal of Geophysical Research, v. 94, p. 4333–4359.

Marty, K., 1994, Finite strain analysis of the Chugach accretionary complex: Implications for tectonic processes, Richards Highway transect, Alaska [master's thesis]: New Orleans, Louisiana, University of New Orleans, 88 p.

Moore, J.C., Byrne, T., Plumley, P.W., Reid, M., Gibbons, H., and Coe, R.S., 1983, Paleogene evolution of the Kodiak Islands, Alaska: Consequences of ridge-trench interaction in a more southerly latitude: Tectonics, v. 2, p. 265–293.

Nilsen, T.H., and Zuffa, G.G., 1982, The Chugach terrane: A Cretaceous trench-fill deposit, southern Alaska, *in* Leggett, J.K., ed., Trench-forearc geology: Sedimentation and tectonics on modern and ancient active plate margins: Geological Society [London] Special Publication No. 10, p. 213–227.

Nokleberg, W.J., Plafker, G., Lull, J.S., Wallace, W.K., and Winkler, G.R., 1989, Structural analysis of the southern Peninsular, southern Wrangellia, and northern Chugach terranes along the Trans-Alaska Crustal Transect, northern Chugach Mountains, Alaska: Journal of Geophysical Research, v. 94, p. 4297–4320.

Oldow, J.S., Bally, A.W., Avé Lallemant, H.G., and Leeman, W.P., 1989, Phanerozoic evolution of the North American Cordillera; United States and Canada, *in* Bally, A.W., and Palmer, A.R., eds., The geology of North America: An overview: Boulder, Colorado, Geological Society of America, Geology of North America, v. A, p. 139–232.

Owen, W.H., 1984, The calculation of a best-fit ellipsoid from elliptical sections on arbitrarily oriented planes: Journal of Structural Geology, v. 6, p. 571–578.

Pavlis, T.L., 1982, Origin and age of the Border Ranges fault of southern Alaska and its bearing on the late Mesozoic tectonic evolution of Alaska: Tectonics, v. 1, p. 343–368.

Pavlis, T.L., and Crouse, G.W., 1989, Late Mesozoic strike-slip movement on the Border Ranges fault system: Evidence from mylonitic schists of the eastern Chugach Mountains, southern Alaska: Journal of Geophysical Research, v. 94, p. 4321–4332.

Pavlis, T.L., and Sisson, V.B., 1995, Tectonic processes during ridge subduction: Chugach metamorphic complex, southern Alaska: Geological Society of America Bulletin, v. 107, p. 1333–1355.

Pavlis, T.L., and Sisson, V.B., 2003, Development of a subhorizontal decoupling horizon in a transpressional system, Chugach metamorphic complex, Alaska: Evidence for rheological stratification of the crust, *in* Sisson, V.B., Roeske, S.M., and Pavlis, T.L., eds., Geology of a transpressional orogen developed during ridge-trench interaction along the North Pacific margin: Boulder, Colorado, Geological Society of America Special Paper 371, p. 191–216 (this volume).

Pavlis, T.L., Monteverde, D.L., Bowman, J.R., Rubenstone, J., and Reason, M., 1988, Early Cretaceous near-trench plutonism in southern Alaska: A tonalite-trondhjemite intrusive complex injected during ductile-thrusting: Tectonics, v. 7, p. 1179–1199.

Plafker, G., Jones, D.L., and Pessagno, E.A., Jr., 1977, A Cretaceous accretionary flysch and mélange terrane along the Gulf of Alaska margin: U.S. Geological Survey Circular 967, p. 76–79.

Plafker, G., Nokleberg, W.J., and Lull, J.S., 1989, Bedrock geology and tectonic evolution of the Wrangellia, Peninsular, and Chugach terranes along the Trans-Alaska Crustal Transect in the Chugach Mountains and southern Copper River basin, Alaska: Journal of Geophysical Research, v. 94, p. 4255–4295.

Plafker, G., Lull, J.S., Nokleberg, W.J., Pessel, G.H., Wallace, W.K., and Winkler, G.R., 1992, Geologic map of the Valdez A-4, B-3, B-4, C-3, C-4, and D-4 quadrangles, northern Chugach Mountains, and southern Copper River Basin, Alaska: U.S. Geological Survey Miscellaneous Investigation Map I-2164, scale 1:125,000.

Plafker, G., Moore, J.C., and Winkler, G.R., 1994, Geology of the southern Alaska margin, *in* Plafker, G., and Berg, H.C., eds., The geology of Alaska: Boulder, Colorado, Geological Society of America, Geology of North America, v. G-1, p. 389–450.

Plumley, P.W., Coe, R.S., and Byrne, T., 1983, Paleomagnetism of the Paleocene Ghost Rocks Formation, Prince William terrane, Alaska: Tectonics, v. 2, p. 295–314.

Ramsay, J.G., and Huber, M.I., 1983, The techniques of modern structural geology volume 1: Strain analysis: London, Academic Press, 307 p.

Roedder, E., 1984, Fluid inclusions: Mineralogical Society of America Reviews in Mineralogy, v. 12, 644 p.

Roeske, S.M., Snee, L.W., and Bunds, M.P., 1993, $^{40}Ar/^{39}Ar$ dates from the Border Ranges fault system, a hydrothermally altered brittle-ductile transition strike-slip shear zone, southern Alaska: Geological Society of America Abstracts with Programs, v. 25, no. 6, p. 419.

Roeske, S.M., Snee, L.W., and Pavlis, T.L., 2003, Dextral slip reactivation of an arc-forearc boundary during Late Cretaceous–Early Eocene oblique convergence in the northern Cordillera, *in* Sisson, V.B., Roeske, S.M., and Pavlis, T.L., eds., Geology of a transpressional orogen developed during ridge-trench interaction along the North Pacific margin: Boulder, Colorado, Geological Society of America Special Paper 371, p. 141–169 (this volume).

Sample, J.C., and Fisher, D.M., 1986, Duplex accretion and underplating in an ancient accretionary complex, Kodiak Islands, Alaska: Geology, v. 14, p. 160–163.

Sample, J.C., and Moore, J.C., 1987, Structural style and kinematics of an underplated slate belt, Kodiak and adjacent islands, Alaska: Geological Society of America Bulletin, v. 99, p. 7–20.

Sander, B., 1930, Gefügekunde der Gesteine: Vienna, Austria, Springer-Verlag, 352 p.

Schiffman, P., and Day, H.W., 1999, Petrological methods for the study of very low-grade metabasites, *in* Frey, M., and Robinson, D., eds., Low-grade metamorphism: Malden, Massachusetts, Blackwell Science, p. 108–142.

Simpson, C., and De Paor, D., 1993, Strain and kinematic analysis in a general shear zone: Journal of Structural Geology, v. 15, p. 233–268.

Sisson, V.B., and Hollister, L.S., 1988, Low-pressure facies series metamorphism in an accretionary sedimentary prism, southern Alaska: Geology, v. 16, p. 358–361.

Sisson, V.B., and Pavlis, T.L., 1993, Geologic consequences of plate reorganization: An example from the Eocene southern Alaska forearc: Geology, v. 21, p. 913–916.

Sisson, V.B., Hollister, L.S., and Onstott, T.C., 1989, Petrologic and age constraints on the origin of a low-pressure/high-temperature metamorphic complex, southern Alaska: Journal of Geophysical Research, v. 94, p. 4392–4410.

Sisson, V.B., Poole, A.R., Harris, N.R., Cooper Burner, H., Pavlis, T.L., Copeland, P., Donelick, R.A., and McLelland, W.C., 2003, Geochemical and geochronologic constraints for genesis of a tonalite-trondhjemite suite and associated mafic intrusive rocks in the eastern Chugach Mountains, Alaska: A record of ridge-transform subduction, *in* Sisson, V.B., Roeske, S.M., and Pavlis, T.L., eds., Geology of a transpressional orogen developed during ridge-trench interaction along the North Pacific margin: Boulder, Colorado, Geological Society of America Special Paper 371, p. 293–326 (this volume).

Smart, K.J., Pavlis, T.L., Sisson, V.B., Roeske, S.M., and Snee, L.W., 1996, The Border Ranges fault system in Glacier Bay National Park, Alaska: Evidence for major late Mesozoic–early Cenozoic dextral strike-slip motion: Canadian Journal of Earth Science, v. 33, p. 1268–1282.

Teyssier, C., Tikoff, B., and Markley, M., 1995, Oblique plate motion and continental tectonics: Geology, v. 23, p. 447–450.

Tikoff, B., and Fossen, H., 1999, Three-dimensional reference deformations and strain facies: Journal of Structural Geology, v. 21, p. 1497–1512.

Tysdal, R.G., and Plafker, G., 1978, Age and continuity of the Valdez Group, southern Alaska: U.S. Geological Survey Bulletin 1457-A, p. 120–131.

Vrolijk, P., 1987, Tectonically driven fluid flow in the Kodiak accretionary complex, Alaska: Geology, v. 15, p. 466–469.

Weinberger, J., and Sisson, V.B., 2003, Pressure and temperature conditions of brittle-ductile vein emplacement in the greenschist facies, Chugach metamorphic complex, Alaska: Evidence from fluid inclusions, *in* Sisson, V.B., Roeske, S.M., and Pavlis, T.L., eds., Geology of a transpressional orogen developed during ridge-trench interaction along the North Pacific margin: Boulder, Colorado, Geological Society of America Special Paper 371, p. 217–235 (this volume).

Winkler, G.R., Silberman, M.L., Grantz, A., Miller, R.J., and MacKevett, E.M., Jr., 1981, Geologic map and summary geochronology of the Valdez quadrangle, southern Alaska: U.S. Geological Survey Open-File Report 80-892-A, scale 1:250,000.

Zhang, Y.-G., and Frantz, J.D., 1987, Determination of the homogenization temperatures and densities of supercritical fluids in the system NaCl-KCl-H_2O using synthetic fluid inclusions: Chemical Geology, v. 64, p. 335–350.

Manuscript Accepted by the Society February 5, 2003

Printed in the USA

Geological Society of America
Special Paper 371
2003

Development of a subhorizontal decoupling horizon in a transpressional system, Chugach metamorphic complex, Alaska: Evidence for rheological stratification of the crust

Terry L. Pavlis
Department of Geology and Geophysics, University of New Orleans, New Orleans, Louisiana 70148, USA

Virginia B. Sisson
Department of Earth Science, MS-126, Rice University, Houston, Texas 77005-1892, USA

ABSTRACT

Field studies east of the Copper River provide an unusual view of the architecture of a ductile strike-slip system where the Chugach metamorphic complex plunges westward from high-grade gneiss to greenschist facies rocks of the Chugach terrane. In this down-plunge view of a transpressional system we observe a three-phase ductile deformational history similar to the Tana River region about 75 km to the east. In the Copper River area, however, spatial variations in structural development provide important insights into syntectonic flow in lithologically homogeneous rocks ranging from high-grade gneiss to greenschist facies phyllite. Early structures (D_1) formed prior to the Chugach metamorphic complex and produced a steeply dipping layering (S_0) and layer-parallel cleavage (S_1) following thrust imbrication and associated folding. Following D_1, the earliest structures of the Chugach metamorphic complex (D_2) were associated with a vertical shortening and horizontal, margin-parallel extension. D_2 produced gently inclined to recumbent subisoclinal to isoclinal folds (F_2) in $S_1 = S_0$ layering with axes parallel to the ~WNW-trending elongation lineation. These structures were in turn overprinted by a third set of structures (D_3) that produced upright, gently plunging folds (F_3) of S_2 that vary spatially. In the high-grade gneissic core of the complex, D_3 structures are homogeneously distributed with tight to subisoclinal mesoscopic folding throughout the core, and no significant spatial variation in fabric development. In contrast, structurally above a transition from schist to gneiss, D_3 fabric development is domainal with bands of intense S_3 foliation development separated by broad regions in which D_3 fabrics are indistinct or absent. Within these D_3 high-strain bands, numerous fabric asymmetries indicate that these bands are zones of distributed dextral shear. These observations suggest that during D_3, upper crustal dextral strike-slip faults passed downward into discrete shear zones, but these shear zones did not penetrate to the base of the crustal column. Instead, the strike-slip motion was dispersed into a broad zone of distributed shear within the high-grade gneisses at depth, and the contact between these two crustal levels was a D_3 decoupling surface.

We propose a tectonic model for the Chugach metamorphic complex that combines effects of highly oblique Kula-Farallon subduction with the coeval interactions of the Kula–Farallon–North American triple junction. Reanalysis of the plate velocities

Pavlis, T.L., and Sisson, V.B., 2003, Development of a subhorizontal decoupling horizon in a transpressional system, Chugach metamorphic complex, Alaska: Evidence for rheological stratification of the crust, *in* Sisson, V.B., Roeske, S.M., and Pavlis, T.L., eds., Geology of a transpressional orogen developed during ridge-trench interaction along the North Pacific margin: Boulder, Colorado, Geological Society of America Special Paper 371, p. 191–216. ©2003 Geological Society of America

together with observations from the Border Ranges fault system immediately inboard of the Chugach metamorphic complex indicate that the Chugach metamorphic complex developed when the Chugach terrane was a forearc sliver. If slip partitioning were complete, this forearc sliver migrated northward faster than the triple junction. Thus, the triple junction moved slowly southward within the forearc sliver, which is consistent with regional geochronological trends. The proposed mid-crustal decoupling horizon formed within this evolving transform system and D_2 and D_3 probably overlapped in time. In this interpretation D_2 primarily records vertical displacement gradients related to mid-crustal decoupling that initiated during supracrustal strike slip along the Border Ranges fault. With prograde metamorphism of the Chugach metamorphic complex, the decoupling horizon rose to higher crustal levels, and strike slip stepped outboard to form D_3 dextral shear zones within the Chugach metamorphic complex.

The inference of mid-crustal decoupling implies a grossly different rheologic behavior across the gneiss transition. We propose that the gneissic front represents a fundamental change in macroscopic flow behavior from homogeneous linear flow in the high-grade gneisses to strain-softening, nonlinear flow in the overlying rocks. This rheological change is interpreted as a consequence of a deformation mechanism switch from diffusion creep in gneiss to crystal plastic mechanism in the schist with an associated rheological transition from linear to nonlinear flow.

Keywords: Chugach metamorphic complex, ridge subduction, plate reorganization, triple junction, forearc sliver, Border Ranges fault, crustal rheology, strike-slip, polyphase-deformation, down-plunge crustal section.

INTRODUCTION

In the last 15 years significant evidence has accumulated that strike-slip systems are not necessarily confined to narrow shear zones that penetrate the lithosphere (Zoback et al., 1987; Oldow et al., 1989, 1990; Teyssier and Tikoff, 1998; Teyssier et al., 2003). Instead, many strike-slip faults apparently pass downward into zones of penetrative flow where the strain is dispersed over a large volume of rock. An important question, however, is the width of the zone of dispersed strain. If a surface fault passes downward into a narrow shear zone, the large-scale kinematic system is effectively rigid-body motion, whereas if this deep shear zone is several tens of kilometers wide, a description of the system is a three-dimensional finite strain problem. Thatcher (1995) described these two end members as microplate and continuum descriptions respectively.

A variety of models have been used to evaluate these effects, including kinematic models of slip partitioning in transpressional systems (e.g., Teyssier et al., 1995), dynamic models to explain the unusual stress-states on active strike-slip faults (e.g., Zoback et al., 1987), and dynamic models of transpressional systems using thin-sheet approximations (e.g., England and Jackson, 1989). An important prediction of many of these models is that as slip on discrete faults in the near surface environment passes downward into distributed flow, the accompanying vertical displacement gradients will produce a subhorizontal decoupling horizon within the lithosphere (e.g., Zoback et al., 1987; Oldow et al., 1989, 1990; Teyssier et al., 2003). This transition has generally been assumed to be coincident with the brittle-ductile transition (e.g., Jones et al., 1994; Teyssier and Tikoff, 1998). Although features of this type have been inferred from geophysical data (e.g., Serpa et al., 1988; Jones et al., 1994), geologic studies generally have not found extensive evidence for this type of feature in the rock record.

In this paper we present evidence that a mid-crustal decoupling horizon developed within the forearc of southern Alaska during a prograde metamorphic event closely tied to ridge subduction and transpressional movements driven by oblique convergence. We emphasize, however, that decoupling did not occur at the brittle-ductile transition, but instead occurred at the petrologic transition from schistose to gneissic rocks; an observation that we ascribe to a rheologic distinction between gneissic and schistose rocks. We develop this conclusion from observations in an unusual down-plunge crustal section of the western termination of the Chugach metamorphic complex, a Mesozoic accretionary complex that was subjected to high-*T*, low-*P* metamorphism when the Kula-Farallon spreading center was subducted along the north Pacific margin during Eocene time. We begin by describing field relationships illustrating a marked difference in the structure of high-grade gneiss at the deepest structural levels relative to higher-level schists. This structural transition occurs across a thin horizon that we mapped as a regional, gently dipping surface. We recognize a system of dextral shear zones above the horizon that disappear downward across the transition into homogeneously distributed deformation below. We conclude that this gently dipping surface is a subhorizontal decoupling horizon that developed along the petrologic schist to gneiss transition. We further propose that this coincidence between the "gneiss front" and the decoupling horizon is consistent with a rheologic transition that would be predicted by a deformation mechanism switch from crystal plastic mechanisms in schists (with nonlinear viscous behavior) to diffusion-dominated flow in gneissic rocks below (with linear viscous behavior).

This study, together with two other papers in this volume (Pavlis et al., Chapter 7, and Roeske et al., Chapter 6) also provides important new constraints on the timing and structural consequences of ridge-trench interactions along this margin. These regional relationships are used to develop a new hypothesis for the development of the Chugach metamorphic complex as a forearc sliver that moved laterally along the margin with the Kula-Farallon triple junction. We infer that this moving reference frame was the primary reason for the unusually high-grade conditions achieved within the complex.

TECTONIC SETTING OF THE CHUGACH METAMORPHIC COMPLEX

The Chugach metamorphic complex is a high-temperature/low-pressure metamorphic belt developed in the forearc of the Alaskan-Aleutian arc (Fig. 1). Metamorphism within the complex is Eocene in age and affected rocks that can be traced laterally into a Cretaceous-Paleogene forearc accretionary complex (Fig. 1) (Hudson and Plafker, 1982). This accretionary assemblage is known locally as the Valdez and Orca Groups, but is commonly referred to on a regional scale as the Chugach and Prince William terranes respectively (Hudson and Plafker, 1982). The metamorphic history of this belt stands in contrast to typical low-*T*/high-*P* assemblages of forearc accretionary complexes with a progressive metamorphic sequence more typical of arc terranes. These rocks record an andalusite-sillimanite, low-*P* field gradient with the highest-grade rocks reaching the low-pressure upper amphibolite facies (~650 °C at ~300 MPa) in a high-grade gneissic core (Sisson et al., 1989).

Significant evidence has accumulated (e.g., Sisson et al., 1989; James et al., 1989; Plafker et al., 1989; Sisson and Pavlis, 1993; Pavlis and Sisson, 1995; Bradley et al., 1993, and this volume, Chapter 1; Haeussler et al., 1995; Harris et al., 1996) that the Chugach metamorphic complex is the highest-temperature part of a belt of forearc metamorphism and granitic magmatism and that these events are the product of an early Cenozoic ridge-trench interaction along the margin. Regional tectonic analyses and paleomagnetic data indicate that the Chugach metamorphic complex formed well south of its present position (e.g., Bol et al.,

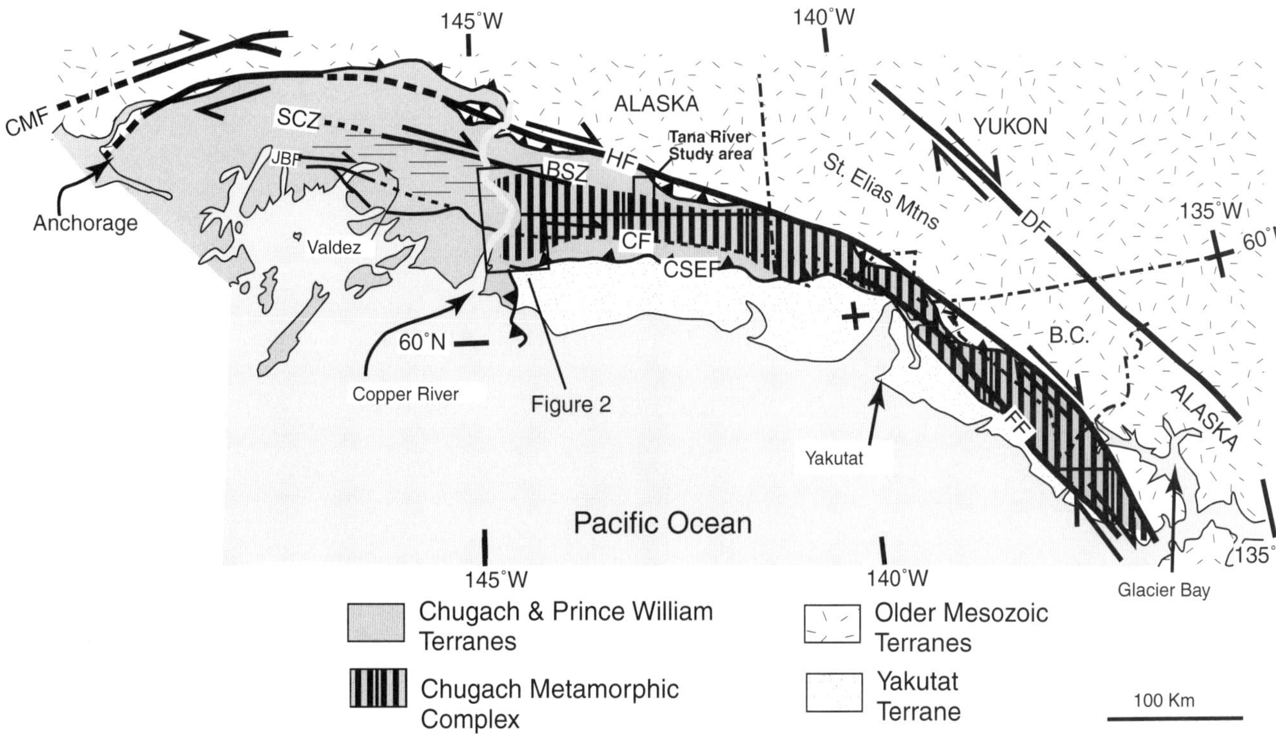

Figure 1. Tectonic map of southern Alaska emphasizing the position of the Chugach metamorphic complex relative to adjacent major tectonic features. Note the position of the Hanagita fault (HF), the Tertiary reactivation of the Border Ranges fault, and the Bremner shear zone (BSZ), which play an important role in the tectonics of the system discussed in this paper. FF—Fairweather fault, CSEF—Chugach–St. Elias fault, CF—Contact fault, HF—Hanagita fault, SCZ—Stuart Creek zone, JBF—Jack Bay fault, CMF—Castle Mountain fault, and DF—Denali fault. After Hudson and Plafker (1982) with modifications from field work by T. Pavlis, V.B. Sisson, and S. Roeske.

1992; Plafker et al., 1994), although the exact position is debated. For example, some of the best paleomagnetic data are from ophiolitic assemblages accreted in the forearc. Some fraction of the northward motion of these fragments may have occurred on an oceanic plate prior to accretion (Kusky and Young, 1999). Nonetheless, it is widely accepted that the rocks were initially accreted during the latest Cretaceous to early Cenozoic in a trench offshore from what is now western British Columbia and were transported northward by margin-parallel strike-slip faults (e.g., Plafker et al., 1994). These dextral faults include the well-known motions on the Denali and Tintina faults (e.g., Plafker et al., 1994). They also include the Border Ranges fault of southern Alaska (Roeske et al., 1992, 1993, this volume, Chapter 6; Smart et al., 1996) as well as the Coast shear zone of southeast Alaska and British Columbia (Andronicos et al., 1999; Chardon et al., 1999). Prior to Middle Eocene time, the ridge that interacted with this trench to form the Chugach metamorphic complex was almost certainly the Kula-Farallon spreading center (e.g., Moore et al., 1983). However, the main period of metamorphism in the Chugach metamorphic complex (ca. 54–48 Ma) is coincident with a north Pacific plate reorganization. The Pacific-Farallon ridge, Kula-Farallon ridge, or both, may have interacted with the margin to form the high-grade assemblages of the Chugach metamorphic complex (Sisson and Pavlis, 1993; Pavlis and Sisson, 1995). Alternatively, a fourth plate may have been present (Bradley et al., this volume, Chapter 1). In our opinion, geologic relationships in the complex are most easily explained by a plate reorganization that involved the Kula plate. This model involves a Kula-Farallon spreading center that migrated northward (ancient coordinates) relative to North America with oblique subduction of the Kula and more direct convergence of Farallon plate (Sisson and Pavlis, 1993; Pavlis and Sisson, 1995).

The Chugach metamorphic complex is developed almost exclusively in a homogeneous assemblage of interbedded graywackes and argillites that are widely recognized as a turbidite sequence that was deposited in the Late Cretaceous to Eocene subduction zone of the northern Cordillera (e.g., Tysdal and Plafker, 1978; Nilsen and Zuffa, 1982). This assemblage locally contains interbedded volcanic rocks and ophiolitic assemblages are present locally (e.g., Plafker et al., 1989). In the eastern Chugach Mountains, however, metavolcanic rocks are almost exclusively restricted to the outboard part of the belt (e.g., Lull and Plafker, 1990) and are rare within the high-grade core of the Chugach metamorphic complex. This relationship is important because although the monotony of the protolith handicaps detailed descriptions of the internal structural geometry of the complex, the homogeneity of the rocks throughout the complex is critical to assessment of structural processes. Specifically, because lithologic variations are insignificant, macroscopic changes in rheology due to rock type are not important in the complex. Thus, any differences in fabric development are exclusively the result of the kinematic and thermal variations during development of the complex.

Our field studies between the Copper River and the Tana River regions (Fig. 1) show that the western ~1/2 of the Chugach metamorphic complex is characterized by a similar structural history. Between ~142° W and just east of Yakutat in Nunatak Fjord, the complex had been investigated only in reconnaissance studies by the U.S. and Canadian geological surveys (Campbell and Dodds, 1982, 1983; Hudson et al., 1977) prior to 1995. Our recent work in the St. Elias Range (Sisson et al., 1998), however, confirms that the structural history recognized in the western Chugach metamorphic complex also continues into the St. Elias Range to at least ~140° W at Mt. Logan. Important differences are seen across strike in the St. Elias Range as the southern margin of the Chugach metamorphic complex has a history more closely akin to that observed to the south near Yakutat (e.g., Poole, 1996; Sisson et al., this volume, Chapter 13). In this paper, we limit our discussion to the better-known, western half of the complex.

GENERAL GEOLOGIC RELATIONSHIPS

Scope of This Study

This paper summarizes results of six summers (1991–1996) of field work in which we attempted to systematically map the western edge of the Chugach metamorphic complex just east of the Copper River where the high-grade core of the complex plunges westward beneath lower-grade rocks (Figs. 1 and 2). This area is large and highly glaciated, and there are no roads in the area. Access was by foot traverses conducted from fly camps established by helicopter, with a few partial days of helicopter assisted spot checks over the entire area of Figure 2. This limitation is important because the accuracy of the mapping in Figure 2 is variable with some areas well constrained (Plates 1 and 2 on CD-ROM accompanying this volume[1]) and others idealized from only reconnaissance work. Nonetheless, our detailed coverage is sufficiently extensive that we are confident the major first-order features have been recognized, at least in areas within a few kilometers of the gneissic core.

Figure 2 is generalized from more detailed mapping on enlarged 1:63,360 topographic maps, aerial photographs, or both. Plates 1 and 2 show details in specific areas produced from field sheets ranged in working scale from 1:12,000 to 1:30,000, although most of our mapping was at ~1:24,000 scale on enlargements of U2 aerial photography. Photogeologic maps were orthocorrected using ERDAS Imagine software from control points picked on photographs and rubber sheet corrected to remove parallax distortions. The total field effort represented by these maps is ~300 person days of field work over the six year period.

Since the rocks are lithologically monotonous, the most distinctive features of the rock structure were revealed by systematic mapping of the form surfaces of major planar structural elements. The compiled digital maps in Plates 1 and 2 (in CD-ROM accompanying this volume) show these basic data

[1]Plates 1 and 2 are also available as GSA Data Repository item 2003126, available on request from Documents Secretary, GSA, P.O. Box 9140, Boulder, CO 80301-9140, USA, editing@geosociety.org, or at www.geosociety.org/pubs/ft2003.htm.

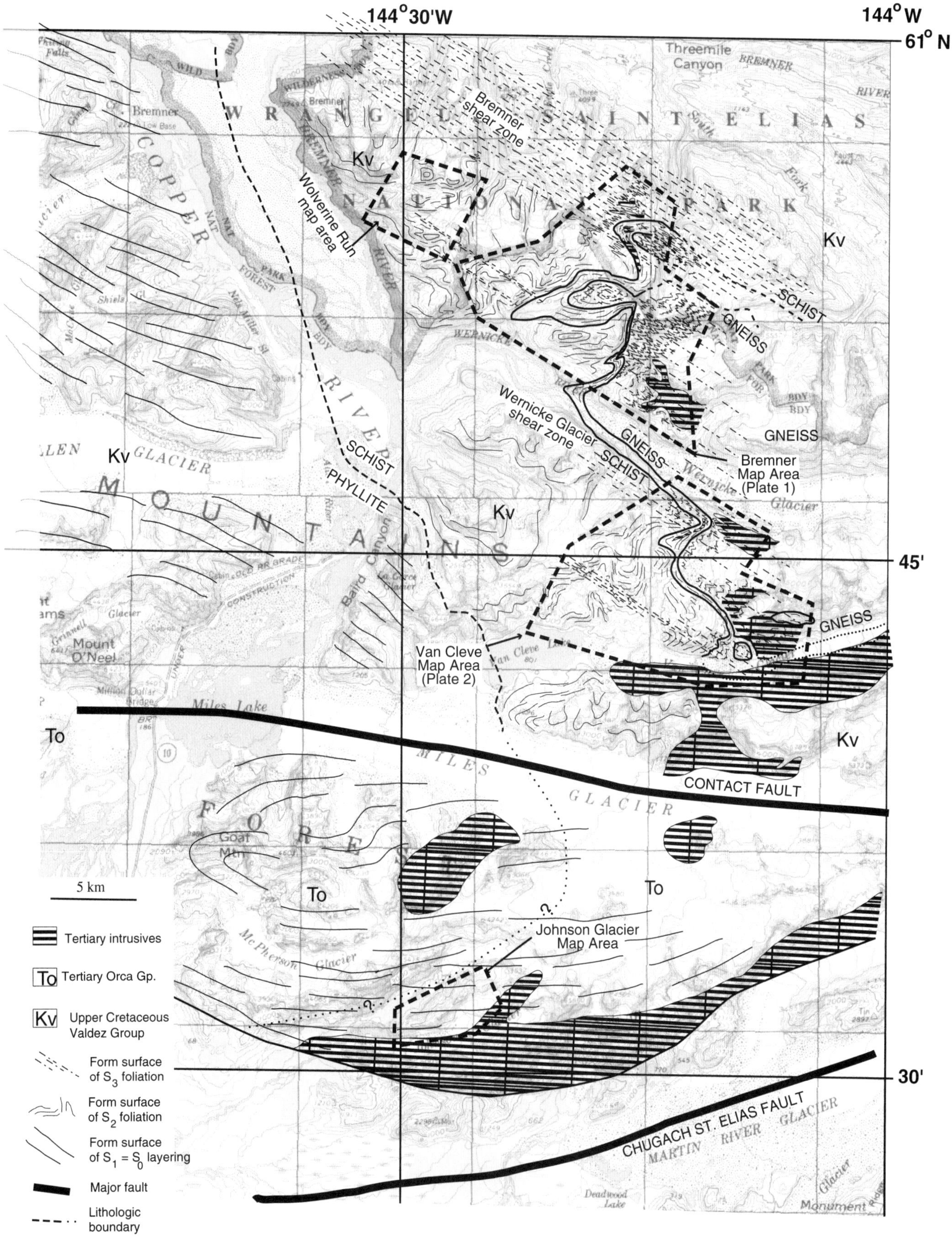

Figure 2. Generalized geologic map of the western termination of the Chugach metamorphic complex. Map is plotted on topographic base of the U.S. Geological Survey Cordova 1:250,000 sheet. Map generalized from our mapping and published reconnaissance mapping of Winkler and Plafker (1993).

as linework placed on distinct layers in the digital map, which should be toggled on and off to display views emphasizing different structural generations.

Lithologic Units and Distribution of Rock Units

The most conspicuous regional feature of the western Chugach metamorphic complex is a down-plunge metamorphic progression from upper amphibolite facies gneiss to lower greenschist facies phyllite. This metamorphic assemblage is intruded by a series of tonalitic intrusives that are variably deformed due to emplacement through different phases of the syn-metamorphic flow. Intrusive rocks, however, are not voluminous in the region and account for <10% of the area shown in Figure 2.

Previous reconnaissance studies of the region distinguished three metamorphic units within the complex that were easily recognized both on the ground and from the air (e.g., Hudson and Plafker, 1982; Winkler and Plafker, 1993): (1) a phyllite unit characterized by black to dark gray weathering well-layered/foliated rock; (2) a schist unit recognized by brown (rusty) weathering color and flaggy exposures; and (3) a gneiss unit characterized by brown to gray weathering and relatively rounded exposures. These units remain useful divisions for gross characteristics of the complex, but these divisions can produce misleading conclusions without information on fabric relationships. Thus, we developed both lithologic and structural subdivisions within these general lithologic divisions.

The phyllite to schist transition is poorly exposed in the study area because it occurs almost exclusively beneath Holocene cover of the Copper River canyon. We know little about this transition. The schist to gneiss transition, however, is well exposed throughout the area and is the most dramatic feature within the western termination of the Chugach metamorphic complex. The boundary is easily recognized as a distinctive gneissic "front" that separates enveloping high-grade schists from gneissic rocks in the core of the complex (Figs. 2 and 3; Plates 1 and 2). In our fieldwork, we were able to trace this boundary in much greater detail than in previous studies (Fig. 2). The schist-gneiss transition is a gently west-dipping surface throughout the area with the schists always structurally above the gneiss. We were also generally able to trace two distinct boundaries within this schist-gneiss transition to define a "transitional gneiss" unit shown in Figure 2 and Plates 1 and 2. The top of the schist-gneiss transition was mapped at the first appearance downward of quartzofeldspathic segregations. These segregations invariably appear first in meta-argillite units rather than metagraywacke and become increasingly abundant through the transitional gneiss. The base of the transitional gneiss unit was defined by the full conversion of the rock to gneiss. This criterion allows routine mapping of the structural level where continuity of primary lithologic layering is destroyed by development of complex quartzofeldspathic vein/segregation networks. Typically, this conversion is a combination of structural overprinting effects and full development of gneissic segregations in both metagraywacke and argillite layers. The coincidence of the schist-gneiss transition with structural variations is a critical observation relevant to this paper (see below).

We recognized a similar gneissic transition in the Tana River region (Pavlis and Sisson, 1995). In that region, however, the transition occurred across strike and the narrow width of the zone could be attributed to structural complications. In both areas, the transitional gneiss unit has a comparable structural thickness of 50–100 m. A unique feature of the Copper River area, however, is that the boundary is a very low-angle contact as well as a structural discontinuity separating rocks with distinct structural styles.

STRUCTURAL OBSERVATIONS

Early Deformation Prior to Development of the Chugach Metamorphic Complex: Premetamorphic Architecture of the Complex

Throughout the study area the rocks have a common early ductile deformational history (D_1) that predated the metamorphism of the Chugach metamorphic complex. Within the Chugach metamorphic complex, younger greenschist to amphibolite facies fabrics have largely transposed D_1 fabrics. However, fold interference patterns were recognized sporadically throughout the area, demonstrating continuity of D_1 structure into the high-grade rocks (e.g., Pavlis and Sisson, 1995). Observations outside the Chugach metamorphic complex along the Richardson Highway (e.g., Nokleberg et al., 1989; Pavlis et al., this volume, Chapter 7) as well as in the Tana River area (Pavlis and Sisson, 1995) demonstrate that D_1 was accompanied by lower to middle greenschist facies metamorphism. Pressure solution appears to have been the dominant D_1 deformation mechanism. The most conspicuous D_1 structures are a penetrative, layer-parallel cleavage (S_1) axial planar to mesoscopic to macroscopic sub-isoclinal to isoclinal folds (F_1) in sedimentary layering (S_0). Although mesoscopic F_1 folds are relatively abundant in thinly bedded units, macroscopic F_1 folds are only sporadically developed. F_1 folds are generally limited to upright to steeply inclined, south-vergent anticlinal hinges separating large domains of north-facing strata. This suggests that the macroscopic D_1 structure in the area is dominated by imbricate thrusting with a cleavage superimposed on a thrust-related homocline (e.g., Pavlis and Sisson, 1995).

Within the Chugach metamorphic complex, D_1 structure is seen best within the Wolverine Run map area (Fig. 2) where relatively low strains allow routine determination of stratigraphic younging (see detailed map in Weinberger and Sisson, this volume, Chapter 9). Throughout this area, F_2 folding is intense with subisoclinal folds producing a general north dip to layering with an associated gently north-dipping S_2 foliation (Figs. 3A and 3D). These folds are downward facing and north vergent. That is, the F_2 folds are antiformal synclines and synformal anticlines with no observed cases of south vergence. This observation is difficult to reconcile if D_1 produced large-

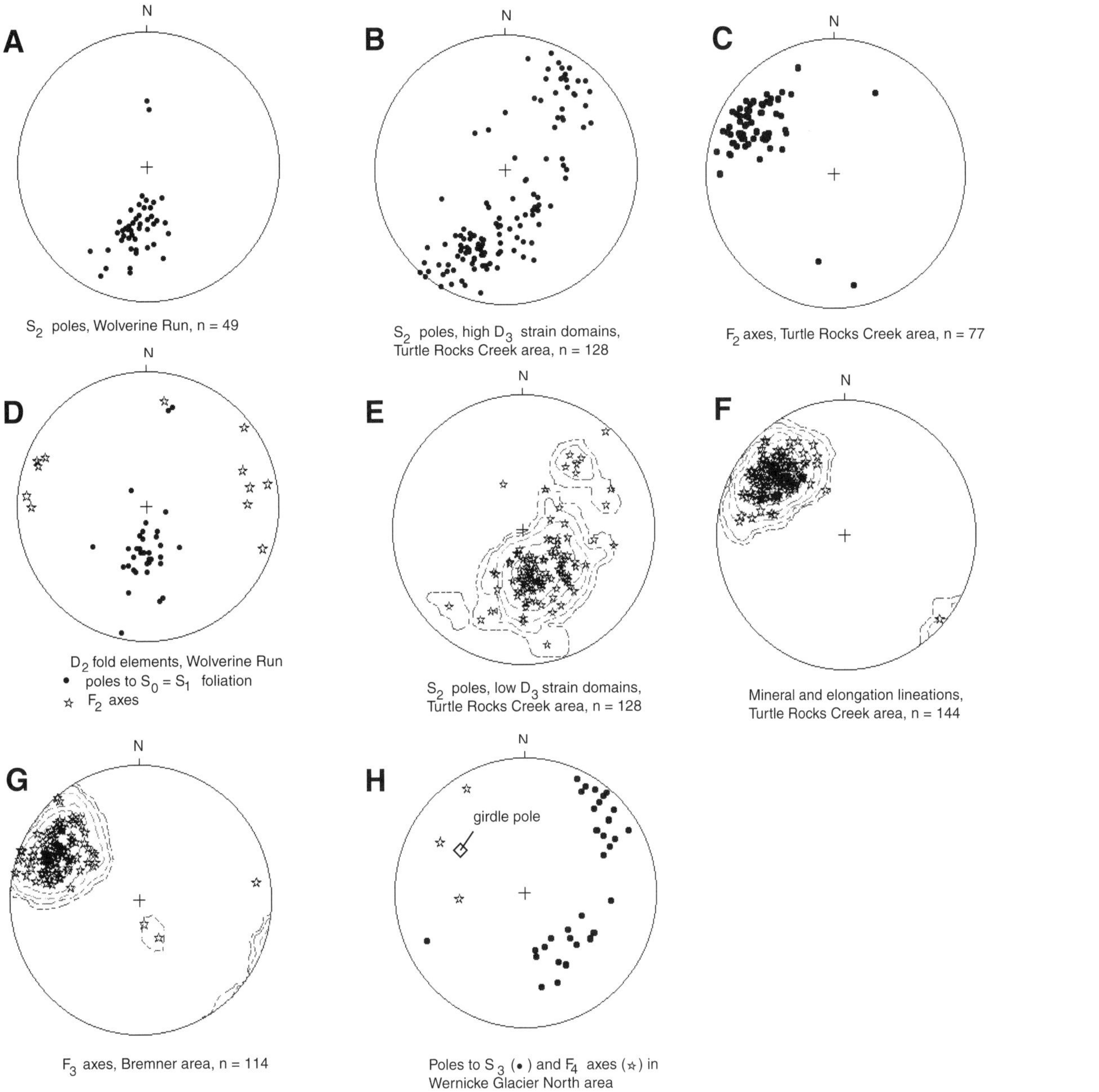

Figure 3 (*on this and following page*). Equal-area stereographic projection diagrams summarizing orientation data from the Bremner map area.

scale folds with south-facing strata. Thus, we suggest that D_1 was dominated by thrusting with dominantly north-facing strata in stacked thrust sheets.

Although the details of D_1 are important regionally and have implications for subduction zone processes (e.g., Fisher and Byrne, 1987, 1992; Sample and Moore, 1987), the primary significance of D_1 in the Chugach metamorphic complex is as an event that produced the primary structural architecture that preceded the main metamorphism. It is impossible to fully assess this geometry because there are no unequivocal indicators of paleohorizontal at the close of D_1. Nonetheless, the field data strongly suggest that at the close of D_1, the dominant structure of the Valdez Group in the Chugach metamorphic complex was a relatively simple north-dipping homocline formed by steeply dipping imbricated and folded strata (S_0) with a layer-parallel cleavage (S_1). This conclusion is also supported by our observations throughout the areas to the south, west, and north of the western termination of the Chugach metamorphic complex. In all these areas, layering dips steeply and is overprinted by structural fabrics related to the Chugach metamor-

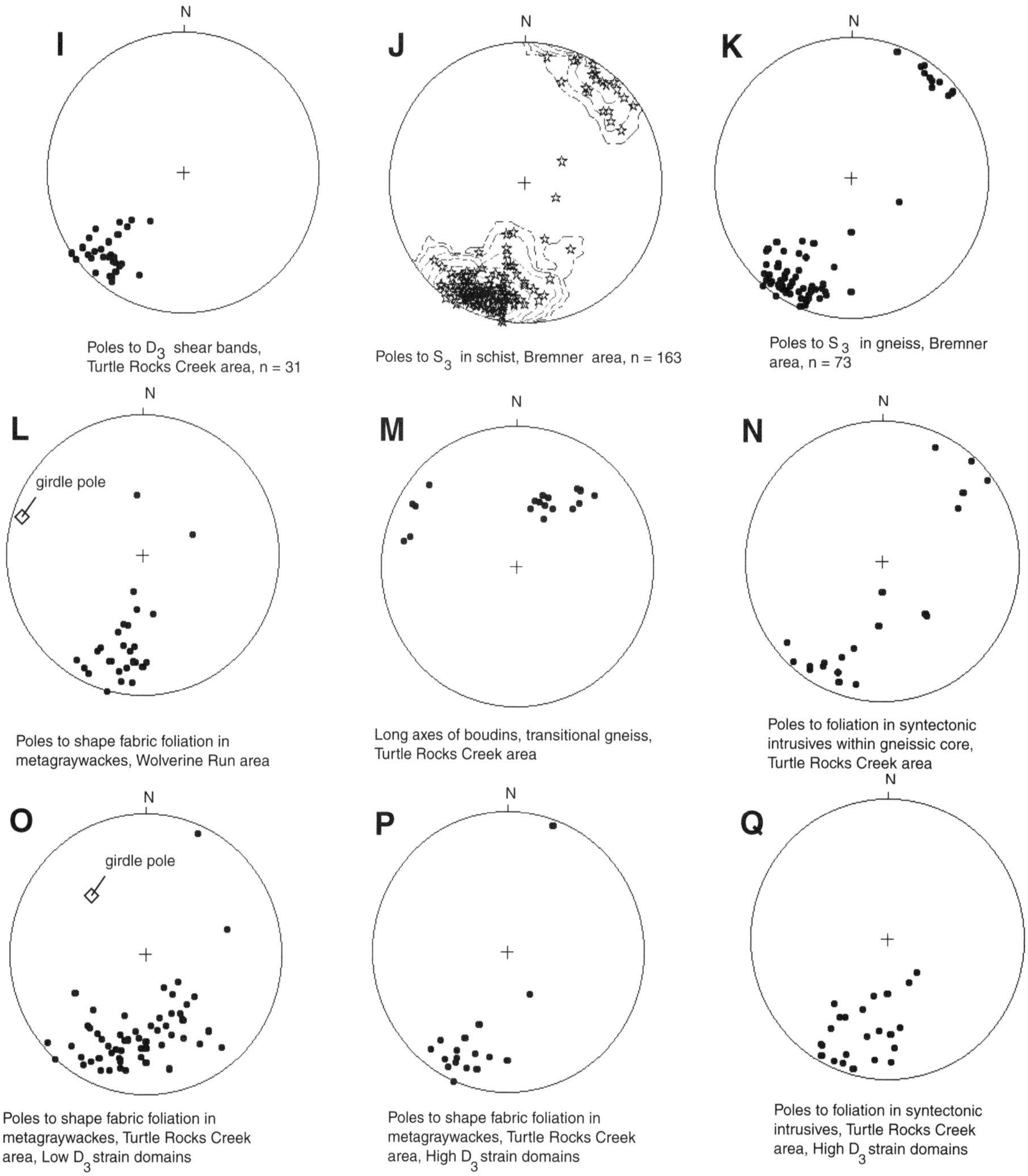

Figure 3 (*continued*).

phic complex. We recognize that overprints well outside the Chugach metamorphic complex along the Richardson Highway include steeply dipping fabrics that overprint D_1 fabrics (e.g., Nokleberg et al., 1989) and these overprints may be partially responsible for this regional pattern. Nonetheless, the details of these overprints along the Richardson Highway are compatible with the assumption of steep dips at the close of D_1 for the area immediately west of the Chugach metamorphic complex (Pavlis et al., this volume, Chapter 7). Moreover, because of the unique downplunge view of the structure, the steep dips west of the Copper River, together with the observed structural geometry within the schists, are difficult to reconcile unless bedding and the subparallel S_1 cleavage were steeply dipping at the close of D_1.

Overprinting and Structural Sequence

During prograde metamorphism of the Chugach metamorphic complex, the rocks were subjected to three kinematically distinct deformational events. Two of these events (D_2 and D_3) are well documented along the north flank of the Chugach metamorphic complex in the Tana River region (Pavlis and Sisson, 1995). The first of these (D_2) involved a near vertical shortening that took the steeply dipping fabrics from D_1 and produced a subhorizontal LS-tectonite fabric with a foliation (S_2) axial planar to gently inclined to recumbent folds (F_2) with axes parallel to a prominent WNW trending elongation lineation (Figs. 3 and 4). The planar fabric generated during this event (S_2) was in turn folded into a series of horizontal to gently plunging, upright to steeply inclined folds during the last ductile deformational event (D_3).

In the Copper River area, this same sequence is present, but D_2 and D_3 strain variations have produced structural variations that lead us to differentiate a series of structural subdomains. A fourth deformational event (D_4) is recognized locally (Plate 1) but the regional significance of this event is unknown.

To clarify the continuity of structure between this region and the Tana River region we retain the terminology from that area (Pavlis and Sisson, 1995). Thus, as in the Tana River region, we define structural domains that are approximately equivalent to the regional map units of Hudson and Plafker (1982); gneissic rocks represent Domain 3 and enveloping schists represent Domain 2. In this description, Domain 1 would represent rocks outside the mapped area of Plates 1 and 2 to the west of the Copper River and to the north of the complex.

Domain 2 in the Copper River area is distinguished by predominance of D_2 structural fabrics with local development of NW trending bands of prominent D_3 structures interpreted as D_3 shear zones. On a macroscopic scale, Domain 2 represents a schist envelope structurally overlying Domain 3, the gneissic core of the complex, with a prominent foliation (S_2) defining the form surface of the envelope (Figure 2; Plate 2). Throughout Domain 2, the prominent continuous cleavage is generally a S_2 foliation; even in areas of high D_3 strain, and S_2 foliation is not completely transposed. In Domain 3, S_2 is also generally recognizable, but is conspicuously folded by mesoscopic F_3 folds. These variations are described below in the context of spatial

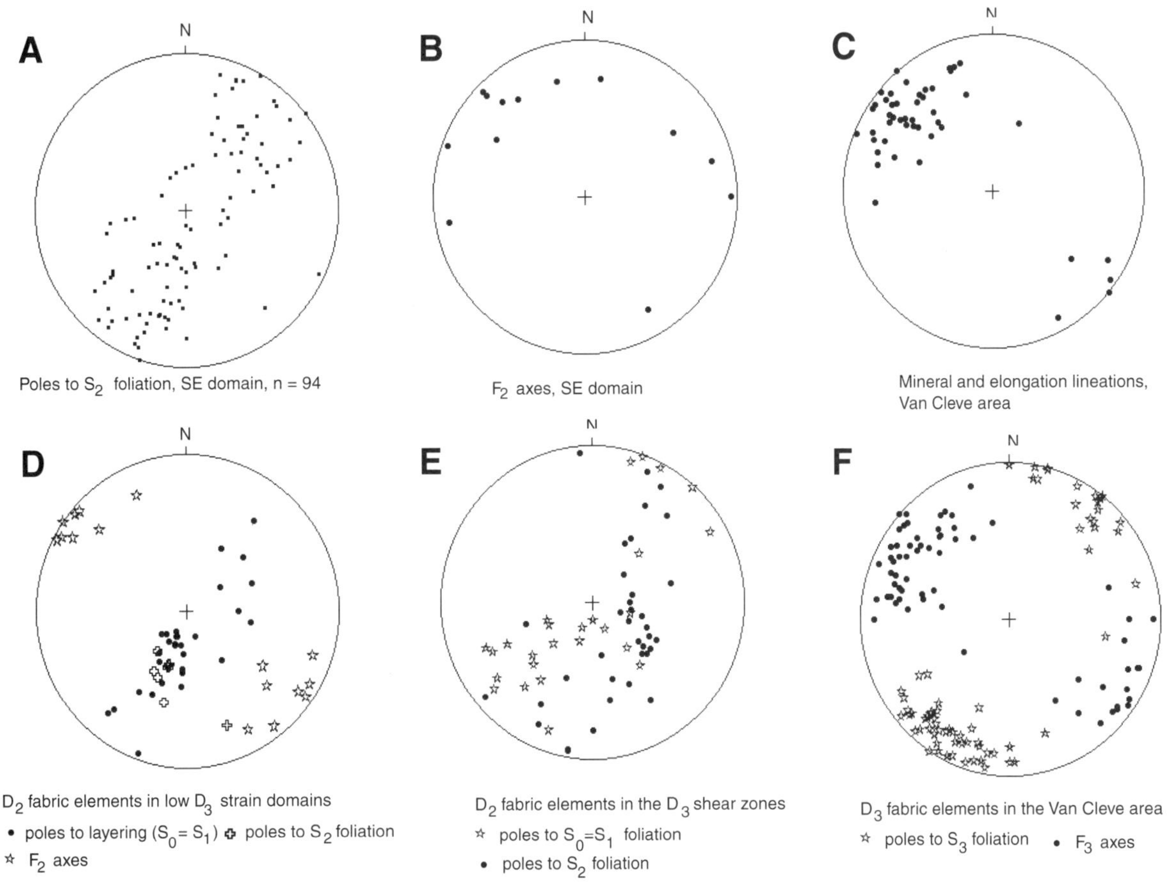

Figure 4. Equal-area stereographic projection diagrams summarizing orientation data from the Van Cleve map area.

variations seen in three distinct map areas outlined in Figure 2 and Plates 1 and 2.

Northern Section: Bremner River to Wernicke Glacier

The area between the Wernicke and Bremner drainages (Fig. 2) contains excellent exposures on a large glacially sculpted plateau above the tree line that provide extensive accessible outcrops. This area also contains abundant low-variance assemblages for metamorphic thermobarometry as well as deformed conglomerates that allowed local estimates of finite strain within the complex. Thus, much of our work in this region concentrated on this area.

Macroscopic structures in this segment are characterized by two major structural domains: (1) a southern domain in which D_3 structures are virtually absent from the schist envelope other than a few small shear zones, but are prominent in underlying gneiss (referred to as the Echo Lake domain); and (2) a northern domain in which D_3 structures are dominant in both the schist and gneiss—this domain is the most significant dextral shear zone within the mapped area (referred to as the Bremner shear zone).

Wernicke Glacier to the Bremner Shear Zone (Echo Lake Domain)

The schist-gneiss transition is readily traceable as a low-angle contact as it emerges from dense vegetative cover along the Wernicke River. To the north, the contact traces out a complex map pattern northward to Transition Glacier (Plate 1). This complex map pattern, however, is almost entirely due to steep topography intersecting the low-angle surface. The large-scale geometry of the contact is a broad antiformal arch plunging WNW subparallel to mesoscopic fold axes (Plate 1 and Figure 3). This entire segment of the schist-gneiss transition also lacks any smaller-scale structural relief from F_3 folding (Plate 1). Instead, the schist envelope is characterized by remarkably homogeneous D_2 structures at scales of several km. A few 50–100 m wide D_3 shear zones occur as distinct domains of steeply dipping foliation and tight F_3 folding. In general, however, the entire schist envelope is characterized by gently NW dipping S_2 foliation axial planar to subisoclinal-isoclinal F_2 folds (Plate 1, Figs. 3C–3D) with an associated gently NW plunging elongation lineation (Fig. 3F).

The large-scale structure of the schist in this segment stands in stark contrast to the gneiss exposed directly below the schist-gneiss transition. Here, mesoscopic close to sub-isoclinal F_3 folds are the most prominent feature in the rock. The transition from the schist envelope to the gneiss with conspicuous F_3 folds occurs over a vertical dimension of <100–150 m throughout this segment and occurs exclusively within the "transitional gneiss" unit. Figure 5A illustrates this relationship where the transition is well exposed on a large cliff face above Goat Wallow Glacier, but the map pattern in Plate 1 also shows the pattern on a larger scale.

F_2 folds become successively tighter as the schist-gneiss transition is approached. Typically, layering is largely transposed into the S_2 foliation within 50–100 meters above the gneiss transition, and transposition is nearly complete within the transitional gneiss (e.g., compare form surfaces of $S_1 = S_0$ and S_2 in Plate 1). These observations imply increasing D_2 strain toward the schist-gneiss contact, a relationship supported by finite-strain measurements (see below). D_2 shear sense indicators, however, are rare. The only indicators of D_2 shear sense observed were a series of asymmetric boudins in deformed dikes and quartz veins (Fig. 6C) along the north side of Transition Glacier (Plate 1). The asymmetries of these boudinaged rocks indicate a top to the east sense of shear. This area is, however, within 500 m of the Bremner shear zone and has a distinctive D_3 overprint that has rotated S_2 to moderate dips. Thus, it is possible the fabric asymmetries are D_3 structures because when F_3 folds rotated S_2 to steeper dips a superimposed D_3 dextral shear would produce an asymmetry indistinguishable from D_2 top to the east shear on these moderately north-dipping surfaces.

Although D_3 structures are relatively homogeneous in gneissic rocks throughout the mapped area of Plate 1, an important feature occurs just north of the Wernicke Glacier (Plate 1). Here, the homogeneous S_3 foliation typical of the gneiss is deformed into a set of upright, steeply northwest plunging folds (F_4). These folds are largely recognized at map scale where they appear as changes in the surface trace of the S_3 form surface (SE corner, Plate 1), but a few mesoscopic structures were also observed. Where observed, these folds are not associated with a new foliation, but do appear to be related to mesoscopic domainal shear zones that locally contain gneissic segregations parallel to the shear zones. Because these structures are limited to a small area, they are probably not significant. They probably represent a local, minor perturbation in the general flow that produced D_3.

Bremner Shear Zone

At Transition Glacier (Plate 1), the deformation changes northward into a major D_3 shear zone, referred to here as the Bremner shear zone after Bremner River, the closest formally named geographic feature. This zone extends northwestward to at least the Copper River (Fig. 2) as observed on our overflights and a few helicopter stops. It appears to continue farther westward as a shear zone into lower-grade rocks where it produces a prominent topographic lineament along the Tiekel River and Stuart Creek. Along the Richardson Highway this lineament is the boundary between two distinct structural domains, and Pavlis et al. (this volume, Chapter 7) interpret this "Stuart Creek shear zone" as the western extension of the Bremner shear zone.

The Bremner shear zone is a conspicuous structure within the schist envelope but becomes cryptic when traced along strike into the gneiss transition where the structure disappears into gneiss that is indistinguishable from gneissic rocks across strike to the south. From south to north, the shear zone is first

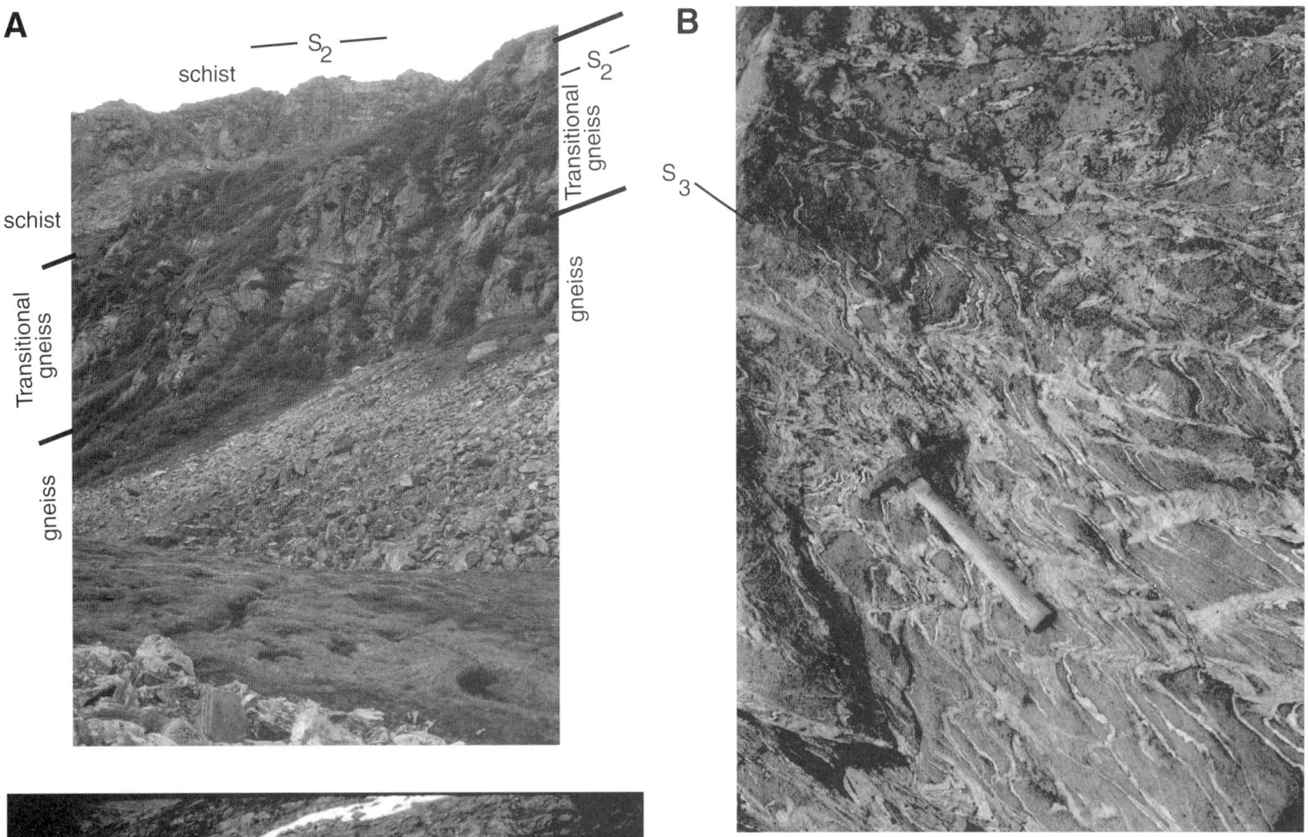

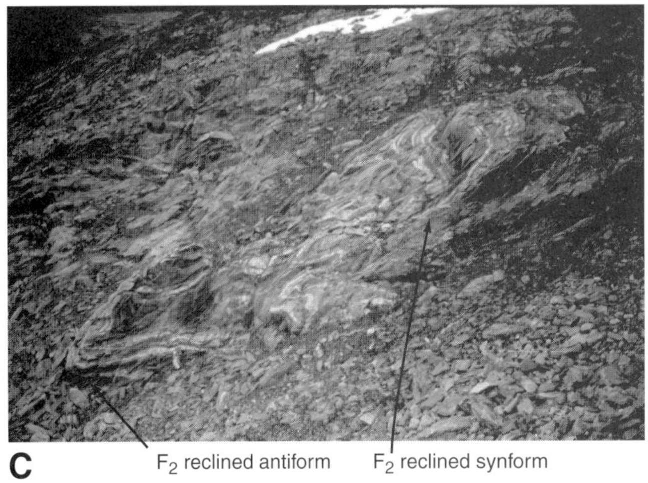

Figure 5. Field views illustrating the nature of the schist-gneiss transition. A: Cliff face west of Goat Wallow Glacier with the gneiss transition on the slope, the homoclinal S_2 foliation within the area, and sharp transition to the gneiss. Height of the cliff face is approximately 800 m. B: Field photo of an exposure of gneiss taken close to the site where the photo A was taken. Note the prominent F_3 folds and disruption of primary layering by development of gneissic banding. C: Field photo showing a typical exposure of schist ~200 m above the gneiss transition illustrating the contrast in structural style from the gneiss. Note the nearly recumbent F_2 subisoclinal folds in layering that are typical of the schist envelope and the absence of F_3 folds or crosscutting S_3 foliation. For scale there is a person just to the right of the arrow pointing to the F_2 antiform.

recognized as an open, upright F_3 folding of the S_2 foliation that contrasts with the homoclinal S_2 foliation in the schist envelope to the south. These folds get successively tighter to the north, and within a cross-strike distance of ~1 km, F_3 folds become subisoclinal with a transposition of layering into S_3 conspicuous at scales of tens of meters. Associated with this tightening of F_3 folds is a development of a domainal mica foliation (S_3) and a prominent gently NW plunging elongation and mineral lineation (Fig 3). The lineation is an elongation lineation as demonstrated by boudinage of early andalusite with development of fibrous sillimanite within the boudin necks (Fig. 6A). Both asymmetry of these fibers as well as mesoscopic shear bands indicate dextral shear sense throughout the zone (Figs. 6A and 6B).

The map pattern of the transitional gneiss unit is complex as it approaches the Bremner shear zone. Part of this complexity is topographic (Plate 1), but it is also due to structural relief as well as blurring of the contact by syn-tectonic plutonism. The principal structural feature is a large, open antiform in the schist-gneiss transition (see cross section A–A′, Plate 1), which deflects this contact westward across Turtle Rocks creek. However, unlike similar structures to the south of Wernicke Glacier (see below), the gneiss transition does not continue northward

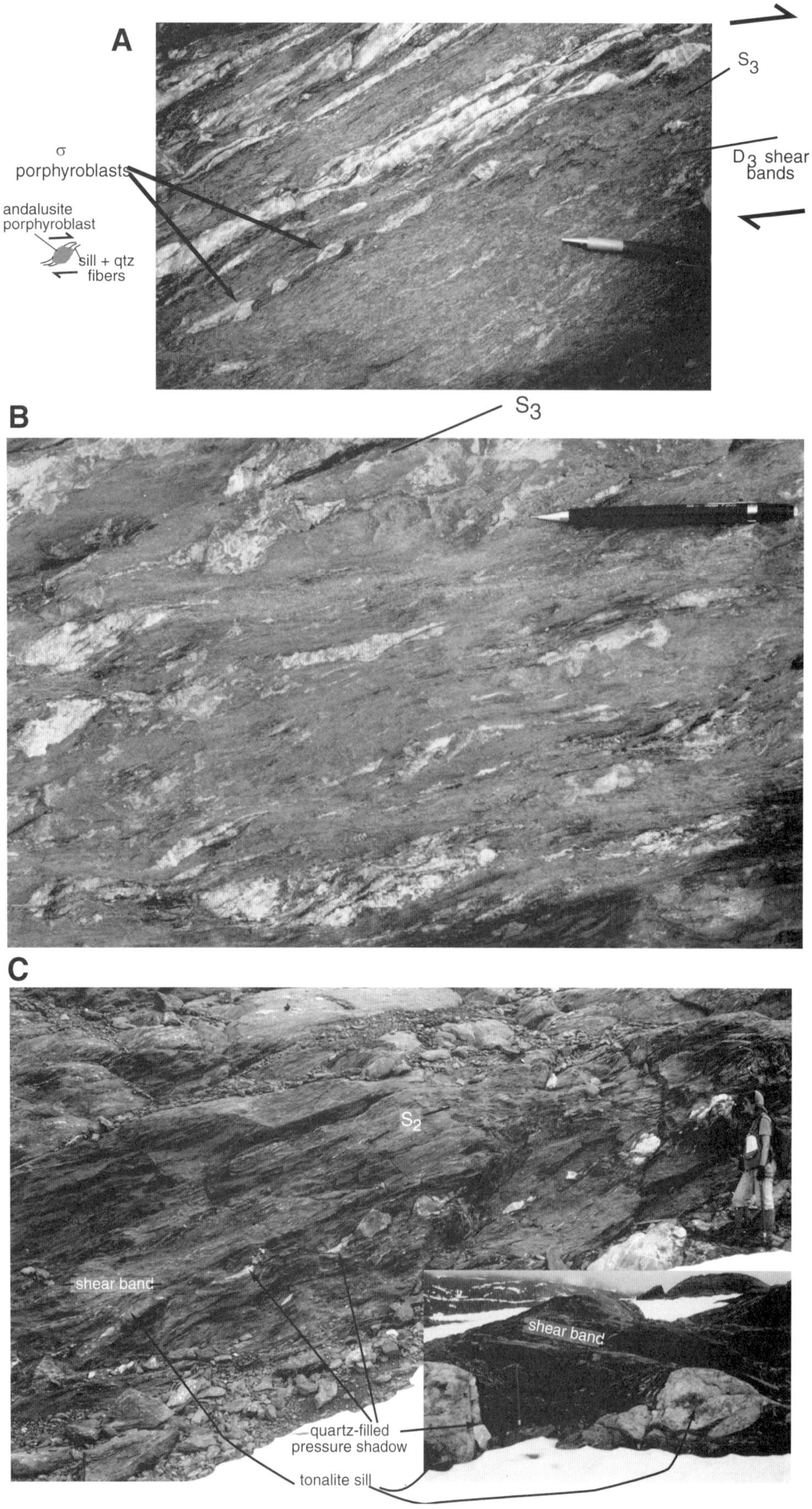

Figure 6. Fabric asymmetries indicating shear sense for D_2 and D_3 in the Chugach metamorphic complex. A: Boudinage of early (syn-D_2) andalusite with fibrous sillimanite pressure shadows showing stretching parallel to the subhorizontal, D_3 lineation indicates the lineation is a stretching lineation. Asymmetric pressure shadows indicating dextral shear sense for the Bremner shear zone. View is of a flat surface approximately along the lineation and perpendicular to foliation. B: Mesoscopic shear bands and asymmetric pressure shadows in porphyroblasts illustrating dextral shear in the Bremner shear zone. Pencil is parallel to shear bands and the label shows obliquity of foliation relative to the shear bands indicating dextral shear (view is a flat surface ~perpendicular to S_3 foliation). C: Large-scale asymmetric boudinage of a Ti_2 sill with associated quartz-filled pressure shadows (inset) lower right shows details of boudins) along S_2 foliation suggesting D_2 top to the east shear sense. View is looking north on a surface ~parallel to lineation and perpendicular to S_2, which dips moderately away from the view.

through the Bremner shear zone. Instead, it swings sharply to the east into parallelism with the shear zone just north of Turtle Rocks Creek (Plate 1). Thus, at least within the mapped area, the Bremner shear zone bounds the gneissic rocks on the northern side. In addition, other structural complications occur where the shear zone passes downward into the gneiss. In this area, transitional gneiss along Turtle Rocks Creek shows complex, convolute folds (e.g., details of the S_2 form surface, Plate 1) that indicate either an additional folding phase or, more likely, D_3 sheath folding superimposed on older fabrics. Farther to the east (structurally downward), however, the gneiss becomes structurally indistinguishable from gneisses in other areas. Thus, the complications in the transitional gneiss are either a high-strain effect within the transitional gneiss, a product of the emplacement of voluminous plutonic rocks nearby (Plate 1), or both.

The Bremner shear zone is also intruded by multiple generations of syn-tectonic tonalitic plutons. One of the most significant of these plutons is a large plutonic sheet exposed along the schist-gneiss transition just north of Transition Glacier and continuing to the east bank of Turtle Rocks Creek (Plate 1). However, this large pluton is only one of innumerable sill-like sheets along S_2 foliation that comprise a significant fraction of the exposures on the ridge north of Turtle Rocks Creek (Plate 1). All of these plutons contain a prominent foliation parallel to S_2 as well as a subhorizontal lineation. This LS fabric is also overprinted by D_3 shear-related fabrics (Plate 1). This plutonic complex is interpreted as syn-D_2 because (1) it contains a conspicuous D_2 fabric (Plate 1 and Fig. 3) indicating it experienced significant D_2 flow, yet that fabric is less conspicuous than in adjacent rocks; (2) here, as well as in the Tana River region (Pavlis and Sisson, 1995), plutons of this generation crosscut metamorphosed felsic porphyries that are similar to Tertiary dikes outside the Chugach metamorphic complex—an observation implying these plutonic bodies are younger bodies emplaced during development of the Chugach metamorphic complex; and (3) most importantly, these plutonic bodies show no evidence of chilled margins or contact aureole development at their margins implying they were emplaced after most of the heating had occurred in the Chugach metamorphic complex, and are, therefore, syn-D_2. Finally, these syn-D_2 plutons are also cut by even younger plutons. These include both syn-D_3 plutons (Ti_{3a}), which contain a prominent foliation parallel to S_3, and even younger late to post-tectonic plutons (Ti_{3b}) that have no recognizable foliation.

The plutonic assemblages along the Bremner shear zone demonstrate that plutonism occurred continuously throughout the D_2-D_3 progression in a manner similar to that recognized by Pavlis and Sisson (1995), Harris et al. (1996) and Sisson et al. (this volume, Chapter 13) elsewhere in the Chugach metamorphic complex. In contrast to the Tana River area, the Bremner River area has only rare late generation hypabyssal dikes (Ti_4 of Pavlis and Sisson, 1995, and Harris et al., 1996). It also contains large undeformed discordant plutons (Ti_{3b}), which are not abundant in the Tana River region.

Southern Section: Wernicke Glacier to the Chugach St. Elias Fault

The southernmost exposure of the high-grade gneissic core of the Chugach metamorphic complex is exposed in the area between the Wernicke and Van Cleve Glaciers (Fig. 2). Metamorphism and plutonism related to the Chugach metamorphic complex extends to the south to the Chugach-St. Elias fault (Miller et al., 1984). Immediately south of Van Cleve Glacier, the metamorphic assemblage is engulfed by a large pluton (Fig. 2). Thus, we know little about the structural history in that area other than limited information from screens within the pluton and from spot observations just south of Van Cleve Lake (Fig. 2, Plate 2).

Farther south, across the Miles Glacier (Fig. 2), Paleogene metamorphism or Early Cenozoic metamorphism was developed in a different assemblage, the Early Cenozoic Orca Group. This large segment of the Chugach metamorphic complex has only been examined by reconnaissance work (Miller et al., 1984; Winkler and Plafker, 1993). We conducted a detailed study along the southernmost edge of this assemblage in the Johnson Glacier area (Fig. 2) and have conducted limited aerial reconnaissance of the bulk of the area. From aerial observation and available data from reconnaissance studies, we conclude that the structures are relatively homoclinal between Miles Glacier and Johnson Glacier. Thus, our observations at Johnson Glacier are probably diagnostic of this large area.

Van Cleve Map Area

Between the Wernicke and Van Cleve Glaciers (Fig. 2) the schist-gneiss transition is well exposed in an area of modest topographic relief relative to adjacent areas. The area contains the southern limit of the gneissic core of the complex, and is characterized by slightly lower metamorphic pressures than areas north of Wernicke Glacier. That is, garnet is absent and many of the highest-grade rocks are andalusite-cordierite gneisses (Sisson et al., 1989, 1997).

As in the Bremner River area (Plate 1), schists above the gneiss transition (Domain 2, Plate 2) are dominated by relatively homogeneous D_2 fabrics with narrow bands of intense D_3 fabric development. Throughout most of the schist envelope, S_2 foliation dips gently to the NNE and is associated with gently inclined, close to isoclinal F_2 folds. In areas of low D_3 strain, these F_2 structures are approximately cylindrical (Fig. 4D). F_2 axes plunge gently WNW subparallel to a conspicuous elongation lineation (Fig. 4C). In areas well above the schist-gneiss transition, these homogeneous fabrics are locally cut by mappable bands of D_3 fabric development that represent distinct, steeply dipping shear zones. Typically these zones are marked by an internal zone where D_3 fabric development is intense such that S_2 foliation is nearly transposed by close to subisoclinal, upright, gently northwest plunging F_3 folds (Figs. 4E and 4F). Lineations are well developed within these zones, but are subparallel to the gently WNW plunging lineations in adjacent D_2

dominant domains (Fig. 4C). The boundaries to these zones are generally relatively sharp (Plate 2) with a transition zone <10% of the total width of the zone. This boundary zone is characterized by more open fold geometries that tighten toward the interior of the zone. Shear sense indicators are prominent within all of these zones and consistently indicate a dextral sense of shear parallel to the elongation lineation.

The most prominent of the D_3 shear zones is ~1 km wide and occurs just south of the Wernicke Glacier (Fig. 2). This shear zone is informally referred to here as the Wernicke Glacier shear zone and is particularly significant because it produces a left stepping offset of the schist-gneiss transition across the Wernicke Glacier (Fig. 2). The left-stepping offset is initially confusing in light of mesoscopic evidence within the shear zone for a dextral shear sense; e.g., abundant asymmetric boudins, pressure shadows in boudinaged andalusite, and fabric asymmetries. This problem is easily resolved, however, by consideration of the three-dimensional geometry of this down-plunge section. If the slip vector plunged westward, dextral-oblique motion would produce north-side-up displacement across the zone. This north-side-up component would produce an apparent sinistral shift provided that the slip vector plunged more steeply than the gently northwest dipping schist-gneiss contact, and that this shift was not obliterated by crustal flow or thermal re-equilibration at the gneiss-transition. This inference agrees with two observations. First, elongation lineations (Fig. 4C) plunge westward throughout the shear zone and are subparallel to the intersection line between the shear zone and the gneiss contact. Thus, although the maximum finite stretching direction does not necessarily equate to the slip vector (e.g., Tikoff and Teyssier, 1994), the observed geometry is consistent with a west-plunging slip vector. Second, there is an apparent increase in metamorphic pressures north of the Wernicke Glacier (Sisson et al., 1989, 1997) indicated by the absence of garnet and the occurrence of cordierite-andalusite gneiss to the south in the Van Cleve area. Thus, the north side of the Wernicke Glacier shear zone has been more deeply exhumed than to the south, which is consistent with dextral oblique, north-side-up movement on the Wernicke Glacier shear zone.

Like the Bremner shear zone, the Wernicke Glacier shear zone is also spatially associated with a plutonic complex localized near the schist-gneiss transition. This observation could be coincidental, because one of the largest syn-D_2 plutons recognized during our studies is in the Goat Wallow Glacier area (Plate 1), an area with virtually no D_3 overprints in the schist envelope. Nonetheless, because both the Wernicke Glacier shear zone and Bremner shear zone have large syn-D_3 plutons, their occurrence in the shear zones may not be coincidental. That is, these shear zones may have developed at these positions due to crustal weakening in the vicinity of the plutonic complex in the manner suggested by Hollister and Crawford (1986) or Pavlis (1996).

The distinctive feature of the Van Cleve map area is a spatial variation in the geometry of the schist to gneiss transition. Immediately south of the Wernicke Glacier shear zone (Fig. 2, Plate 2), the schist-gneiss transition is a nearly homoclinal, gently WNW dipping surface. It can be continuously traced for ~4 km with no significant structural relief (Plate 2, Fig. 4D). However, ~3–4 km north of Van Cleve Glacier (along Draco Glacier, Plate 2), the schist-gneiss contact is warped into a series of macroscopic upright folds with amplitudes of ~1 km. These folds produce both a macroscopic zigzag pattern to the schist-gneiss contact and an isolated oblong body of schist and transitional gneiss surrounded by gneiss; a geometry produced by a large doubly plunging synform (Plate 2). These macroscopic folds are F_3 structures. The curvature of the fold axes is shown in stereonet plots (Fig. 4F) by occurrence of both southeast and northwest plunges to the fold axes. In addition, scatter of S_2 poles (Fig. 4A) along a northeast striking, vertical great circle suggests an average, approximately horizontal fold axis. These large folds are also developed along strike from a significant D_3 shear zone (Plate 2) suggesting a direct association with this shear zone.

The local structural relief along the schist-gneiss transition has important implications for the structural history. Specifically, because the schist-gneiss transition is almost certainly a temperature-controlled boundary, the structural relief along the boundary indicates a lateral variation in thermal structure, a later deformation of an early formed thermal boundary, or both. Our tentative conclusion is that both effects produced the observed structure. For example, the relief across the Wernicke Glacier shear zone is almost certainly synchronous with prograde metamorphism because early andalusite was boudinaged and then overgrown by sillimanite within the shear zone. In addition, peak metamorphic pressure appears to be higher north of this zone. In contrast, the folding in the southern part of the area may be associated with late folding of the contact during peak or falling temperature conditions. Nonetheless, even in this case, the macroscopic folding is spatially associated with a conspicuous shear zone at higher structural levels to the west (Plate 2) and the structures disappear at depth into the homogeneously deformed gneiss. Thus, these observations do not alter our premise of a major change in structural behavior across the gneiss transition.

Johnson Glacier Area

This small area displays a markedly different structural history from all other exposures we have examined in the western Chugach metamorphic complex. The area lies immediately north of an elongate Tertiary plutonic complex that extends eastward for ~50 km. The plutonic contact is covered by a glacier in the study area, but map pattern and general structural relationships suggest the plutonic complex is a large, subvertical sheet.

Metamorphic rocks north of Johnson Glacier are dominantly metagraywacke, partially converted to biotite-garnet gneiss in close proximity to the pluton. The metamorphic grade rapidly decreases northward to upper greenschist facies phyllite. Structurally, these metasedimentary rocks are characterized by two major fabrics (Fig. 7): an early layer-parallel foliation (S_1) and an intense second phase (D_2) foliation (S_2) with an associated mineral lineation. S_2 is ~EW striking and steeply north dipping to vertical throughout the area (Fig. 7) and is axial planar to numerous vertical to steeply plunging folds in $S_1 = S_0$. The mineral lineation is subparallel to fold axes (Fig. 7) and is in part an intersection lineation, but boudinage structures suggest the lineation is also a finite stretching direction. Shear sense indicators are not conspicuous, and the kinematic significance of these fabrics is not well established. The markedly different geometry, overprinting relationships, and thermal history of this region relative to adjacent areas implies a major discontinuity across the region: presumably, the Contact fault, which parallels the Miles Glacier and separates the Orca and Valdez Groups (Fig. 2).

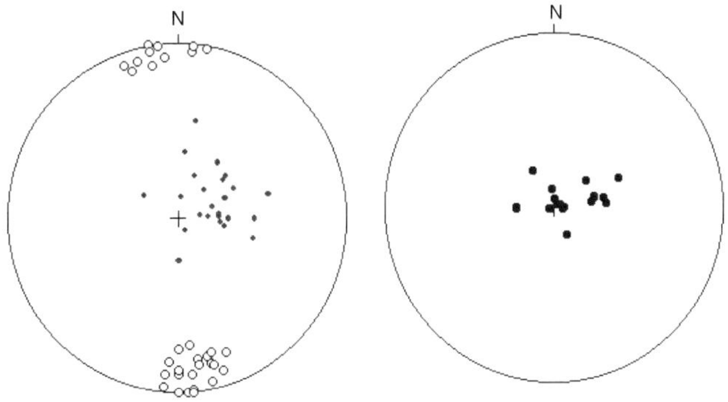

Poles to S_2 foliation (○) and mineral/elongation lineation (•) Johnson Glacier area (Orca Group)

F_2 axes, Johnson Glacier area

Figure 7. Equal-area stereographic projection diagrams summarizing orientation data from the Johnson Glacier map area.

Finite Strain Data

Pavlis and Sisson (1995) used finite strain data obtained from deformed conglomerates, sandstones, and plutonic rocks in the Tana River region to examine the effects of strain superposition in fabric development within the Chugach metamorphic complex. In that study, we were able to show the kinematics of the D_2-D_3 structural succession through a combination of analyzing finite strain in metasedimentary rocks and syntectonic plutons. Our data from the syntectonic plutons also allowed modeling of the strain superposition. In this study, we expanded that database using the same methods reported in Pavlis and Sisson (1995). The finite strain data are presented on a logarithmic Flinn plot (Fig. 8A). Several observations are apparent:

1. In comparison to the Tana River region, metaconglomerates from the Bremner River area show much greater scatter in ellipsoid shape, although strain magnitudes are comparable.

2. Measurements of shapes of argillite rip-up clasts in the Bremner area are sparse, but dominated by apparent flattening ellipsoid shapes. These contrast markedly with the Tana River area where similar rocks show constrictional ellipsoid shapes.

3. Xenolith shape fabrics in three plutonic bodies within the Bremner River region all show apparent constrictional ellipsoids, including two bodies that we interpreted as syn-D_3 plutons. Note that one of the syn-D_3 plutons as well as the syn-D_2 plutons are located within the Bremner shear zone, but the third pluton was located within the gneiss and possessed a pronounced foliation parallel to S_3. This observation contrasts with the Tana River region where syn-D_2 plutons show constrictional strains, but syn-D_3 plutons show apparent flattening strains.

4. The data are relatively sparse from the Bremner River area, particularly given the size of this area relative to the Tana River region. Nonetheless, the data do provide important regional constraints when considered in a spatial context that we can compare to the Tana River data.

Observations 1 and 2 together imply there is significantly greater strain heterogeneity in Bremner River region, which is not surprising given the spatial variations in structural style. This strain heterogeneity is not random but instead is correlated to structural domains. For example, all but one of the metaconglomerates and argillite rip-up clasts from the Wolverine Run area (Fig. 2) display apparent flattening shape fabrics. In this area D_3 strain is cryptic to non-existent, and D_2 structures are dominant. In contrast, the sites that show apparent constriction are all located within, or near, significant D_3 shear zones, and all are <1 km above the gneiss transition. Thus, we conclude that the apparent constrictional strains are probably a result of D_2-D_3 strain superposition, whereas D_2 flattening was dominant where D_3 strain is low.

This conclusion is also consistent with the finite strain modeling results we reported for the Tana River region (Pavlis and Sisson, 1995). These models provide a qualitative basis to explain the observations in the Bremner River area. Specifically, in the Tana River region, we were able to use shape fabrics in syn-D_3 plutons to restore D_3 strains on the limbs of macroscopic D_2 folds. This restoration shifted the observed apparent constrictional shapes into the flattening field indicating D_2 produced flattening strains. This flattening was explained by the inference that D_2 involved regional, ~plane strain with subhorizontal, EW stretching and vertical shortening in combination with localized flexural-flow-related shear on the limbs of macroscopic folds (e.g., Fig. 7C in Pavlis and Sisson, 1995). In the Bremner River area, we were unable to find syn-D_3 plutons to provide the same constraint, but the structural style observations suggest an analogous, qualitative similarity. That is, areas of apparent flattening are where D_3 strains are low and these areas the strains are dominated by a combination of regional flow (vertical shortening and EW elongation) plus subhorizontal NS stretching related to flexural flow on the limbs of F_2 folds. Constrictional strains, in contrast, represent either a bulk flow effect, or more likely, the effect of finite-strain superposition of D_2 and D_3 strains.

Results from three syntectonic intrusive rocks are also broadly consistent with this conclusion. Nonetheless, all three

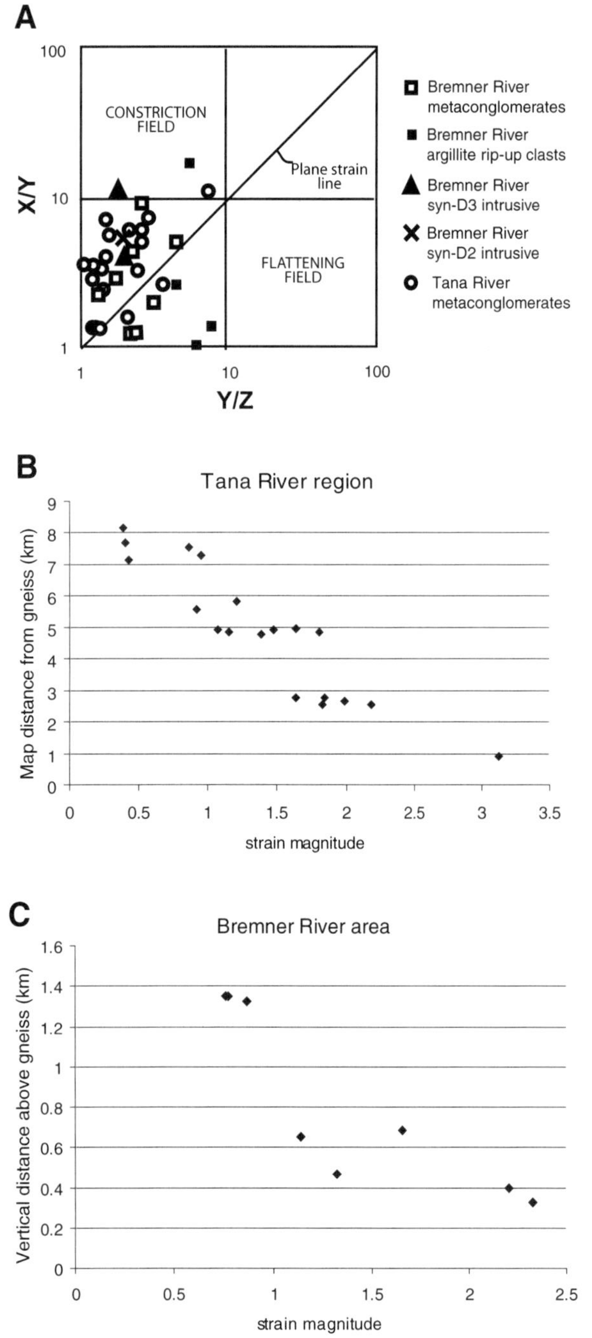

$$\text{strain magnitude}(\varepsilon_s) = \frac{1}{\sqrt{3}}\left[(\varepsilon_1-\varepsilon_2)^2+(\varepsilon_2-\varepsilon_3)^2+(\varepsilon_3-\varepsilon_1)^2\right]^{\frac{1}{2}}$$

Where: $\varepsilon = \ln$ (stretch) = natural strain [Nadai, 1963]

Figure 8. Summary of finite strain data from the Bremner River area in comparison to finite strain data from the Tana River area reported by Pavlis and Sisson (1995). A: Flinn diagram summarizing finite strain ellipsoid shapes determined from metaconglomerates and syntectonic intrusives in the both the Bremner region and Tana River area. B: Strain magnitude vs. map distance from the schist-gneiss transition in the Tana River area showing systematic increase in strain toward the transition. C: Strain magnitude vs. vertical distance above the schist-gneiss transition in the Bremner region showing increased strain toward the transition analogous to the Tana River region.

plutonic bodies, including the syn-D_3 plutons, show apparent constriction, which raises questions on the nature of D_3 flow. Specifically, in the syn-D_2 pluton (Fig. 8A), constriction can be explained by D_2-D_3 strain superposition analogous to that in the metaconglomerates. In the syn-D_3 plutons, however, this conclusion is not necessarily appropriate. This result could be a consequence of an inherited shape fabric (e.g., igneous flow fabric) overprinted by D_3 strain. Nonetheless, we suggest that it is more likely that our field criteria underestimated the relative age of the plutons, and the syn-D_3 plutons are actually late D_2. That is, we distinguished syn-D_2 versus syn-D_3 plutons on the basis of fabric geometries relative to adjacent rocks. If the pluton possessed a single, continuous cleavage subparallel to S_3 foliation in adjacent rocks, we assumed the pluton was syn-D_3. Alternatively, if the pluton was emplaced late during D_2 motions or if D_2 and D_3 were broadly synchronous, this criterion could systematically underestimate the relative emplacement age. We tentatively conclude that the apparent syn-D_3 plutons are actually late D_2 and the apparent constrictional strains are the product of D_2-D_3 superpositions. Nonetheless, more data are ultimately needed to evaluate this hypothesis.

The most important result of finite strain measurements from both the Copper River region and Tana River region is a spatial variation of strain magnitudes with structural position relative to the schist-gneiss transition (Fig. 8). In both the Tana River and Copper River regions, the finite strain magnitude, calculated as Nadai's (1963) strain parameter, increases systematically with decreasing distance to the transition. This conclusion is qualitatively obvious in the field by inspection of metaconglomerates, but is also evident in the finite strain data (Fig. 8). Thus, regardless of the interpretation of the D_2 to D_3 strain progression, strain magnitude increases rapidly towards the gneiss transition, implying this boundary is a fundamental structure within the crustal column.

DISCUSSION

Interpretation of the Kinematic Development of the Chugach Metamorphic Complex

Summary and Interpretation of the Kinematics of D_2 and D_3

The general kinematic history for the Chugach metamorphic complex recognized in the Tana River region by Pavlis and Sisson (1995) can be traced westward to the Copper River region. In the Copper River region, however, these high-temperature D_2 and D_3 structures plunge westward beneath lower-grade rocks with a different, but related, kinematic history (e.g., see Pavlis et al., this volume, Chapter 7). To the north of Miles Glacier (Fig. 2), the D_2 to D_3 progression is associated with peak metamorphism with D_2 preceding the metamorphic temperature peak and D_3 synchronous with it. In this area, D_2 was associated with a steeply south-plunging to vertical shortening axis with an associated subhorizontal, WNW trending stretching axis. This geometry is similar to the Tana River region, although lineations in the Tana River region plunge to the west whereas in this region lineations plunge uniformly to the west-northwest or northwest. This variation is probably a regional relationship related to strain variations within the down-plunge transition to low-grade rocks.

Measured finite strains in the Tana River region are predominantly constrictional (Pavlis and Sisson, 1995), which seems different from the Copper River area where both flattening and constrictional strains are observed (Fig. 8). However, the constrictional strains in the Tana River region are the result of a local, relatively unusual, situation where most of the strain data were obtained from a section where F_2 folding was largely absent and D_2 occurred primarily through layer-parallel shortening of steeply dipping layers (Pavlis and Sisson, 1995). In both the Tana River and the Copper River areas, sites with conspicuous F_2 folds both show apparent flattening strains on fold limbs and constrictional strains in fold closures, provided D_3 overprints are minor or can be restored. This implies that the bulk D_2 kinematics are similar in the two areas with a regional ~plane strain imposed on initially steeply dipping layers producing folds with axes parallel to the elongation direction. Flattening strain occurred on fold limbs due to a combination of the regional flow plus flexural-flow-related shear on the limbs of F_2 folds (e.g., as inferred by Pavlis and Sisson, 1995). Many of the observed constrictional strains from the Copper River area are probably the product of D_2-D_3 strain superposition, but absence of syn-tectonic plutons to constrain the strain path leaves some doubt on this interpretation. Thus, additional work would be needed to clarify these uncertainties. In any case, however, the strain magnitude calculated from these data indicates a marked strain increase towards the schist-gneiss transition implying it is a significant structural feature.

D_2 fabrics display evidence of top to the east shear along the north side of Transition Glacier (Plate 1, Fig. 6C). Moreover, in the Bremner River area the macroscopic structure hints at top to the east shear. Specifically, if we assume the gneiss transition is subparallel to a shear zone boundary, the ~10°–15° west dip of this contact (B–B′, Plate 1) is shallower than the 20°–25° WNW dip of S_2 (e.g., between Echo Lake and Frozen Toe Lake, Plate 1); an asymmetry suggesting top to the east shear. There are, however, no equivalent shear sense indicators in the Van Cleve Glacier area, suggesting spatial variation in shear along the boundary. We suggest this variation is primarily the product of a variable shear sense or "general shear" condition along most of the boundary. More importantly, however, the flow-field diverged strongly from simple shear (e.g., flattening or constriction, depending on F_2 folding) and thus, classic shear sense indicators probably failed to develop, even at high shear strains.

As in the Tana River region, the gently dipping to flat-lying foliation generated during D_2 was overprinted by D_3 structures that include upright folds in S_2 and shear bands showing evidence of dextral strike slip. The important distinction, however, is the manner in which the deformation was distributed during D_3. In the gneissic core, D_3 generated a relatively homogeneous fabric indistinguishable from the Van Cleve Glacier to the northern edge of the gneissic core in Turtle Rocks Creek. In contrast, D_3 structures are strongly domainal in the schist envelope that surrounds the gneissic core, and these domains of intense D_3 fabric development are associated with fabric asymmetries indicative of dextral shear. These domains are separated by a narrow zone marked primarily by the "transitional gneiss" (Fig. 2, Plates 1 and 2), in which along-strike variations are intermediate in character. Note, however, that this structural transition is not strictly limited to the transitional gneiss as D_3 structures "smear" across the boundary in the form of mesoscopic folds and distributed strain up to 100–200 m above the transitional gneiss contact (e.g., the southern edge of the Bremner shear zone, Plate 1). Thus, like most real world systems, the transitions responsible for the variations in structure across the gneiss transition are gradational, yet even with the addition of the overlying schists, the transition is remarkably narrow.

We conclude that the D_3 fabric domains within the schist envelope are D_3 dextral shear zones that continued westward outside the area influenced by metamorphism. We present evidence elsewhere (Pavlis et al., this volume, Chapter 7) that these dextral shear zones continue westward in lower-grade (greenschist facies) rocks as a broad band of fabric overprints to at least the longitude of Valdez (Fig. 1). There is a large gap in our database between the Copper River and the area examined by Pavlis et al. (this volume, Chapter 7). Nonetheless, these two areas display very similar kinematic histories despite major differences in macroscopic scale structure.

Interpretation of the Macroscopic Significance of Chugach Metamorphic Complex Structure and Broader Implications

The fundamental variation in the structural style of D_3 with structural level represents, in our view, the most important first order observation from our field effort in the Copper River region. The structures imply that within the gneissic core, D_3 flow occurred as a uniform, laminar flow distributed relatively homogeneously throughout the core. At higher structural levels, however, the flow associated with the same D_3 event was partitioned into distinct shear zones. The boundary between these two flow domains occurs primarily within the transitional gneiss, but locally occurs over a wider zone.

These relationships imply that the boundary between the schist envelope and the gneissic core was both an important petrologic boundary and a major kinematic boundary separating two distinct structural levels with different flow regimes. That is, the horizontal displacement field in the overlying schists was grossly different from displacements within the gneiss. Thus, these two flow regimes must have set up a strong vertical displacement gradient across what is now the gneiss transition.

Teyssier et al. (2003) recently proposed a theoretical basis for this process. They modeled the effects of rigid-body slip passing downward into distributed flow through a narrow transition that they referred to as an "attachment zone" (inset, Fig. 9). These models illustrate the effects of the vertical displacement gradients between the two crustal levels and predict a subhorizontal foliation within the "attachment zone." This prediction is remarkably similar to what we observe along the schist-gneiss transition in the Copper River region. Interestingly, these models also predict steep foliation domains within the "attachment zone," but these steep domains are located between upper crustal blocks rather than beneath upper crustal shear zones (Fig. 9). It is possible that features like the Bremner and Wernicke Glacier shear zones (Fig. 2) are equivalent to these steep-dip domains rather than the deep-seated base of strike-slip shear zones. This hypothesis cannot be evaluated from available data, however, and additional field work is needed west of the Copper River to determine if these shear zone pass laterally into faults or disappear at higher structural levels.

Our field observations together with theoretical models give rise to a fundamental question of the connection between the D_2 and D_3 structures. Since Sander (1930) laid down the basic concepts of fabric analysis, a fundamental assumption in structural analysis has been that fabric overprinting is a relative age indicator. In the last 60 years it has become clear that this relative age relationship is not necessarily applicable to absolute time because many tectonic processes are time-transgressive. Nonetheless, in most cases, these diachronous processes have been ascribed to large-scale processes such as events that migrate laterally along an orogen with time. In the Chugach metamorphic complex, our observations raise the specter that deformational processes can also be diachronous vertically within the crustal column. In particular, we recognize two distinct large-scale scenarios for the Chugach metamorphic complex, but in one of these D_2 and D_3 are broadly synchronous within a general wrench system (e.g., Fig. 9).

One possibility for the D_2 to D_3 progression in the Chugach metamorphic complex is that it is related to large-scale kinematic variations related to the triple junction interactions (e.g., Sisson and Pavlis, 1993; Pavlis and Sisson, 1995). In this scenario, the D_2 to D_3 progression was linked to triple junction migration with D_2 recording vertical shortening as the triple junction approached and D_3 recording dextral shear following passage of the triple junction when the plate kinematics changed.

This scenario remains a reasonable working model for the Chugach metamorphic complex, but the down-plunge view of the Copper River region suggests an alternative scenario. Specifically, if D_3 flow was partitioned within the crustal column, it is possible that the displacement gradients set up across the flow boundary—the gneiss transition—could have generated D_2 type structures that were all part of the same deformational event represented by the D_3 system (e.g., the model of Teyssier et al., 2003; Fig. 9). If this scenario is true, then observed fabric overprints may have a complex connection to an evolving thermal-mechanical system. For example, the general motion during D_3 was apparently dominated by dextral shear. In the schist envelope, this motion would produce domains approximating rigid-body motion bounded by narrow shear zones that take up the bulk of the horizontal displacement gradient. At depth, however, that same bulk motion occurs as a broad zone of dextral shear, and as a result, vertical displacements gradients would separate the rigid blocks above from the distributed flow below (Fig. 9). These vertical displacement gradients would produce subhorizontal shear parallel to the slip vector of the dextral shear zones. The sense of shear, however, could be variable in space and time across this boundary (e.g., Teyssier et al., 2003). For example, if the strike-slip shear zones activated at different times, or multiple shear zones were active, the shear sense across the subhorizontal surface could vary across strike, in time, or both. In any case, this process would result in the development of a subhorizontal decoupling horizon within the crust that separated these two flow domains.

In the Chugach metamorphic complex, this process could have been time transgressive vertically. As temperatures rose within the complex during prograde metamorphism, the gneiss transition—the site of the boundary—would also rise to higher levels in the crust with the increased temperature. Thus, rocks now below the gneiss transition may have passed through the decoupling surface as the gneiss transition rose to higher levels in the crust. This leads to the intriguing possibility that the observed D_2 to D_3 progression in the gneissic rocks could represent the rocks passing through the decoupling zone as the gneiss transition rose to successively higher levels in the crust during prograde heating. This latter effect would be consistent with the observed metamorphic history where peak conditions accompany D_3. It would also mean, however, that D_2 and D_3 were manifestations of the same event and that event was diachronous both horizontally and vertically.

It is difficult to distinguish these alternatives from local geology alone, but regional tectonic considerations provide important clues. Several observations are more easily explained if D_2 was fundamentally an earlier event and D_3 was superimposed by a regional change in kinematics. First, since the regional tectonic framework involves a triple junction interaction, this is an ideal site for a major change in kinematics; e.g., as the triple junction passes, the margin experiences a major change in subduction slip vector. This presumably would translate into changes in deformation kinematics as proposed by Sisson and Pavlis (1993). Secondly, both the observed macroscopic folding of the gneiss transition in the southern edge of the Van Cleve area as well as the offset in the gneiss contact along the Wernicke Glacier shear zone (Fig. 2) suggest that D_3 was established after the gneiss transition had developed at a relatively constant structural level in the crust. This qualitative

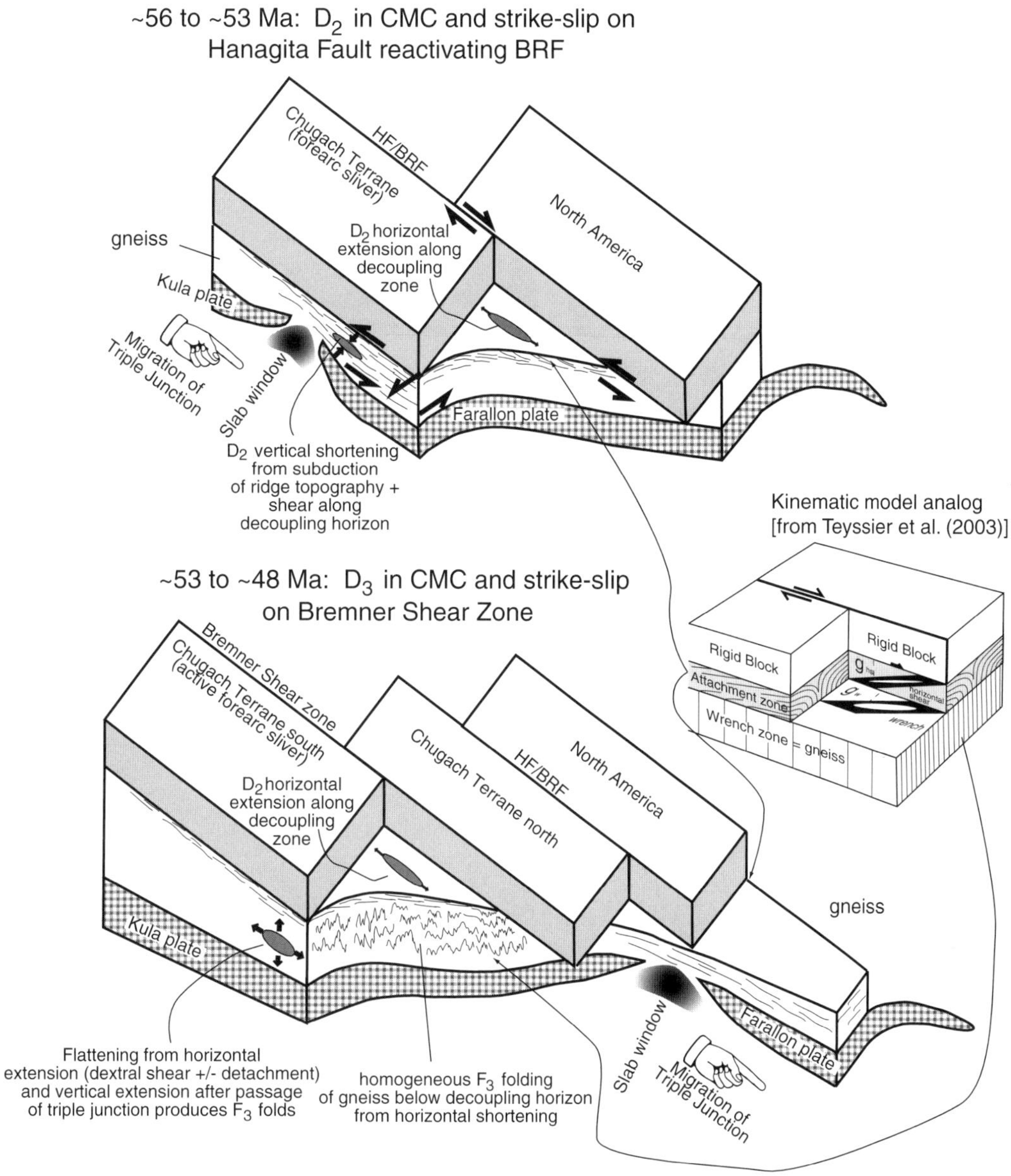

Figure 9. Cartoon summarizing our conceptual model for the D_2 to D_3 succession. Model assumes a plate interaction model summarized in Figure 12 with a south (upper left to lower right) migration of the triple junction relative to forearc sliver bounded on the northeast (upper right) by the Border Ranges fault (BRF)/Hanagita fault (HF) system at early stages (top) and by the both the Hanagita fault and Bremner shear zone (bottom) at a later time. Inset (right center) shows a kinematic model for the development of the decoupling zone (from Teyssier et al., 2003). Figure is vertically exaggerated to illustrate the inferred deep structure, and no attempt is made to show the finite width of the ductile strike-slip systems at depth. See text for a more detailed discussion.

reasoning could be flawed in a system as thermally and rheologically dynamic as this one, yet this simpler view is difficult to dismiss. Finally, finite strain observations from syn-D_3 plutons in the Tana River area (Pavlis and Sisson, 1995) indicate that although D_3 was associated with dextral shear, the bulk D_3 strain was a flattening event rather than a plane strain. In a distributed shear system, bulk flattening is indicative of a transpressional system (e.g., Tikoff and Teyssier, 1994). Thus, D_3 was probably not a plane-strain, simple-shear event associated with a dextral fault, but instead D_3 was probably a transpressional event. Thus, D_3 probably included not only horizontal shortening related to dextral shear, but also additional displacements perpendicular to strike to accommodate the contractional component to the deformation. If lower and upper structural levels were decoupled, these contractional displacements should be different across the decoupling horizon and recognizable from the orientation of kinematic axes. In the Copper River region, however, D_2 and D_3 fabric axes are all coaxial and do not fit with this relationship. Conversely, transpressional systems are commonly slip partitioned (e.g., Fitch, 1972; Jarrard, 1987) into zones of strike-slip shear and zones of thrusting. When partitioning occurs, the motion vectors are partitioned as well (e.g., Jarrard, 1987), such that plate motion between a forearc sliver and an underthrusting plate are effectively parallel to the strike-slip system. In this case, a crustal decoupling horizon would show the same flow direction as the strike-slip system, as seen in the Chugach metamorphic complex.

Interestingly, there is some evidence that D_3 in the Chugach metamorphic complex was associated with slip partitioning. First, the shear zones in the Chugach metamorphic complex as well as the backstop to the accretionary complex, the Border Ranges fault, were all active strike-slip systems during the history of the Chugach metamorphic complex (Roeske et al., this volume, Chapter 6). At the same time, however, a subduction thrust system was present below the system. Thus, this thrust system could have taken up all the thrust component of the deformation while the strike-slip systems took up the lateral component. Second, although we have only seen the southern edge of the Chugach metamorphic complex in the Johnson Glacier area and in the St. Elias range, both of these areas show very different structural fabrics from the bulk of the Chugach metamorphic complex. These rocks show a steep (Johnson Glacier) to moderately north dipping foliation (St. Elias region) with a down-dip lineation; a structural geometry consistent with a ductile thrust system. Thus, the southern edge of the complex may have taken up D_3 thrusting at the same time as the D_3 dextral shear zones took up the strike-slip component, effectively producing complete slip partitioning.

Despite difficulties in distinguishing a unique scenario for the D_2-D_3 progression within the Chugach metamorphic complex, this does not alter our fundamental observation that during D_3 the deformation was strongly partitioned at different crustal levels. Our conclusion is that the schist envelope is an "orogenic float" in a sense similar to that of Oldow et al. (1990). In the upper crustal "lid," D_3 is marked by motion of rigid body blocks bounded by narrow shear zones. This deformation was partitioned into strike-slip systems (e.g., the shear zones of the Bremner River area, the Wernicke Glacier shear zone, and the Hanagita fault) and thrust systems (e.g., the megathrust itself beneath the complex and a ductile thrust along the southern edge of the metamorphic complex). Below, in the gneissic rocks, D_3 occurred through distributed flow in a broad zone encompassing the entire gneissic complex. Thus, the "lid" moves relatively independently from the "core," but both are linked by the same general kinematic system. In this view of the crustal structure, the gneiss transition represents a decoupling horizon between the upper crustal domain displaying microplate behavior and a lower crustal domain of distributed, homogeneous deformation.

Regional Tectonic Implications and Applications to Problems of North Pacific Plate Motion

Several large-magnitude Early Tertiary strike-slip systems are known in the northern Cordillera, and there is general agreement that these fault systems are a major factor in large northward latitudinal displacements of outer Cordilleran terranes inferred from paleomagnetic data (e.g., Hillhouse and Coe, 1994). A puzzling problem, however, has been the evidence that rocks as young as Paleocene–early Eocene in the Prince William terrane have been displaced as much as 2000 km northward in the last 55–60 Ma, while coeval rocks immediately inboard on the Peninsular terrane show little, if any, latitudinal displacement (e.g., Hillhouse et al., 1985; Stamatakos et al., 1989; Hillhouse and Coe, 1994). We note here, however, that the results of this study as well as two related studies (Pavlis et al., this volume, Chapter 7; Roeske et al., this volume, Chapter 6) may resolve this problem by a combination of large-magnitude strike slip along the Hanagita fault (Roeske et al., this volume, Chapter 6) and previously unrecognized strike-slip systems within the Chugach terrane.

In this study the most significant strike-slip structure appears to be the Bremner shear zone and its probable correlative, the Stuart Creek shear zone (Fig. 1). The Bremner shear zone is a 2–3 km wide zone of intense ductile deformation surrounded by a broader zone of lower strain (Fig. 2). Strains estimated from xenoliths in syn-tectonic plutons (Fig. 8) are high (maximum ellipticities as high as 20), which is consistent with D_3 fabric transposition observed within the zone. If these strain estimates are typical of the zone, the offset across the zone is modest—a few tens of km based on shear strains of 5–10 across this 3 km wide zone. This structure appears to connect to the Stuart Creek shear zone recognized in lower-grade rocks ~75 km to the west along the Richardson Highway (Pavlis et al., this volume, Chapter 7). The Stuart Creek zone is marked by a broad (~20 km) zone of distributed strain and a narrow, poorly exposed contact that presumably represents the high-strain interior of the shear zone. Displacement across the Stuart Creek zone is more difficult to constrain because the interior of the shear zone is poorly exposed and may include brittle faults. Nonetheless, it is a significant crustal feature. Thus, we interpret these two features together as a significant dextral strike-slip system developed within the forearc region.

Regionally, dextral offset on the Bremner shear zone is probably less significant than the Hanagita fault system; the Late Cretaceous–Early Eocene structure that reactivated the Border Ranges fault (Fig. 1) as a dextral strike-slip zone (Smart et al., 1996; Roeske et al., 1992, 1993, this volume, Chapter 6). This structure was active during the early phases of development of the Chugach metamorphic complex (ca. 54 Ma). It probably shut down during the waning phases of metamorphism at ca. 51 Ma (Roeske et al., this volume, Chapter 6). Thus, D_2 in the Chugach metamorphic complex was presumably occurring while the Hanagita system was active, and D_3

occurred either at the same time or slightly later when part of the strike-slip motion was taken up within the Chugach metamorphic complex. This chronology implies a connection between these events and that the Chugach metamorphic complex was fundamentally associated with both triple junction interactions and the coeval motion of forearc slivers.

The mechanism for development of the dextral shear systems is tied to the triple junction interaction that ultimately generated the Chugach metamorphic complex, yet that process also raises regional questions of the connection between plate interaction and the dextral shear. All north Pacific plate models agree that prior to ca. 56 Ma, the Kula-Farallon plate boundary was the most likely ridge system interacting with the North America subduction zone to generate the Chugach metamorphic complex (e.g., Engebretson et al., 1985; Stock and Molnar, 1988). The resulting triple junction (Fig. 10) had a general northwest to southeast (present coordinates) migration based on both plate motions and age progressions of forearc intrusives (e.g., Sisson and Pavlis, 1993; Bradley et al., 1993; Haeussler et al., 1995). Following an Eocene plate reorganization, however, it is neither known which plates interacted with the North American margin nor how the triple junction migrated along the margin. Three scenarios have been suggested for this late Eocene triple junction: (1) a Kula-Pacific model calls on the demise of the Kula spreading center (e.g., Byrne, 1979; Stock and Molnar, 1988), and in this case a southward migrating Farallon–Pacific–North

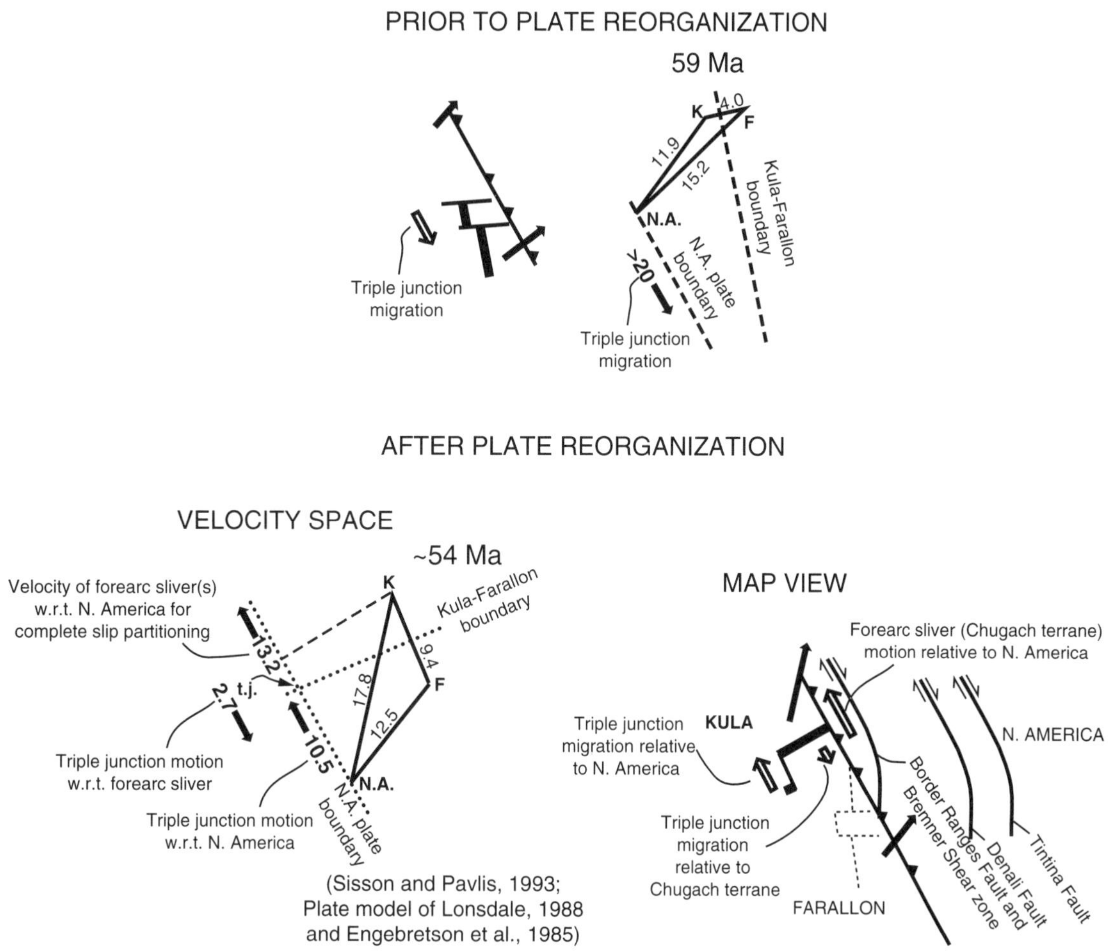

Figure 10. Plate motion model for development of the Chugach metamorphic complex assuming interaction between North America, Kula, and Farallon plates with the addition of a microplate (forearc sliver) between North America and the subduction zone. Figure is a modification of a figure in Pavlis and Sisson (1995) and uses the same linear velocities from that study. Note that in this model if the forearc sliver moves with the full lateral component of the Kula–North American motion, it moves northward faster than the triple junction; hence, the triple junction migrates southward on the forearc sliver. The lateral component presumably was taken up primarily on the Hanagita/Border Ranges fault system, but part of the motion may also have been accommodated on the better-known Denali and Tintina fault systems. See text for implications of this result.

American triple junction was consumed; (2) a Kula-Farallon model calls on continued rapid spreading of the Kula-Pacific ridge (e.g., Engebretson et al., 1985; Lonsdale, 1988) and the resultant Kula-Farallon-Pacific triple junction would have migrated rapidly northward along the margin (e.g., Sisson and Pavlis, 1993); and (3) a Resurrection plate model (Bradley et al., this volume, Chapter 1), which invokes a fourth north Pacific plate emanating from a ridge-ridge-ridge triple junction on the Kula-Farallon spreading center—because this plate is completely subducted, there are no constraints on its possible plate kinematics.

In our initial studies of the Chugach metamorphic complex, we were attracted to the Kula-Farallon model because the northward backtracking of the Kula–Farallon–North America triple junction provided an explanation for both the thermal history of the Chugach metamorphic complex and the D_2-D_3 structural history within the complex (Sisson and Pavlis, 1993; Pavlis and Sisson, 1995). As more data have accumulated, evidence for the northward migration of the triple junction has been lacking. Bradley et al. (1993) and Haeussler et al. (1995) present conclusive evidence for southeastward (modern coordinates) migration of plutonic ages in the forearc. Poole (1996) and Sisson et al. (this volume, Chapter 13) obtained geochronological data from the Yakutat region indicating a ca. 54 Ma pluton outboard from younger (ca. 48 Ma) plutonic bodies, but these bodies were separated by the Neogene Fairweather fault. Given the large magnitude of dextral slip along the Fairweather fault (Plafker et al., 1994), this observation implies either extremely rapid northward migration of the triple junction, involvement of transform offsets, or both (e.g., Poole, 1996, and Sisson et al., this volume, Chapter 13). In any case, the younger ages east of the Fairweather fault imply the triple junction continued migrating southward within what is now the Chugach metamorphic complex from ca. 54 to 48 Ma.

This conflict between the plate model and the age progressions in the forearc belt can be resolved if the Chugach terrane was a forearc sliver during development of the Chugach metamorphic complex. Forearc slivers move with a different relative motion than adjacent plates (e.g., Jarrard, 1987). In perfectly partitioned oblique-convergence, the strike-slip fault bounding the sliver takes up all of the oblique component of convergence, and the forearc sliver moves parallel to the margin at this transcurrent velocity with pure dip slip beneath the forearc sliver (e.g., Jarrard, 1987). In normal oblique convergence, this process is well understood (e.g., Jarrard, 1987). However, if forearc slivering occurs during triple junction interaction, the system is considerably more complex because another plate is present in addition to the two converging plates and their associated forearc sliver (Fig. 10). It is the specifics of this interaction associated with the Chugach metamorphic complex, however, that provide insight to the history.

If we consider the specific velocities of the Kula–Farallon–North American system after the plate reorganization and add the evidence that the Chugach metamorphic complex was moving as a forearc sliver, the three-plates and the microplate can be analyzed as a unit (Fig. 10). The velocity triangles from the linear velocities show the inferred northward motion of the triple junction relative to North America reported by Sisson and Pavlis (1993). Addition of an inferred forearc sliver has a simple, but significant, result. If the forearc sliver moved laterally at the full lateral component of Kula–North American motion, the forearc sliver actually moved northward faster than the triple junction (Fig. 10). As a result, the triple junction would migrate southward along the forearc sliver, even though the triple junction was migrating northward relative to North America (Fig. 10).

This analysis is critical to the understanding of the Chugach metamorphic complex because it resolves a long-standing problem of why the high-temperature metamorphism was localized in the Chugach metamorphic complex. We previously suggested a two-phase heating due to backtracking of the triple junction (Sisson and Pavlis, 1993; Pavlis and Sisson, 1995). Figure 9 shows a much simpler mechanism. We propose that the forearc sliver, which was already actively moving along the margin through strike slip on the Hanagita fault system (Roeske et al., this volume, Chapter 6) was accelerated by increasing subduction obliquity during the Eocene plate reorganization. The new triple junction migrated northward with, but more slowly than, the forearc sliver. As a result, the triple junction motion effectively slowed to a very low velocity (<30 km/m.y.), and that slow migration allowed greater heating in this area relative to adjacent segments of the margin.

This velocity analysis also has important implications for the structural history of the Chugach metamorphic complex and the role of triple junction processes. We previously suggested that the D_2 to D_3 progression recorded the passage of the triple junction and the change from subduction of the Kula to Farallon plates. In this new analysis, the opposite plate subduction history is indicated; i.e., Farallon subduction followed by Kula subduction as the triple junction migrated southward along the forearc sliver.

In order to reconcile this plate model, we propose a tectonic model illustrated in Figure 9. The model begins as the new spreading ridge emerges from the plate reorganization, and the Border Ranges fault (Hanagita fault system) takes up the dextral convergence component as the forearc sliver is driven northward (upper half, Fig. 9). During plate reorganization, the triple junction may have jumped northward or migrated rapidly to a position northwest of the Chugach metamorphic complex because older plutons occur well to the southwest (Sisson et al., this volume, Chapter 13). As the Border Ranges system takes up the oblique component, the triple junction migrates slowly southward (upper left to lower right, Fig. 9) relative to the forearc sliver bounded by the trench and the Border Ranges fault. We infer that the slowly migrating triple junction produces the main prograde metamorphism in the Chugach metamorphic complex, although earlier heating may have occurred during earlier triple junction passes (e.g., as suggested by Sisson and Pavlis, 1993). As heating develops high-grade metamorphic assemblages at depth, the forearc sliver decouples along the gneiss-schist interface. At the same time as the upper crustal rocks above the gneiss transition move as a rigid body, the gneissic rocks at depth flow homogeneously in distributed dextral shear. The contact between these flow domains develops a sharp vertical displacement gradient (Fig. 9). We suggest that the combination of this decoupling, together with vertical shortening produced by approach of more buoyant rocks along the ridge, is the main process responsible for D_2 deformation within the western Chugach metamorphic complex. Specifically, subhorizontal shear along the interface produces a steeply dipping shortening axis, and when this shortening is combined with additional shortening from subduction of ridge topography, the observed vertical D_2 shortening is produced.

Ultimately, the triple junction migrated past the Copper River region to the SE (lower half, Fig. 9). At this stage, we infer that the complex went through a kinematic switch that led to development of most of the observed D_3 structure. We infer this timing because D_3 was coincident with the metamorphic peak in the Chugach metamorphic complex, and thus, this event must have occurred during and immediately following the passage of the triple junction. This conclusion also implies late development of the Bremner shear zone, after significant slip had accumulated on the Hanagita/Border Ranges system (e.g., Roeske et al., this volume, Chapter 6). Thus, we show this change in position of the zone of active slip to include motion on the D_3 shear zones of the Chugach metamorphic complex and the Hanagita System (Fig. 9). In this scenario, D_3 in the gneiss is produced by a combination of distributed dextral shear and vertical extension within the gneiss. The vertical extension is not required to account for the observed F_3 folding within the gneiss. Instead, it is suggested by apparent flattening strains observed by Pavlis and Sisson (1995) within the gneiss in the Tana Glacier area. We assume that vertical extension may have developed as the inverse of the process that produced the D_2 vertical shortening. That is, as the triple junction passed, the previously thinned accretionary complex would have to readjust to the increasingly older, and deeper, lithosphere being subducted. As a result, the flow would have to respond by vertical thickening.

This model (Fig. 9) is a hypothesis that requires further testing, yet it provides a reasonable explanation for most features of the Chugach metamorphic complex. As more geochronological data accumulate, the details will undoubtedly evolve, but we believe the basic tectonic framework will remain. That is, this is a system in which (1) the process is closely linked to triple junction interaction; (2) the upper and lower crust are decoupled at the gneiss transition; and (3) the D_2 to D_3 progression is associated with variations in upper crustal versus lower crustal kinematics.

Our model also has important implications for interpretations of crustal architecture from geophysical data collected in areas dominated by strike-slip tectonics (Fig. 9). Many previous studies have suggested that subhorizontal detachments should exist, or do exist, along major strike-slip systems. These studies generally assumed that decoupling would occur at the brittle-ductile transition (e.g., Jones et al., 1994). Our observations in the Chugach metamorphic complex imply the gneiss transition may be the more fundamental boundary; e.g., seismic reflection would probably readily image the low-angle D_2 fabrics but the low-angle fabrics would be exclusively in amphibolite facies rocks, well below the brittle-ductile transition.

Implication for Rheology of the Lithosphere

Several features, including association between D_2 and D_3 in the Chugach metamorphic complex, the distribution of D_3 structures, and the grossly different behavior of the rocks at different crustal levels have important implications for the rheology of the lithosphere. Our observations imply a markedly different mechanical behavior across the gneiss transition. We suggest that the difference in behavior reflects a change in rheology from nonlinear viscous behavior in the schist to a linear-viscous behavior in the gneiss.

This conclusion arises from qualitative consideration of the behavior of linear versus nonlinear fluids. In linear viscous materials, viscosity is a constant; i.e., stress is directly proportional to strain rate with viscosity as the proportionality constant. In visco-plastic behavior, there is a nonlinear stress-strain rate relationship in which apparent viscosity decreases with increasing stress; i.e., the effective viscosity is much higher at low stress than high stress. This type of material is characterized by a macroscopic behavior in which small-scale perturbations produce marked strain localization in the form of shear zones or other "banded perturbations" (e.g., Cobbold, 1977a, 1977b). In contrast, linear viscous materials typically would produce distributed flow because the material flows at all stress levels with a linear dependence of stress and strain-rate. Thus, unless stress gradients are steep, the finite strain accumulates in a broader zone than it would in a visco-plastic fluid. Qualitatively, this relationship is often described as the distinction between standard pipe flow of a linear fluid (e.g., water) and a nonlinear fluid characterized by plug flow; i.e., material extruded from the center of the pipe is virtually undeformed. Applying this qualitative reasoning to the Chugach metamorphic complex, we infer that the distributed flow in the gneiss probably reflects a macroscopic linear viscous behavior. In contrast, the behavior of the schist with strain localization in distinct shear zones is indicative of nonlinear flow.

Although this conclusion arises primarily from the structural behavior of the rocks in the Chugach metamorphic complex, it is important to realize that theoretical studies of the mechanism for gneissic segregations provide independent evidence for the same conclusion. The mechanism for the formation of gneissic segregations, or other types of "metamorphic differentiation" features is a long-standing problem in geology (e.g., Turner, 1941). In high-grade metamorphic rocks, gneisses may result from in situ melting and the leucocratic bands are partial melts (e.g., Ashworth, 1985). In the Chugach metamorphic complex, however, this mechanism seems highly unlikely at the gneissic front since these rocks show neither anatectic textures (e.g., restite development) nor are metamorphic temperatures sufficient for anatexis (Sisson et al., 1989). The leucocratic bands might also represent injected magmas from an external source; i.e., injection migmatites.

Similar observations and the occurrence of metamorphic segregations at lower-grade metamorphic conditions have led geochemists to infer that gneissic segregations are the product of fluid-assisted diffusion (Fisher, 1973; Wintsch, 1975, 1978; Robin, 1979; Sawyer and Robin, 1986; Dewers and Ortoleva, 1990). Typically, this process has been ascribed to the effects of differential diffusion rates of mobile constituents within micas or along mica grain boundaries (e.g., Robin, 1979); i.e., the mica acts as a "diffusional pipe" to accelerate diffusion. This process can be viewed as a high-temperature pressure-solution process in which the fluid assisting the diffusion is saturated in quartzofeldspathic components (e.g., Wintsch, 1978). Mass transfer of these components allows continuous deformation of the rock (e.g., Robin, 1979; Dewers and Ortoleva, 1990). This is a particularly attractive model for the Chugach metamorphic complex as veins are ubiquitous throughout the lower-grade parts of the complex. In addition, they show a distinctive pattern with increasing metamorphic grade: Quartz + calcite give way to quartz veins within the greenschist facies, followed by quartz + aluminosilicate or quartz + cordierite veins in amphibolite facies rocks. Thus, at even higher grades, addition of calcium and sodium to this fluid could produce quartzofeldspathic veins analogous to the quartzofeldspathic segregations in the gneiss.

Regardless of the details of the process that ultimately led to the gneissic segregations, all of these processes fundamentally are diffusional mechanisms. Diffusion mechanisms are important from the perspective of rock rheology because a diffusion mechanism leads to a linear viscous rheology (e.g., Elliot, 1970; Rutter, 1983). In contrast, most middle crustal rocks are thought to be dominated by crystal plastic deformation mechanisms, and experiments show that crystal plastic deformation leads to power-law flow (e.g., Turcotte and Schubert, 1982; Kirby, 1983) except for Harper-Dorn creep (e.g., Wang et al., 1994). Thus, development of gneissic banding should also be associated with a deformation mechanism switch from crystal-plastic mechanisms to diffusional mechanisms. Accordingly, there should be a change in rheology from linear viscous flow in gneiss to power-law, nonlinear flow in lower-grade rocks; a prediction consistent with the observed structural behavior in the Chugach metamorphic complex.

Additional studies are needed on the details of the processes leading to the gneissic banding in the Chugach metamorphic complex before this hypothesis can be tested fully. Nonetheless, our conclusions have important implications for worldwide studies of the behavior of the continental crust. The characteristic rock of high-grade metamorphic terranes is gneiss, and in Archean rocks high-grade tonalitic gneisses constitute a large percentage of the crust. We hypothesize that in most active orogenic belts the lower crust would be composed of deforming gneisses. Thus, to model crustal deformation, the appropriate rheology for this lower crustal material would be a linear viscous rheology. More importantly, this lower crustal material is overlain by low- to medium-grade rocks dominated by crystal-plastic deformation mechanisms and a nonlinear rheology. The boundary between these different crustal layers might vary in thickness and rheologic contrast, particularly in more typical areas where crustal heterogeneity would play a greater role. Future mechanical models of crustal behavior should consider the affects of this rheological stratification.

ACKNOWLEDGMENTS

This work was supported by NSF grants NSF EAR-9307169 and EAR-9018717 to Pavlis and EAR-9020391 and EAR-9304062 and National Geographic Society Grant 6010-97 to Sisson. Christian Teyssier, Cees Van Staal, and Sarah Roeske provided insightful reviews. Rob Holcombe's program, Georient, was used in stereonet plotting. We are grateful to Walt Woodrow and Jan Jorgensen of ERA helicopters for the working with us to make the challenging logistics of this area easy, and to Danny Rosencrans at Wrangell-St. Elias National Park for assistance with logistics and the NPS permitting process. We also appreciate field assistance from John Bowman, Nancy Harris, Anne Poole, and Holly Cooper-Burner.

REFERENCES CITED

Andronicos, C.L., Hollister, L.S., Davidson, C., and Chardon, D., 1999, Kinematics and tectonic significance of transpressive structures within the Coast Plutonic Complex, British Columbia: Journal of Structural Geology, v. 21, p. 229–230.

Ashworth, J.R., 1985, Introduction, *in* Ashworth, J.R., ed., Migmatites: Glasgow, Blackie, p. 1–35.

Bol, A., Coe, R.S., Gromme, C.S., and Hillhouse, J.W., 1992, Paleomagnetism of the Resurrection Peninsular, Alaska—Implications for the tectonics of southern Alaska and the Kula-Farallon Ridge: Journal of Geophysical Research, v. 97, p. 17213–17232.

Bradley, D.C., Haeussler, P., and Kusky, T.M., 1993, Timing of early Tertiary ridge subduction in southern Alaska: U.S. Geological Survey Bulletin 2068, p. 13–22.

Bradley, D., Kusky, T., Haeussler, P., Goldfarb, R., Miller, M., Dumoulin, J., Nelson, S.W., and Karl, S., 2003, Geologic signature of early Tertiary ridge subduction in Alaska *in* Sisson, V.B., Roeske, S.M., and Pavlis, T.L., eds., Geology of a transpressional orogen developed during ridge-trench interaction along the North Pacific margin: Boulder, Colorado, Geological Society of America Special Paper 371, p. 19–49 (this volume).

Byrne, T., 1979, Late Paleocene demise of the Kula-Pacific spreading center: Geology, v. 7, p. 341–344.

Campbell, R.B., and Dodds, C.J., 1982, Geology of the Kluane, Mt. St. Elias, and Dezadeash map areas, Yukon Territory: Geological Survey of Canada Open Files nos. 829, 830, 831, 4 map sheets, scale 1:250,000.

Campbell, R.B., and Dodds, C.J., 1983, Geology of the Tatshenshini River and Yakutat map areas, Yukon Territory: Geological Survey of Canada Open Files nos. 926, 927, 1:125,000, 4 sheets.

Chardon, D., Andronicos, C.L., and Hollister, L.S., 1999, Large-scale transpressive shear zone patterns and displacements within magmatic arcs; the Coast plutonic complex, British Columbia: Tectonics, v. 18, p. 278–292.

Cobbold, P.R., 1977a, Description and origin of banded perturbations, I: Regional strain, local perturbations and deformation bands: Canadian Journal of Earth Science, v. 14, p. 1721–1731.

Cobbold, P.R., 1977b, Description and origin of banded perturbations, II: Rheology and the growth of banded perturbations: Canadian Journal of Earth Science, v. 14, p. 2510–2523.

Dewers, T., and Ortoleva, P., 1990, Geochemical self-organization III: A mechano-chemical model of metamorphic differentiation: American Journal of Science, v. 290, p. 473–521.

Elliot, D., 1970, Diffusion flow laws in metamorphic rocks: Geological Society of America Bulletin, v. 84, p. 2645–2664.

Engebretson, D.C., Cox, A., and Gordon, R.G., 1985, Relative motions between oceanic and continental plates in the Pacific basin: Boulder, Colorado, Geological Society of America Special Paper 206, 59 p.

England, P., and Jackson, J., 1989, Active deformation of the continents: Annual Reviews of Earth and Planetary Science, v. 17, p. 197–226.

Fisher, D.M., and Byrne, T., 1987, Structural evolution of underthrust sediments, Kodiak Islands, Alaska: Tectonics, v. 6, p. 775–793.

Fisher, D.M., and Byrne, T., 1992, Strain variations in an ancient accretionary complex: Implications for forearc evolution: Tectonics, v. 11, p. 330–347.

Fisher, G.W., 1973, The application of ionic equilibria to metamorphic differentiation: An example: Contributions to Mineralogy and Petrology, v. 29, p. 91–103.

Fitch, T.J., 1972, Plate convergence, strike-slip faults, and internal deformation adjacent to southeast Asia and the western Pacific: Journal of Geophysical Research, v. 77, p. 4432–4460.

Haeussler, P.J., Bradley, D., Goldfarb, R., Snee, L., and Taylor, C., 1995, Link between ridge subduction and gold mineralization in southern Alaska: Geology, v. 23, p. 995–998.

Harris, N.R., Sisson, V.B., Wright, J.E., and Pavlis, T.L., 1996, Evidence for Eocene mafic underplating during forearc intrusive activity, eastern Chugach mountains, Alaska: Geology, v. 24, p. 263–266.

Hillhouse, J.W., and Coe, R.S., 1994, Paleomagnetic data from Alaska, *in* Plafker, G., and Berg, H.C., eds., The geology of Alaska: Boulder, Colorado, Geological Society of America, Geology of North America, v. G-1, p. 797–812.

Hillhouse, J.W., Gromme, C.S., and Csejtey, B., Jr., 1985, Tectonic implications of paleomagnetic poles from lower Tertiary volcanic rocks, south-central Alaska: Journal of Geophysical Research, v. 14, p. 12,523–12,535.

Hollister, L.S., and Crawford, M.L., 1986, Melt-enhanced deformation: A major tectonic process: Geology, v. 14, p. 558–561.

Hudson, T.L., and Plafker, G., 1982, Paleogene metamorphism of an accretionary flysch terrane, eastern Gulf of Alaska: Geological Society of America Bulletin, v. 93, p. 1280–1290.

Hudson, T.L., Plafker, G., and Turner, D.L., 1977, Metamorphic rocks of the Yakutat–St. Elias area, south-central Alaska: U.S. Geological Survey Journal of Research, v. 1, p. 125–136.

James, T.S., Hollister, L., and Morgan, W.J., 1989, Thermal modeling of the Chugach metamorphic complex: Journal of Geophysical Research, v. 94, p. 4411–4423.

Jarrard, R.D., 1987, Relations among subduction parameters: Reviews of Geophysics, v. 24, p. 217–284.

Jones, D.L., Gaymer, R., Wang, C., Mc Evilly, T.V., and Lomax, A., 1994, Neogene tranpressive evolution of the California Coast Ranges: Tectonics, v. 13, p. 561–574.

Kirby, S.H., 1983, Rheology of the lithosphere: Reviews of Geophysics and Space Physics, v. 21, p. 1458–1487.

Kusky, T.M., and Young, C.P., 1999, Emplacement of the Resurrection Peninsula ophiolite in the forearc during a ridge-trench encounter southern Alaska: Journal of Geophysical Research, v. 104, p. 29,025–29,054.

Lonsdale, P., 1988, Paleogene history of the Kula plate: Offshore evidence and onshore implications: Geological Society of America Bulletin, v. 100, p. 733–754.

Lull, J.S., and Plafker, G., 1990, Geochemistry and paleotectonic implications of metabasaltic rocks in the Valdez Group, southern Alaska, *in* Dover, J.H., and Galloway, J.P., eds., Geologic studies in Alaska by the U.S. Geological Survey, 1989: U.S. Geological Survey Bulletin 1946, p. 29–38.

Miller, M.L., Dumoulin, J.A., and Nelson, S.N., 1984, A transect of metamorphic rocks along the Copper River, Cordova and Valdez quadrangles, Alaska: U.S. Geological Survey Circular 939, p. 52–57.

Moore, J.C., Byrne, T., Plumley, P.W., Reid, M., Gibbons, H., and Coe, R.S., 1983, Paleogene evolution of the Kodiak Islands, Alaska: Consequences of ridge-trench interaction in a more southerly latitude: Tectonics, v. 2, p. 265–293.

Nadai, A., 1963, Theory of Flow and Fracture of Solids: New York, McGraw Hill, 705 p.

Nilsen, T.H., and Zuffa, G.G., 1982, The Chugach terrane: A Cretaceous trench-fill deposit, southern Alaska, *in* Leggett, J.K., ed., Trench-forearc geology: Sedimentation and tectonics on modern and ancient active plate margins: Geological Society [London] Special Publication No. 10, p. 213–227.

Nokleberg, W.J., Plafker, G., Lull, J.S., Wallace, W.K., and Winkler, G.R., 1989, Structural analysis of the southern Peninsular, southern Wrangellia, and northern Chugach terranes along the Trans-Alaska Crustal Transect, northern Chugach Mountains, Alaska: Journal of Geophysical Research, v. 94, p. 4297–4320.

Oldow, J.S., Bally, A.W., Avé Lallemant, H.G., and Leeman, W.P., 1989, Phanerozoic evolution of the North American Cordillera; United States and Canada, *in* Bally, A.W., and Palmer, A.R., eds., The geology of North America: An overview: Boulder, Colorado, Geological Society of America, Geology of North America, v. A, p. 139–232.

Oldow, J.S., Bally, A.W., and Avé Lallemant, H.G., 1990, Transpression, orogenic float, and lithospheric balance: Geology, v. 18, p. 991–994.

Pavlis, T.L., 1996, Fabric development in syn-tectonic intrusive sheets as a consequence of melt-dominated flow and thermal softening of the crust: Tectonophysics, v. 253, p. 1–31.

Pavlis, T.L., and Sisson, V.B., 1995, Tectonic processes during ridge subduction: Chugach metamorphic complex, southern Alaska: Geological Society of America Bulletin, v. 107, p. 1333–1355.

Pavlis, T.L., Marty, K., and Sisson, V.B., 2003, Constrictional flow within the Eocene forearc of southern Alaska: An effect of dextral shear during ridge subduction, *in* Sisson, V.B., Roeske, S.M., and Pavlis, T.L., eds., Geology of a transpressional orogen developed during ridge-trench interaction along the North Pacific margin: Boulder, Colorado, Geological Society of America Special Paper 371, p. 171–190 (this volume).

Plafker, G., Nokleberg, W.J., and Lull, J.S., 1989, Bedrock geology and tectonic evolution of the Wrangellia, Peninsular, and Chugach terranes along the Trans-Alaska Crustal Transect in the Chugach Mountains and southern Copper River basin, Alaska: Journal of Geophysical Research, v. 94, p. 4255–4295.

Plafker, G., Moore, J.C., and Winkler, G.R., 1994, Geology of the southern Alaska margin, *in* Plafker, G., and Berg, H.C., eds., The geology of Alaska: Boulder, Colorado, Geological Society of America, Geology of North America, v. G-1, p. 389–450.

Poole, A.R., 1996, Age and geochemical constraints on ridge subduction for igneous rocks of the eastern Chugach Mountains, Alaska [M.S. thesis]: Houston, Texas, Rice University, 83 p.

Robin, P.-Y., 1979, Theory of metamorphic segregation and related processes: Geochimica et Cosmochimica Acta, v. 43, p. 1586–1600.

Roeske, S.M., Pavlis, T.L., Snee, L.W., and Sisson, V.B., 1992, $^{40}Ar/^{39}Ar$ isotopic ages from the Wrangellia-Alexander terrane along the Border Ranges fault system in the eastern Chugach Mountains and Glacier Bay, Alaska: U.S. Geological Survey Bulletin 1999, p. 180–195.

Roeske, S.M., Snee, L.W., and Bunds, M.P., 1993, $^{40}Ar/^{39}Ar$ dates from the Border Ranges fault system, a hydrothermally altered brittle-ductile transition strike-slip shear zone, southern Alaska: Geological Society of America Abstracts with Programs, v. 25, no. 6, p. 419.

Roeske, S.M., Snee, L.W., and Pavlis, T.L., 2003, Dextral slip reactivation of an arc-forearc boundary during Late Cretaceous–Early Eocene oblique convergence in the northern Cordillera, *in* Sisson, V.B., Roeske, S.M., and Pavlis, T.L., eds., Geology of a transpressional orogen developed during ridge-trench interaction along the North Pacific margin: Boulder, Colorado, Geological Society of America Special Paper 371, p. 141–169 (this volume).

Rutter, E.H., 1983, Pressure solution in nature, theory and experiment: Geological Society [London] Journal, v. 140, p. 725–740.

Sample, J.C., and Moore, J.C., 1987, Structural style and kinematics of an underplated slate belt, Kodiak and adjacent islands, Alaska: Geological Society of America Bulletin, v. 99, p. 7–20.

Sander, B., 1930, Gefügekunde der Gesteine: Vienna, Austria, Springer Verlag, 352 p.

Sawyer, E.W., and Robin, P.-Y., 1986, The subsolidus segregation of layer-parallel quartz-feldspar veins in greenschist to upper amphibolite facies metasediments: Journal of Metamorphic Geology, v. 4, p. 237–260.

Serpa, L., de Voogd, B., Wright, L., Willemin, J., Oliver, J., Hauser, E., and Troxel, B., 1988, Structure of the central Death Valley pull-apart basin and vicinity from COCORP profiles in the southern Great Basin: Geological Society of America Bulletin, v. 100, p. 1437–1450.

Sisson, V.B., and Pavlis, T.L., 1993, Geologic consequences of plate reorganization: An example from the Eocene southern Alaska forearc: Geology, v. 21, p. 913–916.

Sisson, V.B., Hollister, L.S., and Onstott, T.C., 1989, Petrologic and age constraints on the origin of a low-pressure/high-temperature metamorphic complex, southern Alaska: Journal of Geophysical Research, v. 94, p. 4392–4410.

Sisson, V.B., Cooper, H., Pavlis, T.L., and Poole, A., 1997, Contrasting metamorphic histories within the Chugach metamorphic complex: Differential uplift related to transpression on the Fairweather fault, Alaska: Geological Society of America Abstracts with Programs, v. 29, no. 6, p. 277.

Sisson, V.B., Pavlis, T.P., and Stüwe, K., 1998, Along-strike differences in the forearc associated with Eocene triple-junction interactions in the Mt. St. Elias–Mt. Logan region: American Geophysical Union Eos Abstracts, v. 79, p. 913.

Sisson, V.B., Poole, A.R., Harris, N.R., Cooper Burner, H., Pavlis, T.L., Copeland, P., Donelick, R.A., and McLelland, W.C., 2003, Geochemical and geochronologic constraints for genesis of a tonalite-trondhjemite suite and associated mafic intrusive rocks in the eastern Chugach Mountains, Alaska: A record of ridge-transform subduction, *in* Sisson, V.B., Roeske, S.M., and Pavlis, T.L., eds., Geology of a transpressional orogen developed during ridge-trench interaction along the North Pacific margin: Boulder, Colorado, Geological Society of America Special Paper 371, p. 293–326 (this volume).

Smart, K.J., Pavlis, T.L., Sisson, V.B., Roeske, S.M., and Snee, L.W., 1996, The Border Ranges fault system in Glacier Bay National Park, Alaska: Evidence for major late Mesozoic–early Cenozoic dextral strike-slip motion: Canadian Journal of Earth Science, v. 33, p. 1268–1282.

Stamatakos, J.A., Kodama, K.P., Pavlis, T.L., and Vittorio, L., 1989, Paleomagnetism of Cretaceous and Paleocene sedimentary rocks across the Castle Mountains fault, south-central Alaska, *in* Hillhouse, J.W., ed., Deep structure and past kinematics of accreted terranes: American Geophysical Union Geophysical Monograph Series, v. 50, p. 151–177.

Stock, J., and Molnar, P., 1988, Uncertainties and implications of the Late Cretaceous and Tertiary position of North America relative to the Farallon, Kula, and Pacific plates: Tectonics, v. 7, p. 1339–1384.

Teyssier, C., and Tikoff, B., 1998, Strike-slip partitioning transpresion of the San Andreas fault system: a lithospheric-scale approach, *in* Holdsworth, R.E., et al., eds., Continental transpressional and transtensional tectonics: Geological Society [London] Special Publication No. 135, p. 143–158.

Teyssier, C., Tikoff, B., and Markley, M., 1995, Oblique plate motion and continental tectonics: Geology, v. 23, p. 447–450.

Teyssier, C., Tikoff, B., and Weber, J., 2003, Attachment between brittle and ductile crust at wrenching plate margins: European Geophysical Society Special Publication (in press).

Thatcher, W., 1995, Microplate versus continuum descriptions of active tectonic deformation: Journal of Geophysical Research, v. 100, p. 3885–3894.

Tikoff, B., and Teyssier, C., 1994, Strain modeling of displacement partitioning in transpressional orogens: Journal of Structural Geology, v. 16, p. 1575–1588.

Turcotte, D.L., and Schubert, G., 1982, Geodynamics: Applications of continuum physics to geological problems: New York, J. Wiley and Sons, 450 p.

Turner, F.J., 1941, The development of pseudo-stratification by metamorphic differentiation in the schists of Otago, New Zealand: American Journal of Science, v. 239, p. 1–16.

Tysdal, R.G., and Plafker, G., 1978, Age and continuity of the Valdez Group, southern Alaska: U.S. Geological Survey Bulletin 1457-A, p. 120–131.

Wang, J.N., Hobbs, B.E., Ord, A., Shimamoto, T., and Toriumi, M., 1994, Newtonian dislocation creep in quartzites: Implications for the rheology of the lower crust: Science, v. 265, p. 1204–1205.

Weinberger, J., and Sisson, V.B., 2003, Pressure and temperature conditions of brittle-ductile vein emplacement in the greenschist facies, Chugach metamorphic complex, Alaska: Evidence from fluid inclusions, *in* Sisson, V.B., Roeske, S.M., and Pavlis, T.L., eds., Geology of a transpressional orogen developed during ridge-trench interaction along the North Pacific margin: Boulder, Colorado, Geological Society of America Special Paper 371, p. 217–235 (this volume).

Winkler, G.R., and Plafker, G., 1993, Geologic map of the Cordova and Middleton Island quadrangles, southern Alaska: U.S. Geological Survey Miscellaneous Investigations Map I-1984, scale 1:250,000.

Wintsch, R.P., 1975, Feldspathization as a result of deformation: Geological Society of America Bulletin, v. 86, p. 35–38.

Wintsch, R.P., 1978, A chemical approach to the preferred orientation of mica: Geological Society of America Bulletin, v. 89, p. 1715–1718.

Zoback, M.D., Zoback, M.L., Mount, V.S., Suppe, J., Eaton, J.P., Healy, J.H., Oppenheimer, D.H., Reasenberg, P.A., Jones, L.M., Raleigh, C.B., Wong, I.G., Scotti, O., and Wentworth, C.M., 1987, New evidence on the state of stress of the San Andreas fault system: Science, v. 238, p. 1105–1111.

Manuscript Accepted by the Society February 5, 2003

Printed in the USA

Geological Society of America
Special Paper 371
2003

Pressure and temperature conditions of brittle-ductile vein emplacement in the greenschist facies, Chugach metamorphic complex, Alaska: Evidence from fluid inclusions

Jill Weinberger*
Virginia B. Sisson
Department of Earth Science, MS-126, Rice University, Houston, Texas 77005-1892, USA

ABSTRACT

Fluid inclusions within the quartz veins cutting greenschist facies metasedimentary rocks of the Valdez group provide insight into the changes in deformational style displayed in two vein systems from the northern Chugach metamorphic complex. Six samples were analyzed in order to determine temperature and pressure of vein emplacement. Metamorphic temperature estimates from chlorite geothermometry range from 325 to 350 °C. This information combined with fluid inclusion geobarometry yields a high paleogeothermal gradient ranging from 58 to 35 °C/km for the accretionary prism. This higher-than-normal geothermal gradient represents elevated temperatures associated with Eocene subduction of the Kula-Farallon ridge. Isochores calculated for each vein show little temperature variation (less than 40 °C), and large pressure variations (up to 200 MPa). Maximum pressure differences within one vein set are as much as 150 MPa. In addition, the temperature of early vein emplacement was close to the brittle–ductile transition, which is consistent with textures of veins that vary from brecciated to folded. Late stage hydrocarbon-bearing inclusions are present in the samples, and may indicate long-distance, post-peak metamorphic, fluid flow through this region. The source for the hydrocarbon-bearing fluids is unknown and poses many interesting questions about regional fluid flow distances and directions after peak metamorphism.

Keywords: fluid inclusions, geothermometry, geothermal gradient, ridge subduction, Chugach metamorphic complex, greenschist facies, geothermobarometry, vein emplacement.

INTRODUCTION

The study of fluid inclusions in relation to specific structural problems has made many advancements in recent years (e.g., O'Hara and Haak, 1992; Hodgkins and Stewart, 1994; Shelton et al., 1996; Boullier, 1999; Lespinasse, 1999; Eichhubl and Boles, 2000). In the upper crust, deformation is dominated by brittle deformation; but both ductile and pseudo-ductile viscous behavior also occurs in some folds. Recently, Gratier et al. (1999) and Gray et al. (1999) proposed a complex interplay between these two styles of deformation as they form at different strain rates with fracturing occurring rapidly and viscous creep at longer time scales. Typically, these relations are only investigated experimentally (e.g., Hirth and Tullis, 1989; Logan et al., 1992). In naturally deformed rocks, variables such as fluid pressure, fluid composition, lithology and grain size affect the deformation processes (e.g., Newman and Mitra, 1994; Ohlmacher and Aydin, 1997; Wyder and Mullis, 1998; Gray et al., 1999; Bruhn et al., 2000). One of the best ways to estimate paleofluid pressure, composition, and migration in rocks is to use fluid inclusions. These record the composition of

*Current address: Scripps Institution of Oceanography, University of California, San Diego, La Jolla, California 92093-0208, USA; jlweinb@ucsd.edu.

Weinberger, J., and Sisson, V.B., 2003, Pressure and temperature conditions of brittle-ductile vein emplacement in the greenschist facies, Chugach metamorphic complex, Alaska: Evidence from fluid inclusions, *in* Sisson, V.B., Roeske, S.M., and Pavlis, T.L., eds., Geology of a transpressional orogen developed during ridge-trench interaction along the North Pacific margin: Boulder, Colorado, Geological Society of America Special Paper 371, p. 217–235. ©2003 Geological Society of America paleofluids, orientation of microcracks, and most importantly, fluid density, which is related to pressure and temperature. This information can then be used to document paleohydrology of accretionary margins (e.g., Vrolijk, 1987; Fisher et al., 1995; Underwood et al., 1999; Sakaguchi, 1999a; Lewis et al., 2000).

The northwestern terminus of the Chugach metamorphic complex (Fig. 1) is composed of greenschist facies metasedimentary rocks infused with quartz veins emplaced along joints, faults, and shear zones. The pervasive nature of the quartz veins shows that large quantities of fluids were released from water-rich sediments that were heated and overthrust during ridge collision and subduction (e.g., Sisson and Hollister, 1988; Sisson and Pavlis, 1993; Pavlis and Sisson, 1995).

This study examined fluid inclusions in these quartz veins in the Wolverine Run Glacier region to determine the composition, pressure, and temperature of the fluids present at the time of vein formation. Assuming that the fluids trapped are representative of those present during metamorphism, and that the inclusions have retained a constant volume and composition since entrapment, fluid inclusion data can be used to help explain observations of differences in emplacement textures and deformation between sets of veins. In particular, we investigated the deformational styles seen in two outcrops using data gathered from microthermometric, electron microprobe, and ultraviolet (UV) microscopy studies on fluid inclusions.

REGIONAL GEOLOGY

The Chugach terrane is an elongate belt of mainly Mesozoic and Cenozoic accretionary complex rocks approximately 100 km wide and 2200 km long, which runs from the Sanak Islands to the Chatham Strait in southeastern Alaska (Jones et al., 1987; Plafker et al., 1989; Plafker et al., 1994). The accretionary complex is bounded to the north by the Border Ranges fault, and extends across the Contact fault into younger accretionary prism rocks of the Prince William Terrane. The belt is primarily composed of Maastrichtian age continentally derived sediments of the Valdez Group (Plafker et al., 1977). The metasedimentary rocks are composed of graywacke, siltstone, and argillaceous mudstone, with rare pebble conglomerate (e.g., Dumoulin, 1987). Turbidite sedimentary features have been mostly obscured by later deformation in the Chugach metamorphic complex but are preserved south and west of the complex (Plafker et al., 1994). Farmer et al. (1993) proposed that the source for the metasediments was the Coast Mountains in British Columbia. Structurally overlying the Valdez Group is a mélange complex that, together with the Valdez Group, represents the entire late Mesozoic accretionary prism (e.g., Plafker et al., 1989). The Valdez group is between the Wrangellia Terrane to the north and the Prince William Terrane and Yakutat Terrane to the south (Plafker et al., 1977; Plafker et al., 1994).

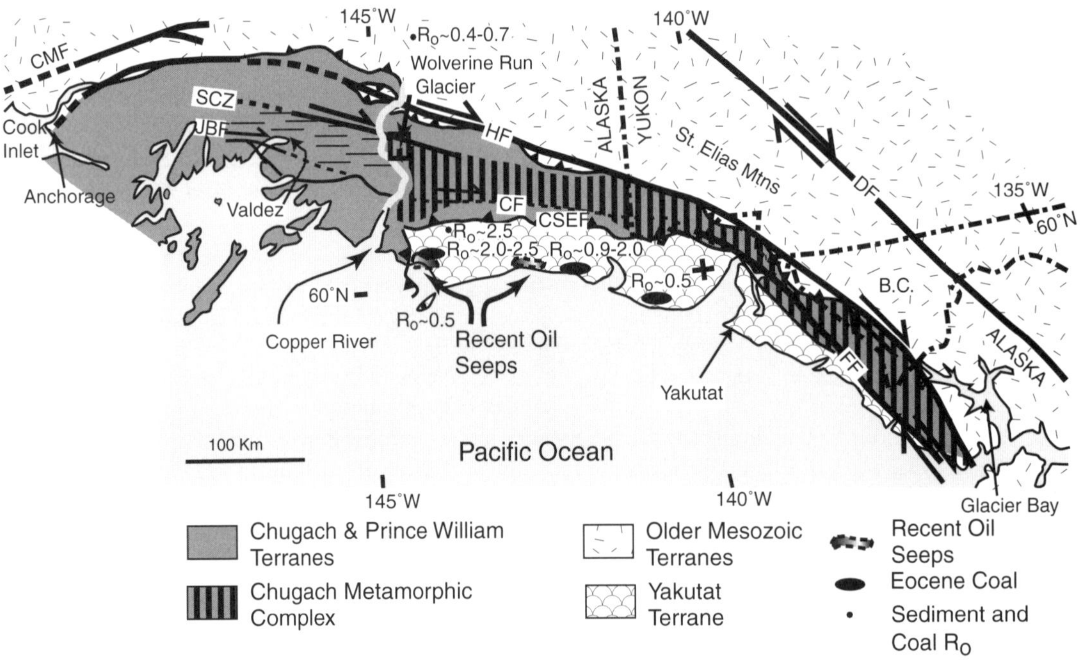

Figure 1. General tectonic setting for the Wolverine Run Glacier study area. Inset box near Copper River shows approximate location of Figure 2. Also shown are locations of oil seeps analyzed by Bence et al. (1996), coal vitrinite reflectance (R_O) from Plafker (1987), and sediment vitrinite reflectance from Armentrout (1981) and Johnsson et al. (1999). Map is adapted from Winkler and Plafker (1993) and Plafker (1987) with additonal mapping by Sisson, Pavlis, Roeske, Marty, and Bowman. Abbreviations for various faults: FF—Fairweather fault, CSEF—Chugach–St. Elias fault, CF—Contact fault, HF—Hanagita fault, SCZ—Stuart Creek zone, JBF—Jack Bay fault, CMF—Castle Mountain fault, and DF—Denali fault.

Within the Chugach terrane is a series of higher-temperature metamorphic rocks, the Chugach metamorphic complex. This metamorphic complex records an intricate series of unusual events in the Eocene (Hudson et al., 1979; Hudson and Plafker, 1982; Sisson et al., 1989). The Andean-type margin, active since the Late Cretaceous, was disrupted by Eocene plate reorganization (Sisson and Pavlis, 1993; Pavlis and Sisson, 1995, this volume, Chapter 8). Metamorphic grade in the region varies from sub-greenschist facies to amphibolite facies. High-grade rocks are generally restricted to the contact aureoles of plutons to the west and increased regional metamorphic grade to the east (Moore et al., 1983; Plafker et al., 1989; Sisson and Pavlis, 1993; Bradley et al., this volume, Chapter 1). Accordingly, metamorphic temperatures and pressures vary within the region with peak temperature reaching ~670 °C at approximately 300 MPa (Sisson et al., 1989).

The Chugach metamorphic complex experienced three periods of deformation during prograde metamorphism (e.g., Pavlis and Sisson, 1995, this volume, Chapter 8). The first deformation (D_1) formed in response to accretionary processes. The second deformation (D_2) was a near vertical shortening that produced a subhorizontal LS tectonite fabric. The third deformation (D_3) folded the foliation into horizontal to gently plunging, inclined folds. Relatively low strain in the northwestern terminus of the Chugach metamorphic complex (Wolverine Run Glacier region) makes it possible to identify stratigraphic younging and, in turn, identify the F_2 folds as antiformal synclines and synformal anticlines; all are north-vergent structures (Fig. 2). Pavlis and Sisson (this volume, Chapter 8) provide a more comprehensive discussion of D_2 and D_3 deformation in this study area.

Peak metamorphism occurred in the western Chugach metamorphic complex between 58 and 56 Ma synchronous with felsic intrusive rocks of the Sanak-Baranof belt (Hudson et al., 1979; Hudson and Plafker, 1982; Sisson et al., 1989; Haeussler et al., 1995; Harris et al., 1996; Sisson et al., this volume, Chapter 13). Younger intrusive rocks range in composition from felsic to mafic and were emplaced between 37 and 34 Ma (Nelson et al., 1985; Goldfarb et al., 1986; Bradley et al., this volume, Chapter 1). The presence of high-temperature/low-pressure minerals, the common occurrence of gold mineralization (Haeussler et al., 1995, this volume, Chapter 5), and the presence of a forearc plutonic suite emplaced during a period of continued subduction (Harris et al., 1996; Sisson et al., this volume, Chapter 13), indicate that an additional heat source affected the region. Many models attempt to explain the anomalous high heat flow conditions within the Chugach metamorphic complex. Whereas these models disagree on the exact sequence of events, most postulate that subduction of ridge and transform segments resulted in the high-temperature, low-pressure mineral assemblage seen today (e.g., Pavlis and Sisson, this volume, Chapter 8; Sisson et al., this volume, Chapter 13; Bradley et al., this volume, Chapter 1).

An extensive region of greenschist facies rocks of the Valdez Group occurs from west of Valdez to just east of the Copper River. This region is characterized by development of chlorite and biotite (Miller et al., 1984; Sisson et al., 1989). Metamorphism of the Valdez Group to lower greenschist facies introduced a regionally extensive network of quartz veins (Goldfarb et al., 1986; Stüwe, 1986; Vrolijk, 1987; Sisson et al., 1989; Fisher and Brantley, 1992; Fisher et al., 1995; Clark et al., 1995; Brantley et al., 1997). Veins in this network formed during infiltration of multiple generations of hot fluids. These veins display changes in deformational style with age. Elsewhere, late quartz veins are associated with gold mineralization (e.g., Stüwe, 1986; Goldfarb et al., 1986; Haeussler et al., 1995, this volume, Chapter 5). The early quartz veins provide the focus of our study.

METHODS

Vein Samples

Within the study area, there are five vein generations, with up to two present in a single outcrop (Figs. 3 and 4). In this region, all veins consist mostly of quartz with some chlorite at their edges. Elsewhere in the Chugach metamorphic complex, veins emplaced in greenschist facies rocks often include calcite. The earliest vein generation was emplaced syn-D_1 to pre-D_2. The early veins are generally thin, though it is difficult to assess their original size due to later deformation. This earliest vein generation is boudinaged, indicating that the veins experienced ductile deformation during D_2 while undergoing layer-parallel extension (loc. 93ASn33, Fig. 2; Fig. 3B). Deformation associated with subsequent generations of veins changed emplacement style dramatically. In metagraywackes, the second-generation veins are linear networks of variable size that crosscut foliation (loc. 93ASn39, Fig. 2; Fig. 3D). In contrast, second-generation veins in metaargillite are aligned parallel to D_2 foliation and have a blobby appearance. In one metagraywacke outcrop, the second-generation veins display a transition from brittle to ductile characteristics (loc. 93ASn35, Fig. 2; Fig. 4A–C). In the western portion of the outcrop, the vein is a linear network and the outcrop has a brecciated texture (Figs. 4A and 4B). In the eastern end of the outcrop (Figs. 4A and 4C), the veins are sigmoidal en echelon tension arrays crudely parallel to D_2 foliation. This texture is indicative of mixed brittle-ductile behavior and shows progressive simple shear with vein opening during ductile shortening. The third vein generation consists of thin linear veins that parallel D_3 axial planes (loc. 93ASn30, Fig. 2; Fig. 4D). Dominated by brittle deformation, fourth-generation veins contain rounded to angular brecciated clasts of the host rock and are thicker than any of the earlier vein generations (loc. 93ASn33, Fig. 2; Fig. 3C). Rare fifth-generation veins occur parallel to late N-S striking joints. These veins have vuggy quartz and commonly show an alteration halo in which biotite is altered to chlorite. The fifth-generation veins were not sampled for this study. The first three vein generations were emplaced pre- to syn-peak metamorphic temperatures. The occurrence of retrograde chlorite around some of the veins and crosscutting

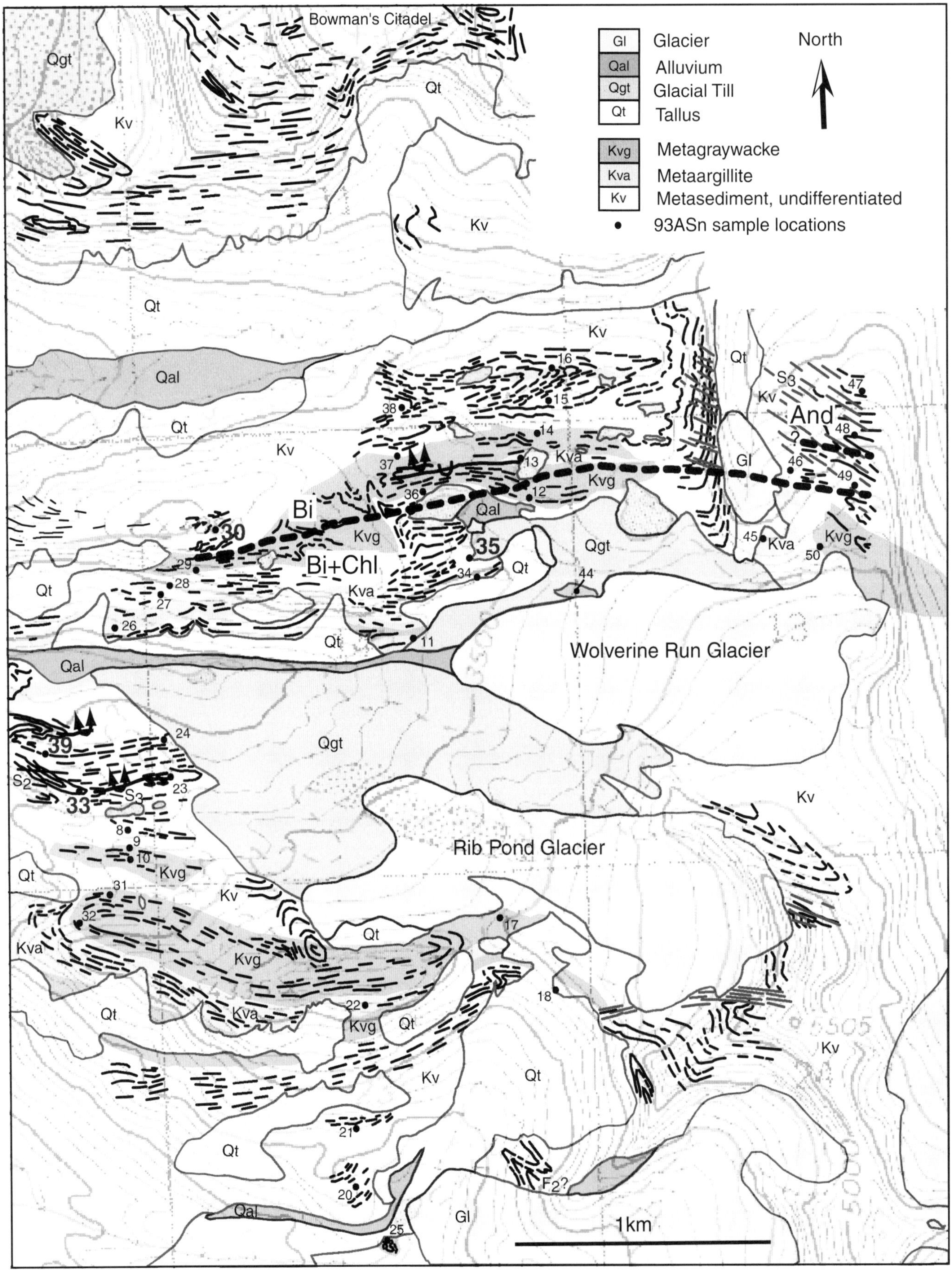
Bowman's Citadel
Gl Glacier
Qal Alluvium
Qgt Glacial Till
Qt Tallus
Kvg Metagraywacke
Kva Metaargillite
Kv Metasediment, undifferentiated
93ASn sample locations
North
Wolverine Run Glacier
Rib Pond Glacier
Bi
Bi+Chl
And
F2?
1km

Figure 2. Geologic, topographic, and sample location map of the Wolverine Run Glacier study area. Mapping was at ~1:24,000 scale on enlargements of U2 aerial photography. Geologic maps were orthocorrected using ERDAS Imagine software from control points picked on photographs and rubber sheet corrected to remove major parallax distortions. Topographic map is enlarged from the 1:63,360 Cordova D_2 quadrangle. The dashed isograd lines indicate distribution of mineral assemblages in metapelitic lithologies with occurrence of biotite (Bi), andalusite (And), and biotite + chlorite (Bi + Chl). Large numbers indicate sample for quartz veins shown in Figures 3 and 4. Thin lines are foliation trends. In some locations, S_3 is shown as thin red lines. Overturned folds shown with double arrows. In many locations, the interbeds of metagraywacke and metaargillite interbedded at a fine scale not distinguishable at this scale and are shown as undifferentiated. The Stewart Creek fault zone and Bremner shear zone are parallel to the S_3 foliation (see Pavlis et al., this volume, Chapter 7; Pavlis and Sisson, this volume, Chapter 8).

nature of the veins suggest that the fourth and fifth generation were emplaced after peak metamorphism.

Two outcrops within 2 km of each other in the Wolverine Run Glacier region (Fig. 2) were sampled for this study (Table 1). Four samples from one 3 m long outcrop (93ASn35) represent the entire range of deformational styles within the region; two of these samples come from a single vein set that had both ductile and brittle deformational features (Fig. 4). These samples were separated by ~1 m. The second outcrop occurs in a D_3 synform with interbedded metaargillite and metagraywacke; samples collected here (93ASn33) represent the first and fourth vein generations.

Analysis of Fluid Inclusions

We used a U.S. Geological Survey heating-freezing stage made by FLUID Inc. for microthermometric measurements. Fluid inclusions ranged in size from less than 1 to 20 µm, with most falling between 5 and 10 µm. We avoided those inclusions that appeared to have necked down, or re-equilibrated to lower temperatures and pressures after entrapment. Stage calibration used synthetic CO_2 inclusions (–56.6 °C) as well as synthetic H_2O inclusions (0 °C and 374.1 °C). We typically made measurements of final ice melting temperature ($T_{M, ICE}$) and homogenization temperature (T_H). Due to the small size and low salinity of many fluid inclusions, the observation of first ice melting is difficult, but was recorded if possible. $T_{M, ICE}$ measurements were recorded using cycling technique described by Roedder (1984) and Goldstein and Reynolds (1994) at heating rates of 0.5 °C/min. We used faster rates of 1–5 °C/min to measure T_H. The precision of $T_{M, ICE}$ varied from ±0.1 to ±0.3 °C and the precision of T_H around ±2 °C; both were determined by repeat analysis of individual fluid inclusions.

Examination of doubly polished thin sections under a UV light source indicates the presence of aromatic hydrocarbons and other fluorescent material (e.g., Burruss, 1981; McDonald and Skilbeck, 1996). We employ an Olympus microscope, equipped with a UV (355 nm) light source and filters to study the fluid inclusions. Some of the hydrocarbon-bearing fluid inclusions were easy to identify from their light, yellow-brown to chocolate brown color in plain light. In other cases, we made sure to determine that the fluorescent material was within the section and not due to the presence of epoxy cement along cracks or dust accumulated on the surface. We made qualitative measurements of hydrocarbon abundance and relations to other inclusions. Fluorescence color is an indicator of hydrocarbon maturity (e.g., Henry and Donovan, 1984; Hagemann and Hollerbach, 1985; McLimans, 1987; Stasiuk and Snowden, 1997). Yellow fluorescing hydrocarbons have longer chains than blue fluorescing hydrocarbons and are therefore less mature (McLimans, 1987; Stasiuk and Snowden, 1997). The assessment of maturity is purely qualitative as it is difficult to compare oils from different basins and regions (Burruss, 1991).

Mineral Chemistry

Characterization of chlorite was carried out by electron microprobe analyses conducted with a Cameca SX-50 at Rice University. Analyses utilized a 20 s counting time and a 5 µm beam to minimize decomposition during analyses. Operating conditions were 15 kV accelerating potential and 15 nA sample current. We used both natural and synthetic calibration standards. Data reduction used correction methods of Pouchou and Pichoir (1985). We attempted to use cathodoluminescence (CL) of quartz to classify textural zones related to generations of fluid inclusions as suggested by Boiron et al. (1992). The quartz in these veins had a uniform CL response; thus, this technique was not employed.

RESULTS

Fluid Inclusion Analysis

Composition and textural relations vary both between and within individual samples. Two major textural types of fluid inclusions are (1) those found along healed micro-fractures and (2) those found as separate inclusions either distinct from all other inclusions or in clusters. The latter individual inclusions do not cross grain boundaries. We infer that these inclusions formed early and designated them as primary inclusions for this study. Fluid inclusions found along healed microfractures cross grain boundaries; we designated these as secondary inclusions. Both generations of inclusions have a variety of compositions, although no primary inclusions contain hydrocarbons. None of the inclusions had textures indicative of fluid loss (e.g., Vityk and Bodnar, 1995; Audétat and Günther, 1999). There are four types of fluid compositions:

Type 1 Inclusions

Two-phase aqueous inclusions are present in all samples. They range in size from 2 to 15 µm. These aqueous inclusions occur as both primary and secondary inclusions. Shapes range

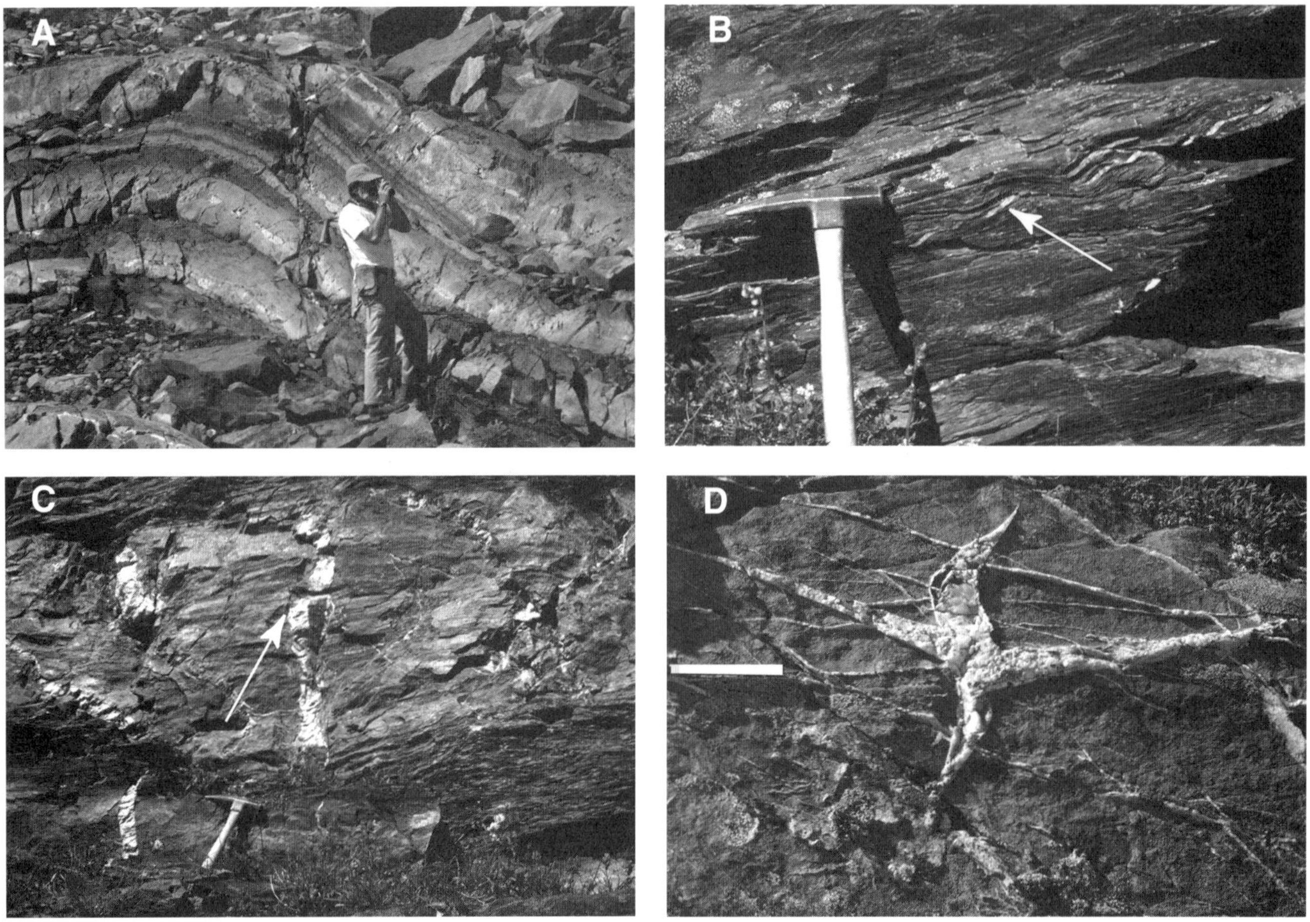

Figure 3. Outcrop photographs from locations 93ASn33 and 93ASn39. In all photographs, hammer handle is 22 cm. A: D_3 synform in interbedded metaargillite and metagraywacke at 93ASn33. B: Early quartz veins (sample 93ASn33a). Arrow points to boudinaged and deformed D_2 quartz vein in metaargillite. Perpendicular to this exposure, the quartz vein appears folded. C: Late post-D_3 quartz veins (sample 93ASn33b). Arrow points to brittle late breccia vein in metagraywacke. Note that the vein does not extend into metaargillite at the bottom of the outcrop. D: Quartz vein network at 93ASn39. The white box approximates the hammer that was in a shadow. These veins have dilational jogs and are approximately parallel to S_3 foliation.

from round to oblong with primary inclusions tending to be round and secondary inclusions tending to be elongate and lobate.

Type 2 Inclusions

One to two-phase liquid, higher hydrocarbon–bearing inclusions are present in all samples as secondary inclusions. The hydrocarbon-bearing inclusions commonly occur on grain boundaries and along healed micro-fractures. They range in color from colorless to light, yellow brown to a deeper chocolate brown and in size from <1 to 6 μm. Shapes of liquid hydrocarbon inclusions vary from round to elongate and irregular. Many appear to have necked down.

Type 3 Inclusions

Two- or three-phase aqueous plus liquid hydrocarbon (both methane and higher hydrocarbon) inclusions occur as secondary fluid inclusions. These have clear central bubbles with slightly clouded outer rims and range in size from 2 to 4 μm. Inclusions are circular to oblong in shape and do not appear to have necked down.

Type 4 Inclusions

Two-phase aqueous inclusions with a CO_2 vapor phase occur in one sample (93ASn33b). They are much less common than the either type 1 or 2 inclusions. These inclusions occurred as a small cluster similar to other primary inclusions and are not found along healed fissures within the vein. These inclusions range in size from 4 to 10 μm and are rounded to oval in shape. We observed only one cluster of such inclusions.

$T_{M, ICE}$ and Fluid Inclusion Salinity

Characteristics of all the inclusions are presented in Table 1. As is often the case in fluid inclusion studies, the range

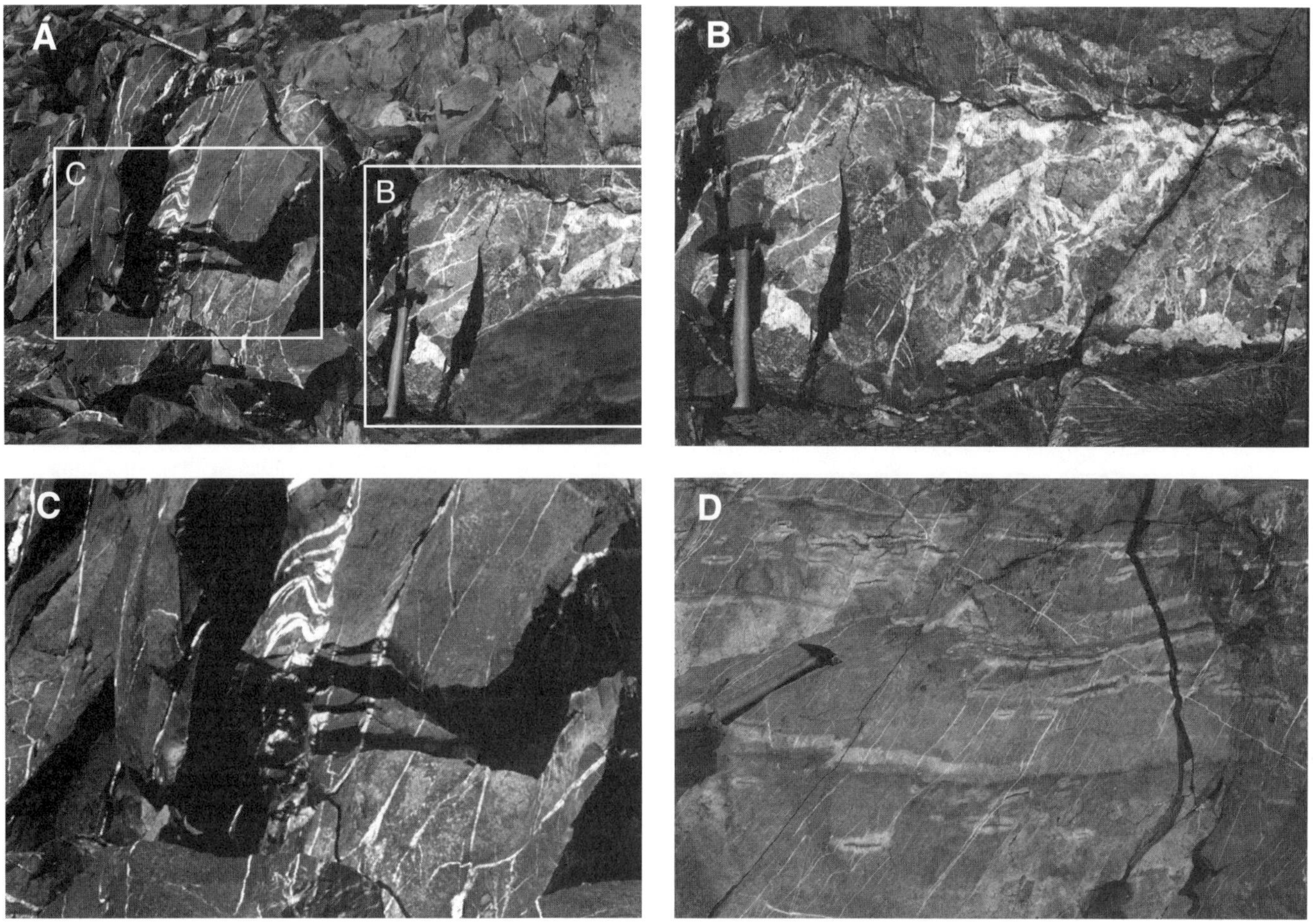

Figure 4. Outcrop photographs from locations 93ASn35 and 93ASn30. In all photographs, hammer handle is 22 cm. A: Quartz vein network showing both brecciated and ductile fabrics within the same vein set. Boxes show locations for B and C. Close-up of quartz vein with breccia texture (sample 93ASn35a). Note angular blocks of metagraywacke within the vein system. C: Close-up of quartz vein tips with ductile, folded appearance (sample 93ASn35b). In the field, it was possible to trace the quartz veins between B and C through the area in the shadow. D: Thin linear veins in metagraywacke from 93ASn30 that crosscut both lithologic layering and torpedo-shaped mineral zones in calc-silicate pods. These veins parallel S_3 foliation. These veins are similar in appearance to those sampled at 93ASn35c.

in melting temperatures ($T_{M, ICE}$) is smaller than that for homogenization temperatures. Equilibrium melting temperatures range from –1.1 to –17.9 °C for type 1 inclusions, corresponding to 2–21 wt% NaCl equivalent (calculated using equations of Zhang and Frantz, 1987). Within individual samples, the salinity range is generally within 5 to 10 wt% except for sample 93ASn33a, which has a variation from 3 to 19 wt% NaCl equivalent (Fig. 5). This sample has a sparse number of relatively small fluid inclusions (most under 1 μm long) making it difficult to obtain reliable observations. The few low-salinity inclusions in this sample reside in one small cluster near the edge of the vein, whereas the high-salinity inclusions all occur away from the edge. The low-salinity inclusions are similar in appearance, composition and density to those in the late vein from this location. Thus, these inclusions may relate to a late stage fluid infiltration. The rest of the samples lack obvious or systematic changes in the salinity between different vein generations. In addition, there are no indications for either mixing of fluids or cooling of fluids during vein formation. It is not known what is the cause of the salinity variations. Oxygen isotope results from the central Chugach metamorphic complex indicate that there were at least two episodes of metamorphic fluid infiltration; these fluids have exchanged with felsic magmas (Bowman et al., this volume, Chapter 10). The salinity probably reflects variable amounts of water and salts produced during dehydration of clays. The variation in salinity indicates that, for most veins, a restricted fluid composition was available for each episode of vein formation despite the wide range in T_H (see below). In general, the earlier veins have higher salinity than the later veins. These salinity values are similar to those of greenschist facies quartz-calcite veins (0 to 8 wt% NaCl equivalent) from the Richardson Highway (Pavlis et al., this volume, Chapter 7) as well as those for gold-bearing quartz veins (0 to 7 wt% NaCl equivalent) in the Valdez district (Goldfarb et al., 1986).

TABLE 1. SUMMARY OF FLUID INCLUSION COMPOSITIONS

Sample	Deformational style	Relative age	Inclusion types*	Average T_M	Average T_H
93ASn33a	ductile: boundinaged pre-D_2 to syn-D_2	Early	1 and 2	−10 to −12	160 to 170
93ASn33b	brittle: brecciated syn-D_3 to post-D_3	Late	1, 2, 3, and 4	−6 to −8	180 to 190
93ASn35a	brittle: brecciated syn-D_2 in metagreywacke	Early	1, 2, and 3	−6 to −8	260 to 270
93ASn35b	Ductile: folded tip of 35a syn-D_2	Early	1 and 2	−6 to −8	210 to 220
93ASn35c	brittle: thin veins in metagraywacke, syn-D_3	Middle	1 and 2	−6 to −8	210 to 220
93ASn35d	brittle: thick breccia on left lateral post-D_3 fault	Late	1 and 2	−8 to −10	180 to 190

*Type 1 is H_2O-rich, Type 2 is hydrocarbon, Type 3 is CH_4 and/or CH_4 + H_2O, and Type 4 is CO_2-rich. For Type 1 inclusions, T_M—melting temperature and T_H—homogenization temperature, both in °C.

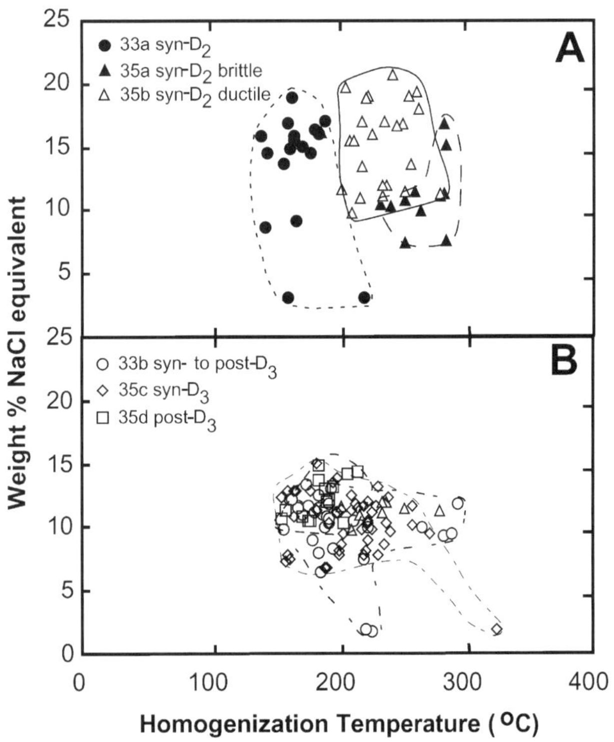

Figure 5. Salinity (wt% NaCl equivalent) versus homogenization temperature (°C) for type 1 aqueous fluid inclusions. Circled regions delineate each fluid inclusion population. A: Early syn-D_2 veins. B: Late syn- to post-D_3 veins.

T_H Measurements

Homogenization results for type 1 inclusions are given in Figure 6. In all cases, aqueous fluids homogenized to the liquid phase. Due to the variation in fluid inclusion composition, homogenization temperatures ranged from −113 °C in a type 2 hydrocarbon-bearing inclusion, to 320 °C in a type 1 aqueous inclusion. The range of T_H both between and within veins is important as we use T_H to calculate isochores for the various samples. We assumed that inclusions with homogenization temperatures less than −100 °C were pure methane for the purposes of generating and comparing pressures to those derived from the aqueous inclusions. This approximation, however, is not valid in all cases. We calculated isochores using the MacFlinCor program of Brown and Hagemann (1995). We used the temperature of the most abundant fluid inclusions estimated from histograms to calculate representative isochores. Data from aqueous, type 1 inclusions in each vein analyzed fall into a cluster on a plot of salinity versus homogenization temperature (Fig. 5). Isochores calculated for type 4 CO_2-rich inclusions do not pass near peak metamorphic conditions as determined from peak metamorphic mineral assemblages (Fig. 7). In contrast, the isochores for the type 3 methane-rich inclusions fall slightly below the peak metamorphic conditions.

UV Analyses

We observed three types of liquid hydrocarbon fluid inclusions. The first is a two-phase liquid hydrocarbon-bearing inclusion on microcracks that cross grain boundaries; we assume these formed in a late crack-seal deformation. A second type of two-phase hydrocarbon inclusion occurs on grain boundaries. Both of these textures indicate that the liquid hydrocarbon-bearing inclusions form later than the other fluid types. A third type of fluorescent inclusions is large (5 to 7 μm) three-phase inclusions with a vapor bubble that ranges from yellow to chocolate brown in color. We observed this type on grain boundaries.

Within the hydrocarbon-bearing inclusions fluorescence color ranges from light blue to yellow. Those that fluoresce yellow and white tend to be brighter than those that fluoresce light blue. The light blue fluorescent inclusions have high refractive indices and do not have vapor bubbles at room temperature. Rather, vapor bubbles for these inclusions homogenize to liquid at temperatures ranging from −113 to −60 °C. Yellow fluorescent inclusions, on the other hand, have low refractive indices and vapor bubbles at room temperature. However, accurate measurements of homogenization temperatures could not obtained for these inclusions because of their small size.

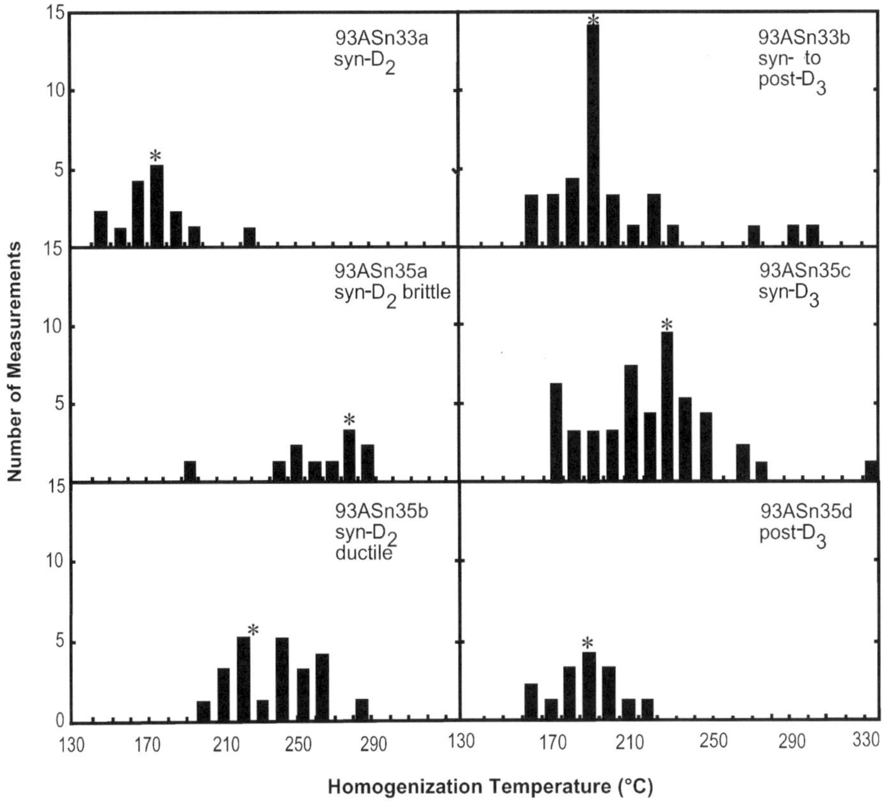

Figure 6. Histograms showing distribution of homogenization temperatures (°C) for aqueous, type 1 fluid inclusions. The asterisk shows the T_H used to calculate isochores for Figure 7.

Chlorite Geothermometry

Table 2 gives the results of representative analyses of chlorite from several quartz veins. The compositional variation of tetrahedral Al in chlorite is a function of the metamorphic temperature (e.g., Cathelineau and Nieva, 1985; Cathelineau, 1988). Reviews of chlorite geothermometry pointed out that empirical chlorite geothermometers are not universally applicable, as compositional variables depend on the mineral assemblage and bulk rock protolith; quartz and a fluid phase must also be in equilibrium (De Caritat et al., 1991; Zane et al., 1996; Merriman and Peacor, 1999; Schmidt et al., 1999). Recent work in high-pressure terrains (Bröcker and Day, 1995; Smith et al., 1999) and low-pressure terrains (Bevins et al., 1991; Merriman et al., 1995; Xie et al., 1997; Smith and Yardley, 1999) yield reasonable agreement between calculated chlorite temperatures and mineral assemblage data. Temperature uncertainty is on the order of ±30 °C based upon analytical uncertainty. Cathelineau (1988) calibrated an empirical chlorite geothermometer for feldspathic sandstone in equilibrium with an aqueous fluid, similar to the situation for the greenschist facies Chugach metamorphic complex.

In the absence of any other reliable geothermometer for these vein samples, we estimated the temperature for chlorite formation using the calibration of Cathelineau (1988). These temperatures must be regarded with a degree of caution as the chlorite in these veins is only in equilibrium with quartz. The calculated temperatures range from 325 to 350 °C. The results from the chlorite geothermometry are consistent with temperatures estimated from the occurrence of chlorite + biotite assemblage in the metaargillites and the actinolite + epidote assemblage in calcareous metagraywackes (Fig. 7). The values are low, however, compared to the temperature estimate for the occurrence of andalusite in the northeastern corner of the Wolverine Run Glacier region (Fig. 2).

DISCUSSION

Vein Emplacement

Our estimates for *P-T* conditions of vein formation gives information on the relation of veining to bulk rock metamorphism and deformation. First, we note that there is little variation in paleotemperature derived from chlorite geothermometry between the two locations. This small paleotemperature variation is not sufficient to explain the observed differences in deformational styles within or between the veins. Second, we assume that the paleodepth for the region was close to 10 km defined by the presence of andalusite + biotite. Figure 7 shows the calculated isochores for both sample locations along with the temperature range calculated by chlorite geothermometry

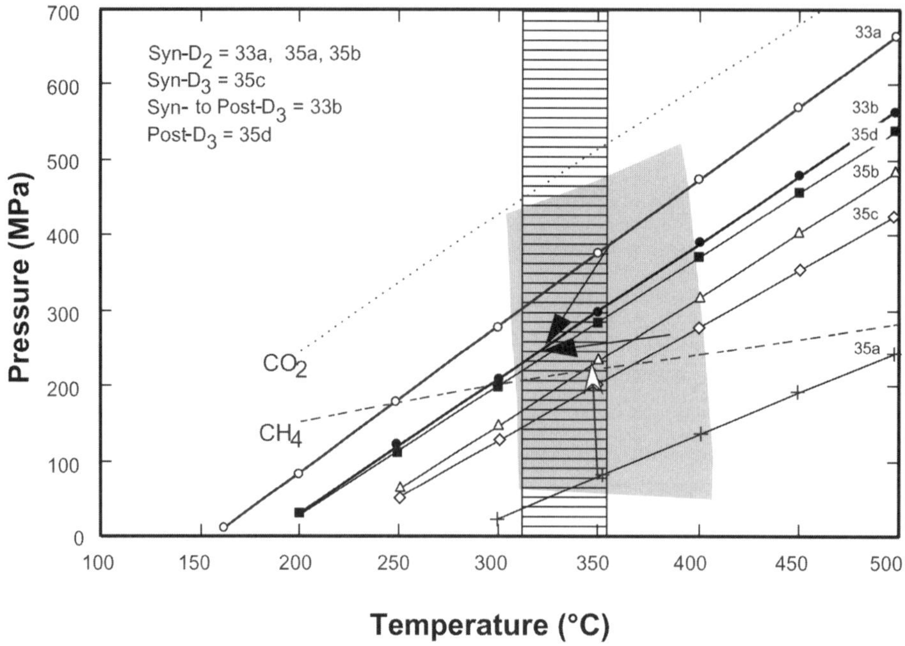

Figure 7. Pressure-temperature plot showing representative isochores for fluid inclusions. Aqueous, type 1 isochores as solid lines with thick lines for 93ASn33 and thin lines for 93ASn35. The filled symbols indicate the syn- to post-D_3 veins; the open symbols and crosses indicate syn-D_2 to syn-D_3 veins. CO_2-rich, type 3 inclusions—dotted line; CH_4-rich inclusions—dashed line. The ruled region corresponds to the temperatures derived from chlorite geothermometry. The shaded region shows the peak metamorphic conditions derived from the coexistence of biotite + chlorite in the metaargillite, epidote + actinolite in the calcareous metagraywacke, and lack of lawsonite in metabasaltic lithologies. The arrow with an open head shows changes in pressure during formation of brittle–ductile vein set. The two arrows with solid heads show possible *P-T* paths between selected isochores for cooling from early syn-D_2 to late post-D_3 quartz veins. Many other *P-T* paths are possible between the different vein sets. See text for discussion of *P-T* paths between brittle and ductile deformation.

and estimated from adjacent mineral assemblages. The isochore diagram reveals several trends both between the two vein sets and within each vein set.

Syn- to Post-D_3 Veins

First, note the similarity of isochores for inclusions for late vein inclusions at the two locations (compare 93ASn33b and 93ASn35d). These two isochores show a maximum pressure difference of approximately 10 MPa within the temperature range estimated from chlorite geothermometry. There is a variation in the salinity range between the two veins (2–13 wt% NaCl equivalent versus 8–15 wt% NaCl equivalent). Salinity differences may reflect different local conditions as these two vein sets are separated by ~2 km. The similarity between their isochores indicates that the two regions experienced D_3 deformation and vein emplacement at similar pressures and temperatures (Fig. 7).

Syn-D_2 Veins (Southern Location)

Although the two locations reached similar pressure and temperature conditions, the isochores for earlier veins at each outcrop indicate that they took different paths to get there. Isochores for early veins from the two locations are significantly different (compare 93ASn35b and 93ASn33a). The early vein at the southern location (sample 93ASn33a) underwent ductile deformation after emplacement and is the only one from this study to display microstructural characteristics of ductile deformation with grain size reduction. In order to reach the temperature and pressures dictated by the isochore of 33b, the vein system at location 93ASn33 must have had a decrease in temperature and/or pressure (see Fig. 7). We infer that initially the rock was at 350–400 °C as the vein is ductily deformed (93ASn33a). Then temperature decreased to at least ~325–350 °C as the later veins (93ASn33b) have brittle textures; this assumes that the ductile to brittle transition is close 350 °C (Stöckhert et al., 1999). A decrease in pressure from approximately 380 MPa to 280 MPa, with a corresponding decrease in temperature from ~400 to ~350 °C is one path that would explain the change in deformational style. Vrolijk (1987) and Brantley et al. (1997) document changes in pressure of up to 150 MPa within vein systems from Kodiak Island (~1000 km to the west in the Chugach terrane); our inferred pressure drop is in the same range. Also, note that both the Kodiak Island and Wolverine

TABLE 2: REPRESENTATIVE CHLORITE ANALYSES AND GEOTHERMOMETRY

Sample	93ASn33a	93ASn33b	93ASn35d
SiO_2	25.01	26.079	28.29
Al_2O_3	21.29	21.47	20.02
TiO_2	0.07	0.070	0.23
FeO	24.19	23.807	25.29
MgO	13.32	14.941	12.36
MnO	0.38	0.489	0.31
CaO	0.03	0.092	0.12
Na_2O	0.04	0.050	0.10
K_2O	0.05	0.033	0.12
F	0.00	0.044	0.12
Cl	0.00	0.020	0.04
H_2O	10.99	11.366	11.28
Total	95.36	98.46	98.28
Structural formula <36 O>			
Si	5.460	5.491	5.506
Al (Total)	5.477	5.327	5.244
Al iv	1.270	1.288	1.247
Fe	4.417	4.192	4.176
Mg	4.335	4.689	4.596
Mn	0.070	0.086	0.067
Ca	0.007	0.021	0.082
Na	0.015	0.021	0.030
K	0.014	0.009	0.010
Temperature (°C)	347	353	340

Note: Temperature calculated according to Cathelineau (1988).

Run Glacier regions formed in accretionary prisms probably subjected to either triple junction interaction and/or ridge subduction. The changing paleofluid conditions can be explained by either exhumation of the host rock, cooling, change in fluid pressure or a combination of these three factors. The late stage vein underwent brittle deformation during retrograde cooling either from subduction cooling, exhumation, or both.

One puzzling aspect of the veins from southern outcrop 93ASn33 is the occurrence of type 4 CO_2-bearing inclusions. These inclusions occur as a single cluster far from other inclusions, and have the steepest isochore for sample 93ASn33b. The isochore intersects the upper limit of the estimated temperature range from chlorite geothermometry at a pressure that is 200 MPa above the pressure for the type 1 aqueous inclusions. This pressure (~500 MPa) is much higher than that estimated from the occurrence of andalusite in the region. We infer that these inclusions are therefore probably not primary despite their appearance. They may contain a significant quantity of either methane or nitrogen as their T_{M,CO_2} was –62 °C, well below that for pure CO_2 (–56.6 °C). In addition, there are secondary type 3 methane-rich inclusions in this sample. The possible higher methane content of these type 4 inclusions may indicate that they are related to the late stage influx of liquid hydrocarbon and methane-rich fluids (see as follows).

Syn- to Post-D_2 Veins (Northern Location)

Unlike the early vein from the southern location (93ASn33), isochores calculated for the earliest vein from northern location (sample 93ASn35) probably indicate a change in either pressure or temperature conditions, but not both. Since all the early veins (93ASn35a–c) have shallower $\delta P/\delta T$ slopes than isochores from 35d, the possible *P-T* paths must either record an increase in pressure or decrease in temperature to reach the *P-T* conditions of the late veins (93ASn35d). We believe that a decrease in temperature is more likely. In the same way that cooling and exhumation could explain the shift in isochore positions of southern region, the shift in isochore positions from samples 93ASn35b and 93ASn35c to 93ASn35d could be a result of a similar change in *P-T* conditions. We infer that cooling occurred between the emplacement of these two veins and the last generation. In reality, there could be several possible paths that connect the two isochores, and our data do not sufficiently constrain this path. Our preferred interpretation implies that the early veins in the southern location were emplaced before the early veins in the north. This observation fits the field observations that the early veins at 93ASn33 are deformed by D_2 and that the early veins at 93ASn35 were emplaced syn-D_2.

We interpret that both the thin, folded vein (sample 93ASn35b) and thin linear veins (sample 93ASn35c) were emplaced close to lithostatic pressure as their isochores pass close to the peak metamorphic conditions defined by host rock mineral assemblages. Since the veins are hosted by different protoliths, their appearance (folded versus linear) may be related to the strength properties of the respective host rocks. The difference in appearance may also be related to relative timing of the vein emplacement. It is difficult to determine if the foliation in the graywacke host (93ASn35c) formed during D_2 or D_3; thus, this vein may have been emplaced syn to post D_2 or syn to post-D_3 deformation.

Somewhat more surprising is the difference between the isochores for samples from the same vein (compare 93ASn35b to 93ASn35a). From field relationships, these two samples were emplaced synchronously; yet, the peaks of their homogenization temperature histograms differ by as much as 60 °C (Table 1). Since these two samples are part of a single vein set, we assume that the fluid inclusions in these veins were trapped at the same temperature of ~340 °C. Using this assumption and the difference in homogenization temperature yields maximum pressure differences of up to 140 MPa between these two samples. While the pressure differences recorded here exceed those of Vrolijk (1987), they fall within the limits of other pressure differentials found between inclusions in similar investigations (e.g., Sibson et al., 1988; Meere, 1995). This maximum pressure differential of 140 MPa could be related to the difference between the limits of hydrostatic and lithostatic pressure conditions. At a depth of 10 km, there would be a difference of approximately 170 MPa between lithostatic and hydrostatic values (Sibson et al., 1988). Thus, the change in deformational style across outcrop 93ASn35 may represent a change from near hydrostatic to near-lithostatic pressure conditions related to tectonic and/or deformational processes. Theories of emplacement mechanisms during deformation processes include catastrophic vein opening and fluid influx (e.g., Bruhn et al., 1994). In the vein set at site 93ASn35, one end probably was emplaced by hydrofacturing causing it to display common characteristics associated with brittle deformation. Textural features of 35b are almost explosive and include large angular pieces of the host rock contained within the vein as well as jagged edges to the vein. In addition, the amount of fluid influx may have been greater during this episode as the vein is thicker here. We base this on quartz solubility as in aqueous fluids solubility is affected by changes in temperature, salinity, and pressure (Newton and Manning, 2000). The observed changes in temperature, pressure, and salinity recorded by these veins do not affect quartz solubility, as it changes by ~0.02 mol/kg H_2O, calculated using empirical equations of Shibue (1996). Thus, we assume vein thickness is proportional to the amount of silica-saturated fluid present. After the initial fluid influx, pressure may have decreased to near hydrostatic as recorded by the fluid inclusion isochores. The other end of the vein probably records deformation mechanisms such as ductile creep at near-lithostatic conditions.

Evidence from other fluid inclusion studies (Bruhn et al., 1994; Robert et al., 1995; Lewis et al., 2000; Eichhubl and Boles, 2000) indicates that the cyclic behavior is akin to the fault-valve mechanism described by Sibson (1990) or breaching of sealed rocks in magmatic epithermal ore deposits (Fournier, 1999). At any time, different parts of a basin or fault zone may be at different stages in the fluid-pressure cycle (Hunt, 1990; Evans and Chester, 1995; Ortoleva et al., 1995; Bruhn et al., 2000). Field evidence shows that such hydraulically isolated compartments are typically up to tens of meters wide and hundreds of meters long. Thus, this vein system may not be in a fault zone but solely record local effects of changing fluid-pressure.

Another explanation for the observed difference in entrapment conditions concerns emplacement near the brittle/plastic transition temperature. Recently, Stöckhert et al. (1999) proposed that brittle plastic transition occurs at temperatures between 310 and 350 °C, close to that estimated here. Therefore, changes of rock temperature during vein formation may result in vein textures varying from brittle to ductile. A possible scenario was that the initial fluid infiltrating the rock was at or above the host rock temperature (≥350 °C) and the initial vein formed in the ductile regime. As increasingly more fluid infiltrated, the fluid cooled as well as the host rock to below the brittle–ductile transition (≤350 °C); thus, this portion of the vein was emplaced in the brittle regime. This interpretation involving only a temperature change does not fit with isochores for 93ASn35b to 93ASn35a.

Two other variables may affect the transition between ductile to brittle process: strain rate and protolith. Protolith variability is not significant in this outcrop since the vein system occurs in one unit. Also, we believe that strain rate did not vary appreciably during the emplacement of this single vein set. Thus, both temperature and pressure changes probably influenced the textural development of this vein system. Since we cannot determine which end of the vein system was emplaced first, we cannot uniquely determine the mechanism with the available data.

These observations are summarized in Figure 8, and we suggest the following sequence of fluid events. Prior to and during the D_2 deformation and metamorphism, devolatilization of clays in the accretionary prism results in silica saturated aqueous fluids with moderate salinity. As the region undergoes greenschist facies metamorphism, these fluids precipitate as the first generation of veins in the ductile regime (Fig. 8A). Then, during and after D_2 deformation, the region undergoes a brief period with change in fluid pressure and/or temperature that results in both ductile and brittle deformation presumably related to variation in fluid pressure at conditions near the brittle-ductile transition (Fig. 8B). Subsequent cooling introduced lower-salinity aqueous fluids generated at deeper levels in the accretionary prism; these were entrapped in late post-D_3 veins. Finally, late type 2, 3 and 4 fluids are trapped in these veins as secondary fluid inclusions (Fig. 8D).

Paleogeothermal Gradient

Chlorite analyses yield similar paleotemperature for the two outcrops. Average temperatures for vein samples from site 93ASn33 are close to 350 °C whereas those from site 93ASn35 are closer to 340 °C. These results are statistically indistinguishable and not sufficient to cause the differing deformational styles of veins seen in the two sites. If we accept these values, the paleotemperature difference between the two outcrops is small; yet, it follows a regional gradient with temperature increasing towards the northeast as suggested by andalusite bearing rocks at the northeast corner of this region (Fig. 2). The peak metamorphic pressure can be defined from the intersection of the isochores for the early veins with the metamorphic temperatures; this yields 250 to 300 MPa. If we assume an average density of 2.6 to 3.0 gm/cm^3 for a crustal column, this yields a paleodepth equivalent of 6 to 10 km. This paleodepth estimate is similar that estimated for greenschist facies conditions on the Richardson Highway approximately 80 km to the west (Pavlis et al., this volume, Chapter 7) and amphibolite facies conditions elsewhere in the Chugach metamorphic complex (Sisson et al., 1989). The pressure and temperature estimates correspond to a range of paleogeothermal gradients of 58 to 35 °C/km. Note this is only an approximation as geothermal gradients in subduction zones are non-linear due to changes thermal conductivity and fluid flow. Our estimated paleogeothermal gradient is similar to

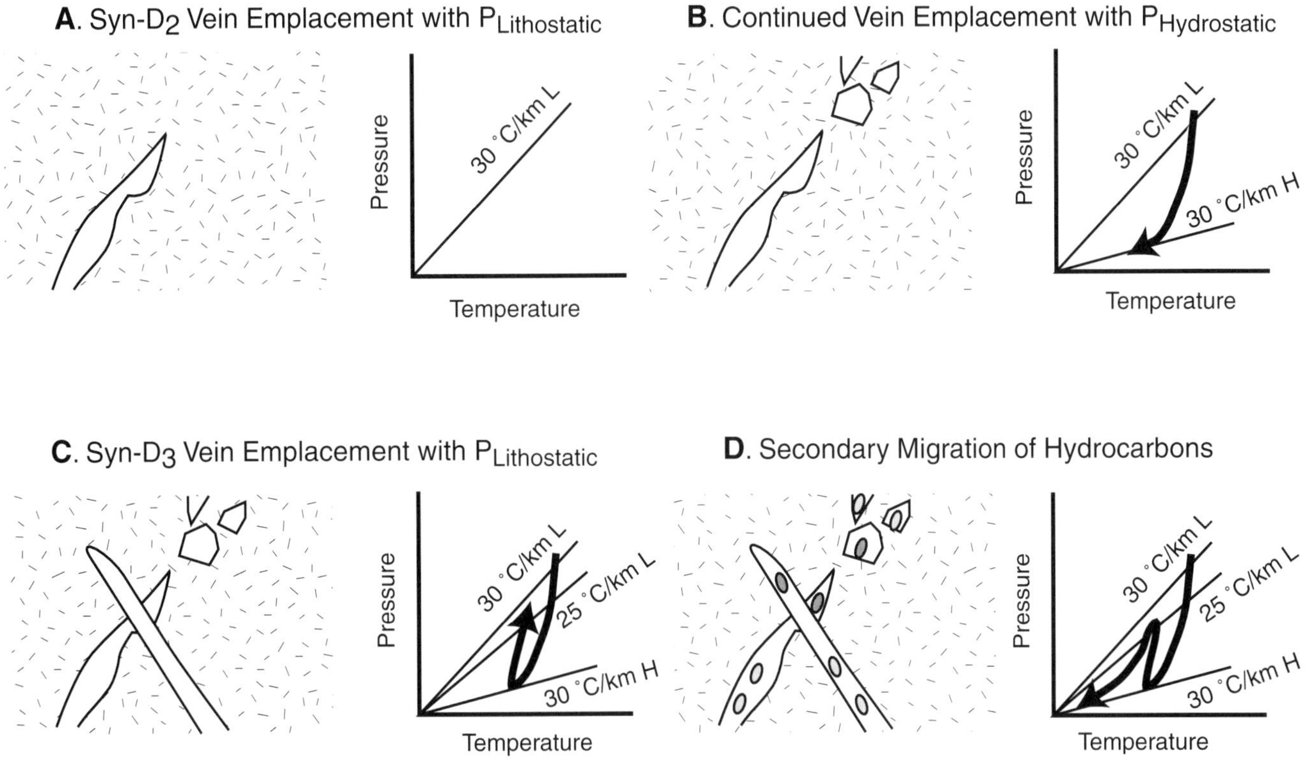

Figure 8. Schematic diagram showing sequence of vein growth, deformation, and fluid entrapment. Also shown are generalized *P-T* conditions indicated by fluid inclusion isochores for each event. Different UV fluorescence in late hydrocarbon fluid inclusions is indicated by light gray and dark gray fluid inclusions. A: Veins are emplaced pre- to syn-D_2 at relatively high paleogeothermal gradients at suprahydrostatic pressure. These veins are then folded and deformed by D_3 as seen in vein 93ASn33a. B: During D_2, pressure fluctuates between higher and lower values resulting in both brittle and ductile veins at 93ASn35 (see isochores for 93ASn35a and 93ASn35b on Fig. 7). C: During D_3, veins form parallel to S_3 axial planes at slightly lower paleogeothermal gradients as seen in 93ASn35c. After D_3, veins are emplaced at even lower paleogeothermal conditions (not shown) as documented in 93ASn33b and 93ASn35d. D: After vein emplacement, late hydrocarbon fluids are trapped as secondary fluid inclusions in all quartz veins.

the Richardson Highway (Pavlis et al., this volume, Chapter 7). This paleogeothermal gradient is higher than would be expected for "normal" accretionary prisms with subduction of cool oceanic crust. An alternative model involves subduction of hotter oceanic crust associated with ridge subduction. Our estimated range in paleogeothermal gradient, however, is lower than the shallow domain of the Chile Rise, a site of present-day ridge subduction (in excess of 70 °C/km; Cande et al., 1987) as well as for relict ridge subduction during Miocene time for the Franciscan complex (~60–120 °C/km; Underwood et al., 1999). It overlaps with the range estimated for the ridge subduction event in the Cretaceous and Tertiary Shimanto accretionary prism (40–50 °C/km; Laughland and Underwood, 1993; Sakaguchi, 1996, 1999a, 1999b).

Hydrocarbon Source and Migration

The timing and origin of petroleum-bearing fluid inclusions in sedimentary basins has been successfully determined by combining fluid inclusion thermometry with UV spectroscopy (e.g., Burruss, 1981; Bodnar, 1990; Orange et al., 1993; Wilkinson et al., 1998). The presence of hydrocarbon-bearing fluid inclusions in these samples was a surprise as no previous studies of greenschist facies metamorphic rocks have reported secondary hydrocarbon fluid inclusions. These inclusions are located along grain boundaries, and in some cases seem to define grain boundaries, indicating that the grain boundary acted as a plane of weakness along which fluids flowed. Hydrocarbon-bearing inclusions also display typical crack seal textures that cross grain boundaries providing additional indications of possible cyclic fluid flow (Fisher et al., 1995). Textural relations show that all hydrocarbon-bearing inclusions are secondary inclusions. The late stage hydrocarbon-bearing fluids trapped in these samples, therefore, do not represent unusually high-temperature preservation of hydrocarbons trapped in situ during greenschist facies metamorphism. Instead, the hydrocarbons must have migrated from outside the field region.

There are two known possible sources of organic-rich material that could be linked to the hydrocarbons: (1) black shales from Cretaceous flysch in the Copper River basin to the north and (2) Tertiary rocks of the Yakutat Terrane to the south. The Copper River basin is an unlikely source as Kirschner (1994) reports only minor oil and gas shows from several exploratory wells in the Copper River basin, about 80 km to the north. In addition, the vitrinite reflectance (R_0) for organic matter in these wells is only 0.4 to 0.7 (Johnsson et al., 1999). Oil generation begins above an R_0 of ~0.56. Thus, this region would produce immature oils with yellow fluorescence. The blue to yellow UV fluorescence of the observed secondary hydrocarbons suggests that source beds had a higher R_0 value than this (A.E. Bence, 1997, personal commun.). Furthermore, seismic refraction data indicate that the Copper River basin is presently 2–3 km deep (Goodwin et al., 1989; Fuis et al., 1991; Ambos et al., 1995) suggesting depths too shallow for large-scale hydrocarbon generation. Finally, the most significant problem is structural as the Border Ranges fault, a major east-west fault system, places a large slab of crystalline basement between the Copper River Basin and the study area. This basement slab would presumably act as a barrier to any north-south hydrocarbon migration (Fig. 9). Fuis and Plafker (1991) present an alternative interpretation of seismic data across southern Alaska with the Jurassic Talkeetna Formation of the Peninsular terrain underthrust beneath the Chugach terrain along a south-dipping thrust. If this interpretation is valid, this may be a pathway for deep migration of hydrocarbon-rich fluids from Copper River basin equivalents. Other interpretations (e.g., Plafker, 1987; Fuis and Plafker, 1991; Grantz et al., 1991), however, place the Chugach terrain underneath the Peninsular terrain, which would not allow migration of the hydrocarbons from the Copper River Basin into the study area region.

The second possible source is hydrocarbon from the Yakutat terrane rocks that were thrust beneath the study region in late Miocene to Recent time (Plafker et al., 1994). These hydrocarbons include a known oil field area near Katalla immediately south of the study area, coal deposits from the same region in the Yakutat terrane, and organic-rich sediments, both on and off shore.

Hydrocarbons could have been generated from Eocene to Oligocene coal deposits that crop out in the northern Yakutat terrane. These coals have variable vitrinite reflectance (R_0) ranging from ~0.5% in the east to ~2.5% and possibly higher near Katalla (Fig. 1; Plafker, 1987). Some of the coals are anthracite related to localized heating from adjacent hypabyssal dikes (A.E. Bence, 1999, personal commun.). Hydrocarbons in the six samples studied are probably mature with a high API gravity index from their light blue fluorescence and high refractive index, which means they are more mature than most of the coals (A.E. Bence, 1997, personal commun.). Although some inclusions are less mature, based on fluorescence, we suggest that most could not have come from the coal deposits to the south. In addition, most coals tend to produce gas rather than oils. The long distance required for flow of these hydrocarbons to the Wolverine Run Glacier region precludes the coals being the source of hydrocarbons. If there were deeply buried coals or organic-rich sediments, then they may be a possible source of hydrocarbons.

Another potential source is the younger Late Oligocene to Quaternary sedimentary rocks of the Yakutat terrane. However, samples from outcrops, dredge hauls and drill core indicate they are thermally immature and low in kerogen (Armentrout, 1981; Mull and Nelson, 1986; Plafker, 1987; Johnsson et al., 1999). Armentrout (1981) reports up to 0.75 wt% C in some of the lower Yakataga Formation. Both Armentrout (1981) and Plafker (1987) estimate the maturity as R_0 ~0.5–0.6%, which is too low to be the source for the observed hydrocarbon fluid inclusions. In the offshore, a calcite dredge sample is depleted in ^{13}C; this may indicate oxidation of Eocene methane (Barnes et al., 1982). Mull and Nelson (1986) and Johnsson et al. (1999) indicate a region of over mature sedimentary rocks with R_0 ranging from 2.0 to 3.6 north of the Bering Glacier and south of the Contact fault. So, more deeply buried sediments of the Yakutat Terrane could be a source for the observed hydrocarbons. A limited amount of maturity data from sediment samples from wells in the Yakutat terrane shows both small decreases and increases of R_0 with depth (see Johnsson et al., 1999).

A Yakutat source for the hydrocarbons could be the same source as for the oil seeps south of the field area (Fig. 1). The oil seeps are a manifestation of a hydrocarbon system. One shallow field at Katalla was developed and produced hydrocarbons in the early 20th century (reported in Plafker, 1971). Bence et al. (1996) examined these seeps in an attempt to distinguish them from oil spilled in the *Exxon Valdez* disaster. Maturity level for these and other oil seeps is moderate (Plafker, 1987; Bence et al., 1996). Thus, these may be appropriate maturity levels for the hydrocarbons found in these samples.

Thus, the question of where these hydrocarbons originated remains unanswered. Additional studies, such as geochemical "fingerprinting," are needed to identify the source of these hydrocarbons. Long distance migration of hydrocarbons is possible along faults (Hunt, 1990; Anderson et al., 1994; Grauls and Baleix, 1994; Ge and Garven, 1994; Garven, 1995; Fleming et al., 1998; Bruhn et al., 2000). Long distance migration of aqueous fluids has been inferred for several accretionary wedges including Peru (Dia et al., 1993) and Barbados (e.g., Moore and Vrolijk, 1992). Bruhn et al. (2000) identified aqueous fluids in the Castle Mountain fault system ~150 km to the west. Anderson et al. (1994) suggest that hydrocarbon migration from over-pressured shales in the Gulf of Mexico may result in locally higher temperatures and pressures along a fault, with the fault zone periodically releasing hydrocarbons to shallower levels. A similar mechanism may have operated along late stage fractures through the southern Alaska margin from a slightly more mature source than those for the Katalla region. That is, as the Yakutat terrane rocks were underthrust beneath the margin, the hydrocarbons could have been released to the hanging wall. The initial accretion of the Yakutat terrane may have occurred as early as 30 Ma (Plafker, 1987) with underthrusting continuing for at least the last 7.1 m.y. (Bruns, 1985). Fission track geochronology indicates a change in the cooling rate at ca. 8 Ma for gneisses about 15 km to the south (Sisson et al., this volume, Chapter 13). This

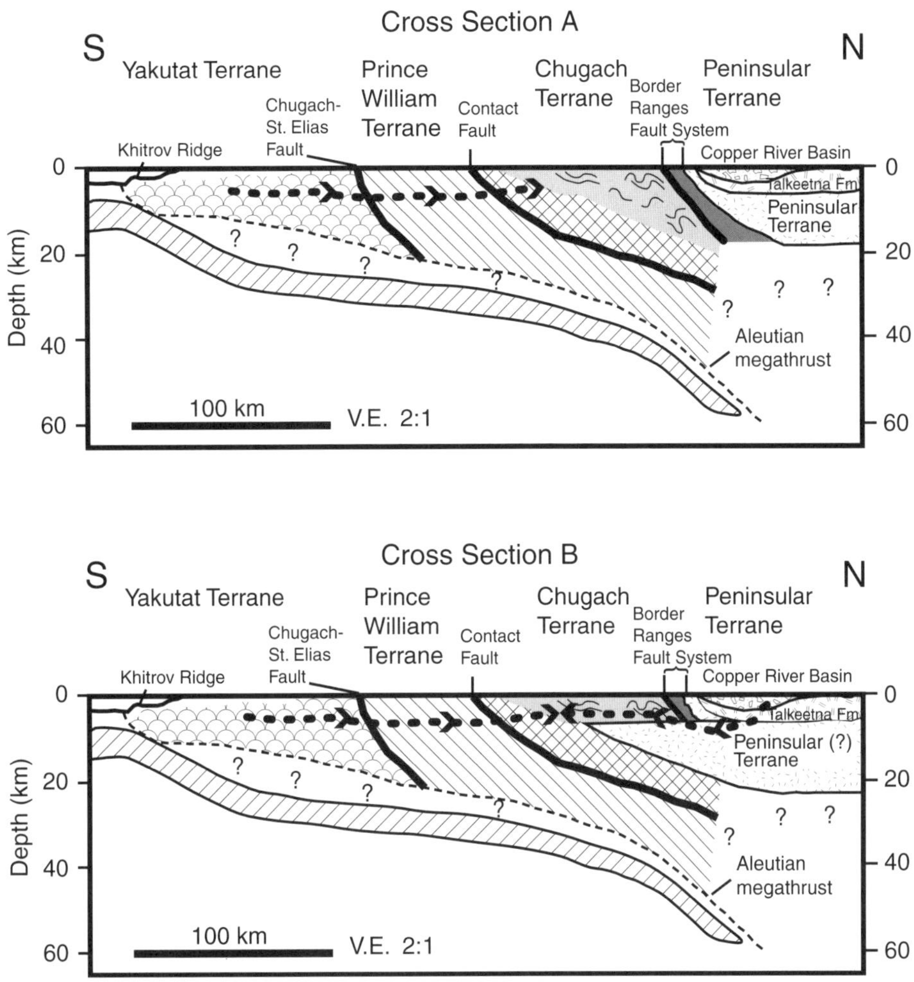

Figure 9. Two schematic cross sections from continental shelf-break in the Gulf of Alaska to the Copper River Basin constructed from geophysical and geological data from Grantz et al. (1991), Fuis et al. (1991), and Fuis and Plafker (1991), Winkler and Plafker (1993), Brocher et al. (1994) and Fuis (1998). In both cross sections, the heavy dashed lines with arrows indicate possible migration paths for hydrocarbons either from the organic matter and/or coals in the Yakutat Terrane or from the Copper River Basin into the Wolverine Run Glacier region. The Pacific and Yakutat plates are welded together and possibly subducting together underneath the Aleutian megathrust (thin dashed line). The patterns for the Chugach, Yakutat, and Peninsular Terranes are the same as Figure 1. The Pacific plate is shown with left-dipping diagonal lines. The Prince William terrane low-grade graywackes in right-dipping diagonal lines. A possible fragment of underplated Kula plate is shown in crosshatched pattern. The McHugh Formation is shown as dark gray. The Talkeetna Formation underlies the Copper River basin. Vertical exaggeration is 2:1. A: In this interpretation, the Border Ranges fault system extends beneath the Peninsular terrane. This may create a barrier to late flow of hydrocarbon-rich fluids sourced from the Copper River Basin. B: An alternative model places the Peninsular terrane beneath the Chugach terrane (see Fuis and Plafker, 1991). The geophysical data cannot be used to distinguish between these two possible cross sections.

change in cooling rate may correlate with underthrusting of the Yakutat terrane that would lead to migration of the hydrocarbons into the Wolverine Run Glacier region.

CONCLUSIONS

This study adds another piece to the puzzle related to fluid history and deformation for a portion of the western Chugach metamorphic complex. The fluid entrapment conditions derived from chlorite geothermometry and mineral assemblages combined with fluid inclusion isochores yield calculated paleogeothermal gradients from 58 to 35 °C/km. This gradient is higher than that expected for accretionary margins with subduction of cold oceanic crust. High geothermal gradients may reflect subduction of a ridge and/or creation of a slab window similar to the Miocene event in the Franciscan Complex and the Cretaceous and Tertiary events in the Shimanto accretionary prism (Laughland and Underwood, 1993; Underwood et al., 1999; Sakaguchi, 1999a, 1999b; Lewis et al., 2000). In this study area, two locations separated by 2 km underwent early deformation at disparate pressure and temperature conditions. Ductile structures such as boudinage seen in early veins indicate that vein emplacement was at temperatures and/or pressures higher than subsequent periods of vein emplacement. At one outcrop, there is a vein set that has textures that range from thin folded veins to thick brecciated veins. One explanation for the differences in these textures was a change from near-lithostatic to near-hydrostatic pressure conditions. Another explanation is that changes in temperature near the brittle to ductile transition caused the textural change. It is also possible that both mechanisms operated simultaneously to produce this vein set. By the last period of vein emplacement both outcrops were at the same pressure-temperature conditions.

These late thick veins may have remained open longer and could have developed a positive feedback loop in which the longer the vein stayed open, the more fluids flow through it, which in turn keeps the vein open even longer. Had this feedback loop not been interrupted, it may have lead to the development of a brittle ductile shear zone, and eventually to faulting (e.g., Clark et al., 1995). The earlier thinner veins did not remain open as long, and therefore did not have the time to develop the feedback loop.

The latest fluids to flow through the area were not restricted to the latest veins. Rather, late stage liquid hydrocarbon-bearing inclusions are found in all veins as well as matrix quartz, indicating that the hydrocarbon-rich fluids pervasively invaded the region. There is little evidence to suggest that these late stage fluids originated from the immediate field area. In fact, the high pressure and temperature associated with greenschist facies metamorphism prohibit preservation of original liquid hydrocarbons within the greenschist facies Wolverine Run Glacier region. Instead, the fluids must have migrated from some distance away, and may be related to the source beds of oil seeps or coals found to the south of the field area (Fig. 1). However, until geochemical "fingerprinting" is obtained, neither precise nor accurate determinations of the origins of these hydrocarbons are possible. We tentatively suggest that the fluid flow was either (1) migration of hydrocarbons from a cryptic source up to 100 km away, similar to basin scale fluid flow described for shallow level rift environments (Fleming et al., 1998; Sytla et al., 1998) and thrust terranes (Ge and Garven, 1994; Garven, 1995) or (2) migration from the underplated Yakutat terrane.

ACKNOWLEDGMENTS

Rice University Undergraduate Scholars Program and NSF EAR-9304062 and EAR-9725661 provided funding for this research. We would like to thank Terry Pavlis and John Bowman for their assistance in the field, Milton Pierson for his help on the microprobe, Ted Bence for information on the origin and maturity of southern Alaskan oil seeps, and Elizabeth Johnson for providing useful comments. Dan Orange and Michael Underwood provided very helpful reviews.

REFERENCES CITED

Ambos, E.L., Mooney, W.D., and Fuis, G.S., 1995, Seismic refraction measurements within the Peninsular terrane, south-central Alaska: Journal of Geophysical Research, v. 100, p. 4079–4095.

Anderson, R.N., Flemings, P., Losh, S., Austim, J., and Woodham, R., 1994, Gulf of Mexico fault drilled seen as oil, gas migration pathway: Oil and Gas Journal, v. 94, p. 97–104.

Armentrout, J.M., 1981, Distribution of organic carbon in the Gulf of Alaska Neogene and Holocene sedimentary record, *in* Thiede, J., and Suess, E., eds., Coastal upwelling, its sedimentary record, Part B: Sedimentary records of ancient coastal upwelling: New York, New York, Plenum Press, p. 421–452.

Audétat, A., and Günther, D., 1999, Mobility and H_2O loss from fluid inclusions in natural quartz crystals: Contributions to Mineralogy and Petrology, v. 137, p. 1–14.

Barnes, I., Plafker, G., White, L.D., and Armstrong, A.K., 1982, Potential natural gas in the Gulf of Alaska indicated by calcite depleted in carbon-13, *in* Coonrad, W.L., ed., The U.S. Geological Survey in Alaska: Accomplishments in 1980: U.S. Geological Survey Circular 844, p. 143–146.

Bence, A.E., Kvenvolden, K.A., and Kennicutt, M.C., 1996, Organic geochemistry applied to environmental assessments of Prince William Sound, Alaska, after the *Exxon Valdez* oil spill—A review: Organic Geochemistry, v. 24, p. 7–42.

Bevins, R.E., Robinson, D., and Rowbotham, G., 1991, Compositional variations in mafic phyllosilicates from regional low-grade metabasites and application of the chlorite geothermometer: Journal of Metamorphic Geology, v. 9, p. 711–721.

Bodnar, R.J., 1990, Petroleum migration in the Miocene Monterey Formation, California, USA: Constraints from fluid inclusion studies: Mineralogical Magazine, v. 54, p. 295–304.

Boiron, M.C., Essarraj, S., Sellier, E., Cathelineau, M., Lespinasse, M., and Poty, B., 1992, Identification of fluid inclusions in relation to their host microstructural domains in quartz by cathodoluminescence: Geochimica et Cosmochimica Acta, v. 56, p. 175–185.

Boullier, A.-M., 1999, Fluid inclusions: Tectonic indicators: Journal of Structural Geology, v. 21, p. 1229–1235.

Bowman, J.R., Sisson, V.B., Valley, J.W., and Pavlis, T.L., 2003, Oxygen isotope constraints on fluid infiltration associated with high-temperature–low-pressure metamorphism (Chugach metamorphic complex) within the Eocene southern Alaska forearc, *in* Sisson, V.B., Roeske, S.M., and Pavlis, T.L., eds, Geology of a transpressional orogen developed during ridge-trench interaction along the North Pacific margin: Boulder, Colorado, Geological Society of America Special Paper 371, p. 237–252 (this volume).

Bradley, D., Kusky, T., Haeussler, P., Goldfarb, R., Miller, M., Dumoulin, J., Nelson, S.W., and Karl, S., 2003, Geologic signature of early Tertiary ridge subduction in Alaska *in* Sisson, V.B., Roeske, S.M., and Pavlis, T.L., eds, Geology of a transpressional orogen developed during ridge-trench interaction along the North Pacific margin: Boulder, Colorado, Geological Society of America Special Paper 371, p. 19–49 (this volume).

Brantley, S.L., Fisher, D.M., Deines, P., Clark, M.B., and Myers, G., 1997, Segregation veins: Evidence for the deformation and dewatering of a low-grade metapelite, *in* Holness, M.B., ed., Deformation-enhanced fluid transport in the Earth's crust and mantle: Mineralogical Society Series 8, p. 267–288.

Brocher, T.M., Fuis, G.S., Fisher, M.A., Plafker, G., Moses, M.J., Taber, J.J., and Christensen, N.I., 1994, Mapping the megathrust beneath the northern Gulf of Alaska using wide-angle seismic data: Journal of Geophysical Research, v. 99, p. 11,663–11,685.

Bröcker, M., and Day, H.W., 1995, Low-grade blueschist facies metamorphism of metagreywackes, Franciscan Complex, northern California: Journal of Metamorphic Geology, v. 13, p. 61–78.

Brown, P.E., and Hagemann, S.G., 1995, MacFlinCor and its application to fluids in Archean lode-gold deposits: Geochimica et Cosmochimica Acta, v. 59, p. 3943–3952.

Bruhn, L.R., Parry, W.T., Yonkee, W.A., and Thompson, T., 1994, Fracturing and hydrothermal alteration in normal fault zones: Pure and Applied Geophysics, v. 142, p. 609–644.

Bruhn, L.R., Parry, W.T., and Bunds, M.P., 2000, Tectonics, fluid migration, and fluid pressure in a deformed forearc basin, Cook Inlet, Alaska: Geological Society of America Bulletin, v. 112, p. 550–563.

Bruns, T.R., 1985, Tectonics of the Yakutat Block, an allochthonous terrane in the northern Gulf of Alaska: U.S. Geological Survey Open-File Report 85-13, 112 p.

Burruss, R.C., 1981, Hydrocarbon fluid inclusions in studies of sedimentary diagenesis, *in* Hollister, L.S., and Crawford, M.L., eds., Short course in fluid inclusions: Applications to petrology: Mineralogical Association of Canada Short Course Handbook, v. 6, p. 138–156.

Burruss, R.C., 1991, Practical aspects of fluorescence microscopy of petroleum fluid inclusions, *in* Barker, C.E., and Kopp, O.C., eds., Luminescence microscopy and spectroscopy: Tulsa, Oklahoma, SEPM (Society for Sedimentary Geology) Short Course 25, p. 1–7.

Cande, S.C., Leslie, R.B., Parra, J.C., and Hobart, M., 1987, Interaction between the Chile Rise and Chile Trench: Geophysical and geothermal evidence: Journal of Geophysical Research, v. 92, p. 495–520.

Cathelineau, M., 1988, Cation site occupancy in chlorites and illites as a function of temperature: Clay Mineralogy, v. 23, p. 471–485.

Cathelineau, M., and Nieva, D., 1985, A chlorite solid solution geothermometer: The Los Azufres (Mexico) geothermal system: Contributions to Mineralogy and Petrology, v. 91, p. 235–244.

Clark, M.B., Brantley, S.L., and Fisher, D.M., 1995, Power-law vein thickness distributions and positive feedback in vein growth: Geology, v. 23, p. 975–978.

De Caritat, P., Hutcheon, I., and Walshe, J.L., 1991, Chlorite geothermometry: A review: Clays and Clay Minerals, v. 41, p. 219–239.

Dia, A.N., Aqulina, L., Boulègue, J., Bourgois, J., Suess, E., and Torres, M., 1993, Origin of fluids and related barite deposits at vent sites along the Peru convergent margin: Geology, v. 21, p. 1099–1102.

Dumoulin, J.A., 1987, Sandstone composition of the Valdez and Orca Groups, Prince William Sound, Alaska: U.S. Geological Survey Bulletin 1774, 37 p.

Eichhubl, P., and Boles, J.R., 2000, Rates of fluid flow in fault systems—Evidence for episodic rapid fluid flow in the Miocene Monterey Formation, Coastal California: American Journal of Science, v. 300, p. 571–600.

Evans, J.P., and Chester, F.M., 1995, Fluid-rock interaction in faults of the San Andreas system: Inferences from San Gabriel fault rock geochemistry and microstructures: Journal of Geophysical Research, v. 100, p. 13,007–13,020.

Farmer, G.L., Ayuso, R., and Plafker, G., 1993, A Coast Mountains provenance for the Valdez and Orca groups, southern Alaska, based on Nd, Sr, and Pb isotopic evidence: Earth and Planetary Science Letters, v. 116, p. 9–21.

Fisher, D.M., and Brantley, S.L., 1992, Models of quartz overgrowth and vein formation: Deformation and episodic flow in an ancient subduction zone: Journal of Geophysical Research, v. 97, p. 20,043–20,061.

Fisher, D.M., Brantley, S.L., Everett, M., and Dzvonik, J., 1995, Cyclic fluid flow through a regionally extensive fracture network within the Kodiak accretionary prism: Journal of Geophysical Research, v. 100, p. 12,881–12,894.

Fleming, C.G., Couples, G.D., and Haszeldine, R.S., 1998, Thermal effects of fluid flow in steep fault zones, *in* Jones, G., et al., eds., Faulting, fault sealing, and fluid flow in hydrocarbon reservoirs: Geological Society [London] Special Publication No. 147, p. 217–229.

Fournier, R.O., 1999, Hydrothermal processes related to movement of fluid from plastic into brittle rock in the magmatic-epithermal environment: Economic Geology, v. 94, p. 1193–1212.

Fuis, G.S., 1998, West margin of North America—A synthesis of recent seismic transects: Tectonophysics, v. 288, p. 265–292.

Fuis, G.S., and Plafker, G., 1991, Evolution deep structure along the Trans-Alaska Crustal Transect, Chugach Mountains and Copper River Basin: Journal of Geophysical Research, v. 96, p. 4229–4253.

Fuis, G.S., Ambos, E.L., Mooney, W.D., Christensen, N.I., and Geist, E., 1991, Crustal structure of accreted terranes in southern Alaska, Chugach Mountains and Copper River Basin, from seismic refraction results: Journal of Geophysical Research, v. 96, p. 4187–4227.

Garven, G., 1995, Continental-scale groundwater flow and geologic processes: Annual Reviews of Earth and Planetary Sciences, v. 24, p. 89–117.

Ge, S., and Garven, G., 1994, A theoretical model for thrust-induced deep groundwater expulsion with application to the Canadian Rocky Mountains: Journal of Geophysical Research, v. 99, p. 13,851–13,868.

Goldfarb, R.J., Leach, D.L., Miller, M.L., and Pickthorn, W.J., 1986, Geology, metamorphic setting, and genetic constraints of epigenetic lode-gold mineralization within the Cretaceous Valdez Group, south-central Alaska, *in* Keppie, J.D., et al., eds., Turbidite hosted gold deposits: Geological Association of Canada Special Paper 32, p. 87–105.

Goldstein, R.H., and Reynolds, T.J., 1994, Systematics of fluid inclusions in diagenetic minerals: Tulsa, Oklahoma, SEPM (Society for Sedimentary Geology) Short Course 31, 199 p.

Goodwin, E., Fuis, G.S., Nokleberg, W.J., and Ambos, E.L., 1989, The crustal structure of the Wrangellia terrane along the Tok Highway, eastern-southern Alaska: Journal of Geophysical Research, v. 94, p. 16,037–19,049.

Grantz, A., Moore, T.E., and Roeske, S.M., 1991, Centennial Continent-Ocean Transect A-3: Gulf of Alaska to Arctic Ocean: Boulder, Colorado, Geological Society of America, Geology of North America 15, v. TRA-A3, 72 p., 3 sheets, scale 1:500,000.

Gratier, J.-P., Renard, F., and Labume, P., 1999, How pressure solution creep and fracturing processes interact in the upper crust to make it behave in both a brittle and viscous manner: Journal of Structural Geology, v. 21, p. 1189–1197.

Grauls, D.J., and Baleix, J.M., 1994, Role of overpressured and in situ stresses in fault controlled hydrocarbon migration: a case study: Marine and Petroleum Geochemistry, v. 11, p. 734–742.

Gray, D.R., Janssen, C., and Vapnik, Y., 1999, Deformation character and paleo-fluid flow across a wrench fault within a Paleozoic subduction-accretion system: Waratah fault zone, southeastern Australia: Journal of Structural Geology, v. 21, p. 191–214.

Haeussler, P.J., Bradley, D., Goldfarb, R.J., Snee, L., and Taylor, C., 1995, Link between ridge subduction and gold mineralization in southern Alaska: Geology, v. 23, p. 995–998.

Haeussler, P.J., Bradley, D.C., and Goldfarb, R.J., 2003, Brittle deformation along the Gulf of Alaska margin in response to Paleocene-Eocene triple junction migration, *in* Sisson, V.B., Roeske, S.M., and Pavlis, T.L., eds, Geology of a transpressional orogen developed during ridge-trench interaction along the North Pacific margin: Boulder, Colorado, Geological Society of America Special Paper 371, p. 119–140 (this volume).

Hagemann, H.W., and Hollerbach, A., 1985, The fluorescent behavior of crude oils with respect to their thermal maturation and degradation: Organic Geochemistry, v. 10, p. 473–480.

Harris, N.R., Sisson, V.B., Wright, J., and Pavlis, T.L., 1996, Evidence for Eocene mafic underplating during forearc intrusive activity, eastern Chugach Mountains, Alaska: Geology, v. 24, p. 263–266.

Henry, M.E., and Donovan, T.J., 1984, Luminescence properties and chemical composition of crude oils: U.S. Geological Survey Open-File Report 84-385, 38 p.

Hirth, G., and Tullis, J., 1989, The effects of pressure and porosity on the micromechanics of the brittle-ductile transition in quartzite: Journal of Geophysical Research, v. 94, p. 17,825–17,838.

Hodgkins, M.A., and Stewart, K.G., 1994, The use of fluid inclusions to constrain fault zone pressure, temperature, and kinematic history: an example from the Alpi Apuane, Italy: Journal of Structural Geology, v. 16, p. 85–96.

Hudson, T., and Plafker, G., 1982, Paleogene metamorphism of an accretionary flysch terrain, eastern Gulf of Alaska: Geological Society of America Bulletin, v. 93, p. 1280–1290.

Hudson, T., Plafker, G., and Peterman, Z., 1979, Paleogene anatexis along the Gulf of Alaska margin: Geology, v. 7, p. 573–577.

Hunt, J.M., 1990, Generation and migration of hydrocarbons from abnormally pressured fluid compartments: American Association of Petroleum Geologists Bulletin, v. 74, p. 1–12.

Johnsson, M.J., Evans, K.R., and Marshall, H.A., 1999, Thermal maturity of sedimentary rocks in Alaska; digital resources: U.S. Geological Survey DDS-0054, one CD-ROM.

Jones, D.L., Silberling, N.J., Coney, P.J., and Monger, J.W.H., 1987, Lithotectonic terrane map of Alaska (west of the 141st meridian): U.S. Geological Survey Miscellaneous Field Studies Map MF-1847A, scale 1:2,500,000.

Kirschner, C.E., 1994, Interior basins of Alaska, *in* Plafker, G., and Berg, H.C., eds., The geology of Alaska: Boulder, Colorado, Geological Society of America, Geology of North America, v. G-1, p. 469–493.

Laughland, M.M., and Underwood, M.B., 1993, Vitrinite reflectance and estimates of paleotemperature within the Upper Shimanto Group, Muroto Peninsula, Shikoku, Japan, *in* Underwood, M.B., ed., Thermal evolution of the Tertiary Shimanto belt, southwest Japan: Boulder, Colorado, Geological Society of America Special Paper 273, p. 25–44.

Lespinasse, M., 1999, Are fluid inclusion planes useful in structural geology?: Journal of Structural Geology, v. 21, p. 1237–1243.

Lewis, J.C., Byrne, T.B., Pasteris, J.D., London, D., and Morgan, G.B., IV, 2000, Early Tertiary fluid flow and pressure-temperature conditions in the Shimanto accretionary complex of southwest Japan: Constraints from fluid inclusions: Journal of Metamorphic Geology, v. 18, p. 319–335.

Logan, J.M., Dengo, C.A., Higgs, N.G., and Wang, Z.Z., 1992, Fabrics of experimental fault zones: Their development and relationship to mechanical behavior, *in* Evans, B., and Wong, T.-F., eds., Fault mechanics and transport properties: San Diego, Academic Press, p. 33–67.

McDonald, S.J., and Skilbeck, C.G., 1996, Authigenic fluid inclusions in lithic sandstone: A case study from the Permo-Triassic Gunnedah Basin, New South Wales: Australian Journal of Earth Sciences, v. 43, p. 217–228.

McLimans, R.K., 1987, The application of fluid inclusions to migration of oil and diagenesis in petroleum reservoirs: Applied Geochemistry, v. 2, p. 585–603.

Meere, P.A., 1995, High and low density fluids in a quartz vein from the Irish Variscides: Journal of Structural Geology, v. 17, p. 435–446.

Merriman, R.J., and Peacor, D.R., 1999, Very low-grade metapelites: Mineralogy, microfabrics, and measuring reaction progress, *in* Frey, M., and Robinson, D., eds., Low-grade metamorphism: Oxford, Blackwell Scientific, p. 10–60.

Merriman, R.J., Roberts, B., Peacor, D.R., and Hirons, S.R., 1995, Strain-related differences in the crystal growth of white mica and chlorite: A TEM and XRD study of the development of metapelitic microfabrics in the Southern Uplands thrust terrane, Scotland: Journal of Metamorphic Geology, v. 13, p. 559–576.

Miller, M.L., Dumoulin, J.A., and Nelson, S.W., 1984, A transect of metamorphic rocks along the Copper River, Cordova and Valdez quadrangles: U.S. Geological Survey Circular 939, p. 52–57.

Moore, J.C., and Vrolijk, P., 1992, Fluids in accretionary prism: Reviews of Geophysics, v. 30, p. 113–135.

Moore, J.C., Byrne, T., Plumley, P.W., Reid, M., Gibbons, H., and Coe, R.S., 1983, Paleogene evolution of the Kodiak Islands, Alaska: Consequences of ridge-trench interaction in a more southerly latitude: Tectonics, v. 2, p. 265–293.

Mull, C.G., and Nelson, S.W., 1986, Anomalous thermal maturity data from the Orca Group (Paleocene and Eocene), Katalla–Kayak Island area, *in* Brasch-Winkler, S., and Reed, K.M., eds., Geologic studies in Alaska by the U.S. Geological Survey during 1985: U.S. Geological Survey Circular 978, p. 50–55.

Nelson, S.W., Dumoulin, J., and Miller, M.L., 1985, Geologic map of the Chugach National Forest, Alaska: U.S. Geological Survey Miscellaneous Field Studies Map MF-1645B, 16 p., 1 plate, scale 1:250,000.

Newman, J., and Mitra, G., 1994, Fluid-induced deformation and recrystallization of dolomite at low temperatures along a natural fault zone, Mountain City window, Tennessee: Geological Society of America Bulletin, v. 106, p. 1267–1280.

Newton, R.C., and Manning, C.E., 2000, Quartz solubility in H_2O-NaCl and H_2O-CO_2 solutions at deep crust-upper mantle pressures and temperatures: 2–15 kbar and 500–900 °C: Geochimica et Cosmochimica Acta, v. 64, p. 2993–3005.

O'Hara, K., and Haak, A., 1992, A fluid inclusion study of pressure and salinity variations in the footwall of the rector branch thrust, North Carolina, USA: Journal of Structural Geology, v. 14, p. 579–589.

Ohlmacher, G.C., and Aydin, A., 1997, Mechanics of vein, fault, and solution surface formation in the Appalachian Valley and Ridge, northeast Tennessee, USA: Implications for fault friction, state of stress and fluid pressure: Journal of Structural Geology, v. 19, p. 927–944.

Orange, D.L., Geddes, D.S., and Moore, J.C., 1993, Structural and fluid evolution of a young accretionary complex; the Hoh rock assemblage of the western Olympic Peninsula, Washington: Geological Society of America Bulletin, v. 105, p. 1053–1075.

Ortoleva, P., Al-Shaieb, Z., and Puckette, J., 1995, Genesis and dynamics of basin compartments and seals: American Journal of Science, v. 295, p. 345–427.

Pavlis, T.L., and Sisson, V.B., 1995, Structural history of the Chugach metamorphic complex in the Tana River region, eastern Alaska: A record of Eocene ridge subduction: Geological Society of America Bulletin, v. 107, p. 1333–1355.

Pavlis, T.L., and Sisson, V.B., 2003, Development of a subhorizontal decoupling horizon in a transpressional system, Chugach metamorphic complex, Alaska: Evidence for rheological stratification of the crust, *in* Sisson, V.B., Roeske, S.M., and Pavlis, T.L., eds, Geology of a transpressional orogen developed during ridge-trench interaction along the North Pacific margin: Boulder, Colorado, Geological Society of America Special Paper 371, p. 191–216 (this volume).

Pavlis, T.L., Marty, K., and Sisson, V.B., 2003, Constrictional flow within the Eocene forearc of southern Alaska: An effect of dextral shear during ridge subduction, *in* Sisson, V.B., Roeske, S.M., and Pavlis, T.L., eds, Geology of a transpressional orogen developed during ridge-trench interaction along the North Pacific margin: Boulder, Colorado, Geological Society of America Special Paper 371, p. 171–190 (this volume).

Plafker, G., 1971, Pacific margin Tertiary basin, *in* Cram, I.H., ed., Petroleum provinces in North America: American Association of Petroleum Geologists Memoir 15, p. 120–135.

Plafker, G., 1987, Regional geology and petroleum potential of the northern Gulf of Alaska continental margin, *in* Scholl, D.W., et al., eds., Geology and resource potential of the continental margin of western North America and adjacent ocean basins: Houston, Texas, Circum-Pacific Council for Energy and Mineral Resources, Earth Science Series, v. 6, p. 229–268.

Plafker, G., Jones, D.L., and Pessagno, E.A., Jr., 1977, A Cretaceous accretionary flysch and melange terrane along the Gulf of Alaska margin, *in* Blean, K.M., ed., The United States Geological Survey in Alaska: Accomplishments during 1976: U.S. Geological Survey Circular 751-B, p. B41–B43.

Plafker, G., Nokleberg, W.J., and Lull, J.S., 1989, Bedrock geology and tectonic evolution of the Wrangellia, Peninsular, and Chugach terranes along the Trans-Alaska Crustal Transect in the Chugach Mountains and southern Copper River Basin, Alaska: Journal of Geophysical Research, v. 94, p. 4255–4295.

Plafker, G., Moore, J.C., and Winkler, G.R., 1994, Geology of the southern Alaska margin, *in* Plafker, G., and Berg, H.C., eds., The geology of Alaska: Boulder, Colorado, Geological Society of America, Geology of North America, v. G-1, p. 451–467.

Pouchou, J.L., and Pichoir, F., 1985, PAP procedure for improved quantitative microanalysis, *in* Armstrong, J.T., ed., Microbeam analysis: San Francisco, California, San Francisco Press, p. 88–97.

Robert, F., Boullier, A.-M., and Firaous, K., 1995, Gold-quartz veins in metamorphic terrains and their bearing on the role of fluids in faulting: Journal of Geophysical Research, v. 100, p. 12,861–12,879.

Roedder, E., 1984, Fluid inclusions: Reviews in Mineralogy 12, 644 p.

Sakaguchi, A., 1996, High paleothermal gradient with ridge subduction beneath the Cretaceous Shimanto accretionary prism, southwest Japan: Geology, v. 24, p. 795–798.

Sakaguchi, A., 1999a, Thermal maturity in the Shimanto accretionary prism, southwest Japan, with the thermal change of the subducting slab: Fluid inclusion and vitrinite reflectance study: Earth and Planetary Science Letters, v. 173, p. 61–74.

Sakaguchi, A., 1999b, Thermal structure and paleo-heat flow in the Shimanto accretionary prism, southwest Japan: The Island Arc, v. 8, p. 350–372.

Schmidt, D., Livi, K.J.T., and Frey, M., 1999, Reaction progress in chloritic material: An electron microbeam study of the Taveyanne graywacke, Switzerland: Journal of Metamorphic Geology, v. 17, p. 229–241.

Shelton, K.L., Underwood, M.B., Burstein, I.B., Haeussler, G.T., and Howell, D.G., 1996, Stable-isotope and fluid-inclusion studies of hydrothermal quartz and calcite veins from the Kandik thrust belt of east-central Alaska: Implications for thermotectonic history and terrane analysis, *in* Johnsson, M.J., and Howell, D.G., eds., Thermal evolution of sedimentary basins in Alaska: U.S. Geological Survey Bulletin B2142, p. 111–131.

Shibue, Y., 1996, Empirical expressions of quartz solubility in H_2O, H_2O+CO_2, and H_2O+NaCl fluids: Geochemical Journal, v. 30, p. 339–354.

Sibson, R.H., 1990, Conditions for fault valve behavior, *in* Knipe, R.H., and Rutter, E.H., eds., Deformation mechanisms, rheology, and tectonics: Geological Society [London] Special Publication No. 54, p. 15–28.

Sibson, R.H., Robert, F., and Poulsen, K.H., 1988, High angle reverse faults, fluid pressure cycling, and mesothermal gold-quartz deposits: Geology, v. 16, p. 551.

Sisson, V.B., and Hollister, L.S., 1988, Low-pressure facies series metamorphism in an accretionary sedimentary prism, southern Alaska: Geology, v. 16, p. 358–361.

Sisson, V.B., and Pavlis, T.L., 1993, Geologic consequences of plate reorganization: An example from the Eocene southern Alaska forearc: Geology, v. 21, p. 913–916.

Sisson, V.B., Hollister, L.S., and Onstott, T.C., 1989, Petrologic and age constraints on the origin of a low-pressure/high-temperature metamorphic complex, southern Alaska: Journal of Geophysical Research, v. 94, p. 4392–4410.

Sisson, V.B., Poole, A.R., Harris, N.R., Cooper Burner, H., Pavlis, T.L., Copeland, P., Donelick, R.A., and McLelland, W.C., 2003, Geochemical and geochronologic constraints for genesis of a tonalite-trondhjemite suite and associated mafic intrusive rocks in the eastern Chugach Mountains, Alaska: A record of ridge-transform subduction, *in* Sisson, V.B., Roeske, S.M., and Pavlis, T.L., eds, Geology of a transpressional orogen developed during ridge-trench interaction along the North Pacific margin: Boulder, Colorado, Geological Society of America Special Paper 371, p. 293–326 (this volume).

Smith, C.S., Sisson, V.B., Avé Lallemant, H.G., and Copeland, P., 1999, Two contrasting *P-T-t* paths in the Villa de Cura blueschist belt, Venezuela: A record of ~97 Ma subduction in the Caribbean: Geological Society of America Bulletin, v. 111, p. 831–848.

Smith, M.P., and Yardley, B.W.D., 1999, Fluid evolution during metamorphism of the Otago Schist, New Zealand: (I) Evidence from fluid inclusions: Journal of Metamorphic Geology, v. 17, p. 173–186.

Stasiuk, L.D., and Snowden, L.R., 1997, Fluorescence microscopy of synthetic and natural hydrocarbon fluid inclusions: Crude oil chemistry, density and application to petroleum migration: Applied Geochemistry, v. 12, p. 229–241.

Stöckhert, B., Brix, M.R., Kleinschrodt, R., Hurford, A.J., and Wirth, R., 1999, Thermochronometry and microstructures of quartz—A comparison with experimental flow laws and predictions on the temperature of the brittle-plastic transition: Journal of Structural Geology, v. 21, p. 351–360.

Stüwe, K., 1986, Structural evolution of the Port Wells mining district, Prince William Sound, south-central Alaska: Implications for the origin of the gold lodes: Mineralium Deposita, v. 21, p. 288–295.

Sytla, O., Pedersen, J.I., and Hamborg, M., 1998, On the vertical and lateral distribution of hydrocarbon migration velocities during secondary migration, *in* Parnell, J., ed., Dating and duration of fluid flow and fluid-rock interaction: Geological Society [London] Special Publication No. 144, p. 221–232.

Underwood, M.B., Shelton, K.L., McLaughlin, R.J., Laughland, M.M., and Solomon, R.M., 1999, Middle Miocene paleotemperature anomalies within the Franciscan Complex of northern California: Thermo-tectonic responses near the Mendocino triple junction: Geological Society of America Bulletin, v. 111, p. 1448–1467.

Vityk, M.O., and Bodnar, R.J., 1995, Textural evolution of synthetic fluid inclusions in quartz during reequilibration, with application to tectonic reconstruction: Contributions to Mineralogy and Petrology, v. 121, p. 309–323.

Vrolijk, P., 1987, Tectonically driven fluid flow in the Kodiak accretionary complex, Alaska: Geology, v. 15, p. 466–469.

Wilkinson, J.J., Lonergan, L., Fairs, T., and Herrington, R.J., 1998, Fluid inclusion constraints on conditions and timing of hydrocarbon migration and quartz cementation in Brent Group reservoir sandstones, Columba Terrace, northern North Sea, *in* Parnell, J., ed., Dating and duration of fluid flow and fluid-rock interaction: Geological Society [London] Special Publication No. 144, p. 69–90.

Winkler, G.R., and Plafker, G., 1993, Geologic map of the Cordova and Middleton Island quadrangles, southern Alaska: U.S. Geological Survey Miscellaneous Investigations Series Map I-1984, scale 1:250,000.

Wyder, R.F., and Mullis, J., 1998, Fluid impregnation and development of fault breccias in the Tavetsch basement rocks (Sedrun, Central Swiss Alps): Tectonophysics, v. 294, p. 89–107.

Xie, X., Byerly, G.R., and Ferrell, R.E., Jr., 1997, IIb trioctahedral chlorite from the Barberton greenstone belt: Crystal structure and rock composition constraints with implications to geothermometry: Contributions to Mineralogy and Petrology, v. 126, p. 275–291.

Zane, A., Sassi, F.P., and Sassi, R., 1996, New data on chlorite as petrogenetic indicator mineral in metamorphic rocks: Plinius, v. 16, p. 212–213.

Zhang, Y.-G., and Frantz, J.D., 1987, Determination of the homogenization temperatures and densities of supercritical fluids in the system NaCl-KCl-H_2O using synthetic fluid inclusions: Chemical Geology, v. 64, p. 335–350.

Manuscript Accepted by the Society February 5, 2003

Printed in the USA

Geological Society of America
Special Paper 371
2003

Oxygen isotope constraints on fluid infiltration associated with high-temperature–low-pressure metamorphism (Chugach metamorphic complex) within the Eocene southern Alaska forearc

John R. Bowman*
Department of Geology and Geophysics, University of Utah, Salt Lake City, Utah 84112, USA

Virginia B. Sisson
Department of Earth Science, MS-126, Rice University, Houston, Texas 77005-1892, USA

John W. Valley
Department of Geology and Geophysics, University of Wisconsin, Madison, Wisconsin 53706, USA

Terry L. Pavlis
Department of Geology and Geophysics, University of New Orleans, New Orleans, Louisiana 70148, USA

ABSTRACT

The Chugach metamorphic complex is a high-temperature–low-pressure metamorphic belt developed in the Eocene forearc of southern Alaska. A transect along Tana Glacier shows $^{18}O/^{16}O$ depletion of quartz with increasing metamorphic grade in both rock matrix and associated veins. These $^{18}O/^{16}O$ depletions are too large to result from closed system increase in metamorphic grade (temperature). Therefore, this indicates exchange of the metasedimentary rocks with an external reservoir of lower $\delta^{18}O$, hot aqueous fluid and/or felsic magma, at some stage(s) in the thermal and deformational history of the Chugach metamorphic complex. The concordant decrease in $\delta^{18}O$ values of quartz from both rock matrix and associated veins indicates that the fluid infiltration occurred prior to or during formation of the quartz veins. In turn, this fluid infiltration must have preceded the deformation of the quartz veins. The metasediments of the Chugach metamorphic complex experienced at least two stages of fluid infiltration and associated $^{18}O/^{16}O$ depletion. The first stage occurred prior to or during the second deformation (D_2), which was near the thermal peak for intermediate-grade zones (andalusite, staurolite). The second stage occurred after D_2 but prior to the third deformation (D_3) and before the thermal peak of metamorphism for the high-grade core (sillimanite zone schists and gneiss).

It is unlikely that this pattern of $^{18}O/^{16}O$ depletion is the result of a single, geometrically simple fluid flow system operating at the scale of the exposed metamorphic section at Tana Glacier and at the peak of metamorphism. Evidence against such simple flow scenarios includes (1) structural evidence for multiple phases of deformation; (2) textural relationships among deformation fabrics, multiple generations of low-$\delta^{18}O$ quartz veins, and metamorphic index minerals; and (3) constraints on the

*jrbowman@mines.utah.edu

Bowman, J.R., Sisson, V.B., Valley, J.W., and Pavlis, T.L., 2003, Oxygen isotope constraints on fluid infiltration associated with high-temperature–low-pressure metamorphism (Chugach metamorphic complex) within the Eocene southern Alaska forearc, *in* Sisson, V.B., Roeske, S.M., and Pavlis, T.L., eds., Geology of a transpressional orogen developed during ridge-trench interaction along the North Pacific margin: Boulder, Colorado, Geological Society of America Special Paper 371, p. 237–252.

patterns of isotopic shifts in flow systems based on transport theory. Instead, evidence suggests that the δ¹⁸O-distance profile is more likely the result of more than one generation of fluid infiltration occurring at different times, places, and orientations during the prograde metamorphic and structural development of the Tana Glacier section. General oxygen isotope data sets such as those presented here for the Tana Glacier area and those previously published for other areas may not uniquely constrain the flow geometry responsible for the depletion patterns. These data sets also need to be integrated with structural and petrological studies in order to constrain the timing of fluid flow with respect to deformation and thermal history of the metamorphic belt.

The general lowering of $\delta^{18}O$ in the intermediate- to high-grade rocks of the Chugach metamorphic complex was caused by fluid infiltration (recorded by injection of quartz veins) prior to the thermal peak of metamorphism. These fluids were most likely exchanged with hot igneous intrusions. Thus, the detailed structural, thermal, and oxygen isotope record of these rocks indicates that magmatic heating is most likely responsible for this episode of high-temperature, low-pressure, Buchan metamorphism.

Keywords: fluid infiltration, ridge subduction, Chugach metamorphic complex, oxygen isotope transport, open system, $^{18}O/^{16}O$ depletion, quartz veins, Buchan metamorphism.

INTRODUCTION

Stable isotope studies of a number of Buchan and Barrovian metamorphic belts document a significant decrease in $\delta^{18}O$ values of whole rock and constituent minerals with increase in metamorphic grade (Garlick and Epstein, 1967; Rye et al., 1976; Wickham and Taylor, 1985; Criss and Fleck, 1986; Baker et al., 1989; Stern et al., 1992; Baker and Matthews, 1995; Cartwright et al., 1995; Oliver et al., 1998). In all of these areas, $^{18}O/^{16}O$ depletions exceed the effects of prograde dehydration reactions; these depletion trends require the infiltration of external, relatively lower-$\delta^{18}O$ fluids into these metamorphic terranes at some point in their metamorphic history. However, definitive evidence constraining the timing of fluid infiltration with respect to deformation history and the thermal peak of metamorphism is generally lacking.

The Chugach metamorphic complex is a belt of high-temperature/low-pressure (Buchan-style) metasedimentary rocks (principally argillite and graywacke protoliths) located within the Eocene forearc of southern Alaska in the eastern Chugach Mountains. Previous studies (Hudson et al., 1979; Hudson and Plafker, 1982; Sisson et al., 1989; Sisson and Pavlis, 1993) have shown that this amphibolite facies metamorphic belt is associated in space and time with subduction and intrusion of felsic magmas into the forearc, a thermally anomalous situation. Sisson et al. (1989) and Sisson and Pavlis (1993) have proposed that the heating responsible for the Chugach metamorphic complex results from intrusion of these felsic magmas, whose generation in turn occurred at deeper levels of the crust and is related to plate reorganization events and ridge subduction in the Eocene. These authors also proposed that multiple generations of quartz veins throughout much of the Chugach metamorphic complex—in places making up as much as 5 vol% of an outcrop—may indicate active fluid infiltration of the complex, and that this infiltration may have played some role in the heating of the complex.

Field structural studies by Pavlis and Sisson (1995, this volume, Chapter 8) have documented in detail the deformation history of the Chugach metamorphic complex. This framework clearly constrains the timing of formation of abundant quartz veins found in the Chugach metamorphic complex with respect to deformation stages (see below). As a result, there is an unusually good opportunity to utilize isotopic evidence for fluid infiltration (from the $\delta^{18}O$ values of quartz veins and host rocks) to constrain clearly the timing of any fluid infiltration in the Chugach metamorphic complex within its thermotectonic evolution.

Our objectives for this study were to evaluate oxygen isotope evidence for fluid infiltration and to constrain its timing. We focused on a transect along the Tana Glacier and Tana River in the central Chugach metamorphic complex. Specifically, we used oxygen isotope analyses of quartz separates from rock matrix and associated quartz veins to document progressive $^{18}O/^{16}O$ depletion with increasing grade. This $^{18}O/^{16}O$ depletion is in excess of that expected from closed-system effects, and indicates that metasediments were infiltrated by fluids relatively depleted in $^{18}O/^{16}O$. We also present textural evidence for timing of vein formation with respect to major deformation stages and the thermal peak of metamorphism. This in turn constrains the timing of the $^{18}O/^{16}O$ depletion of the rock matrix.

REGIONAL GEOLOGY

The Chugach metamorphic complex is developed in a relatively homogeneous sequence of interbedded metagraywackes and metaargillites of the Late Cretaceous Valdez Group (Hudson et al., 1979; Plafker et al., 1989). The Valdez Group is part of the Chugach Terrane of southern Alaska, which is bounded structurally on the north and south by the Border Ranges and Contact faults, respectively (Fig. 1). The Chugach metamorphic complex is a linear belt of high-temperature/low-pressure, amphibolite facies metamorphic rocks developed parallel to these faults and to the trend of the Chugach Terrane.

A well-exposed transect of the Chugach metamorphic complex occurs along the east edge of the Tana Glacier and Tana River (Fig. 1). Figure 2 is a simplified map of this Tana Glacier section showing locations of samples used for isotopic analysis, metamorphic isograds, and a summary of the deformational features in the section (see below). Field, structural, petrologic, and geochronologic studies in this area, augmented by observations in other areas of the Chugach metamorphic complex, form the basis for the following summary (Table 1) of the deformational and metamorphic history of the complex (Sisson and Hollister, 1988; Sisson et al., 1989; Sisson and Pavlis, 1993; Pavlis and Sisson, 1995, this volume, Chapter 8; Harris et al., 1996).

Ductile Deformation

The Tana Glacier area underwent three phases of ductile deformation (Sisson and Pavlis, 1993; Pavlis and Sisson, 1995) (Table 1; Fig. 2). The first phase (D_1) involved formation of a layer-parallel fabric (S_1), which strikes approximately E-W and dips steeply north. This fabric is thought to reflect the regional deformation processes associated with accretion of the Chugach Terrane. It developed well before the passage of the Kula–Farallon–North American triple junction—and before any thermal effects from that passage (Sisson and Pavlis, 1993). This deformation is observed in the Slide Pond region about 15 km northeast (Pavlis and Sisson, 1995). The second deformation phase (D_2) involved development of shallow-dipping ("flat") foliation (S_2) and E-W trending stretching and lineation (L_2). Finite strain increases to the south (i.e., up-grade). Sisson and Pavlis (1993) interpret D_2 to record vertical crustal shortening and horizontal, orogen-parallel (e.g., approximately E-W) elongation. In the northern Tana Glacier transect, there is only layer-parallel shortening and no conspicuous F_2 folds; this region is identified as Domain 2 (Fig. 2).

Andalusite-bearing quartz veins in the andalusite and staurolite zones are deformed by D_2 (see below), so thermal peak in these intermediate-grade zones (lower amphibolite facies) is pre- to syn-D_2. However, static overgrowth of sillimanite on D_2 fabric indicates that D_2 developed prior to peak metamorphism—is pre-metamorphic peak—at higher grades (upper amphibolite facies sillimanite schist and gneiss) in the Chugach metamorphic complex. Hence, thermal peak of metamorphism was not synchronous in time for the various zones of the amphibolite facies that define the Chugach metamorphic complex itself (Table 1). $^{40}Ar/^{39}Ar$ dating (Sisson et al., 1989) indicates that the metamorphic peak in the Chugach metamorphic complex occurred prior to 52 Ma (age of hornblende Ar blocking temperature), and that the complex cooled rapidly between 52 and 50 Ma to the biotite Ar blocking temperature. This rapid cooling is also inferred from U/Pb zircon and fission track geochronology on associated intrusives near Wernicke Glacier, west of this region (Sisson et al., this volume, Chapter 13).

The third deformation phase (D_3) involved production of a spaced crenulation cleavage (S_3) by folding of S_2 fabric into steeply inclined and shallowly plunging folds (F_3), with fold axes striking approximately E-W. Domain 3 along the Tana Glacier transect begins where prominent F_2 folds are evident as well as D_3 structures. The intensity of D_3 deformation increases to the south (i.e., up-grade) across the schist-gneiss transition zone (Fig. 2; also see below), culminating in a penetrative S_3 foliation within the gneiss. This has been identified as Domain 4. D_3 appears to

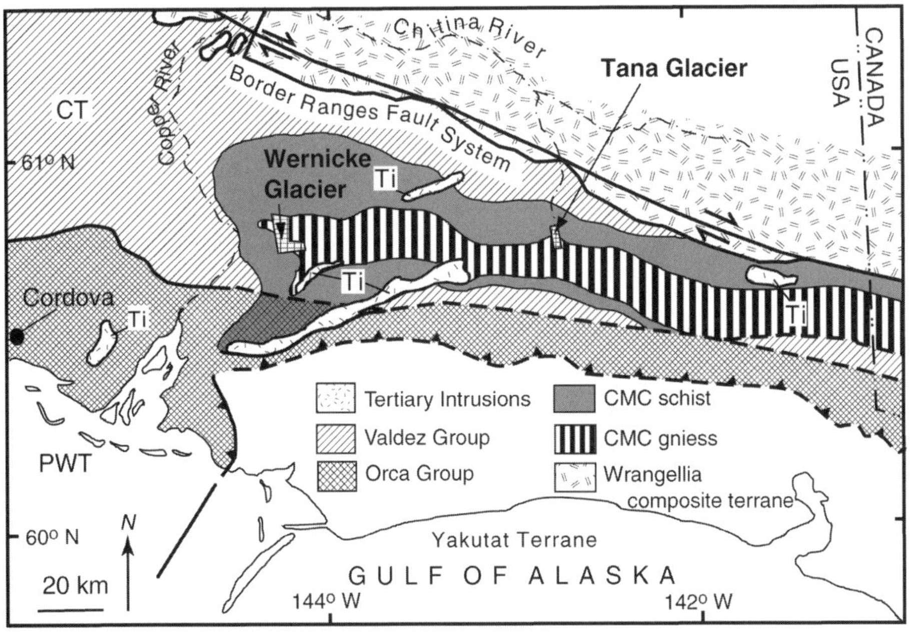

Figure 1. Regional geologic map of the Chugach metamorphic complex within the Chugach Terrane, modified from Hudson and Plafker (1982). Locations of the Tana Glacier and Wernicke Glacier areas within the Chugach metamorphic complex are noted. Abbreviations used are Ti—Tertiary intrusion, CT—Chugach terrane, PWT—Prince William terrane, and CMC—Chugach metamorphic complex.

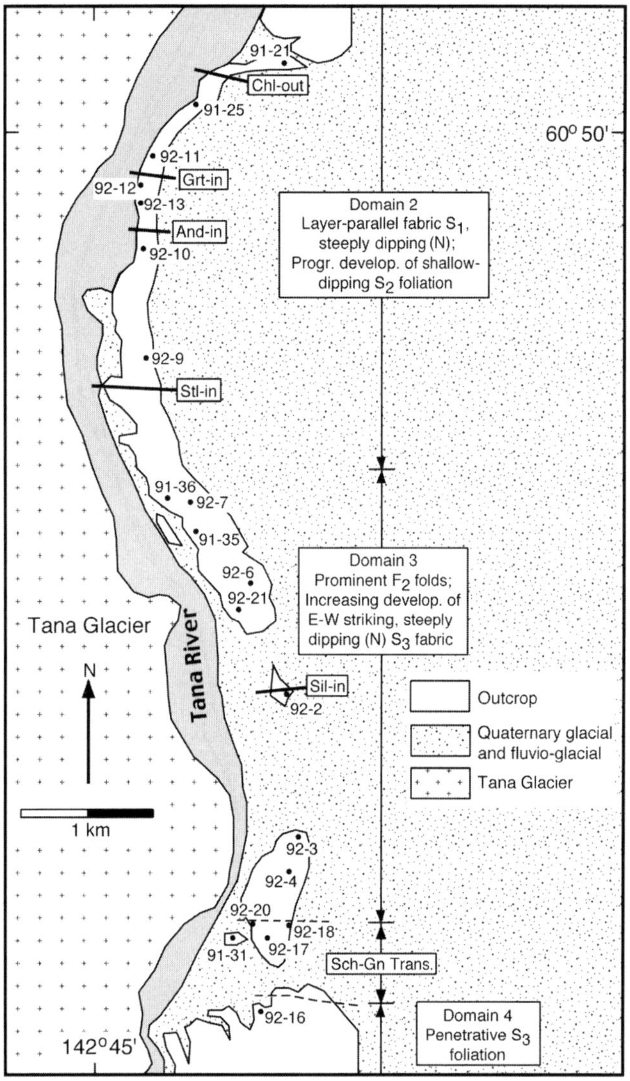

Figure 2. Simplified map of the Tana Glacier section showing locations of metamorphic isograds, samples used in this study, and domains of deformation in the Chugach metamorphic complex. Abbreviations: And—andalusite; Chl—chlorite; Grt—garnet; Sil—sillimanite; Stl—staurolite; Sch-Gn Trans.—schist-gneiss transition. The dashed lines on the south end of the map mark the schist-gneiss transition zone. Adapted from Pavlis and Sisson (1995).

record a distinct subhorizontal shortening event. It is synchronous with, but ceases before the conclusion of felsic magma intrusion (see below) and peak metamorphism in the high-grade gneiss core of the Chugach metamorphic complex (Table 1).

Metamorphic Petrology

Metamorphic grade increases from greenschist facies (Bi zone) to upper amphibolite facies (Sil zone) southward over a lateral distance of <7 km. The observed sequence of isograds southward is chlorite (Chl)-out, garnet (Grt)-in, andalusite (And)-in, staurolite (Stl)-in, sillimanite/fibrolite (Sil)-in, and Stl-out (Fig. 2). The initial amphibolite facies index mineral (garnet) appears to be superimposed on D_1 fabric, indicating that amphibolite facies conditions followed earlier regional development of greenschist facies and D_1. Accompanying the increase in grade is an increase in grain size of the schists, and ultimately, development of gneiss in the high-grade core of the Chugach metamorphic complex (Fig. 2). Garnet-biotite Fe/Mg exchange geothermometry and garnet-aluminosilicate-plagioclase (GASP) geobarometry indicate temperatures of approximately 450 °C in the upper Bi zone at the north end of the Tana Glacier section, and maximum conditions of 650 °C and 250 to 300 MPa in the high-grade gneiss core at its southern end (Sisson et al., 1989; V.B. Sisson, 2002, personal commun.).

The transition from schist to gneiss (Fig. 2) occurs over a relatively short interval where felsic sills of various sizes intrude along foliation and bedding planes and a gneissic fabric (S_3) is developed. The transition from schist to gneiss is mapped as complete in the field when metaargillite can no longer be distinguished from metagraywacke. There is outcrop evidence for fine-scale injection of felsic (tonalite) magma along foliation/bedding planes. Magma injection clearly plays a role in the formation of the high-grade gneiss core at Tana Glacier, but recent field work by V.B. Sisson indicates that some of the gneiss development is probably the result of pressure solution/diffusion mechanisms as well. In contrast, "in situ" melting is probably a subordinate means of formation of these gneisses. Mineralogically, this schist-to-gneiss transition involves increases in the amounts of quartz and plagioclase and a decrease in the amount of biotite.

The development of gneiss is spatially related to igneous intrusions within the Chugach metamorphic complex. The amount of igneous intrusion varies throughout the complex, locally reaching 10–20 vol% in the gneiss core. Within the Tana Glacier section shown in Figure 2, igneous intrusions are principally sills and dikes of modest to small thickness (≤2 m). There is more extensive igneous intrusion, and correspondingly more extensive development of associated migmatitic gneiss, in the Wernicke Glacier area in the western part of the Chugach metamorphic complex (map location in Fig. 1).

In general, the orientation of isograd surfaces throughout the Chugach metamorphic complex is poorly known; however, in the Tana Glacier area, the intermediate- to high-grade isograds are steeply dipping. Isograd sequence and petrologic geothermometry suggest a horizontal temperature change of approximately 200 °C over <7 km. This change could be the result either of a large horizontal temperature gradient or of significant crustal thinning. As limited geobarometry indicates little or only modest increase in pressure over the Tana Glacier section, significant crustal thinning (in the vertical component) is unlikely; hence, the isograd sequence suggests the existence of significant horizontal thermal gradients associated with peak metamorphism in the Chugach metamorphic complex. It is unknown if structural effects such as D_3 oblique dextral shear or a D_3 anticlinorium may have enhanced the apparent thermal gradient.

TABLE 1. RELATIONSHIPS IN THE EOCENE BETWEEN DEFORMATIONAL EVENTS, QUARTZ VEINS, METAMORPHIC THERMAL PEAK, AND IGNEOUS INTRUSION IN THE CHUGACH METAMORPHIC COMPLEX, ALASKA

Deformation stage	Generation of quartz vein*	Metamorphic thermal peak	Igneous intrusion
	(1) Pre-D_1		
D_1			Ti1 (syn-D_1)
	(2) Post-D_1 to **pre-D_2**		
D_2	(3) Pre-D_2 to **syn-D_2**	Pre-D_2 to syn-D_2 for intermediate grades (And, St zones)	Ti2 (syn-D_2)
	(4) Post-D_2 to **pre-D_3**		
D_3	(5) Syn- D_3 to late-D_3	Syn-D_3 to post-D_3 for high grades (Sil zone, Gneiss)	Ti3a (syn-D_3) Ti3b (late-D_3 to post-D_3)
			Ti4 (post-D_3)

Note: And—Andalusite, St—staurolite, Sil—sillimanite.

*The three most commonly occurring generations of quartz veins (2, 3, and 4) are discussed in the text using the shortened titles highlighted in **bold type**.

Quartz Veins

Quartz veins occur throughout the Chugach metamorphic complex; veins are ubiquitous at low to intermediate grades but less so in the highest-grade gneisses. In the former, veins are locally abundant, reaching 5 vol% in individual outcrops. Crosscutting relationships between metamorphic structures and veins indicate that multiple generations of quartz veins exist within the Chugach metamorphic complex (see also Weinberger and Sisson, this volume, Chapter 9; Pavlis et al., this volume, Chapter 7). Sometimes two generations are present in single outcrops. Veins are usually deformed to some extent. In their present, deformed state, quartz veins vary in thickness from a few millimeters to as much as 5–10 cm. Highly deformed quartz veins are usually isoclinally folded and often boudinaged, so their original length and thickness are not known. Relatively undeformed veins have lengths varying from a few centimeters to the lengths of outcrops (meters).

Although the veins are dominantly quartz, plagioclase is usually present, and its abundance increases at higher metamorphic grades. Detailed vein mineralogy varies with grade. Greenschist facies veins often contain calcite in addition to quartz. Veins within the amphibolite facies metasediments sometimes contain aluminosilicate minerals. With increasing grade, veins can contain andalusite (And), And + sillimanite (Sil), and Sil. Veins in gneiss were not observed to contain aluminosilicate minerals.

The timing of vein formation bears on the timing for observed shifts in the $\delta^{18}O$ values of the metasediments (see discussion below). Accordingly, we utilize crosscutting relationships to establish the age of veins relative to the three major phases of deformation and thermal peak of metamorphism (Table 1). This analysis shows that multiple generations of veins occur in the Chugach metamorphic complex, most of which formed prior to D_3 deformation and prior to the attainment of peak temperatures (approximately 650 °C) in the high-grade core (sillimanite zone gneiss) of the complex. Hence, several stages of vein formation—and any episodes of fluid infiltration associated with these stages of vein formation—occurred prior to D_3 and the thermal peak of metamorphism in the high-grade core.

Veins occur as early as D_1 associated with accretion of flysch (Valdez Group) but are found only in the lowest grade (chlorite, biotite zones) in the Tana Glacier transect. Figure 3A illustrates an outcrop in the north end of the field area that has been affected by D_2; this outcrop contains two generations of quartz veins. The earlier set of veins (I) in this outcrop is parallel to F_1 foliation and compositional banding, but is now deformed within shallow-dipping and overturned, isoclinal folds. These folds are the result of D_2; hence, these veins are post-D_1 and pre-D_2. This vein generation is common in low-grade rocks and is referred to in subsequent discussions as pre-D_2. The later set of quartz veins (II) in Figure 3A is undeformed and aligned parallel to the axial planes of the D_2 folds and is interpreted to be syn- to post-D_2.

There is additional evidence for the development of quartz veins before D_2. Figure 3B shows a syn-D_2 igneous sill at site 92-18 that contains an inclusion of schist with quartz veins that do not extend into the sill. The quartz veins must then be pre- to syn-D_2. Figure 3C shows "blobby," deformed quartz veins containing pink andalusite at site 91-36 in the staurolite zone (Fig. 2). These veins are deformed subparallel to the flat-lying F_2 foliation, and are early- to syn-D_2. This generation of quartz veins is common in intermediate-grade zones and will be referred to as syn-D_2 in subsequent discussion. The presence of andalusite in these syn-D_2 veins (Fig. 3D) indicates that in these intermediate-grade zones (andalusite, staurolite) of the Chugach metamorphic complex, thermal peak ($T \geq 450$ to 475 °C) is also achieved prior to the conclusion of D_2 deformation (Table 1).

In the high-grade schists (sillimanite zone), the schist-gneiss transition zone, and gneiss, veins developed prior to D_3 (Table 1). In some cases, textural evidence indicates that these veins are post-D_2. Figure 3E illustrates folded quartz veins in an outcrop of schist (metaargillite) near site 91-31 within the transition zone to gneiss. These quartz veins are concordant with F_2 foliation. The

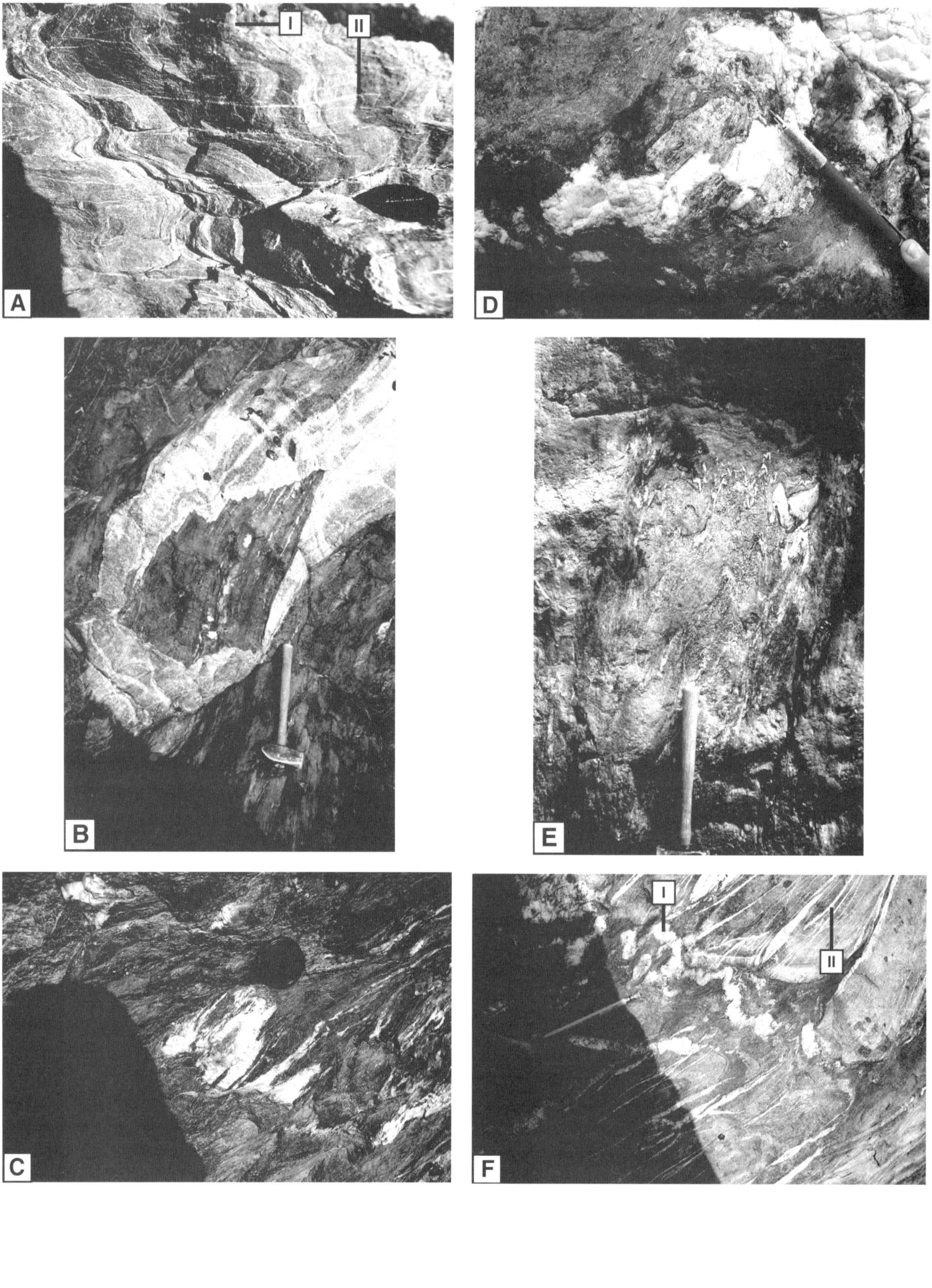
I
II
A
D
B
E
I
II
C
F

Figure 3. A: Photograph of an outcrop in the north end of the field area that has been affected by phase two (D_2) deformation; this outcrop contains two generations of quartz veins. The earlier set of veins (I) in this outcrop is parallel to F_1 foliation and compositional banding, and is now deformed within overturned, isoclinal fold forms characterized by approximately E-W axial planes that are shallowly dipping to the south. These folds are the result of D_2 deformation; hence, these veins are pre-D_2. The later set of quartz veins (II) is undeformed and aligned parallel to the axial planes of the D_2 folds and is interpreted to be syn- to post-D_2. B: Photograph of an igneous sill at site 92-18 (Fig. 2), interpreted to be syn-D_2 (Ti2). Note that the margins of this sill have developed a foliation parallel to the F_2 foliation in the wall rocks. This sill contains an inclusion of schist, which in turn contains deformed quartz veins that do not extend across the sill. The quartz veins must then be pre- to syn-D_2. C: Photograph of "blobby," deformed quartz veins up to 4 cm thick, containing pink andalusite at site 91-36 in the staurolite zone (Fig. 2). These veins are deformed subparallel to the flat-lying F_2 foliation and are interpreted to be early- to syn-D_2. D: Close-up view of large grains of pink andalusite in the deformed, syn-D_2 veins of Figure 3C. E: Photograph of folded quartz veins developed in an outcrop of schist (meta-argillite) near site 91-31 within the transition zone to gneiss (Fig. 2). These quartz veins are concordant with F_2 foliation. The isoclinal folds defined by the deformed F_2 foliation and quartz veins are concordant with the steeply dipping F_3 foliation developed in the schist. This generation of veins is interpreted to be post-D_2 and pre-D_3. F: Photograph illustrates development of two generations of quartz veins in an outcrop at site 92-17 (Fig. 2). The earlier generation of highly deformed quartz veins (I) is analogous to those shown in Figure 3E and is post-D_2 and pre-D_3. The later set of quartz veins (II) is undeformed and aligned parallel to the axial planes of the D_3 folds and is interpreted to be syn- (late) to post-D_3. (A color version of this figure is on the CD-ROM accompanying this volume.)

isoclinal folds defined by the deformed F_2 foliation and quartz veins are parallel to the steeply dipping F_3 foliation developed in the schist. This generation of veins is interpreted to be post-D_2 and pre-D_3 and will be termed pre-D_3 in subsequent discussions. Figure 3F illustrates development of two generations of quartz veins in a nearby outcrop (site 92-17). The earlier generation of highly deformed quartz veins (I) is analogous to those shown in Figure 3E and is post-D_2 and pre-D_3. The later set of quartz veins (II) is undeformed and aligned parallel to the axial planes of the D_3 folds and is interpreted to be late- to post-D_3.

Igneous Intrusions

Tertiary felsic igneous activity (referred to as Ti1 through Ti4; Table 1) overlaps the metamorphic and deformation history of the Chugach metamorphic complex and has been subdivided into four generations with respect to the three main phases of ductile deformation in the complex (Sisson and Pavlis, 1993; Pavlis and Sisson, 1995; Harris et al., 1996; Sisson et al., this volume, Chapter 13). Igneous intrusions in the Tana Glacier section include dikes, sills, and discrete plutons. The first phase of intrusives are pre-metamorphic and not considered in this study. Dikes and sills intruded into the middle- to upper-amphibolite facies schists that are synchronous with D_2 are referred to as Ti2 intrusions. A third phase of intrusive (Ti3) dikes, sills, and plutons are interpreted to be syn- to post-D_3 and synchronous with the thermal peak of metamorphism in the high-grade core (sillimanite zone, gneiss) of the Chugach metamorphic complex. The latest phase of intrusive felsic dikes (Ti4) occurs parallel to a north-south joint set and is largely undeformed; these are interpreted to be post-D_3 and peak of metamorphism.

Deformational and Thermal History of the Chugach Metamorphic Complex

Sisson and Hollister (1988) proposed a two-stage heating process to be responsible for the thermally anomalous development of greenschist and amphibolite facies (e.g., the Chugach metamorphic complex itself) metamorphism in this forearc setting (Sisson and Pavlis, 1993). The initial greenshist facies metamorphism is associated with the regional development of D_1 and D_2. Temperatures of 300–400 °C are higher than would be expected from conduction in a forearc system (James et al., 1989) for the approximately 10 km depth inferred for the Tana Glacier section from metamorphic zoning sequence and geobarometry. This may reflect the effects of fluid advection via the quartz vein systems that characterize the D_2 phyllite/schists (Sisson et al., 1989). Fluid flux and associated advective heating have also been proposed to be responsible for prehnite-pumpellyite facies metamorphism in correlative rocks on Kodiak Island by Vrolijk et al. (1988). However, recent thermal modeling by Molnar and England (1995), which incorporates the effects of frictional heating similar to that observed in Taiwan, suggest that temperatures of 300–400 °C at approximately 10 km depth may not be anomalous for a forearc.

Greenschist facies metamorphism was followed by significant felsic magmatism (Ti2 and Ti3 igneous intrusions) and accompanying increase in D_2 strain and later D_3. This magmatism is thought to result from the thermal effects of the passage of the Kula–Farallon–North American triple junction (Sisson and Pavlis, 1993). It is considered to be the immediate heat source responsible for ultimately producing the high-*T*/low-*P*, amphibolite facies schists and gneissic core of the Chugach metamorphic complex. However, the detailed crosscutting relationships among deformation fabrics, quartz veins, and metamorphic index minerals indicate that maximum temperatures (thermal peak) are not achieved simultaneously across grade within the amphibolite facies Chugach metamorphic complex (Table 1). Because andalusite-bearing quartz veins are deformed by D_2, the thermal peak within the intermediate-grade schists ($T \geq 450$ to 475 °C) occurs prior to the conclusion of D_2 deformation—pre- to syn-D_2. Within the high-grade schist (sillimanite zone) and gneiss core where D_3 has deformed quartz veins and produced a penetrative fabric, sillimanite is undeformed and randomly oriented, statically overprinting the D_3 fabric. Hence, thermal peak in the highest-grade sillimanite schist and gneiss ($T \geq 650$ °C) occurs late- to post-D_3.

ANALYTICAL RESULTS

Analytical Methods

One of the objectives of this study was to evaluate isotopic evidence for fluid infiltration, including the extent of isotopic communication between external fluids—magma and aqueous fluid—and the rock matrix, so we emphasized isotopic analysis of quartz from the rock matrix (matrix quartz) and from veins within the same hand sample (vein quartz). Samples used for extraction of quartz from the rock matrix avoided quartz veins; quartz veins were removed and prepared separately. Quartz separates for oxygen isotope analysis were prepared with standard grinding, magnetic separation, and heavy liquid techniques. Quartz separates were treated with cold HF to remove feldspar; this treatment has no measurable effect on the $\delta^{18}O$ value of the quartz. Oxygen isotope analyses of a few quartz separates and the small group of whole-rock analyses were made with conventional silicate fluorination techniques utilizing BrF_5 (Clayton and Mayeda, 1963). Estimated analytical precision for these samples is ±0.2‰, and these analyses are reported to one decimal place in Table 2. The rest of the quartz separates were analyzed with a laser-aided fluorination system (Spicuzza et al., 1998) at the University of Wisconsin. Analytical precision is evaluated through replicate analysis of UWG-2, an internal lab garnet standard (Valley et al., 1995) and is ±0.05‰; these analyses are reported to two decimal places. All isotopic compositions are reported in standard δ notation relative to standard mean ocean water (SMOW) (Craig, 1961).

Oxygen Isotope Compositions of Matrix Quartz and Whole Rock

Table 2 gives the measured $\delta^{18}O$ values of quartz separates from the bulk rock matrix (matrix quartz) and for a few whole rock samples from metaargillites and metagraywackes. Figure 4 shows $\delta^{18}O$ values of matrix quartz plotted as a function of location within the Tana Glacier transect (Fig. 2). In the lowest-grade (Chl and Bi zones) rocks, the $\delta^{18}O$ values of matrix quartz are in the range 14.60 to 15.10‰ in metaargillite. Up-grade through Domain 2 in which only D_1 folds are developed and only pre-D_2 veins are present (to 2500 m), the $\delta^{18}O$ values of quartz in both metaargillite and metagraywacke progressively decrease from 15.10‰ (Chl zone) to values as low as 13.6‰. Over this interval, there is no systematic difference in the $\delta^{18}O$ values of quartz from either lithology. Starting in the staurolite zone (3500 m), where syn-D_2 quartz veins are present and D_3 fabric is superimposed on these rocks (Domain 3), the $\delta^{18}O$ values of matrix quartz from both metaargillite and metagraywacke decrease further to values between 12.3 and 13.3‰. Over the interval of section characterized by full development of D_3 fabric (3500 to 4500 m), the $\delta^{18}O$ values of matrix quartz have some variability (almost 1‰), but $\delta^{18}O$ values do not change systematically up-grade, or with degree of development of D_3 fabric. Further up-grade, through the sillimanite zone, the schist-gneiss transition, and gneiss, D_3 fabric is progressively superimposed on these rocks. However, the $\delta^{18}O$ values of matrix quartz in both metaargillite and metagraywacke remain approximately constant in the range 12.60 to 13.10‰ over this entire interval (Fig. 4); there is no further $^{18}O/^{16}O$ depletion upon transition from schist to gneiss. Thus, the decrease in $\delta^{18}O$ value of matrix quartz is not linear with either metamorphic grade or deformation history in the Tana Glacier area. Matrix quartz from both lithologies is depleted in $^{18}O/^{16}O$ to the maximum observed amount by the middle of the staurolite zone (3500 m) and by the conclusion of D_2 deformation.

A more detailed sampling traverse was taken through a section of interbedded metaargillites and metagraywackes in the garnet zone at site 92-12 (1120 m). The $\delta^{18}O$ values of matrix quartz from the two protoliths vary by 0.38 and 0.15‰ (Table 2). Within the higher-grade schists (upper staurolite and sillimanite zones) and into the gneiss, the $\delta^{18}O$ values of matrix quartz from the metagraywackes tend to be slightly higher (excluding sample 92-7D at 3500 m) than the metaargillite (Fig. 4; Table 2). Within the schist-gneiss transition at site 92-17 (6765 m), there are still significant variations (≤0.4‰) in the $\delta^{18}O$ values of quartz within both protoliths.

The amount of $^{18}O/^{16}O$ depletion recorded by matrix quartz in the Tana Glacier section is up to 2.50‰. Limited data indicate that the decrease in $\delta^{18}O$ values of matrix quartz up-grade in the Chugach metamorphic complex is also mirrored in the $\delta^{18}O$ values of whole-rock samples (Table 2). Depletion relative to the original $\delta^{18}O$ values of the protolith may be even greater, as Goldfarb et al. (1986) report $\delta^{18}O$ values from 13.9 to 17.0‰ for quartz from gold-quartz veins in lower-grade Valdez Group metasediments in the Port Valdez mining district.

Oxygen Isotope Compositions of Vein Quartz

The $\delta^{18}O$ values of vein quartz (Table 2) also decrease up metamorphic grade and through the deformation history by somewhat greater amounts (a maximum of 3.6‰) than recorded by matrix quartz (Fig. 5). Pre-D_2 quartz veins are progressively depleted in $^{18}O/^{16}O$ up-grade (to 2500 m), from $\delta^{18}O$ values as high as 15.3‰ to values as low as 13.65‰. Syn-D_2 quartz veins from the mid-staurolite zone (3500 to 4500 m) have still lower $\delta^{18}O$ values ranging from 13.25‰ down to 12.75‰ that do not vary with position or grade. Within the D_3 domain (staurolite and sillimanite zones) where development of D_3 fabric is not yet pervasive (3700 to 6500 m), post-D_2/pre-D_3 quartz veins have similar $\delta^{18}O$ values. Hence, there is no systematic change in $\delta^{18}O$ values of either syn-D_2 veins or pre-D_3 veins over the interval 3500 to 6500 m within the Tana Glacier section. Within the schist-gneiss transition and gneiss where D_3 fabric becomes fully developed, the $\delta^{18}O$ values of the pre-D_3 quartz veins decrease further, to $\delta^{18}O$ values as low as 11.75‰.

For the Tana Glacier transect as a whole, successively younger generations of quartz veins have lower $\delta^{18}O$ values. This trend generally holds for individual outcrops and hand

TABLE 2. OXYGEN ISOTOPE COMPOSITIONS OF QUARTZ SEPARATES AND WHOLE ROCK SAMPLES

Sample	Zone*	Location† (m)	Protolith§	Matrix phase	$\delta^{18}O$ value	$\delta^{18}O$ value, vein quartz	Vein age#
Metamorphic Rocks							
91-21A	Chl	220	arg	quartz	14.89	14.95	pre/syn-D_1
"		"	"			15.30	post-D_1
"		"	"	wr	12.5		
92-11B	Bi	940	arg	quartz	14.72		post-D_1
92-12A	Grt	1120	gw	quartz	14.81	14.55	post-D_1
92-12D	"	"	arg	quartz	14.92	14.65	post-D_1
92-12E	"	"	gw	quartz	14.80	14.56	post-D_1
92-12F	"	"	gw	quartz	14.66	14.10	post-D_1
92-12H	"	"	arg	quartz	14.54	14.75	syn/post-D_1
92-13C	"	1280	gw	quartz	14.24	14.38	pre/syn-D_1
92-10A	And	1615	arg	quartz	14.26		no veins
92-10B	"	"	gw			13.74	syn/post-D_1; pre-D_2
92-10D	"	"	gw	quartz	13.72	13.87	syn/post-D_1; pre-D_2
92-10E	"	"	arg	quartz	14.11	13.72	syn/post-D_1; pre-D_2
92-9A	"	2425	arg	quartz	13.71		syn/post-D_1; pre-D_2
92-9B	"	"	gw	quartz	13.78		syn/post-D_1; pre-D_2
91-36	Stl	3485	arg			13.30	pre/syn-D_2
92-7B	"	3515	arg	quartz	13.20	13.00	pre/syn-D_2
92-7D	"	"	gw	quartz	12.30	12.79	pre/syn-D_2
91-35B	"	3735	gw	quartz	13.25	12.99	pre/syn-D_2
"		"	"			12.71	late/post-D_2
92-6A	"	4000	arg	quartz	12.98		no veins
92-6B	"	"	arg	quartz	12.58	12.30	post-D_2; pre-D_3
92-21A	"	4325	arg	quartz	12.90		pre/syn-D_2
92-21B	"	"	arg	quartz	12.80		pre/syn-D_2
"			"	wr	11.0		
92-21C	"	"	gw	quartz	13.09		
92-21D	"	"	arg	quartz	12.69		
92-21F	"	"	gw	quartz	13.18	13.26	syn/post-D_2
92-21G	"	"	gw	quartz	13.24	12.98	syn/post-D_2
92-2A	Sil	4940	arg	quartz	13.38	12.82	pre-D_3
"	"		"	wr	11.2		
92-3A	"	6015	gw			13.00	
92-3C	"	"	arg	quartz	12.7		no veins
92-4A	"	6265	arg	quartz	12.9		
92-4B	"	"	gw	quartz	13.02	13.00	pre-D_3
92-17A	sch/gn	6765	gw			11.97	post-D_2; pre-D_3
92-17B	"	"	gw	quartz	12.66	12.30	pre-D_3
92-17C	"	"	arg			12.51	pre-D_3
92-17D	"	"	arg	quartz	13.04		pre-D_3
92-17E	"	"	arg	quartz	12.62	12.35	pre-D_3
91-31B	"	6780	gw	quartz	13.05	11.88	post-D_2; pre-D_3
91-32	"	6800	arg	quartz	12.76	11.84	post-D_2; pre-D_3
92-20A	"	6675	arg?	quartz	12.87		
92-20B	"	"	gw	quartz	12.99	12.39	post-D_2; pre-D_3
92-20D	"	"	gw	quartz	12.90	12.38	
92-20E	"	"	arg	quartz	12.71	11.75	post-D_2; pre-D_3
92-16A	Gn	7310	arg?	quartz	12.84	11.95	post-D_2; pre-D_3
92-16B	"	"	?			12.04	
Igneous Rocks							
89-ASN-87C	Trondhj dike			quartz	10.81		Ti2
91-ASN-61A	Granite dike			quartz	12.55		Ti2
89-ASN-87E	Granite dike			quartz	10.56		Ti3
91-ASN-20	Granite dike			quartz	12.03		Ti3
91-AHA-42	Tonalite pluton			quartz	10.35		Ti3
91-ASN-64	Granite dike			quartz	10.78		Ti3

*Metamorphic zone abbreviations: And—andalusite; Bi—biotite; Chl—chlorite; Grt—garnet; Stl—staurolite; Sil—sillimanite; gn—gneiss; sch/gn—schist-gneiss transition zone.

†Sample location is measured (in meters) from the north end of the map area (Fig. 2).

§Protolith and other abbreviations: arg—argillite; gw—greywacke; trondhj—trondhjemite; wr—whole rock.

#Age of vein or igneous intrusion relative to the major phases of deformation, D_1, D_2, and D_3; see text.

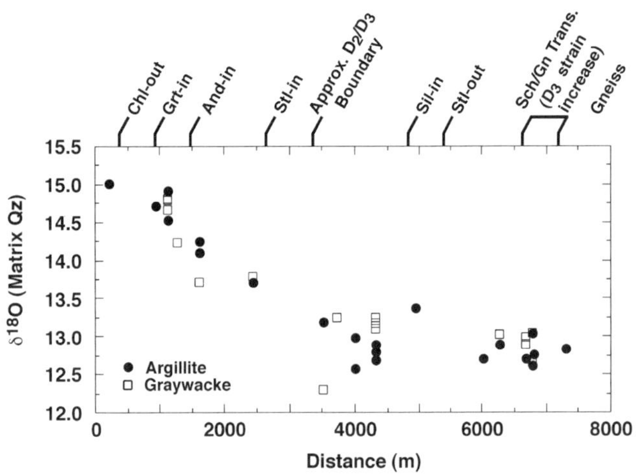

Figure 4. Plot of the $\delta^{18}O$ values of matrix quartz versus position along the Tana Glacier section. Sample location is measured southward from the north end of the map area (Fig. 2). Locations of metamorphic isograds, the schist-gneiss transition zone (Sch/Gn Trans.), gneiss, and the approximate boundary between the areas affected by D_2 and D_3 phases of deformation shown for reference. Symbols used are argillite (solid circles) and graywacke (open squares).

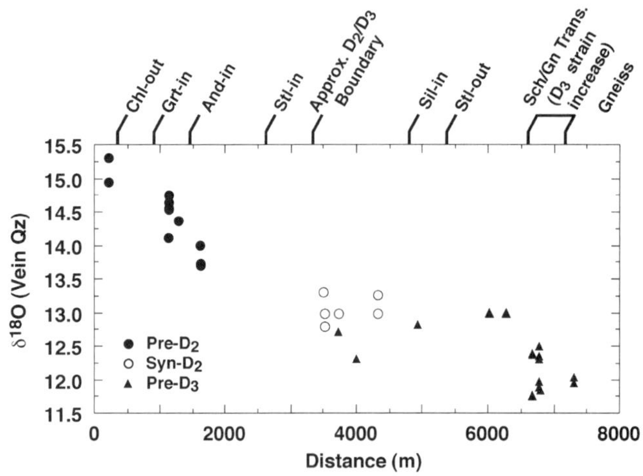

Figure 5. Plot of the $\delta^{18}O$ values of vein quartz versus position along the Tana Glacier section. Vein quartz is distinguished by the age of the vein with respect to the principal stages of deformation experienced in the area: pre-D_2 (solid circles); syn-D_2 (open circles); and pre-D_3 (solid triangles). Sample location is measured southward from the north end of the map area. Locations of metamorphic isograds, the schist-gneiss transition zone (Sch/Gn Trans.), gneiss, and the approximate boundary between the areas affected by D_2 and D_3 phases of deformation shown for reference.

samples as well. Table 2 contains two samples (91-21, 91-35B) and two outcrops (92-12, 92-17) where two generations of quartz veins were separated and analyzed isotopically. In all but the lowest-grade sample (91-21, which contains pre-D_1 veins), the $\delta^{18}O$ values of the younger veins are somewhat lower (by up to 0.54‰) than the $\delta^{18}O$ values of the older veins.

Figure 6 illustrates well the parallel decrease in $\delta^{18}O$ values of both matrix and vein quartz from the same hand samples through the deformation history and with increase in metamorphic grade. Through all metamorphic grades up to the schist-gneiss transition and through both D_2 and D_3 stages of deformation, the pre-D_2 and syn-D_2 generations of quartz veins in both metaargillite and metagraywacke protoliths closely approach oxygen isotope exchange equilibrium with matrix quartz (the one-to-one line in Figure 6). In Domain 4, where D_3 fabric is developed fully, the pre-D_3 quartz veins have significantly lower $\delta^{18}O$ values than matrix quartz by as much as 1.17‰.

Oxygen Isotope Compositions of Igneous Quartz

The $\delta^{18}O$ values of quartz from six igneous rocks sampled in the Tana Glacier section are also compiled in Table 2. Quartz separates from two pre- to syn-metamorphic Ti2 igneous rocks have $\delta^{18}O$ values of 10.81 and 12.55. The latter is equivalent to the $\delta^{18}O$ values of matrix quartz from the highest-grade schists and gneisses. The $\delta^{18}O$ values of quartz from four syn- to post-metamorphic Ti3 igneous rocks range from 10.35 to 12.03‰. These $\delta^{18}O$ values are lower than any recorded by matrix quartz from the metasediments, but overlap the range of $\delta^{18}O$ values recorded by pre-D_3 quartz veins from the highest-grade schists and gneisses in the Tana River section.

DISCUSSION

Closed-System Temperature Effects

Up-grade in the Tana Glacier transect, the $\delta^{18}O$ values of matrix quartz decrease by amounts reaching 2.50‰. However, part of this decrease must result from the increase in metamorphic grade. This is because the oxygen isotope fractionation factors between quartz and the other minerals decrease with increase in temperature, and the prograde reactions will change the proportions of quartz in the rock. The amount of this decrease during closed-system, prograde metamorphism can be calculated through oxygen isotope mass balance (Chamberlain et al., 1990; Kohn and Valley, 1994), using measured changes in mineral modes with metamorphic grade and for the increase in temperature, 450 to 650 °C, recorded by garnet-biotite geothermometry across the Tana Glacier transect.

Measured mineral modes in selected samples from both low (biotite zone) and high grade (gneiss) are given in Table 3. Oxygen isotope fractionation factors for the mineral pairs were taken from the compilation of internally consistent fractionation factors published by Kohn and Valley (1998a, 1998b). The results of the mass-balance calculations are also shown in Table 3. These oxygen isotope mass-balance calculations assume that minerals maintain oxygen isotope exchange equilibrium. The mass-balance calculations predict that the $\delta^{18}O$

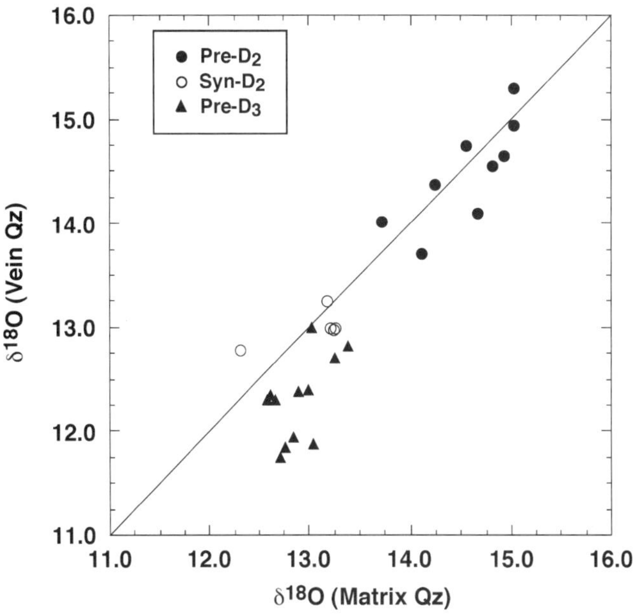

Figure 6. Plot of the $\delta^{18}O$ values of quartz from veins and quartz from the adjacent rock matrix (vein-matrix pairs at the hand specimen scale). These vein-matrix pairs are distinguished by the age of the vein with respect to the principal stages of deformation experienced in the area: pre-D_2 (solid circles); syn-D_2 (open circles); and pre-D_3 (solid triangles). Solid line (slope = 1) denotes oxygen isotope exchange equilibrium between vein and matrix quartz.

value of quartz in the metaargillite should decrease by 1.2‰ with an increase in temperature from 450 °C (biotite zone) to 650 °C (gneiss). The groups of samples representing the low- and high-grade metaargillite in Table 3 were selected to have the greatest difference in biotite content, and hence these mass-balance calculations will yield the *maximum* decrease in $\delta^{18}O$ value of quartz. Given observed variations in mineral proportions and uncertainties in fractionation factors, the mass-balance calculations predict that the $\delta^{18}O$ value of matrix quartz should decrease between 0.9 and 1.2‰ up-grade in response to closed-system mineral (dehydration) reactions and to the accompanying increase in temperature (Fig. 7A).

Exchange of the Chugach Metamorphic Complex Metasediments with an External Oxygen Reservoir

The observed 2.50‰ depletion in $\delta^{18}O$ values experienced by the matrix quartz up-grade in the Tana Glacier section (and the 3.6‰ depletion in vein quartz) is significantly greater than can be explained by closed-system dehydration reactions and an increase in temperature (Fig. 7A). This large decrease in $\delta^{18}O$ value of the metasediments with increasing temperature indicates progressive exchange of these metasediments with an external, relatively lower $\delta^{18}O$ reservoir of oxygen; either one or more generations of magma (represented by the Ti2 and Ti3 igneous rocks), aqueous fluid that equilibrated with these igneous reservoirs, or both.

TABLE 3. MEASURED MINERAL MODES FROM THE TANA GLACIER SECTION WITH CALCULATED CHANGES IN THE $\delta^{18}O$ VALUES OF MINERALS

	Low Grade			High Grade (gneiss)				
Phase	Mode* (Vol. Fr.)	Mol Fr†	Oxygen Mol Fr.	Mode* (Vol. Fr.)	Mol Fr. †	Oxygen Mol Fr.	Initial $\delta^{18}O$†† (at 450 °C)	Final $\delta^{18}O$ §§ (at 650 °C)
Quartz	0.292	0.682	0.313	0.350	0.716	0.372	14.89	13.70
Plagioclase§	0.392	0.207	0.380	0.436	0.201	0.418	12.78	12.41
Sillimanite	–	–	–	0.012	0.011	0.014	–	11.41
Biotite#	0.264	0.092	0.253	0.181	0.055	0.172	9.91	10.64
Chlorite	0.045	0.011	0.047	–	–	–	9.91	–
Garnet	–	–	–	0.014	0.006	0.018	–	10.44
Opaques**	0.007	0.008	0.008	0.006	0.006	0.006	2.86	6.31
Whole rock							12.50	12.50

*Modes (volume fraction) were measured by electron microprobe analysis (EMPA) of 2500 points in polished thin sections on a grid pattern determined by the average grain size in the sample. Low-grade argillite is the average of two samples; high-grade gneiss is the average of three samples.

†Molar volumes of solid solution minerals based on typical EMPA results from Sisson et al. (1989) and this study: biotite (Ann50), chlorite (Chm60), garnet (Alm75), plagioclase (An20).

§Some K-feldspar included in high-grade gneiss.

#Includes minor muscovite.

**Opaques are principally ilmenite, pyrite, and graphite (at low grade).

††These calculated $\delta^{18}O$ values are based on the measured whole-rock value of 12.5 per mil for sample 91-21A (see Table 2).

§§As above, but the whole-rock $\delta^{18}O$ value is adjusted for the effect of removal of the water released by dehydration of hydrous silicates.

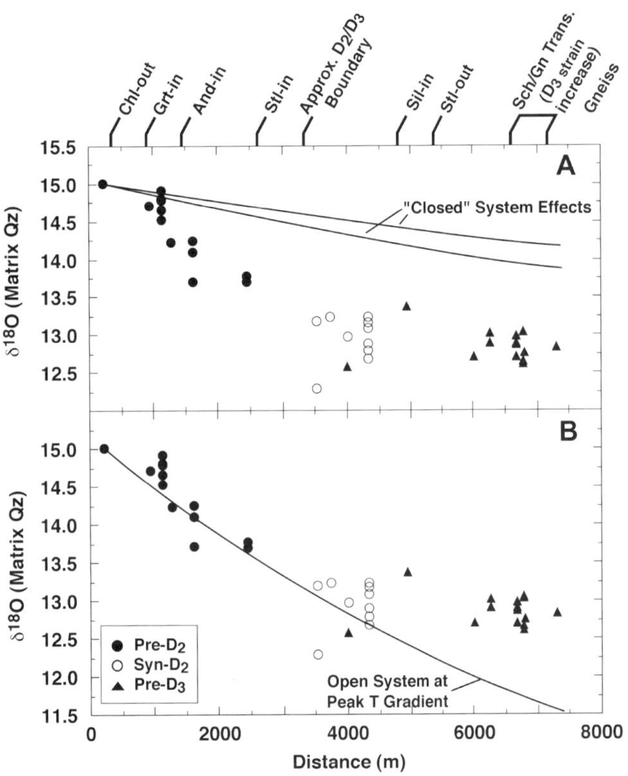

Figure 7. Plot of the $\delta^{18}O$ values of matrix quartz versus position along the Tana Glacier section. Note that the matrix quartz is defined in terms of the age of the principal generation of quartz veins within the rock matrix to facilitate comparison to the trends with grade defined by vein quartz illustrated in Figure 5: pre-D_2 (solid circles); syn-D_2 (open circles); and pre-D_3 (solid triangles). A: Two calculated "closed-system" trends of $^{18}O/^{16}O$ depletion for matrix quartz are superimposed. These curves represent the maximum $^{18}O/^{16}O$ depletion that would be produced in the metasedimentary rocks by closed-system, prograde metamorphic reactions (see text for further discussion of these calculations). The maximum calculated depletion varies from 0.9 to 1.2‰, depending on bulk composition (observed variation in mineral modes). B: Superimposed is a calculated, "open system" curve marking the progressive decrease in $\delta^{18}O$ value that would be experienced by matrix quartz for simple equilibrium exchange one-dimensional fluid flow—either up- or down-*T*—along the peak metamorphic temperature gradient, and within the water-dominated segment of the flow path (curve labeled "Open System at Peak T Gradient").

The schists experience significant $^{18}O/^{16}O$ depletions kilometers away from exposed igneous rock. Further, the $\delta^{18}O$ values of matrix quartz from the highest-grade schists and gneisses overlap only partially with the $\delta^{18}O$ values of quartz in the igneous rock samples (10.35 to 12.55, Table 2). Hence, complete exchange directly between magma and metasediment on a large scale (e.g., >100s of meters to kilometers) in the Tana Glacier section is unlikely. The large-scale $^{18}O/^{16}O$ depletions experienced by the schists and gneisses more likely occurred by exchange with infiltrating fluids; these fluids in turn may have initially equilibrated with the igneous reservoirs. The parallel decrease in $\delta^{18}O$ values of quartz veins and matrix (host rock) quartz (Fig. 6) suggests that the fluids responsible for lowering $\delta^{18}O$ values in the rock matrix are also responsible for the precipitation of quartz in these veins. Regardless, the scale of isotope exchange between the rock matrix and infiltrating fluid is at least that of the typical hand specimen, 5 to 10 cm, used to prepare the matrix quartz separates. Spacing of quartz veins ranges from 1 to 10 cm or more, generally compatible with the scale of isotopic exchange documented by these samples.

Implications for Fluid Flow Geometry and Timing of Fluid Infiltration

The section exposed at Tana Glacier has exchanged with an external, lower-$\delta^{18}O$ reservoir (hot aqueous fluid and/or magma) during one or more episodes of fluid infiltration and magma injection. However, it appears unlikely that the decrease in $\delta^{18}O$ values up-grade at Tana Glacier was ever the result of a single, coherent and geometrically simple fluid flow system operating at the scale of the exposed metamorphic section and operating at the peak of metamorphism. Evidence against such simple flow scenarios includes structural data and deformation history, quartz vein systems and fluid inclusion microthermometry data, and constraints on the patterns of isotopic shifts in flow systems based on transport theory.

Deformation History

The existence of multiple phases of intense deformation in the Tana Glacier area (Pavlis and Sisson, 1995) indicates that different parts of this section were subjected to geometrically distinct stress fields at different times (Fig. 2). The flow geometry would probably have changed with change in orientation of the differential stress regime, as the stress field and resulting deformation would likely partition fluid flow (Etheridge et al., 1983; Oliver, 1996; Ord and Oliver, 1997). Given that different portions of the Tana Glacier transect were subjected to different phases of deformation, it seems unlikely that there would ever exist a single, coherent stress regime and permeability field—and therefore a single, coherent flow system—that could characterize this entire area for any significant period of time. Hence, it seems unlikely from a mechanical and hydrological point of view that a single, coherent flow regime of simple geometry produced the entire measured $^{18}O/^{16}O$ depletion pattern in the Tana Glacier section in one contemporaneous event.

Significance of Quartz Veins

The existence of multiple generations of quartz veins within the Tana Glacier section is direct field evidence for multiple generations of silica mobility (on at least the local, outcrop scale) and perhaps, for multiple episodes of larger-scale fluid flow. Most important, the timing of formation of quartz veins constrains the timing of $^{18}O/^{16}O$ depletion—and fluid infiltration—in the Tana Glacier section with respect to major phases of deformation and metamorphic peak temperature.

The complicated fluid history is further manifested by the presence of multiple generations of fluid inclusions within these quartz vein systems (Xu et al., 1994). In many samples, there is an early generation of fluid inclusions crosscut by a later generation with late low-salinity brines. They report three types of early fluid inclusion compositions ranging from low-salinity brines, mixed CO_2-H_2O solutions, to CO_2-rich fluids. The fluid composition changes from low-salinity brines at low grade to CO_2-rich fluids in the high-grade schists. Above the schist to gneiss transition, the quartz veins have rare, early H_2O-rich fluids with 5–10% CO_2 along with abundant, late low-salinity brines. The late, low-salinity brines were trapped during cooling after peak metamorphic conditions. The abundance of late fluids varies along the Tana Glacier transect.

Both the quartz from the rock matrix and the vein quartz exhibit significant, parallel decreases in their $\delta^{18}O$ values with increasing metamorphic grade that require infiltration and isotopic exchange of the rock matrix with relatively lower-$\delta^{18}O$ fluids. However through all metamorphic grades short of the schist-gneiss transition, vein and matrix quartz are in or closely approach isotopic exchange equilibrium (Fig. 6), a condition normally associated with closed-system formation of quartz veins. The combination of these two characteristics requires either that the quartz veins formed in closed systems (on the scale of outcrops within individual lithologic units) *subsequent* to fluid infiltration and $^{18}O/^{16}O$ depletion of the rock matrix, or that the veins formed in open systems *during* the infiltration of lower-$\delta^{18}O$ fluids and therefore *contemporaneously* with the $^{18}O/^{16}O$ depletion of the rock matrix. In the open system case, the same fluids responsible for the veins are responsible for $^{18}O/^{16}O$ depletion of the rock matrix, and the rock matrix is largely equilibrated (e.g., becomes lower in $\delta^{18}O$) with this infiltrating fluid. Regardless of whether veins formed under open- or closed-system circumstances, the event(s) of fluid infiltration responsible for the decrease in $^{18}O/^{16}O$ in the rock matrix must have occurred prior to, or during the time(s) of quartz vein formation.

Quartz veins (post-D_2/pre-D_3) in the schist-gneiss transition zone and gneiss are systematically depleted in $^{18}O/^{16}O$ by as much as 1.1‰ with respect to matrix quartz. A straight-forward interpretation of these disequilibrium vein-matrix pairs is that the veins formed when even lower-$\delta^{18}O$ fluid infiltrated the schist-gneiss transition and gneiss after D_2 and prior to or during the early stages of D_3. The incomplete isotopic exchange between these vein fluids and the adjacent rock matrix (on the cm scale) suggests that the fluid infiltration was focused along fractures now occupied by veins, limited in flux and/or duration for this section of the Chugach metamorphic complex, and that any subsequent heating and/or cooling of this section was relatively rapid. $^{40}Ar/^{39}Ar$ geochronology indicating rapid cooling (two million years) from the thermal peak of metamorphism for the Chugach metamorphic complex down through the biotite closure temperature is compatible with such a scenario (Sisson et. al, 1989).

Regardless of the mode of vein formation, the fluid infiltration and $^{18}O/^{16}O$ depletion experienced by the metasediments in the Tana Glacier transect must have preceded the phase of deformation in which the quartz veins were deformed. Field observations discussed above (e.g., crosscutting relationships between veins and major structural fabrics developed in these metasediments) establish the timing of vein formation—and fluid infiltration—with respect to deformation phases. These relationships indicate that $^{18}O/^{16}O$-depleted quartz veins ($\delta^{18}O \leq 13.0‰$) formed prior to or during the early stages of D_2 in schists from the andalusite zone to the sillimanite zone. Therefore, fluid infiltration occurred prior to or during the early stages of D_2 deformation. Some of these $^{18}O/^{16}O$-depleted veins contain andalusite, indicating that the thermal peak also occurred prior to or during D_2 within the intermediate-grade section of the Chugach metamorphic complex. Therefore, fluid infiltration also occurred at or just prior to the thermal peak for these intermediate-grade rocks; both fluid infiltration and thermal peak occurred prior to or during D_2 deformation within the intermediate-grade section of the Chugach metamorphic complex.

Quartz veins with still lower $\delta^{18}O$ values (many lower than 12.0‰) formed after D_2 and prior to the D_3 deformation within the high-grade schist-gneiss transition zone and gneiss (pre-D_3 veins, Fig. 5). Therefore, the fluid infiltration responsible for this observed $^{18}O/^{16}O$ depletion occurred after D_2 but prior to D_3 deformation. Within the high-grade schist (sillimanite zone) and gneiss core, undeformed and randomly oriented sillimanite overprints the D_3 fabric, indicating that the thermal peak in these high-grade regions is achieved after D_3. Hence, fluid infiltration not only occurred prior to D_3 deformation but also before the thermal peak (approximately 650 °C) was achieved within the high-grade core (sillimanite zone and gneiss) of the Chugach metamorphic complex.

In summary, the combination of oxygen isotope evidence with crosscutting relationships among deformation fabrics, quartz veins, and metamorphic index minerals indicates that the metasediments of the Chugach metamorphic complex experienced at least two stages of fluid infiltration and associated $^{18}O/^{16}O$ depletion. The first stage occurred prior to or during D_2 deformation and near (at or before) the thermal peak for intermediate-grade zones (andalusite, staurolite). The second stage occurred after D_2 but prior to D_3 deformation and before the thermal peak of metamorphism for the high-grade core (sillimanite zone schists and gneiss) of the Chugach metamorphic complex.

Constraints to Flow Geometry from Patterns of Oxygen Isotopic Shifts

Matrix quartz in the rocks of the Tana Glacier section are depleted to fairly uniform $\delta^{18}O$ values of +12.5 to +13.0‰ over a wide lateral expanse and range of metamorphic grade (mid-staurolite up-grade to the gneiss core). Rocks whose $\delta^{18}O$ values have been lowered to constant values over a significant range of metamorphic grade are unlikely to be developed by a geometrically simple up- or down-*T* flow system. Basic considerations from transport theory illustrate why this is the case.

For advection-dominated flow systems and equilibrium exchange, rocks will experience significant shifts in $\delta^{18}O$ value over significant lateral distances only in the water-dominated segment of a flow path (Baumgartner and Rumble, 1988; Bowman and Willett, 1991; Dipple and Ferry, 1992). For flow along temperature gradients, the $\delta^{18}O$ values of rock (and constituent minerals) within the water-dominated segment of the flow path must change as a function of the temperature change (Dipple and Ferry, 1992; Bowman et al., 1994). These changes in $\delta^{18}O$ values will be a function only of temperature change because the $\delta^{18}O$ value of the fluid will be constant within the water-dominated segment of the flow path (Bowman et al., 1994). Hence, the spatial gradient in $\delta^{18}O$ values ($d[\delta^{18}O]/dx$) preserved by rock (and constituent minerals) within the water-dominated segment of a flow path will be a function only of the spatial gradient in temperature, dT/dx, that exists along this segment of the flowpath during fluid flow. Note that if these $\delta^{18}O$ values are uniform for any significant lateral distance along the water-dominated segment of the flowpath, this section of the flowpath must be isothermal.

The Tana Glacier section has experienced significant $^{18}O/^{16}O$ depletions (up to 2.5‰) over much of its length (Figs. 4 and 5). Thus, much of the Tana Glacier section lies within the water-dominated segments of the flow system(s) responsible for this $^{18}O/^{16}O$ depletion. Field and petrographic observations indicate that isogradic surfaces are sub-vertical with little or only modest pressure increase within the Tana Glacier transect. Therefore, the measured $\delta^{18}O$-distance profile is approximately normal to isotherms and subparallel to both the peak metamorphic field gradient and to any proposed up- or down-T flow path of simple geometry.

The field gradient preserved by the peak metamorphic sequence in the Tana Glacier section records temperature of approximately 450 °C at the garnet isograd and about 650 °C within the schist-gneiss transition. Extrapolating this gradient down grade yields a T = 430 °C at the location of the lowest-grade sample in the Tana Glacier transect. If one-dimensional fluid flow occurred normal to these isothermal surfaces and along this peak temperature gradient (i.e., simple up- or down-T flow), then the $\delta^{18}O$ values of quartz must decrease—in the direction of temperature increase—by 3.6‰. This decrease corresponds to the decrease in the quartz-water fractionation factor (Clayton et al., 1989) over this temperature change (430 to 650 °C) and is shown by the "open system" curve superimposed on the measured $\delta^{18}O$-distance profile at Tana Glacier in Figure 7B. However, the observed decrease in the $\delta^{18}O$ values of matrix quartz from the lowest- to highest-grade rocks is less than 2.5‰. For such simple one-dimensional flow scenarios then, the matrix quartz from the highest-grade schists and gneisses should have $\delta^{18}O$ values more than 1‰ lower than measured. Most critically, the change in grade from the mid-staurolite zone (about 4000 m) to gneiss (≥7300 m) would require a 1.2‰ decrease in the $\delta^{18}O$ value of matrix quartz (Fig. 7B). However, there is no observed systematic decrease in $\delta^{18}O$ values across this interval, which, according to transport theory, requires that this interval be near-isothermal for a flowpath oriented parallel to this section (e.g., normal to the isograds). Such near-isothermal fluid flow is incompatible with the peak metamorphic field gradient preserved across this same interval. This higher-grade portion of the measured $\delta^{18}O$-distance profile is actually more compatible with the calculated, closed-system profile (Fig. 7A). This suggests that the higher-grade segment of the Tana Glacier transect was nearly a closed system isotopically—and experienced relatively little fluid infiltration—at peak metamorphic conditions. This conclusion is supported by the isotopic equilibrium between vein and matrix quartz in samples from the staurolite through sillimanite zones.

In light of these incompatibilities, the observed patterns of $^{18}O/^{16}O$ depletion in the Tana section suggest either that: (1) significant fluid infiltration and isotopic depletion occurred *prior* to attainment of peak metamorphism within the high-grade core of the Chugach metamorphic complex, produced by a geometrically simple flow system characterized by a much lower lateral temperature gradient; or (2) the observed pattern of decreasing $\delta^{18}O$ is the integrated composite of several generations of two- or three-dimensional flow systems operating at different times, in different areas, and in different orientations (perhaps including sub-vertical flow at the exposed level of the Tana Glacier section) during the *prograde* metamorphic and structural development of the Tana section; that is, before attainment of peak metamorphism within the high-grade core of the Chugach metamorphic complex. The textural relationships between quartz veins, deformation stages, and index minerals (sillimanite), which suggest multiple generations of fluid flow prior to the thermal peak (650 °C) of metamorphism within the high-grade core of the Chugach metamorphic complex, support the latter alternative.

Implications for Metamorphic Heating

The parallel decrease in $\delta^{18}O$ of matrix and vein quartz indicates that fluid infiltration produced quartz veins and caused the general lowering of $\delta^{18}O$ values in progressively higher-grade metamorphic rocks of the Chugach metamorphic complex. The isotope data, combined with crosscutting relationships among quartz veins, deformation stages, igneous intrusions, and metamorphic index minerals indicate that there are at least two fluid infiltration events that overlap in time with igneous activity but precede the thermal peak of metamorphism in the high-grade core of the Chugach metamorphic complex. Oxygen isotope evidence (particularly the lower $\delta^{18}O$ values of the pre-D_3 veins) further indicates that these infiltrating fluids likely exchanged with regional igneous reservoirs emplaced during this time (Ti2 and at deeper levels, Ti3). These fluids can advect into the high-grade core more quickly than heat will be conducted there from these intrusions. As a result, the quartz veins produced by these fluids formed prior to the thermal peak of metamorphism in the high-grade core of this section of the Chugach metamorphic complex. Thus, the detailed structural, thermal, and oxygen isotope record of these metamorphic rocks indicates that igneous activity is likely responsible for these fluid infiltration events and is also likely responsible for the metamorphic heating in this Buchan terrane, reinforcing the previous conclusions of Sisson and Hollister (1988) and Sisson and Pavlis (1993).

Comparison with Other Metamorphic Terranes

The systematic decrease in $\delta^{18}O$ values of metasediments that correlate with increase in metamorphic grade in the Tana Glacier section is similar to that observed in metapelites from a number of regional metamorphic terranes (Garlick and Epstein, 1967; Rye et al., 1976; Wickham and Taylor, 1985; Criss and Fleck, 1986; Cartwright et al., 1995; Oliver et al., 1998). In particular, the $\delta^{18}O$-grade profile in the Trois Seigneurs Massif (Wickham and Taylor, 1985) is quite similar to that for the Tana Glacier section; significant decrease in $\delta^{18}O$ from low to medium grade, uniformly $^{18}O/^{16}O$-depleted rocks at higher grades into a gneiss core. In all of these areas, the decreases in $\delta^{18}O$ values are sufficiently large to require infiltration of low-$\delta^{18}O$ fluids at some point in the metamorphic history of these terranes. None of these previous studies have definitive evidence constraining the timing of fluid infiltration with respect to deformation history and thermal peak(s) of metamorphism, the geometry of flow, and whether or not multiple infiltration events have occurred in these terranes.

Some studies propose or imply that the $^{18}O/^{16}O$ depletion is the result of fluid infiltration involving a hydrothermal system on the scale of the metamorphic terrane and operating at or near the thermal peak of metamorphism (Cartwright et al., 1995; Oliver et al., 1998). These studies have then applied one-dimensional flow models and used the peak metamorphic field gradient to estimate time-integrated fluid flux (q_{TIFF}) for the terrane. However, if the complex relationships between multiple episodes of deformation, fluid infiltration, and thermal peak of metamorphism documented within the Chugach metamorphic complex are representative, it is unlikely that the observed low $\delta^{18}O$ values in any of these metamorphic terranes were developed in a single flow event of simple geometry at the peak of metamorphism. For example, the existence of uniformly low-$\delta^{18}O$ rocks over significant change of grade in the Trois Seigneurs Massif (Wickham and Taylor, 1985) precludes simple up- or down-T fluid flow along the peak metamorphic field gradient, as is demonstrated in this paper for the same situation in the Chugach metamorphic complex. In another example, Oliver et al. (1998) measured $^{18}O/^{16}O$ depletions upgrade in the Adelaide fold belt, Australia, that are too large to result from simple one-dimensional fluid flow over the lateral temperature gradient defined by the peak metamorphic field gradient; a higher temperature gradient was needed to produce the observed decrease in $\delta^{18}O$ values.

This study of the Chugach metamorphic complex documents multiple episodes of fluid infiltration that occur during the prograde thermal and deformational history of the complex and certainly prior to the achievement of maximum temperature (650 °C) in the highest-grade core of the complex. Other studies suggest that the bulk of fluid infiltration occurs prior to the thermal peak of metamorphism. Bickle and Chapman (1990) document short (<1 m) transport distances of Sr isotopes at carbonate-pelite lithologic contacts in the Trois Seigneurs, Pyrenees metamorphic belt. They conclude that these short distances are not compatible with the large (kilometers) oxygen transport distances necessary to account for the regional (5 km) scale $^{18}O/^{16}O$ depletion and homogenization that is documented in the Trois Seigneurs terrane by Wickham and Taylor (1985). They suggest that these differences can be explained by fluid infiltration in the brittle regime well prior to the peak of metamorphism, at lower-temperature conditions where kinetic dispersion could effectively decouple the mobility of Sr and O isotopes.

Several isotope studies of regional metamorphic terranes have also inferred multiple episodes of fluid infiltration (Baker et al., 1989; Baker and Matthews, 1995). Baker et al. (1989) present isotopic evidence for multiple episodes of fluid infiltration during Miocene Barrovian metamorphism in Naxos—an initial episode during the prograde segment of the evolution of the metamorphic belt and another episode at or after the peak of metamorphism.

General oxygen isotope data sets such as those presented here for the Tana Glacier area and others previously published for such places as Trois Seigneurs (Wickham and Taylor, 1985) may not uniquely constrain the flow geometry responsible for the depletion patterns. These data sets also need to be integrated with structural and petrological studies in order to constrain the timing of fluid flow with respect to deformation and thermal history of the metamorphic belt.

ACKNOWLEDGMENTS

Financial support for this study was provided by NSF grants EAR-9205218 and EAR-9903274 to J.R. Bowman, NSF grant EAR-9628260 to J.W. Valley, and DOE grant FG02-93ER14389 to J.W. Valley. We thank Mike Spicuzza for his expert support in the laser silicate stable isotope laboratory at the University of Wisconsin and Wes Christensen for mineral separations at the University of Utah. We also appreciate constructive reviews by Steve Dunn and Peter Nabelek.

REFERENCES CITED

Baker, J., and Matthews, A., 1995, The stable isotopic evolution of a metamorphic complex, Naxos, Greece: Contributions to Mineralogy and Petrology, v. 120, p. 391–403.

Baker, J., Bickle, M.J., Buick, I.S., Holland, T.J.B., and Matthews, A., 1989, Isotopic and petrological evidence for the infiltration of water-rich fluids during the Miocene M2 metamorphism on Naxos, Greece: Geochimica et Cosmochimica Acta, v. 53, p. 2037–2050.

Baumgartner, L.P., and Rumble, D., III, 1988, Transport of stable isotopes, I: Development of a continuum theory for stable isotope transport: Contributions to Mineralogy and Petrology, v. 98, p. 417–430.

Bickle, M.J., and Chapman, H.J., 1990, Strontium and oxygen isotope decoupling in the Hercynian Trois Seigneurs Massif, Pyrenees—Evidence for fluid circulation in a brittle regime: Contributions to Mineralogy and Petrology, v. 104, p. 332–347.

Bowman, J.R., and Willett, S.D., 1991, Spatial patterns of oxygen isotope exchange during one-dimensional fluid infiltration: Geophysical Research Letters, v. 18, p. 971–974.

Bowman, J.R., Willett, S.D., and Cook, S.J., 1994, Oxygen isotope transport and exchange during fluid flow: One-dimensional models and applications: American Journal of Science, v. 294, p. 1–55.

Cartwright, I., Vry, J., and Sandiford, M., 1995, Changes in the stable isotope ratios of metapelites and marbles during regional metamorphism, Mount Lofty Ranges, South Australia—Implications for crustal scale fluid flow: Contributions to Mineralogy and Petrology, v. 120, p. 292–310.

Chamberlain, C.P., Ferry, J.M., and Rumble, D., 1990, The effect of net transfer reactions on the isotopic composition of minerals: Contributions to Mineralogy and Petrology, v. 105, p. 322–336.

Clayton, R.N., and Mayeda, T.K., 1963, The use of bromine pentafluoride in the extraction of oxygen from oxides and silicates for isotopic analysis: Geochimica et Cosmochimica Acta, v. 17, p. 43–52.

Clayton, R.N., Goldsmith, J.R., and Mayeda, T.K., 1989, Oxygen isotope fractionation in quartz, albite, anorthite, and calcite: Geochimica et Cosmochimica Acta, v. 53, p. 725–733.

Craig, H., 1961, Standard for reporting concentrations of deuterium and oxygen-18 in natural waters: Science, v. 133, p. 1833–1834.

Criss, R.E., and Fleck, R.J., 1986, Petrogenesis, geochronology, and hydrothermal systems of the northern Idaho Batholith and adjacent areas based on $^{18}O/^{16}O$, D/H, $^{87}Sr/^{86}Sr$, and $^{40}Ar/^{39}Ar$ studies: U.S. Geological Survey Professional Paper 1436, p. 95–137.

Dipple, G.M., and Ferry, J.M., 1992, Fluid flow and stable isotopic alteration in rocks at elevated temperatures with applications to metamorphism: Geochimica et Cosmochimica Acta, v. 56, p. 3539–3550.

Etheridge, M.A., Wall, V.J., and Vernon, 1983, The role of the fluid phase during regional metamorphism and deformation: Journal of Metamorphic Geology, v. 1, p. 205–226.

Garlick, G.D., and Epstein, S., 1967, Oxygen isotope ratios in coexisting minerals of regionally metamorphosed rocks: Geochimica et Cosmochimica Acta, v. 31, p. 181–214.

Goldfarb, R.J., Leach, D.L., Miller, M.L., and Pickthorn, W.J., 1986, Geology, metamorphic setting, and genetic constraints of epigenetic lode-gold mineralization within the Cretaceous Valdez Group, south-central Alaska, *in* Keppie, J.D., et al., eds., Turbidite hosted gold deposits: Geological Association of Canada Special Paper 32, p. 87–105.

Harris, N.R., Sisson, V.B., Wright, J.E., and Pavlis, T.L., 1996, Evidence for Eocene mafic underplating during forearc intrusive activity, eastern Chugach Mountains, Alaska: Geology, v. 24, p. 263–266.

Hudson, T.G., and Plafker, G., 1982, Paleogene metamorphism in an accretionary flysch terrane, eastern Gulf of Alaska: Geological Society of America Bulletin, v. 93, p. 1280–1290.

Hudson, T.G., Plafker, G., and Peterman, Z.E., 1979, Paleogene anatexis along the Gulf of Alaska margin: Geology, v. 7, p. 573–577.

James, T.S., Hollister, L.S., and Morgan, W.J., 1989, Thermal modeling of the Chugach metamorphic complex: Journal of Geophysical Research, v. 94, p. 4392–4410.

Kohn, M.J., and Valley, J.W., 1994, Oxygen isotope constraints on metamorphic fluid flow, Townshend Dam, Vermont, USA: Geochimica et Cosmochimica Acta, v. 58, p. 5551–5566.

Kohn, M.J., and Valley, J.W., 1998a, Obtaining equilibrium oxygen isotope fractionations from rocks: Theory and examples: Contributions to Mineralogy and Petrology, v. 132, p. 209–224.

Kohn, M.J., and Valley, J.W., 1998b, Effects of cation substitutions in garnet and pyroxene on equilibrium oxygen isotope fractionations: Journal of Metamorphic Geology, v. 16, p. 625–639.

Molnar, P., and England, P., 1995, Temperatures in zones of steady-state underthrusting of young oceanic lithosphere: Earth and Planetary Science Letters, v. 131, p. 57–70.

Oliver, N.H.S., 1996, Review and classification of structural controls on fluid flow during regional metamorphism: Journal of Metamorphic Geology, v. 11, p. 477–492.

Oliver, N.H.S., Dipple, G.M., Cartwright, I., and Schiller, J., 1998, Fluid flow and metasomatism in the genesis of the amphibolite-facies, pelite-hosted Kanmantoo copper deposit, South Australia: American Journal of Science, v. 298, p. 181–218.

Ord, A., and Oliver, N.H.S., 1997, Mechanical controls on fluid flow during regional metamorphism—Some numerical models: Journal of Metamorphic Geology, v. 15, p. 345–360.

Pavlis, T.L., and Sisson, V.B., 1995, Structural history of the Chugach metamorphic complex in the Tana River region, eastern Chugach Mountains, Alaska: A record of tectonic processes during Eocene subduction of the Kula-Farallon spreading center: Geological Society of America Bulletin, v. 107, p. 1333–1355.

Pavlis, T.L., and Sisson, V.B., 2003, Development of a subhorizontal decoupling horizon in a transpressional system, Chugach metamorphic complex, Alaska: Evidence for rheological stratification of the crust, *in* Sisson, V.B., Roeske, S.M., and Pavlis, T.L., eds, Geology of a transpressional orogen developed during ridge-trench interaction along the North Pacific margin: Boulder, Colorado, Geological Society of America Special Paper 371, p. 191–216 (this volume).

Pavlis, T.L., Marty, K., and Sisson, V.B., 2003, Constrictional flow within the Eocene forearc of southern Alaska: An effect of dextral shear during ridge subduction, *in* Sisson, V.B., Roeske, S.M., and Pavlis, T.L., eds, Geology of a transpressional orogen developed during ridge-trench interaction along the North Pacific margin: Boulder, Colorado, Geological Society of America Special Paper 371, p. 171–190 (this volume).

Plafker, G., Nokleberg, W.P., Lull, J.S., Pessel, G.H., Wallace, W.K., and Winkler, G.R., 1989, Bedrock geology and tectonic evolution of the Wrangellia, Peninsular, and Chugach terranes along the Trans-Alaska Crustal Transect in the Chugach Mountains and southern Copper River Basin: Journal of Geophysical Research, v. 94, p. 4255–4295.

Rye, R.O., Schuiling, R.D., Rye, D.M., and Jansen, J.B.H., 1976, Carbon, hydrogen, and oxygen isotope studies of the regional metamorphic complex at Naxos, Greece: Geochimica et Cosmochimica Acta, v. 40, p. 1031–1049.

Sisson, V.B., and Hollister, L.S., 1988, Low-pressure facies series metamorphism in an accretionary sedimentary prism, southern Alaska: Geology, v. 16, p. 358–361.

Sisson, V.B., and Pavlis, T.L., 1993, Geologic consequences of plate reorganization: An example from the Eocene southern Alaska forearc: Geology, v. 21, p. 913–916.

Sisson, V.B., Hollister, L.S., and Onstott, T.C., 1989, Petrologic constraints on the origin of a low-pressure/high-temperature complex, eastern Chugach Mountains, Alaska: Journal of Geophysical Research, v. 94, p. 4392–4410.

Sisson, V.B., Poole, A.R., Harris, N.R., Cooper Burner, H., Pavlis, T.L., Copeland, P., Donelick, R.A., and McLelland, W.C., 2003, Geochemical and geochronologic constraints for genesis of a tonalite-trondhjemite suite and associated mafic intrusive rocks in the eastern Chugach Mountains, Alaska: A record of ridge-transform subduction, *in* Sisson, V.B., Roeske, S.M., and Pavlis, T.L., eds, Geology of a transpressional orogen developed during ridge-trench interaction along the North Pacific margin: Boulder, Colorado, Geological Society of America Special Paper 371, p. 293–326 (this volume)

Spicuzza, M.J., Valley, J.W., Kohn, M.J., Girard, J.P., and Fouillac, A.M., 1998, The rapid heating, defocused beam technique—A CO_2-laser-based method for highly precise and accurate determination of $\delta^{18}O$ values of quartz: Chemical Geology (Isotope Geoscience), v. 144, p. 195–203.

Stern, L.A., Chamberlain, C.P., Barnett, D.E., and Ferry, J.M., 1992, Stable isotope evidence for regional-scale fluid migration in a Barrovian metamorphic terrane, Vermont, USA: Contributions to Mineralogy and Petrology, v. 112, p. 475–489.

Valley, J.W., Kitchen, N., Kohn, M.J., Niendorf, C.R., and Spicuzza, M.J., 1995, UWG-2, A garnet standard for oxygen isotope ratio—Strategies for high precision and accuracy with laser heating: Geochimica et Cosmochimica Acta, v. 59, p. 5223–5231.

Vrolijk, P., Myers, G., and Moore, J.C., 1988, Warm fluid migration along tectonic melanges in the Kodiak accretionary complex, Alaska: Journal of Geophysical Research, v. 93, p. 10,313–10,324.

Weinberger, J., and Sisson, V.B., 2003, Pressure and temperature conditions of brittle-ductile vein emplacement in the greenschist facies, Chugach metamorphic complex, Alaska: Evidence from fluid inclusions, *in* Sisson, V.B., Roeske, S.M., and Pavlis, T.L., eds, Geology of a transpressional orogen developed during ridge-trench interaction along the North Pacific margin: Boulder, Colorado, Geological Society of America Special Paper 371, p. 217–235 (this volume).

Wickham, S.M., and Taylor, H.P., Jr., 1985, Stable isotopic evidence for large-scale seawater infiltration in a regional metamorphic terrane; the Trois Seigneurs Massif, Pyrenees, France: Contributions to Mineralogy and Petrology, v. 91, p. 122–137.

Xu, P., Sisson, V.B., Bowman, J.B., and Leeman, W.P., 1994, Fluid evolution of a low-*P*, high-*T* metamorphic terrrane, Chugach Mtns, southern Alaska: Geological Society of America Abstracts with Programs, v. 26, no. 7, p. 226.

Manuscript Accepted by the Society February 5, 2003

Geological Society of America
Special Paper 371
2003

Metamorphism within the Chugach accretionary complex on southern Baranof Island, southeastern Alaska

Cathy L. Zumsteg*
Glen R. Himmelberg
101 Geological Sciences, University of Missouri, Columbia, Missouri 65211, USA

Susan M. Karl
Peter J. Haeussler
U.S. Geological Survey, 4200 University Drive, Anchorage, Alaska 99508, USA

ABSTRACT

On Baranof Island, southeastern Alaska, we identify four metamorphic events that affect rocks associated with the Chugach accretionary complex. This study focuses on the M1 and M4 metamorphic events. Mesozoic schists, gneisses, and migmatitic gneisses exposed near the Kasnyku pluton on central Baranof Island represent the M1 metamorphic rocks. These rocks underwent amphibolite facies metamorphism. Calculated temperatures and pressures range from about 620 to 780 °C and 5.5 to 6.6 kbar and are compatible with the observed metamorphic mineral assemblages.

The M4 metamorphism affected rocks of the Sitka Graywacke on southern Baranof Island, producing extensive biotite and garnet zones as well as andalusite and sillimanite zones at the contacts of the Crawfish Inlet and Redfish Bay plutons. Calculated M4 temperatures and pressures from the andalusite and sillimanite zones range from 575 to 755 °C and 3.4 to 6.9 kbar. These results fall within the sillimanite stability field, at pressures higher than andalusite stability. These results may indicate the M4 metamorphic event occurred along a *P-T* path along which the equilibration of aluminosilicate–garnet–plagioclase–quartz did not occur or was not maintained. This interpretation is supported by the occurrence of andalusite and sillimanite within the same sample. We propose the data reflect a clockwise *P-T* path with peak M4 metamorphism of the sillimanite-bearing samples adjacent to the intrusions at an approximate depth of 15 to 20 km, followed by rapid uplift without reequilibration of garnet–plagioclase–aluminosilicate–quartz.

The large extent of the biotite zone, and possibly the garnet zone, suggests that an additional heat source must have existed to regionally metamorphose these rocks during the M4 event. We suggest the M4 regional thermal metamorphism and intrusion of the Crawfish Inlet and Redfish Bay plutons were synchronous and the result of heat flux from a slab window beneath the accretionary complex at that time. If our conclusions regarding the effect of the slab window are correct, the style of metamorphism is different from the Chugach metamorphic complex, which is clearly linked to a slab window. Therefore, our findings would suggest that there is no distinct metamorphic signature for slab window effects.

Keywords: Alaska, aluminosilicates, Baranof Island, Chugach terrane, metamorphism, Sitka Graywacke.

*clz0f3@mizzou.edu

Zumsteg, C.L., Himmelberg, G.R., Karl, S.M., and Haeussler, P.J., 2003, Metamorphism within the Chugach accretionary complex on southern Baranof Island, southeastern Alaska, *in* Sisson, V.B., Roeske, S.M., and Pavlis, T.L., eds., Geology of a transpressional orogen developed during ridge-trench interaction along the North Pacific margin: Boulder, Colorado, Geological Society of America Special Paper 371, p. 253–267.

INTRODUCTION

Metamorphic rocks exposed within the Chugach accretionary complex on Baranof Island (Fig. 1) occur in proximity to Paleocene-Eocene intrusions that are the easternmost expression of the Sanak-Baranof plutonic belt of Hudson (1979). The Sanak-Baranof plutonic belt intrudes a Mesozoic and Cenozoic accretionary complex, which lies between the Border Ranges fault and the current margin of southern Alaska (Hudson, 1979; Plafker et al., 1977). Plutons in this belt yield progressively younger ages from 61 Ma on the west at Sanak Island, to 50 Ma on the east at Baranof Island (Bradley et al., 1993; Bradley et al., 2000). Metamorphic rocks that range in grade from sub-greenschist to amphibolite facies occur discontinuously throughout the Sanak-Baranof plutonic belt. Some of the higher-grade metamorphic rocks, such as those on Baranof Island, have been interpreted to be a product of thermal metamorphism by the intrusions (Reifenstuhl, 1983; Loney and Brew, 1987). Others, such as the Chugach metamorphic complex, a large-scale high-temperature, low-pressure assemblage of rocks, are not directly attributed to contact metamorphism (Sisson and Hollister, 1988; Sisson et al., 1989; Sisson and Pavlis, 1993).

Typical thermal models of accretionary complexes do not readily explain the generation of heat necessary to produce the Sanak-Baranof plutonic belt, nor the high-grade metamorphism within the Chugach metamorphic complex. Thus, several workers have proposed that the ultimate source of heat was related to the subduction of young, hot oceanic crust at trench-ridge-trench triple junction as it migrated along the continental margin of Alaska (Marshak and Karig, 1977; Helwig and Emmett,

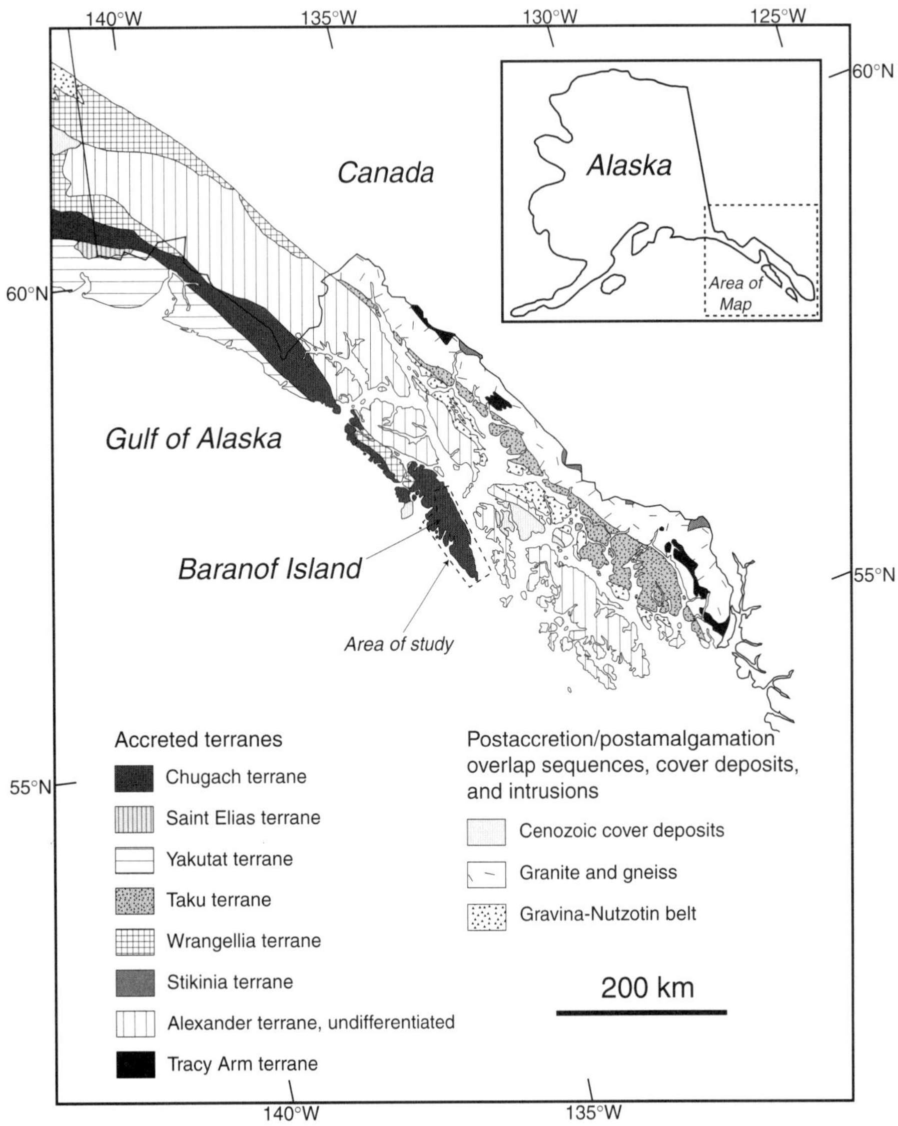

Figure 1. Map showing the study area on Baranof Island; terranes of southeastern Alaska; and postaccretion/postamalgamation overlap sequences, cover deposits, and intrusions (modified from Silberling et al., 1994).

1981; Moore et al., 1983; Sisson and Hollister, 1988; Sisson et al., 1989; Bradley et al., 1993; Sisson and Pavlis, 1993; Bradley et al., 2000). The subducted ridge may have been the Kula-Farallon spreading ridge (Marshak and Karig, 1977), or perhaps another spreading ridge (Bradley et al., 1993; Haeussler et al., 2000). Subduction of an oceanic spreading center would develop a slab window below the accretionary complex capable of elevating the thermal conditions (Dickinson and Snyder, 1979; DeLong et al., 1979; Thorkelson and Taylor, 1989; Thorkelson, 1996).

Loney et al. (1975) summarized the geology of Baranof Island and Loney and Brew (1987) described the relationships of the metamorphic rocks to the Crawfish Inlet and Redfish Bay intrusions on southern Baranof Island. This report extends their work to include the metamorphic rocks associated with the Kasnyku pluton from central Baranof Island as well as additional sampling on southern Baranof Island. In this study we provide and compare thermobarometric data to metamorphic conditions implied by mineral assemblages and established petrogenetic grids in order to place constraints on the thermal evolution of the eastern part of the Chugach accretionary complex. The mineral assemblages and thermobarometric data are evaluated to determine to what extent the metamorphism was related to the intrusions and if the metamorphism is consistent with slab window effects as proposed for other areas of the Chugach accretionary complex. The data and interpretations are then integrated and compared with those reported for other areas of the Sanak-Baranof plutonic belt.

GEOLOGIC SETTING

The metamorphic rocks of the study area (Fig. 2) that occur in proximity to Tertiary plutons of the Sanak-Baranof belt on Baranof Island are composed of Mesozoic semischist, gneiss, and migmatitic gneiss (Mzs, Mzm); Cretaceous flysch and mélange facies rocks of the Chugach accretionary complex (KJs, KJsh, KJss, KJk, KJkh, KJkm); and Triassic metasedimentary and metavolcanic rocks (Trsv) (Plafker et al., 1977; Decker, 1980; Johnson and Karl, 1985; Gehrels and Berg, 1994). In the study area the flysch facies is the Jurassic-Cretaceous Sitka Graywacke (Berg and Hinckley, 1963), which is composed dominantly of turbidites with subordinate mafic volcanic rocks (Berg and Hinckley, 1963; Loney et al., 1975; Johnson and Karl, 1985). The mélange facies rocks include Jurassic-Cretaceous Khaz Formation of the Kelp Bay Group. (Karl, 1982; Johnson and Karl, 1985). The Khaz Formation is a slatey argillite with lenses of tuff, graywacke, and chert (Karl, 1982; Johnson and Karl, 1985).

Plutons in the study area are from north to south: the Kasnyku (referred to as the Baranof Lake pluton by Bradley et al., 1993), Indigo Lake, Crawfish Inlet, Gut Bay, and Redfish Bay plutons (Loney et al., 1975) (Fig. 2). The composition of these plutons ranges from hornblende and biotite granodiorite, to tonalite and quartz diorite. The Kasnyku, Indigo Lake, Crawfish Inlet, and Redfish Bay plutons are considered to be part of the Sanak-Baranof plutonic belt. Conventional K-Ar dating of biotite, muscovite, and/or hornblende of these plutons yielded cooling ages of 48.3 to 39.6 Ma (Loney et al., 1967; Reifenstuhl, 1983; Bradley et al., 1993). U-Pb dating of zircon from the Crawfish Inlet pluton yielded an age of 50.1 Ma (Brew et al., 1991) and from the Redfish Bay pluton, 51.4 Ma (R. Friedman, 2000, personal commun.). The Gut Bay pluton is younger; K-Ar dating of biotite and hornblende yielded cooling ages from 31.5 to 24.3 Ma (Loney et al., 1967). The Kasnyku pluton intrudes the Mesozoic metamorphic rocks and the Triassic metasedimetary and metavolcanic rocks of the Chugach accretionary complex mélange. The Crawfish Inlet and Redfish Bay plutons for the most part intrude the Sitka Graywacke.

Loney et al. (1975) recognized two metamorphic events in the Chugach accretionary complex on Baranof Island. The first event was a low-grade regional metamorphism under prehnite-pumpellyite and lower greenschist facies conditions. The second event was primarily a regional thermal event, which they attributed to intrusion of the Crawfish Inlet and Redfish Bay plutons. The regional thermal event overprinted the low-grade regional event on southern Baranof Island and produced metamorphic rocks that range in grade from the biotite zone to sillimanite zone (Loney et al., 1975; Loney and Brew, 1987).

On the basis of our geologic mapping, petrography, and dating of metamorphic minerals, we recognize a minimum of four metamorphic events. The earliest recognized event (M1) was a regional metamorphic amphibolite facies event represented by the Mesozoic schist, gneiss, and migmatitic gneiss unit (Mzs, Mzm) exposed around the Kasnyku pluton on the east side of Baranof Island. The metamorphic fabric of the gneiss and schist, and the contact between the high-grade metamorphic rocks and the lower-grade Khaz mélange, are truncated by the intrusion. These rocks may be correlative with similar rocks on Duffield Peninsula on northern Baranof Island.

The second event (M2) was a regional greenschist facies metamorphism that produced phyllite, marble, and greenstone (Trsv) exposed between Red Bluff Bay and Patterson Bay on the east side of Baranof Island. $^{40}Ar/^{39}Ar$ dating of white mica indicates an age of 155 Ma (L. Snee, 2003, personal commun.). Metamorphic event M3 is equivalent to the prehnite-pumpellyite facies metamorphism recognized by Loney et al. (1975). K-Ar dating of metamorphic actinolite and sericite (collected from Chichagof Island) indicates a Cretaceous age (106 to 91 Ma) (Decker et al., 1980) and provides the only isotopic determination of this event. This metamorphism affected the Sitka Graywacke and the Kelp Bay Group of the Chugach accretionary complex and is attributed to the accretionary event. Primary sedimentary structures and textures are preserved where the M3 metamorphism has not been substantially overprinted by later metamorphic events. Metamorphic event M4 is a regional event that affected the Sitka Graywacke between Patterson Bay and Port Alexander on southern Baranof Island. The metamorphic grade ranges from biotite to garnet zone over most of the area,

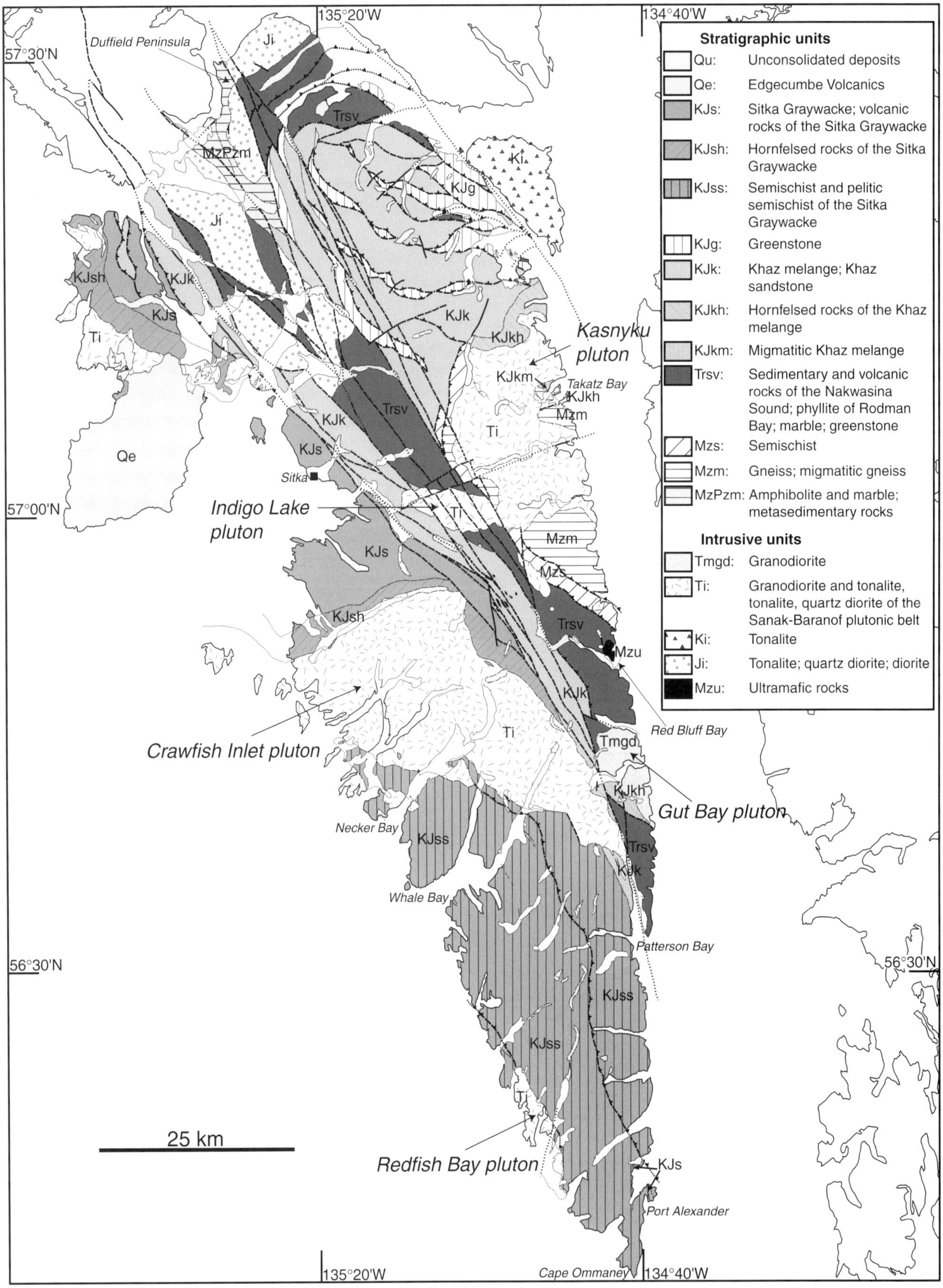

Figure 2. Geologic map showing the stratigraphic and intrusive units of Baranof Island.

but increases to andalusite and sillimanite zone near contacts with the Crawfish Inlet and Redfish Bay plutons.

MINERAL ASSOCIATIONS AND METAMORPHIC ZONES

This study focuses on the metamorphic rocks that surround the Kasnyku, Crawfish Inlet, and Redfish Bay plutons (M1 and M4). Mineral associations were determined by examination of more than 200 thin sections of samples collected from more than 150 locations (Fig. 3). Most samples were collected in association with field mapping during the summer of 1999 along the coastline and fjords of southern Baranof Island from Takatz Bay on the eastern coast, south to Cape Ommaney at the southern extent of the island, and north to Necker Bay on the western coast. The fjords provided excellent traverses deep into the interior of Baranof Island across the metamorphic zones and plutonic contacts. Other samples from the interior of the island were previously collected by Glen Himmelberg and Robert Loney using helicopter support.

Mineral associations from the Mesozoic pelitic and semipelitic schists and gneisses around the Kasnyku pluton (Fig. 3), and from inclusions within the pluton, are most commonly quartz–plagioclase–muscovite–biotite–garnet±staurolite±sillimanite. Although a garnet zone and a sillimanite zone could be inferred on the basis of the absence or presence of sillimanite, respectively, the regional distribution of sillimanite, as far as 6 km from the pluton contact, and consideration of protolith compositional effects precludes recognition of isograds and zones. Texturally the schist and gneiss are completely recrystallized with granoblastic quartz and feldspar and the foliation defined by micas and mineral segregations. Garnet porphyroblasts range from anhedral to euhedral and staurolite is subhedral. Sillimanite occurs as fine needles and mats.

The distribution of mineral associations supports our interpretation based on field relationships that the Mesozoic metamorphic rocks in the area of the Kasnyku pluton are a result of a high-grade regional metamorphism that predates the intrusion of the pluton. Rare quartz–plagioclase–muscovite–chlorite–biotite–garnet associations with hornfelsic texture occur near the contact with the pluton and are considered to represent thermal recrystallization of the preexisting sillimanite-grade, regionally metamorphosed rocks.

From Patterson Bay south to Port Alexander, mineral associations allow distinction of biotite, garnet, andalusite, and sillimanite zones (Fig. 3). Staurolite occurs in a few samples in the andalusite zone allowing recognition of an andalusite-staurolite subzone and an andalusite-staurolite–absent subzone. However, the limited number of samples containing staurolite precludes mapping these subzones.

Some samples with abundant, well-formed andalusite also contain minor amounts of sillimanite needles and mats. Similarly, some samples in the sillimanite zone also contain rare andalusite. In one sample (91GH12D) relict grains of sillimanite are enclosed in andalusite porphyroblasts. In all other samples, however, there is no textural relationship between the two polymorphs to suggest the nature of their co-occurrence within the same sample. Possible explanations for the co-occurrence of andalusite and sillimanite are discussed below.

The biotite zone is regionally extensive, occurring between the southern extension of the Crawfish Inlet pluton and the Redfish Bay pluton. The garnet isograd is well defined around the Redfish Bay pluton, but irregular about the Crawfish Inlet pluton. The andalusite and sillimanite isograds are also poorly defined. We attribute the scattered occurrence of the higher-grade index minerals and the resultant difficulty in mapping continuous isograds to the compositional heterogeneity of the protoliths. The metamorphic zones on the north side of the Crawfish Inlet pluton are markedly narrower than those on the south side. Classification of the metamorphic zones is in relative agreement with the zones presented by Loney and Brew (1987). The major exception is that we did not identify cordierite in any of our samples; thus, we have not recognized the andalusite-cordierite and garnet-cordierite zones of Loney and Brew (1987).

Textures of the metamorphic rocks range from incompletely recrystallized metagraywacke in the biotite zone to texturally mature schists in the andalusite and sillimanite zones. Relict clasts of quartz and feldspar are common in biotite zone rocks but rare in the garnet zone. Primary sedimentary structures are generally preserved in the biotite zone, and less commonly in the garnet zone. The andalusite and sillimanite zones are characterized by recrystallized mosaics of quartz and feldspar with no relict clasts. Foliation in the schists is defined by one or more of the minerals graphite, chlorite, biotite, and muscovite. Garnet occurs as subhedral to euhedral porphyroblasts. Andalusite commonly occurs as chiastolite porphyroblasts and sillimanite occurs as fibrolite, fine needles, and mats. In the higher-grade rocks, retrograde alteration is evident as chlorite and sericite.

MINERAL CHEMISTRY

Electron microprobe analyses of thirteen polished thin sections were completed on the JEOL-733 electron microprobe equipped with Advanced Microbeam automation at Washington University, St. Louis, Missouri. Sample selection was based on appropriate metamorphic mineral association, evidence of textural equilibrium with minor retrograde alteration, and distribution around the plutons. Standards used include natural mineral silicates and oxides. Analyses were obtained with a 15 kV accelerating voltage, a 30 nA beam current and a beam diameter of 1 μm for garnet, 5 μm for plagioclase, and 10 μm for biotite. X-ray matrix corrections were done with a modified Armstrong (1988) CITZAF routine incorporated into the software. Chemical compositions of garnet, biotite, and plagioclase used in thermobarometric calculations are given in Tables 1 through 3.

Analyzed garnet porphyroblasts range in size from approximately 0.2 to 2.8 mm and were subhedral to euhedral. The largest and most euhedral garnets in each thin section were selected for

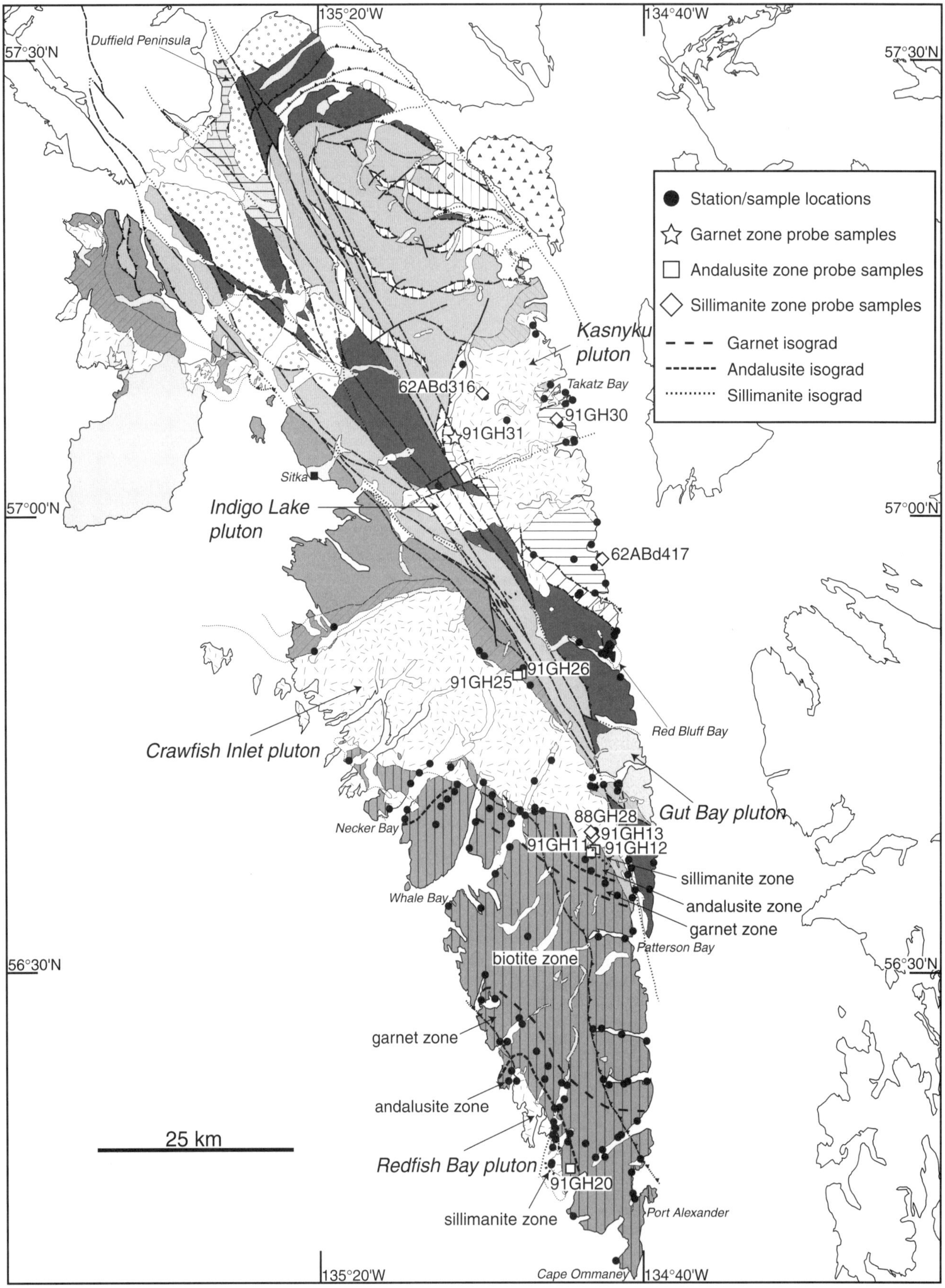

Figure 3. Map showing the station and sample locations; probe sample locations; and metamorphic zones south of the Crawfish Inlet pluton and surrounding the Redfish Bay pluton on Baranof Island. See Figure 2 for description of the stratigraphic and intrusive units.

TABLE 1. REPRESENTATIVE ANALYSES OF GARNET

Sample	62ABd316 c	62ABd316 r	62ABd417 c	62ABd417 r	88GH28A c	88GH28A r	91GH11C c	91GH11C r	91GH12A c
SiO_2	37.42	37.08	37.35	37.34	37.24	37.73	36.56	36.95	36.90
Al_2O_3	21.65	21.31	22.10	21.99	21.21	21.41	22.26	22.36	21.41
TiO_2	0.01	0.02	0.03	n.d.	0.13	0.12	0.04	0.02	0.08
FeO*	29.84	32.25	31.94	31.57	32.16	33.11	31.43	31.99	27.25
MnO	5.36	4.53	3.13	5.25	3.04	3.78	4.62	4.86	9.27
MgO	3.04	2.86	3.90	2.74	3.80	2.69	2.53	2.42	2.49
CaO	2.32	1.38	1.35	1.22	1.59	1.40	1.96	1.82	1.62
Σ	99.64	99.44	99.80	100.11	99.17	100.24	99.40	100.42	99.02
Structural formula on a basis of 12 oxygens									
Si	3.00	3.00	2.98	2.99	3.00	3.02	2.95	2.96	3.00
Al^{IV}	0.00	0.00	0.02	0.01	0.00	0.00	0.05	0.04	0.00
Al^{VI}	2.05	2.03	2.06	2.07	2.01	2.02	2.07	2.07	2.04
Ti	0.00	0.00	0.00	n.d.	0.01	0.01	0.00	0.00	0.00
Fe^{2+}	2.00	2.18	2.13	2.12	2.17	2.22	2.12	2.14	1.85
Mn	0.36	0.31	0.21	0.36	0.21	0.26	0.32	0.33	0.64
Mg	0.36	0.35	0.46	0.33	0.46	0.32	0.30	0.29	0.30
Ca	0.20	0.12	0.12	0.11	0.14	0.12	0.17	0.16	0.14
Σ	7.97	7.99	7.98	7.99	8.00	7.97	7.98	7.99	7.97
Xalm	0.68	0.74	0.73	0.73	0.73	0.76	0.73	0.73	0.63
Xprp	0.12	0.12	0.16	0.11	0.15	0.11	0.10	0.10	0.10
Xgrs	0.07	0.04	0.04	0.04	0.05	0.04	0.06	0.05	0.05
Xsps	0.12	0.10	0.07	0.12	0.07	0.09	0.11	0.11	0.22
Fe/(Fe+Mg)	0.85	0.86	0.82	0.87	0.83	0.87	0.88	0.88	0.86

Sample	91GH12A r	91GH12C c	91GH12C r	91GH13D c	91GH13D r	91GH20D h	91GH20D r	91GH25C c	91GH25C r
SiO_2	37.41	38.13	38.07	36.23	37.65	37.27	37.36	37.07	37.21
Al_2O_3	21.89	22.03	21.96	20.82	21.83	21.54	21.87	21.66	21.58
TiO_2	0.04	0.02	0.02	2.31	n.d.	0.11	0.01	0.01	n.d.
FeO*	28.61	28.98	30.33	28.52	28.34	27.36	28.17	31.63	31.13
MnO	8.14	7.99	7.14	8.06	8.56	9.29	8.45	6.07	6.64
MgO	2.25	2.47	2.29	2.84	2.37	2.66	2.63	2.42	2.22
CaO	1.52	1.62	1.53	1.46	1.46	1.59	1.53	1.38	1.17
Σ	99.86	101.24	101.34	100.24	100.21	99.82	100.02	100.24	99.95
Structural formula on a basis of 12 oxygens									
Si	3.00	3.02	3.02	2.91	3.01	3.00	3.00	2.98	3.00
Al^{IV}	0.00	0.00	0.00	0.09	0.00	0.00	0.00	0.02	0.00
Al^{VI}	2.08	2.05	2.05	1.89	2.06	2.04	2.06	2.04	2.05
Ti	0.00	0.00	0.00	0.14	n.d.	0.01	0.00	0.00	n.d.
Fe^{2+}	1.92	1.92	2.01	1.92	1.90	1.84	1.89	2.13	2.10
Mn	0.55	0.54	0.48	0.55	0.58	0.63	0.57	0.41	0.45
Mg	0.27	0.29	0.27	0.34	0.28	0.32	0.31	0.29	0.27
Ca	0.13	0.14	0.13	0.13	0.13	0.14	0.13	0.12	0.10
Σ	7.95	7.96	7.96	7.97	7.96	7.98	7.96	7.99	7.97
Xalm	0.67	0.66	0.70	0.65	0.66	0.63	0.65	0.72	0.72
Xprp	0.09	0.10	0.09	0.12	0.10	0.11	0.11	0.10	0.09
Xgrs	0.05	0.05	0.04	0.04	0.04	0.05	0.04	0.04	0.03
Xsps	0.19	0.19	0.17	0.19	0.20	0.22	0.20	0.14	0.15
Fe/(Fe+Mg)	0.88	0.87	0.88	0.85	0.87	0.85	0.86	0.88	0.89

Note: c—core; r—rim; h—hook at rim; n.d.—not detected.
*Total iron calculated as FeO.

(*continued*)

TABLE 1. REPRESENTATIVE ANALYSES OF GARNET (*continued*)

Sample	91GH26A c	91GH26A r	91GH26B c	91GH26B r	91GH30D c	91GH30D r	91GH31A c	91GH31A r
SiO_2	37.06	37.16	37.15	37.31	37.31	37.15	37.49	37.29
Al_2O_3	21.66	21.49	21.00	21.74	21.88	21.94	21.36	21.95
TiO_2	n.d.	n.d.	0.03	n.d.	0.02	n.d.	0.03	0.05
FeO*	27.06	27.41	28.39	29.90	31.29	30.46	27.49	31.62
MnO	10.40	10.01	8.69	8.05	4.89	7.15	6.90	3.12
MgO	2.29	1.87	2.31	2.10	3.36	2.50	2.09	3.02
CaO	1.50	1.38	1.77	1.62	1.46	1.43	4.48	2.99
Σ	99.97	99.32	99.34	100.72	100.21	100.63	99.84	100.04
Structural formula on a basis of 12 oxygens								
Si	2.99	3.01	3.01	2.99	2.98	2.98	3.01	2.98
Al^{IV}	0.01	0.00	0.00	0.01	0.02	0.02	0.00	0.02
Al^{VI}	2.05	2.06	2.01	2.04	2.04	2.05	2.20	2.04
Ti	n.d.	n.d.	0.00	n.d.	0.00	n.d.	0.00	0.00
Fe^{2+}	1.82	1.86	1.93	2.00	2.09	2.04	1.85	2.11
Mn	0.71	0.69	0.60	0.55	0.33	0.49	0.47	0.21
Mg	0.28	0.23	0.28	0.25	0.40	0.30	0.25	0.36
Ca	0.13	0.12	0.15	0.14	0.13	0.12	0.39	0.26
Σ	7.99	7.97	7.98	7.98	7.99	8.00	8.17	7.98
Xalm	0.62	0.64	0.65	0.68	0.71	0.69	0.63	0.72
Xprp	0.10	0.08	0.09	0.09	0.14	0.10	0.08	0.12
Xgrs	0.04	0.04	0.05	0.05	0.04	0.04	0.13	0.09
Xsps	0.24	0.24	0.20	0.19	0.11	0.17	0.16	0.07
Fe/(Fe+Mg)	0.87	0.89	0.87	0.89	0.84	0.87	0.88	0.85

Note: c—core; r—rim; n.d.—not detected.
*Total iron calculated as FeO.

microprobe analysis. The analyzed garnets are free of chlorite alteration and analyses were made on points away from contact with biotite to eliminate the effect of localized Fe-Mg exchange reactions. Representative garnet compositional zoning profiles are illustrated in Figure 4. The garnets from the andalusite- and sillimanite-bearing samples exhibit similar compositional zoning profiles. These profiles suggest that the garnets were homogeneous at peak temperatures, with diffusion processes modifying the rims upon cooling. The garnet compositional zoning profile from the garnet–chlorite–biotite assemblage on northern Baranof Island records growth-zoning conditions.

Matrix biotite was analyzed. To minimize possible local Fe-Mg exchange effects, the grains selected for analysis were free of chloritization and not in contact with garnet. The analyzed biotite was chemically unzoned.

Plagioclase grains near analyzed garnet porphyroblasts were selected for analysis. A minimum of three plagioclase grains per sample, with four to five core and rim points per grain, were analyzed. Plagioclase in sillimanite zone samples shows essentially no zoning, and there is no significant point to point or grain to grain variation in composition within a sample. The range in An content of plagioclase among the five sillimanite zone samples is only 0.30 to 0.33. Some plagioclase grains in andalusite-bearing samples do show zoning; in all cases, the rims are more anorthitic than the cores. Rim compositions from grain to grain within a given sample are generally quite consistent. Even between samples from the andalusite zone of the Crawfish Inlet pluton, the anorthite content of plagioclase rims ranges only from about 0.31 to 0.36. The sample from near the Redfish Bay pluton, however, has near homogeneous plagioclase core and rim compositions of about $An_{24–26}$.

GEOTHERMOBAROMETRY

Procedures and Results

Temperatures and pressures under which samples of Baranof Island schists and gneisses were metamorphosed were calculated from microprobe analyses and experimentally and empirically calibrated equilibria. Temperature determinations are based on garnet-biotite Fe-Mg exchange equilibrium as calibrated by Ferry and Spear (1978). Pressure determinations are based on the garnet–aluminumosilicate–quartz–plagioclase equilibrium (GASP) as formulated by Newton and Haselton (1981). This calibration was used so that the equilibria involving andalusite and sillimanite would be internally consistent.

TABLE 2. REPRESENTATIVE ANALYSES OF BIOTITE

Sample	62ABd316	62ABd417	88GH28A	91GH11C	91GH12A	91GH12C	91GH13D	91GH20D	91GH25C	91GH26A	91GH26B	91GH30D	91GH31A
SiO_2	35.99	34.98	35.22	34.70	34.56	34.73	34.93	36.28	35.02	35.42	34.60	34.93	35.97
Al_2O_3	21.03	20.28	20.53	20.43	20.35	20.49	20.73	19.75	20.33	20.34	20.40	20.37	17.82
TiO_2	1.64	2.55	2.19	1.44	1.60	2.02	2.30	1.75	1.82	1.67	1.40	2.56	1.53
FeO*	18.33	19.78	20.23	20.29	20.16	20.75	19.89	18.87	20.67	19.71	19.92	19.36	19.33
MnO	0.06	0.17	0.13	0.04	0.22	0.13	0.28	0.15	0.14	0.17	0.13	0.16	0.11
MgO	10.25	8.75	9.19	9.62	9.41	8.81	9.00	10.40	9.20	9.52	9.29	9.32	11.57
CaO	n.d.	n.d.	n.d.	n.d.	0.01	n.d.	0.01	n.d.	n.d.	n.d.	n.d.	n.d.	n.d
Na_2O	0.27	0.21	0.12	0.18	0.20	0.19	0.24	0.12	0.09	0.08	0.11	0.15	0.12
K_2O	8.50	8.77	8.99	8.18	8.32	8.58	8.45	9.13	8.39	8.92	8.96	8.80	8.49
Σ	96.07	95.49	96.60	94.88	94.83	95.70	95.83	96.45	95.66	95.83	94.81	95.65	94.94
Structural formula on a basis of 11 oxygens													
Si	2.68	2.65	2.64	2.64	2.64	2.64	2.63	2.71	2.65	2.67	2.65	2.64	2.73
Al^{IV}	1.32	1.35	1.36	1.36	1.36	1.36	1.37	1.29	1.35	1.33	1.35	1.36	1.27
Al^{VI}	0.52	0.46	0.46	0.48	0.47	0.47	0.48	0.44	0.47	0.48	0.49	0.45	0.33
Ti	0.09	0.15	0.12	0.08	0.09	0.12	0.13	0.10	0.10	0.10	0.08	0.15	0.09
Fe^{2+}	1.14	1.25	1.27	1.29	1.29	1.32	1.25	1.18	1.31	1.24	1.27	1.22	1.23
Mn	0.00	0.01	0.01	0.00	0.01	0.01	0.02	0.01	0.01	0.01	0.01	0.01	0.01
Mg	1.14	0.99	1.03	1.09	1.07	1.00	1.01	1.16	1.04	1.07	1.06	1.05	1.31
Ca	n.d.	n.d.	n.d.	n.d.	0.00	n.d.	0.00	n.d.	n.d.	n.d.	n.d.	n.d.	n.d.
Na	0.04	0.03	0.02	0.03	0.03	0.03	0.03	0.02	0.01	0.01	0.02	0.02	0.02
K	0.81	0.85	0.86	0.80	0.81	0.83	0.81	0.87	0.81	0.86	0.88	0.85	0.82
Σ	7.74	7.74	7.77	7.77	7.77	7.78	7.73	7.78	7.75	7.77	7.81	7.75	7.81
Fe/(Fe+Mg)	0.50	0.56	0.55	0.54	0.55	0.57	0.55	0.50	0.56	0.54	0.55	0.54	0.48

Note: n.d.—not detected.
*Total iron calculated as FeO.

TABLE 3. REPRESENTATIVE ANALYSES OF PLAGIOCLASE

Sample	62ABd316 r	62ABd417 r	88GH28A r	91GH11C r	91GH12A r	91GH12C r	91GH13D r	91GH20D r	91GH25C r	91GH26A r	91GH26B r	91GH30D r
SiO_2	60.83	60.56	60.66	59.47	59.90	59.86	59.91	62.16	60.09	60.17	59.38	60.74
Al_2O_3	25.21	25.24	25.68	25.93	25.61	26.30	25.77	24.03	25.96	25.77	26.14	25.22
FeO*	0.08	0.02	0.21	0.14	0.09	0.14	0.11	0.05	0.17	0.05	0.12	0.04
CaO	6.20	6.34	6.35	7.10	6.76	7.01	6.72	5.08	6.55	6.75	7.17	6.19
Na_2O	8.09	7.94	7.67	7.26	7.63	7.15	7.34	8.77	7.56	7.73	7.43	7.99
K_2O	0.07	0.07	0.09	0.07	0.09	0.08	0.09	0.08	0.11	0.08	0.07	0.08
BaO	0.01	0.01	0.02	0.03	0.03	0.03	0.04	0.01	0.01	0.03	0.01	0.03
Σ	100.49	100.18	100.68	100.00	100.11	100.57	99.98	100.18	100.45	100.58	100.32	100.29
Structural formula on a basis of 8 oxygens												
Si	2.69	2.69	2.68	2.65	2.66	2.65	2.67	2.75	2.66	2.66	2.64	2.69
Al^{IV}	1.31	1.31	1.32	1.35	1.34	1.35	1.33	1.25	1.34	1.34	1.36	1.31
Al^{VI}	0.01	0.01	0.02	0.01	0.01	0.02	0.02	0.00	0.02	0.01	0.01	0.01
Fe^{2+}	0.00	0.00	0.01	0.01	0.00	0.01	0.00	0.00	0.01	0.00	0.00	0.00
Ca	0.29	0.30	0.30	0.34	0.32	0.33	0.32	0.24	0.31	0.32	0.34	0.29
Na	0.69	0.68	0.66	0.63	0.66	0.61	0.63	0.75	0.65	0.66	0.64	0.69
K	0.00	0.00	0.01	0.00	0.01	0.00	0.01	0.00	0.01	0.00	0.00	0.00
Ba	0.00	0.00	0.00	0.00	0.00	0.00	0.00	0.00	0.00	0.00	0.00	0.00
Σ	4.99	4.99	5.00	4.99	5.00	4.97	4.98	4.99	5.00	4.99	4.99	4.99
Xan	0.30	0.31	0.31	0.35	0.32	0.35	0.33	0.24	0.32	0.33	0.35	0.30
Xab	0.70	0.69	0.68	0.65	0.67	0.65	0.66	0.76	0.67	0.67	0.65	0.70
Xor	0.00	0.00	0.01	0.00	0.01	0.00	0.01	0.00	0.01	0.00	0.00	0.00

Note: r—rim; n.d.—not detected.
*Total iron calculated as FeO.

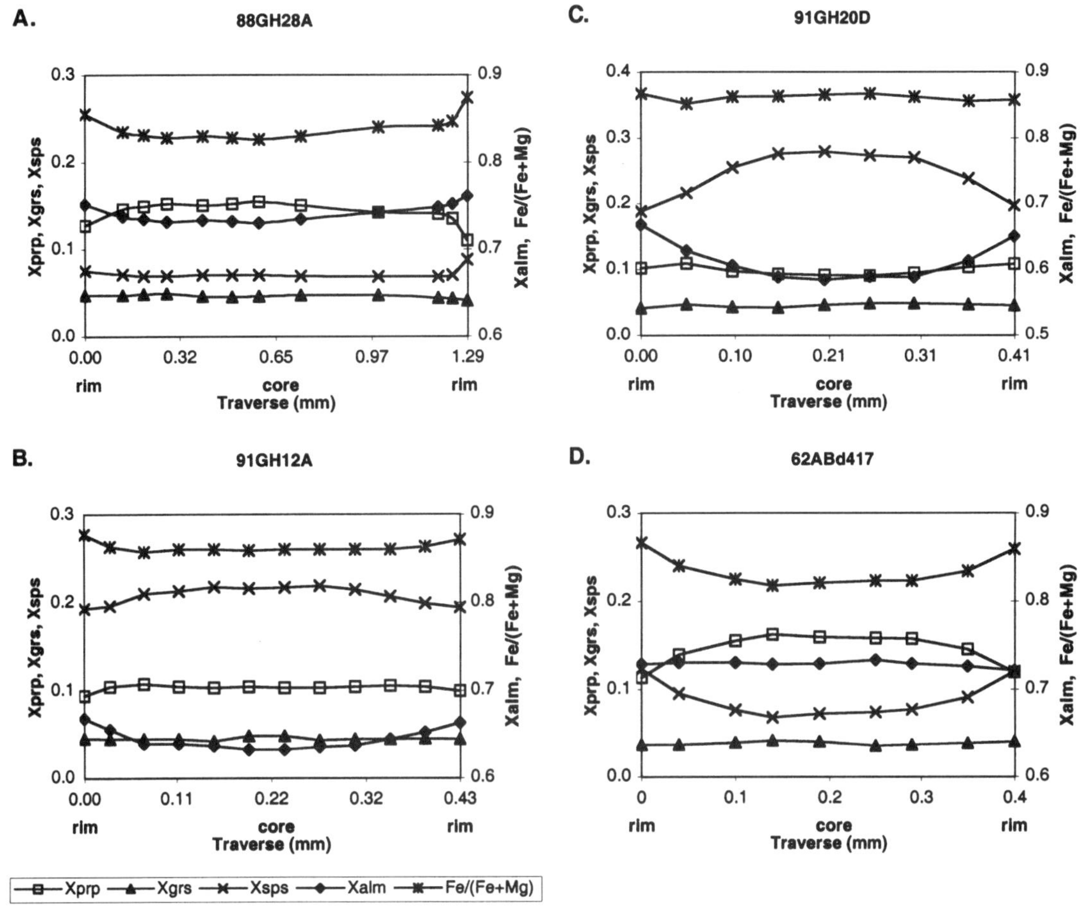

Figure 4. Representative garnet compositional zoning profiles. A: M4 sillimanite zone rocks associated with the Crawfish Inlet pluton. B: M4 andalusite zone rocks associated with the Crawfish Inlet pluton. C: M4 andalusite zone rocks associated with the Redfish Bay pluton. D: M1 sillimanite-bearing rocks.

The later calibration of the GASP geobarometer by Koziol and Newton (1988) was determined for kyanite and sillimanite, but not for andalusite. Pressures calculated using the Koziol and Newton (1988) geobarometer for sillimanite-bearing samples are about 1 to 1.5 kbar higher than obtained by the Newton and Haselton (1981) calibration. Activity-composition relationships used are those proposed by Hodges and Crowley (1985).

Temperature and pressure were calculated for andalusite- and sillimanite-bearing samples in an attempt to recover the peak conditions of the M1 and M4 metamorphic events. Because of the flat zoning profiles, suggesting chemical homogenization at peak temperatures, garnet core compositions were used in all calculations. Averaged plagioclase rim compositions, which showed no core to rim zoning or zoning toward more An-rich rims, were used. It was then assumed that the garnet and plagioclase compositions used were in equilibrium with matrix biotite, andalusite or sillimanite, and quartz. However, as will be presented in the following sections, the results may or may not reflect true equilibrium peak conditions. Therefore, referral to "peak" conditions in this section may not necessarily reflect true equilibrium metamorphic peak temperatures and pressures, but rather describes the highest metamorphic conditions calculated from the mineral data.

The results of the temperature and pressure calculations are shown in Figure 5 along with the aluminosilicate stability fields of Holdaway (1971). Absolute error in calculated temperature and pressure associated with calibration uncertainties can be quite large (± several kbar and more than 100 °C, Hodges and Crowley, 1985; Hodges and McKenna, 1987). The Koziol and Newton (1988) calibration of GASP reduces the uncertainty of calculated pressure by a factor of two, but as noted above the pressures are increased by 1 to 1.5 kbar. Errors are also associated with activity-composition relations; calculated temperatures and, thus, pressures are reduced by about 20 °C and 300 bar, respectively, if the Ferry and Spear (1978) ideal solution model for garnet and biotite are used in the Fe-Mg exchange

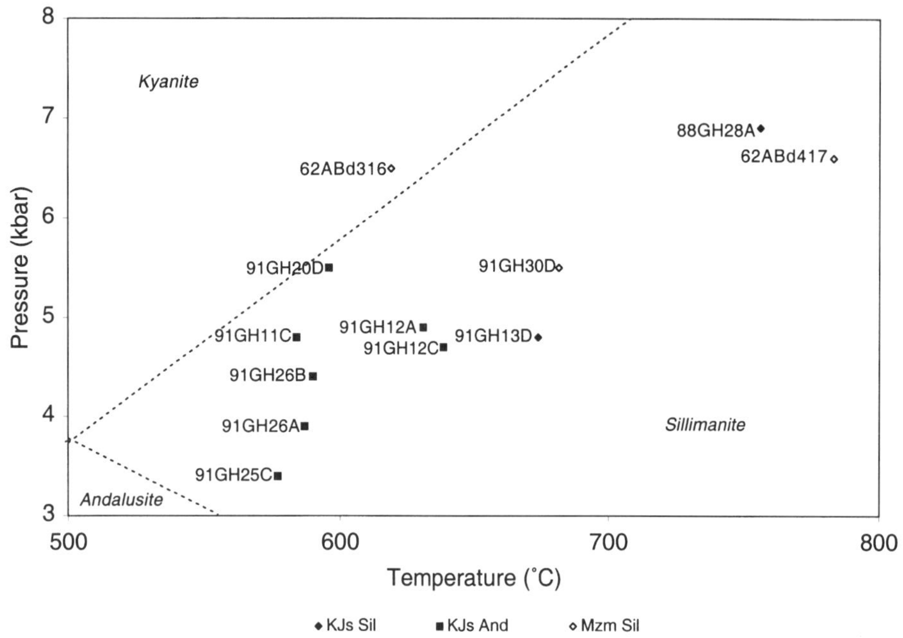

Figure 5. Calculated pressures and temperatures shown with respect to the aluminosilicate stability fields of Holdaway (1971) for the M4 andalusite- and sillimanite-bearing samples within the Sitka Graywacke, and the M1 sillimanite-bearing samples from the Mesozoic rocks. KJs—Sitka Graywacke; Mzm—Mesozoic metamorphic rocks; Sil—sillimanite; And—andalusite.

thermometer. The relative precision of the temperature and pressure determinations due to uncertainties in microprobe analyses is about 15 °C and 500 bar (1σ).

Interpretation of Results

Calculated temperatures and pressures for three samples of the M1 Mesozoic gneiss and schist in the area of the Kasnyku pluton range from about 620 to 780 °C and 5.5 to 6.6 kbar. All are sillimanite-bearing samples and, with one exception, they plot in the sillimanite stability field (Fig. 5). The exception plots in the kyanite stability field, just above the sillimanite-kyanite univariant curve, and can be attributed to analytical, calibration, or activity-composition error. Considering that andalusite was not observed in any samples of the gneiss and schist in this area, the calculated temperatures and pressures may be reasonable approximations of the conditions of M1 regional metamorphism.

The calculated temperatures and pressures for M4 metamorphic rocks near and between the Crawfish Inlet and Redfish Bay plutons range from 575 to 755 °C and 3.4 to 6.9 kbar. All samples, sillimanite- and andalusite-bearing, plot in the sillimanite stability field and all but two are at pressures significantly higher than the maximum pressure for andalusite stability as indicated by Holdaway's (1971) aluminosilicate triple point (3.76 kbar). The discrepancy between the determined andalusite stability field and the calculated temperature and pressure for andalusite-bearing samples implies that (1) the calculated temperatures and pressures are incorrect because of large errors in calibration of thermobarometers, activity-composition relations, or analytical error; or (2) the metamorphism followed a *P-T* path along which equilibration of aluminosilicate–garnet–plagioclase–quartz did not occur or was not maintained for all samples of the *P-T* array.

We consider the first alternative unlikely, mainly because numerous other studies have shown that the calibrations and activity-composition relations used in this study commonly yield geologically reasonable temperatures and pressures (e.g., Hodges and Spear, 1982; Essene, 1989; Spear, 1989, 1993; Sisson et al., 1989; Himmelberg et al., 1991). Likewise, repeat analysis of garnet, plagioclase, and biotite in several samples yielded results within the statistical precision mentioned above. Low mole fractions of grossular and anorthite have been demonstrated to increase the uncertainty in pressure calculations regardless of the thermodynamic databases, activity models, etc. (Todd, 1998). Todd's (1998) results suggest caution should be applied when the product of X_{grs} and X_{an} is <0.05, which may lead to pressure errors >3 kbar. All of the analyzed samples yield $X_{grs}X_{an}$ products ranging from 0.01 to 0.02, which may contribute to the unreasonable pressures.

We consider non-equilibrium of the GASP minerals the most reasonable explanation. This is supported by the fact that some of the samples that contain andalusite also contain sillimanite needles and mats, and some sillimanite zone samples contain minor amounts of andalusite. It is unlikely that these samples with two aluminosilicate polymorphs reflect univariant equilibrium. Instead, one of the polymorphs must exist outside its stability field and thus is not part of the stable assemblage. Even where only one polymorph is present, it too may not have reached chemical equilibrium with the associated plagioclase, garnet, and biotite.

We propose three limiting *P-T* paths for creation of the determined *P-T* array and the occurrence of aluminosilicates; the first two assume that the temperature and pressure calculated for the sillimanite zone samples approximate the peak conditions. (1) The metamorphism followed a counterclockwise andalusite-sillimanite *P-T* path where garnet–plagioclase–quartz equilibrated with sillimanite, and andalusite persisted beyond its stability field; (2) the metamorphism followed a clockwise *P-T* path where garnet–plagioclase–quartz equilibrated with sillimanite, but did not reequilibrate with andalusite during cooling and decompression; or (3) the metamorphism followed a *P-T* path where aluminosilicate–garnet–plagioclase–quartz did not equilibrate in any of the samples, and thus calculations yield incorrect temperatures and pressures.

The proposed counterclockwise *P-T* path assumes that the calculated high temperatures and moderate pressures of sillimanite zone samples approximate the peak metamorphic conditions. The *P-T* path would involve intrusion and initial metamorphism at shallow depth, consistent with the andalusite stability field, followed by crustal thickening and continued heating to the peak conditions at higher pressure in the sillimanite stability field. Persistence of andalusite can be attributed to sluggish reaction rates due to small free energy differences between phases (Kerrick, 1987, 1990), which is supported by the common occurrence of both andalusite and sillimanite in many of the samples. Crustal thickening could be accomplished by thrusting, or ballooning of magmas at shallower levels similar to that proposed by Brown and Walker (1993) and Brown (1996). Although counterclockwise *P-T* paths are believed to be common for andalusite-sillimanite metamorphic field gradients (Spear, 1993), there is no independent evidence on Baranof Island to support crustal thickening following intrusion. Furthermore, thrust thickening of the crust should produce a regional perturbation of the thermal structure resulting in regional metamorphism (England and Richardson, 1977; England and Thompson, 1984).

The second proposed *P-T* path also considers that the calculated temperatures and pressures of sillimanite zone samples approximate peak metamorphic conditions. In this case, the peak conditions would have been achieved by burial of the accretionary complex to depths of ~15–20 km followed by intrusion and associated metamorphism at that depth. Thermal models of subduction zones predict that a steady-state thermal gradient (10 °C/km) in the accretionary wedge at this depth would result in low temperatures of ~150–200 °C, which would be consistent with the M3 prehnite-pumpellyite grade of metamorphism of the area prior to intrusion. Andalusite could form along this path as a result of decompression and cooling following peak conditions. The occurrence of relict sillimanite enclosed in andalusite in sample 91GH12D lends some support to this interpretation. Two-dimensional thermal modeling (e.g., Quottoon pluton, British Columbia, Harrison and Clarke, 1979) illustrates that cooling accompanied by uplift and erosion causes rocks at different distances from the pluton to not only achieve different thermal maxima at markedly different times, but the thermal maxima could also be achieved at significantly different pressures (Spear, 1993). Qualitative application of this model to the accretionary complex metamorphic rocks lends some support to the proposed clockwise *P-T* path with the sillimanite-bearing schists near the contact equilibrating at higher pressures than the andalusite-bearing schists further from the contact equilibrating at lower pressures. However, the calculated temperatures and pressures for andalusite bearing samples that are outside the andalusite stability field indicate that if a *P-T* path similar to this one was followed, equilibrium of andalusite with garnet-plagioclase was not obtained.

The third proposed *P-T* path suggests that the calculated temperatures and pressures are incorrect largely because of non-equilibration of aluminosilicate–garnet–plagioclase–quartz and that the *P-T* path was largely isobaric and compatible with the limiting pressure of the andalusite stability field. This interpretation is supported by the common occurrence of andalusite and sillimanite in the same sample, suggesting sluggish and incomplete reactions. Although the calculated temperatures, as determined using the calculated pressures, for andalusite-bearing samples are too high relative to the andalusite stability field, the temperature calculations yield reasonable, compatible results if constrained by lower pressures of the andalusite stability field. This suggests that the garnet-biotite Fe-Mg exchange reaction may have attained equilibrium even though the net-transfer aluminosilicate–garnet–plagioclase–quartz association did not. Independent evidence for pressures limited by the andalusite stability field comes from Al in hornblende from the Crawfish Inlet plutons, which indicate pressures of crystallization of about 3 kbar (Brew et al., 1991). However, these pressures may reflect post-magmatic reequilibration, possibly related to emplacement of the Gut Bay stock, rather than emplacement of the Crawfish Inlet pluton (J. Hammarstrom, 1990, written commun.).

DISCUSSION AND CONCLUSIONS

On the basis of field relations, structure, and distribution of mineral associations we conclude that the Mesozoic gneisses and schists in the area of the Kasnyku pluton are a result of a regional amphibolite facies metamorphic event (M1) and that they are not a product of thermal metamorphism in response to intrusion of the Kasnyku pluton. This group of metamorphic rocks is tentatively correlated with the Mesozoic-Paleozoic gneisses and schists exposed on Duffield Peninsula on northern Baranof Island. Intrusion of the Kasnyku pluton recrystallized the gneisses and schists and produced a garnet zone hornfels aureole about 1 km wide or less. Thermobarometry of sillimanite-bearing schists and gneisses yield temperatures ranging from about 620 to 780 °C at pressures of about 5.5 to 6.6 kbar. These temperatures possibly reflect the approximate conditions of peak M1 metamorphism.

On southern Baranof Island, in the metamorphosed Sitka Graywacke, sillimanite, andalusite, garnet, and biotite zones occur progressively outward from the contacts with the Crawfish Inlet and Redfish Bay plutons. Although the zones are not well defined, in part because of protolith differences, and because both andalusite and sillimanite occur in many samples, the spatial relationship indicates a genetic relationship between the plutons and the peak metamorphism of the highest-grade zones. Calculated pressures and temperatures for andalusite- and sillimanite-bearing schists yield a *P-T* array that falls in the sillimanite stability field and, for the most part, at pressures higher than the maximum pressure of stability of andalusite. One possible explanation is that the data reflect a clockwise *P-T* path with peak metamorphism of the sillimanite-bearing samples resulting from intrusion of the Crawfish Inlet and Redfish Bay plutons at a depth of about 15 to 20 kilometers, followed by rapid uplift and erosion into the andalusite stability field without complete reequilibration of the aluminosilicate–garnet–plagioclase–quartz net-transfer reaction. Alternatively, all calculated temperatures and pressures could be incorrect due to nonequilbration of the aluminosilicate–garnet–plagioclase–quartz association. Aluminum in hornblende in the Crawfish Inlet pluton suggest pressures of about 3 kbar (Brew et al., 1991), but there is some question whether this pressure reflects shallow emplacement or post-magmatic reequilibration, possibly related to the emplacement of the nearby Gut Bay stock (J. Hammarstrom, 1990, written commun.). Recalculating temperatures to a pressure of 3 kbar yields values for andalusite- and sillimanite-bearing samples ranging from about 575 to 630 °C and 665 to 740 °C, respectively, which is consistent with the experimentally determined stability fields for andalusite and sillimanite.

Although the mineral zones appear to be spatially related to the plutons, the garnet isograd occurs about 6 km outward from the southern contact of the Crawfish Inlet pluton and about 12 km outward from the Redfish Bay pluton. In addition, the biotite zone extends over a region of about 25 km between the two plutons. Thus, it would appear that a heat source other than only the plutons must have existed to regionally metamorphose the rocks to biotite and, possibly, garnet grade. Loney et al. (1975) and Loney and Brew (1987) attributed the regional occurrence of biotite zone rocks to connection of the plutons at depth, but we found no evidence such as small intrusions or numerous dikes or sills to support this suggestion.

Several workers (Sisson et al., 1989; Sisson and Pavlis, 1993) have proposed that the high-temperature/low-pressure metamorphism of the Chugach metamorphic complex, elsewhere in the Chugach accretionary complex, was related to enhanced heat flux from a slab window associated with the migration of a trench-ridge-trench triple junction (Marshak and Karig, 1977). Brown (1998) has argued a similar cause for the high-temperature/low-pressure metamorphism of the Ryoke and Abukuma metamorphic belts in Japan, and Bradley et al. (1993, 2000) have argued that the origin of the Sanak-Baranof plutons is also related to the subduction of an oceanic spreading center. Thus we suggest, too, that the M4 regional thermal metamorphism of the Sitka Graywacke and the intrusion of the Crawfish Inlet and Redfish Bay plutons were largely synchronous and both a result of heat flux from a slab window beneath the accretionary complex. Enhanced heat flux from the plutons resulted in development of the high-temperature sillimanite and andalusite zones and possibly the broad garnet zones.

The low- to moderate-*P* and high-*T* metamorphic rocks on Baranof Island are different from those of the Chugach metamorphic complex. The Chugach metamorphic complex consists of an elongate (>200 km) high-grade core (approximately 675 °C and 3 kbar) surrounded by successively lower-grade assemblages without any direct spatial relationship to plutons (Sisson et al., 1989; Sisson and Pavlis, 1993), whereas on southern Baranof Island, M4 high-grade schists occur adjacent to plutons and grade outward to regional M4 garnet and biotite zone schists. The difference in distribution of metamorphic assemblages, peak metamorphic conditions and *P-T* paths of the Sitka Graywacke on southern Baranof Island and the Chugach metamorphic complex to the north (Sisson et al., 1989; Sisson and Pavlis, 1993) illustrates that, if the metamorphism of these two areas is related to high heat flow from a slab window, there is no distinct metamorphic signature associated with this origin. Similar conclusions were drawn by Brown (1998) in his study of the effects of slab window metamorphism in the Japanese Islands. As pointed out by Brown (1998), however, ridge-trench interactions generate distinctive packages of igneous, metamorphic, structural, and sedimentation effects.

ACKNOWLEDGMENTS

The U.S. Geological Survey and University of Missouri—Columbia provided support for this project. Thanks go to Maria Luisa Crawford and Harold Stowell for their helpful reviews of the original manuscript and to Dan Kremser for his assistance on the use of the electron microprobe at Washington University, St. Louis. Sarah Roeske provided field collaboration for the dated northern Baranof sample. Frank and Linda Churchill of the MV Venture and the crew of the MV Raven's Fire provided safe and enjoyable passage through Alaskan waters.

REFERENCES CITED

Armstrong, J.T., 1988, Quantitative analysis of silicate and oxide minerals: Comparison of Monte-Carlo, ZAF, and Phi-Rho-Z procedures, *in* Newbury, D.E., ed., Microbeam analysis: San Francisco, California, San Francisco Press, p. 239–246.

Berg, H.C., and Hinckley, D.W., 1963, Reconnaissance geology of northern Baranof Island, Alaska: U.S. Geological Survey Bulletin 1141-O, p. O1–O24.

Bradley, D.C., Haeussler, P.J., and Kusky, T.M., 1993, Timing of early Tertiary ridge subduction in southern Alaska, *in* Dusel-Bacon, C., and Till, A.B., eds., Geologic studies in Alaska by the U.S. Geological Survey, 1992: U.S. Geological Survey Bulletin 2068, p. 163–177.

Bradley, D.C., Parrish, R., Clendenen, W., Lux, D., Layer, P., Heizler, M., and Donley, T.D., 2000, Tertiary ridge subduction in southern Alaska, *in* Kelley, K.D., and Gough, L.P., eds., Geologic studies in Alaska by the U.S. Geological Survey, 1998: U.S. Geological Survey Professional Paper 1615, p. 5–20.

Brew, D.A., Hammarstrom, J.M., Himmelberg, G.R., Wooden, J.L., Loney, R.A., and Karl, S.M., 1991, Crawfish Inlet pluton, Baranof Island, southeastern Alaska—A north-tilted Eocene body or an untilted enigma?: Geological Society of America Abstracts with Programs, v. 23, no. 2, p. 8.

Brown, E.H., 1996, High-pressure metamorphism caused by magma loading in Fiordland, New Zealand: Journal of Metamorphic Geology, v. 14, p. 441–452.

Brown, E.H., and Walker, H.W., 1993, A magma-loading model for Barrovian metamorphism in the southeast Coast Plutonic Complex, British Columbia and Washington: Geological Society of America Bulletin, v. 105, p. 479–500.

Brown, M., 1998, Ridge-trench interactions and high-*T*–low-*P* metamorphism, with particular reference to the Cretaceous evolution of the Japanese Islands, *in* Treloar, P.J., and O'Brien, P., eds., What drives metamorphism and metamorphic reactions?: Geological Society [London] Special Publication No. 138, p. 137–169.

Decker, J.E., Jr., 1980, Geology of a Cretaceous subduction complex, western Chichagof Island, southeastern Alaska [Ph.D. thesis]: Stanford, California, Stanford University, 135 p.

Decker, J.E., Wilson, F.H., and Turner, D.L., 1980, Mid-Cretaceous subduction event in southeastern Alaska: Geological Society of America Abstracts with Programs, v. 12, p. 103.

DeLong, S.E., Schwarz, W.M., and Anderson, R.N., 1979, Thermal effects of ridge subduction: Earth and Planetary Science Letters, v. 44, p. 239–246.

Dickinson, W.R., and Snyder, W.S., 1979, Geometry of subducted slabs related to San Andreas transform: Journal of Geology, v. 87, p. 609–627.

England, P.C., and Richardson, S.W., 1977, The influence of erosion upon the mineral facies of rocks from different metamorphic environments: Geological Society [London] Journal, v. 134, p. 201–213.

England, P.C., and Thompson, A.B., 1984, Pressure-temperature-time paths of regional metamorphism, I: Heat transfer during the evolution of regions of thickened continental crust: Journal of Petrology, v. 25, p. 894–928.

Essene, E.J., 1989, The current status of thermobarometry in metamorphic rocks, *in* Daly, J.S., et al., eds., The evolution of metamorphic belts: Geological Society [London] Special Publication No. 43, p. 1–44.

Ferry, J.M., and Spear, F.S., 1978, Experimental calibration of the partitioning of Fe and Mg between biotite and garnet: Contributions to Mineralogy and Petrology, v. 66, p. 113–117.

Gehrels, G.E., and Berg, H.C., 1994, Geology of southeastern Alaska, *in* Plafker, G., and Berg, H.C., eds., The geology of Alaska: Boulder, Colorado, Geological Society of America, Geology of North America, v. G-1, p. 451–467.

Haeussler, P.J., Bradley, D., and Miller, M., 2000, Life and death of the Resurrection plate: evidence for an additional plate in the NE Pacific in Paleocene-Eocene time: Geological Society of America Abstracts with Programs, v. 32, no. 7, p. A-382.

Harrison, T.M., and Clarke, G.K.C., 1979, A model of the thermal effects of igneous intrusion and uplift as applied to the Quottoon pluton, British Columbia: Canadian Journal of Earth Sciences, v. 16, p. 411–420.

Helwig, J., and Emmet, P., 1981, Structure of the Early Tertiary Orca Group in Prince William Sound and some implications for the plate tectonic history of southern Alaska: Journal of the Alaska Geological Society, v. 1, p. 12–35.

Himmelberg, G.R., Brew, D.A., and Ford, A.B., 1991, Development of inverted metamorphic isograds in the western metamorphic belt, Juneau, Alaska: Journal of Metamorphic Geology, v. 9, p. 165–180.

Hodges, K.V., and Crowley, P.D., 1985, Error estimation in empirical geothermometry and geobarometry for pelitic systems: American Mineralogist, v. 70, p. 702–709.

Hodges, K.V., and McKenna, L.W., 1987, Realistic propogation of uncertainties in geologic thermobarometry: American Mineralogist, v. 72, p. 671–680.

Hodges, K.V., and Spear, F.S., 1982, Geothermometry, geobarometry, and the Al_2SiO_5 triple point at Mt. Moosilauke, New Hampshire: American Mineralogist, v. 67, p. 1118–1134.

Holdaway, M.J., 1971, Stability of andalusite and the aluminum silicate phase diagram: American Journal of Science, v. 271, p. 97–131.

Hudson, T., 1979, Calc-alkaline plutonism along the Pacific rim of southern Alaska: U.S. Geological Survey Open-File Report 79-953, 31 p.

Johnson, B.R., and Karl, S.M., 1985, Geologic map of western Chichagof and Yakobi Islands, southeastern Alaska: U.S. Geological Survey Miscellaneous Investigations Series Map I-1506, scale 1:250,000.

Karl, S.M., 1982, Geochemical and depositional environments of Upper Mesozoic radiolarian cherts from the northeastern Pacific rim and from Pacific DSDP cores [Ph.D. thesis]: Stanford, California, Stanford University, 245 p.

Kerrick, D.M., 1987, Fibrolite in contact aureoles of Donegal, Ireland: American Mineralogist, v. 72, p. 240–254.

Kerrick, D.M., 1990, The Al_2SiO_5 polymorphs: Reviews in Mineralogy, v. 22, 406 p.

Koziol, A.M., and Newton, R.C., 1988, Redetermination of the anorthite breakdown reaction and improvement of the plagioclase-garnet-Al_2SiO_5-quartz geobarometer: American Mineralogist, v. 73, p. 216–223.

Loney, R.A., and Brew, D.A., 1987, Regional thermal metamorphism and deformation of the Sitka Graywacke, southern Baranof Island, southeastern Alaska: U.S. Geological Survey Bulletin 1779, 17 p.

Loney, R.A., Brew, D.A., and Lanphere, M.A., 1967, Post-Paleozoic radiometric ages and their relevance to fault movements, northern southeastern Alaska: Geological Society of America Bulletin, v. 78, p. 511–526.

Loney, R.A., Brew, D.A., Muffler, L.J.P., and Pomeroy, J.S., 1975, Reconnaissance geology of Chichagof, Baranof, and Kruzof Islands, southeastern Alaska: Geological Survey Professional Paper 792, 105 p.

Marshak, R.S., and Karig, D.E., 1977, Triple junctions as a cause for anomalously near-trench igneous activity between the trench and volcanic arc: Geology, v. 5, p. 233–236.

Moore, J.C., Byrne, T., Plumley, P.W., Reid, M., Gibbons, H., and Coe, R.S., 1983, Paleogene evolution of the Kodiak Islands, Alaska: Consequences of ridge-trench interaction in a more southerly latitude: Tectonics, v. 2, p. 265–293.

Newton, R.C., and Haselton, H.T., 1981, Thermodynamics of the garnet-plagioclase-Al_2SiO_5-quartz geobarometer, *in* Newton, R.C., et al., eds., Thermodynamics of minerals and melts: Advances in Physical Geochemistry, v. 1, p. 131–147.

Plafker, G., Jones, D.L., and Pessagno, Jr., E.A., 1977, A Cretaceous accretionary flysch and mélange terrane along the Gulf of Alaska margin, *in* Blean, K.M., ed., The United States Geological Survey in Alaska: Accomplishments during 1976: U.S. Geological Survey Circular 751-B, p. B41–B43.

Reifenstuhl, R.R., 1983, A geological and geophysical study of the Goddard hot springs area, Baranof Island, southeast Alaska [M.S. thesis]: Fairbanks, University of Alaska, 113 p., 1 sheet, scale 1:250,000.

Silberling, N.J., Jones, D.L., Monger, J.W.H., Coney, P.J., Berg, H.C., and Plafker, G., 1994, Lithotectonic terrane map of Alaska and adjacent parts of Canada, *in* Plafker, G., and Berg, H.C., eds., The geology of Alaska: Boulder, Colorado, Geological Society of America, Geology of North America, v. G-1, plate 3, scale 1:2,500,000.

Sisson, V.B., and Hollister, L.S., 1988, Low-pressure facies series metamorphism in an accretionary sedimentary prism, southern Alaska: Geology, v. 16, p. 358–361.

Sisson, V.B., and Pavlis, T.L., 1993, Geologic consequences of plate reorganization: An example from the Eocene southern Alaska forearc: Geology, v. 21, p. 913–916.

Sisson, V.B., Hollister, L.S., and Onstott, T.C., 1989, Petrologic and age constraints on the origin of a low-pressure/high-temperature metamorphic complex, southern Alaska: Journal of Geophysical Research, v. 94, p. 4392–4410.

Spear, F.S., 1989, Relative thermobarometry and metamorphic *P-T* paths, *in* Daly, J.S., et al., eds., Evolution of metamorphic belts: Geological Society [London] Special Publication No. 43, p. 63–81.

Spear, F.S., 1993, Metamorphic phase equilibria and pressure-temperature-time paths: Mineralogical Society of America Monograph 1, 799 p.

Thorkelson, D.J., 1996, Subduction of diverging plates and the principles of slab window formation: Tectonophysics, v. 255, p. 47–63.

Thorkelson, D.J., and Taylor, R.P., 1989, Cordilleran slab windows: Geology, v. 17, p. 833–836.

Todd, C.S., 1998, Limits on the precision of geobarometry at low grossular and anorthite content: American Mineralogist, v. 83, p. 1161–1167.

Manuscript Accepted by the Society February 5, 2003

Geological Society of America
Special Paper 371
2003

Controls on intrusion of near-trench magmas of the Sanak-Baranof belt, Alaska, during Paleogene ridge subduction, and consequences for forearc evolution

Timothy M. Kusky*
Department of Earth and Atmospheric Sciences, St. Louis University, St. Louis, Missouri 63103, USA

Dwight Bradley
U.S. Geological Survey, 4200 University Drive, Anchorage, Alaska 99508, USA

D. Thomas Donley
P.O. Box 243834, Anchorage, Alaska 99508, USA

David Rowley
Department of Geophysical Sciences, The University of Chicago, Chicago, Illinois 60637, USA

Peter J. Haeussler
U.S. Geological Survey, 4200 University Drive, Anchorage, Alaska 99508, USA

ABSTRACT

A belt of Paleogene near-trench plutons known as the Sanak-Baranof belt intruded the southern Alaska convergent margin. A compilation of isotopic ages of these plutons shows that they range in age from 61 Ma in the west to ca. 50 Ma in the east. This migrating pulse of magmatism along the continental margin is consistent with North Pacific plate reconstructions that suggests the plutons were generated by migration of a trench-ridge-trench triple junction along the margin. On the Kenai Peninsula the regional lower greenschist metamorphic grade of the turbiditic host rocks, texture of the plutons, contact-metamorphic assemblage, and isotopic and fluid inclusion studies suggest that the plutons were emplaced at pressures of 1.5–3.0 kbars (5.2–10.5 km) into a part of the accretionary wedge with an ambient temperature of 210–300 °C. The presence of kyanite, garnet, and cordierite megacrysts in the plutons indicates that the melts were generated at a depth greater than 20 km and minimum temperature of 650 °C. These megacrysts are probably xenocrystic remnants of a restitic or contact metamorphic phase entrained by the melt during intrusion. However, it is also possible that they are primary magmatic phases crystallized from the peraluminous melt.

Plutons of the Sanak-Baranof belt serve as time and strain markers separating kinematic regimes that predate and postdate ridge subduction. Pre-ridge subduction structures are interpreted to be related to the interaction between the leading oceanic plate and the Chugach terrane. These include regional thrust faults, NE-striking map-scale folds with associated axial planar foliation, type-1 mélanges, and an array

*kusky@eas.slu.edu

Kusky, T.M., Bradley, D., Donley, D.T., Rowley, D., and Haeussler, P., 2003, Controls on intrusion of near-trench magmas of the Sanak-Baranof Belt, Alaska, during Paleogene ridge subduction, and consequences for forearc evolution, *in* Sisson, V.B., Roeske, S.M., and Pavlis, T.L., eds., Geology of a transpressional orogen developed during ridge-trench interaction along the North Pacific margin: Boulder, Colorado, Geological Society of America Special Paper 371, p. 269–292.

of faults within the contact aureole indicating shortening largely accommodated by layer-parallel extension. Syn-ridge subduction features include the plutons, dikes, and ductile shear zones within contact aureoles with syn-kinematic metamorphic mineral growth and foliation development. Many of the studied plutons have sheeted margins and appear to have intruded along extensional jogs in margin-parallel strike-slip faults, whereas others form significant angles with the main faults and may have been influenced by minor faults of other orientations. Some of the plutons of the Sanak-Baranof belt have their long axes oriented parallel to faults of an orthorhombic fault set, implying that these faults may have provided a conduit for magma emplacement. This orthorhombic set of late faults is interpreted to have initially formed during the ridge subduction event, and continued to be active for a short time after passage of the triple junction. ENE-striking dextral faults of this orthorhombic fault system exhibit mutually crosscutting relationships with Eocene dikes related to ridge subduction, and mineralized strike-slip and normal faults of this system have yielded $^{40}Ar/^{39}Ar$ ages identical to near-trench intrusives related to ridge subduction. Movement on the orthorhombic fault system accommodated exhumation of deeper levels of the southern Alaska accretionary wedge, which is interpreted as a critical taper adjustment to subduction of younger oceanic lithosphere during ridge subduction. These faults therefore accommodate both deformation of the wedge and assisted emplacement of near-trench plutons. Structures that crosscut the plutons and aureoles include the orthorhombic fault set and dextral strike-slip faults, reflecting a new kinematic regime established after ridge subduction, during underthrusting of the trailing oceanic plate with new dextral-oblique convergence vectors with the overriding plate. The observation that the orthorhombic fault set both cuts and is cut by Eocene intrusives demonstrates the importance of these faults for magma emplacement in the forearc.

A younger, ca. 35 Ma suite of plutons intrudes the Chugach terrane in the Prince William Sound region, and their intrusion geometry was strongly influenced by pre-existing faults developed during ridge subduction. The generation of these plutons may be related to the sudden northward migration of the triple junction at ca. 40–33 Ma, as the ridge was being subducted nearly parallel to the trench during this interval. These younger plutons are used to provide additional constraints on the structural evolution of the wedge. Late- to post-ridge subduction fabrics include a pressure solution cleavage and additional movement on the orthorhombic fault system. After triple junction migration, subduction of the trailing oceanic plate involved a significant component of dextral transpression and northward translation of the Chugach terrane. This change in kinematics is recorded by very late gouge-filled dextral faults in the late structures of the accretionary prism.

Keywords: Sanak-Baranof belt, Alaska, near-trench magmas, forearc, ridge subduction, Kenai Peninsula, xenocryst.

INTRODUCTION

Forearc regions are normally relatively "cold" places under tectonic compression, yet a few contemporary forearc accretionary prisms contain numerous granitoid intrusions (Pitcher, 1983). Ridge subduction has been invoked to explain the presence of near-trench magmas in several forearc regions (e.g., Marshak and Karig, 1977; Dixon and Farrar, 1980; Bradley et al., 1993; Sisson and Pavlis 1993) as well as anomalous deformation, thermal signatures, and landward structural vergence in accretionary prisms (e.g., Hibbard and Karig, 1990; Lallemand et al., 1992; Kusky et al., 1997a, 1997b).

Near-trench magmatic rocks appear in most cases to represent a mixture of MORB components from the subducted ridge contaminated by a sedimentary component of the overriding accretionary prism (Moore et al., 1983; Hibbard and Karig, 1990; Harris et al., 1996; Maeda and Kagami, 1996; Lytwyn et al., 2000; Sisson et al., this volume, Chapter 13), although an arc-like component has been recognized in the Woodlark near-trench magmas (Perfit et al., 1987; Johnson et al., 1997). Metamorphic signatures associated with ridge subduction are known to include high-temperature, low-pressure metamorphic facies as documented in the Chugach metamorphic complex (Hudson and Plafker, 1982; Hudson et al., 1979; James et al., 1989; Sisson and

Hollister, 1988; Sisson et al., 1989). Structural signatures of the subduction of ridges have received slightly less attention with most data derived from active examples, simply because it is difficult to relate specific structures in ancient accretionary prisms to ridge subduction events (Vogt et al., 1976; Furlong et al., 1989; Thorkelson and Taylor, 1989; Hibbard and Karig, 1990; von Huene and Lallemand, 1990; Lallemand et al., 1992; Kusky et al., 1997a, 1997b; Kusky and Young, 1999).

Here, we document aspects of the structural history of an accretionary prism before, during, and after ridge subduction. We use plutons that are believed to have been generated by passage of the triple junction (see Bradley et al., 1993, this volume, Chapter 1) as a time marker separating different kinematic regimes to evaluate the structural history before, during, and after ridge subduction. We also examine faults associated with gold-quartz mineralization formed during ridge subduction. The strain history recorded by the plutons is somewhat different from the gold-quartz veins/faults (Haeussler et al. this volume). The plutons can also be utilized as strain markers injected into the wedge between two different kinematic regimes, including (a) interactions of the "oceanic plate A"—North America plates prior to ridge subduction, and (b) interaction of the "oceanic plate B"—North America plates after ridge subduction. We refer to the leading, eastern oceanic plate that was subducted before passage of the triple junction as oceanic plate A, and the western or trailing oceanic plate that was subducted after ridge subduction as oceanic plate B. This is because the record of a similar-age ridge subduction event in the Olympic Mountains (Wells et al., 1984; Babcock et al., 1992, 1994) raises some uncertainty as to whether the ridge subduction record in southern Alaska is a product of subduction of the Kula-Farallon ridge or a ridge between the Kula and another smaller plate to the east (Bradley et al., 1993, this volume, Chapter 1), colloquially named the Resurrection plate by Haeussler et al. (2000). Understanding the kinematic framework that predates and postdates pluton emplacement, on local scales, can thus be used to infer the larger-scale change in kinematics associated with passage of the triple junction. In the following sections, we document structures that predate and postdate the ca. 56.0 ± 0.3 Ma (U-Pb on monazite, Bradley et al., 2000) Nuka, Tustumena–Harris Bay, and Aialik plutons, and we relate these to subduction of oceanic plates A and B, respectively. We do the same for a suite of ca. 35 Ma plutons that intrude the accretionary prism in the Prince William Sound area and offer a model for the tectonic origin of these Late Eocene plutons that also relates to ridge subduction. We also document and discuss mechanisms of emplacement of these plutons, as the intrusion mechanisms are directly related to the mechanics of ridge subduction.

REGIONAL GEOLOGY

Southern Alaska is composed of a series of accreted terranes representing relict Paleozoic, Mesozoic, and Cenozoic arc-trench systems, oceanic plateaus, and flysch basins (Plafker et al., 1989). The Chugach–Prince William composite terrane (Fig. 1) is over 2200 km long and extends from Baranof Island in southeastern Alaska westward to Valdez and then southwestward to the Sanak and Shumagin Islands. It consists of deformed Triassic-Cretaceous mélange with exotic blocks of Permian limestone, Upper Cretaceous–Tertiary flysch with minor tholeiitic volcanic rocks, and a thin discontinuous band of dismembered ophiolites (Nelson et al., 1985; Bol et al., 1992; Nelson and Nelson, 1993; Crowe et al., 1992; Kusky et al., 1997a, 1997b; Kusky and Young, 1999).

Triassic-Cretaceous mélange in the Chugach terrane is known as the McHugh Complex in the Chugach Mountains (Clark, 1972; Bradley and Kusky, 1990, 1992; Kusky et al., 1997b; Kusky and Bradley, 1999). The McHugh Complex has been regionally metamorphosed to prehnite-pumpellyite facies (Clark, 1972; Kusky et al., 1997a), consistent with pressures of 3–8 kbars and burial to 11–30 km at relatively low temperatures. A second component of the Chugach terrane consists of deformed Cretaceous flysch known as the Valdez Group (Tysdal and Plafker, 1978; Nilsen and Zuffa, 1982; Kusky et al., 1997b) metamorphosed to lower-greenschist grade (Bradley et al., 1999). Hawley (1992) and Fisher and Brantley (1992) suggest that there has been a large-scale mobilization of many elements because sericite and chlorite occur in quartz/calcite veins. They suggest a minimum temperature of 200 °C for the deformation and alteration since albite precipitated in equilibrium with sericite and is not stable below that temperature. Goldfarb et al. (1986), Borden et al. (1992) and Taylor et al. (1994) report isotopic and fluid inclusion data from ca. 55 Ma quartz veins in the region showing that the veins (and presumably the related plutons) were emplaced at 1.5–3.0 kbars (5.2–10.5 km) at temperatures of 210–300 °C.

Paleogene flysch in the Prince William terrane is known as the Orca Group (Moffit, 1954; Plafker, 1987; Kveton, 1989). Scarce fossils indicate that portions of the Valdez Group is Late Cretaceous (Maestrichtian) in age (Jones and Clark, 1973; Bradley et al., 1999) while the Orca Group ranges in age from Paleocene to early Eocene (Winkler, 1976; Tysdal and Case, 1979; Tysdal et al., 1977). The Valdez and Orca metasedimentary rocks contain lode gold deposits that occur in quartz dikes and veins as free metals or combined with sulfides (Tuck, 1933; Silberman et al., 1981; Goldfarb et al., 1986; Stüwe, 1986; Haeussler et al., 2000, this volume). An anomalous high-temperature, low-pressure metamorphic belt has been well documented in the eastern Chugach Mountains (Pavlis et al., 1988; Sisson et al., 1989, this volume, Chapter 13).

Early Tertiary near-trench intrusive rocks are widespread in the Chugach accretionary wedge. They belong to what Hudson (1983) named the Sanak-Baranof belt, a 2200 km long belt including biotite-muscovite granodiorites, leucotonalites, trondhjemites, and dikes ranging in composition from basalt to rhyolite (Marshak and Karig, 1977; Hill, 1979; Hill et al., 1981; Barker et al., 1992; Bradley and Kusky, 1992). The plutons, plugs, dikes, and sills lie far outboard of an evidently

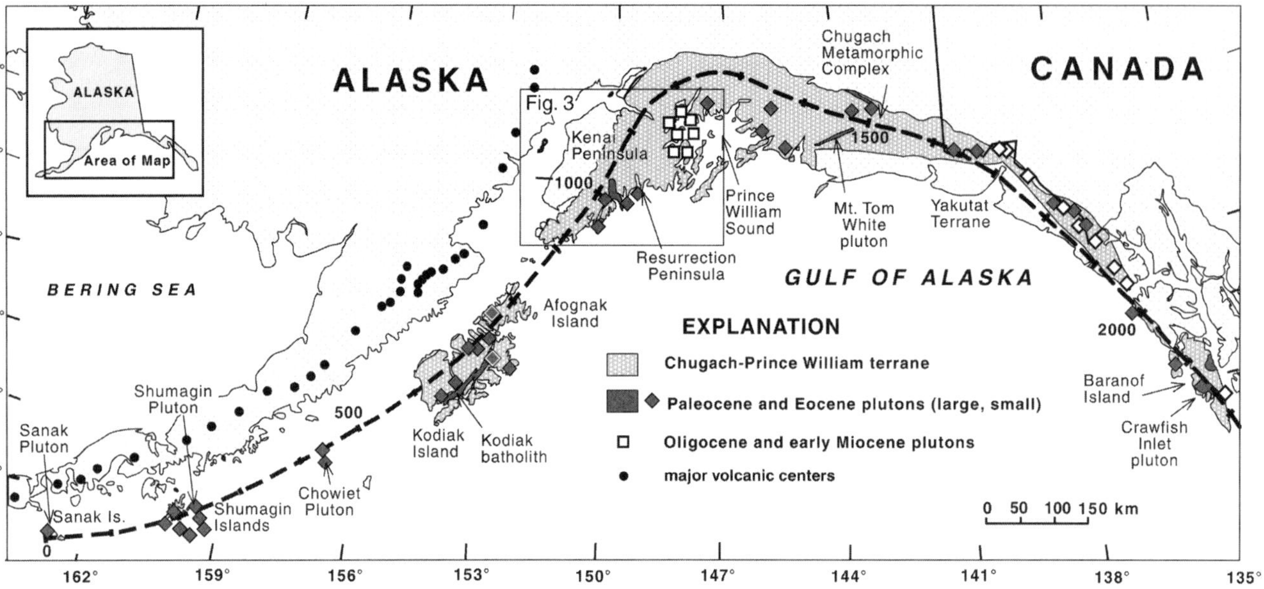

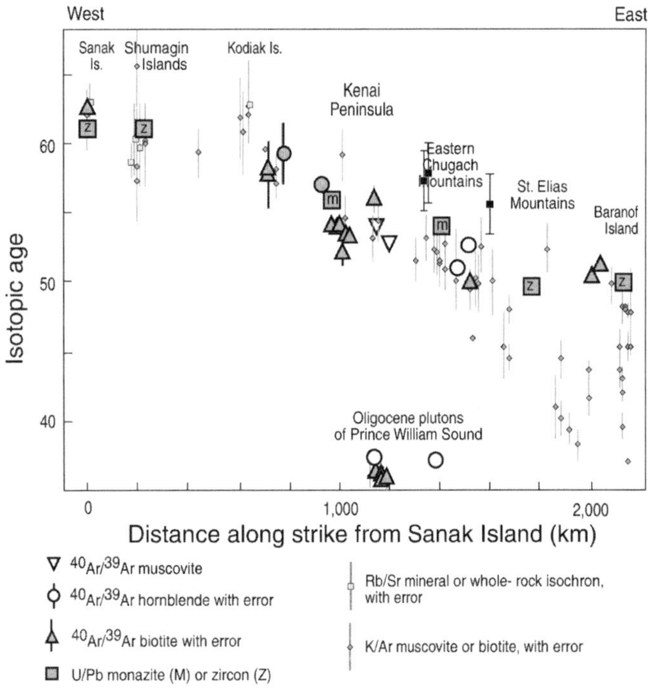

Figure 1. A: Map of Chugach terrane showing location of near-trench plutons of Sanak-Baranof belt. B: Plot showing the age of the plutons vs. strike distance from Sanak Island (from Bradley et al., 2000).

coeval magmatic arc (Fig. 1). They intrude trench deposits and ophiolite fragments that had only been deposited and offscraped a few million years earlier (Kusky and Young, 1999). The age of near-trench magmatism decreases from west to east around the Gulf of Alaska, from 61 Ma at Sanak and the Shumagins to ca. 50 Ma at Baranof Island (Bradley et al., 1993, 2000). One interpretation of the near-trench magmatism is that it marks the site of subduction of an oceanic ridge. The age progression (Fig. 1B), then, is interpreted to track the migration of the triple junction (Fig. 2) (Bradley et al., 1993, 2000).

GRANITE OF THE HARDING ICEFIELD REGION

A remote area on the outer Kenai Peninsula between the Harding Icefield and the Pacific Ocean hosts several large plutons and numerous small stocks and plugs of the Sanak-

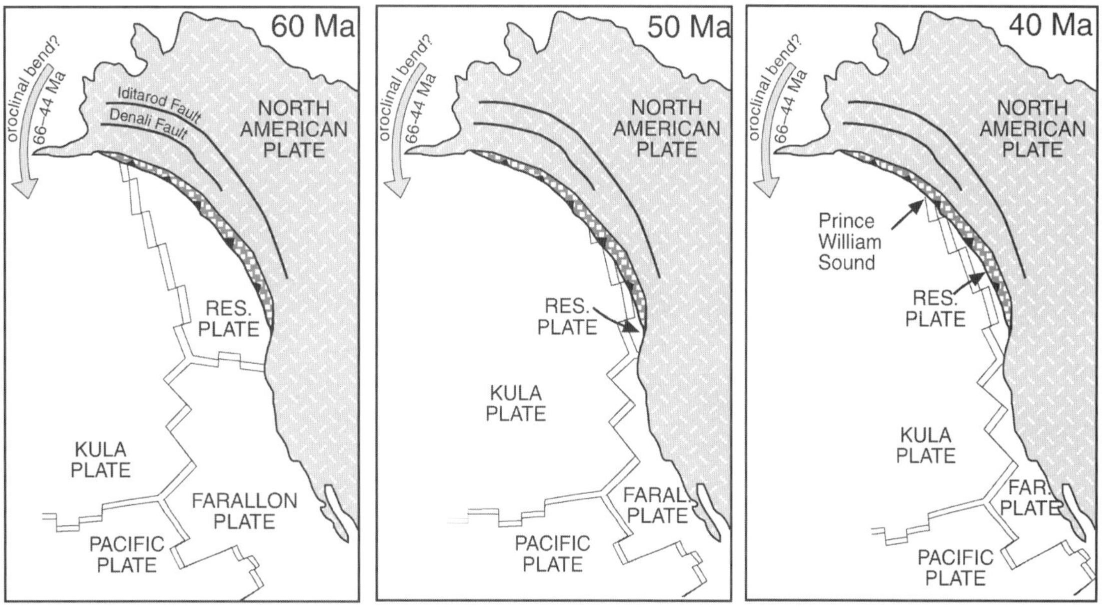

Figure 2. Schematic plate reconstructions, showing the Kula–Resurrection (Oceanic plate B)–North America triple junction moving southward along the margin from 60 to 50 Ma, then jumping back north by 40 Ma in response to the early Eocene plate reorganization. See Bradley et al. (this volume, Chapter 1) for alternative plate reconstructions. In each of the time intervals, the entire Alaskan orocline is shown as partly straightened out, and the Chugach–Prince William terrane is restored by ~820 km southward along the margin. Reconstruction at 60 Ma is from Bradley et al. (this volume, Chapter 1).

Baranof belt (Figs. 1 and 3). Martin et al. (1915) were the first to map the rocks of the Kenai Peninsula, followed by Cowan and Boss (1978), Magoon et al. (1976), Tysdal and Case (1979), and Bradley et al. (1999). Martin et al. (1915) suggested that because the plutons truncate folded sedimentary rocks thought to be Jurassic to Cretaceous, they are probably Tertiary. Tysdal and Case (1979) named these plutons the "granite of the Harding Icefield region" and suggested that the granite forms a batholith that extends into the Seldovia quadrangle to the west and northward ~50 km, outcropping discontinuously as nunataks in the Harding Icefield (Figs. 4 and 5A). Bradley et al. (1999) mapped the granitoids and surrounding rocks in the Seldovia quadrangle, documenting the southern and western extent of the granitic rocks. Bradley and Wilson (2000) traced the pluton into the Kenai quadrangle. Compositions vary from tonalite near the contacts through granite in the pluton interior (Bradley et al., this volume, Chapter 1). Tysdal and Case (1979) report that the contacts between the batholith and country rock are steeply dipping, and dominantly bedding parallel, although some contacts are discordant (Fig. 4).

The Nuka, Aialik, Harris Bay, and Tustumena plutons are the main plutonic components of the "granite of the Harding Icefield" (Fig. 4) and together form one of the largest (>2500 km^2) near-trench igneous complexes in the Sanak-Baranof belt, and apparently in the world. The Kodiak Batholith on Kodiak Island is longer, but much thinner, making it smaller in area (1650 square km) than the plutonic complex of the Harding Icefield. The Nuka and Aialik plutons have broadly ovoidal shapes that are elongate nearly parallel to the grain of accreted rocks, but the Tustumena–Harris Bay complex forms a significant angle with earlier structures (Fig. 4).

The Nuka pluton crops out from Nuka Bay to Thunder Bay (Figs. 3 and 4). The Nuka pluton has yielded an U/Pb (zircon and monazite) age of 56 ± 0.5 Ma (Bradley et al., 2000), and a $^{40}Ar/^{39}Ar$ age spectrum that is a plateau corresponding to an isochron age of 54.2 ± 0.08 Ma (Bradley et al., 2000). The Nuka pluton is a buff-rust weathering massive, coarse to finely crystalline, granitic rock with abundant euhedral plagioclase feldspar and quartz, lesser and varying amounts of potassium feldspar, muscovite, chlorite, and a few percent of mafic minerals including hornblende and biotite. Near the contact with the country rocks, the Nuka pluton is a fine-grained, well-foliated tonalite that forms numerous tabular dikes and sills crosscutting the country rock (Fig. 5B). This is in sharp contrast to the coarse-grained, orthoclase-rich, granodiorite in the center of the pluton. Metasedimentary and gabbroic xenoliths form minor but ubiquitous components of the Nuka pluton (Fig. 5C) and large xenocrysts of kyanite are present (Fig. 5D). The pluton exhibits a moderately strong magmatic foliation (defined by long axes of plagioclase and rarely hornblende, and aligned biotite) and a moderately to strongly developed tectonic foliation (defined by biotite and recrystallized quartz ribbons). In most places in the

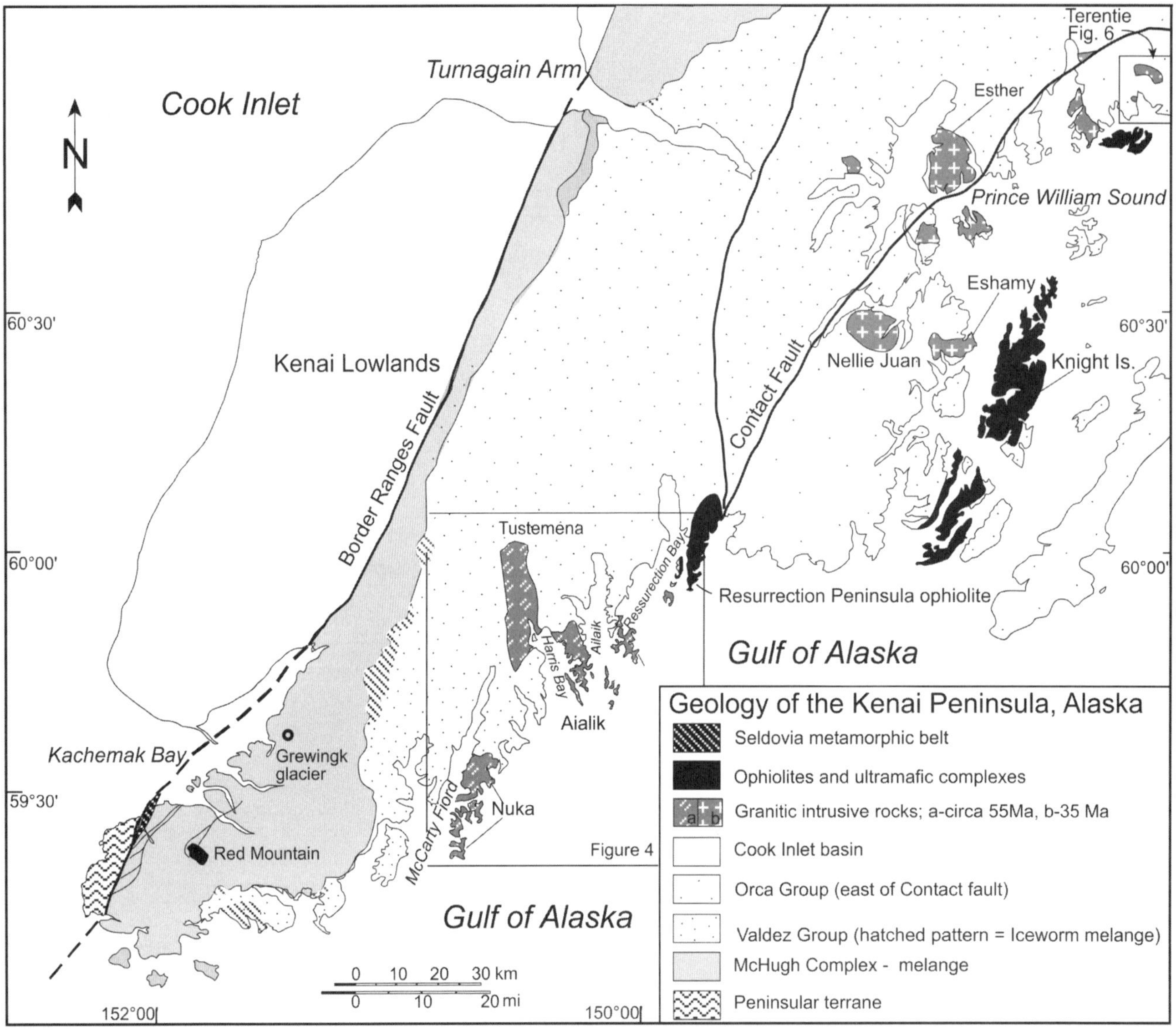

Figure 3. Simplified geologic map of the Kenai Peninsula, showing the setting of the Nuka, Aialik, Harris Bay, Terentiev, and related plutons.

Nuka pluton, the coaxial magmatic and tectonic foliation strike parallel to the long axis of the pluton, are steeply dipping, and contain a gently north-plunging lineation (Fig. 4).

The Tustumena–Harris Bay pluton forms a continuous S-shaped intrusive complex, with a sharp bend centered on the head of Harris Bay in Northwestern Lagoon (Fig. 4). The Tustumena–Harris Bay pluton is spectacularly exposed on small nunataks, and in the fiords of Northwestern Lagoon (Fig. 4), but is largely buried by ice. Contacts shown on the map (Fig. 4) are based on detailed mapping that include many small nunataks, too small to show on the map at this scale (see http://wrgis.wr.usgs.gov/open-file/of99-18/). The pluton has yielded a $^{40}Ar/^{39}Ar$ plateau age on biotite of 53.23 ± 1.14 Ma (Bradley et al., 2000), whereas a biotite separate from the Harris Bay lobe analyzed at University of Alaska, Fairbanks, yielded three plateau ages: 51.1 ± 0.5, 53.5 ± 0.5, and 49.6 ± 1.7 Ma (Bradley et al., 2000). The variation may be due to the presence of a small amount of excess argon. The Tustumena–Harris Bay pluton is a medium- to coarse-grained granodiorite containing biotite, muscovite, rare hornblende, and xenocrystic kyanite (Donley et al., 1995). Tysdal and Case (1979) reported subordinate granitic and tonalitic phases. The pluton locally displays a tectonic foliation that also affects aplitic dikes (Fig. 5E). The granodioritic rocks in the area of the bend in NW Lagoon are much more strongly deformed than typical granites of the Harding Icefield region, and they display both magmatic and crystal-plastic

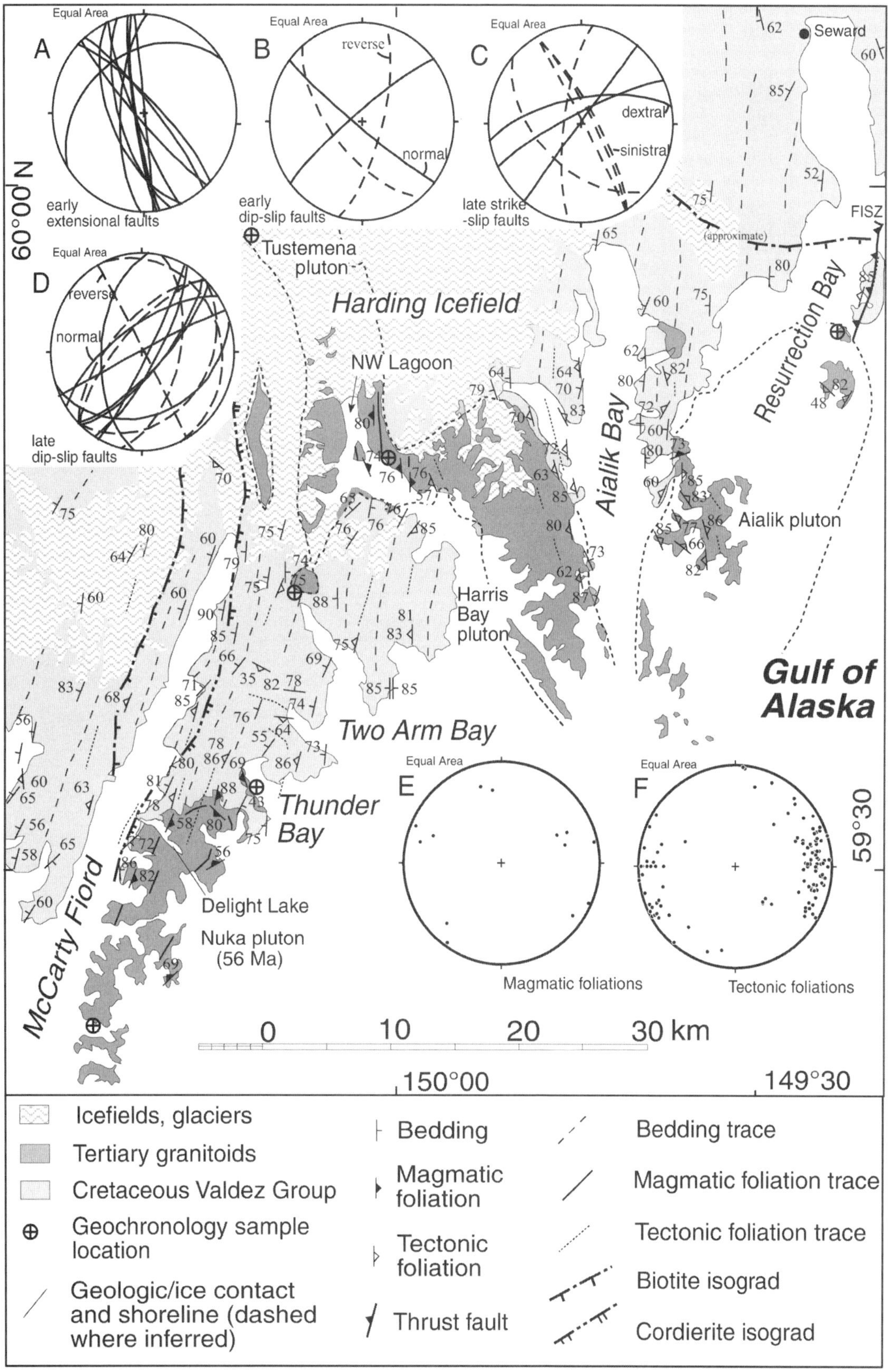

Figure 4. Structural map of the Nuka–Aialik–Harris Bay plutons, showing relationships between magmatic and tectonic foliations. The western half of the map is simplified from Bradley et al. (1999), and the eastern half is based on the map of Tysdal and Case (1979). Both maps were supplemented with additional mapping and detailed field observations, and contacts between plutons and country rocks are largely interpreted based on these observations. Insets are lower-hemisphere, equal-area projections showing the orientations of brittle fractures in and around the Nuka and Aialik plutons. Early faults predate pluton emplacement, late faults postdate pluton emplacement. A: Early orogen-parallel extension fractures. B: Early high-angle dip-slip faults (reverse dashed, normal solid). C: Late strike-slip faults (sinistral dashed, dextral solid). D: Late dip-slip faults (normal solid, reverse dashed). E: Magmatic foliations. F: Tectonic foliations.

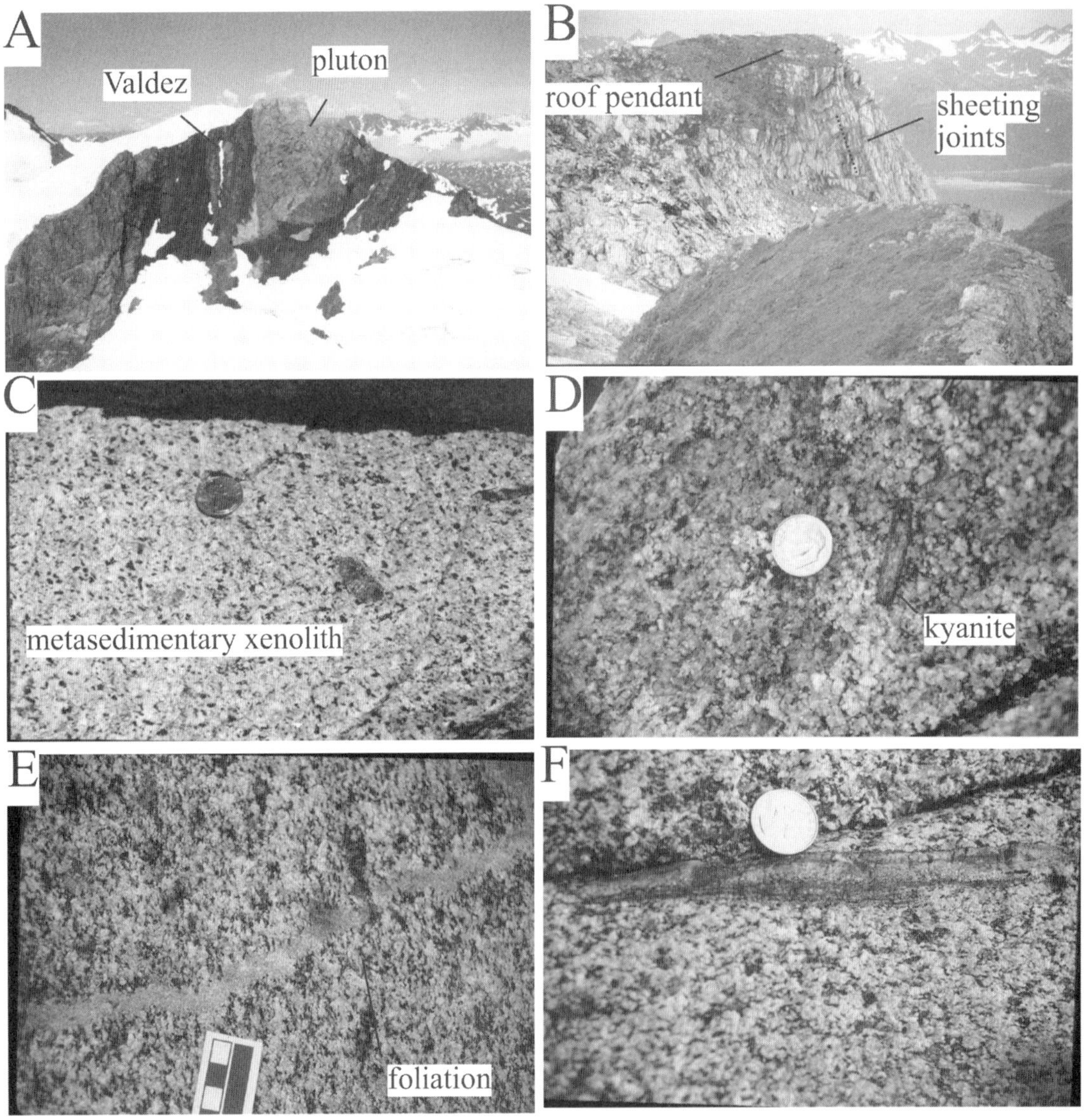

Figure 5. Field photographs. A: Contact of Tustumena pluton and Valdez Group exposed on Nunatak in the Harding Icefield. B: Nuka pluton (Delight Lake in background) showing vertical sheeting joints and roof pendant of Valdez Group. C: Metasedimentary xenolith in Aialik pluton. D: Kyanite xenocryst in Nuka pluton. E: Foliated granodiorite and aplite dike in Tustumena–Harris Bay pluton, from shear zone in Harris Bay. F: Foliated granodiorite and metasedimentary xenolith from Tustumena–Harris Bay pluton in Harris Bay. G: Dikes related to plutons. H: Tertiary dike cutting foliation of Iceworm mélange. I: Foliated dike. J: Apophysis from dike that cuts foliation but is apparently folded. K: Faulted dike from Grewingk Glacier. L: Plug of granodiorite from Crow Pass (Turnagain Arm) showing large stoped block.

deformation fabrics (Figs. 5E and 5F). These foliations are parallel to the long axis of the Tustumena lobe of the pluton.

The Aialik pluton outcrops discontinuously around the mouth of Resurrection Bay and Aialik Bay, and has an unknown contact relationship with the Harris Bay pluton (Fig. 4). A biotite separate from the northern part of the Aialik pluton on Hive Island yielded a $^{40}Ar/^{39}Ar$ plateau age of 53.4 ± 0.4 Ma (Bradley et al., 2000). The Aialik pluton is a massive rust-buff weathering, medium- to coarse-grained tonalite to granodiorite that has abundant shallowly dipping sheeting joints. Mineral percentages range from 30 to 45% plagioclase, 10–30% orthoclase, 20–30% quartz, 3–10% biotite, 3–7% amphibole, 1–4% muscovite, 0–5% chlorite, epidote, apatite, zircon, kyanite, garnet, sillimanite, cordierite, and opaques. It is nearly identical to the Nuka and Tustumena–Harris Bay plutons in mineral percentages and magmatic and tectonic fabrics, and it too contains gabbroic xenoliths that are elongate parallel to the magmatic foliation. It also contains metasedimentary xenoliths and megacrystic aluminosilicates

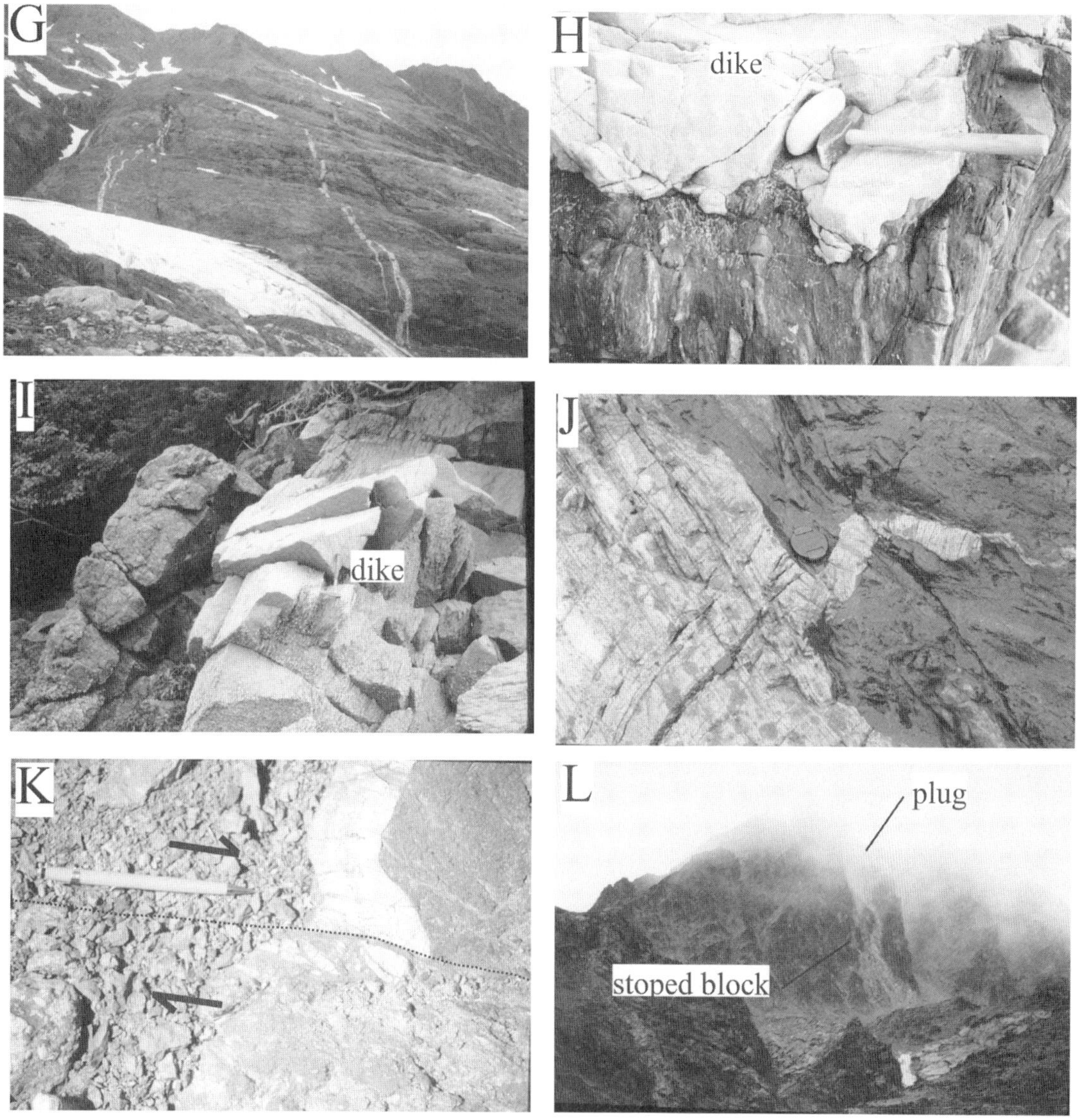

Figure 5 (*continued*).

(kyanite, cordierite), particularly along its margins. The only notable difference is that the Aialik pluton is locally very biotite-rich, with large (~2 cm), strongly aligned books of mica composing up to 20% of the rock. Both the Nuka and Aialik plutons contain several small, late stage, tourmaline-bearing pegmatites.

Between the Nuka and Tustumena–Harris Bay plutons, there are several smaller outcroppings of granitic rocks in Two Arm Bay (Paguna stock), Thunder Bay, and west of the area at Chernof Glacier (Figs. 1, 3, and 4; Bradley et al., 2000). The Chernof stock has yielded a $^{40}Ar/^{39}Ar$ biotite age of 54.2 ± 1.14 Ma, whereas the sill in Thunder Bay has a biotite $^{40}Ar/^{39}Ar$ plateau age of 53.7 ± 0.1 Ma (Bradley et al., 2000). Although these outcrops are not visibly part of the Nuka, Tustumena–Harris Bay, or Aialik plutons, their identical mineralogy, overlapping closure dates on biotite (Bradley et al., 2000), close proximity to much larger outcrops of granodiorite, and the fact that the bedrock separating the outcrops are thin ridges of high-grade hornfels suggest that they may be part of the same system of Paleogene plutons.

In all of the granitoids of the Harding Icefield region, plagioclase forms well-zoned framework grains that have a range of An content from 25 to 50%. The cores of the plagioclase have a patchy appearance due to later albitization. Many of the larger plagioclase grains, near the contact with country rock, show evidence of post-crystallization deformation including fractured grains, deformation twinning, offset albite twinning and rarely undulose extinction. Orthoclase, near the contact with country rock, rarely occurs as individual crystals but rather occurs as an orthoclase-quartz granophyric intergrowth between the plagioclase. Away from the contacts, orthoclase forms small euhedral crystals that in many cases are mantled around the much larger plagioclase laths. Quartz resides interstitially between the feldspars and mafic minerals. It is commonly deformed and/or recrystallized near the margins of the pluton and has a strong polycrystalline habit. Biotite occurs as large, partly euhedral mats and as a secondary mineral after amphibole, kyanite, and cordierite. High-Mg amphibole occurs as small anhedral crystals that show evidence of retrogression to biotite and chlorite. Muscovite is rarely primary, as it is secondary sericite after plagioclase and kyanite. Epidote occurs as small, zoned, euhedral to anhedral crystals, mantled in biotite and as an overgrowth on hornblende, that are found predominantly near the contact with the country rock. Kyanite forms large (>2.5 cm) megacrysts near the contact with the country rock, particularly in association with xenoliths (mostly metasedimentary). The kyanite has a reaction rim of andalusite, sillimanite, muscovite, biotite and chlorite (Donley et al., 1995). Additional work is needed on this assemblage to determine the reactions and *P-T-t* path followed by the kyanite crystals from their source to their final emplacement. Cordierite is found near the contact with the country rock as well, but typically forms much smaller, anhedral crystals found almost exclusively in the cores of large biotite crystals. The kyanite and cordierite phenocrysts have not been observed in the same sample, or to occur in any spatial or temporal relationship. Garnet is present in minor amounts in the plutons, but is more common in the dikes associated with the plutons.

Origin of Aluminosilicate Xenocrysts

Garnet and cordierite crystals are not uncommon in peraluminous granites. However, kyanite megacrysts are very rare in granitic magmas. Examples of granitic rocks bearing other aluminosilicate megacrysts are known from several places, including the Blue Mountains of northeastern Oregon (Johnson et al., 1997), and in the Grenville province (Perreault and Martignole, 1988). There are at least three possible origins for the large kyanite and other aluminosilicate megacrysts in the plutons of the Sanak-Baranof belt. First, they could be part of a restitic phase plucked from the wall rock in the granitic source region, and carried upward by the melt (e.g., Vernon, 1983). Partial melting of pelitic source rocks has been shown in several cases to generate trapped crystals of K-feldspar, garnet, cordierite, biotite, kyanite, spinel, sillimanite, and other minerals (e.g., Couturié, 1969; Tracy and Robinson, 1983; Clark and Lyons, 1986; Barbey, 1991; Hibbard, 1995). The reaction rims around the kyanite may have effectively isolated the kyanite xenocrysts from the melt so that they did not re-react with the melt to produce phases stable at lower pressures and temperatures. Reactions between magma and restitic xenocrysts are known to produce biotite-rich reaction rims around the inclusions (e.g., Couturié, 1969). In a similar way, the kyanite megacrysts might represent pieces of a contact metamorphic phase, plucked from the wall rocks during the magmas ascent and entrapped in the marginal phases of the plutons. This second process has been documented in Hercynian granites from the Central Iberian Massif (Ugidos, 1990; Recio et al., 1992; Ugidos and Recio, 1993).

The kyanite megacrysts in the Nuka and Aialik plutons could alternatively be a magmatic phase, because they occur only as large single megacrysts that have aspect ratios of up to 17:1, and none of the xenoliths spatially associated with the kyanites have kyanite as a constituent mineral. The kyanite has margaritic-kaolinitic exsolution-alteration along cleavage traces. In any case, the presence of the kyanite megacrysts indicate that the aluminous melt was generated at a depth exceeding 20 km at a minimum temperature of 650 °C (e.g., Essene, 1986; Spear, 1993; Hibbard, 1995). It is likely that the kyanite crystals were initially formed by heat from mafic intrusions generated from the slab window, which metamorphosed and partially melted the base of the accretionary prism. The hybrid magmas then rose to higher crustal levels, where they cooled and crystallized.

Magmatic and Tectonic Foliations in the Granite of Harding Icefield

Structural studies of the Nuka, Tustumena–Harris Bay, and Aialik plutons were undertaken to understand the mechanisms of intrusion of such large volumes of melt into an accretionary prism and to elucidate the effects of the incorporation of large plutonic bodies into the accretionary wedge. Emphasis was placed on determining relationships between fabrics of magmatic (pre-complete crystallization) and tectonic (hypersolidus, or crystal-plastic) origin, and establishing the relative timing of pluton emplacement and fabric development through examination of the geometrical relationships between foliations and porphyroblast growth in the contact aureoles of the plutons (e.g., Hutton, 1988; Dell'Angelo and Tullis, 1988; Paterson and Tobish, 1992). Relationships between plutons and regional structures were also examined. The sequence of crosscutting relationships within these plutons and their contact aureoles is used to determine the kinematic regime in the wedge before, during, and after pluton emplacement. Since the plutons are interpreted to be a consequence of ridge subduction, this information may be related to plate convergence parameters and wedge mechanics before, during, and after passage of the triple junction.

The plutons intrude the Cretaceous Valdez Group, which in the area is a thick sequence of turbiditic sandstone and shale with some minor granule conglomerate and calcareous siltstone (Dumoulin, 1987, 1988; Bradley et al., 1999; Kusky and Young, 1999). The Valdez Formation is locally transposed into a mélange (Iceworm mélange) (Kusky et al., 1997b) consisting of blocks of graywacke in a phacoidally cleaved argillite matrix. This mélange fabric is associated with the Chugach Bay thrust (Kusky et al., 1997b). In the area of study, the Valdez Formation turbidites are dominantly foliated, gray to black, variably silt-rich slate and phyllite interbedded with thin (1–6 cm) siltstone beds. The slates and sandstones are moderately recrystallized from regional metamorphism generally to lower greenschist facies, with muscovite, chlorite, quartz, albite, and minor calcite making up the bulk of the new minerals (Goldfarb et al., 1986; Gibbons, 1988; Bradley et al., 1999).

A regional contact aureole surrounds the Nuka, Tustumena–Harris Bay, and Aialik plutons. It includes a zone ranging from 5 to 25 km wide (Fig. 4), of lower greenschist to lower amphibolite facies meta-turbidites. The area between the Nuka pluton, Tustumena pluton, and the Aialik pluton (Fig. 4) is a large area of generally greenschist through amphibolite facies metamorphic rocks formed as a coalesced contact aureole from the numerous intrusions in this area. We refer to this area as the Kenai metamorphic complex, and note that it is similar to but at lower grade than the Chugach metamorphic complex in the northern Chugach Mountains (Sisson et al., 1989). The thickness and extent of this aureole in such a shallow crustal level suggests that the region may be underlain by numerous other plutons just below the surface, forming the regional contact aureole. On the northwest side of the Nuka pluton near Delight Lake (Fig. 4), the meta-sandstone and pelite have a purple color and a distinct lumpy appearance due to foliation-parallel biotite and cordierite porphyroblasts. Cordierite is surrounded by white mica that define a foliation-parallel stretching lineation, plunging gently north in most places, but with steep plunges in others. The cordierite has undergone syn-growth deformation as shown by asymmetric porphyroclast tails and inclusion patterns, as well as a strong polycrystalline habit. Within 100–1600 meters of the contact, cordierite is generally no longer present, but the hornfels are still biotite-rich, with minor muscovite, epidote, and calcite. The next 300–2600 meters is marked by a rapid decline in the percentage of biotite, and the rocks take on a blocky rusty weathering appearance and are green on fresh surfaces, reflecting the assemblage quartz, plagioclase, muscovite, chlorite, epidote, and calcite. The remaining contact aureole is a gradual transition to lower grades of regional metamorphism.

Figure 4 shows foliation orientations within the Nuka, Aialik, and Tustumena–Harris Bay plutons and the surrounding Valdez formation. The magmatic foliation is parallel to the intrusive contacts and generally trends ENE (Fig. 4). In many places along the margins of the Nuka pluton, the granodiorite forms steeply dipping magmatic sheets defined by the magmatic foliation (parallel to traces of magmatic foliation on Fig. 3, and shown as photo in Fig. 4B), but on the NE side of the pluton these sheets become subhorizontal. The majority of magmatic and tectonic foliations in the Nuka and Aialik plutons are subparallel to the intrusive contacts and regional structural trend (Fig. 4), although at the NE and SW ends of the plutons, the magmatic foliation and pluton contacts are discordant with the regional tectonic foliation. Dikes that stem from the northeastern contact of the Nuka pluton are oriented both parallel and perpendicular to the regional cleavage; those that are perpendicular to the cleavage are locally openly to tightly folded, with axial surfaces striking NE and dipping NW. The magmatic foliations have a variable orientation (perhaps indicating magmatic processes or a variable strain field; Paterson et al., 1989a, Paterson and Miller, 1998) whereas the tectonic foliations have a more consistent orientation (parallel to the long axis of the plutons).

Near the contacts, the magmatic foliations are not visible due to later tectonic overprinting, except at the NE and SW ends of the plutons, where the magmatic and tectonic foliations are at high angles to each other (Fig. 4). The tectonic foliation is defined by euhedral to kinked books of biotite and attenuated, polycrystalline recrystallized quartz ribbons. Along the NW margin of the Nuka pluton, in McCarty Fiord, the magmatic foliation is clearly parallel with the intrusive margin, whereas along the NE margin of the pluton the magmatic and tectonic foliations are discordant. Lineations defined by aligned hornblende and biotite are developed in some zones, and generally plunge to the north. The magmatic and tectonic fabrics tend to be parallel along the NE/SW-trending concordant margins of the plutons and are at high angles to one another along the NW/SE-trending discordant contacts. This discordance implies that the plastic deformation fabrics cut the magmatic deformation fabrics, and postdate them by some time.

Early Paleogene Seldovia Dike Swarm

The Kenai Peninsula is cut by locally dense, discrete swarms of dikes (Fig. 5G), showing a calc-alkaline differentiation trend and ranging in composition from basalt to rhyolite, with dacite being the most abundant dike composition (Lytwyn et al., 2000; Bradley et al., this volume, Chapter 1). The dikes and ca. 55 Ma plutons have overlapping compositions, but the dikes have a greater range in composition than the plutons (Bradley et al., this volume, Chapter 1). The basalts and basaltic andesites show the most primitive "MORB-like" REE trends (Bradley et al., this volume, Chapter 1), whereas the more evolved dacites and rhyolites show HREE depletion and LREE enrichment, consistent with mixing of melted accretionary prism material with the MORB source magmas (e.g., Harris et al., 1996; Lytwyn et al., 2000; Sisson et al., this volume, Chapter 13). Interestingly, some of the dikes have xenocrystic kyanite and/or garnet, suggesting genesis at depths of 20 km or more. The dikes and plutons are interpreted to be comagmatic based on overlapping isotopic ages and chemistry (Bradley et al., this volume, Chapter 1). The dikes cut mélange fabrics of the McHugh Complex (Fig. 5H) and foliations in the Valdez Group, showing that they are relatively late structural features. A few of the dikes are foliated and possibly folded, but most cut all the fabric elements in the accreted metasediments (Figs. 5I and 5J). On the lower Kenai Peninsula, and east of Anchorage, some of the dikes are spatially related to near-trench plutons. Several dikes have yielded $^{40}Ar/^{39}Ar$ ages of 57–53 Ma (Bradley et al., 2000, 1999), suggesting that the entire swarm may be related to passage of the triple junction that generated the plutons of the Sanak-Baranof belt.

Most of the dikes on the lower Kenai Peninsula strike east-southeast, with minor differences between the basaltic and andesitic dikes (east-west strikes), and the dacite-rhyolite dikes (east-southeast/west-northwest strikes). Two dikes with anomalous Au concentrations strike southeast-northwest (Fig. 6). The significance of these differences is unclear, but could be related to counterclockwise rotations of the stress field during intrusion of progressively differentiated dikes, perhaps with slight difference in time of intrusion. Perhaps more significantly, most dikes indicate extension at about 15° to the regional strike, implying that the stress field operative during dike emplacement differed from the stress field during development of the main Cretaceous structural fabric in the Chugach terrane (Kusky et al., 1997a, 1997b; Kusky and Bradley, 1999; Haeussler et al., this volume, Chapter 5).

Figure 6 presents a regional compilation of dike orientations from the Chugach terrane. In the northern Kenai Peninsula and the Prince William Sound area (Tana Glacier, Valdez, Port Wells), the dikes show about the same orientation with respect to the structural grain, but are more N-S in orientation, suggesting that they have been rotated in the hinge of the southern Alaska orocline (Fig. 6). A paleomagnetic study of the dikes that was aimed at quantifying these rotations (A. Bol, 1992, personal commun.) was inconclusive. Dikes on the western limb of the orocline (Matanuska-Knik area, Eagle River area, Girdwood, Hope-Sunrise District) also strike about N-S, but lie at an acute angle to the local grain. In southeastern Alaska, dikes strike NW, subparallel to the margin and the local structural grain. Except in southeast Alaska, the net effect of dike emplacement was orogen-parallel extension; the cumulative thickness of the dikes suggests perhaps 1% extension was accommodated by dike intrusion.

Circa 35 Ma Plutons in the Prince William Sound Region

A younger, ca. 35–30 Ma suite of plutons intrudes the Chugach–Prince William terranes in the Prince William Sound area (Fig. 3). Plutons of this group include the Terentiev (29.2 ± 0.3 Ma, $^{40}Ar/^{39}Ar$ on microcline; Nelson et al., 1999), Esther (35.5 Ma, K/Ar—biotite; L. Snee, personal commun., 1992), Nellie Juan (36.1 Ma, K/Ar—biotite; L. Snee, personal commun., 1992), and Eshamy bodies (Fig. 2). The Miners Bay pluton (Fig. 3) has yielded $^{40}Ar/^{39}Ar$ age on biotite of 38.6 ± 0.6 to 41 Ma (Nelson et al., 1999), but a dioritic phase of this pluton yielded an $^{40}Ar/^{39}Ar$ age of 59.1 ± 0.1 Ma (Nelson et al., 1999), suggesting that this pluton may be part of the older Sanak-Baranof belt, and be equivalent to the Nuka, Aialik, and other plutons discussed above. These ca. 35 Ma plutons have to-date defied compelling tectonic explanation. We suggest the following scenario as a tentative explanation for their origin. At 40–33 Ma, the ridge was being subducted nearly parallel to the continental margin (see Bradley et al., Fig. 16, this volume, Chapter 1). A small change in the orientation of the ridge in response to the early Eocene plate reorganization (Sisson and

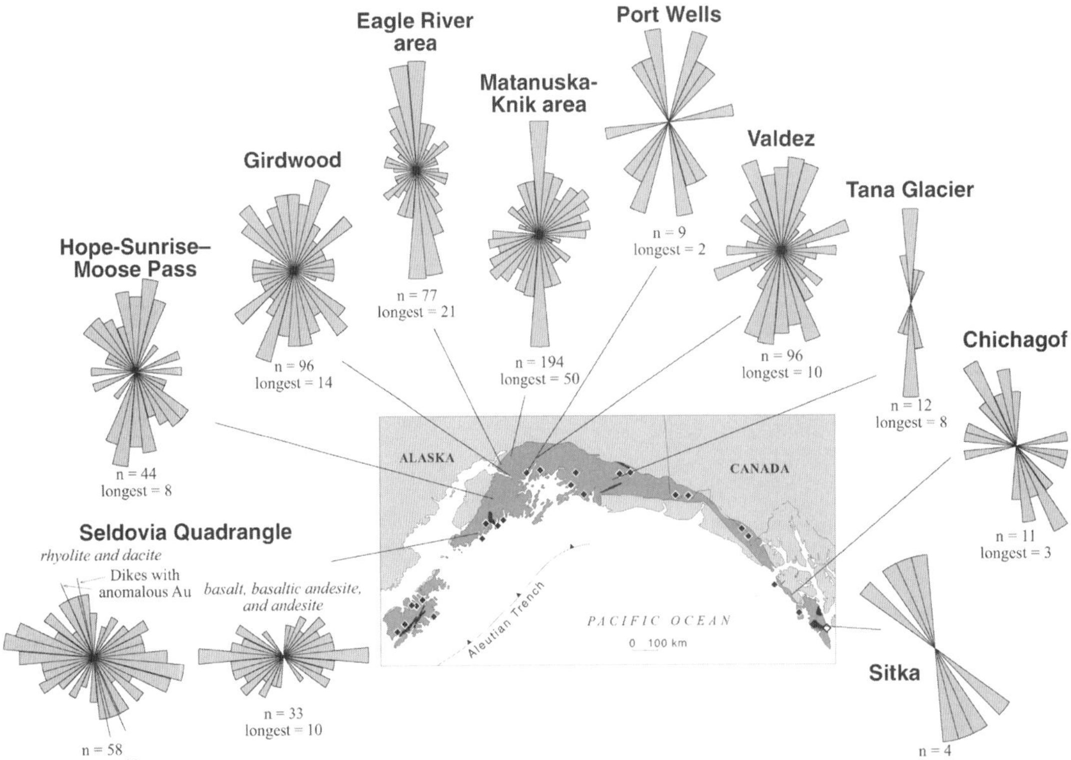

Figure 6. Map of Chugach terrane showing orientation of dikes of Paleocene-Eocene age (after Haeussler et al., this volume).

Pavlis, 1993) would cause it to migrate instantaneously northward, intersecting the continental margin where it curved significantly in southern Alaska (Fig. 2C). The sudden re-appearance of the ridge and slab window in the Prince William Sound area caused a second generation of melting and near-trench magma generation, explaining the ca. 35 Ma plutonism. After this, the rest of the ridge would be subducted in a relatively short time (depending on the geometry of ridge-transform offsets) along western North America. For the purposes of this work, we use the ca. 35 Ma plutons as spatial and temporal markers for the development of structures in the wedge. By determining which structures predate, postdate, and are synchronous with the 55 and 35 Ma plutons, we have two additional constraints on the structural evolution of the accretionary wedge as a whole.

Terentiev Pluton

The Terentiev pluton is a medium- to fine-grained, locally well-foliated, two-mica and amphibole-bearing granite (Nelson et al., 1999). It has yielded a $^{40}Ar/^{39}Ar$ isochron age of 29.2 ± 0.3 Ma (Nelson et al., 1999), interpreted as an exhumation and cooling age from intrusion near 34 Ma. The Terentiev, Granite Cove, and Cedar Bay plutons are enriched in alkalis and silica compared to other ca. 35 Ma plutons in Prince William Sound, consistent with derivation from melting of evolved rocks of the accretionary prism (Nelson et al., 1999). The Terentiev is finer-grained and has more amphibole near its margins and is more quartz-rich at higher structural levels. It has a mottled phase toward the center of the pluton that consists of fine-grained amphibole-rich granite with interstitial pods and zones of coarser-grained two-mica granite. The core of the pluton is a medium-grained two-mica granite with minor and varying amounts of amphibole. Mineral percentages are approximately 20–35% plagioclase, 15–35% orthoclase, 10–25% quartz, 3–7% biotite, 1–4% muscovite, 1–8% amphibole, 1–5% sulfides, and 1–4% opaques, garnet, secondary epidote, and unknowns.

The Terentiev pluton intrudes the Orca Group (Fig. 7), forming a contact aureole of albite-epidote hornfels facies (Nelson et al., 1999), with rare staurolite (Haeussler and Nelson, 1993). In the area of Figure 7, the Orca Group includes units of massive sandstone, turbidites, conglomerate, siltstone, shale, and chaotic deposits interpreted as slump deposits (Nelson et al., 1999). Minor volcanic rocks (Fig. 7) include volcanogenic mudstone with hyaloclastite. These are interbedded with purple to green calcareous shale, thin-bedded pelagic limestone, and tectonically disrupted graywacke-shale units (Nelson et al., 1999; Haeussler and Nelson, 1993).

Magmatic foliations are defined by aligned plagioclase crystals and locally by aligned amphibole crystals. Several separate internal lobes and circular patterns defined by the magmatic foliation can be distinguished within the magma chamber (Fig. 7). Magmatic folds defined by folded magmatic layering are visible both on the outcrop and map scales. These magmatic folds tend to be parallel or subparallel to the margins. The great vertical relief in this pluton (topographically lower in the southeast, higher in the northwest; Fig. 7) revealed a lower zone characterized by lobate magmatic foliations, a higher zone with relatively parallel or tabular foliation traces, and a highest zone containing many blocks of country rock just below a roof pendant.

Pre-emplacement fabrics in the aureole of the Terentiev pluton consist of annealed bedding-parallel faults and ductile mélange fabrics preserved in the biotite-grade hornfels. The first stages of syn-emplacement deformation are represented by pods to tabular zones of intrusion breccia consisting of angular clasts of granite in a very fine-grained green groundmass. The breccia groundmass is geochemically identical to the granite (S.W. Nelson, 1993, personal commun.). To the south of Terentiev Lake a large mass of similar intrusion breccia associated with the Granite Cove granitic stock (Fig. 7) has been dated at 29.2 ± 0.1 Ma ($^{40}Ar/^{39}Ar$ plateau age on K-spar: Nelson et al., 1999). Tabular zones of intrusion breccia in the Terentiev pluton are parallel to several other structures, including early epidote/quartz veins, later iron oxide/quartz veins, relatively late NW-striking sinistral faults, and an EW-striking normal fault set. These faults are parallel to, and may be part of the late orthorhombic fault set that Kusky et al. (1997a) related to ridge subduction.

The long axis of the Terentiev pluton is parallel to NW-striking sinistral faults in the surrounding Orca Group (Figs. 7 and 8). These faults are probably part of the ca. 55 Ma regional set of orthorhombic faults (Fig. 8F) that Kusky et al. (1997b) related to late stages of Paleogene ridge subduction. The Terentiev pluton seems therefore to have intruded at ca. 35–30 Ma along extinct fault sets formed during ridge subduction at ca. 55 Ma, soon after this part of the Orca Group was accreted. The magmas likely used the old faults as structurally weak planes to intrude along, forming intrusive contacts where faults previously existed. Several NW-striking faults also cut the Terentiev pluton. Slickensides on the dominant NNW-striking sinistral fault set show changes in orientation with time. Early motions along these faults (preserved in the epidote/quartz veins) had a significant dip-slip component whereas younger FeO/quartz veins record a progressive trend towards strictly strike-slip motion (plunges on the last set are approximately 1°–4°). These relationships are similar to those of the 55 Ma ridge-subduction related orthorhombic fault set at Turnagain Arm.

The Terentiev pluton is also cut by NE-striking dextral fault zones, which are the youngest structures recognized, and postdate all plutonism. These are typically several centimeters to meters wide zones of fault gouge, with several tens of centimeters to meters (or more) displacement. Haeussler and Nelson (1993) note that several ca. 35–30 Ma plutons in the Prince William Sound area are cut by very late dextral faults including the Contact fault.

The magmatic and tectonic structures present within the Terentiev pluton and its associated contact aureole can be attributed to the subduction of the western plate (oceanic plate B), which was the "trailing" plate of a migrating, subducting spreading ridge. Pre-emplacement fabrics, though not thoroughly studied in this area, are consistent with subduction

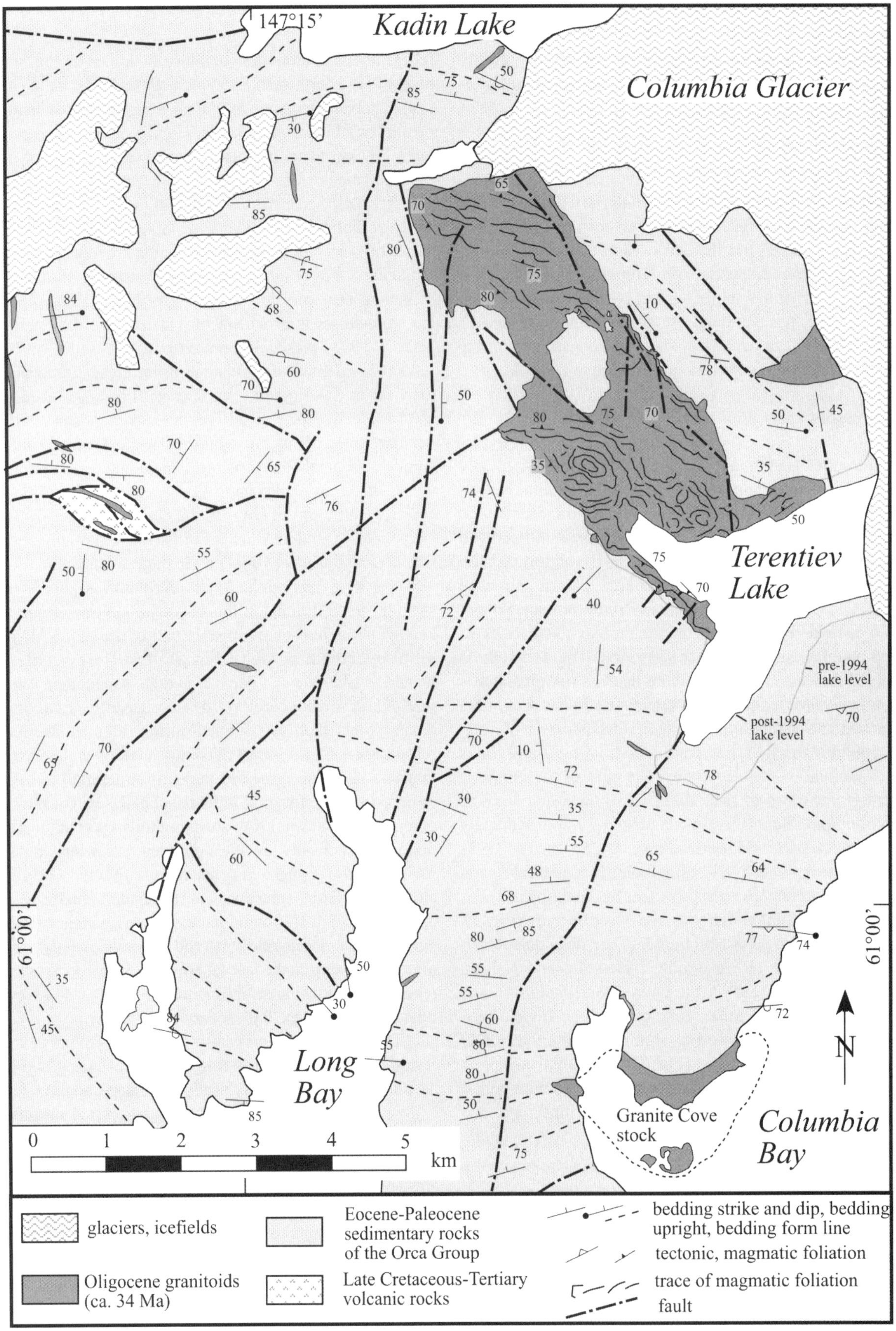

Figure 7. Map of the ca. 30 Ma Terentiev pluton and surrounding Orca Group (compiled from Nelson et al. (1999), and 1:6000 scale mapping of magmatic foliations by authors). Map emphasizes relationships between magmatic and tectonic foliations.

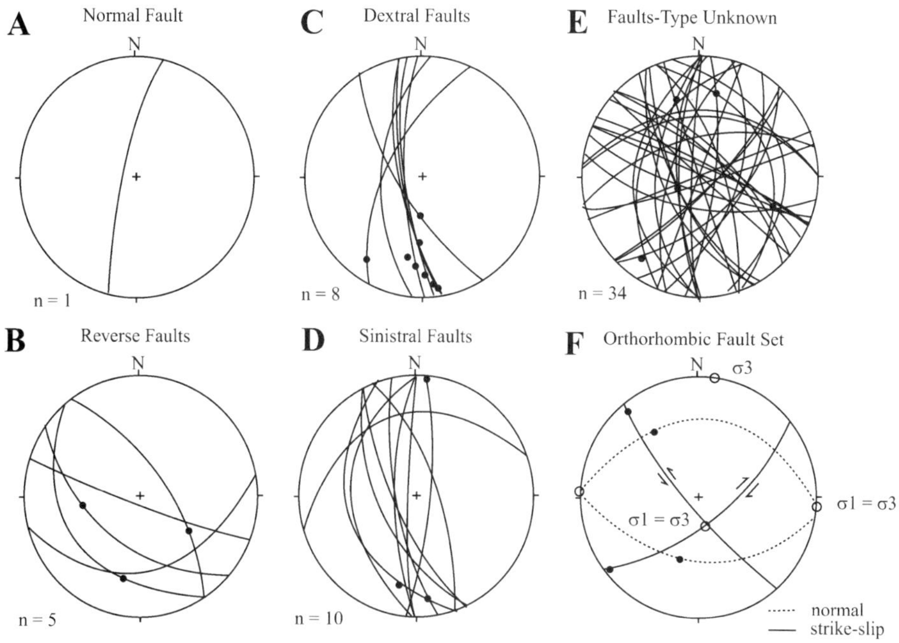

Figure 8. Structural data (after Haeussler and Nelson, 1993) from the Columbia Glacier–Terentiev pluton area, showing the orientation of normal (A), reverse (B), dextral (C), sinistral (D), and unknown faults (E). F: Geometry of Paleogene orthorhombic fault set from the Turnagain Arm area (from Kusky et al., 1997b). Note that Terentiev pluton is elongate parallel to the sinistral faults of the regional orthorhombic set, and to sinistral and thrust faults in the study area.

kinematics interpreted from along strike (Kusky and Bradley, 1999) and include evidence for layer-parallel extension. Syn-emplacement fabrics are apparently restricted to the plutonic body. These fabrics include magmatic foliations that define magmatic lobes and folds near the contact with the country rock. The contact zones are characterized by abundant angular metasedimentary xenoliths that typically have their long axes parallel to the contact. In some cases there is a strongly bimodal size distribution of the xenoliths, in which case the small ones have their long axes parallel to much larger xenoliths, which in turn have their long axes parallel to the contact with the country rock. Late-stage emplacement related structures through post-emplacement fabrics include intrusion breccia dikes, high- and lower-temperature late magmatic veining, and brittle faults. It is possible that the intrusion breccia is related to a late magmatic (Granite Cove stock?), highly volatile explosive stage. The dominant near field mechanism for magmatic emplacement for the Terentiev pluton appears to be diking and thermal shattering forming stoped blocks near the magma/host rock contact. The origin of the lobes defined by the magmatic foliations is less clear; they could be magmatic diapirs, or they could represent complex folds formed in a crystal-liquid mush due to strain being concentrated in the crystallizing magma chamber.

Eshamy and Nellie Juan Plutons

The ca. 35 Ma Eshamy pluton (Fig. 3) consists of a series of thin dikes parallel to regional orogenic strike, but perpendicular to the long axis of the pluton. It is a multi-pulse pluton composed of a series of individual sheets or dikes (e.g., Hibbard and Watters, 1985), with a sequence of crosscutting relationships indicating a magmatic series from gabbro to quartz-feldspar porphyry to biotite-hornblende granodiorite to gabbro to granite. It is a fairly shallow-level intrusion, based on spectacular fracturing and stoping along the pluton–wall rock boundary. Xenoliths are especially abundant along the pluton's southwestern end. Based on our reconnaissance observations, we conclude that the Eshamy pluton was probably never a large body of magma, but represents a magma pathway that was active for many batches of melt, with at least two generations of gabbro generated at depth (from the slab window?), explaining the repetition of compositional pulses. This distinction is important, because it suggests that the Eshamy was never a single batch of magma, but represents a time series of dikes, a couple of meters wide at a time, striking parallel to regional strike, then cut by a NE cleavage. The Eshamy pluton bears many similarities to other multiple-pulse, sheet-like intrusions including the Calamity Peak Complex in the Black Hills (Duke et al., 1988). In contrast, the Nellie Juan pluton is extremely homogenous, with only a weak transitional magmatic/solid state foliation parallel to regional strike.

MAGMATIC EMPLACEMENT MECHANISMS

The style of magmatic emplacement during ridge subduction can be evaluated through interpretation of field relationships, the structure of the plutons, mineralogy, and metamorphic reactions (e.g., Castro, 1987). The Nuka and Aialik plutons form a series of granitic-to-granodioritic sheets that strike approximately 020° along their long sides, and have a variety of dips, predominantly steeply northwest. The magmatic layers, however, curve along with the pluton contacts, forming discordant contacts with country rocks along their short NW-striking margins. Most contacts are relatively planar and utilize pre-existing anisotropies and fabrics within the Valdez Group. The Tustumena–Harris Bay pluton strikes NNW and shows stronger magmatic deformation fabrics than the other plutons, suggesting that its intrusion is largely controlled by a NNW striking shear zone. Isotopic studies of pluton-related quartz veins (Goldfarb et al., 1986; Borden et al., 1992; Taylor et al., 1994) indicate emplacement at 5.2–10.5 km depth and 210–300 °C. The regional metamorphism of the Valdez group is lower greenschist facies, and the contact aureole is cordierite-biotite (amphibolite) grade, also indicating that the plutons were emplaced in a shallow (approximately 3–25 km for lower greenschist facies) and cool (~250–450 °C) part of the accretionary prism. Since the plutons have megacrystic kyanite apparently concentrated along their contacts, the peraluminous melt must have been generated at a depth of at least 20 km at a minimum temperature of 650 °C, in the PT space where the stability field of kyanite is above the granitic solidus (e.g., Essene, 1986; Hibbard, 1995). Therefore the melt must have migrated upward to its emplacement level of less than 5.2–10.5 km from at least this depth, entraining the kyanite and other xenocrystic phases along its intrusion path. If the kyanite crystals were in contact with the melt, then the magma must have risen at a rapid rate that would prevent the kyanite from recrystallizing to a more stable aluminosilicate or phyllosilicate. However, the reaction rims around the kyanite crystals effectively isolated them from reacting with the melt.

Relationships of Plutons to Regional Structures

The intrusion of many granitic bodies is triggered by some kind of tectonic activity, and understanding specific triggers is one of the fundamental goals of many granite studies (e.g., Leake, 1990; Petford et al., 2000). The large plutons of the Sanak-Baranof belt on the Kenai Peninsula and Kodiak Island are predominantly oriented slightly oblique to the main terrane-bounding faults and large-scale regional strike. They are generally parallel to local foliations and bedding traces. In particular, the Kodiak batholith, the composite Tustumena–Harris Bay plutons, and the large pluton near Mt. Tom White in the Bearing Glacier quadrangle (eastern Chugach Mountains; Fig. 1) are all elongated about 15°–45° counterclockwise from the main bounding faults of the Chugach terrane (Fig. 9). The Nuka and Aialik plutons are elongate parallel to regional strike, and are associated with regional scale lineaments parallel to or coincident with known faults (Fig. 3).

Many plutons are emplaced into dilational bends in strike-slip fault systems (e.g., Hutton, 1987, 1988; Hutton et al., 1990; Hutton and Reavey, 1992; Tobish and Cruden, 1995; Saint Blanquat et al., 1998). Examples include the Main Donegal granite (Hutton, 1982), the Mount Peyton pluton of Newfoundland (Kusky et al., 1987), the Doctor's Flat pluton, Victoria, Australia (Morand, 1992), and many Sierra Nevadan plutons (e.g., Tikoff and Saint Blanquat, 1997; Tikoff et al., 1999). The Nuka and Aialik plutons exhibit sheets and dikes along their contacts, as predicted for plutons whose intrusion is controlled by space opened by active faulting (Yoshinobu et al., 1998). The Chugach terrane experienced a change in kinematics from predominantly head-on convergence to dextral transpression with the passage of the triple junction, and this change formed dextral strike-slip faults in the forearc that are coeval and slightly younger than an orthorhombic fault set formed during ridge subduction (Kusky et al., 1997b). Therefore, a fault-controlled pull-apart mechanism may have controlled the intrusion of the Nuka and Aialik plutons at 55 Ma, which are situated along possible extensional jogs along traces of one of the numerous orogen-parallel strike-slip faults, including the Contact fault (Fig. 9). This mechanism does not, however, explain the Tustumena–Harris Bay pluton, whose long axis is oblique to regional strike.

The long axis of the Tustumena–Harris Bay pluton is parallel to NW-striking sinistral late faults of the ca. 55 Ma orthorhombic fault set on the Kenai Peninsula (Fig. 9; Kusky et al., 1997b). We suggest that a fault of the orthorhombic set may have provided a structural weakness along which magmas of the Tustumena–Harris Bay pluton intruded, largely replacing the fault with an intrusive contact. The area of the sharp bend in the Tustumena–Harris Bay pluton preserves an unusually strong NW-striking tectonic foliation (Fig. 5E) that we interpret as related to this shear zone.

Tikoff and Teyssier (1992) propose an alternative pluton emplacement model for transpressional arcs that may have applications to the Sanak-Baranof belt. In their model, plutons are emplaced into en echelon P-shear tensional bridges, forming elongate plutons oriented approximately 15° from the main orogenic strike. P-shears (Logan et al., 1979) form at low angles to the main transpressional structures in rheologically simple, brittle systems. R-shears may connect individual P-shears, and movement on the crustal scale shear system creates extensional jogs into which magmas may be emplaced. However, it is unknown how applicable this general model may be to rheologically complex systems like the southern Alaska forearc, nor is it known how these P-shears and related fractures may connect with ductile structures at depth. In the model of Tikoff and Teyssier (1992), the spacing and the amount of overlap between individual P-shears determines the shape of the individual plutons. P-shears commonly have a large overlap area resulting in elongate plutons oriented with their long walls parallel to the P-shears (Tikoff and Teyssier, 1992). Additionally, the largest strains should be on the margins of the plutons along the P-shears, and most foliations in the plutons should be nearly vertical and parallel to the P-shears (Tikoff and Teyssier, 1992).

Some of the field relationships in the Sanak-Baranof belt are apparently explained by this alternative model. The orientations of the many plutons are parallel to the hypothetical P-shears, and the pluton emplacement mechanisms point to fracture-controlled intrusion. In the case of the southern Kenai Peninsula, however, it is difficult to identify specific structures that may be P-shears or R-shears, whereas most pluton-related structures correlate with the main dextral strike-slip faults or the orthorhombic fault set. However, other P-shears may be obscured by intrusion by the

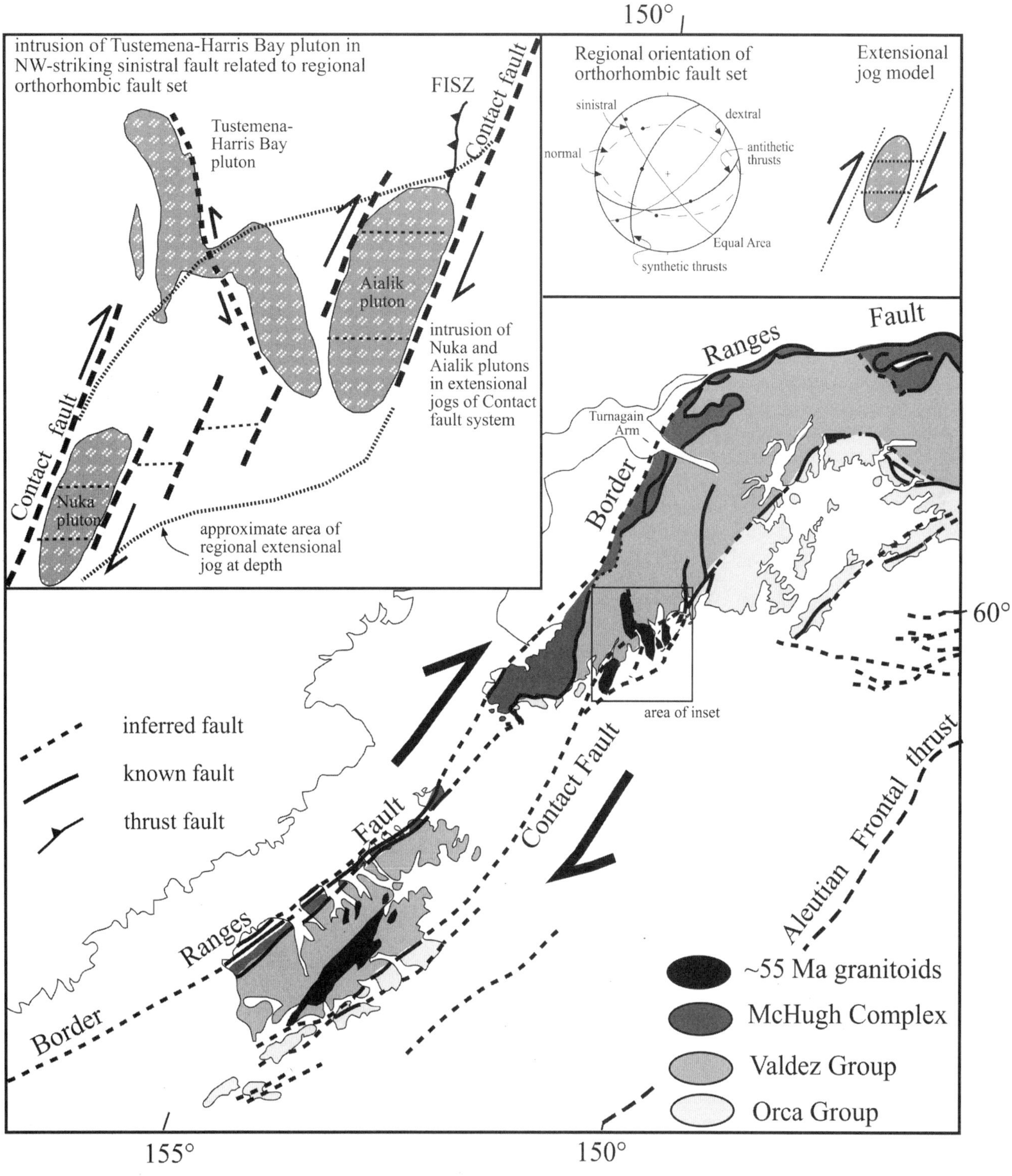

Figure 9. Map showing relationship of plutons to orthorhombic fault set. The Nuka and Aialik plutons are elongate parallel to orogen-parallel strike-slip faults and are interpreted to have intruded into dilational jogs during movement on these faults. The Tustumena–Harris Bay pluton is parallel to and deformed by NW-striking sinistral faults of the ridge-subduction-related orthorhombic fault system, and probably used these faults as a structural weakness to ease intrusion into the wedge.

granites, i.e., plutonism lagged behind faulting, and the plutons merely used pre-exisiting faults as planes of weakness to intrude along. Other faults may be covered by glaciers, outwash, or may be only exposed at the bottoms of deeply incised fiords. The lack of identifiable P-shears may also be partly attributable to the fact that the region has only been mapped at the reconnaissance scale of 1:250,000. Nonetheless, we note the well-exposed shear zone in the Tustumena pluton may continue along its margin to the north, and that several of the deep fiords that mark the boundaries of the plutons are parallel to the predicted P-shear direction (e.g., McCarty Fiord, Harris Bay, Aialik Bay, and Resurrection Bay). In addition, the southern margin of the Tustumena pluton is abruptly terminated along a glaciated valley parallel to the predicted R-shear direction.

One variation of the fault-related intrusion mechanism is preserved on Hive Island and Ragged Island in Resurrection Bay. Here, the northwestern side of the Aialik pluton appears to have been emplaced along a thrust fault (Fox Island shear zone; FISZ in Fig. 4) that formed during emplacement of the Resurrection Peninsula ophiolite between 57 and 53.4 Ma as a consequence of ridge subduction (Kusky and Young, 1999). This thrust fault may have intersected one of the deep-seated shear systems at depth, and magma migrated along the fault. Such migration of magmas along an active rapidly moving thrust, with the consequent melt-enhanced deformation (e.g., Karlstrom et al., 1993; Yoshinobu et al., 1998) could explain the rapid emplacement of the Resurrection Peninsula ophiolite.

Use of the circa 56 Ma Plutons as Strain Markers, to Estimate Plate Kinematic Parameters Before, During, and After Ridge Subduction

We use the ca. 56 Ma plutons of the Sanak-Baranof belt as time and strain markers, separating kinematic regimes that predate, postdate, and were synchronous with ridge subduction. The suite of 35–30 Ma plutons in Prince William Sound provide an additional time and strain marker that can be used in the same context. Any structures related to ridge subduction will be late features (Paleogene), superimposed on fabrics related to early accretion, but perhaps contemporaneous with structures resulting from coastwise strike slip and/or orocline formation. Structures associated with magma emplacement (for the ca. 56 Ma plutons and dikes) are regarded as related to ridge subduction, whereas the chronology of other structures is established by crosscutting relationships. We have placed considerable emphasis on establishing which structures cut the 56 Ma plutons, and which cut or are cut by the ca. 35 Ma plutons.

We emphasize the sequence of development of late structures, paying particular attention to the relative timing of deformation and near-trench magmatism. Hypersolidus fabrics produced by magmatic flow and deformation of hot plutons are differentiated from crystal-plastic fabrics, which are entirely post-crystallization (e.g., Marre, 1986; Paterson and Tobish, 1988; Paterson et al., 1989a, 1989b, 1991).

Relative Timing of Pluton Emplacement and Fabric Development

Pre–circa 56 Ma Pluton Emplacement Structures

Structures and fabrics associated with Late Cretaceous subduction present in the Valdez Group includes regional scale, northeast-trending isoclinal folds; bedding and cleavage are generally parallel to one another (Kusky et al., 1997b). The mélange of Iceworm Peak is a type-1 mélange that forms a regionally mappable unit along the contact between the McHugh Complex and the Valdez Group. Kinematic analysis of the fabric elements that are cut by ridge-subduction related dikes led Kusky et al. (1997b) to suggest that the pre–triple junction interaction contraction and convergence was close to perpendicular to the paleo-trench, during subduction of oceanic plate A (Farallon or Resurrection) beneath the Chugach terrane.

Minor faults related to accretion and subduction of the oceanic plate A are also preserved in the contact aureole of the Nuka, Tustumena–Harris Bay, and Aialik plutons. The dominant style of deformation was orogen-parallel extension (Figs. 4A and 4B). The orientations of these extensional slip surfaces are consistent throughout the contact aureole (345°, 85°W) and reflect bedding-parallel extension perpendicular to the maximum compression direction. These extensional slip surfaces also cut the fabric of the Iceworm mélange and are therefore post-Maastrichtian (age of Valdez protolith to Iceworm mélange) and older than 57 Ma (age of crosscutting dike) (Kusky et al., 1997a). There are other early dip-slip faults that are at a higher angle to bedding that are cut by dikes and therefore also predate ridge subduction. The pre-ridge subduction faults reflect E-W contraction and N-S extension, perhaps formed during early dewatering and volume loss during early accretion (e.g., Kusky and Bradley, 1999).

Syn–circa 56 Ma Pluton Emplacement Structures

Structures associated with emplacement of the Nuka, Tustumena–Harris Bay, and Aialik plutons and subduction of the spreading center include brittle/ductile faults, foliation development and porphyroblast growth in contact aureoles, and brittle dikes. Dikes associated with the emplacement of the plutons (in close proximity to the pluton margins) have been boudinaged and folded. Dikes oriented subparallel to the pluton margins are boudinaged, whereas dikes oriented at a high angle to the shear direction are ptygmatically folded. This relationship suggests that the plutons expanded or ballooned during intrusion, superimposing a local strain upon the regional strain field (e.g., Bateman, 1984, 1985; Paterson, 1988; Paterson and Vernon, 1995). The fact that not only the dikes but also the lithified graywackes are ductilely deformed (post-mélange and post-bedding-parallel brittle faults) suggests that the cold accretionary prism must have experienced a significant localized thermal softening event, consistent with the cordierite-biotite (amphibolite) metamorphic grade of the aureole. It is unlikely that these fabrics are post emplacement since the rheological contrast between the low-grade Valdez Group and the metamorphic aureole would have resulted in a strain partitioning and localization of deformation at the margin of the aureole and not ductile shear zone formation within the aureole.

The plutons have a strong tectonic fabric at their margins and a weakly developed tectonic foliation away from their margins. This is because the margins, having cooled faster than the interior of the plutons, were able to preserve tectonic fabrics, whereas the more liquid interior of the pluton did not preserve fabrics related to this phase of deformation. A melt that is greater than 55% solidified experiences a dramatic increase in viscosity due to crystal locking and will support a shear stress (Arzi, 1978; Scaillet et al., 1997). The mostly crystalline melt could therefore have foliations related to tectonic/syn-emplacement strains. This tectonic/syn-emplacement foliation is also developed within later-stage, syn-emplacement, crosscutting dikes as well (Fig. 5J). This strain is the result of both regional and local emplacement stresses (e.g., Yoshinobu et al., 1998).

Post–circa 56 Ma Pluton Emplacement Structures

Post-emplacement brittle fabric patterns are more complex because, in some instances, the later faults have reactivated older faults. In an effort to distinguish between reactivation of pre-existing faults and the creation of new fracture surfaces, careful selection of brittle faults for measurement was made. The bulk of the orientations in Figures 4C and 4D reflect faults mapped in the Nuka, Tustumena–Harris Bay and Aialik plutons and those in the contact aureoles that were observed to crosscut the pluton or dikes, or in a few cases earlier brittle to ductile features. There is a set of conjugate strike-slip faults but most faults are high-angle normal faults and lower-angle reverse/thrust faults. Post ridge subduction deformation is marked by an orthorhombic fault system (three dip-slip and two strike-slip). The dip-slip faults have orientations of N40°E, 80°E for the normal faults, N20°E, 10°E and N40°E, 70°W for the reverse faults while the strike-slip faults have orientations of 335°, 85°E for sinistral faults and N60°E, 75°NW for the dextral faults. These are similar to the syn-ridge subduction orthorhombic fault set described from the Turnagain Arm area by Bradley and Kusky (1990) and Kusky et al. (1997b), although rotated slightly. The presence of these faults in the plutons shows that the stress system that formed the orthorhombic fault set persisted though the late-ridge subduction related plutonism, and that the faults and plutons are spatially and temporally related.

These late faults, formed during subduction of oceanic plate B (Kula) indicate a significant component of dextral transpression, associated with orogen-perpendicular extension that is probably related to exhumation and critical taper recovery following ridge subduction (Kusky et al., 1997b).

Additional Constraints on Structural History of the Wedge from circa 35 Ma Plutons

The Terentiev pluton intruded the Orca Group along a set of fractures that is similar in orientation to both the late faults and the orthorhombic fault set described from the Valdez Group by Kusky et al. (1997b). Emplacement along this pre-existing weakness may explain the anomalous NW elongation of the Terentiev pluton. The Terentiev pluton is cut by numerous NE-striking dextral strike-slip faults, showing that dextral transpression continued past 30 Ma.

CONCLUSIONS

Physical evidence from the Nuka, Tustumena–Harris Bay, and Aialik granitic complex and surrounding rocks has implications for ridge subduction and intrusion mechanisms of near-trench magmas associated with ridge subduction. The granites of the Harding Icefield region were intruded into a fault network within the accretionary prism and these faults include orogen-parallel strike-slip faults as well as an orthorhombic fault set including NW and NE striking strike-slip faults. Xenocrystic kyanite and garnet indicate that the magmas were generated above 650 °C at pressures exceeding 5.5 kbars (>20 km; Fig. 10). Metamorphic mineral assemblages, isotopic and fluid inclusion studies indicate that the plutons intruded at 5.2–10.5 km depth, into wall rocks at 210–300 °C. Several of the plutons appear to have intruded along NW-striking sinistral faults of the orthorhombic fault set, formed at an angle to the main regional dextral shear system developed as a consequence of ridge subduction (Fig. 10).

Structures present in the contact aureoles of the Nuka, Tustumena–Harris Bay, and Aialik plutons that can be attributed to deformation prior to the subduction of the oceanic ridge include regional NE-striking thrusts, folds, and zones of ductile to brittle layer-parallel extension (mélange foliation and bedding-parallel normal/listric faults). There are also two conjugate sets of larger, brittle dip-slip faults that may occur at higher structural levels in the accretionary prism.

Syn–56 Ma pluton emplacement deformation includes strains related to intrusion and expansion of plutons and intrusion of swarms of dikes throughout the region. The dikes generally strike at a high angle to regional bedding and cleavage and reflect a component of orogen-parallel extension at a high angle to the main orogen-parallel strike-slip faults. Intrusion of most of the magmas related to ridge subduction seems therefore to be controlled in some way by the large-scale orogen-parallel faults. Plutons intruded into extensional jogs along the main faults, oblique fracture systems related to the orthorhombic fault set, and dikes intruded at high angles to the faults and reflect extension in the same direction.

Post–56 Ma pluton emplacement deformation includes high-angle normal and lower-angle reverse/thrust faults. The normal faults strike at approximately 040° and dip near vertical. The reverse faults have a similar strike but dip shallowly to the southeast and less commonly steeply to the northwest. There is also a less common and well-developed conjugate set of strike-slip faults. Together, these late fault sets possess orthorhombic symmetry, and are remarkably similar in orientation to the late ridge subduction related orthorhombic fault system documented from the Turnagain Arm area (Kusky et al., 1997b). Overall,

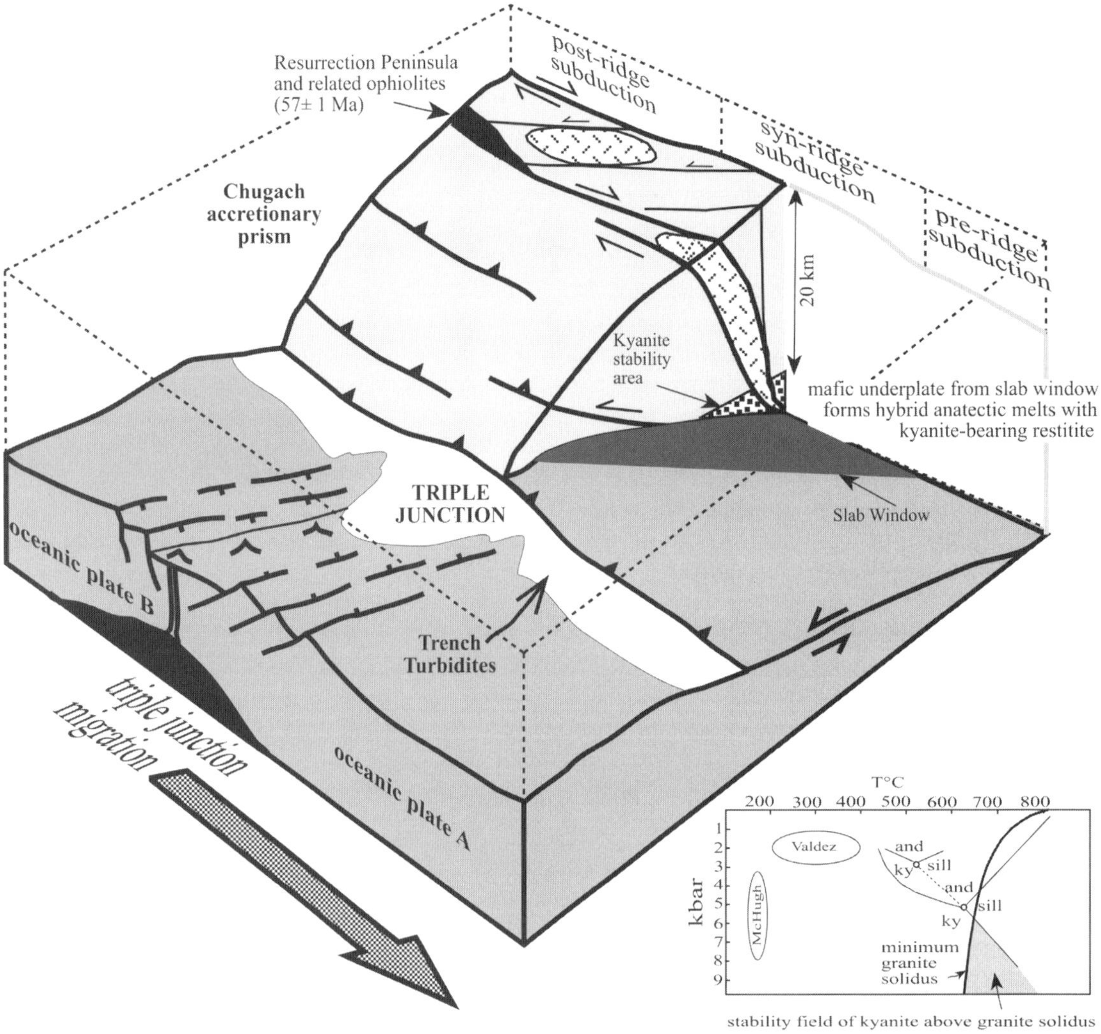

Figure 10. Three-dimensional sketch indicating the relative position and timing of ridge subduction, formation of the slab window, generation of the Nuka-Aialik and Harris Bay plutons, and activation of the late out-of-sequence thrust faults including the Fox Island shear zone.

this orthorhombic fault set accommodated dextral oblique transpression, with $\sigma_1 = \sigma_2$ about WNW-ESE axes, along with extensional faulting in the upper part of the wedge accommodating exhumation in response to recovery of critical taper after ridge subduction (Kusky et al., 1997b).

The suite of 35–30 Ma plutons in Prince William Sound is interpreted to have formed in response to the ridge migrating back northward in response to the early Eocene plate reorganization (Fig. 2). These plutons add an additional temporal strain marker that is used to calibrate the structural evolution of the wedge. Post-30 Ma pluton emplacement deformation includes margin-parallel dextral strike-slip faulting, probably associated with the northward translation of the Chugach terrane with respect to North America.

Implications for Forearc Evolution and Crustal Growth

Recognizing the Sanak-Baranof belt as a product of ridge subduction has several implications for forearc evolution and the interpretation of linear belts of plutons in ancient mountain belts. Forearcs are not necessarily places characterized exclusively by high-*P*, low-*T* metamorphic series and a lack of plutonism, but may contain high-*T*, low-*P* metamorphism in association with belts of plutonic rocks if the forearc was affected by ridge subduction. Similarly, belts of magmatic rocks in ancient mountain belts may not necessarily represent individual arc terranes, but could be a paired arc/forearc system that experienced a ridge subduction event. The record of ridge subduction events may vary considerably in individual examples depending on the plate geometry and rates of triple junction and slab window migration. However, some of the hallmark signatures of ridge subduction in forearcs include along strike diachronous intrusion of tonalitic, trondhjemitic, granodiorite, to granitic plutons, high-temperature metamorphism, and diachronous gold mineralization, and belts of anomalous complex faulting. We have shown in this contribution how the near-trench magmatic rocks may serve as diachronous temporal and strain markers that separate kinematic regimes in the forearc that represent interactions of the leading and trailing subducting plates with the overriding plate. We have also provided several examples of how faults can control pluton emplacement, both as inactive zones of structural weakness, and as active dilational bends.

Structural, thermal, and magmatic aspects of the Chugach terrane are similar to the geology of Archean greenstone-granodiorite terranes (Pavlis et al., 1988; Kusky, 1989; Barker et al., 1992; Kusky and Polat, 1999). In both, deformation is locally melt-dominated, and plutons follow a low-K series from diorite to trondhjemite (Pavlis et al., 1988). Metamorphism is of a high-temperature low-pressure series. Ridge subduction was most likely an important process in the Archean, when the total number of plates was higher, and the number of ridge-trench encounters was probably greater (Kusky and Polat, 1999). The southern Alaska margin then may serve as a relatively "modern" example of processes that were likely to have been important in Archean forearc evolution and continental growth.

ACKNOWLEDGMENTS

The work was funded by National Science Foundation grants EAR 9304647 and 9706699 awarded to T. Kusky and by the Alaska Mineral Resource Assessment Program of the U.S. Geological Survey. Numerous people participated in the field program, including Alison Till, Charles Young, and Shawn Dolan. Lorainne Fan compiled the map of magmatic foliations for the Terentiev pluton. This manuscript has benefited from discussions and reviews by C.J. Northrup, S. Roeske, Bill McLelland, A. Yoshinobu, and V. Sisson.

REFERENCES CITED

Arzi, A.A., 1978, Critical phenomena in the rheology of partially melted rocks: Tectonophysics, v. 44, p. 173–184.

Babcock, R.S., Burmester, R.F., Engebretson, D.C., Warnock, A., and Clark, K.P., 1992, A rifted margin origin for the Crescent basalts and related rocks in northern Coast Range volcanic province, Washington and British Columbia: Journal of Geophysical Research, v. 97, p. 6799–6821.

Babcock, R.S., Suczek, C.A., and Engebretson, D.C., 1994, The Crescent "terrane," Olympic Peninsula, southern Vancouver Island: Washington Division of Geology and Earth Resources Bulletin, v. 80, p. 141–157.

Barbey, P., 1991, Restites in migmatites and autochthonous granites: Their main features and their genesis, *in* Didier, J., and Barbarin, B., eds., Enclaves and granite petrology: Amsterdam, Elsevier, p. 479–491.

Barker, F., Farmer, G.L., Ayuso, R.A., Plafker, G., and Lull, J.S., 1992, The 50 Ma granodiorite of the eastern Gulf of Alaska: Melting in an accretionary prism in the forearc: Journal of Geophysical Research, v. 97, p. 6757–6778.

Bateman, R., 1984, On the role of diapirism in the segregation, ascent, and final emplacement of granitoids: Tectonophysics, v. 110, p. 211–231.

Bateman, R., 1985, Aureole deformation by flattening around a diapir during in situ ballooning: The Cannibal Creek granite: Journal of Geology, v. 93, p. 293–310.

Bol, A.J., and Coe, R.S., Gromme, C.S., and Hillhouse, J.W., 1992, Paleomagnetism of the Resurrection Peninsula, Alaska: Implications for the tectonics of southern Alaska and the Kula-Farallon ridge: Journal of Geophysical Research, v. 97, p. 17,213–17,232.

Borden, J.C., Goldfarb, R.J., Gent, C.A., Buruss, R.C., and Roushey, B.H., 1992, Geochemistry of lode-gold deposits, Nuka Bay district, southern Kenai Peninsula: U.S. Geological Survey Bulletin 2041, p. 13–22.

Bradley, D.C., and Kusky, T.M., 1990, Kinematics of late faults along Turnagain Arm, Cretaceous accretionary complex, south-central Alaska, *in* Dover, J., and Galloway, J., eds., Geologic studies in Alaska by the U.S. Geological Survey, 1989: U.S. Geological Survey Bulletin 1946, p. 3–10.

Bradley, D.C., and Kusky, T.M., 1992, Deformation history of the McHugh accretionary complex, Seldovia quadrangle, south-central Alaska, *in* Bradley, D.C., and Ford, A.B., eds., Geologic studies in Alaska by the U.S. Geological Survey during 1990: U.S. Geological Survey Bulletin 1999, p. 17–32.

Bradley, D.C., and Wilson, F.H., 2000, Reconnaissance bedrock geology of the southeastern Kenai quadrangle, Alaska: U.S. Geological Survey Professional Paper 1615, p. 59–63.

Bradley, D.C., Haeussler, P.J., and Kusky, T.M., 1993, Timing of early Tertiary ridge subduction in southern Alaska, *in* Dusel-Bacon, C., and Till, A.B., eds., Geologic studies in Alaska by the U.S. Geologic Survey, 1992: U.S. Geological Survey Bulletin 2068, p. 163–177.

Bradley, D.C., Kusky, T.M., Haeussler, P., Karl, S.M., and Donley, D.T., 1999, Geologic map of the Seldovia quadrangle: U.S. Geological Survey Open-File Report 99-18, scale 1:250,000, with marginal notes. Also available as an Internet publication: http://wrgis.wr.usgs.gov/open-file/of99-18/ (March 2003).

Bradley, D.C., Parrish, R., Clendenen, W., Lux, D., Layer, P., Heizler, M., and Donley, D.T., 2000, New geochronological evidence for the timing of early Tertiary ridge subduction in southern Alaska: U.S. Geological Survey Professional Paper 1615, p. 5–21.

Bradley, D., Kusky, T., Haeussler, P., Goldfarb, R., Miller, M., Dumoulin, J., Nelson, S.W., and Karl, S., 2003, Geologic signature of early Tertiary ridge subduction in Alaska *in* Sisson, V.B., Roeske, S.M., and Pavlis, T.L., eds., Geology of a transpressional orogen developed during ridge-trench interaction along the North Pacific margin: Boulder, Colorado, Geological Society of America Special Paper 371, p. 19–49 (this volume).

Castro, A., 1987, On granitoid emplacement and related structures: A review: Geologische Rundschau, v. 76, p. 101–124.

Clark, R.G., Jr., and Lyons, J.B., 1986, Petrogenesis of the Kinsman intrusive suite: Peraluminous granitoids of western New Hampshire: Journal of Petrology, v. 27, p. 1365–1393.

Clark, S.H.B., 1972, The McHugh Complex of south-central Alaska: U.S. Geological Survey Bulletin 1372-D, p. D1–D11.

Couturié, J.P., 1969, Sur l'anteriorite du granite porphyroide de la Margeride par rapport au granite a cordierite du Valey (Massif Central Francais): Compte Rendu Academie Science (Paris), v. 269D, p. 2298–2300.

Cowan, D.S., and Boss, R.F., 1978, Tectonic framework of the southwestern Kenai Peninsula, Alaska: Geological Society of America Bulletin, v. 89, p. 155–158.

Crowe, D.E., Nelson, S.W., Brown, P.E., Shanks, W.C., III, and Valley, J.W., 1992, Geology and geochemistry of volcanogenic massive sulfide deposits and related igneous rocks, Prince William Sound, south-central Alaska: Economic Geology, v. 87, p. 1722–1746.

Dell'Angelo, L.N., and Tullis, J., 1988, Experimental deformation of partially melted granitic aggregates: Journal of Metamorphic Geology, v. 6, p. 495–515.

Dixon, J., and Farrar, E., 1980, Ridge subduction, eduction, and Neogene tectonics of southwestern North America: Tectonophysics, v. 67, p. 81–99.

Donley, D.T., Kusky, T.M., and Bradley, D.C., 1995, Emplacement of the Tertiary Nuka, Aialik, and related near-trench plutons, Chugach accretionary wedge, Alaska: Geological Society of America Abstracts with Programs, v. 27, no. 5, p. 15.

Duke, E.F., Redden, J.A., and Papike, J.J., 1988, Calamity Peak layered granite-pegmatite complex, Black Hills, South Dakota, Part I: Structure and emplacement: Geological Society of America Bulletin, v. 100, p. 825–840.

Dumoulin, J.A., 1987, Sandstone composition of the Valdez and Orca Groups, Prince William Sound, Alaska: U.S. Geological Survey Bulletin 1774, 37 p.

Dumoulin, J.A., 1988, Sandstone petrographic evidence and the Chugach–Prince William Terrane boundary in southern Alaska: Geology, v. 16, p. 456–460.

Essene, E., 1986, Geologic thermometry and barometry, *in* Ferry, J., ed., Characterization of metamorphism through mineral equilibria: Mineralogical Society of America Reviews in Mineralogy, v. 10, p. 153–206.

Fisher, D.M., and Brantley, S.L., 1992, Models of quartz overgrowth and vein formation: deformation and episodic fluid flow in an ancient asubduction zone: Journal of Geophysical Research, v. 97, p. 20,043–20,061.

Furlong, K.P., Hugo, W.D., and Zandt, G., 1989, Geometry and evolution of the San Andreas fault system in northern California: Journal of Geophysical Research, v. 94, p. 3100–3100.

Gibbons, H., 1988, Microstructures and metamorphism in an accretionary prism in Prince William Sound, Alaska [M.S. thesis]: Santa Cruz, University of California, 191 p.

Goldfarb, R.J., Leach, D.L., Miller, M.L., and Pickerthorn, W.J., 1986, Geology, metamorphic setting, and genetic constraints of epigenetic lode-gold mineralization within the Cretaceous Valdez Group, south-central Alaska, *in* Keppie, J.D., et al., eds., Turbidite hosted gold deposits: Geological Association of Canada Special Paper 32, p. 87–105.

Haeussler, P.J., and Nelson, S.W., 1993, Structural evolution of the Chugach–Prince William Terrane at the hinge of the orocline in Prince William Sound, and implications for ore deposits, *in* Dusel-Bacon, C., and Till, A.B., eds., Geologic studies in Alaska by the U.S. Geological Survey in 1992: U.S. Geological Survey Bulletin 2068, p. 143–162.

Haeussler, P.J., Bradley, D.C., Miller, M.L., and Wells, R., 2000, Life and death of the Resurrection plate: Evidence for an additional plate in the NE Pacific in Paleocene-Eocene time: Geological Society of America Abstracts with Programs, v. 32, no. 7, p. A-382.

Haeussler, P.J., Bradley, D.C., and Goldfarb, R.J., 2003, Brittle deformation along the Gulf of Alaska margin in Response to Paleocene-Eocene triple junction migration, *in* Sisson, V.B., Roeske, S.M., and Pavlis, T.L., eds., Geology of a transpressional orogen developed during ridge-trench interaction along the North Pacific margin: Boulder, Colorado, Geological Society of America Special Paper 371, p. 119–140 (this volume).

Harris, N.R., Sisson, V.B., Wright, J.E., and Pavlis, T.L., 1996, Evidence for Eocene mafic underplating during forearc intrusive activity, eastern Chugach Mountains, Alaska: Geology, v. 24, p. 263–266.

Hawley, B.W., 1992, Structural, metamorphic, and geochemical study of the Seldovia Bay fault, Alaska: A relict Cretaceous subduction zone [Ph.D. thesis]: Salt Lake City, University of Utah, 132 p.

Hibbard, J.P., and Karig, D.E., 1990, Structural and magmatic responses to spreading ridge subduction: An example from southwest Japan: Tectonics, v. 9, p. 207–230.

Hibbard, M.J., 1995, Petrography to petrogenesis: Englewood Cliffs, New Jersey, Prentice-Hall, 587 p.

Hibbard, M.J., and Watters, R.J., 1985, Fracturing and diking in incompletely crystallized granitic plutons: Lithos, v. 18, p. 1–12.

Hill, M.D., 1979, Volcanic and plutonic rocks of the Kodiak-Shumagin shelf, Alaska: Subduction deposits and near-trench magmatism [Ph.D. thesis]: Santa Cruz, University of California, 274 p.

Hill, M.D., Morris, J., and Whelan, J., 1981, Hybrid granodiorites intruding the accretionary prism, Kodiak, Shumagin, and Sanuk Islands, southwest Alaska: Journal of Geophysical Research, v. 86, p. 10,569–10,590.

Hudson, T., 1983, Calc-alkaline plutonism along the Pacific rim of southern Alaska, *in* Roddick, J.A., ed., Circum-Pacific plutonic terranes: Boulder, Colorado, Geological Society of America Memoir 159, p. 159–169.

Hudson, T., and Plafker, G., 1982, Paleogene metamorphism in an accretionary flysch terrane, eastern Gulf of Alaska: Geological Society of America Bulletin, v. 98, p. 265–279.

Hudson, T., Plafker, G., and Peterman, Z.E., 1979, Paleogene anatexis along the Gulf of Alaska margin: Geology, v. 7, p. 573–577.

Hutton, D.W.H., 1982, A tectonic model for the emplacement of the Main Donegal granite, NW Ireland: Geological Society [London] Journal, v. 139, p. 615–631.

Hutton, D.H.W., 1987, Strike-slip terranes and a model for the evolution of the British and Irish Caledonides: Geological Magazine, v. 124, p. 405–425.

Hutton, D.H.W., 1988, Granite emplacement mechanisms and tectonic controls: Inferences from deformation studies: Transactions of the Royal Society of Edinburgh, Earth Sciences, v. 79, p. 245–255.

Hutton, D.H.W., and Reavy, R.J., 1992, Strike-slip tectonics and granite petrogenesis: Tectonics, v. 11, p. 960–967.

Hutton, D.W.H., Dempster, T.J., Brown, P.E., and Becker, S.D., 1990, A new mechanism of granite emplacement: Intrusion in active shear zones: Nature, v. 343, p. 452–455.

James, T.S., Hollister, L.S., and Morgan, W.J., 1989, Thermal modelling of the Chugach metamorphic complex: Journal of Geophysical Research, v. 94, p. 4411–4423.

Johnson, K., Barnes, C., and Miller, C., 1997, Petrology, geochemistry, and genesis of high-Al tonalite and trondhjemites of the Cornucopia stock, Blue Mountains, northeastern Oregon: Journal of Petrology, v. 38, p. 1585–1611.

Jones, D.L., and Clark, S.H.B., 1973, Upper Cretaceous (Maestrichtian) fossils from the Kenai-Chugach Mountains, Kodiak and Shumagin Islands, southern Alaska: U.S. Geological Survey Journal of Research, v. 1, p. 125–136.

Karlstrom, K., Miller, C., Kingsbury, J., and Wooden, J., 1993, Pluton emplacement along an active ductile thrust zone, Piute Mountains, southeastern California: Interaction between deformational and solidification processes: Geological Society of America Bulletin, v. 105, p. 213–230.

Kusky, T.M., 1989, Accretion of the Archean Slave province: Geology, v. 17, p. 63–67.

Kusky, T.M., and Bradley, D.C., 1999, Kinematics of mélange fabrics: Examples and applications from the McHugh Complex, Kenai Peninsula, Alaska: Journal of Structural Geology, v. 21, p. 1773–1796.

Kusky, T.M., and Polat, A., 1999, Growth of granite-greenstone terranes at convergent margins and stabilization of Archean cratons: Tectonophysics, v. 305, p. 43–73.

Kusky, T.M., and Young, C., 1999, Emplacement of the Resurrection Peninsula ophiolite in the southern Alaska forearc during a ridge-trench encounter: Journal of Geophysical Research, v. 104, p. 29,025–29,054.

Kusky, T.M., Kidd, W.S.F., and Bradley, D.C., 1987, Displacement history of the Northern Arm fault and its bearing on the post-Taconic evolution of north-central Newfoundland: Journal of Geodynamics, v. 7, p. 105–133.

Kusky, T.M., Bradley, D.C., Haeussler, P., and Karl, S., 1997a, Controls on accretion of flysch and mélange belts at convergent margins: Evidence from The Chugach Bay thrust and Iceworm mélange, Chugach Terrane, Alaska: Tectonics, v. 16, p. 855–878.

Kusky, T.M., Bradley, D.C., and Haeussler, 1997b, Progressive deformation of the Chugach accretionary complex, Alaska, during a Paleogene ridge-trench encounter: Journal of Structural Geology, v. 19, p. 139–157.

Kveton, K.J., 1989, Structure, thermochronology, provenance, and tectonic history of the Paleogene Orca Group in southwestern Prince William Sound, Alaska [Ph.D. thesis]: Seattle, University of Washington, 173 p.

Lallemand, S.E., Malavielle, J., and Calassou, S., 1992, Effects of oceanic ridge subduction on accretionary wedges: Experimental modeling and marine observations: Tectonics, v. 11, p. 1301–1313.

Leake, B.E., 1990, Granite magmas: Their sources, initiation, and consequences of emplacement: Geological Society [London] Journal, v. 147, p. 579–589.

Logan, J.M., Friedman, M., Higgs, N., Dengo, C., and Shimamoto, T., 1979, Experimental studies of simulated gouge and their application to studies of natural fault zones: U.S. Geological Survey Open-File Report 79-1239, p. 305–343.

Lytwyn, J., Gilbert, S., Casey, J., and Kusky, T.M., 2000, Geochemistry of near-trench intrusives associated with ridge subduction, Seldovia quadrangle, southern Alaska: Journal of Geophysical Research, v. 105, p. 27,957–27,978.

Maeda, J., and Kagami, H., 1996, Interaction of a spreading ridge and an accretionary prism: Implications from MORB magmatism in the Hidaka magmatic zone, Hokkaido, Japan: Geology, v. 24, p. 31–34.

Magoon, L.B., Adkison, W.L., and Egbert, R.M., 1976, Map showing geology, wildcat wells, Tertiary plant fossil localities, K-Ar age dates, and petroleum operations, Cook Inlet area, Alaska: U.S. Geological Survey Miscellaneous Investigations Series Map I-1019, scale 1:250,000.

Marre, J., 1986, The structural analysis of granitic rocks: New York, Elsevier, 123 p.

Marshak, R.S., and Karig, D.E., 1977, Triple junctions as a cause for anomalously near-trench igneous activity between the trench and volcanic arc: Geology, v. 5, p. 233–236.

Martin, G.C., Johnson, B.L., and Grant, U.S., 1915, Geology and mineral resources of Kenai Peninsula, Alaska: U.S. Geological Survey Bulletin 587, 243 p.

Moffit, F.H., 1954, Geology of the Prince William Sound region, Alaska: U.S. Geological Survey Bulletin 989-E, p. 225–310.

Moore, J.C., Byrne, T., Plumley, P.W., Reid, M., Gibbons, H., and Coe, R.S., 1983, Paleogene evolution of the Kodiak Islands, Alaska: Consequences of ridge-trench interaction in a more southerly latitude: Tectonics, v. 2, p. 265–293.

Morand, V.J., 1992, Pluton emplacement in a strike slip shear zone: The Doctors Flat pluton, Victoria, Australia: Journal of Structural Geology, v. 14, p. 205–213.

Nelson, S.W., and Nelson, M.S., 1993, Geochemistry of ophiolitic rocks from Knight Island, Prince William Sound, Alaska, *in* Bacon, C.R., and Till, A.B., eds., Geologic studies in Alaska by the U.S. Geological Survey in 1992: U.S. Geological Survey Bulletin 2068, p. 130–142.

Nelson, S.W., Dumoulin, J.A., and Miler, M.L., 1985, Geologic map of the Chugach National Forest, Alaska: U.S. Geological Survey Miscellaneous Field Studies Map MF-1645-B, scale 1:250,000.

Nelson, S.W., Miller, M.L., Haeussler, P.J., Snee, L.W., Phillips, P.J., and Huber, C., 1999, Preliminary geologic map of the Chugach National Forest Special Study Area, Alaska: U.S. Geological Survey Open-File Report 99-362, scale 1:63,000.

Nilsen, T.H., and Zuffa, G.G., 1982, The Chugach Terrane, a Cretaceous trench-fill deposit, southern Alaska, *in* Leggett, K.J., ed., Trench forearc geology: Cambridge, Massachusetts, Blackwell, p. 213–227.

Paterson, S.R., 1988, Cannibal Creek granite: Post-tectonic "ballooning" intrusion, or pre-tectonic piercement diapir?: Journal of Geology, v. 96, p. 730–736.

Paterson, S.R., and Miller, R.B., 1998, Mid-crustal magmatic sheets in the Cascades Mountains, Washington: Implications for magma ascent: Journal of Structural Geology, v. 20, p. 1435–1363.

Paterson, S.R., and Tobish, O.T., 1988, Using pluton ages to date regional deformations: Problems with commonly used criteria: Geology, v. 16, p. 1108–1111.

Paterson, S.R., and Tobish, O.T., 1992, Rates of processes in magmatic arcs: Implications for the timing and nature of pluton emplacement and wall rock deformation: Journal of Structural Geology, v. 14, p. 291–300.

Paterson, S.R., and Vernon, R., 1995, Bursting the bubble of ballooning plutons; a return to nested diapirs emplaced by multiple processes: Geological Society of America Bulletin, v. 107, p. 1356–1380.

Paterson, S.R., Tobish, O.T., and Vernon, R.H., 1989a, Penrose conference report: Criteria for establishing the relative timing of pluton emplacement and regional deformation: Geology, v. 17, p. 475.

Paterson, S.R., Vernon, R.H., and Tobisch, O.T., 1989b, A review of criteria for the identification of magmatic and tectonic foliations in granitoids: Journal of Structural Geology, v. 11, p. 349–363.

Paterson, S.R., Tobish, O.T., and Vernon, R.H., 1991, Emplacement and deformation of granitoids during volcanic arc construction in the foothills terrane, central Sierra Nevada, California: Tectonophysics, v. 191, p. 89–110.

Pavlis, T.L., Monteverde, D.H., Bowman, J.R., Rubenstone, J.L., and Reason, M.D., 1988, Early Cretaceous near-trench plutonism in southern Alaska: A tonalite-trondjhemite intrusive complex injected during ductile thrusting along the Border Ranges fault: Tectonics, v. 7, p. 1179–1199.

Perfit, M., Langmuir, C., Baekisapa, M., Chappell, B., Johnson, R., Staudigel, H., and Taylor, S., 1987, Geochemistry and petrology of volcanic rocks from the Woodlark basin: Addressing questions of ridge subduction, *in* Taylor, B., and Exon, N., eds., Marine geology, geophysics, and geochemistry of the Woodlark Basin–Solomon Islands: Houston, Texas, Circum-Pacific Council for Energy and Mineral Resources, Earth Science Series, v. 7, p. 113–154.

Perreault, S., and Martignole, J., 1988, High-temperature cordierite migmatites in the northeastern Grenville province: Journal of Metamorphic Geology, v. 6, p. 673–696.

Petford, M., Cruden, A.R., McCraffrey, K.J.W., and Vigneresse, J.-L., 2000, Granite magma formation, transport and emplacement in the Earth's crust: Nature, v. 408, p. 669–673.

Pitcher, W.S., 1983, Granite type and tectonic environment, *in* Hsü, K., ed., Mountain building processes: London, Academic Press, p. 19–40.

Plafker, G., 1987, Regional geology and petroleum potential of the northern Gulf of Alaska continental margin, *in* Scholl, D.W., et al., eds., Geology and resource potential of the continental margin of western North America and adjacent ocean basins: Houston, Texas, Circum-Pacific Council for Energy and Mineral Resources, Earth Science Series, v. 6, p. 229–268.

Plafker, G., Nokleberg, W.J., and Lull, J.S., 1989, Bedrock geology and tectonic evolution of the Wrangellia, Peninsular, and Chugach terranes along the Trans-Alaska Crustal Transect in the Chugach Mountains and southern Copper River Basin, Alaska: Journal of Geophysical Research, v. 94, p. 4255–4295.

Recio, C., Fallick, A.E., and Ugidos, J.M., 1992, A stable isotopic ($\delta^{18}O$, δD) study of the late-Hercynian granites and their host rocks in the Central Iberian Massif, Spain: Transactions of the Royal Society of Edinburgh, Earth Sciences, v. 83, p. 247–257.

Saint Blanquat, M., Tikoff, B., Teyssier, C., and Vigneresse, J.L., 1998, Transpressional kinematic and magmatic arc, *in* Holdsworth, R.E., et al., eds., Continental transpressional and transtensional tectonics: Geological Society [London] Special Publication No. 135, p. 327–340.

Scaillet, B.Holts, F., and Pichavant, M., 1997, Rheological properties of granitic magmas in their crystallization range, *in* Bouchez, J.J., ed., Granite: From segregation to melt to emplacement fabrics: Dordrecht, Netherlands, Kluwer Academic Publishers, p. 11–29.

Silberman, M.L., Mitchell, P.A., and O'Neil, J.R., 1981, Isotopic data bearing on the origin and age of the epithermal lode gold deposits in the Hope-Sunrise mining district, northern Kenai Peninsula, Alaska: U.S. Geological Survey Circular 823-B, p. B81–B84.

Sisson, V.B., and Hollister, L.S., 1988, Low-pressure facies series metamorphism in an accretionary sedimentary prism, southern Alaska: Geology, v. 16, p. 358–361.

Sisson, V.B., and Pavlis, T.L., 1993, Geologic consequences of plate reorganization: An example from the Eocene southern Alaska forearc: Geology, v. 21, p. 913–916.

Sisson, V.B., Hollister, L.S., and Onstott, T.C., 1989, Petrologic and age constraints on the origin of a low-pressure/high-temperature metamorphic complex, southern Alaska: Journal of Geophysical Research, v. 94, p. 4392–4410.

Sisson, V.B., Poole, A.R., Harris, N.R., Cooper Burner, H., Pavlis, T.L., Copeland, P., Donelick, R.A., and McLelland, W.C., 2003, Geochemical and geochronologic constraints for genesis of a tonalite-trondhjemite suite and associated mafic intrusive rocks in the eastern Chugach Mountains, Alaska: A record of ridge-transform subduction, *in* Sisson, V.B., Roeske, S.M., and Pavlis, T.L., eds., Geology of a transpressional orogen developed during ridge-trench interaction along the North Pacific margin: Boulder, Colorado, Geological Society of America Special Paper 371, p. 293–326 (this volume).

Spear, F.S., 1993, Metamorphic phase equilibria and pressure-temperature-time paths: Mineralogical Society of America Monograph Series 1, 799 p.

Stüwe, K., 1986, Structural evolution of the Port Wells mining district, Prince William Sound, south-central Alaska: Implications for the origin of the gold lodes: Mineralium Deposita, v. 21, p. 288–295.

Taylor, C.D., Goldfarb, R.J., Snee, L.W., Ghent, C.A., Karl, S.M., and Haeussler, P.J., 1994, New age data for gold deposits and granites, Chichagof mining district, southeast Alaska: Evidence for a common origin: Geological Society of America Abstracts with Programs, v. 26, no. 7, p. A140.

Thorkelson, D.J., and Taylor, R.P., 1989, Cordilleran slab windows: Geology, v. 17, p. 833–836.

Tikoff, B., and Saint Blanquat, M., 1997, Transpressional shearing and strike-slip partitioning in the Late Cretaceous Sierra Nevada magmatic arc, California: Tectonics, v. 16, p. 442–459.

Tikoff, B., and Teyssier, C., 1992, Crustal-scale, en-echelon "P-shear" tensional bridges: A possible solution tothe batholithic room problem: Geology, v. 20, p. 927–930.

Tikoff, B., Saint Blanquat, M., and Teyssier, C., 1999, Translation and the resolution of the pluton space problem: Journal of Structural Geology, v. 21, p. 1109–1117.

Tobish, O.T., and Cruden, A.R., 1995, Fracture controlled magma conduits in an oblique convergent continental magmatic arc: Geology, v. 23, p. 941–944.

Tracy, R.J., and Robinson, P., 1983, Acadian migmatite types in pelitic rocks of central Massachusetts, *in* Atherton, M.P., and Gribble, C.D., eds., Migmatites, melting, and metamorphism: Nantwich, UK, Shiva Press, p. 163–173.

Tuck, R., 1933, The Moose Pass–Hope district, Kenai Peninsula, Alaska: U.S. Geological Survey Bulletin 849-I, p. 460–527.

Tysdal, R.G., and Case, J.E., 1979, Geologic map of the Seward and Blying Sound quadrangles, Alaska: U.S. Geological Survey Map I-1150, scale 1:250,000.

Tysdal, R.G., and Plafker, G., 1978, Age and continuity of the Valdez Group, southern Alaska, *in* Sohl, N.F., and Wright, W.B., eds., Changes in stratigraphic nomenclature by the U.S. Geological Survey, 1977: U.S. Geological Survey Bulletin 1457-A, p. A120–A124.

Tysdal, R.G., Case, J.E., Winkler, G.R., and Clark, S.H.B., 1977, Sheeted dikes, gabbro, and pillow basalt in flysch of coastal southern Alaska: Geology, v. 5, p. 377–383.

Ugidos, J.M., 1990, Granites as a paradigm of genetic processes of granitic rocks: I-types vs. S-types, *in* Dallmayer, R.D., and Martinez Garcia, E., eds., Pre-Mesozoic geology of Iberia: Berlin, Springer Verlag, p. 189–206.

Ugidos, J.M., and Recio, C., 1993, Origin of cordierite-bearing granites by assimilation in the Central Iberian Massif (CIM), Spain: Chemical Geology, v. 103, p. 27–43.

Vernon, R.H., 1983, Restite, xenoliths, and micrgranitoids enclaves in granites: Journal of the Royal Society of New South Wales, v. 116, p. 77–103.

Vogt, P.R., Lowrie, A., Bracey, D.R., and Hey, R.N., 1976, Subduction of aseismic oceanic ridges: Effects on shape, seismicity, and other characteristics of consuming plate boundaries: Boulder, Colorado, Geological Society of America Special Paper 172, 59 p.

von Huene, R., and Lallemand, S., 1990, Tectonic erosion along the Japan and Peru convergent margins: Geological Society of America Bulletin, v. 102, p. 704–720.

Yoshinobu, A.S., Okaya, D.A., and Paterson, S.R., 1998, Modeling the thermal evolution of fault-controlled magma emplacement models: Implications for the solidification of granitoid plutons: Journal of Structural Geology, v. 20, p. 1205–1218.

Wells, R.E., Engebretson, D.C., Snavely, P.D., and Coe, R.S., 1984, Cenozoic plate motions and the volcano-tectonic evolution of western Oregon and Washington: Tectonics, v. 3, p. 274–294.

Winkler, H.G.F., 1976, Petrogenesis of metamorphic rocks: New York, Springer-Verlag, 334 p.

Manuscript Accepted by the Society February 5, 2003

Printed in the USA

Geological Society of America
Special Paper 371
2003

Geochemical and geochronologic constraints for genesis of a tonalite-trondhjemite suite and associated mafic intrusive rocks in the eastern Chugach Mountains, Alaska: A record of ridge-transform subduction

Virginia B. Sisson*
Anne R. Poole*
Nancy R. Harris*
Department of Earth Science, MS-126, Rice University, Houston, Texas 77005-1892, USA

Holly Cooper Burner
Terry L. Pavlis
Department of Geology, University of New Orleans, New Orleans, Louisiana 70148, USA

Peter Copeland
Department of Geosciences, University of Houston, Houston, Texas 77204-3399, USA

Raymond A. Donelick
Department of Earth Science, MS-126, Rice University, Houston, Texas 77005-1892, USA

William C. McLelland
Department of Geological Sciences, University of Idaho, Moscow, Idaho 83844-3022, USA

ABSTRACT

Igneous rocks in the Chugach metamorphic complex of southern Alaska form part of the Sanak-Baranof belt, a series of forearc plutons believed to have been formed by subduction of a spreading ridge. The intrusives are predominantly tonalite-trondhjemite-granodiorite suite. There are some associated mafic plutonics close to the eastern end of the belt near Yakutat and Glacier Bay National Park. Trace element and Nd-Sr isotopic data suggest that tonalitic magmas were derived from the mixing of two sources: melted accretionary wedge sedimentary rocks and mafic material underplated at the base of the wedge as the ridge was subducted. The mafic material could include metabasaltic amphibolites at Nunatak Fjord that have a mid-ocean ridge basalt–composition protolith.

Early Tertiary ridge migration along the Alaskan margin was generally toward the east. New U-Pb and $^{40}Ar/^{39}Ar$ geochronology data indicate synchronous magmatism at 54 Ma in the western Chugach metamorphic complex and Nunatak Fjord region. There is also a younger 49 Ma intrusive at Nunatak Fjord. The Nunatak Fjord region is subdivided into two blocks: (1) Chugach block has young amphibole cooling ages of 20 Ma, whereas (2) Boundary-Fairweather block has older cooling ages of ca. 53 Ma, which are contemporaneous with an U-Pb age on an adjacent tonalitic pluton.

*E-mail: Sisson—jinnys@rice.edu. Present addresses: Poole—National Park Service, Intermountain GIS Center, University of New Mexico, Bandelier Hall West, Rm. 215, Albuquerque, NM 87131, USA; Harris—ChevronTexaco Overseas Petroleum, 4800 Fournace Place, BOB W-1044, Bellaire, TX 77401, USA

Sisson, V.B., Poole, A.R., Harris, N.R., Cooper Burner, H., Pavlis, T.L., Copeland, P., Donelick, R.A., and McLelland, W.C., 2003, Geochemical and geochronologic constraints for genesis of a tonalite-trondhjemite suite and associated mafic intrusive rocks in the eastern Chugach Mountains, Alaska: A record of ridge-transform subduction, *in* Sisson, V.B., Roeske, S.M., and Pavlis, T.L., eds., Geology of a transpressional orogen developed during ridge-trench interaction along the North Pacific margin: Boulder, Colorado, Geological Society of America Special Paper 371, p. 293–326.

This shows that within the Chugach metamorphic complex, ridge subduction was probably complicated by subduction of a transform, causing the locus of magmatism to appear to jump ~130 km eastward from Van Cleve Glacier to Nunatak Fjord near Yakutat at 54 Ma. In addition, forearc slivers record different emplacement, thermal, and exhumation processes along the margin.

Keywords: gabbro, mid-ocean ridge basalt, granite, granodiorite, tonalite, trondhjemite, dacite, gabbronorite, pyroxenite

INTRODUCTION

The interaction of a spreading ridge and a subduction zone is a poorly understood process, but one known to create several unusual geological features (e.g., Sisson, 1994; Pavlis et al., 1995). Most notably, subduction of young, relatively hot oceanic ridge material may produce magmas of distinct compositions in both the arc and forearc of a convergent margin. Marshak and Karig (1977) were the first to propose that forearc magmatism was a consequence of subduction of a spreading ridge in southern Alaska. Other examples of young plutons found near the trench occur in some active ridge subduction encounters, such as Chile Trench (Nazca-Antarctic-South American triple junction) where intrusive bodies and hot springs occur more than 100 km seaward of the modern volcanic arc (Forsythe et al., 1986; Guivel et al., 1999). In the Woodlark triple junction region, near the Solomon Islands, submarine volcanoes have been found in the forearc (Taylor and Exon, 1987). Other modern and ancient occurrences of this phenomenon include the Eocence-Miocene magmatism in the Shimanto accretionary prism of southwest Japan (Miyake, 1985; Hibbard and Karig, 1990a; Osozawa, 1992; Kiminami et al., 1993; Maeda and Kagami, 1996), Cretaceous magmatism in southwest Japan (Kinoshita, 1995, 1999, 2002), recent magmatism in Antarctic Peninsula (Hole and Larter, 1993; Hole et al., 1994), and Pliocene magmatism in Baja and southern California (Storey et al., 1989; Cole and Basu, 1995).

As a ridge is subducted, it may open a window in the downgoing slab (e.g., Thorkelson, 1996; Johnston and Thorkelson, 1997). This could have several effects on the overlying wedge, causing it to differ from a typical accretionary prism. First, the higher temperature over the slab window would cause metamorphism in what is normally a refrigerated system. Second, mafic material upwelling through the slab window might be incorporated at the base of the prism. This "underplated" material could have the composition of depleted mantle or mid-ocean ridge basalt (MORB). Third, the slab window could induce magmatism in the forearc, either by melting of accretionary wedge sediments, or from magmas derived from the mafic underplating, or both. If the slab window were to migrate along the margin, it might create a string of such intrusions through time. In addition, ridge-trench collision results in a variety of magmatic processes that cannot be characterized by a single set of geochemical features. For example, magmas at the Chile margin triple junction range from andesitic to dacitic, with isotopic ratios showing a mafic origin—possibly even a mantle component (Kaeding et al., 1990). Above the subducted ridge there is a gap in the associated island arc volcanics. To the south, volcanics associated with rejuvenation of magmatism are adakitic or may have signatures of ocean island basalt (e.g., Kay et al., 1993; Gorring and Kay, 2001; D'Orazio et al., 2001). Mafic dikes associated with ridge subduction occur in the Shimanto accretionary prism of Japan and have MORB compositions (Hibbard and Karig, 1990b). Magmas at the Antarctic Peninsula (Hole and Larter, 1993; Hole et al., 1994) and at Baja California (Storey et al., 1989) are primarily alkali basalt, geochemically indistinguishable from ocean island and continental alkali basalt.

The Sanak-Baranof plutonic belt defined by Hudson (1979) is often cited as an example of ancient ridge subduction. This series of early Cenozoic felsic forearc plutons extends 2200 km from Sanak Island near the Aleutian Peninsula to Baranof Island in the southeast (Fig. 1). Similar forearc felsic magmatism also occurred on southern Vancouver Island (Groome et al., this volume, Chapter 14). During the early Tertiary, in the North Pacific, the diverging Kula and Farallon plates were subducted underneath the North American plate. The Kula plate has since

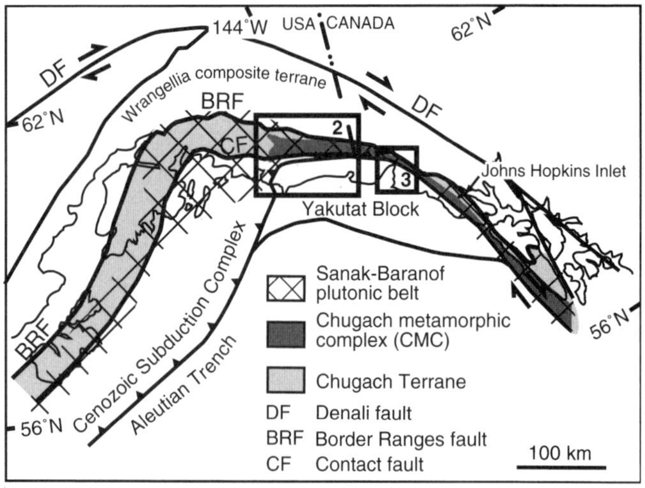

Figure 1. Regional tectonic setting of the Chugach metamorphic complex (adapted from Plafker et al., 1989). The two boxes indicate regions shown in Figure 2—western Chugach metamorphic complex, and Figure 3—Yakutat Bay and Nunatak Fjord region. Wrangellia composite terrane represents the combined Peninsular, Alexander, and Wrangellia terranes.

been almost entirely subducted, except for one small piece south of the Aleutian Islands (Byrne, 1979; Engebretson et al., 1985; Stock and Molnar, 1988; Lonsdale, 1988; Atwater, 1989; Bunge and Grand, 2000). The triple junction migrated to the east from latest Cretaceous to Eocene time as the two ocean plates were subducted from 61 to 50 Ma (Sisson and Hollister, 1988; Bradley et al., 1993; Bradley et al., this volume, Chapter 1).

The Sanak-Baranof plutons are partly derived from melted accretionary wedge sediments, with some mafic component (Hudson et al., 1979; Hill et al., 1981; Barker et al., 1992; Hudson, 1994; Lytwyn et al., 2000; Bradley et al., this volume, Chapter 1). Harris et al. (1996), using Sr and Nd isotopic analyses and rare-earth-element geochemistry, found that magmas of the western Chugach metamorphic complex lay on a mixing trend between the host metasedimentary rock and a mantle-derived end member. This study presents new trace and major element and whole rock isotope data from Nunatak Fjord and Van Cleve Glacier regions to expand the previous work. New U-Pb, $^{40}Ar/^{39}Ar$, and fission track geochronologic data from the Chugach metamorphic complex further constrain the timing of migration of the triple junction through this section of the Sanak-Baranof plutonic belt. In addition, U/Pb geochronologic data are presented for one pluton that crosscuts both the Chugach metamorphic complex and the Border Ranges fault in Glacier Bay.

GEOLOGIC SETTING

The Chugach metamorphic complex is a region of high-temperature, low-pressure metamorphism and felsic magmatism located in the forearc of the southern Alaskan margin. The Chugach metamorphic complex also contains the most voluminous plutonic assemblages of the Sanak-Baranof belt. This study concentrated on two areas: the western end of the complex near the Copper River and the region ~130 km to the east near Nunatak Fjord near Yakutat Bay (areas 2 and 3 respectively in Fig. 1). These rocks have not been described in detail in the literature and regional descriptions do not do justice to the highly varied nature of the rock types and metamorphic grade. Thus, the regional geology of these areas is presented below (Figs. 2 and 3).

If the ridge subduction hypothesis is applicable for this relatively small area, two features are expected. First, if magmatism corresponded with the passage of the triple junction through the region, plutonism should be time transgressive. To help test this hypothesis, we also present geochronologic data for several intrusives as well as a pluton from Glacier Bay National Park. Second, mafic material underplated through the slab window may have acted as a source for intrusive rocks in the Chugach metamorphic complex; if so, this should be evident through trace element and isotopic geochemical data.

Western Chugach Metamorphic Complex Region

Harris et al. (1996) previously examined the ridge subduction hypothesis for the western Chugach metamorphic complex. The Chugach metamorphic complex, shown in Figure 2, lies mainly in the Chugach Terrane, between the Contact and Border Ranges fault systems. Here, the accretionary prism comprises predominantly island-arc derived argillites and graywackes of the Cretaceous Valdez Group (Nilsen and Zuffa, 1982; Farmer et al., 1993). It has been proposed that exhumation of the Coast Plutonic complex of British Columbia is the

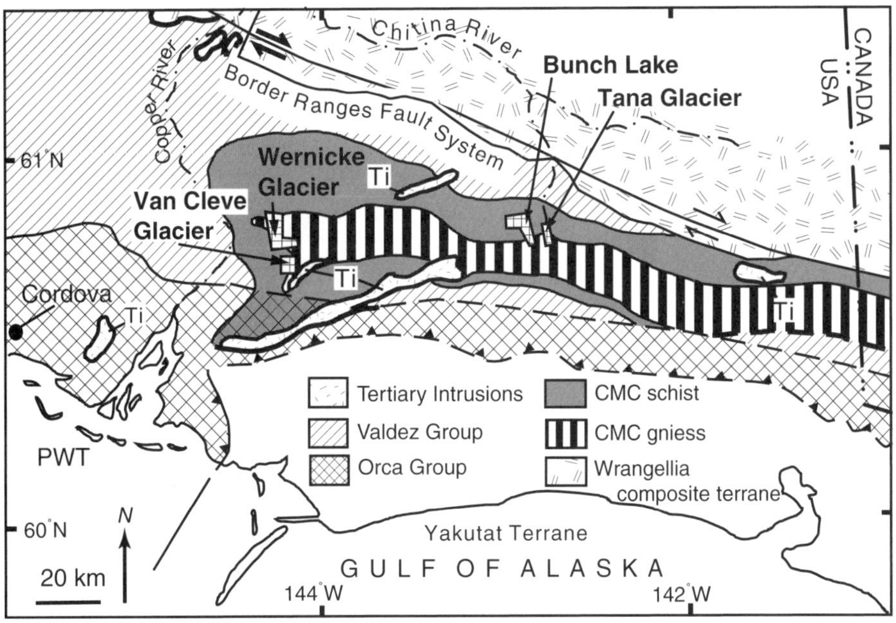

Figure 2. Bedrock geologic map of the western Chugach metamorphic complex showing major Tertiary intrusive rocks (Ti). Adapted from Hudson and Plafker (1982). Boxed regions show locations of felsic intrusives described in Harris et al. (1996) and this study.

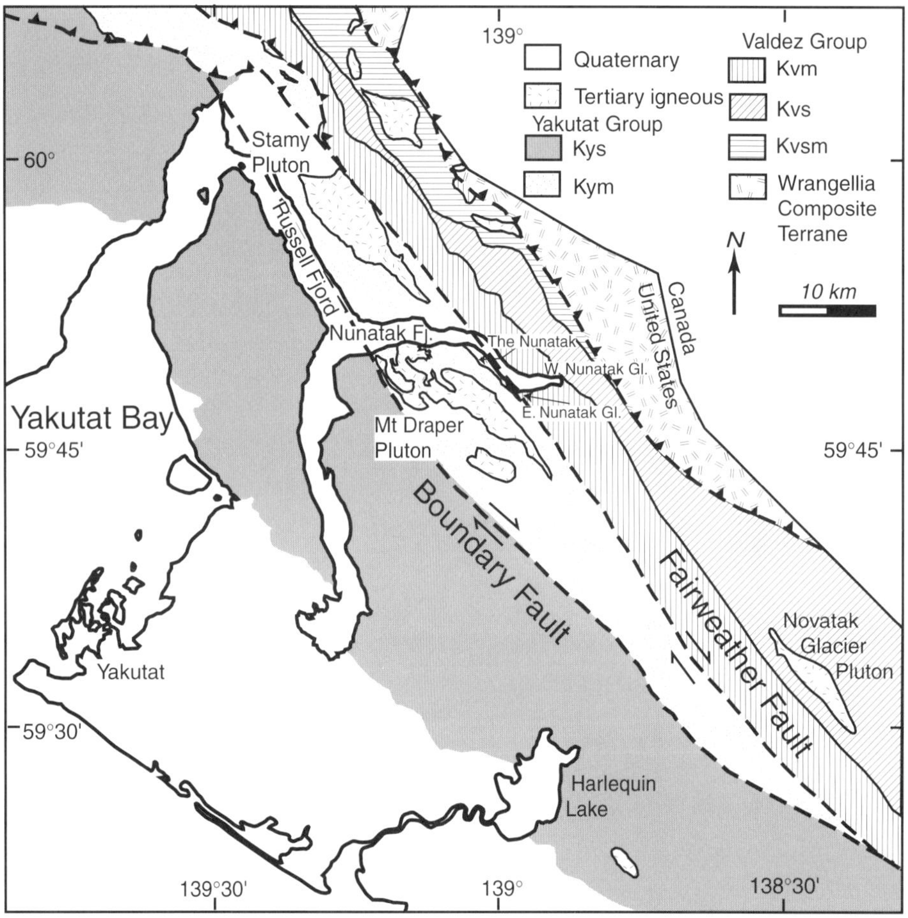

Figure 3. Bedrock geologic map of the Yakutat Bay and Nunatak Fjord region. Adapted from Hudson et al. (1977a, 1977b), Plafker (1999, personal commun.), and unpublished mapping by the authors. Kvm—Cretaceous Valdez Group metavolcanic, Kvs—Cretaceous Valdez Group schist, Kvsm—Cretaceous Valdez Group interbedded metavolcanic and schist, Kym—Cretaceous Yakutat Group mélange that may be correlative with Valdez and Sitka Formations, Kys—Cretaceous Yakutat Group.

provenance for the Valdez Group (e.g., Plafker, 1987; Farmer et al., 1993; Plafker et al., 1994; Sample and Reid, this volume, Chapter 2). The Valdez Group sediments were metamorphosed to greenschist facies during the early Eocene and later to amphibolite-facies temperatures (Hudson and Plafker, 1982; Sisson et al., 1989). Maximum metamorphic conditions in the migmatitic core of the Chugach metamorphic complex have been estimated at 250–350 MPa and 600–650 °C (Sisson et al., 1989). Four phases of deformation, three ductile (D_{1-3}) and one brittle (D_4), have been identified in the western and central portion of the complex (Pavlis and Sisson, 1995; Pavlis and Sisson, this volume, Chapter 8). Igneous rocks occur as plutons varying in size from small stocks to an 80 km long body and as dikes and sills, from 10 cm to several meters in width. These rocks range in composition from trondhjemite to tonalite (most common), granodiorite, and granite (Fig. 4).

Igneous activity in this part of the Chugach metamorphic complex can be divided into four episodes (Ti_{1-4}), determined by crosscutting relationships of the intrusions to each other and to the four phases of deformation. The first three intrusive phases are pre- to synmetamorphic and the final phase is post-metamorphic. The first phase (Ti_1) intruded pre-D_2 and is now represented by metamorphosed dikes that are foliated and boudinaged in low- to medium-grade metasediments. The second-phase (Ti_2) sills, dikes, and plutons were intruded syn-D_2 into middle to upper-amphibolite facies schist prior to the peak metamorphism. Injection of the third-phase (Ti_3) dikes and plutons continued during and after D_3 (also see Pavlis and Sisson, this volume, Chapter 8). The latest intrusions (Ti_4) are hypabyssal dikes intruded parallel to the D_4 joint set. Of the four generations, Ti_2 and Ti_3 bodies are the most abundant in volume and number, while Ti_4 is limited to thin dikes, more abundant near Bunch Lake and Tana Glacier and rare in the western Chugach metamorphic complex near Van Cleve Glacier and Wernicke Glacier (Fig. 2; Harris et al., 1996). In this study, we present new data for the Van Cleve Glacier region as well as supplementary data to Harris et al. (1996) for Bunch Lake, Tana Glacier, and Wernicke Glacier regions. Since this is a reconnaissance study, generally only one sample per igneous body was acquired.

Nunatak Fjord region

Igneous and metamorphic rocks of the Chugach metamorphic complex in the Yakutat-St. Elias area were first studied by

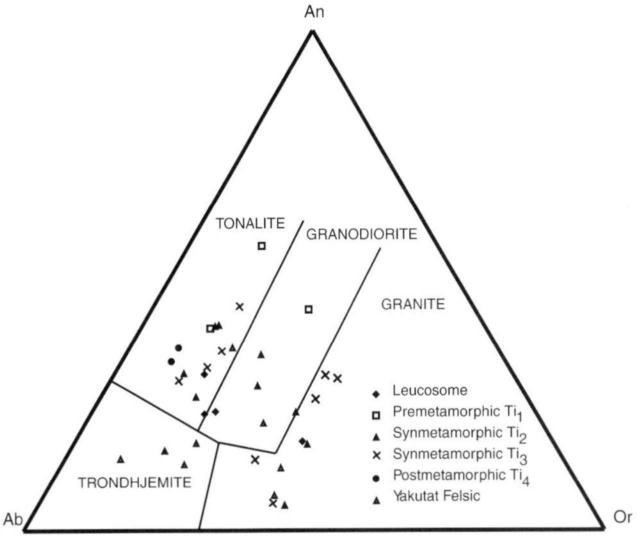

Figure 4. Normative albite (Ab), anorthite (An), orthoclase (Or) classification of felsic rocks in the Chugach metamorphic complex. Fields from Barker (1979). See text for explanation of Ti_1–Ti_4.

Hudson et al. (1977a, 1977b) and have received little attention since then. Specific areas examined for this study were Nunatak Fjord region and the Novatak Glacier pluton (Fig. 3). In this region, the Yakutat terrane, a microcontinental block, collided with the margin in late Neogene time (Plafker et al., 1994). The region is cut by several large Neogene high-angle faults; the most significant is the Fairweather fault and its offshoot, the Boundary fault. The region to the northeast of the Fairweather fault is referred to as the Chugach block and includes the eastern end of Nunatak Fjord and the Novatak Glacier area. The western part of Nunatak Fjord lies in the sliver between the Fairweather and Boundary faults, known as the Boundary-Fairweather block (Hudson et al., 1977a). Post-Eocene dextral strike slip along the Fairweather fault has been estimated at up to 600 km (Plafker et al., 1994). Rock type, metamorphic grade, and deformational history vary dramatically between the two blocks. Consequently, the geology of the Nunatak Fjord area is best understood when treated as two separate regions.

Chugach Block

The country rock in the Chugach block from the Fairweather fault trace to approximately 2 km up East Nunatak Glacier (in its 1994 location) consists almost entirely of amphibolite schist and gneiss ranging from epidote-amphibolite to upper amphibolite facies (Kvm). These are interlayered with plagioclase ± quartz layers and veins of varying sizes. Some of these veins are pegmatitic in texture and contain garnet, muscovite, and biotite; others have virtually pure feldspar. Some fine-grained felsic dikes several meters wide crosscut the amphibolite (e.g., 95APo25a). Within the amphibolite, grains of chalcopyrite and pyrrhotite as well as pods of diopside and zoisite occur. Some of the more schistose sections of the amphibolite are fine grained with randomly oriented blades of hornblende. Foliation in the amphibolite decreases to the northeast, to the extent that at the easternmost point reached on the traverse, it appears to be almost plutonic. Thus, part of this mafic complex may represent metamorphosed mafic plutonic rocks. Farther east, approximately 4 km from the Fairweather fault, the amphibolite is interleaved with a metagraywacke (unit Kvs) that is similar to metagraywacke in the western Chugach metamorphic complex. The fact that amphibolite is interleaved with the metasediments rather than crosscutting suggests that its protolith was volcanic tuffs, volcaniclastic sediments, or concordant sills. Ductile strains are high, however, and thus, concordance could be due to transposition of layering. Nonetheless, the amphibolites are similar to rocks described as metabasalts by Lull and Plafker (1990). Other metabasaltic and metadiabasic rocks are found along the southern margin of the Chugach terrane in a 3–4 km thick sequence (Winkler and Plafker, 1981; Lull and Plafker, 1990; Crowe et al., 1992). Similar amphibolites also occur along the Mt. St. Elias ridge between Mt. St. Elias and the Seward Glacier, as well as on the northern edge of the Bagley Icefield (Campbell and Dodds, 1978; Sisson et al., 1998). Both Lull and Plafker (1990) and Crowe et al. (1992) propose that most of the metabasalts fall into the field of island arc tholeiites, remnants of an ancient intra-oceanic arc. In contrast, amphibolites from the Yakutat region have MORB compositions.

The Novatak Glacier pluton lies within metagraywacke and metaargillite (Kvsm) of the Chugach block, southeast of Nunatak Fjord. Like most Chugach metamorphic complex intrusives, it is tonalitic in composition, and had been dated by K-Ar amphibole at 62.5 Ma (Hudson et al., 1977a). Although the center of the pluton shows little or no foliation, arms of intrusive material extending from the main pluton do show a strong alignment of biotite and amphibole that parallels the foliation of the surrounding country rock of garnet-bearing metagraywacke and metaargillite. Next to the pluton, the metagraywacke and tonalite mingle, forming an injection migmatite. This gneissic texture does not extend far from the intrusive contact. There are also some dikes of garnet- and muscovite-bearing pegmatite up to several meters wide that may be related to the Novatak Glacier pluton. As at Nunatak Fjord, the metagraywacke is interleaved with and contains boudinaged amphibolite (Kvm) along a gradational contact into amphibolite along the Fairweather fault.

Rocks of the Chugach block have a different deformational history from the western Chugach metamorphic complex. Amphibolites, metasedimentary schist, and gneiss record three phases of ductile deformation, but this deformation is kinematically distinct. The rock exhibits two continuous cleavages in which the older cleavage (S_1) is virtually transposed by the younger cleavage (S_2) in upright to steeply inclined, steeply to moderately north plunging, subisoclinal to isoclinal F_2 folds with S_2 axial planar to F_2 folds. D_3 is domainal and is represented by a more WNW striking foliation.

Boundary-Fairweather Block

In the Boundary-Fairweather block along Nunatak Fjord, the country rock is a weakly metamorphosed turbidite sequence of the Cretaceous Yakutat Group (Kym), consisting of graywacke, argillite, and mélange, possibly correlative with the Valdez Group to the west. The metamorphic grade is lower greenschist facies except in narrow (100 m) contact aureoles. Intrusions include tonalite, granodiorite, and granite similar to those in other parts of the Chugach metamorphic complex along with two-mica and garnet-bearing granite, diorite, gabbro, and in the Fairweather fault zone itself, pyroxenite and gabbro. Most of the rocks on the northern shore of Nunatak Fjord are severely deformed by the Fairweather fault. Samples for geochronology were taken from the two-mica granite (known as the Mt. Draper pluton) and from a nearby diorite.

The Boundary-Fairweather block records only a two-phase deformational history that varies within the block and is grossly different from the adjacent Chugach block. Along Nunatak Fjord, only one deformational event is recognizable and the rocks are weakly metamorphosed slates. The dominant feature in the area is the Mt. Draper pluton. This pluton exhibits a weak foliation striking NE with steep dips. However, there is a discordant contact between the pluton and the surrounding country rock, and foliation is probably an igneous flow foliation. To the west of the Mt. Draper pluton is a small body of diorite with only 300 m exposed. The metamorphic contact aureole around the diorite is 10 m thick with static growth of garnet + staurolite. The contact aureole with the Mt. Draper pluton only has static growth of biotite within 50 m of the contact.

The southern contact of Mt. Draper pluton contrasts markedly with that seen along Nunatak Fjord. Metamorphic grade is clearly higher with garnet + hornblende in amphibolites and garnet-mica schists in a broad contact aureole. These rocks display two ductile fabrics. Initial layering (S_0) parallels foliation (S_1) that is dominantly steeply dipping along north-south strikes associated with D_1. D_2 structures fold D_1 about an axis (F_2) that trends north-south and is steeply plunging. D_2 produced a predominantly north-south–striking foliation and also dips steeply north.

Between Mt. Draper and West Nunatak Glacier is a fault bounded slice of predominantly igneous rock that is referred to as "The Nunatak," a geographic term from the early twentieth century, when it was surrounded by West Nunatak Glacier. The Nunatak is characterized by intrusives with complex crosscutting relationships. Despite the complex crosscutting relationships, there is an igneous foliation (Sig) that has been folded. Fold axes are steeply plunging, but it is unclear if these structures relate to those in the adjacent blocks. Just to the southwest of "The Nunatak" is a small isolated outcrop, which has coarse-grained pyroxenite (94APo25a) that grades into medium-grained gabbro (94APo25b). A fine-grained gabbroic dike (94APo25c) crosscuts both.

About 40 km farther to the south in the Boundary-Fairweather block, there are highly deformed tonalite sills and pegmatites. These felsic intrusives were not analyzed for this study. They contain muscovite, biotite, and garnet. In this region, metamorphism is clearly on a regional scale and not limited to contact aureoles. In the host mica schists, kyanite is replaced by sillimanite, indicating much higher pressures than along the shores of Nunatak Fjord.

In line with the interpretation of Hudson et al. (1977b), we interpret the Boundary-Fairweather block as an oblique crustal section that may be far-traveled relative to the adjacent rocks of the Chugach block, northeast of the Fairweather fault. The dramatic variation in metamorphic grade and structure across the block implies upper crustal levels are now exposed as weakly metamorphosed slates along Nunatak Fjord and the high-grade schists in the southern portion of the block represent much deeper levels. Thus, the Mt. Draper pluton is now exposed in an oblique view from its roof and shallow margin in the north to its floor and deeper margin in the south. The northward tilting of the Boundary-Fairweather block probably was produced by Neogene motion on the Fairweather and Boundary faults. Moreover, this motion has also translated the Boundary-Fairweather block a significant distance northward relative to the Chugach block.

Glacier Bay Mafic Intrusive

The Border Ranges fault system in Johns Hopkins Inlet of Glacier Bay National Park and Preserve separates the relatively unmetamorphosed Valdez Group from the crystalline rocks of the Alexander terrane (e.g., Smart et al., 1996). Within the ~8 km wide Border Ranges shear zone is an undeformed small gabbroic plug (previously described as a diorite by Decker and Plafker, 1982). This body crosscuts ductily deformed Mesozoic metasedimentary rocks. The gabbro includes thin screens of gneissic amphibolite. It crosscuts a ductile mylonitic foliation formed during greenschist to subgreenschist facies conditions (Smart et al., 1996). The gabbroic plug is similar in size and mineralogy to the gabbro west of the "The Nunatak" in Nunatak Fjord. It is possible that this unnamed gabbro is related to a suite of layered gabbroic bodies in the Fairweather Range (e.g., Loney and Himmelberg, 1983; Foley et al., 1997). There are also some small gabbronorite bodies on Yakobi and Chicagof Islands (e.g., Himmelberg et al., 1987). Bradley et al. (this volume, Chapter 1) report biotite $^{40}Ar/^{39}Ar$ ages of 50.5 to 50.2 Ma for a gabbronorite body on Chicagof Island. At this time, the geochronologic data are too sparse to correlate the different mafic bodies.

Both the amphibolite screens and gabbro are crosscut by two 1–2 m wide felsic porphyry dikes. These dikes may be correlative with an undeformed biotite tonalite at the western edge of the Border Ranges shear zone in Johns Hopkins Inlet. This tonalite body has a $^{40}Ar/^{39}Ar$ biotite total gas age of 33.1 ± 0.3 Ma (Smart et el., 1996). Goldfarb (1997) reports a similar $^{40}Ar/^{39}Ar$ biotite age of 30 Ma for a pegmatite that crosscuts the La Perouse–Crillon gabbro body in the Fairweather Range.

Intrusive History

The different deformational histories of the sub regions in the Nunatak Fjord and Glacier Bay, together with a general lack of crosscutting relationships among intrusives, makes the classification of the felsic intrusives from these areas into the Ti_{1-4} system used in the Chugach metamorphic complex difficult. Unlike the western Chugach metamorphic complex, there appear to be only one or two clearly definable generations of magmatism that are both synmetamorphic. We did not recognize any pre-metamorphic intrusives in either Nunatak Fjord or Glacier Bay. We found only one late hypabyssal dike in Novayak Nunatak.

ANALYTICAL TECHNIQUES

Samples were analyzed for major and trace elements at several laboratories: 1994 and 1995 samples were done by X-ray fluorescence spectrometry (XRF) and inductively coupled plasma–mass spectrometry (ICP-MS) at Washington State University and instrumental neutron activation analysis (INAA) at the Oregon State University Radiation Center, 1991 samples by XRF, ICP, and INAA at Activation Laboratories, and 1989 samples by ICP at Rice University and INAA at Oregon State University Radiation Center. Sample locations as well as analytical techniques used are given in Appendix 1. Nd, Sr, and U/Pb isotope data for the 94APo and 95APo samples were determined using well-established techniques on a Finnigan MAT 262 multi-collector mass spectrometer previously at Rice University. Analytical uncertainties for $^{87}Sr/^{86}Sr$ and $^{144}Nd/^{143}Nd$ are ±0.000015 and ±0.000023 respectively. U/Pb isotope data for 89GP19b were done using similar techniques on a Finnigan MAT 261 multi-collector mass spectrometer at University of California at Santa Barbara (e.g., McClelland and Mattinson, 1996). Mineral separates for U/Pb geochronology were prepared using standard techniques; these minerals were nonmagnetic at 1.8 amps with a 15° forward and a 2°–7° side tilt on a Franz magnetic separator. They were subsequently size-fractioned and handpicked to a greater than 99% purity. Three or four size fractions of zircon were processed for each sample. In one sample of a synmetamorphic Ti_2 sill (94APo36) near Van Cleve Glacier, fractions of monazite and xenotime were also analyzed. Most of the zircon data give discordant ages and suggest an inherited component in these plutons.

Mineral chemistry and catholoduminescence were determined using a Cameca SX-50 electron microprobe at Rice University and University of California, Davis. Operating conditions were 15 kV and 15 nA for all minerals, except for 10 nA sample current for feldspar. The beam diameter was 5 μm for most minerals, again except for 10 μm used for feldspar. Calibration was done on natural and synthetic mineral and oxide standards, and all corrections were performed using PAP reduction procedures (Pouchou and Pichoir, 1987).

Mineral separates for $^{40}Ar/^{39}Ar$ analyses were prepared by standard techniques. All separates were >99% pure. These were irradiated at Texas A&M University for approximately 30 hours. Analyses were performed by the standard step-heating technique on a MAP 215-50 rare-gas mass spectrometer in the Argon Thermochronology Laboratory at the University of Houston. Neutron fluence was monitored with a standard mineral (Fish Canyon Tuff sanidine, 27.9 Ma; Steven et al., 1967; Hurford and Hammerschmidt, 1985; Cebula et al., 1986). Interfering reactions were accounted for by analyzing optical grade CaF_2 and synthetic glass with a K_2O content of 18 wt%.

Apatite and zircon separates from one sample (91ASn105) near Goat Wallow Glacier were prepared for fission track geochronological analysis. These were analyzed using the external detector method after irradiation in the Texas A&M University Radiation Center. Both numbers of tracks as well as track length distributions were measured. Fission track lengths were measured using Cf^{252}-irradiated grain mounts (e.g., Donelick and Miller, 1991). Apatite fission track age and length data were grouped into kinetic classes using the calibrated parameter (Dpar), the mean etch pit diameter parallel to the crystallographic c-axis (Donelick, 1993; Burtner et al., 1994; Donelick, 1995; Carlson et al., 1999; Ketcham et al., 1999). These data were then used to model a thermal history using AFTSolve (©1996–2000 Donelick Analytical, Inc., and Richard A. Ketcham), which implements various laboratory calibrations for fission tracks in apatite in response to heating and/or cooling histories. Full details concerning these calibrations and the various uses of AFTSolve are available in Carlson et al. (1999), Donelick et al. (1999), Ketcham et al. (1999), and Ketcham et al. (2000). The models use fission track ages and track length distributions to calculate a series of 30,000 randomly generated temperature histories. These randomly generated temperature histories were constrained by independently known information such as the intrusive age and present-day temperature. A temperature history was deemed acceptable when both the model fission track age and the model fission track length distribution matched their measured counterparts with a level of confidence of 0.05 or greater. A temperature history was deemed good when both the model fission track age and length distribution matched their measured counterparts with a level of confidence of 0.50 or greater.

RESULTS

Petrography and Mineral Chemistry

The Sanak-Baranof igneous rocks consist of tonalite, granodiorite, and trondhjemite (Hudson, 1979; Hill et al., 1981; Barker et al., 1992); less common are granitic pegmatites, diorite, and gabbro (Fig. 4). Some of the bodies also have sparse mafic enclaves. The grain size in the gabbro and pegmatite is generally larger than the other rock types. Most of the tonalite and trondjhemite are hypidiomorphic granular and exhibit seriate grain-size distribution. Some samples are porphyritic with euhedral phenocrysts of plagioclase that commonly have altered cores of sericite ± epidote. Common minerals include amphibole and biotite as mafic phases, with occasional muscovite and garnet with interstitial micrographic quartz-plagioclase intergrowths. The mafic plutonic rocks also have clinopyroxene except for the gabbro from Johns Hopkins Inlet, which has both clino- and orthopyroxene. Some pegmatites have tourmaline and garnet in vermicular intergrowths with quartz and plagioclase. Other accessory minerals include zircon, apatite, titanite, Fe-Ti oxides and rare monazite, xenotime, and allanite. Minor interstitial K-feldspar is present in most felsic rocks. Secondary minerals include chlorite, epidote, and sericite. The earlier sills and plutonic bodies have plastically deformed quartz and contain a mica foliation. Feldspar rim composition ranges from An_{24-32} in tonalite to An_{54} in gabbro (Table 1). Some feldspars exhibit oscillatory zoning with cores generally being more calcic. The gabbro from Glacier Bay has cores of both clinopyroxene ($Wo_{41}En_{44}Fs_{15}$) and orthopyroxene ($Wo_4En_{60}Fs_{36}$) in amphibole. In both felsic and mafic intrusives, amphibole is not zoned and is magnesio- or tschermakitic-hornblende with a few actinolitic hornblende compositions (Table 2). Compositional variations in the amphibole are controlled by tshermakitic substitution.

Major Elements

This suite of intrusive rocks spans a wide range of SiO_2 content; mafic samples (gabbro and pyroxenite) range from 49 to 52 wt%, felsic intrusives such as biotite ± hornblende tonalite range from 60 to 67 wt%, two mica tonalite ranges from 58 to 72 wt%, and garnet-bearing tonalite/trondhjemite ranges from 74 to 77 wt% (Table 3). Among the tonalitic rocks, hornblende-bearing samples have lower SiO_2 contents averaging 65 wt% compared to an average of 70 wt% for two mica and garnet-bearing tonalite. There is nearly a complete overlap among the major and trace element chemistry between all the different generations of felsic intrusives. Several thin leucosomes or felsic segregations in gneissic rocks also have similar chemistry to the pre- to synmetamorphic Ti_{1-3} intrusives. In addition, the felsic rocks from Nunatak Fjord are similar to those from the western and central Chugach metamorphic complex. The majority of the felsic rocks fall in the magnesian and calcic to calc-alkaline fields defined by Frost et al. (2001) similar to other Cordilleran granitoids. They also have lower values of $Na_2O + K_2O - CaO$ than Archean tonalitic suites (e.g., Frost et al., 2001). In all felsic units, TiO_2, Al_2O_3, total Fe, MgO,

TABLE 1. REPRESENTATIVE ANALYSES OF FELDSPAR

Sample:	94APo47	94APo49	95APo1	95APo13	95APo35a	95APo44e
Location:	Van Cleve	Van Cleve	Novatak	Novatak	Nunatak	Nunatak
Rock type:	tonalite	tonalite	tonalite	tonalite	tonalite	gabbro
SiO_2	60.00	57.70	60.43	60.20	58.74	54.27
Al_2O_3	25.36	26.21	23.69	24.38	23.87	29.30
FeO	0.12	0.15	0.13	0.81	0.51	0.22
CaO	6.79	7.66	4.89	5.62	6.71	11.18
Na_2O	7.70	7.08	8.34	8.07	7.79	5.20
K_2O	0.21	0.10	0.27	0.14	0.19	0.08
Sr_2O	0.05	0.07	0.11	0.06	0.09	0.11
BaO	0.04	0.03	0.02	0.05	0.05	0.02
Total	100.27	99.00	97.89	99.33	97.95	100.38
Cations per 8 Oxygen						
Si	2.669	2.607	2.740	2.700	2.656	2.443
Al	1.330	1.396	1.267	1.289	1.326	1.555
Fe	0.004	0.005	0.005	0.027	0.017	0.007
Ca	0.323	0.371	0.238	0.270	0.325	0.539
Na	0.664	0.620	0.733	0.702	0.683	0.454
K	0.012	0.006	0.016	0.008	0.011	0.004
Sum	5.002	5.005	4.998	4.996	5.019	5.004
Ternary Mol %						
An	32.4	37.2	24.1	27.5	31.9	54.0
Ab	66.4	62.2	74.3	71.6	67.0	45.5
Or	1.2	0.6	1.6	0.8	1.1	0.5

TABLE 2. REPRESENTATIVE ANALYSES OF AMPHIBOLE

Sample:	94APo47	94APo49	95APo1	95APo13	95APo44e	95APo25	95APo35a	89GP19b
Location:	Van Cleve	Van Cleve	Novayak	Novayak	Nunatak	Nunatak	Nunatak	Glacier Bay
Rock type:	tonalite	tonalite	tonalite	tonalite	tonalite	gabbro	gabbro	gabbro
Amphibole name:	act-hbl	tr	tsch-hbl	tsch-hbl	hbl	hbl	tsch-hbl	hbl
SiO_2	50.59	54.20	41.93	43.49	47.82	50.43	42.39	45.39
TiO_2	0.40	0.05	0.55	0.51	1.35	0.50	1.09	1.78
Al_2O_3	5.53	3.60	13.35	14.39	10.25	6.23	11.50	8.24
FeO	16.46	7.43	18.97	17.67	15.10	11.05	18.85	19.48
MnO	0.66	0.21	0.44	0.51	0.31	0.25	0.38	0.24
MgO	11.78	19.34	6.94	7.52	10.84	16.08	8.53	10.71
CaO	11.48	11.44	11.24	11.06	10.49	11.89	11.34	10.69
Na_2O	0.60	0.41	1.21	1.37	1.35	0.71	1.11	1.57
K_2O	0.39	0.08	1.38	0.93	0.45	0.28	0.99	0.29
F	0.05	0.01	0.14	0.18	0.12	0.04	0.08	0.06
Cl	0.02	0.01	0.01	0.01	0.03	0.02	0.01	0.01
H_2O*	2.03	2.12	1.87	1.94	2.07	2.09	1.91	1.89
O=F,Cl	0.03	0.04	0.08	0.06	0.01	0.02	0.06	0.03
Total	100.32	99.74	98.21	99.75	100.44	100.17	98.70	100.32
Cations per 23 Oxygen								
Si	7.366	7.520	6.439	6.479	6.909	7.159	6.440	7.104
Al^{iv}	0.634	0.480	1.561	1.521	1.091	0.841	1.560	0.896
Al^{vi}	0.314	0.108	0.855	1.006	0.655	0.201	0.499	0.625
Ti	0.044	0.005	0.063	0.058	0.147	0.053	0.124	0.210
Fe^{3+}	0.407	0.833	0.249	0.299	0.433	0.669	0.600	0.000
Fe^{2+}	1.597	0.029	2.187	1.903	1.391	0.643	1.794	2.550
Mn	0.081	0.025	0.057	0.065	0.038	0.030	0.049	1.421
Mg	2.557	4.000	1.588	1.670	2.335	3.404	1.933	0.056
Ca	1.791	1.701	1.849	1.765	1.624	1.809	1.846	1.794
Na	0.171	0.111	0.361	0.395	0.377	0.195	0.328	0.477
K	0.072	0.015	0.270	0.177	0.082	0.050	0.192	0.057
F	0.024	0.034	0.086	0.067	0.004	0.018	0.056	0.029
Cl	0.004	0.003	0.003	0.003	0.003	0.006	0.009	0.002
OH*	1.972	1.963	1.911	1.930	1.993	1.977	1.936	1.968
Sum	17.03	16.83	17.48	17.34	17.08	17.05	17.37	17.19
(Ca+Na)	1.962	1.812	2.000	2.000	2.000	2.000	2.000	2.000
Na	0.171	0.111	0.151	0.235	0.376	0.191	0.154	0.206
(Na+K)	0.072	0.015	0.480	0.337	0.084	0.055	0.366	0.328
$Mg/(Mg+Fe^{2+})$	0.616	0.993	0.421	0.468	0.627	0.841	0.519	0.021
$Fe^{3+}/(Fe^{3+}+Al^{vi})$	0.564	0.885	0.226	0.229	0.398	0.769	0.546	0.000
Geothermobarometry[†]								
H&Z	not valid	not valid	820	880	490	not valid	640	370
H et al.			890	950	510		690	380
J&R			680	720	390		520	300
S			850	900	530		680	420
A&S			890	830	490		630	320
T (°C)	614	627	642	715	720	700	706	750

Notes: act—actinolite, hbl—hornblende, tr—tremolite, tsch—tschermakite.

*H_2O and OH is calculated by difference.

[†]Pressure calculated using calibrations of H&Z—Hammarstrom and Zen (1986), H et al.—Hollister et al. (1987), J&R—Johnson and Rutherford (1989), S—Schmidt (1992), A&S—Anderson and Smith (1995). Temperature calculated using Holland and Blundy (1994).

TABLE 3. WHOLE-ROCK MAJOR AND TRACE ELEMENT ANALYSES

Sample:	89ASn62c	91AHa31	91ASn61b	91ASn63	91ASn83	91ASn93	94APo49	89ASn87c
Location:	Tana	Tana	Tana	Tana	Tana	Goat Wallow	Van Cleve	Tana
Mineralogy:	bt-ms	bt-ms	bt-ms	bt-ms	bt-hbl	hbl-bt	px-bt	bt-ms
Rock type:	leucosome	leucosome	leucosome	leucosome	Ti_1	Ti_1	Ti_1 or Ti_2	Ti_2
SiO_2	70.00	58.38	70.52	68.96	64.22	60.18	66.57	71.89
Al_2O_3	14.26	23.73	14.72	15.61	17.93	13.38	10.39	15.03
TiO_2	0.450	0.290	0.360	0.500	0.290	0.250	0.127	0.190
FeO*	3.72	3.11	3.19	3.35	3.26	6.16	4.20	1.41
MnO	0.060	0.130	0.050	0.050	0.040	0.120	0.113	0.040
MgO	1.65	1.14	1.27	1.19	1.80	7.62	8.77	0.46
CaO	2.74	2.80	2.74	3.86	5.42	5.81	5.43	2.19
Na_2O	3.47	3.66	3.36	3.56	3.57	1.64	1.34	4.92
K_2O	1.68	4.76	1.82	1.46	1.28	0.96	1.92	1.68
P_2O_5	0.22	0.14	0.18	0.20	0.14	0.04	0.041	0.12
LOI	1.33	1.68	1.03	0.77	0.81	2.85		0.58
Total	99.58	99.82	99.24	99.51	98.76	99.01	98.90	98.51
Sc	9	10	8	8	7	23	25	3
V		32	58	34	26	180	148	
Cr	41	7	41	5	0	190	588	<0.5
Ni							175	
Cu							13	
Nb	9.0					5.8	3.6	5.0
Y	16	36	24	12	10	18	20	6
Zr	94	159	142	172	99	99	43	75
Ba	913	2501	935	776	446	385	394	685
Sr	310	409	339	315	277	214	179	412
Rb	50	78	52	58	43	31	40	46
Cs	1.9	1.8	2.1	1.9	1.7	1.2	3.76	2.6
Pb							13	
Hf	2.60	5.40	3.60	3.60	2.10	1.60	1.36	2.60
Ta		1.80	3.00	3.00	2.20	1.90	0.45	
Th	4.20	12.00	6.90	3.10	3.10	6.50	2.57	2.70
U	2.40	6.50	2.80	1.90	2.00	4.00	3.08	1.50
La	16.2	35.4	20.4	14.1	8.1	7.7	9.8	7.7
Ce	32	68	42	28	16	17	23.7	15
Pr								
Nd	15	30	21	13	8	9	14.5	8
Sm	3.1	6.5	4	2.2	1.6	2.4	3.25	1.5
Eu	1.09	1.17	1.19	0.83	0.51	0.67	0.52	0.51
Gd	2.7	5.2	3.6	2.6	2.2	2.2		1.3
Tb	0.5	0.9	0.6	0.4	0.3	0.3	0.5	0.2
Dy								
Ho								
Er								
Tm								
Yb	1.48	3.07	2.01	0.88	0.86	1.45	2.11	0.64
Lu	0.2	0.49	0.26	0.14	0.12	0.2	0.33	0.07
B		19.9	4.7	5.2	2.6	4.9		

Note: See text for intrusive classification scheme. Mineralogy includes phases other than Qz-Ab-An-Or with abbreviations after Kretz (1983) with ms—muscovite, bt—biotite, px—pyroxene, hbl—hornblende, pl—plagioclase, and grt—garnet.

*Total Fe as FeO.

(*continued*)

TABLE 3. WHOLE-ROCK MAJOR AND TRACE ELEMENT ANALYSES (*continued*)

Sample:	91AHa32	91AHa42	91AHa50	91ASn72b	91AHa6b	91AHa12	94APO36
Location:	Tana	W of Tana	S of Tana	Tana	Wernicke	Wernicke	Van Cleve
Mineralogy:	bt-ms	bt	bt-hbl	bt-hbl	bt-hbl	bt-grt	bt
Rock type:	Ti_2	Ti_2	Ti_2	Ti_2	Ti_2	Ti_2	Ti_2
SiO_2	69.03	68.26	67.66	63.65	64.80	66.85	74.26
Al_2O_3	14.67	15.59	14.01	17.54	16.72	15.52	15.15
TiO_2	0.440	0.440	0.690	0.320	0.310	0.470	0.348
FeO*	3.70	3.21	5.05	3.45	3.43	5.08	2.31
MnO	0.100	0.060	0.080	0.050	0.060	0.110	0.038
MgO	1.20	0.86	1.66	1.77	2.72	1.98	0.92
CaO	0.77	4.00	3.86	5.48	5.45	4.03	3.24
Na_2O	3.56	3.94	3.08	3.52	3.39	2.58	3.87
K_2O	4.12	1.16	2.52	1.34	1.38	2.06	1.57
P_2O_5	0.12	0.16	0.18	0.12	0.14	0.20	0.09
LOI	0.89	0.74	1.43	1.10	1.02	1.25	
Total	98.6	98.42	100.22	98.34	99.42	100.13	101.79
Sc	5	13	13	9	10	12	7
V	8	18	72	24	48	54	36
Cr	5	7	35	2	43	28	28
Ni							9
Cu							1
Nb							10.9
Y	10	28	34	10	16	28	13.9
Zr	100	206	281	124	116	185	146
Ba	609	631	1206	506	502	733	549
Sr	102	219	186	297	201	228	224
Rb	97	40	74	38	44	62	46
Cs	5.8	1.8	2.5	2.5	2	2.5	2.4
Pb							15.60
Hf	2.10	4.20	6.10	2.00	2.80	4.20	4.25
Ta	3.30	3.70	3.50	2.40	3.80	2.70	0.53
Th	1.70	6.50	10.00	3.00	4.70	6.10	7.55
U	2.40	2.70	2.60	1.80	1.90	3.30	3.21
La	4.8	16.8	28.8	7.9	11.3	14.8	19.74
Ce	9	32	55	16	22	28	34.83
Pr							3.66
Nd	6	15	28	7	11	12	13.63
Sm	1.2	3.2	4.7	1.7	2.5	2.6	2.80
Eu	0.18	0.82	0.78	0.54	0.54	0.77	0.88
Gd	1.1	3.2	4.8	2.2	2.4	2.5	2.30
Tb	0.2	0.6	0.8	0.3	0.5	0.5	0.39
Dy							2.34
Ho							0.49
Er							1.47
Tm							0.23
Yb	0.63	1.96	2.54	0.91	1.59	2.28	1.48
Lu	0.07	0.3	0.38	0.13	0.26	0.36	0.26
B	7.0	3.5	4.8	8.0	2.9	6.7	

Note: See text for intrusive classification scheme. Mineralogy includes phases other than Qz-Ab-An-Or with abbreviations after Kretz (1983) with ms—muscovite, bt—biotite, px—pyroxene, hbl—hornblende, pl—plagioclase, and grt—garnet.

*Total Fe as FeO.

(*continued*)

TABLE 3. WHOLE ROCK MAJOR AND TRACE ELEMENT ANALYSES (*continued*)

Sample:	94APo39	89ASn69	91ASn64	91AHa26	91AHa27	91AHa2	91AHa3
Location:	Van Cleve	Tana	Tana	Tana	Tana	Wernicke	Wernicke
Mineralogy:	bt-mu	Ms-grt	bt-ms	bt	bt-hbl	bt-ms	bt
Rock type:	Ti_2	Ti_3	Ti_3	Ti_4	Ti_4	Ti_3	Ti_3
SiO_2	71.91	77.01	67.68	67.75	66.00	68.74	60.99
Al_2O_3	14.96	13.75	16.48	17.33	17.11	16.28	18.12
TiO_2	0.275	0.070	0.200	0.160	0.150	0.410	0.270
FeO*	2.39	0.66	2.15	2.18	2.17	2.91	4.03
MnO	0.058	0.050	0.040	0.030	0.040	0.050	0.090
MgO	0.94	0.24	0.85	1.90	1.83	1.33	3.46
CaO	2.76	0.74	4.04	4.51	4.97	4.30	6.47
Na_2O	2.69	3.16	4.34	4.34	4.21	3.73	3.18
K_2O	3.28	3.34	1.20	0.76	0.92	1.60	1.70
P_2O_5	0.152	0.14	0.14	0.10	0.10	0.22	0.10
LOI		0.96	0.57	1.26	0.83	0.95	2.33
Total	99.41	100.12	97.69	100.32	98.33	100.52	100.74
Sc	11	4	4	5	5	7	12
V	32		12	30	24	42	72
Cr	44	<0.5	5	45	42	16	38
Ni	22						
Cu	10						
Nb	6.6	6.0		2.4		7.9	
Y	12	10	10	4	8	20	16
Zr	104	51	134	48	104	243	104
Ba	811	485	414	432	471	546	460
Sr	150	90	417	327	294	345	283
Rb	75	64	27	23	25	47	49
Cs	2.45	1.8	1.5	1.1	0.9	1.8	1.4
Pb	13						
Hf	3.04	3.20	2.40	1.80	1.70	2.80	1.80
Ta	0.8		3.70	2.40	2.60	3.70	2.80
Th	4.85	2.50	1.90	1.10	1.00	0.60	2.70
U	1.49	3.60	1.30	0.90	1.30	2.40	1.60
La	11.5	6.3	6.6	2.6	2.5	5.3	9
Ce	21.5	14	13	5	5	11	17
Pr							
Nd	9.7	7	6	3	3	6	7
Sm	2.08	1.6	1.1	0.62	0.67	1.9	1.5
Eu	0.61	0.27	0.48	0.29	0.27	1.04	0.43
Gd		2.3	1.4	0.5	0.6	2.4	1.6
Tb	0.29	0.4	0.2	0.1	0.1	0.5	0.3
Dy							
Ho							
Er							
Tm							
Yb	1.19	1	0.6	0.34	0.33	1.81	1.13
Lu	0.17	0.12	0.07	0.04	0.04	0.26	0.18
B			5.1	2.9			19.6

Note: See text for intrusive classification scheme. Mineralogy includes phases other than Qz-Ab-An-Or with abbreviations after Kretz (1983) with ms—muscovite, bt—biotite, px—pyroxene, hbl—hornblende, pl—plagioclase, and grt—garnet.

*Total Fe as FeO.

(*continued*)

TABLE 3. WHOLE ROCK MAJOR AND TRACE ELEMENT ANALYSES (*continued*)

Sample:	91AHa14c	94APo40	94APo44a	94APO47	94ASn52	94APo11b	94APO11c
Location:	Wernicke	Van Cleve	Van Cleve	Van Cleve	Van Cleve	Mt. Draper	Mt. Draper
Mineralogy:	bt-ms	bt-mu	bt-hbl	bt-hbl	bt	bt-mu	bt-ms-grt
Rock type:	Ti_3	Ti_3	Ti_3	Ti_3	Ti3	Ti	Ti
SiO_2	71.69	71.33	69.96	69.40	69.00	71.34	75.58
Al_2O_3	14.76	14.74	15.14	16.18	15.33	15.91	15.93
TiO_2	0.280	0.399	0.347	0.453	0.524	0.123	0.033
FeO*	2.52	3.05	3.08	3.54	3.37	1.66	0.74
MnO	0.090	0.052	0.056	0.066	0.068	0.030	0.015
MgO	1.06	0.88	1.21	1.82	1.68	0.34	0.21
CaO	1.83	3.40	3.80	4.58	3.22	1.71	1.89
Na_2O	3.63	1.68	2.07	3.50	2.25	3.68	5.15
K_2O	3.10	3.26	3.37	1.67	3.51	4.08	2.33
P_2O_5	0.14	0.137	0.099	0.12	0.195	0.085	0.06
LOI	0.86						
Total	99.96	98.92	99.13	101.33	99.14	98.96	101.93
Sc	8	10	7	11	14	1	2
V	44	35	39	65	62	6	0
Cr	28	38	45	51	62	20	15
Ni		19	24	20	23	11	11
Cu		21	5	19	6	7	74
Nb		6.4	8.0	12.4	7.4	7.7	11.5
Y	22	10	23	24.1	11	3	1.9
Zr	106	157	134	137	134	60	41
Ba	1147	829	906	640	760	945	503
Sr	223	142	121	149	359	199	251
Rb	77	57	70	61	79	112	68
Cs	3.8	1.84	1.9	2.3	4.55	1.66	1.1
Pb		12	7	8.02	16	22	15.48
Hf	2.80	4.26	3.98	3.87	3.77	2.6	1.91
Ta	4.70	0.5	0.69	0.74	0.63	0.64	0.64
Th	5.30	8.39	6.94	4.37	7.61	4.13	0.85
U	8.30	1.81	1.89	3.95	2.67	1.93	2.18
La	15	24.2	13	11.98	20.5	10.9	4.83
Ce	29	37.9	24.9	22.26	44.5	21.9	8.32
Pr				2.52			0.96
Nd	14	16.1	11.9	10.20	20.3	11.8	3.91
Sm	3.1	2.85	2.65	2.84	4.32	2.28	1.20
Eu	0.79	0.68	0.63	0.71	1.08	0.4	0.37
Gd	2.9			3.14			0.82
Tb	0.6	0.29	0.57	0.59	0.43	0.13	0.09
Dy				3.94			0.38
Ho				0.85			0.06
Er				2.63			0.13
Tm				0.38			0.02
Yb	2.3	0.96	2.5	2.59	1.7	0.29	0.13
Lu	0.34	0.14	0.38	0.43	0.22	0.04	0.02
B	6.2						

Note: See text for intrusive classification scheme. Mineralogy includes phases other than Qz-Ab-An-Or with abbreviations after Kretz (1983) with ms—muscovite, bt—biotite, px—pyroxene, hbl—hornblende, pl—plagioclase, and grt—garnet.

*Total Fe as FeO.

(*continued*)

TABLE 3. WHOLE ROCK MAJOR AND TRACE ELEMENT ANALYSES (*continued*)

Sample:	94APo24	95APO21	95APO25a	95APO41b	95APO13	94APO25a	94APO25b
Location:	Mt. Draper	Mt. Draper	Nunatak	Mt. Stamy	Novayak	Nunatak	Nunatak
Mineralogy:	mu-bt	bt-ms	bt-hbl	Bt	bt-hbl	Pl-hbl	hbl-px
Rock type:	Ti	Ti	Ti dike	Ti dike	Ti	Pyroxenite	gabbro
SiO_2	73.88	72.10	75.33	71.72	64.19	51.88	50.41
Al_2O_3	14.88	16.88	15.79	15.34	16.24	6.76	20.95
TiO_2	0.002	0.158	0.135	0.431	0.890	0.832	0.247
FeO*	0.76	1.69	0.90	2.84	6.13	11.06	5.62
MnO	0.006	0.026	0.017	0.051	0.107	0.237	0.115
MgO	0.02	0.50	0.45	0.93	2.97	14.83	8.29
CaO	1.05	2.61	2.05	2.88	4.94	12.24	11.73
Na_2O	4.34	5.09	6.11	3.74	3.43	0.78	2.37
K_2O	4.61	2.45	1.21	3.35	1.91	0.26	0.28
P_2O_5	0.010	0.08	0.08	0.12	0.21	0.04	0.04
LOI							
Total	99.56	101.58	102.07	101.40	101.02	98.92	100.05
Sc	4	2	4	10	19	58	31
V	4	2	11	24	105	489	180
Cr	20	31	33	41	77	823	403
Ni	11	10	12	11	13	111	53
Cu	4	1	51	19	17	127	9
Nb	13.9	9.7	4.4	11.0	13.7	6.8	5.1
Y	2	2.5	2.7	22.8	18.5	27.3	8.3
Zr	31	86	94	166	186	43	35
Ba	27	748	304	1325	1233	44	96
Sr	23	313	548	207	248	42	319
Rb	195	72	21	71	62	2	8
Cs	2.51	1.2	0.4	2.7	1.2	0.1	0.2
Pb	35	14.35	4.45	22.86	10.12	1.08	2.87
Hf	3.18	2.91	2.52	5.48	4.25	1.66	0.72
Ta	1.86	0.57	0.11	0.48	0.50	0.20	0.10
Th	2.66	4.43	1.53	9.59	1.95	0.52	0.50
U	1.36	1.88	0.72	3.39	0.70	0.22	0.22
La	1.3	14.98	11.63	27.46	19.30	5.76	4.14
Ce	3.3	27.79	20.95	51.41	35.19	16.51	7.93
Pr		3.08	2.22	5.73	3.96	2.57	1.06
Nd	2.9	11.79	8.47	22.66	16.31	12.84	4.36
Sm	1.39	2.37	1.69	5.34	3.88	4.06	1.26
Eu	0.14	0.62	0.51	0.96	1.14	0.67	0.63
Gd		1.31	1.04	4.54	3.67	4.22	1.25
Tb	0.16	0.14	0.14	0.73	0.63	0.79	0.23
Dy		0.56	0.62	4.29	3.69	5.08	1.52
Ho		0.08	0.09	0.85	0.74	1.06	0.32
Er		0.18	0.20	2.37	1.97	3.03	0.95
Tm		0.03	0.03	0.32	0.27	0.42	0.14
Yb	0.35	0.14	0.14	2.05	1.56	2.55	0.83
Lu		0.02	0.02	0.34	0.25	0.39	0.14
B							

Note: See text for intrusive classification scheme. Mineralogy includes phases other than Qz-Ab-An-Or with abbreviations after Kretz (1983) with ms—muscovite, bt—biotite, px—pyroxene, hbl—hornblende, pl—plagioclase, and grt—garnet.

*Total Fe as FeO.

(*continued*)

TABLE 3. WHOLE ROCK MAJOR AND TRACE ELEMENT ANALYSES (*continued*)

Sample:	94APo25c	95APO30	94APO9	94APo19b	94APO27	95APO25b	95APO47
Location:	Nunatak	Nunatak	Nunatak	Nunatak	Nunatak	Nunatak	Nunatak
Mineralogy:	hbl-pl	deformed	hbl-pl	hbl-px-pl	hbl-pl	hbl-pl	hbl-pl
Rock type:	gabbro dike	metagabbro	diorite	amphibolite	amphibolite	amphibolite	amphibolite
SiO_2	49.05	50.35	49.64	50.19	47.99	50.41	69.40
Al_2O_3	19.07	17.88	19.14	14.32	14.64	20.95	16.18
TiO_2	1.374	1.513	2.068	1.084	1.685	0.247	0.453
FeO*	9.78	10.64	9.57	10.37	13.64	5.62	3.54
MnO	0.181	0.225	0.157	0.200	0.233	0.115	0.066
MgO	5.53	5.83	10.80	7.64	10.71	11.73	4.58
CaO	10.35	10.23	3.94	12.45	7.21	8.29	1.82
Na_2O	0.38	2.48	3.54	0.19	2.75	2.37	3.50
K_2O	2.67	0.27	0.39	2.39	0.23	0.28	1.67
P_2O_5	0.247	0.16	0.20	0.094	0.14	0.04	0.12
LOI							
Total	98.63	99.58	99.45	98.93	99.23	100.05	101.33
Sc	11	44	27	48	42	43	46
V	32	284	283	307	417	398	411
Cr	44	241	46	285	172	414	119
Ni	22	71	34	91	77	50	45
Cu	10	117	179	100	188	234	269
Nb	6.6	4.3	13.7	3.2	1.8	8.8	6.2
Y	12	40.0	27.9	25	56.9	34.9	39.0
Zr	104	89	123	65	95	106	84
Ba	811	49	148	83	36	87	84
Sr	150	245	346	151	94	144	97
Rb	75	5	8	4	5	3	4
Cs	0.31	0.4	0.49		0.14	0.03	0.02
Pb	13	1.60	0.92	1	0.49	0.68	0.52
Hf	2.06	2.36	3.38	2.15	2.73	2.94	2.39
Ta	0.27	0.12	0.65	0.24	0.15	0.55	0.37
Th	4	0.08	0.64	1	0.16	0.64	0.39
U	2.6	0.27	0.34	1.38	0.09	0.20	0.12
La	9	5.16	9.95	3.3	2.68	9.57	6.39
Ce	29.6	10.17	23.33	9.5	7.88	21.63	15.34
Pr		1.89	3.29		1.45	2.99	2.22
Nd	20	10.51	16.03	6.5	8.88	14.57	10.88
Sm	4.65	3.93	4.88	2.44	4.25	4.64	3.77
Eu	1.31	1.65	1.75	0.92	1.66	1.67	1.31
Gd		5.25	5.25		6.5	5.2	4.75
Tb	0.84	1.01	0.93	0.5	1.39	0.99	0.97
Dy		6.73	5.69		9.77	6.42	6.61
Ho		1.47	1.12		2.18	1.36	1.47
Er		4.30	3.02		6.52	3.79	4.46
Tm		0.59	0.4		0.93	0.52	0.62
Yb	3.63	3.57	2.41	2.02	5.88	3.17	4.08
Lu	0.53	0.55	0.37	0.04	0.94	0.5	0.63
B							

Note: See text for intrusive classification scheme. Mineralogy includes phases other than Qz-Ab-An-Or with abbreviations after Kretz (1983) with ms—muscovite, bt—biotite, px—pyroxene, hbl—hornblende, pl—plagioclase, and grt—garnet.

*Total Fe as FeO.

MnO, CaO, P_2O_5, Zr, Y, Sc, and Nb show broad decreases with increasing SiO_2 (Fig. 5). Na_2O and K_2O both increase with increasing SiO_2. In the Al_2O_3-SiO_2 plot, some of the samples fall in the high-Al tonalite-trondhjemite field defined by Drummond and Defant (1990). Also, on the K_2O-SiO_2 plot, some of the samples overlap compositional fields for melts derived from partial melting of amphibolites or of metagraywackes. There are no distinct trends for Sr or Rb with SiO_2 content. There are two samples (leucosome 91AHa31 and a pre-metamorphic dike 89ASn87c) that do not follow these general trends. The leucosome has a geochemical signature more akin to a metaargillite with high Al_2O_3. Overall, these trends are broad, as each sample is from a different intrusion, each with its own fractionation trend. In addition, most of the samples are from thin dikes and sills that may have different sources as well as evolutionary histories. Additional sampling from larger bodies might yield tightly constrained trends for these individual intrusives but are not likely to change the general major element trends of this family of intrusives.

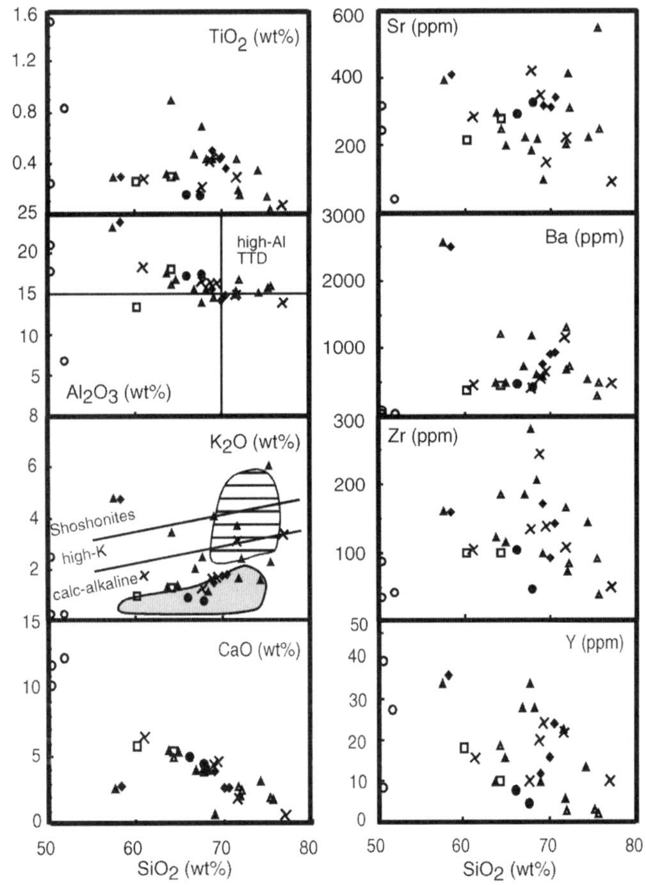

Figure 5. Selected major and trace element compositional variations of igneous rocks from the Chugach metamorphic complex. Symbols the same as in Figure 4 with the addition of open circles for mafic plutonic rocks from Nunatak Fjord region. The fields on the K_2O variation diagram show compositions of water-undersaturated partial melts for two possible protoliths; gray region represents partial melts derived from amphibolites, and ruled region shows partial melts derived from metagraywackes. Data from Beard and Lofgren (1991), Wolf and Wyllie (1994), Patiño Douce and Beard (1996), and Montel and Vielzeuf (1997).

The gabbro from near "The Nunatak" in Nunatak Fjord does not fall along these trends for Al_2O_3, P_2O_5, Ba, Rb, Sr, Y, Zr, V, or Nb. Other elements in the gabbro such as MnO, total Fe, Na_2O, CaO, TiO_2, and K_2O fall on similar trends to the felsic rocks. These are obviously not related by fractional crystallization to the felsic rocks on these plots.

Most of the silicic suite is peraluminous as defined by Shand (1951). These have the assemblage biotite + muscovite ± garnet, which is characteristic of peraluminous, S-type intrusives. The felsic rocks span a wide range in the alumina saturation index (A/CNK) from 0.94 to 1.46, where A/CNK = molar $Al_2O_3/(CaO + Na_2O + K_2O)$. Most of the samples with muscovite have A/CNK below 1.1, the value for S-type intrusives as defined by White et al. (1986). The modified alumina saturation index (A*) of Patiño Douce (1992), defined as A* = (A/CNK) × mol% ($Na_2O + K_2O$), has been proposed to distinguish muscovite-bearing from hornblende-bearing samples. The A* value for the tonalitic suite ranges between 2.3 and 11.3 with a median of 6.0. A value of A* of ~6.2 distinguishes hornblende-bearing samples from muscovite-bearing samples. Many muscovite-bearing samples also contain garnet.

Trace Elements

The tonalitic rocks have characteristics of both high-Al and low-Al suites, according to the criteria of Barker (1979) and Drummond and Defant (1990). The post-metamorphic Ti_4 dikes are all high-Al with high Sr (>300 ppm), low Rb/Sr (<0.1), low Y (<15 ppm), and low Nb (<10 ppm). Some of the other high-Al tonalites do not fit these criteria as they have high Y despite having all the rest of the appropriate geochemical characteristics. The first generation, pre-metamorphic Ti_1 magmas are low-Al with low Sr (<300 ppm), high Y (>15 ppm), and high Rb/Sr (>0.15). The intermediate-age intrusive rocks span a range from high- to low-Al characteristics. There is complete overlap in the trace element geochemistry between the western Chugach metamorphic complex and the Nunatak Fjord region (Fig. 5).

The shaded fields on Figure 6 give rare earth element (REE) data for Valdez Group metasedimentary rocks and igneous rocks from the western part of the complex. As noted by Harris et al. (1996), metasedimentary rocks are light REE-enriched and overlap the top of the range of the synmetamorphic Ti_{1-3} rocks. Post-metamorphic Ti_4 dikes fall at the other extreme, low in all REEs. They are particularly depleted in heavy rare earths, even relative to MORB, implying a garnet residuum.

The REE profiles for felsic intrusives in the Nunatak Fjord area fall into two categories that are not related to the specific fault block in which they occur. Samples from the Novatak Glacier pluton in the Chugach block and the Mt. Stamy pluton in the

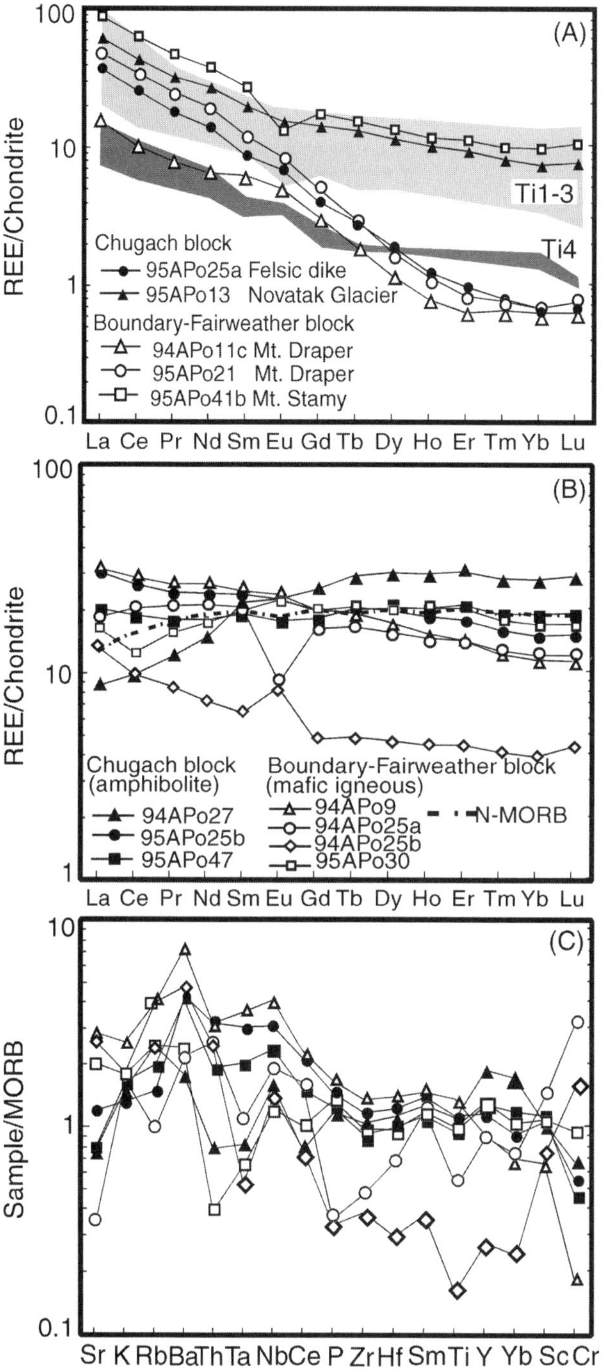

Figure 6. A: Chondrite-normalized REE profiles for felsic rocks of the Nunatak Fjord/Yakutat Bay region. Rocks from the Chugach block are represented by solid symbols; those from the Boundary-Fairweather block by open ones. Light-shaded region shows REE profiles for pre- to synmetamorphic Ti_{1-3} from the Chugach metamorphic complex, and dark-shaded region shows post-metamorphic Ti_4 trends (data from Harris et al., 1996, and this study). B: Chondrite-normalized REE profiles for mafic rocks in the Nunatak Fjord region. Chugach block amphibolites represented by solid symbols; mafic intrusives from the Boundary-Fairweather block by open ones. The profile for N-MORB is shown by a dashed line (Hofmann, 1988). Chondrite normalization values from Anders and Ebihara (1982). C: MORB-normalized spider diagram for mafic rocks of the Nunatak Fjord region. Chugach block amphibolites shown as solid symbols; mafic intrusives from the Boundary-Fairweather block are shown by open ones. Normalization values from Pearce (1983).

Boundary-Fairweather block have profiles similar to the pre- to synmetamorphic Ti_{1-3} rocks. In contrast, the Mt. Draper pluton in the Boundary-Fairweather block and a felsic dike intruded into the Chugach block amphibolites are both highly depleted in heavy rare earths, and thus, they are more akin to the post-metamorphic Ti_4s of the western Chugach metamorphic complex. Note, though, some of their light rare earth element (LREE) signatures overlap that of the pre- to synmetamorphic Ti_{1-3} field.

REE data for mafic rocks of the Nunatak Fjord region are given in Figure 6B. The amphibolites (solid symbols) and mafic plutonic rocks (open symbols) from the Boundary block have profiles similar to the range of N-MORB to E-MORB—relatively flat or slightly depleted to slightly enriched in light rare earths. These flat profiles are also similar to arc tholeiites. The gabbro (94APo25a and 94APo25b) diverges from this trend with large Eu anomalies. The gabbro is much more depleted in heavy rare earth elements (HREEs) than all other mafic rocks. When the samples are shown normalized to MORB (Fig. 6C), they again divide into two distinct groups. "The Nunatak" gabbros show much more variation than the amphibolite samples, particularly among the immobile elements Th through Yb.

Poole (1997) found that Chugach block amphibolites plot in fields for basalt on tectonic discrimination plots after Pearce (1980, 1982) and Shervais (1982). Similar to results of Lull and Plafker (1990), these basalts are transitional between island arc tholeiite and MORB, and they are neither island arc basalt (IAB) nor within-plate basalt (WPB).

Sr and Nd Isotopes

Figure 7 shows ε_{Nd} versus $^{87}Sr/^{86}Sr$ data for metasedimentary rocks, leucosomes (or felsic segregations), and felsic intrusive rocks from the western Chugach metamorphic complex and Nunatak Fjord region (Table 4). Also shown are mafic plutonic rocks from the Nunatak Fjord region and igneous rocks of the Prince William terrane west of the Copper River (data from Barker et al., 1992). Metasedimentary rocks from both high- and low-grade Chugach metamorphic complex samples lie in a field of mainly negative ε_{Nd} (+0.7 to –7.6) and relatively high initial $^{87}Sr/^{86}Sr$ (0.70529 to 0.70654). These values overlap with those reported by Farmer et al. (1993) as well as the field for leucosomes. Post-metamorphic Ti_4 dikes, on the other hand, plot in the region of mantle-derived rocks, with positive ε_{Nd} (+5.8 to +6.7) and low initial $^{87}Sr/^{86}Sr$ (0.70372 to 0.70389). Some of the mafic plutonic rocks from Nunatak Fjord plot in the same region. One

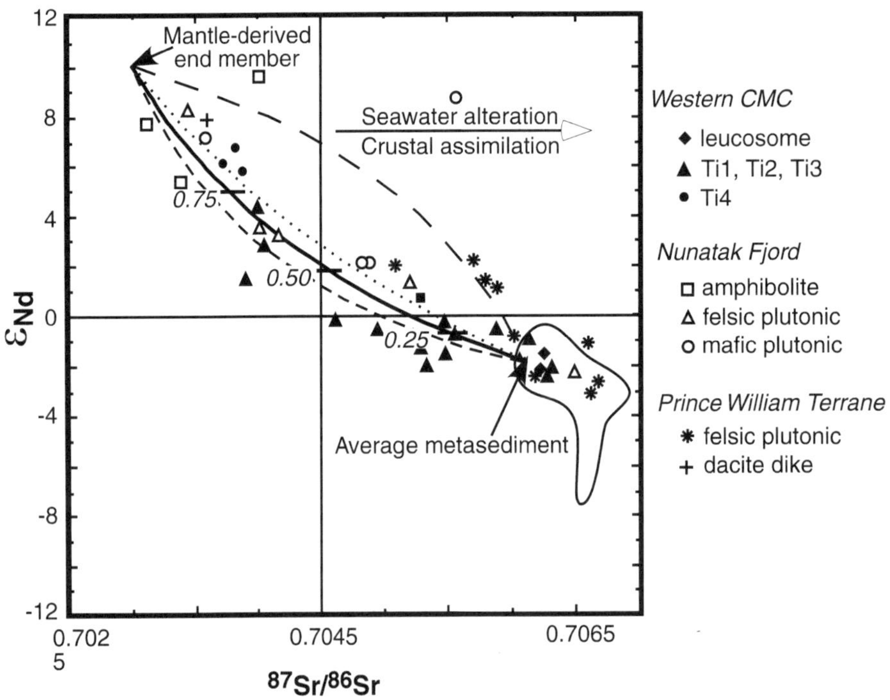

Figure 7. Plot of ε_{Nd} versus initial $^{87}Sr/^{86}Sr$ for metasedimentary rocks, leucosomes, felsic intrusives of the western Chugach metamorphic complex (black symbols), and for amphibolites, felsic intrusives, and mafic intrusives of the Nunatak Fjord/Yakutat Bay region (open symbols) and Prince William terrane (crosses). The analytical error is less than the symbol size. The circled region encloses metasedimentary rocks with an arrow pointing toward an average value. The open-ended arrow shows a theoretical trend for alteration by either seawater or crustal assimilation. Also shown is a simple, two-component mixing curve between an average of the Chugach complex metasedimentary rocks and a mantle-derived component (solid line). Numbers along curve are weight fraction of MORB added to an average Chugach metamorphic complex metasedimentary rock. Alternative two-component mixing curves for MORB-metasediment, and depleted mantle-metasediment produced by melting less than 100% of the metasediment using the Sr distribution coefficient for plagioclase (long dashes for 10% metasediment melt, dotted line for 50% metasediment melt, and medium dashes for 90% metasediment melt). Data for the western Chugach metamorphic complex are from Harris et al. (1996) and this study. Data for Prince William terrane are from Barker et al. (1992). The metasedimentary data are from Harris et al. (1996) and Farmer et al. (1993).

sample of mafic plutonic rock (95APo30) has high positive ε_{Nd} as well as high initial $^{87}Sr/^{86}Sr$. Since this sample is both deformed and altered, the initial $^{87}Sr/^{86}Sr$ value is probably altered by hydrothermal fluid alteration similar to basalt from the Prince William terrane (Barker et al., 1992). Pre- to synmetamorphic Ti_{1-3} rocks are grouped together because no trend is evident between them. These plot along a trend between the metasediments and the late dikes. The felsic plutonic rocks from the Nunatak Fjord region overlap with the values for the western Chugach metamorphic complex. Also note that the felsic plutonic rocks from the Prince William terrane have similar values.

Conditions of Emplacement

Several of the samples have the necessary assemblage for the Al-in-hornblende geobarometer (e.g., Hammarstrom and Zen, 1986; Hollister et al., 1987). Anderson and Smith (1995) proposed that this geobarometer should be calibrated for temperature effects. Crystallization temperatures were estimated using the Holland and Blundy (1994) amphibole-plagioclase geothermometer. Calculated temperatures range from below the solidus (614–627 °C) for two samples from the Van Cleve area and close to the solidus (695–720 °C) for samples from the Nunatak Fjord area (Table 2). The low solidus temperatures suggest the tonalites were H_2O-rich. Likewise, the presence of liquidus amphibole indicates high H_2O contents of ≥5 wt%. The presence of pegmatites, quartz veins, and intrusion breccias at some contacts (e.g., base of Mount Draper pluton) also suggest these magmas were H_2O-rich near the solidus. In contrast, the presence of only aplite dikes and lack of pegmatite in the Boundary-Fairweather block may indicate low fluid content in some of the melts. The intrusion pressure could not be calculated for the two samples from

TABLE 4. Sr AND Nd ISOTOPIC DATA

Sample	Location	Rock type	Concentration (ppm)		Atomic ratio $^{87}Rb/^{86}Sr$	Measured ratio $^{87}Sr/^{86}Sr$*	Initial ratio $^{87}Sr/^{86}Sr$	Concentration (ppm)		Atomic ratio $^{147}Sm/^{144}Nd$	Measured ratio $^{143}Nd/^{144}Nd$†	ε_{Nd} (T)§
			Rb	Sr				Sm	Nd			
94APo9	Nunatak Fjord	diorite	8	346	0.07	0.703653	0.70360	4.35	15.83	0.1661	0.512930	7.1
94APo11c	Mount Draper	tonalite	68	251	0.78	0.704611	0.70400	1.16	4.01	0.1745	0.512744	3.5
94APo25a	Nunatak Fjord	gabbro	2	42	0.14	0.704983	0.70488	3.86	13.72	0.1703	0.512677	2.1
94APo25b	Nunatak Fjord	gabbro	8	319	0.07	0.704893	0.70484	1.14	4.52	0.1530	0.512674	2.1
94APo27	Nunatak Fjord	amphibolite	5	94	0.15	0.704163	0.70404	4.11	9.71	0.2556	0.513059	9.6
94APo36	Van Cleve Gl.	tonalite (Ti_2)	46	224	0.59	0.706590	0.70613	2.45	13.21	0.1119	0.512517	–1.0
94APo47	Van Cleve Gl.	tonalite (Ti_3)	61	149	1.18	0.706804	0.70588	3.33	15.10	0.1334	0.512538	–0.6
95APo13	Novatak Glacier	tonalite	62	248	0.72	0.707068	0.70650	3.40	14.44	0.1423	0.512449	–2.3
95APo21	Mount Draper	granite	72	313	0.67	0.704694	0.70417	1.97	10.81	0.1104	0.512725	3.1
95APo25a	Nunatak Fjord	felsic dike	21	548	0.11	0.703514	0.70343	1.63	9.40	0.1049	0.512993	8.3
95APo25b	Nunatak Fjord	amphibolite	3	144	0.06	0.703435	0.70339	4.19	14.73	0.1720	0.512840	5.3
95APo30	Nunatak Fjord	gabbro	5	245	0.06	0.705637	0.70559	3.22	9.24	0.2107	0.513007	8.6
95APo41b	Mount Stamy	leucosome	71	207	0.99	0.705983	0.70521	4.61	21.77	0.1280	0.512632	1.3
95APo47	Nunatak Fjord	amphibolite	4	97	0.12	0.703255	0.70316	3.50	11.62	0.1822	0.512959	7.6

Note: Values used for CHUR (chondritic uniform reservoir) are $^{143}Nd/^{144}Nd = 0.512638$; $^{147}Sm/^{144}Nd = 0.1967$. Decay constants: Sm = 6.54E – 12 yr^{-1}; Rb = 1.42E – 11 yr^{-1}. Sm and Nd concentrations determined by isotope dilution by addition of a mixed $^{149}Sm–^{150}Nd$ spike prior to sample dissolution. Rb and Sr concentrations determined by XRF at Washington State University. An age of 50 Ma was used to calculate initial Sr and ε_{Nd}.

*Corrected for mass fractionation by normalizing to $^{86}Sr/^{88}Sr = 0.1194$.

†Corrected for mass fractionation by normalizing to $^{146}Nd/^{144}Nd = 0.72190$.

§ $\varepsilon_{Nd}(T) = \{[^{143}Nd/^{144}Nd(T)sample/^{143}Nd/^{144}Nd(T)CHUR]-1\} \times 10,000$.

the Van Cleve area as the amphibole composition has probably re-equilibrated below the solidus. Two samples from Novayak Nunatak area yielded high pressures of ~800 MPa, which are close to the estimated pressure from the metamorphic aureole (Burner, 1996, personal commun.). Two samples from Nunatak Fjord yield pressures close to ~550 MPa, close to that estimated for adjacent amphibolites (Sisson et al., 1997).

The pressure of crystallization for the gabbro from Johns Hopkins Inlet in Glacier Bay is ~320 MPa. This is similar to pressures calculated for deformed plutons within the Border Ranges fault zone (Smart et al., 1996).

Geochronology

Samples for U/Pb geochronology were collected from the Novatak Glacier and Mt. Draper plutons in the Nunatak Fjord region and from synmetamorphic Ti_2 and Ti_3 intrusions near Van Cleve Glacier at the western end of the complex (Fig. 2). In addition, one sample from a gabbroic pluton crosscutting the Border Ranges fault zone in Glacier Bay was analyzed. Hornblende from a diorite (from the Boundary-Fairweather block) and an amphibolite were processed for $^{40}Ar/^{39}Ar$ geochronology. In addition, one sample was analyzed by fission track methods to constrain the late-stage cooling history. Sample locations are given in Table 5.

The morphology of the zircon grains ranges from equant to elongate and grains display a variety of internal structures when imaged by cathodoluminescence (CL; Fig. 8). In some of the samples, zircon grains have a dark, subhedral core surrounded by straight-edged oscillatory zoning, suggesting an inherited core overgrown by magmatic zircon (e.g., Fig. 8A). The zoning may also be highly irregular with patchy development. Some grains have no obvious inherited core (Fig. 8C). By comparing the CL with X-ray dot maps of various trace elements, darker areas in CL correspond to higher U concentration, whereas lighter areas are richer in Hf (compare Fig. 8D with 8E).

Monazite from the synmetamorphic Ti_2 sill (94APo36) at Van Cleve Glacier gives an age of ca. 54 Ma (Table 6). Since the currently accepted U-Pb closure temperature for monazite is 725 ± 25 °C (Parrish, 1990; J.E. Wright, 1996, personal commun.), we interpret this to be the crystallization age for the pluton. Xenotime from the same sample yielded a slightly younger age of 49.0 ± 2.6 Ma. Zircon from this sample has an inherited component (Table 6). Due to analytical difficulties as well as the presence of an inherited component (see Fig. 8A), the age of a nearby synmetamorphic Ti_3 intrusion (94APo47) is not well constrained.

Three multigrain fractions from Novatak Glacier (95APo13) define a concordia line with a lower concordia intercept of 49 ± 7.3 Ma and upper concordia intercept of 571 Ma (Fig. 9). Clustering of data points near the lower intercept causes the large uncertainty for the age. We are confident that this lower intercept age is reasonable as it fits with the trend of decreasing age of forearc magmatism documented along the southern Alaskan margin (e.g., Bradley et al., this volume, Chapter 1). In this sample, there were two populations of zircon grains; one was stubby with inherited cores (Fig. 8B) and the other elongate with no inherited core (Fig. 8C). The lower intercept is interpreted to represent the crystallization age of this undeformed sample. We interpret the upper age as an Early Paleozoic inherited component. In contrast, the Mt. Draper pluton (95APo21) plots with a lower concordia intercept of 54 Ma and upper concordia intercept of ca. 2.2 Ga (Fig. 9). These grains had U-rich zones (Figs. 8D and 8E) as well as inherited cores. Again, the lower intercept is interpreted as the crystallization age with the upper intercept defining a Precambrian inherited component. The inherited component ages are similar to those for detrital zircons in metasedimentary rocks of the Taku terrane of the northern Coast Mountains. Gehrels et al. (1991) report Cambrian zircon grains in a quartzite near Port Haughton and Archean zircon grains in quartzites from Tracy and Endicott Arms. Recent studies by Gehrels (2000) and Gehrels and Boghossian (2000) report other occurrences of detrital zircons with similar ages throughout the Coast Mountains from Bella Coola, British Columbia, to Skagway, Alaska. The presence of inherited zircons with ages similar to those in the Coast Mountains supports the hypothesis that this was the source area for

TABLE 5. LOCATION OF SAMPLES ANALYZED FOR GEOCHRONOLOGY

Sample	Method	Latitude	Longitude
94APo9	Ar/Ar hornblende	59°50'29"N	139°10'57"W
94APo27	Ar/Ar hornblende	59°51'01"N	139°01'33"W
94APo36	U/Pb zircon, monazite, and xenotime	60°45'49"N	144°11'23"W
94APo47	U/Pb zircon	60°45'45"N	144°12'43"W
95APo13	U/Pb zircon	59°36'02"N	138°32'32"W
95APo21	U/Pb zircon	59°50'29"N	139°11'01"W
89GP19b	U/Pb zircon	58°54'32"N	136°57'53"W
91ASn105	fission track zircon and apatite	60°51'45"N	144°13'38"W

Figure 8. Cathodoluminescence (CL) and uranium X-ray dot image of zircon grains analyzed for geochronology. Changes in intensity of the CL shown as grayscale changes representative of chemical variations during crystallization of the zircon as well as late fractures. A: Sample 94Apo47 from a synmetamorphic Ti_3 near Van Cleve Glacier (scale bar is 50 μm). B and C: Sample 95AP013 from Novatak Nunatak (scale bar is 50 μm for B and C is a composite of two images both with scale bar of 20 μm). D and E: sample 95APo21 from the Mt. Draper pluton (scale bar is 20 μm for D, and only 10 μm for E). The box in D shows the location for uranium X-ray dot image (E). The lighter colors correspond to higher amounts of U. The darker CL variation seems to correspond to changes in U concentration shown in E. The lighter areas are richer in Hf and possibly other trace elements. No other trace elements were detected in this zircon grain. The U X-ray dot map reflects changes in concentration from 0.0 to 0.05 wt% UO_2.

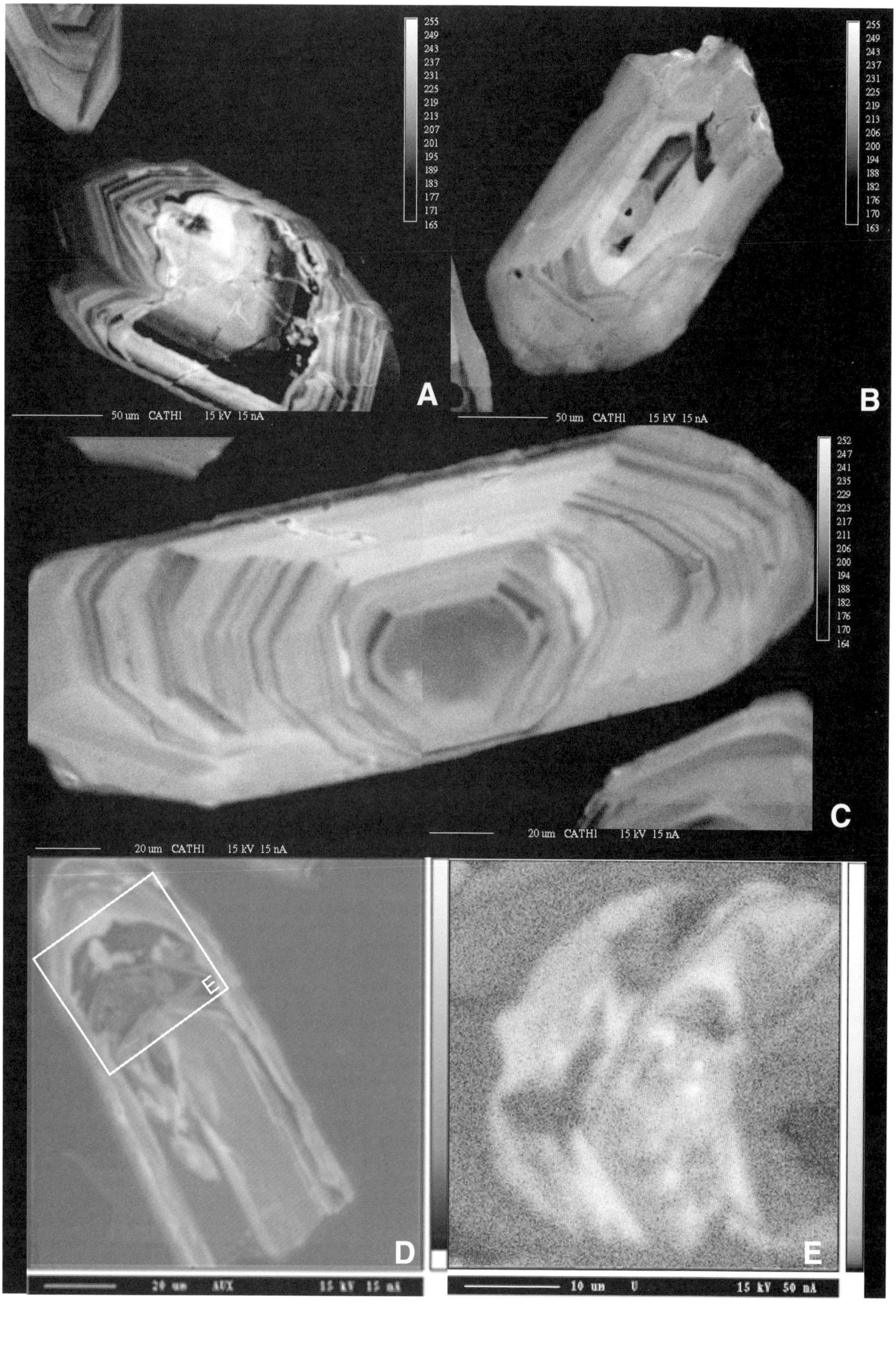
A
50 um CATHl 15 kV 15 nA
B
50 um CATHl 15 kV 15 nA
C
20 um CATHl 15 kV 15 nA
20 um CATHl 15 kV 15 nA
D
E
20 um AUX 15 kV 15 nA
10 um U 15 kV 50 nA

TABLE 6. U-Pb ANALYTICAL DATA AND AGES OF ZIRCON, MONAZITE, AND XENOTIME

Sample[†]	U (ppm)	^{206}Pb[§] (ppm)	Measured ratios[#]			Atomic ratios			Ages (Ma.)[††]		
			$^{206}Pb/^{204}Pb$	$^{207}Pb/^{206}Pb$	$^{208}Pb/^{206}Pb$	$^{206}Pb^*/^{238}U$	$^{207}Pb^*/^{235}U$	$^{207}Pb^*/^{206}Pb^*$	$^{206}Pb^*/^{238}U$	$^{207}Pb^*/^{235}U$	$^{207}Pb^*/^{206}Pb^*$
94APo36 xenotime +100	9482	67.8	30932	0.04747	0.20103	0.00832	0.05390	0.04700	53.4	53.3	49 ± 4.4
94APo36 xenotime 100–150	11363	80.0	17675	0.04783	0.40056	0.00819	0.05310	0.04700	52.6	52.5	49 ± 2.9
94APo36 xenotime 150–210	10570	74.0	8670	0.04863	0.90062	0.00815	0.05273	0.04693	52.3	52.2	46 ± 2.6
94APo36 monazite +100	2497	18.6	1792	0.05405	7.11484	0.00869	0.05490	0.04582	55.8	54.3	—
94APo36 monazite 100–150	2673	19.9	1413	0.05648	6.57194	0.00867	0.05501	0.04604	55.6	54.4	—
94APo36 monazite 150–210	2562	19.0	1270	0.05757	5.98008	0.00861	0.05459	0.04596	55.3	54.0	—
94APo36 zircon 100–150	1426	11.6	5568	0.05262	0.06611	0.00948	0.06536	0.04998	60.9	64.3	194 ± 2.0
94APo36 zircon 150–210	1589	12.9	6453	0.05258	0.08803	0.00942	0.06533	0.05031	60.4	64.3	209 ± 2.1
94APo36 zircon –210	1797	14.2	25964	0.05139	0.07023	0.00920	0.06449	0.05083	59.1	63.5	233 ± 1.6
94APo47 zircon 100–150	1240	10.3	1614	0.05786	0.07626	0.00965	0.06485	0.04876	61.9	63.8	136 ± 3.6
94APo47 zircon 150–210	844	7.3	1415	0.06187	0.08985	0.01007	0.07152	0.05152	64.6	70.1	264 ± 4.2
95APo13 zircon 100–150	438	3.2	2732	0.05374	0.10639	0.00855	0.05700	0.04836	54.9	56.3	117 ± 2.5
95APo13 zircon 150–200	349	2.6	1605	0.05772	0.10944	0.00862	0.05773	0.04856	55.3	57.0	127 ± 3.5
95APo13 zircon <210	88	0.7	126	0.16580	0.38959	0.00943	0.06413	0.04934	60.5	63.1	164 ± 41.7
95APo21 zircon 100–150	3338	25.1	763	0.06796	0.08488	0.00876	0.05879	0.04869	56.2	58.0	133 ± 7.2
95APo21 zircon 150–210	2364	18.2	1185	0.06313	0.07126	0.00898	0.06282	0.05076	57.6	61.9	230 ± 4.7
95APo21 zircon <210	2916	21.9	2207	0.05659	0.06019	0.00874	0.06020	0.04994	56.1	59.4	192 ± 2.7
89GP19b zircon (80)	725	6.0	2342	0.05336	0.44718	0.00652	0.04225	0.04698	41.8	42.0	48 ± 4
89GP19b zircon (100)	537	4.4	1646	0.05608	0.43064	0.00654	0.04241	0.04705	41.9	42.2	52 ± 4
89GP19b zircon (70)	634	5.4	1904	0.05468	0.47887	0.00653	0.04220	0.04686	41.9	42.0	42 ± 5
89GP19b zircon (70)	567	4.8	2015	0.05427	0.47716	0.00653	0.04224	0.04689	41.9	41.9	43 ± 4

Note: 94APo samples analyzed by Poole at Rice University; 89GP19b analyzed by McClelland at the University of California at Santa Barbara.

[†] +100, 100–210, etc., refer to size fractions in mesh.

[§] Denotes radiogenic Pb, corrected for common Pb using the isotopic composition of $^{206}Pb/^{204}Pb$ = 18.5 and 18.6, $^{207}Pb/^{204}Pb$ = 15.6 and 15.5 for samples 95APo samples and 89GP19b, respectively. Sample dissolution and ion exchange chemistry modified from Krogh (1973).

[#] Isotopic compositions corrected for a mass fractionation of 0.11% for 94APo samples and 0.125% for 89GP196. Uncertainties in the $^{208}Pb/^{206}Pb$ and $^{207}Pb/^{206}Pb$ ratios are on the order of 0.02% to 0.05%. $^{206}Pb/^{204}Pb$ errors range from about 0.1% to 10% depending on the ratio (all uncertainties are at the two-sigma level).

[††] Ages calculated using the following constants: decay constants for ^{235}U and ^{238}U = 9.8485E-10 yr^{-1}, and 1.55125E-10 yr^{-1}, respectively; $^{238}U/^{235}U$ = 137.88. Precision on $^{206}Pb^*/^{238}U$ ratios are about 0.2% to 0.3% based on the replicate analysis of two "standard" zircon fractions. Accuracy of the $^{206}Pb^*/^{238}U$ dates, including uncertainties in spike calibration are estimated to be on the order of 0.5%. The two-sigma uncertainties in the $^{207}*Pb/^{206}*Pb$ ages were calculated from the combined uncertainties in mass spectrometry and an assumed uncertainty of ±0.1 in the $^{207}Pb/^{204}Pb$ ratio used for the common correction (Mattinson, 1987) and is consistent with Ludwig (1991). For sample 89GP19b, the $^{207}*Pb/^{206}*Pb$ and $^{206}*Pb/^{238}U$ ratios adjusted assuming a 75% ± 25% efficiency in ^{230}Th exclusion during zircon crystallization (McClelland and Mattinson, 1996).

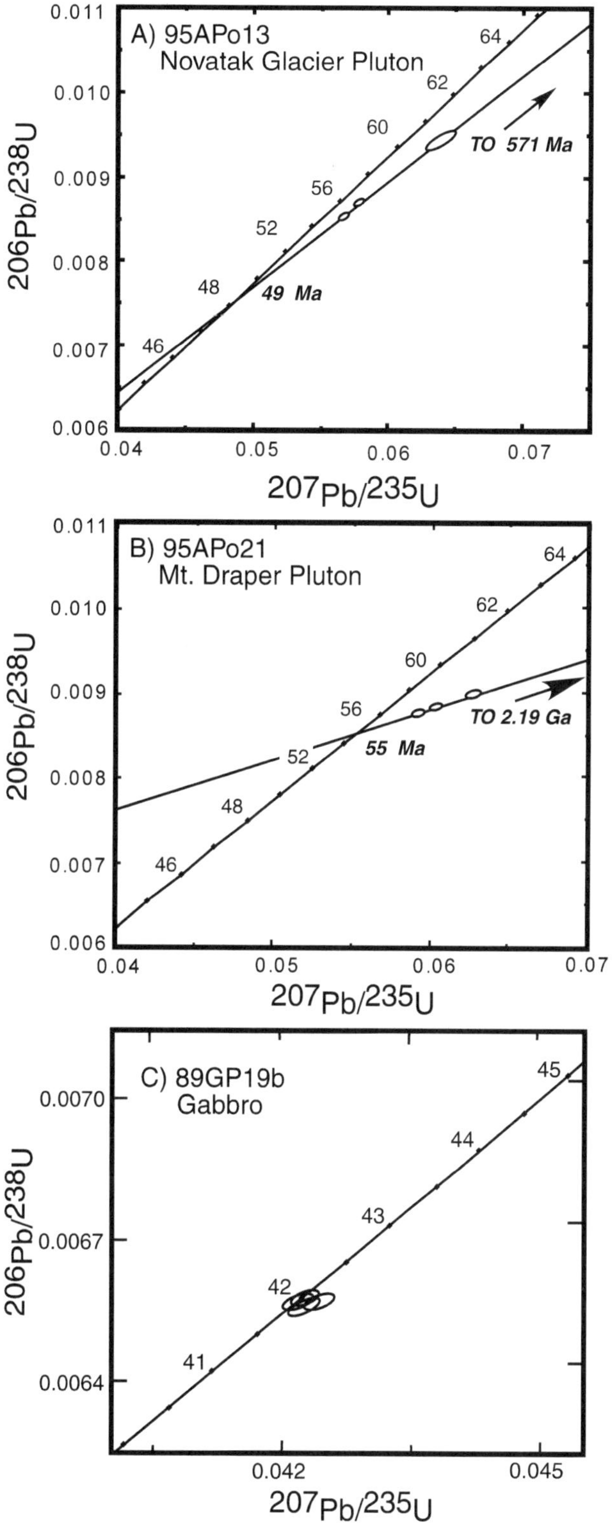

Figure 9. Concordia diagram of zircon fractions from Nunatak Fjord and Glacier Bay. A: Novatak Glacier pluton (95APo13). B: Mt. Draper pluton (95APo21). C: Unnamed pluton from Johns Hopkins Inlet, Glacier Bay (89GP19b). Also shown are arrows pointing toward the upper intercept and error ellipses for each analysis.

the Valdez Group and that these zircons were inherited during the Eocene magmatism.

Four multigrain fractions from the gabbro in Johns Hopkins Inlet (89GP19b) yielded concordant ages of 42 ± 0.1 Ma. This sample has no inherited component. This is substantially younger than the 58.6 ± 5.2 Ma K-Ar age previously determined by Decker and Plafker (1982). The older K-Ar age probably reflects excess Ar incorporated in the orthopyroxene and clinopyroxene cores of hornblende. The hornblende also has a low K_2O percent by weight that would make it difficult to analyze.

$^{40}Ar/^{39}Ar$ geochronology was performed on a hornblende sample from a diorite (94APo9) and two size fractions of hornblende from an amphibolite (94APo27). Two size fractions of hornblende from the Chugach block amphibolite (94APo27) gave indistinguishable results, which we interpret as reflecting cooling through ~500 °C at ca. 20 Ma (Table 7 and Fig. 10). These are interpreted as metamorphic cooling ages. Hornblende from the Boundary block diorite (94APo9) yielded an age of 53.9 ± 0.7 Ma over the entire spectrum and 53.8 ± 0.4 Ma for the plateau (Fig. 10). Although they are located near each other, the timing of emplacement of this sample relative to the Mt. Draper pluton is not known, as no crosscutting relationships were observed between them. Nonetheless, based on having similar U/Pb and $^{40}Ar/^{39}Ar$ dates, they are broadly coeval.

The Goat Wallow tonalite (91ASn105) yielded a zircon fission track age of 38.1 ± 4.9 Ma and an apatite fission track age of 19.3 ± 1.7 Ma (Table 8). The zircon grains have calculated U contents ranging from 800 to 2000 ppm U, higher than most zircons. These results are indistinguishable from data obtained from a large tonalite pluton north of Bremner Glacier with a zircon age of 37.2 ± 3.0 Ma and an apatite age of 19.5 ± 2.9 Ma (C.W. Naesser, 1988, personal commun.). These two plutons are separated by ~40 km. It is also similar to results from an unnamed pluton about 125 km to the east with an apatite age of 20.9 apatite age of 20.9 ± 1.9 Ma and zircon fission track age of 25.2 ± 1.7 Ma (P. O'Sullivan and J. Murphy, 2001, personal commun.). Only 13 track-lengths could be measured in our sample. These have lengths of 9.5 to 15.5 μm with a mean of 13.06 ± 0.44 μm (Fig. 11). The apatite track-length data were modeled using the techniques of Ketcham et al. (1999) and indicate two changes in cooling rate at ca. 20 Ma and 5 Ma with essentially no cooling over this interval.

DISCUSSION

Tonalitic and trondhjemitic rocks of the western Chugach metamorphic complex and Nunatak Fjord region possess characteristics consistent with derivation from a mixture of two source rocks. In addition, this suite has some compositional characteristics similar to those of Archean gray gneisses as well as Phanerozoic high-Al tonalite-trondhjemite suites (Drummond and Defant, 1990; Martin, 1999). Drummond and Defant (1990) suggested that Phanerozoic high-Al tonalite-trondhjemite suites form where geothermal conditions locally mimic those found in

TABLE 7. AR/AR DATA FOR AMPHIBOLE

Step	Age	1σ	°C	%39	Cum. %39	%Rad.	Ca/K
94APo9 (weighted age from steps B-M: 53.8 ± 0.4 Ma; integrated age: 53.9 ± 0.7 Ma)							
A	86.1	5.5	750	14.9315	1.9	13.3	3.8639
B	48.5	1.2	850	8.3196	4.2	45.1	7.4815
C	52.1	0.6	925	8.9411	10.2	73.1	15.4048
D	52.0	0.4	975	8.9297	24.5	81.5	12.9178
E	52.3	0.4	1000	8.9739	29.7	88.8	12.2468
F	56.1	0.6	1025	9.6411	45.5	94.0	11.0989
G	62.4	1.0	1060	10.7536	48.0	94.7	11.2528
H	56.8	0.6	1090	9.7744	53.2	86.9	10.8266
I	54.9	0.5	1120	9.4329	63.9	87.5	11.0138
J	55.8	0.5	1160	9.5998	72.0	89.7	11.3594
K	55.6	0.5	1190	9.5641	80.8	92.4	11.4915
L	51.7	0.4	1250	8.8808	84.0	87.9	11.3765
M	52.3	0.7	1500	8.9804	100.0	59.5	11.8783
94APo27c (weighted age from steps D–L: 19.8 ± 0.4 Ma; integrated age: 20.0 ± 0.7 Ma)							
A	34.2	7.9	700	5.8516	1.8	5.5	6.3149
B	32.2	3.3	800	5.5034	3.2	18.8	3.5727
C	12.6	1.6	900	2.1434	5.6	17.2	12.9108
D	13.5	0.6	970	2.2944	15.1	27.1	17.1627
E	18.7	0.4	1000	3.1831	29.7	51.3	17.6714
F	19.9	0.4	1025	3.8460	41.2	64.7	17.5072
G	19.7	0.6	1050	3.3537	48.2	69.3	17.0735
H	20.2	2.3	1075	3.4374	49.6	59.2	18.5287
I	17.6	0.8	1100	2.9919	54.6	50.3	19.7162
J	19.2	0.4	1125	3.2620	75.1	62.4	19.3866
K	21.7	0.4	1175	3.6927	89.0	54.5	19.8726
L	23.1	0.6	1250	3.9291	96.9	62.2	19.3630
M	39.4	1.5	1500	6.7424	100.0	93.6	18.6219
94APo27f (weighted age from steps E–I: 20.7 ± 0.3 Ma; integrated age: 21.5 ± 1.8 Ma)							
A	49.0	10.3	750	9.8136	1.3	5.8	6.5194
B	36.5	3.2	850	7.2779	2.4	23.6	12.3072
C	20.9	1.6	925	4.1470	4.7	28.4	21.4548
D	16.4	0.4	975	3.2600	14.8	42.9	17.9211
E	19.9	0.5	1000	3.9612	24.0	53.3	18.3787
F	19.4	0.3	1040	3.8469	43.9	71.9	17.8446
G	23.7	0.5	1090	4.7103	51.6	73.6	18.1006
H	21.2	0.3	1120	4.2141	76.0	67.6	18.9085
I	21.0	0.4	1160	4.1655	90.5	61.7	18.7839
J	29.6	0.8	1190	5.8998	94.2	85.3	18.8530
K	17.7	1.1	1250	3.5133	97.8	49.0	18.9254
L	33.3	2.5	1500	6.6500	1000.0	26.1	18.5098

Note: J factor = 0.0003274 for APo9 and APo27c; J factor = 0.002590 for APo 27f

the Archean. This includes where young, relatively hot oceanic crust is subducted. This scenario could be applied to southern Alaska during the Eocene when the Kula-Farallon ridge was subducted. However, other possible sources for high-Al tonalite-trondhjemite suites include partial melting of either lower crustal or volcanic arc basement rocks (Barnes et al., 1996; Atherton and Petford, 1993; Johnson et al., 1997; Smithies and Champion, 2000). To test these hypotheses, partial melting calculations were done using mafic source compositions similar to MORB and mineral-liquid partition coefficients.

Melting of metasedimentary rocks alone could have produced a few of the pre- to synmetamorphic Ti_{1-3} samples, but it could not have been the only source for all magmas (Fig. 7). Barker et al. (1992) proposed that partial melting of the Valdez and Orca sediments could account for most of the magmas in the Prince William terrane. The only exception was dacite dikes that are the products of partial melting of metabasalts in the accretionary prism. The source of heat for partial melting of both the sediments and basalt may have been a ridge transform complex (Barker et al., 1992). In contrast, Harris et al. (1996) modeled the magma mixing using an average of the Chugach metasedimentary rocks (initial $^{87}Sr/^{86}Sr = 0.7061$, $\varepsilon_{Nd} = -1.9$, Sr = 304 ppm, Nd = 22 ppm) and values for MORB (initial $^{87}Sr/^{86}Sr = 0.703$, $\varepsilon_{Nd} = +10$, Sr = 250 ppm, Nd = 8 ppm) to construct a mixing curve shown in Figure 7. From this model, it is predicted that 50–85% of the MORB component is necessary to produce the post-metamorphic Ti_4 magmas with positive ε_{Nd}. Harris et al.'s (1996) isotopic trends are also similar to those recognized elsewhere in the Sanak-Baranof belt. Granodiorite from Sanak, Shumagin, and Kodiak Island has been modeled as the result of mixing of metasediment-derived melts and MORB-like magma (Hill et al., 1981; Reid, 1987). The isotopic results from the Nunatak Fjord amphibolites do not support a model invoking derivation of the felsic magmas solely by partial melting of a MORB-like amphibolite (Fig. 7). The initial Sr and ε_{Nd} of the amphibolite is similar to the late post-metamorphic (Ti_4) intrusives. Thus, these late intrusives may have some component derived from partial melting of amphibolite.

In their model, Harris et al. (1996) assumed that the metasedimentary end member underwent 100% melting. This is clearly not reasonable as partial melting experiments on metasedimentary rocks suggest that only a fraction of the rock (up to ~50%) may partially melt. In order to evaluate this, we used a distribution coefficient of 2.0 for Sr in plagioclase to create approximate mixing curves for variable amounts of partial melts of average Chugach metasediment (Fig. 7). Changing the amount of Sr in an end member changes the shape of the mixing curve dramatically; e.g., compare the case of 10% to 50% partial melt in Figure 7.

These isotopic results show that latest intrusives in the Chugach metamorphic complex (Ti_4) have the largest amount of MORB component. This is also evident in the REE systematics. Depletion in heavy rare earths is indicative of the presence of garnet in the source rock, due to the large variation in partition

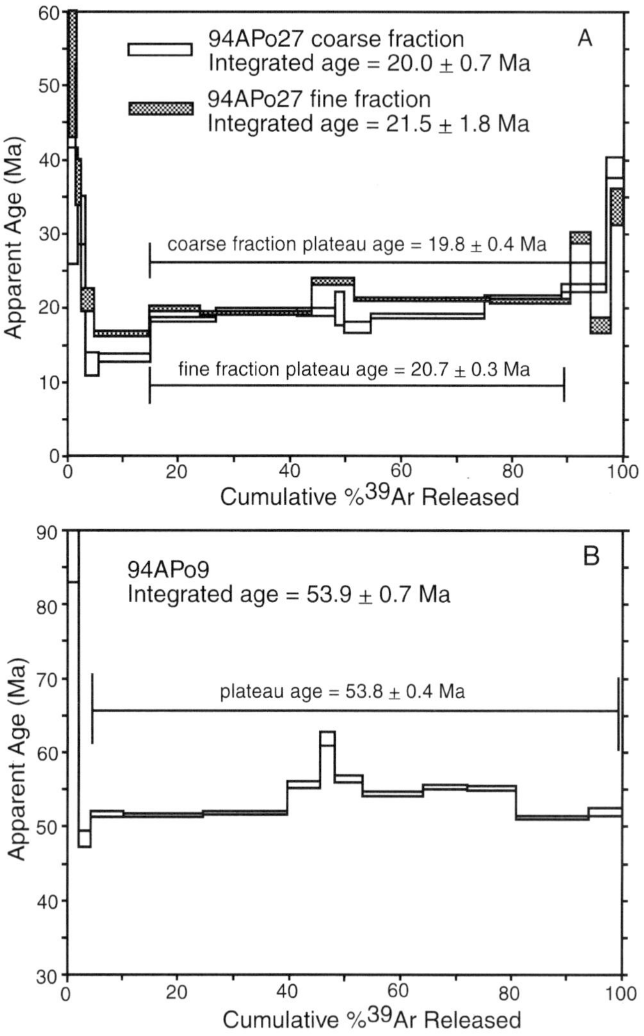

Figure 10. Hornblende $^{40}Ar/^{39}Ar$ incremental step heating spectra from Nunatak Fjord near Yakutat, Alaska. Both integrated age for the entire spectrum and plateau are given. A: Chugach block amphibolite (94APo27) with two size fractions (coarse [open lines] and fine [shaded lines]). B: Diorite in the Boundary-Fairweather block (94APo9).

coefficients between REEs (Rollinson, 1993). Garnet has been observed in the amphibolites; moreover, the concentrations of the lightest rare earths (La, Ce) in the amphibolites are similar to those of the post-metamorphic Ti_4s. We speculate that since the Ti_4s are the latest intrusions, the MORB component increased with time. An alternative hypothesis is that relative abundance of depleted mantle derived material (either previously underplated material or volcanic rocks within the accretionary prism sequence) may vary along strike of the margin and this is the primary factor in generating these magmas. Until more detailed studies are done, it is difficult to ascertain whether these formed from fractional crystallization of depleted primary magma or strictly from partial melting of a depleted mafic source region. Both processes could be involved in formation of these magmas.

The Nunatak Fjord region samples also follow the mixing trend recognized in the western Chugach metamorphic complex. The Novatak Glacier tonalite has a large (perhaps entirely) metasedimentary component, whereas the Mt. Draper granite and tonalite plot close to the post-metamorphic Ti_4 intrusives with 60–70% MORB component. This result is significant because the Mt. Draper pluton is a large, syntectonic felsic intrusive body with garnet and muscovite. This contrasts with the post-metamorphic Ti_4 intrusives, which are limited to dikes that postdate deformation. Since no geochronology exists for post-metamorphic Ti_4 dikes, it could be argued that they are unrelated to the earlier syntectonic intrusions (Ti_1–Ti_3), but there is no doubt that the Mt. Draper pluton is part of the Sanak-Baranof belt. Also, it is interesting that one amphibolite sample from Nunatak Fjord and an associated felsic dike plot close to the hypothetical mantle end-member. Mafic metavolcanic rocks are not common in the western Chugach metamorphic complex; they occur primarily in the southern edge of the complex (e.g., Lull and Plafker, 1990). The presence of extensive metamorphosed mafic metavolcanics, however, together with the higher metamorphic pressures at Nunatak Fjord suggest that these amphibolites may be the exhumed mafic end member that provided the MORB-like source for the felsic magmas.

Using geochemical characteristics, it is possible to correlate the Nunatak Fjord felsic intrusives to the intrusive phases of Harris et al. (1996). Several of the plutons (Novatak Glacier and Mt. Stamy) have geochemical characteristics similar to pre- to synmetamorphic T_{1-3}. Others, including Mt. Draper and a felsic dike, are similar to post-metamorphic Ti_4s. If this correlation is valid, the volume of magmas with a low amount of metasedimentary source material increases eastward. This trend is supported by the observation that late Ti_4 dikes are rare in the western Chugach metamorphic complex and common in the Tana Glacier area (e.g., Harris et al., 1996).

In the Boundary-Fairweather block, two mafic intrusives west of the Mt. Draper pluton (Fig. 3), a diorite (94APo9) and a gabbro (95APo30), have similar trace element compositions. Isotopically, the diorite plots close to the amphibolites; the gabbro has similar ε_{Nd} but altered initial Sr. It is likely that these rocks were derived from or represent the proposed mafic end member.

In contrast, gabbros (samples 94APo25a and 94APo25b) from "The Nunatak" are compositionally different. They may be related to magmas present during ridge-trench interaction as previous studies have found that a variety of mafic magmas were derived from a heterogeneous sub-ridge mantle. For example, Klein and Karsten (1995) and Sturm et al. (2000) found that some segments of present-day Chile ridge near the South American trench had signatures similar to convergent margin magmas and others had N-MORB and E-MORB characteristics. It seems likely that the variable geochemistry of the mafic magmas from Nunatak Fjord reflect a heterogeneous mantle from when the Kula-Farallon ridge interacts with North American continent.

TABLE 8: FISSION TRACK DATA FOR 91ASn105

Mineral	Altitude (feet)	No. of grains	Standard track density (x 10^5)	Spontaneous track density (x 10^5)	Induced track density (x 10^5)	P(χ^2)%	U (ppm)	FT age ±1σ (Ma)	Mean track length (µm)	Standard deviation (µm)
Apatite	4550	14	3.251	0.872	8.738	6.42	122	19.3 ± 1.7	13.06 ± 0.44	1.52
Zircon	4550	5	0.647	11.72	13.38	4.48	1298	38.1 ± 4.9		

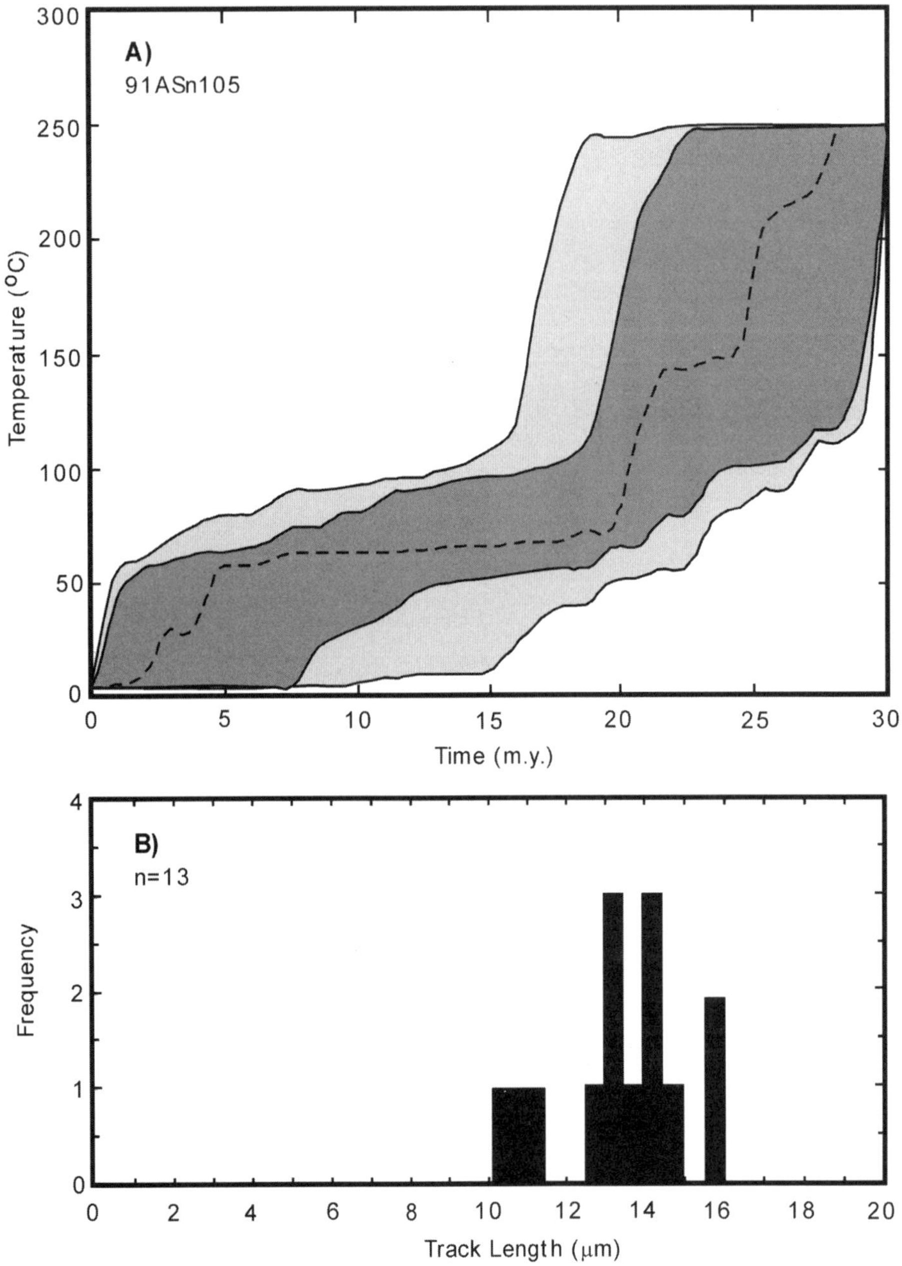

Figure 11. Fission track data and interpretation for apatite from tonalitic gneiss (sample 91ASn105). A: Time-temperature history estimated using methods of Ketcham et al. (1999). The dashed line is the best fit to the data. The dark shaded region shows the good model envelope and the light shaded region outlines the acceptable model envelope. B: Fission track length distribution for 13 tracks.

As described above, the amphibolites from the Nunatak Fjord region appeared to be transitional between island arc tholeiite and MORB. Elsewhere in the Chugach terrane, mafic rocks interbedded with sedimentary rocks have both MORB and primitive island arc affinity (Lull and Plafker, 1990; Crowe et al., 1992). The presence of MORB magmatism in the forearc has been documented in several regions of ridge-trench interaction including Chile (Kaeding et al., 1990; Guivel et al., 1999), western California (Johnson and O'Neil, 1984; Sharma et al., 1991; Cole and Basu, 1995; Leventhal et al., 1995), and in Japanese Shimanto accretionary prism (Hibbard and Karig, 1990b) and Hidaka magmatic zone (Maeda and Kagami, 1996).

Comparison with Igneous Rocks near Cordova and Seldovia

Barker et al. (1992) describe the evolution of felsic igneous rocks in the Prince William terrane near Cordova, about 80 km to the WSW of Van Cleve Glacier region. These felsic intrusives and associated gabbros were emplaced during the last phase of deformation of the Eocene Orca Group. The age constraints for these plutons range from 50 to 53 Ma K-Ar on biotite and hornblende. In general, these bodies are larger than those in the western Chugach metamorphic complex, yet have similar geochemical characteristics. They are peraluminous tonalites and granodiorites with LREE enrichment, high ε_{Nd}, and low $^{87}Sr/^{86}Sr$ ratios. These bodies only have biotite and amphibole in addition to quartz and feldspar; none have garnet and/or muscovite. This mineralogy is consistent with the geochemical data, as none of the Cordova region plutons plot as S-type magmas. The isotopic data overlap with those of both the Orca and Valdez metasedimentary rocks as well as pre- to synmetamorphic Ti_{1-3} from across the Chugach metamorphic complex. Only the McKinley Peak pluton has any indication of a mafic component. A late dacite dike from this region has similar isotopic values to late-stage dikes in the Chugach metamorphic complex. Basalt (N-MORB and E-MORB) interlayered with the Orca Group has interacted with fluids resulting in elevated $^{87}Sr/^{86}Sr$ compared to the postulated MORB end-member. These geochemical characteristics led Barker et al. (1992) to suggest that these plutons were predominantly derived from melted accretionary sediments. They estimated that 65–90% melting of the Orca Group graywacke, and argillite would be needed to produce the REE patterns of the McKinley Peak, Rude River, and Sheep Bay plutons. These plutons do fall on the same mixing trend as magmas of this study to the east; the McKinley Peak pluton in particular may have some mafic component.

Lytwyn et al. (2000) and Bradley et al. (this volume, Chapter 1) describe a suite of mafic to felsic dikes and felsic plutons within the Chugach accretionary prism about 150 km to the west near Seldovia on the Kenai Peninsula. The felsic dikes span a broader compositional range than the felsic plutons. The felsic dikes have very similar major and trace element chemistry to those reported here. Lytwyn et al. (2000) proposed that the wide range in igneous compositions in Seldovia represent mixtures of mafic parental melt with assimilated Chugach terrane sedimentary rock, coupled with fractional crystallization. This inferred parental source becomes more arc-like as the spreading ridge approaches the trench axis. They cannot, however, rule out a MORB-like parental liquid as a source for the Seldovia basaltic dikes. Some of the basaltic andesites have trace element spider diagrams that are similar to enriched or E-MORB (Bradley et al., this volume, Chapter 1), while others are similar to island arc volcanics (Lytwyn et al., 2000). Thus, there can be both island arc and MORB-like compositions in the same ridge system.

Geochronologic Data

The east-to-west, time-transgressive trend of the Sanak-Baranof plutonic belt shows that near-trench magmatism progressed from ca. 61 Ma at Sanak Island to ca. 50 Ma at Baranof Island (Haeussler et al., 1995; Bradley et al., this volume, Chapter 1). Even when the differences in closure temperature for the various dating methods are taken into account, the trend is very general and seems to suggest that there was some variation in the rate of spreading ridge migration or cooling history of the plutons.

The samples from the Chugach terrane—the Van Cleve and Novatak Glacier plutons—are in agreement with the general trend of Haeussler et al. (1995) for the Sanak-Baranof belt. If the spreading ridge moved the ~300 km distance between the Copper River and Nunatak Fjord between 54 Ma and 49 Ma, this results in a migration rate of ~6 cm/yr; this estimate will vary depending on interpretation of geochronologic error. This simple trend, however, is complicated by the plutonic ages from the Boundary-Fairweather block (ca. 54 Ma), which are older than the Novatak pluton. The plutons are now separated by the Neogene Fairweather fault, and the pre-dextral-strike-slip position of this block is critical to the tectonic significance of these ages. Specifically, restoration of the strike-slip motion moves the Boundary-Fairweather block southward toward Baranof Island and thus increases the gap between the two plutons. There is no evidence that there was ever left-lateral movement along the Fairweather fault. One possibility for this age jump is that a ridge segment migrated northwestward along the margin after Eocene plate reorganization (e.g., Sisson and Pavlis, 1993). This hypothesis is allowable if the Boundary-Fairweather block is not displaced >100 km, since the triple junction model predicted northward migration of ~200 km/m.y., which means it takes ~2 m.y. to close the gap and remain within analytical error. Nonetheless, the apparent older age of the Mt. Draper pluton suggests this model may be an oversimplification.

An alternative hypothesis put forth by Sisson and Pavlis (1993) is that the subduction of left- or right-stepping transform faults within the subducting plate could have caused the actual location of the triple junction to jump along the convergent margin, giving the appearance of rapid migration or even backtracking. Shifts such as these may have occurred along the entire length of the Sanak-Baranof belt.

A possible scenario involving the subduction of a transform under the Chugach metamorphic complex is illustrated in Figure 12 to explain occurrence of two 54 Ma plutons. As shown here, subduction of a right-stepping transform fault would create two slab windows, thus allowing for ridge-related magmatism in both regions at 54 Ma. Following Eocene plate reorganization, we infer a history similar to that suggested by Pavlis and Sisson (this volume, Chapter 8). That is, as dextral slip was initiated on the Border Ranges fault, the Chugach terrane was transported laterally along the margin at a higher velocity than the northward rate of the triple junction. As a result, the triple junction migrated relatively slowly southward (eastward in present coordinates) until ca. 48 Ma, when it reached Baranof Island. The magmatism continued until ca. 42 Ma, when the gabbro at Johns Hopkins Inlet, Glacier Bay, was emplaced. This final episode of magmatism may have been contemporaneous with emplacement of gabbronorite in Nunatak Fjord and La Perouse-Crillon layered mafic plutons.

When the tectonic model based on Chugach metamorphic complex geochronology in Figure 12 is compared with the geochemical data, a few trends are seen (cf. Table 9). In the earliest period of magmatism, there is a west to east trend from predominantly felsic plutons with variable amounts of mafic and metasedimentary components in the source region to a mixed mafic to felsic suite with a high percentage of mafic source material in the east. Subsequently, the mafic and felsic intrusives west of the Copper River and felsic intrusives in the western Chugach metamorphic complex have high amounts of mafic source region, whereas to the east there are only felsic intrusives with primarily metasedimentary source characteristics. We speculate that subduction of a ridge segment may cause more mafic material to be underplated below the accretionary prism. This hypothesis is supported by observations elsewhere that if transforms are being subducted, there is no apparent change in the related island arc volcanism. In contrast, in southern Chile, where there is active ridge subduction, there is recent obduction of an ophiolite sequence (e.g., Forsythe et al., 1986). In other regions, the mafic material may be underplated instead of obducted as an ophiolite. Then, after plate reorganization, when the relative positions of transform offsets and ridges change, this will affect the type of material to be partially melted. If the melting occurs in a region that previously had a large amount of mafic material underplated, then the resultant magmas may have a higher percentage of MORB source characteristics. This may account for some of the lateral variation in source rocks in the Chugach metamorphic complex. In support of this hypothesis is the fact that the sequence of magma emplacement is reversed between the blocks in the Nunatak Fjord region: MORB-rich source rocks here formed the first generation of magmatism, while more metasedimentary source-rich intrusions came later. Field evidence in Johns Hopkins Inlet, Glacier Bay, indicates this may be the same sequence since the gabbro is crosscut by felsic dikes. Thus, the intrusives may record different source rocks depending on position of the region relative to the ridge-transform offset before plate reorganization.

The post-intrusive cooling history of the Chugach metamorphic complex is very different from the west near Copper River to the Nunatak Fjord region. In the west, initial rapid cooling to biotite closure temperature (Sisson et al., 1989; Onstott et al., 1989) was followed by a period of slow cooling of ~3 °C/m.y. That is, ~30 m.y. elapsed between biotite Ar and zircon fission track closure temperature to apatite fission track closure temperature at 19 Ma. This cooling history is representative across about 125 km of this belt. This matches the overall topography of the Chugach Mountains and Granite Range, as the highest peaks in this region are all approximately the same elevation. Along strike in the Chugach metamorphic complex, in the Mt. Logan massif, there is an increase in topography and a change in cooling rate in the mid-Miocene at ca. 14 Ma (O'Sullivan and Currie, 1996). This event may be related to initial underplating of the Yakutat terrane at 19 Ma in the west and 14 Ma in the east. This was followed by a period of no cooling between 20 and 5 Ma, suggested by modeling of our apatite fission track data. O'Sullivan and Currie (1996) document rapid cooling in the Pliocene related to surface uplift of the Mt. Logan massif at ca. 5 Ma. Thus, final cooling and surface uplift in the western Chugach metamorphic complex may have been simultaneous with the Mt. Logan massif and a proposed change in relative motion between the Pacific and North American plate (Fitzgerald et al., 1995) and/or resistance to subduction of the Yakutat terrane proposed by Plafker et al. (1992).

In the Nunatak Fjord region, rocks now well east of the Fairweather fault record rapid Eocene cooling with regional correspondence of hornblende K-Ar ages (e.g., Hudson et al., 1977a) with the 49 Ma age of the Novatak Glacier pluton. Near the Fairweather fault, high-temperature cooling ages (~500 °C for hornblende $^{40}Ar/^{39}Ar$) are much younger; the 20 Ma reported here. This region also contains some of the youngest apatite fission track ages in North America at 2–5 Ma (O'Sullivan et al., 1997) suggesting that at least part of this cooling is a Neogene event. We suggest that the simplest explanation is that the young cooling ages along the Fairweather fault imply a regional east tilting of the region east of the Fairweather fault to exhume more deep-seated rocks that lag at temperatures above hornblende closure after initial Eocene cooling. Later Neogene deformation related to the Fairweather fault then uplifted and exhumed the rocks to their present position in the middle Miocene to Recent time.

CONCLUSIONS

Geochemical and isotopic data from the Nunatak Fjord region suggest that the mixing model of Harris et al. (1996) is representative of the entire Chugach metamorphic complex. Previously, the existence of a slab window-derived mafic magma end member was only predicted by geochemistry and by geophysical studies (e.g., Fisher et al., 1989; Fuis et al.,

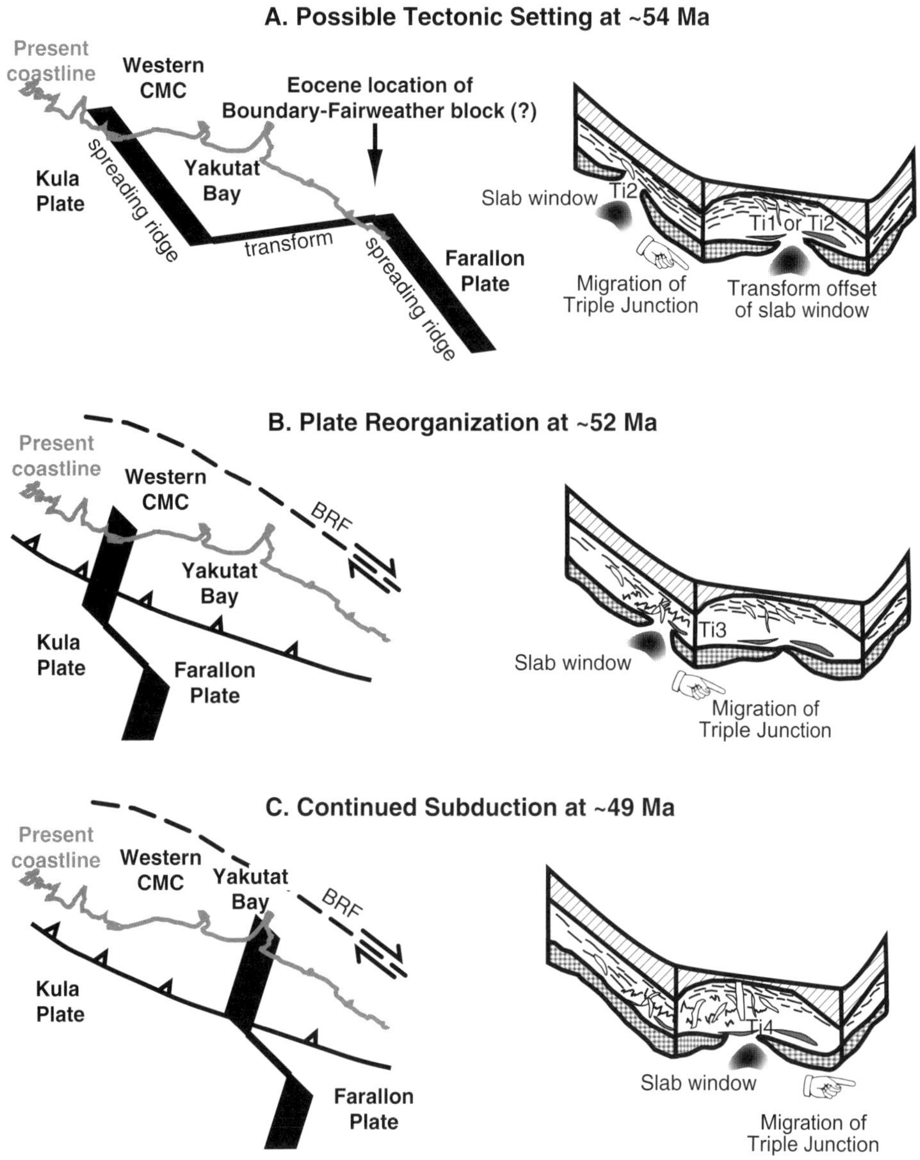

Figure 12. Possible tectonic reconstruction for the Chugach metamorphic complex (CMC) and Sanak-Baranof igneous belt between 54 and 49 Ma involving subduction of a ridge transform. A: Probable position of the subducted ridge at ca. 54 Ma, the time of emplacement of the Van Cleve Glacier pluton (94APo36) in the western Chugach metamorphic complex and of the Mt. Draper pluton (95APo21) and a diorite (94APo9) in the Boundary-Fairweather block at Nunatak Fjord/Yakutat Bay. The location of the Boundary-Fairweather block at 54 Ma is hypothetical because the amount of post-Eocene dextral strike slip along the Fairweather fault at Yakutat Bay is unknown. Subduction of a transform would create a micro-plate with two triple junctions, giving the appearance of a sudden "jump" in ridge-related magmatism within the Sanak-Baranof belt. The schematic block diagram shows positions of slab windows and associated generation of different intrusions (Ti_1 and Ti_2) relative to deformation and metamorphism. This diagram only shows the Chugach metamorphic complex without any forearc slivers that are postulated by Pavlis and Sisson (this volume, Chapter 8). The checked pattern represents subducted oceanic slab with gaps representing slab windows; dark gray areas are mafic underplating above the slab window; the diagonal ruled region is greenschist and lower-grade Valdez Group; the white area with schematic foliation is the amphibolite facies gneiss; and the igneous patterned areas represent different generations of felsic magmas. B: After plate reorganization at ca. 52 Ma, the relative positions of the ridges and transforms are fixed by the plate velocity and therefore undergo an apparent rotation (see Pavlis and Sisson, 1995). Thus, only one locus of magmatism would move down the coast in a systematic fashion. In the schematic block diagram, there is only one slab window that generates synmetamorphic (Ti_3) intrusives. C: Possible setting at ca. 49 Ma, the time of emplacement of the Novatak Glacier pluton in the Chugach block at Nunatak Fjord near Yakutat Bay. The schematic block diagram shows that the slab window has moved south along the coast, generating the latest magmas (Ti_4) only in the southern and eastern portions of the Sanak-Baranof belt. In this scenario, it is possible to partially melt previously underplated depleted mantle material.

TABLE 9. DIFFERENCES IN MAGMATIC HISTORY IN EASTERN CHUGACH MOUNTAINS

West of Copper River	Western Chugach metamorphic complex	Yakutat Chugach Block	Yakutat Boundary–Fairweather Block
Ti_1 0–20% MORB Large plutons	Ti_1 0–75% MORB Small dikes and sills only	Hypabyssal felsic dikes	Granite with 60–100% MORB Related mafic intrusives
	Ti_2 0–75% MORB Dikes and plutons		
	Ti_3 0–75% MOBB Dikes and plutons		
Ti_2 80% MORB Dacite and basaltic dikes	Ti_4 60–100% MORB Small dikes	Ti_2 Tonalite with 0–10% MORB Large plutons	Ti_2 Tonalite with ~40% MORB

1991) that indicated a high-velocity layer underlying the complex. The metabasalt and amphibolite in Nunatak Fjord with MORB affinities confirms that such an end member actually exists within the complex. The fractional amount of mafic and metasedimentary source rock varies from pluton to pluton. The Novatak Glacier pluton, for instance, may have been derived exclusively from the accretionary wedge, whereas the Mt. Draper pluton and the post-metamorphic Ti_4 intrusive bodies in the west isotopically appear to have a large MORB component (≥60%). The amount of MORB component appears to increase toward the east as the relative abundance of post-metamorphic intrusives increases to the east ending at the large Mt. Draper pluton. The relative change of source rocks may be related to the presence and absence of mafic underplated material along the margin.

If the Van Cleve Glacier and Novatak Glacier plutons were emplaced by the same spreading ridge segment, it can be said that the location of the slab window moved from the western end of the complex to the Chugach block part of the Nunatak Fjord region between 54 and 49 Ma, at a rate of ~6 cm/yr. Concordance in the ca. 54 Ma age of the Van Cleve and Mt. Draper plutons, however, imply important complications at this time, including ridge-transform offsets, local ridge backtracking, or both, during this plate reorganization period. Future analysis of structural data throughout the complex as well as further geochronological studies of ridge-related plutons, would help to support or disprove this model.

ACKNOWLEDGMENTS

We would like to thank Jim Wright, Jianping Wen, Michael Quinn, and Mary Anne Brown for their assistance with the U-Pb and whole rock isotope analyses; Terry Spell, University of Houston, for assistance with the $^{40}Ar/^{39}Ar$ dating; Diane Johnson and Charles Knaack at Washington State University for the major and trace element analyses; Liz Johnson for field assistance; Milton Pierson, Sarah Roeske, Jill Weinberger, and Becky Kessler for microprobe assistance; and Bill Leeman for advice. This work was supported by National Science Foundation grants EAR-9304062 and EAR-9020391 with an REU supplement to VBS and EAR-9307169 and EAR-9018717 to TLP, Oregon State University Reactor Sharing Grants to VBS and ARP, Geological Society of America grants to ARP and HCB, and Sigma Xi Grant in Aid of Research 22638 to ARP.

REFERENCES CITED

Anders, E., and Ebihara, M., 1982, Solar system abundances of the elements: Geochimica et Cosmochimica Acta, v. 26, p. 2363–2380.

Anderson, J.L., and Smith, D.R., 1995, The effects of temperature and $f(O_2)$ on the Al-in-hornblende barometer: American Mineralogist, v. 80, p. 549–559.

Atherton, M.P., and Petford, N., 1993, Generation of sodium-rich magmas from newly underplated basaltic crust: Nature, v. 362, p. 144–146.

Atwater, T.M., 1989, Plate tectonic history of the northeast Pacific and western North America, *in* Winterer, E.L., et al., eds., The eastern Pacific Ocean and Hawaii: Boulder, Colorado, Geological Society of America, Geology of North America, v. N, p. 21–72.

Barker, F., 1979, Trondhjemite: Definition, environment and hypotheses of origin, *in* Barker, F., ed., Trondhjemites, dacites, and related rocks: New York, Elsevier Scientific Publishing, p. 1–12.

Barker, F., Farmer, G.L., Ayuso, R., Plafker, G., and Lull, J.S., 1992, The 50 Ma granodiorite of the eastern Gulf of Alaska: Melting in an accretionary prism in the forearc: Journal of Geophysical Research, v. 97, p. 6757–6778.

Barnes, C.G., Peterson, S.W., Kistler, R.W., Murray, R., and Kays, M.A., 1996, Source and tectonics implications of tonalite-trondhjemite magmatism in the Klamath Mountains: Contributions to Mineralogy and Petrology, v. 123, p. 40–60.

Beard, J.S., and Lofgren, G.E., 1991, Dehydration melting and water-saturated melting of basaltic and andesitic greenstones and amphibolites at 1, 3, and 6.9 kb: Journal of Petrology, v. 32, p. 365–401.

Bradley, D.C., Haeussler, P.J., and Kusky, T.M., 1993, Timing of early Tertiary ridge subduction in southern Alaska: U.S. Geological Survey Bulletin 2068, p. 63–177.

Bradley, D., Kusky, T., Haeussler, P., Goldfarb, R., Miller, M., Dumoulin, J., Nelson, S.W., and Karl, S., 2003, Geologic signature of early Tertiary ridge subduction in Alaska *in* Sisson, V.B., Roeske, S.M., and Pavlis, T.L., eds., Geology of a transpressional orogen developed during ridge-trench interaction along the North Pacific margin: Boulder, Colorado, Geological Society of America Special Paper 371, p. 19–49 (this volume).

Bunge, H.P., and Grand, S.P., 2000, Mesozoic plate-motion history below the northeast Pacific ocean from seismic images of the subducted Farallon slab: Nature, v. 405, p. 337–340.

Burtner, R.L., Nigrini, A., and Donelick, R.A., 1994, Thermochronology of Lower Cretaceous source rocks in the Idaho-Wyoming thrust belt: American Association of Petroleum Geologists Bulletin, v. 78, p. 1613–1636.

APPENDIX 1: SAMPLE LOCATIONS AND TECHNIQUES FOR GEOCHEMICAL ANALYSES

Sample	Location	Rock type	Quadrangle	Latitude	Longitude	Techniques
89ASn62c	Tana River	leucosome	BerGl D-5	60°47'17"	142°56'02"	I-R, IN-O
89ASn69	Tana River	Ti_3	BerGl D-5	60°47'09"	142°56'21"	I-R, IN-O
89ASn87c	Tana River	Ti_2	BerGl D-5	60°47'19"	142°43'57"	I-R, IN-O
91AHa26	Tana River	Ti_4	BerGl D-5	60°49'58"	142°44'43"	X-A
91AHa27	Tana River	Ti_4	BerGl D-5	60°46'57"	142°43'56"	X-A
91AHa31	Tana River	leucosome	BerGl D-5	60°47'13"	142°44'01"	X-A
91AHa32	Tana River	Ti_2	BerGl D-5	60°47'13"	142°43'48"	X-A
91AHa36	Tana River	Ti_2	BerGl D-5	60°47'09"	142°56'21"	X-A
91ASn61b	Tana River	leucosome	BerGl D-5	60°46'29"	142°40'27"	X-A
91ASn63	Tana River	leucosome	BerGl D-5	60°46'11"	142°40"05"	X-A
91ASn64	Tana River	Ti_3	BerGl D-5	60°46'22"	142°40'19"	X-A
91ASn72b	Tana River	Ti_2	BerGl D-5	60°49'24"	142°44'52"	X-A
91ASn83	Tana River	Ti_1	BerGl D-5	60°47'19"	142°43'57"	X-A
91AHa42	W of Tana	Ti_2	BerGl D-5	60°49'32"	142°50'03"	X-A
91AHa50	S of Tana	Ti_2	BerGl C-5	60°41'11"	142°43'14"	X-A
91ASn93	Goat Wallow Gl.	Ti_1	Cord D-1	60°50'59"	144°16'42"	X-A
91ASn105	Goat Wallow Gl.	gneiss	Cord D-1	60°51'45"	144°13'38"	not analyzed
91AHa2	Wernicke Gl.	Ti_3	Cord D-1	60°49'16"	144°13'58"	X-A
91AHa3	Wernicke Gl.	Ti_3	Cord D-1	60°49'19"	144°13'31"	X-A
91AHa6b	Wernicke Gl.	Ti_2	Cord D-1	60°49'28"	144°12'44"	X-A
91AHa12	Wernicke Gl.	Ti_2	Cord D-1	60°49'27"	144°14'19"	X-A
91AHa14c	Wernicke Gl.	Ti_3	Cord D-1	60°49'39"	144°14'41"	X-A
94APo36	Van Cleve	Ti_2	Cord D-1	60°45'49"	144°11'24"	I-W; X-W
94APo39	Van Cleve	Ti_2	Cord D-1	60°46'00"	144°11'44"	X-W; IN-O
94APo40	Van Cleve	Ti_3	Cord D-1	60°46'00"	144°13'28"	X-W; IN-O
94APo44a	Van Cleve	Ti	Cord D-1	60°45'26"	144°11'41"	X-W; IN-O
94APo47	Van Cleve	Ti_3	Cord D-1	60°45'05"	144°12'15"	I-W; X-W
94APo49	Van Cleve	Ti_1 or Ti_2	Cord D-1	60°45'35"	144°11'26"	X-W; IN-O
94ASn52	Van Cleve	Ti_3	Cord D-1	60°45'28"	144°09'46"	X-W; IN-O
94APo9	Nunatak Fjord	diorite	Yak D-4	59°50'29"	139°10'57"	I-W; X-W
94APo11b	Mt. Draper	Ti	Yak D-4	59°50'29"	139°10'57"	X-W; IN-O
94APo11c	Mt. Draper	Ti	Yak D-4	59°50'29"	139°10'57"	I-W; X-W
94APo19b	Nunatak Fjord	amphibolite	Yak D-3	59°48'15"?	138°53'25"?	X-W; IN-O
94APo21	Mt. Draper	Ti	Yak D-4	59°50'08"	139°07'53"	X-W; IN-O
94APo25a	Nunatak Fjord	pyroxenite	Yak D-3	59°48'54"	138°55'25"	I-W; X-W
94APo25b	Nunatak Fjord	gabbro	Yak D-3	59°48'54"	138°55'25"	I-W; X-W
94APo25c	Nunatak Fjord	gabbro dike	Yak D-3	59°48'54"	138°55'25"	X-W; IN-O
94APo27	Nunatak Fjord	amphibolite	Yak D-3	59°51'01"	139°01'33"	I-W; X-W
95APo21	Mt. Draper	Ti	Yak D-4	59°50'29"	139°11'02"	I-W; X-W
95APo41b	Mt. Stamy	Ti dike	Yak D-3	59°52'14"	139°06'40"	I-W; X-W
95APo13	Novayak	Ti	Yak C-2	59°36'02"	138°32'32"	I-W; X-W
95APo25a	Nunatak Fjord	Ti dike	Yak D-3	59°48'54"	138°55'25"	I-W; X-W
95APo25b	Nunatak Fjord	amphibolite	Yak D-3	59°48'54"	138°55'25"	I-W; X-W
95APo30	The Nunatak	metagabbro	Yak D-3	59°45'18"	138°55'28"	I-W; X-W
95APo47	Nunatak Fjord	amphibolite	Yak D-3	59°48'52"	138°55'00"	I-W; X-W
89GP19b	Glacier Bay	gabbro	Mt. F D-3	58°54'32"	136°57'53"	not analyzed

Note: Intrusives are classified as in Table 1. 1:63,360 Quadrangles include BerGl—Bering Glacier, Cord—Cordova, Yak—Yakutat, and Mt. F—Mount Fairweather. Techniques include: X-A—X-ray fluorescence spectrometry (XRF) and instrumental neutron activation analysis (INAA) at Activation Labs; I-R—inductively coupled plasma (ICP) at Rice University; X-W—XRF at Washington State University; I-W—inductively coupled plasma-mass spectrometry (ICP-MS) at Washington State University; IN-O—INAA at Oregon State University Reactor Center. B by PGNA at Activation Labs.

Byrne, T., 1979, Late Paleocene demise of the Kula-Pacific spreading center: Geology, v. 7, p. 341–344.

Campbell, R.B., and Dodds, C.J., 1978, Operation St. Elias, Yukon Territory, *in* Current research, part A: Geological Survey of Canada Paper 78-1A, p. 35–41.

Carlson, W.D., Donelick, R.A., and Ketcham, R.A., 1999, Variability of apatite fission track annealing kinetics I: Experimental results: American Mineralogist, v. 84, p. 1213–1223.

Cebula, G.T., Kunk, M.J., Mehnert, H.H., Naeser, C.W., Obradovich, J.D., and Sutter, J.F., 1986, The Fish Canyon Tuff, a potential standard for the $^{40}Ar/^{39}Ar$ and fission-track dating methods [abs.]: Terra Cognita, v. 6, p. 139.

Cole, R.B., and Basu, A.R., 1995, Nd-Sr isotopic geochemistry and tectonics of ridge subduction and middle Cenozoic volcanism in western California: Geological Society of America Bulletin, v. 107, p. 167–179.

Crowe, D.E., Nelson, S.W., Brown, P.E., Shanks, W.C., III, and Valley, J.W., 1992, Geology and geochemistry of volcanogenic massive sulfide deposits and related igneous rocks, Prince William Sound, south-central Alaska: Economic Geology, v. 87, p. 1722–1746.

Decker, J.E., and Plafker, G., 1982, Correlation of rocks in the Tarr Inlet suture zone with the Kelp Bay Group, *in* Coonrad, W.L., ed., The U.S. Geological Survey in Alaska: Accomplishments during 1980: U.S. Geological Survey Circular 844, p. 119–122.

Donelick, R.A., 1993, A method of fission track analysis utilizing bulk chemical etching of apatite: U.S. Patent no. 5,267,274.

Donelick, R.A., 1995, A method of fission track analysis utilizing bulk chemical etching of apatite: Australian Patent no. 658,800.

Donelick, R.A., and Miller, D.S., 1991, Enhanced TINT fission track densities in low spontaneous track density apatites using ^{252}Cf-derived fission fragment tracks: A model and experimental observations: Nuclear Tracks and Radiation Measurements, v. 18, p. 301–307.

Donelick, R.A., Ketcham, R.A., and Carlson, W.D., 1999, Variability of apatite fission track annealing kinetics II: Crystallographic orientation effects: American Mineralogist, v. 84, p. 1224–1234.

D'Orazio, M., Agostini, S., Innocenti, F., Haller, M.J., Manetti, P., and Mazzarini, F., 2001, Slab window-related magmatism from southernmost South America: The late Miocene mafic volcanics from the Estancia Glencross area (~52° S, Argentina-Chile): Lithos, v. 57, p. 67–89.

Drummond, M.S., and Defant, M.J., 1990, A model for trondhjemite-tonalite-dacite genesis and crustal growth via slab melting: Archean to modern comparisons: Journal of Geophysical Research, v. 95, p. 21,503–21,521.

Engebretson, D.C., Cox, A., and Gordon, R.G., 1985, Relative motions between oceanic and continental plates in the Pacific basin: Boulder, Colorado, Geological Society of America Special Paper 206, 59 p.

Farmer, G.L., Ayuso, R., and Plafker, G., 1993, A Coast Mountains provenance for the Valdez and Orca groups, southern Alaska, based on Nd, Sr, and Pb isotopic evidence: Earth and Planetary Science Letters, v. 116, p. 9–21.

Fisher, M.A., Brocher, T.M., Nokleburg, W.J., Plafker, G., and Smith, G.L., 1989, Seismic reflection images of the crust of the northern part of the Chugach transect, Alaska: Results of a survey for the Trans-Alaska Crustal Transect (TACT): Journal of Geophysical Research, v. 94, p. 4424–4440.

Fitzgerald, P.G., Sorkhabi, R.B., Redfield, T.F., and Stump, E., 1995, Uplift and denudation of the central Alaska Range: A case study in the use of apatite fission track thermochronology to determine absolute uplift parameters: Journal of Geophysical Research, v. 100, p. 20,175–20,191.

Foley, J.Y., Light, T.D., Nelson, S.W., and Harris, R.A., 1997, Mineral occurrences associated with mafic-ultramafic and related alkaline complexes in Alaska, *in* Goldfarb, R.J., and Miller, L.D., eds., Mineral deposits of Alaska: Economic Geology Monograph 9, p. 396–449.

Forsythe, R.E., Nelson, E., Carr, M., Kaeding, M., Herve, M., Mpodozis, C., Soffia, J., and Harambour, S., 1986, Pliocene near-trench magmatism in southern Chile: Possible manifestation of ridge collision: Geology, v. 15, p. 413–422.

Frost, B.R., Barnes, C.G., Collins, W.J., Arculus, R.J., Ellis, D.J., and Frost, C.D., 2001, A geochemical classification for granitic rocks: Journal of Petrology, v. 42, p. 2033–2048.

Fuis, G.S., Ambos, E.L., Mooney, W.D., Christensen, N.I., and Geist, E., 1991, Crustal structure of accreted terranes in southern Alaska, Chugach Mountains and Copper River Basin, from seismic refraction results: Journal of Geophysical Research, v. 96, p. 4187–4227.

Gehrels, G.E., 2000, Reconnaissance geology and U-Pb geochronology of the west flank of the Coast Mountains between Juneau and Skagway, southeastern Alaska, *in* Stowell, H.H., and McClelland, W.C., eds., Tectonics of the Coast Mountains, southeastern Alaska and British Columbia: Boulder, Colorado, Geological Society of America Special Paper 343, p. 213–234.

Gehrels, G.E., and Boghossian, N.D., 2000, Reconnaissance geology and U-Pb geochronology of the west flank of the Coast Mountains between Bella Coola and Prince Rupert, coastal British Columbia, *in* Stowell, H.H., and McClelland, W.C., eds., Tectonics of the Coast Mountains, southeastern Alaska and British Columbia: Boulder, Colorado, Geological Society of America Special Paper 343, p. 77–88.

Gehrels, G.E., McClelland, W.C., Samson, S.D., and Patchett, P.J., 1991, U-Pb geochronology of detrital zircons from a continental margin assemblage in the northern Coast Mountains, southeastern Alaska: Canadian Journal of Earth Sciences, v. 28, p. 1285–1300.

Goldfarb, R.J., 1997, Metallogenic evolution of Alaska, *in* Goldfarb, R.J., and Miller, L.D., eds., Mineral deposits of Alaska: Economic Geology Monograph 9, p. 4–34.

Gorring, M.L., and Kay, S.M., 2001, Mantle processes and sources of Neogene slab window magmas from southern Patagonia, Argentina: Journal of Petrology, v. 42, p. 1067–1094.

Groome, W.G., Thorkelson, D.J., Friedman, R.M., Mortensen, J.K., Massey, N.W.D., Marshall, D.D., and Layer, P.W., 2003, Magmatic and tectonic history of the Leech River Complex, Vancouver Island, British Columbia: Evidence for ridge-trench intersection and accretion of the Crescent Terrane, *in* Sisson, V.B., Roeske, S.M., and Pavlis, T.L., eds., Geology of a transpressional orogen developed during ridge-trench interaction along the North Pacific margin: Boulder, Colorado, Geological Society of America Special Paper 371, p. 327–353 (this volume).

Guivel, C., Lagabrielle, Y., Bourgois, J., Maury, R.C., Fourcade, S., Martin, G., and Arnoud, N., 1999, New geochemical constraints for the origin of ridge-subduction-related plutonic and volcanic suites from the Chile triple junction (Taitao Peninsula and Site 862, ODP Leg 141 on the Taitao Ridge): Tectonophysics, v. 311, p. 83–111.

Haeussler, P., Bradley, D., Goldfarb, R., Snee, L.W., and Taylor, C., 1995, Link between ridge subduction and gold mineralization in southern Alaska: Geology, v. 23, p. 995–998.

Hammarstrom, J.M., and Zen, E., 1986, Aluminum in hornblende: An empirical igneous geobarometer: American Mineralogist, v. 71, p. 1297–1313.

Harris, N.R., Sisson, V.B., Wright, J., and Pavlis, T.L., 1996, Evidence for Eocene mafic underplating during forearc intrusive activity, eastern Chugach Mountains, Alaska: Geology, v. 24, p. 263–266.

Hibbard, J.P., and Karig, D.E., 1990a, Alternative plate model for the early Miocene evolution of the southwest Japan margin: Geology, v. 18, p. 170–174.

Hibbard, J.P., and Karig, D.E., 1990b, Structural and magmatic responses to spreading ridge subduction: An example from southwest Japan: Tectonics, v. 9, p. 207–230.

Hill, M., Morris, J., and Whelan, J., 1981, Hybrid granodiorites intruding the accretionary prism, Kodiak, Shumagin, and Sanak Islands, southwest Alaska: Journal of Geophysical Research, v. 86, p. 10,569–10,590.

Himmelberg, G.R., Loney, R.A., and Nabelek, P.I., 1987, Petrogenesis of gabbronorite at Yakobi and Northwest Chichagof islands, Alaska: Geological Society of America Bulletin, v. 98, p. 265–279.

Hofmann, A.W., 1988, Chemical differentiation of the Earth: The relationship between mantle, continental and oceanic crust: Earth and Planetary Science Letters, v. 90, p. 297–314.

Hole, M.J., and Larter, R.D., 1993, Trench-proximal volcanism following ridge crest-trench collision along the Antarctic Peninsula: Tectonics, v. 12, p. 897–910.

Hole, M.J., Saunders, A.D., Rogers, G., and Sykes, M.A., 1994, The relationship between alkaline magmatism, lithospheric extension and slab window formation along continental destructive plate margins, *in* Smellie, J.L., ed., Volcanism associated with extension at consuming plate margins: Geological Society [London] Special Publication No. 81, p. 265–285.

Holland, T., and Blundy, J., 1994, Non-ideal interactions in calcic amphiboles and their bearing on amphibole-plagioclase thermometry: Contributions to Mineralogy and Petrology, v. 116, p. 443–447.

Hollister, L.S., Grissom, G.C., Peters, E.K., Stowell, H.H., and Sisson, V.B., 1987, Confirmation of the empirical correlation of Al in hornblende with pressure of solidification of calc-alkaline plutons: American Mineralogist, v. 72, p. 231–239.

Hudson, T., 1979, Mesozoic plutonic belts of southern Alaska: Geology, v. 7, p. 230–234.

Hudson, T., 1994, Crustal melting events in Alaska, *in* Plafker, G., and Berg, H.C., eds., The geology of Alaska: Boulder, Colorado, Geological Society of America, Geology of North America, v. G-1, p. 657–670.

Hudson, T., and Plafker, G., 1982, Paleogene metamorphism of an accretionary flysch terrane, eastern Gulf of Alaska: Geological Society of America Bulletin, v. 93, p. 1280–1290.

Hudson, T., Plafker, G., and Lanphere, M.A., 1977a, Intrusive rocks of the Yakutat–St. Elias area, south-central Alaska: U.S. Geological Survey Journal of Research, v. 5, p. 155–172.

Hudson, T., Plafker, G., and Lanphere, M.A., 1977b, Metamorphic rocks of the Yakutat–St. Elias area, south-central Alaska: U.S. Geological Survey Journal of Research, v. 5, p. 173–184.

Hudson, T., Plafker, G., and Peterman, Z.E., 1979, Paleogene anatexis along the Gulf of Alaska margin: Geology, v. 4, p. 117–132.

Hurford, A.J., and Hammerschmidt, K., 1985, $^{40}Ar/^{39}Ar$ and K/Ar dating of the Bishop and Fish Canyon Tuffs: Calibration ages for fission-track dating standards: Chemical Geology (Isotope Geoscience Section), v. 58, p. 23–32.

Johnson, C.M., and O'Neil, J.R., 1984, Triple junction magmatism: A geochemical study of Neogene volcanic rocks in western California: Earth and Planetary Science Letters, v. 71, p. 241–262.

Johnson, M.C., and Rutherford, M.J., 1989, Experimental calibration of the aluminum in hornblende geobarometer with application to Long Valley caldera (California) volcanic rocks: Geology, v. 17, p. 837–841.

Johnson, K., Barnes, C.G., and Miller, C.A., 1997, Petrology, geochemistry, and genesis of high-Al tonalite and trondhjemites of the Cornucopia stock, Blue Mountains, northeastern Oregon: Journal of Petrology, v. 38, p. 1585–1611.

Johnston, S.T., and Thorkelson, D.J., 1997, Cocos-Nazca slab window beneath Central America: Earth and Planetary Science Letters, v. 146, p. 465–474.

Kaeding, M., Forsythe, R.D., and Nelson, E.P., 1990, Geochemistry of the Taito ophiolite and near-trench intrusions from the Chile margin triple junction: Journal of South American Earth Sciences, v. 3, p. 161–177.

Kay, S.M., Ramos, V.A., and Marquez, M., 1993, Evidence in Cerro Pampa volcanic rocks for slab-melting prior to ridge-trench collision in South America: Journal of Geology, v. 101, p. 703–714.

Ketcham, R.A., Donelick, R.A., and Carlson, W.D., 1999, Variability of apatite fission track annealing kinetics III: Extrapolation to geological time scales: American Mineralogist, v. 84, p. 1235–1255.

Ketcham, R.A., Donelick, R.A., and Donelick, M.B., 2000, AFTSolve: A program for multi-kinetic modeling of apatite fission-track data: Geological Materials Research, v. 2, no. 1, 32 p.

Kiminami, K., Miyashita, S., and Kawabata, K., 1993, Active ridge-forearc collision and its geological consequences: An example from Late Cretaceous southwest Japan, *in* Komatsu, M., ed., Evolution of the arc crust in southwest Japan: Geological Society of Japan Memoir 42, p. 167–182.

Kinoshita, O., 1995, Migration of igneous activities related to ridge subduction in southwest Japan and the East Asian continental margin from the Mesozoic to the Paleogene: Tectonophysics, v. 245, p. 25–35.

Kinoshita, O., 1999, A migration model of magmatism explaining ridge subduction and its details on a statistical analysis of the granite ages in Cretaceous southwest Japan: The Island Arc, v. 8, p. 181–189.

Kinoshita, O., 2002, Possible manifestations of slab window magmatisms in Cretaceous southwest Japan: Tectonophysics, v. 344, p. 1–13.

Klein, E.M., and Karsten, J.L., 1995, Ocean-ridge basalts with convergent margin geochemical affinities from the Chile Ridge: Nature, v. 374, p. 52–57.

Kretz, R., 1983, Symbols for rock-forming minerals: American Mineralogist, v. 68, p. 227–229.

Krogh, T.E., 1973, A low-contamination method for hydrothermal decomposition of zircon and extraction of U and Pb for isotopic age determinations: Geochimica et Cosmochimica Acta, v. 37, p. 485–494.

Leventhal, J.A., Reid, M.R., Montana, A., and Holden, P., 1995, Mesozoic invasion of crust by MORB-source asthenospheric magmas, U.S. Cordilleran interior: Geology, v. 23, p. 399–402.

Loney, R.A., and Himmelberg, G.R., 1983, Structure of the La Perouse gabbro intrusion, Fairweather Range, southeastern Alaska: Journal of Petrology, v. 24, p. 377–423.

Lonsdale, P., 1988, Paleogene history of the Kula plate: Offshore evidence and onshore implications: Geological Society of America Bulletin, v. 100, p. 733–754.

Ludwig, K.R., 1991, ISOPLOT; a plotting and regression program for radiogenic-isotope data, version 2.53: U.S. Geological Survey, Open File 91-0445, 39 p.

Lull, J.S., and Plafker, G., 1990, Geochemistry and paleotectonic implications of metabasaltic rocks in the Valdez Group, southern Alaska, *in* Dover, J.H., and Galloway, J.P., eds., Geologic studies in Alaska by the U.S. Geological Survey, 1989: U.S. Geological Survey Bulletin 1946, p. 29–38.

Lytwyn, J., Lockhart, S., Casey, J., and Kusky, T., 2000, Geochemistry of near-trench intrusives with ridge subduction, Seldovia quadrangle, southern Alaska: Journal of Geophysical Research, v. 105, p. 27,957–27,978.

Maeda, J., and Kagami, H., 1996, Interaction of a spreading ridge and an accretionary prism: Implications from MORB magmatism in the Hidaka magmatic zone, Hokkaido, Japan: Geology, v. 24, p. 31–34.

Marshak, R.S., and Karig, D.E., 1977, Triple junctions as a cause for anomalous near-trench activity between the trench and volcanic arc: Geology, v. 5, p. 233–236.

Martin, H., 1999, Adakitic magmas: Modern analogues of Archean granitoids: Lithos, v. 46, p. 411–429.

Mattison, J.M., 1987, U-Pb ages of zircons—a basic examination of error propagation: Chemical Geology, Isotope Geoscience section, v. 66, p. 151–162.

McClelland, W.C., and Mattinson, J.M., 1996, Resolving high precision ages from Tertiary plutons with complex zircon systematics: Geochimica et Cosmochimica Acta, v. 60, p. 3955–3965.

Miyake, Y., 1985, MORB-like tholeiites formed within the Miocene forearc basin, southwest Japan: Lithos, v. 5, p. 233–236.

Montel, J.M., and Vielzeuf, D., 1997, Partial melting of greywackes, 2: Compositions of minerals and melts: Contributions to Mineralogy and Petrology, v. 128, p. 176–196.

Nilsen, T.H., and Zuffa, G.G., 1982, The Chugach Terrane, a Cretaceous trench-fill deposit, southern Alaska, *in* Leggett, J.K., ed., Trench-forearc geology: Sedimentation and tectonics of modern and ancient active plate margins: Oxford, UK, Blackwell, p. 213–227.

Onstott, T.C., Sisson, V.B., and Turner, D., 1989, Initial argon in amphiboles from the Chugach Mountains, southern Alaska: Journal of Geophysical Research, v. 94, p. 4361–4372.

Osozawa, S., 1992, Double ridge subduction recorded in the Shimanto accretionary wedge complex, Japan, and plate reconstruction: Geology, v. 20, p. 939–942.

O'Sullivan, P.B., and Currie, L.D., 1996, Thermotectonic history of Mt. Logan, Yukon Territory, Canada: Implications of multiple episodes of middle to late Cenozoic denudation: Earth and Planetary Science Letters, v. 144, p. 251–261.

O'Sullivan, P.B., Plafker, G., and Murphy, J.M., 1997, Apatite fission-track thermotectonic history of crystalline rocks in the northern Saint Elias Mountains, Alaska, *in* Dumoulin, J.A., and Gray, J.E., eds., Geological studies in Alaska by the U.S. Geological Survey in 1995: U.S. Geological Survey Professional Paper1574, p. 283–293.

Parrish, R.R., 1990, U-Pb dating of monazite and its application to geological problems: Canadian Journal of Earth Sciences, v. 27, p. 1431–1450.

Patiño Douce, A.E., 1992, Calculated relationships between activity of alumina and phase assemblages of silica-saturated igneous rocks: petrogenetic implications of magmatic cordierite, garnet, and aluminosilicate: Journal of Volcanology and Geothermal Research, v. 52, p. 43–63.

Patiño Douce, A.E., and Beard, J.E., 1996, Effects of P, $f(O_2)$, and Mg/Fe ratio on dehydration melting of model greywackes: Journal of Petrology, v. 37, p. 999–1024.

Pavlis, T.L., and Sisson, V.B., 1995, Structural history of the Chugach metamorphic complex in the Tana River region, eastern Alaska: A record of Eocene ridge subduction: Geological Society of America Bulletin, v. 107, p. 1333–1355.

Pavlis, T.L., and Sisson, V.B., 2003, Development of a subhorizontal decoupling horizon in a transpressional system, Chugach metamorphic complex, Alaska: Evidence for rheological stratification of the crust, *in* Sisson, V.B., Roeske, S.M., and Pavlis, T.L., eds., Geology of a transpressional orogen developed during ridge-trench interaction along the North Pacific margin: Boulder, Colorado, Geological Society of America Special Paper 371, p. 191–216 (this volume).

Pavlis, T.L., Underwood, M., Sisson, V.B., Serpa, L.F., Prior, D., Marsaglia, K.M., Lewis, S.D., and Byrne, T., 1995, The effects of triple junction interactions at convergent plate margins: Report on the results of the joint JOI/USSAC and GSA Penrose conference: Washington, D.C., Joint Oceanographic Institutions, 44 p.

Pearce, J.A., 1980, Geochemical evidence for the genesis and eruptive setting of lavas from Tethyan ophiolites: Proceedings of the International Ophiolite Symposium, Cyprus, 1979, Institute of Mining Metallurgy, p. 261–272.

Pearce, J.A., 1982, Trace element characteristics of lavas from destructive plate boundaries, *in* Thorpe, R.S., ed., Andesites: New York, Wiley, p. 525–548.

Pearce, J.A., 1983, Role of the sub-continental lithosphere in magma genesis at active continental margins, *in* Hawkesworth, C.J., and Norry, M.J., eds., Continental basalts and mantle xenoliths: Nantwich, UK, Shiva, p. 230–249.

Plafker, G., 1987, Regional geology and petroleum potential of the northern Gulf of Alaska continental margin, *in* Scholl, D.W., et al., eds., Geology and resource potential of the continental margins of western North America and adjacent ocean basins: Houston, Texas, Circum-Pacific Council for Energy and Mineral Resources, Earth Science Series, v. 6, p. 229–268.

Plafker, G., Nokleburg, W.J., and Lull, J.S., 1989, Bedrock geology and tectonic evolution of the Wrangellia, Peninsular, and Chugach terranes along the Trans-Alaska Crustal Transect in the Chugach Mountains and southern Copper River basin, Alaska: Journal of Geophysical Research, v. 94, p. 4255–4295.

Plafker, G., Naeser, C.W., Zimmerman, R.A., Lull, J.S., and Hudson, T., 1992, Cenozoic uplift history of the Mount McKinley area in the central Alaska Range based on fission track dating: U.S. Geological Survey Bulletin 2041, p. 202–212.

Plafker, G., Moore, J.C., and Winkler, G.R., 1994, Geology of the southern Alaska margin, *in* Plafker, G., and Berg, H.C., eds., The geology of Alaska: Boulder, Colorado, Geological Society of America, Geology of North America, v. G-1, p. 389–450.

Poole, A.R., 1997, Age and geochemical constraints on ridge subduction for igneous rocks of the eastern Chugach Mountains, Alaska [M.A. thesis]: Houston, Texas, Rice University, 83 p.

Pouchou, J.L., and Pichoir, F., 1987, PAP procedure for improved quantitative microanalysis, *in* Armstrong, J.T., ed., Microbeam analysis: San Francisco, San Francisco Press, p. 88–97.

Reid, M.R., 1987, Chemical stratification in the crust: Isotopic, trace element, and major element constraints from crustally contaminated lavas and lower crustal xenoliths [Ph.D. thesis]: Cambridge, Massachusetts Institute of Technology, 324 p.

Rollinson, H.R., 1993, Using geochemical data: New York, Wiley, 352 p.

Sample, J.C. and Reid, M.R., 2003, Large-scale, latest Cretaceous uplift along the northeast Pacific Rim: Evidence from sediment volume, sandstone petrography, and Nd isotope signatures of the Kodiak Formation, Kodiak islands, Alaska, *in* Sisson, V.B., Roeske, S.M., and Pavlis, T.L., eds., Geology of a transpressional orogen developed during ridge-trench interaction along the North Pacific margin: Boulder, Colorado, Geological Society of America Special Paper 371, p. 51–70 (this volume).

Schmidt, M.W., 1992, Amphibole composition on tonalite as a function of pressure: an experimental calibration of the Al-in-hornblende barometer: Contributions to Mineralogy and Petrology, v. 110, p. 304–310.

Shand, S.J., 1951, Eruptive rocks (4th edition): New York, Wiley, 488 p.

Sharma, M., Basu, A.R., Cole, R.B., and De Celles, P.G., 1991, Basalt-rhyolite volcanism by MORB-continental crust interaction: Nd, Sr-isotopic evidence from southern San Joaquin basin, California: Contributions to Mineralogy and Petrology, v. 109, p. 159–172.

Shervais, J.W., 1982, Ti-V plots and the petrogenesis of modern and ophiolitic lavas: Earth and Planetary Science Letters, v. 59, p. 101–118.

Sisson, V.B., 1994, Effects of triple junction interactions discussed: Joint Oceanographic Institutions/U.S. Science Advisory Committee Newsletter, v. 7, no. 2, p. 5.

Sisson, V.B., and Hollister, L.S., 1988, Low-pressure facies series metamorphism in an accretionary sedimentary prism, southern Alaska: Geology, v. 16, p. 358–361.

Sisson, V.B., and Pavlis, T.L., 1993, Geologic consequences of plate reorganization: An example from the Eocene southern Alaska forearc: Geology, v. 21, p. 913–916.

Sisson, V.B., Hollister, L.S., and Onstott, T.C., 1989, Petrologic and age constraints on the origin of a low-pressure/high-temperature metamorphic complex, southern Alaska: Journal of Geophysical Research, v. 94, p. 4392–4410.

Sisson, V.B., Cooper, H., Pavlis, T.L., and Poole, A.R., 1997, Contrasting metamorphic histories within the Chugach metamorphic complex: Differential uplift related to transpression on the Fairweather fault, Alaska: Geological Society of America Abstracts with Programs, v. 29, no. 6, p. 277.

Sisson, V.B., Pavlis, T.P., and Stüwe, K., 1998, Along-strike differences in the forearc associated with Eocene triple-junction interactions in the Mt. St. Elias–Mt. Logan region: American Geophysical Union Eos Abstracts, v. 79, p. 913.

Smart, K.S., Pavlis, T.L., Sisson, V.B., Roeske, S.M., and Snee, L.W., 1996, The Border Ranges fault system in Glacier Bay National Park, Alaska: Evidence for major early Cenozoic dextral strike-slip motion: Canadian Journal of Earth Sciences, v. 33, p. 1268–1282.

Smithies, R.H., and Champion, D.C., 2000, The Archaean high-Mg diorite suite: Links to tonalite-trondhjemite-granodiorite magmatism and implications for early Archaean crustal growth: Journal of Petrology, v. 41, p. 1653–1671.

Steven, T.A., Mehnert, H.H., and Obradovich, J.D., 1967, Age of volcanic activity in the San Juan Mountains, Colorado: U.S. Geological Survey Professional Paper 575D, p. 47–55.

Stock, J., and Molnar, P., 1988, Uncertainties and implications of the Late Cretaceous and Tertiary position of North America relative to the Farallon, Kula, and Pacific plates: Tectonics, v. 7, p. 1339–1384.

Storey, M., Rogers, G., Saunders, A.D., and Terrell, D.J., 1989, San Quintin volcanic field, Baja California, Mexico: "Within-plate" magmatism following ridge subduction: Terra Nova, v. 1, p. 195–202.

Sturm, M.E., Klein, E.M., Karsten, J.L., and Karson, J.A., 2000, Evidence for subduction-related contamination of the mantle beneath the southern Chile Ridge: Implications for ambiguous ophiolite compositions, *in* Dilik, Y., et al., eds., Ophiolites and oceanic crust: New insights from field studies and the Ocean Drilling Program: Boulder, Colorado, Geological Society of America Special Paper 349, p. 13–20.

Taylor, B., and Exon, N.F., 1987, An investigation of ridge subduction in the Woodlark-Solomons region: Introduction and overview, *in* Taylor, B., and Exon, N.F., eds., Marine geology, geophysics, and geochemistry of the Woodlark Basin–Solomon Islands: Houston, Texas, Circum-Pacific Council for Energy and Mineral Resources, Earth Science Series, v. 7, p. 1–24.

Thorkelson, D.J., 1996, Subduction of diverging plates and the principles of slab window formation: Tectonophysics, v. 255, p. 47–63.

White, A., Clemens, J.D., Holloway, J.R., Silver, L.T., Chappell, B.W., and Wall, V.J., 1986, S-type granites and their probable absence in southwestern North America: Geology, v. 14, p. 115–118.

Winkler, G.R., and Plafker, G., 1981, Geologic map and cross sections of the Cordova and Middleton Island quadrangles, Alaska: U.S. Geological Survey Open-File Report 81-1164, scale 1:250,000.

Wolf, M.B., and Wyllie, P.J., 1994, Dehydration melting of amphibolite at 10 kb: The effects of temperature and time: Contributions to Mineralogy and Petrology, v. 115, p. 369–383.

Manuscript Accepted by the Society February 5, 2003

Geological Society of America
Special Paper 371
2003

Magmatic and tectonic history of the Leech River Complex, Vancouver Island, British Columbia: Evidence for ridge-trench intersection and accretion of the Crescent Terrane

Wesley G. Groome*
Derek J. Thorkelson
Department of Earth Sciences, Simon Fraser University, Burnaby, British Columbia, V5A 1S6, Canada

Richard M. Friedman
James K. Mortensen
Department of Earth and Ocean Sciences, University of British Columbia, Vancouver, British Columbia, V6T 1Z4, Canada

Nick W.D. Massey
British Columbia Geological Survey Branch, Ministry of Energy and Mines, Victoria, British Columbia, V8W 9N3, Canada

Daniel D. Marshall
Department of Earth Sciences, Simon Fraser University, Burnaby, British Columbia, V5A 1S6, Canada

Paul W. Layer
Geophysical Institute, University of Alaska, Fairbanks, Alaska, 99775, USA

ABSTRACT

The Leech River Complex, part of the Pacific Rim Terrane, is a Cretaceous metasedimentary and metaigneous assemblage on southern Vancouver Island. The Leech River Complex is fault-bounded between the Eocene Metchosin Igneous Complex to the south (part of the Crescent Terrane) and the Paleozoic to Jurassic Wrangel Terrane to the north and provides critical information on the evolution of the central part of the western North American forearc in Cretaceous through Eocene time. Single detrital zircons from the metasedimentary component, known as the Leech River Schist, give U-Pb interpreted ages that range from Precambrian to ca. 103 Ma, indicating a varied source region and a probable Early Cretaceous depositional age. U-Pb geochronology and field investigations indicate at least two magmatic-metamorphic events in the Leech River Complex: one during the Late Cretaceous, and the other during the early Middle Eocene. The peraluminous Jordan River metagranodiorite, a fine-grained biotitic stock and related dikes intruded the Leech River Complex at ca. 88 Ma, during the older magmatic event. Metamorphic pressure-temperature conditions of 525–550 °C and 2–3 kbar are recorded in the contact aureole. The later event occurred during emplacement of the Walker Creek intrusions, a suite of peraluminous tonalite, trondjhemite, and granodiorite dikes that intruded the Complex at ca. 51 Ma and produced a similar metamorphic aureole. Both intrusive suites have

*Current address: Department of Geological Sciences, University of Maine, Orono, ME 04469, USA; wesley.groome@umit.maine.edu.

Groome, W.G., Thorkelson, D.J., Friedman, R.M., Mortensen, J.K., Massey, N.W.D., Marshall, D.D., and Layer, P.W., 2003, Magmatic and tectonic history of the Leech River Complex, Vancouver Island, British Columbia: Evidence for ridge-trench intersection and accretion of the Crescent Terrane, *in* Sisson, V.B., Roeske, S.M., and Pavlis, T.L., eds., Geology of a transpressional orogen developed during ridge-trench interaction along the North Pacific margin: Boulder, Colorado, Geological Society of America Special Paper 371, p. 327–353.

a variably developed schistosity that parallels foliation in the Leech River Schist, and may indicate emplacement during deformation. The Tripp Creek metabasite, a third intrusive unit with an N-MORB geochemical signature, is also interfoliated with the schist and may be the same age as one or both of the peraluminous intrusive suites. Taken together, these magmatic events may be rationalized as products of ridge subduction and slab window formation beneath the Leech River Complex, at the leading edge of the North American plate. Uprise of basaltic melts from the subducting ridge, and consequent anatexis of the forearc metasediments, accounts for the MORB-like and peraluminous intrusions, respectively.

$^{40}Ar/^{39}Ar$ cooling ages for muscovite (ca. 45 Ma) and biotite (ca. 42.5 Ma) indicate rapid cooling of the Leech River Complex in the Middle Eocene. Cooling was probably related to exhumation and denudation of the Complex during underthrusting of the Crescent Terrane along the sinistral-oblique Leech River fault. The Leech River Complex is overlain by undeformed Oligocene Carmanah Group sediments.

Keywords: magmatism, tectonics, slab windows, geochronology, detrital mineral age, geochemistry, major elements, trace elements, Vancouver Island, British Columbia

INTRODUCTION

Forearc accretionary complexes are commonly characterized by low geothermal gradients and a lack of igneous activity (Gill, 1981; Best, 1982; Bucher and Frey, 1994). However, the entry of a spreading ridge into the subduction zone can increase the geothermal gradient, a situation commonly referred to as ridge subduction and slab window formation (Marshak and Karig, 1977; Delong et al., 1979; Dickinson and Snyder, 1979; Forsythe and Nelson, 1985; Sisson and Pavlis, 1993; Thorkelson, 1996). Where this occurs, the temperature of the forearc can be greatly elevated, leading to magmatism and high-temperature metamorphism (Delong et al., 1979; Sisson and Pavlis, 1993). The effects of ridge subduction farther inland, particularly below the arc, are more varied and may require geochemical data to delineate (e.g., Johnston and Thorkelson, 1997). Thus, the presence of magmatic activity within the forearc is arguably one of the key indicators of offshore plate configurations.

Early Tertiary forearc magmatism within the western Cordillera of North America took place in a belt extending from Alaska to Oregon. This belt can be subdivided into two segments, each with a distinct style and tectonic setting. Within a northern segment that extends from southeastern Alaska to southern Vancouver Island, forearc igneous rocks are hosted in accretionary complexes, including the Pacific Rim and Chugach terranes (Hill et al., 1981; Fairchild and Cowan, 1982; Rusmore and Cowan, 1985; Brandon, 1989; Barker et al., 1992; Sisson and Pavlis, 1993; Haeussler et al., 1995; Harris et al., 1996; Groome, 2000). These igneous rocks consist mainly of intermediate to felsic intrusions associated with high-temperature metamorphic assemblages (Hill et al., 1981; Fairchild and Cowan, 1982; Rusmore and Cowan, 1985; Brandon, 1989; Barker et al., 1992; Sisson and Pavlis, 1993; Groome, 2000). Subordinate felsic volcanics and mafic intrusions are present locally (Hill et al., 1981; Groome, 2000). This forearc magmatism has been interpreted to be the result of spreading-ridge subduction (Sisson and Pavlis, 1993; Groome et al., 2000). In the southern segment, which extends from southern Vancouver Island to southern Oregon, Early Tertiary forearc igneous rocks are represented by basaltic rocks of the Crescent Terrane. This terrane was interpreted by Duncan (1982) to be accreted seamounts, but more recent work suggests that it may have formed in a forearc rift basin, which was thrust beneath the continental margin in Eocene time (Wells et al., 1984; Massey, 1986; Babcock et al., 1992).

In this paper, we examine the forearc intrusions in the Leech River Complex, which is part of the Pacific Rim Terrane on southern Vancouver Island (Wheeler and McFeeley, 1991; Fig. 1). The main study area is located in the southeastern part of the Leech River Complex (Fig. 1). We report new results from field and petrographic studies, U-Pb and $^{40}Ar/^{39}Ar$ geochronology, and geochemistry, and define three intrusive suites ranging in age from Late Cretaceous to Middle Eocene. Variations in metamorphic grade are tied to thermal aureoles surrounding the plutons. Cooling ages for micas from the intrusions and their host rocks demonstrate rapid exhumation of the Leech River Complex, and are used to infer the timing of underthrusting of the Crescent Terrane. Subduction of the Kula-Farallon ridge may be linked to Eocene magmatism. Earlier subduction of an unknown spreading ridge is suggested as a possible cause of the Cretaceous igneous activity.

UNITS OF THE PACIFIC RIM TERRANE

The Pacific Rim Terrane is located on western and southern Vancouver Island. It is juxtaposed along faults with Wrangellia to the east and north, and the Crescent Terrane to the west and south. The Pacific Rim Terrane is broadly regarded as an accretionary complex, at least in part, and consists of three principal units: the Pacific Rim Complex, which extends along parts of western Vancouver Island (Brandon, 1989); the Pandora Peak unit, located mainly along southwestern Vancouver Island near Port Renfrew (Rusmore and Cowan, 1985); and the Leech River Complex

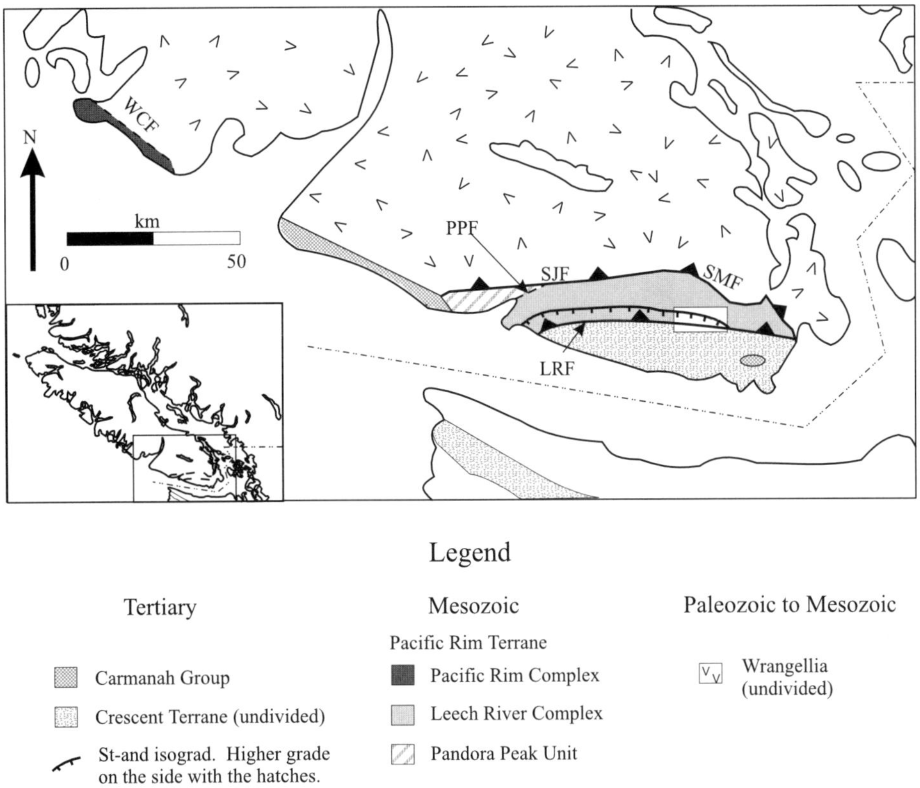

Figure 1. Simplified tectonic assemblage map of southern Vancouver Island (after Wheeler and McFeeley, 1991). Inset: Map of Vancouver Island region; box indicates the region of the main map in the figure. The staurolite isograd is from Read et al., 1991. LRF—Leech River fault; PPF—Pandora Peak fault; SJF—San Juan fault; SMF—Survey Mountain fault; WCF—Westcoast fault. The white box on the main figure outlines the region of Figure 2. St-and—Staurolite-andalusite isograds.

(Fairchild and Cowan, 1982), which extends across the southern end of the island. The basaltic Crescent Terrane is represented by the Metchosin Igneous Complex on the southern tip of Vancouver Island (Massey, 1986), and the Prometheus volcanics to the west of Vancouver Island beneath the Pacific Ocean (Shouldice, 1971). Wrangellia consists mainly of Lower Paleozoic to Mesozoic igneous and sedimentary rocks and was accreted to the continental margin in Jurassic to middle Cretaceous time. Wrangellia apparently served as a buttress against which the Pacific Rim and Crescent Terranes were accreted in Mesozoic to Eocene time. The timing and nature of the accretion of the Pacific Rim Terrane is not well understood and may involve considerable strike-slip faulting (Fairchild and Cowan, 1982; Rusmore and Cowan, 1985; Brandon et al., 1988). Accretion of the Crescent Terrane occurred in the Eocene (Massey, 1986).

The Leech River Complex, the Pandora Peak unit, and the Pacific Rim Complex are all included in the Pacific Rim Terrane (Wheeler and McFeeley, 1991) because of their common tectonic position and their typically disrupted to chaotic sedimentary character. The Leech River Complex differs from the other units by its characteristic foliation and generally higher metamorphic grade (Fairchild and Cowan, 1982; Rusmore and Cowan, 1985).

Pacific Rim Complex

The Pacific Rim Complex is located along the west coast of Vancouver Island. It consists of a Triassic arc sequence (Ucluth Formation) and an unconformably overlying olistostromal mélange (Brandon, 1989). The Ucluth Formation consists of volcanic and sedimentary strata of arc affinity (Brandon, 1989). The overlying olistostromal mélange unit consists of centimeter- to meter-scale angular blocks of chert, graywacke, argillite, and greenstone in a fine-grained matrix of mudstone to siltstone but does not display penetrative deformation (Brandon, 1989). The depositional age of the olistostromal unit is constrained by the presence of Jura-Cretaceous microfossils in the matrix sediments (Brandon, 1989).

Brandon (1989) showed that the Pacific Rim Complex was statically metamorphosed to prehnite-lawsonite grade and interpreted this metamorphism to be Cretaceous in age, based on $^{40}Ar/^{39}Ar$ geochronology on metamorphic minerals. The metamorphic conditions are suggestive of a subduction zone environment. However, Brandon (1989) suggested the rocks of the Pacific Rim Complex were deposited in a marginal basin within submarine debris flows, rather than within an accretionary wedge, based on the presence of blocks of Ucluth Formation greenstone within the mélange. In this scenario, the blueschist metamorphism would be post-deposition.

Pandora Peak Unit

The Pandora Peak unit is fault bounded between the Leech River Complex and Wrangellia near Port Renfrew along the Port Renfrew fault and the San Juan fault, respectively, and is also in fault contact with Wrangellia near Victoria along the Trial Island fault. The Pandora Peak unit consists of black mudstone, terrigenous graywacke, radiolarian ribbon chert, green tuff, metabasaltic greenstone, pebbly mudstone, and local limestone blocks that were deposited during the Late Jurassic to Early Cretaceous (Rusmore and Cowan, 1985).

The Pandora Peak unit underwent blueschist facies metamorphism, with the assemblage of lawsonite-prehnite-calcite being characteristic (Rusmore and Cowan, 1985). The metamorphic assemblage present in the Pandora Peak unit indicates pressure-temperature conditions of about 3 kbar, 175–230 °C. (Rusmore and Cowan, 1985, and references therein).

Leech River Complex

The Leech River Complex is bounded to the north against the Pandora Peak unit and Wrangellia along the Port Renfrew, San Juan, and Survey Mountain faults, and to the south against the Metchosin Igneous Complex (Crescent Terrane) along the Leech River fault (Fairchild and Cowan, 1982; Rusmore and Cowan, 1985). The Leech River Complex consists of pelitic and arenaceous metasediments, and igneous rocks that have been variably metamorphosed and deformed (Fairchild and Cowan, 1982; Groome, 2000). In contrast to the Pandora Peak unit and the Pacific Rim Complex, rocks of the Leech River Complex are characterized by higher-temperature metamorphic assemblages and penetrative foliations.

Metamorphic grade in the Leech River Complex ranges from chlorite zone to staurolite-andalusite zone (Fairchild and Cowan, 1982; Rusmore and Cowan, 1985). Metamorphic grade generally increases towards the southern boundary of the Complex (Fairchild and Cowan, 1982; Rusmore and Cowan, 1985). In the eastern part of the Leech River Complex, increases in metamorphic grade are spatially and, apparently, genetically related to igneous intrusions (Groome et al., 1999). A causal relationship between igneous activity and increased metamorphic grade was also postulated for the western portion of the Complex by Rusmore and Cowan (1985). In the central portion of the Complex, which has not been studied in detail, the controls on metamorphism have not been documented.

Metapelitic rocks are the most abundant rock type in the Leech River Complex. They are typically brown- to black-weathering schist, with an assemblage of biotite-chlorite-white mica-plagioclase-quartz in the greenschist facies regions, and biotite-staurolite-andalusite-plagioclase-muscovite-quartz (± garnet, ± cordierite) in the amphibolite facies regions. The pelitic schist contains discontinuous lenses and patches of white-weathering metasiltite and fine-grained metasandstone; together, these metasediments constitute the Leech River Schist (Muller, 1983). The silty-sandy parts are generally 5–50 cm long and 1–5 cm thick, although some range up to 10 m long and 3 m thick. Their protoliths were probably sublithic to subarkosic arenites on the basis of modal mineralogy and textures in thin section. The metapelite and the coarser layers have been metamorphosed and deformed to such an extent that original sedimentary layering is not preserved. The highly discontinuous nature of the coarse layers is partly due to boudinage and other processes of post-lithification deformation, but a fundamental cause of their ragged shapes and chaotic distribution may be largely synsedimentary in origin, possibly the result of slumping during olistostrome formation. Very similar textures are present in the olistostromal unit of the Pacific Rim Complex, which was not penetratively deformed after lithification (Brandon, 1989).

Metaigneous rocks are also present in the Leech River Complex. Foliated and metamorphosed basaltic layers with small (~1 mm) but conspicuous porphyroclasts of altered plagioclase in the eastern part of the Complex were named the Survey Mountain volcanics by Fairchild (1979). In our study, these rocks were briefly inspected and textures diagnostic of lava or tuffaceous rock were not encountered. Therefore these rocks could be intrusive. If so, they could either be a lower-grade manifestation of the Tripp Creek metabasite or could form a fourth intrusive unit unrelated to the other intrusive units of the Leech River Complex.

INTRUSIVE ROCKS OF THE LEECH RIVER COMPLEX

Three intrusive units are present in the area of detailed study (Fig. 2): the Walker Creek intrusions, the Jordan River metagranodiorite, and the Tripp Creek metabasite. The Walker Creek intrusions are the most widespread intrusive rocks in the study area. They form a tonalite-trondjhemite-granodiorite suite of dikes and sills ranging in size from a few centimeters wide to several meters wide. The Jordan River metagranodiorite crops out as a large stock and several smaller intrusions. The Tripp Creek metabasite consists of several layers of actinolite-rich schist and minor metagabbro. All three intrusive units contain a penetrative foliation. New age determinations, given below, indicate that the Jordan River metagranodiorite is Late Cretaceous, and the Walker Creek intrusions are Eocene. The Tripp Creek metabasite was not dated but is likely to be late Early Cretaceous to Eocene on the basis of similar fabrics and field relations.

Geochemical analyses were obtained for all of the intrusive suites. Samples submitted for analysis were devoid of weathering rinds and were ground in agate. Major oxide analyses were determined by XRF on fused disks at X-Ray Assay Labs, and trace element analyses were carried out by ICP-MS at the University of Saskatchewan. Precision of the major oxides, provided by the laboratory, ranges from 1% to 2% of the reported values. For each trace element, precision was evaluated according to the determination limit provided by the laboratory, and a percentage of the reported element abundance as determined graphically on duplicate analyses (Groome, 2000). Analytical accuracy was evaluated using two CANMET reference materials, MRG-1 and SY-2. The reported values of these standards, when compared to the expected (accepted) abundances, lie within the determined levels of analytical precision.

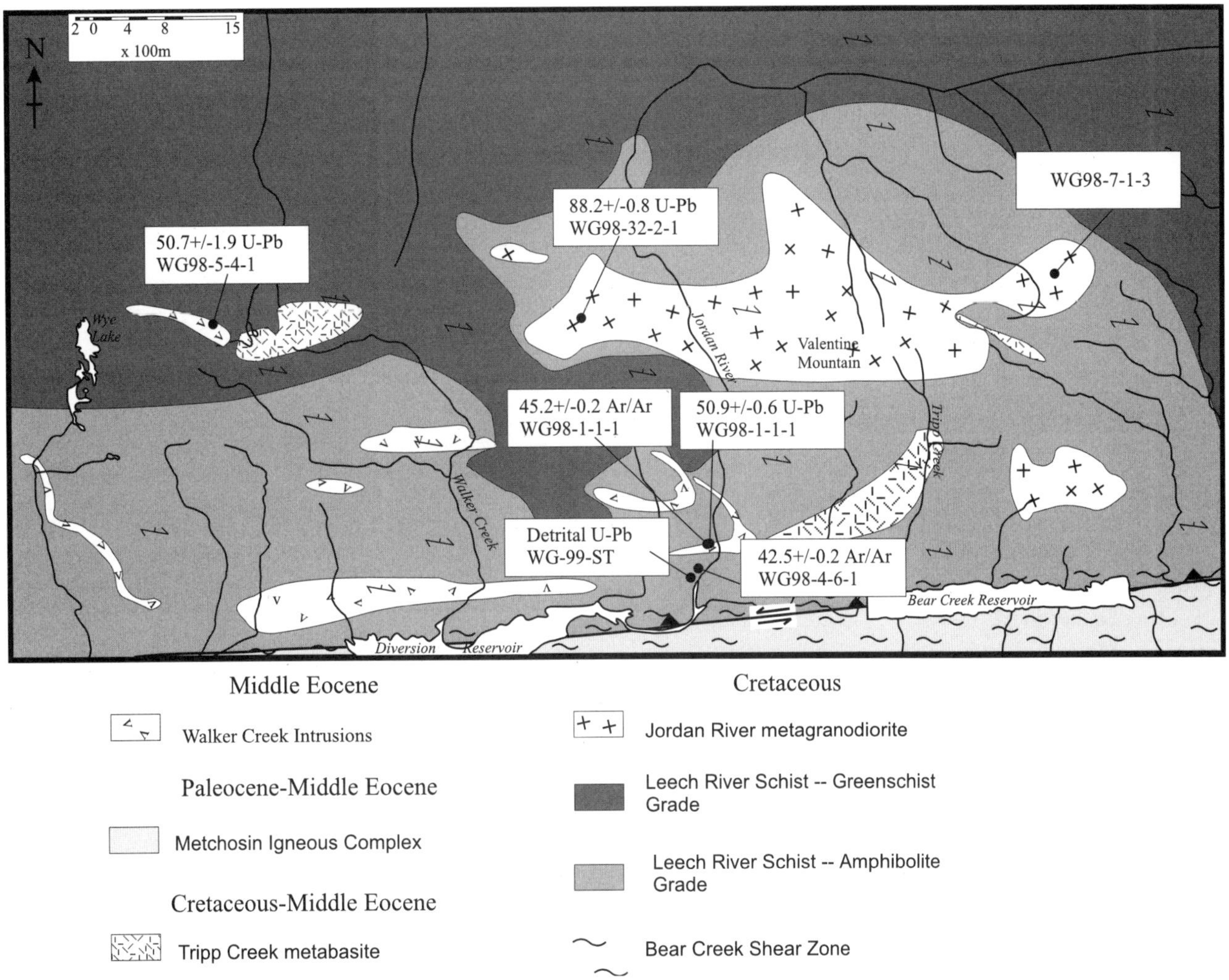

Figure 2. Simplified geologic map of the Jordan River region of the Leech River Complex (modified from Groome, 2000). Sample locations of dated rocks are shown. Foliations are representative and generalized.

Jordan River Metagranodiorite

The Jordan River metagranodiorite comprises a homogenous 6 km long by 2 km wide by 500 m thick stock and approximately ten smaller stocks and dikes. It is a mesocratic biotite ± hornblende granodiorite that is interfoliated with the Leech River Schist along the margins and less deformed towards the interior. The smaller comagmatic dikes and stocks that were found in the study area are interfoliated with the Leech River Schist. These dikes are interpreted to be comagmatic with the Jordan River metagranodiorite based on compositional and structural similarities. Both the main stock and the smaller dikes contain xenoliths of Leech River Schist that preserve penetrative fabrics with orientations independent of the fabric in the intrusion.

In thin section, the Jordan River metagranodiorite has few features preserved that are easily recognizable as primary in origin. Most of the samples appear to have been recrystallized, as indicated by the presence of overgrowths and sutures on most of the grains. The groundmass crystals have a strong preferential orientation, while the larger grains appear to have been rolled during deformation, as indicated by asymmetric pressure shadows. Some plagioclase phenocrysts/porphyroblasts have pressure shadows developed around them, indicative of deformation partitioning during recrystallization. The foliation is defined by the preferential orientation of biotite and of multigrain quartz stringers.

Plagioclase (~An_{30}) is the most abundant rock-forming mineral in the Jordan River metagranodiorite. It constitutes up to 40% of the mode, and is distributed throughout the groundmass and as a phenocryst/porphyroblast phase in most samples. The plagioclase is variably altered to fine-grained clay and muscovite. Quartz and alkali-feldspar are present primarily as groundmass phases in thin section. Some alkali-feldspar crystals have perthitic exsolution present. A few crystals of quartz have a myrmekite texture developed with plagioclase, possibly indicative of simultaneous growth of the two phases.

The primary ferro-magnesian silicate in the Jordan River metagranodiorite is biotite, which generally occurs as well-developed crystals that help to define a foliation. Some of the biotite crystals are pseudomorphed by retrograde chlorite. Hornblende, also present in the Jordan River metagranodiorite, is typically euhedral and partially altered to biotite. In the samples that have no hornblende, it appears that the biotite is primary, as the crystals do not display the morphology of hornblende, and no relict hornblende is present in the cores of the biotite crystals.

Accessory minerals in the Jordan River metagranodiorite include apatite, zircon, titanite, hematite, and other opaques. The Jordan River metagranodiorite is also locally mineralized with chalcopyrite, pyrite, and arsenopyrite. Where observed, this mineralization is disseminated and is spatially associated with alteration zones. These alteration zones are characterized by a reddish iron-oxide staining and a friable character. Abundant quartz veins are also associated with the mineralized zones, implying that the mineralization is the result of post-crystallization hydrothermal activity.

The chemical composition of the metagranodiorite is quite uniform, as indicated by three analyses, two from the main stock and one from a smaller stock. Major element concentrations and CIPW normative mineralogy (Table 1) indicate that the metagranodiorite is slightly peraluminous, sodic, and quartz and hypersthene normative. Trace element abundances normalized to normal mid-ocean ridge basalt (N-MORB) are displayed in Figure 3A. All of these patterns resemble modern continental arcs, with high alkali and alkaline earth concentrations, and relative depletions in Nb and Ta (Fig. 3E). Two samples from the Leech River Schist, shown on Figure 3D, are remarkably similar in chemical composition to the Jordan River intrusions. On discriminant diagrams of Pearce et al. (1984), the Jordan River metagranodiorite samples plot as volcanic arc granitoids (VAG) (Fig. 4).

Tripp Creek Metabasite

The Tripp Creek metabasite comprises at least five lens-like layers up to 20 m thick that are interfoliated with the Leech River Schist, some of which is unusually biotite-rich. It also occurs within cliff faces along the banks of Tripp Creek, as variably foliated, fine- to medium-grained garnet-quartz-plagioclase-actinolite metagabbro that trends into outcrops of fine-grained actinolite-rich schist (Fig. 5). Original contact relations between the Tripp Creek metabasite and the Leech River Schist are obscured by deformation and metamorphism. However, the local igneous texture, and a xenolith of pelitic schist within the metabasite, suggests that the metabasite is intrusive. At some localities, the metabasite is spatially associated with the Walker Creek intrusions. This relation is particularly well expressed along Tripp Creek, where the metabasite crops out within a swarm of closely spaced Walker Creek intrusions. The metabasite is mainly com-

TABLE 1. WHOLE ROCK GEOCHEMISTRY AND CIPW NORMATIVE MINERALOGY

Sample:	98-15-1-1B	98-33-3-2B	98-14-4-1B	98-5-4-2B	98-12-5-1B	98-10-2-2B	98-12-1-1B	98-1-1-1B	98-5-2-3B	98-9-3-2B	98-5-5-3B	99-1-1-1B	99-2-1-1B	98-7-2-2B	98-4-6-1B	Detection Limit
Unit:	KJR	KJR	KJR	eTWC	eTWC	eTWC	eTWC	eTWC	eTWC	eTWC	KeTTC	KeTTC	KeTTC	KLRa	KLRg	
Rock type:	GD	GD	GD	GD	GD	TO	TO	TR	TO	TR	MB	MB	MB	ST	BT	
Northing:	5375326	5374264	5373507	5375781	5372755	5374006	5373016	5373871	5376057	5374291	5375628	5375221	5374232	5375448	5373038	%
Easting:	10430282	10434487	10430772	10425690	10429924	10429687	10428154	10430861	10424497	10430780	10425858	10428031	10431003	10475448	10429581	
Major oxides																
Wt%																
SiO_2	70.91	69.86	70.03	63.10	71.39	70.35	75.79	75.60	75.70	76.98	49.25	46.20	55.10	62.91	65.27	0.01
TiO_2	0.49	0.51	0.48	0.80	0.21	0.27	0.05	0.03	0.01	0.01	1.53	0.91	0.74	0.84	0.74	0.01
Al_2O_3	14.54	15.09	15.02	16.23	16.43	16.95	14.32	14.87	15.18	15.34	14.68	20.30	14.90	19.51	17.56	0.01
Fe_2O_3	3.87	4.10	3.77	5.89	1.55	2.70	1.09	1.32	0.81	0.68	12.30	10.00	9.89	8.05	5.79	0.01
MnO	0.05	0.06	0.06	0.08	0.02	0.04	0.02	0.03	0.03	0.04	0.19	0.15	0.14	0.10	0.10	0.01
MgO	1.73	1.99	1.81	3.96	0.84	0.62	0.14	0.07	0.04	0.03	9.26	7.35	7.52	3.15	2.55	0.01
CaO	3.18	3.15	3.40	4.92	2.80	3.10	1.34	1.52	1.04	0.67	11.52	13.66	8.73	1.21	2.97	0.01
Na_2O	3.96	3.61	3.85	4.00	6.05	4.59	4.40	4.41	4.84	4.15	2.01	2.10	2.50	1.27	2.99	0.01
K_2O	1.50	1.87	1.79	1.36	0.79	1.48	2.94	2.17	2.37	2.07	0.14	0.23	0.56	3.30	2.21	0.01
P_2O_5	0.12	0.11	0.12	0.20	0.06	0.13	0.03	0.07	0.07	0.11	0.12	0.08	0.13	0.16	0.18	0.01
LOI	0.50	0.90	0.35	0.40	0.65	0.65	0.20	0.70	0.65	0.95	0.25	0.25	1.80	1.80	1.80	
SUM	100.35	100.36	100.33	100.55	100.14	100.23	100.11	100.10	100.08	100.06	101.01	100.98	100.21	100.49	100.38	

Note: Rock type abbreviations: GD—granodiorite, TO—tonalite, TR—trondjhemite, ST—staurolite schist, BT—biotite schist, MB—metabasite. Data labels are used on all discriminant diagrams. KJR—Jordan River metagranodiorite; eTWC—Walker Creek intrusions; KLRa—Leech River Schist, amphibolite facies; KLRg—Leech River Schist, greenschist facies; KeTTC—Tripp Creek metabasite.

(continued)

TABLE 1. WHOLE ROCK GEOCHEMISTRY AND CIPW NORMATIVE MINERALOGY (continued)

Sample:	98-15-1-1B	98-33-3-2B	98-14-4-1B	98-5-4-2B	98-12-5-1B	98-10-2-2B	98-12-1-1B	98-1-1-1B	98-5-2-3B
Unit:	KJR	KJR	KJR	eTWC	eTWC	eTWC	eTWC	eTWC	eTWC
Rock type:	GD	GD	GD	GD	GD	TO	TO	TR	TO
Trace elements (ppm)									
Cs	2	2.4	1.7	4.3	1.3	2.3	2.8	2.3	0.82
Ba	680	700	630	480	1300	480	860	570	780
Rb	48	51	48	42	19	41	90	35	73
Th	5.5	4.5	4.2	4.1	1.4	3.5	4.2	2.4	2.2
Pb	10	12	9.8	18	13	12	28	9.1	21
U	1.5	1.5	1.1	2.3	0.47	1.1	1.3	1.2	2.4
K*	12000	16000	15000	11000	6600	12000	24000	18000	20000
Nb	6.7	7	6.7	10	1.5	8.4	11	6.6	9.1
Ta	0.54	0.43	0.45	0.57	0.12	0.63	1.2	0.55	1
La	21	18	18	20	4.6	21	13	13	9.2
Ce	40	35	35	38	8.8	37	24	23	18
Sr	620	350	340	690	950	740	130	350	220
Pr	4.8	4.2	4.2	4.6	1.1	4	2.8	2.5	2.1
Nd	19	16	16	18	4.4	14	11	8.5	7.7
P**	430	210	430	800	200	460	73	200	170
Zr	160	130	120	160	82	190	60	41	33
Hf	4	3.6	3.2	4.1	2.3	4.5	2.3	1.8	1.8
Sm	3.5	3.2	3.2	3.6	0.84	2.5	2.8	1.6	2.1
Eu	0.86	0.83	0.86	1.2	0.31	0.81	0.42	0.41	0.32
Gd	3	2.8	2.8	3.1	0.73	1.8	2.3	1.3	1.8
Tb	0.42	0.38	0.39	0.43	0.09	0.23	0.26	0.16	0.21
Ti***	2900	2500	2900	4600	1300	1600	280	140	43
Dy	2.5	2.3	2.4	2.5	0.48	1.3	1.1	0.87	0.84
Ho	0.51	0.45	0.46	0.45	0.09	0.23	0.14	0.15	0.1
Er	1.5	1.3	1.4	1.2	0.24	0.67	0.29	0.38	0.2
Tm	0.23	0.2	0.19	0.16	0.03	0.09	0.04	0.05	0.03
Y	15	13	13	13	4.1	8	5.2	5.6	4.2
Yb	1.5	1.3	1.3	1.1	0.19	0.62	0.22	0.31	0.12
Lu	0.21	0.18	0.2	0.14	0.03	0.1	0.03	0.03	0.01
V	83	84	84	130	29	4.6	3.1	0.39	0.91
Sc	10	11	10	13	3.8	3.1	1.4	0.83	1.2
Cr	50	37	52	85	26	5.5	11	8.9	20
Ni	18	14	16	82	5.7	2	1.4	0.64	1.4

*K = K_2O * 8300
**P = P_2O_5 * 4360
***Ti = TiO_2 * 6000
**P = P_2O_5 * 4360

(continued)

posed of green, fine-grained, strongly foliated, garnet-quartz-plagioclase-actinolite schist. The foliation is mainly defined by actinolite but is accentuated by preferred orientations of quartz and plagioclase. Garnet typically occupies <2% of the rock, and tends to be restricted to thin (grain-wide) layers.

Three samples of metabasite were analyzed for major and trace elements. On the basis of major elements (Table 1), the unit is basaltic, hypersthene-normative, and metaluminous, with magnesium numbers {100[Mg/(Mg + Fe)]} of approximately 63. Trace element profiles of the samples show nearly flat profiles on an N-MORB-normalized diagram (Fig. 3B). Of particular note is sample WG99-1-1-1B, which is a relatively undeformed metagabbro. The pattern of this sample is quite flat, with moderate enrichments in Cs, Ba, K and Sr, all of which are relatively mobile and may have been enriched during metamorphism. Variations in the profiles of the two other samples can be attributed to the nature of the samples collected. Both samples are actinolite schists and are deformed and metamorphosed to such an extent that original igneous textures are obliterated. The higher degree of deformation and more pervasive metamorphism could have resulted in increased mobility of incompatible elements.

Walker Creek Intrusions

The Walker Creek intrusions consist of a tonalite-trondhjemite-granodiorite suite of at least 100 tabular and lens-like

TABLE 1. WHOLE ROCK GEOCHEMISTRY AND CIPW NORMATIVE MINERALOGY (continued)

Sample:	98-9-3-2B	98-5-5-3B	99-1-1-1B	99-2-1-1B	98-7-2-2B	98-4-6-1B	Detection	ppm	Uncertainty
Unit:	eTWC	KeTTC	KeTTC	KeTTC	KLRa	KLRg	limit		
Rock type:	TR	MB	MB	MB	ST	BT	%		
Trace elements (ppm)									
Cs	0.83	0.04	0.04	1.1	1.8	3.1		0.03	10
Ba	720	23	25	270	820	700		0.17	10
Rb	47	0.78	1.1	17	63	63		0.03	10
Th	1.1	0.37	0.1	2.2	9.1	7.2		0.01	15
Pb	15	3.6	-1	5	14	16		0.05	25
U	1.4	0.13	0.08	1.1	1.9	1.5		0.03	10
K*	17000	1200	1900	4600	27000	18000		83	5
Nb	7.7	4.1	1.9	7.3	12	12		0.50	20
Ta	0.97	0.28	0.13	0.34	0.74	0.74		0.03	10
La	4.3	4.6	2	10	21	22		0.01	10
Ce	8.8	13	6	23	45	45		0.03	5
Sr	130	330	220	290	320	330		0.50	20
Pr	1.1	2	1.1	2.7	5.2	5.4		0.20	10
Nd	3.9	10	5.8	12	20	22		0.03	10
P**	340	490	350	570	560	650		43	5
Zr	30	86	50	110	130	140		0.11	10
Hf	1.7	2.5	1.4	2.8	3.6	3.8		0.10	15
Sm	1.3	3.3	2.1	2.7	4.1	4.5		0.03	10
Eu	0.1	1.1	0.9	0.85	0.97	1.2		0.03	10
Gd	1.6	4	3	2.8	3.9	4.2		0.03	10
Tb	0.26	0.68	0.55	0.47	0.6	0.6		0.03	15
Ti***	31	8800	5500	4400	5000	4800		4.00	10
Dy	1.5	4.3	3.7	3	3.6	3.6		0.03	10
Ho	0.24	0.86	0.76	0.63	0.75	0.7		0.10	20
Er	0.59	2.5	2.3	1.7	2.2	2		0.10	20
Tm	0.07	0.34	0.35	0.26	0.34	0.29		0.03	20
Y	8.1	22	22	18	21	19		0.01	15
Yb	0.43	2.2	2.2	1.7	2.1	1.9		0.10	25
Lu	0.04	0.3	0.36	0.26	0.33	0.29		0.01	10
V	0.34	330	220	170	260	210		0.33	20
Sc	0.69	40	32	25	12	16		0.40	15
Cr	5.4	410	320	410	120	120		1	20
Ni	0.89	180	94	110	57	28		0.10	10

*K = K_2O * 8300
**P = P_2O_5 * 4360
***Ti = TiO_2 * 6000
**P = P_2O_5 * 4360

(continued)

bodies that vary in width from a few cm to several m. Almost all are interfoliated with the enclosing Leech River Schist. The foliation in the intrusions is generally a fine-grained schistosity defined by muscovite and biotite. A few of the intrusions are weakly deformed to undeformed, indicating that magmatism may have outlasted the development of foliation. Many of the intrusions are isolated and separated from one another by tens to hundreds of meters. Others are clustered together, locally forming sheeted dike swarms up to 4 km long and 250 m wide (Fig. 6). The largest single dike is 50 m thick and 1 km long, and belongs to a swarm that crops out along Walker Creek. Many of the Walker Creek intrusions contain abundant metapelitic xenoliths that range from well-preserved clasts with detailed textures to partially digested, featureless ones with indistinct margins (Fig. 7). A thermal aureole similar to the one enclosing the Jordan River metagranodiorite surrounds the areas of concentrated Walker Creek magmatism.

In thin section, Walker Creek intrusions are commonly equigranular, and locally weakly plagioclase-phyric. Groundmass grain size ranges from 0.1 mm to 15 mm, averaging approximately 0.5 to 1.5 mm. Pressure shadows are developed around phenocrysts in some samples, indicative of sub-solidus deformation of these intrusions. A fabric is defined in several samples by the orientation of these pressure shadows, the preferential

TABLE 1. WHOLE ROCK GEOCHEMISTRY AND CIPW NORMATIVE MINERALOGY (continued)

Sample:	98-15-1-1B	98-33-3-2B	98-14-4-1B	98-5-4-2B	98-12-5-1B	98-10-2-2B	98-12-1-1B	98-1-1-1B	98-5-2-3B	98-9-3-2B	98-5-5-3B	99-1-1-1B	99-2-1-1B
Unit:	KJR	KJR	KJR	eTWC	eTWC	eTWC	eTWC	eTWC	eTWC	eTWC	KeTTC	KeTTC	KeTTC
Rock type:	GD	GD	GD	GD	GD	TO	TO	TR	TO	TR	MB	MB	MB
Fe_2O_3/FeO	0.20	0.20	0.20	0.20	0.20	0.20	0.20	0.20	0.20	0.20	0.20	0.20	0.20
Wt% Ap	0.28	0.26	0.28	0.47	0.14	0.31	0.07	0.17	0.16	0.26	0.28	0.19	0.30
Wt% Il	0.93	0.98	0.92	1.52	0.39	0.51	0.09	0.05	0.02	0.01	2.90	1.73	1.42
Wt% Or	8.88	11.06	10.54	8.03	4.67	8.75	17.38	12.84	14.01	12.21	0.84	1.36	3.33
Wt% Ab	33.47	30.55	32.51	33.82	51.19	38.79	37.16	37.29	40.88	35.06	16.95	14.07	21.28
Wt% An	14.99	14.89	16.07	22.27	13.49	14.50	6.46	7.06	4.68	2.58	30.55	45.17	27.95
Wt% C	0.90	1.65	0.86	0.00	0.67	2.48	1.53	2.67	2.94	5.33	0.00	0.00	0.00
Wt% Mt	0.87	0.92	0.85	1.32	0.35	0.61	0.24	0.30	0.18	0.15	2.77	2.21	2.21
Wt% Di	0.00	0.00	0.00	0.66	0.00	0.00	0.00	0.00	0.00	0.00	20.97	17.93	12.08
Wt% Hy	8.56	9.51	8.68	15.91	3.79	4.66	1.70	1.92	1.18	1.01	19.75	0.00	24.65
Wt% Ol	0.00	0.00	0.00	0.00	0.00	0.00	0.00	0.00	0.00	0.00	4.98	15.36	0.00
Wt %Q	31.13	30.20	29.29	16.00	25.32	29.39	35.37	37.70	35.95	43.39	0.00	0.00	6.78
Wt% Ne	0.00	0.00	0.00	0.00	0.00	0.00	0.00	0.00	0.00	0.00	0.00	1.98	0.00
SiO2 Def.	0.00	0.00	0.00	0.00	0.00	0.00	0.00	0.00	0.00	0.00	0.00	0.00	0.00
MIN SUM	100.00	100.00	100.00	100.00	100.00	100.00	100.00	100.00	100.00	100.00	100.00	100.00	100.00
Mol% Al_2O_3	0.14	0.15	0.15	0.16	0.16	0.17	0.14	0.15	0.15	0.15	0.14	0.20	0.15
Mol% Na_2O	0.06	0.06	0.06	0.06	0.10	0.07	0.07	0.07	0.08	0.07	0.03	0.03	0.04
Mol% K_2O	0.02	0.02	0.02	0.01	0.01	0.02	0.03	0.02	0.03	0.02	0.00	0.00	0.01
Mol% CaO	0.06	0.06	0.06	0.09	0.05	0.06	0.02	0.03	0.02	0.01	0.21	0.24	0.16
Mol% A/CNK	1.04	1.10	1.04	0.95	1.03	1.15	1.11	1.20	1.22	1.49	0.60	0.71	0.72
Mol% A/NK	1.79	1.89	1.82	2.01	1.52	1.85	1.37	1.55	1.44	1.69	4.24	5.48	3.16
Alumina class	Per	Per	Per	Met	Per	Per	Per	Per	Per	Per	Met	Met	Met
Rock Mg#	50.81	52.90	52.46	60.93	55.59	34.56	23.09	11.03	10.35	9.35	63.31	63.20	63.99
Silicate Min Mg#	56.96	58.90	58.51	66.96	61.83	39.34	25.32	11.92	10.86	9.56	68.67	67.62	67.89
Na_2O - K_2O<>2	Sod	Pot	Sod	Sod	Sod	Sod	Pot	Sod	Sod	Sod	Pot	Pot	Pot
Plag Ab%	70.34	68.54	68.24	61.71	80.12	73.97	85.93	84.87	90.28	93.53	37.07	29.40	44.69
Plag An%	29.66	31.46	31.76	38.29	19.88	26.03	14.07	15.13	9.72	6.47	62.93	70.60	55.31

Note: Normative minerals: Ab—albite; Il—ilmenite; An—anorthite; Mt—magnetite; Ap—apatite; Ne—nepheline; C—corundum; Ol—olivine; Di—diopside; Or—orthoclase; Hy—hypersthene; Q—quartz. Petrologic data: A/CNK = $Al_2O_3/(CaO + Na_2O + K_2O)$; A/NK = $Al_2O_3/(Na_2O + K_2O)$; Per = peraluminous (A/CNK > 1); Met = metaluminous (A/CNK < 1); Mg# = MgO/(MgO + FeO); Pot = potassic ($Na_2O - K_2O > 2$); Sod = sodic ($Na_2O - K_2O < 2$).

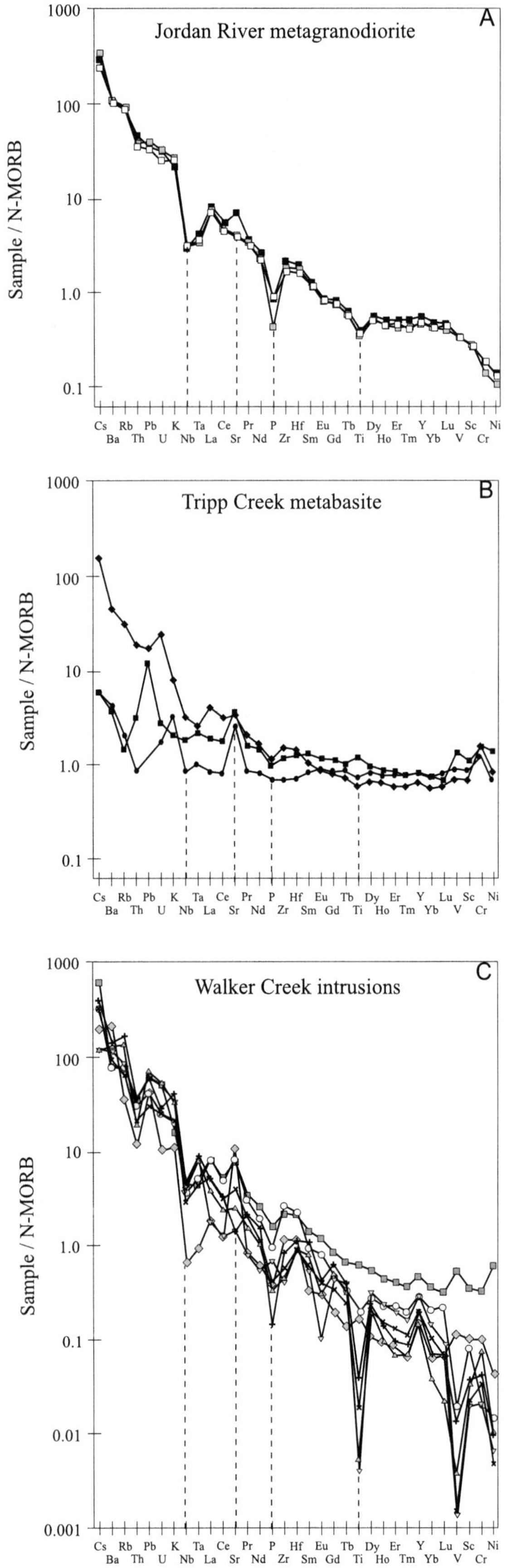

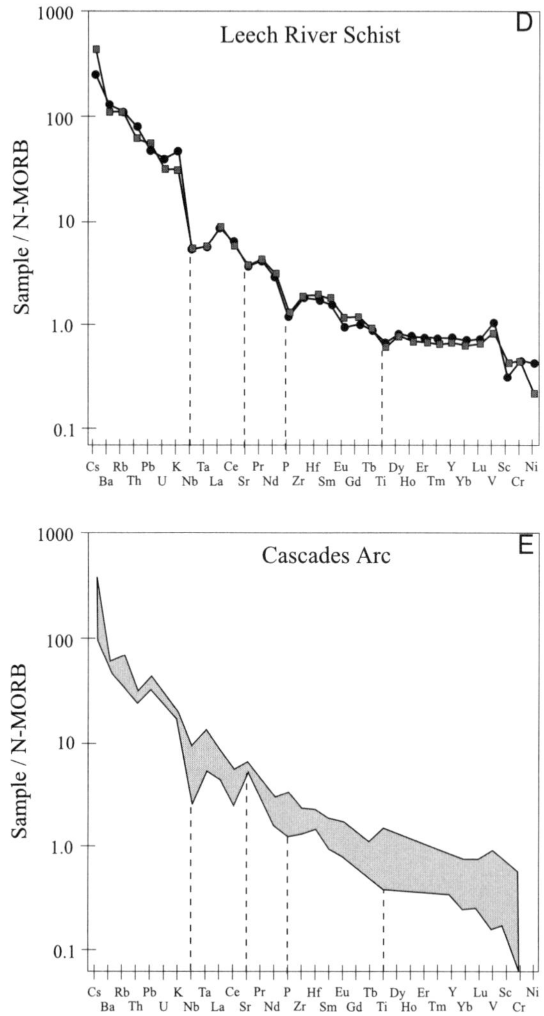

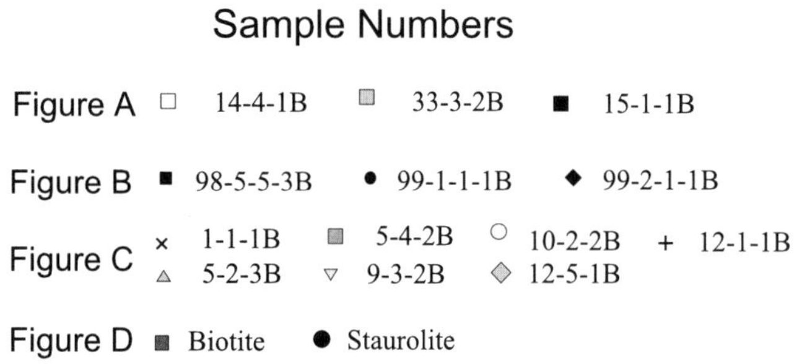

Figure 3. Multi-element geochemical plots from the Leech River Complex. A: Multi-element profile of the Jordan River metagranodiorite. Trace element data are presented in Table 1. N-MORB normalizing factors from Sun and McDonough (1989). B: Multi-element profile of the Tripp Creek metabasite. Trace element data are presented in Table 1. N-MORB normalizing factors from Sun and McDonough (1989). C: Multi-element profile of the Walker Creek intrusions. Trace element data are presented in Table 1. N-MORB normalizing factors from Sun and McDonough (1989). D: Multi-element profile of the Leech River Schist. Trace element data are presented in Table 1. N-MORB normalizing factors from Sun and McDonough (1989). E: Multi-element profile of samples from the Cascades Arc. Data are from Halliday et al., 1983.

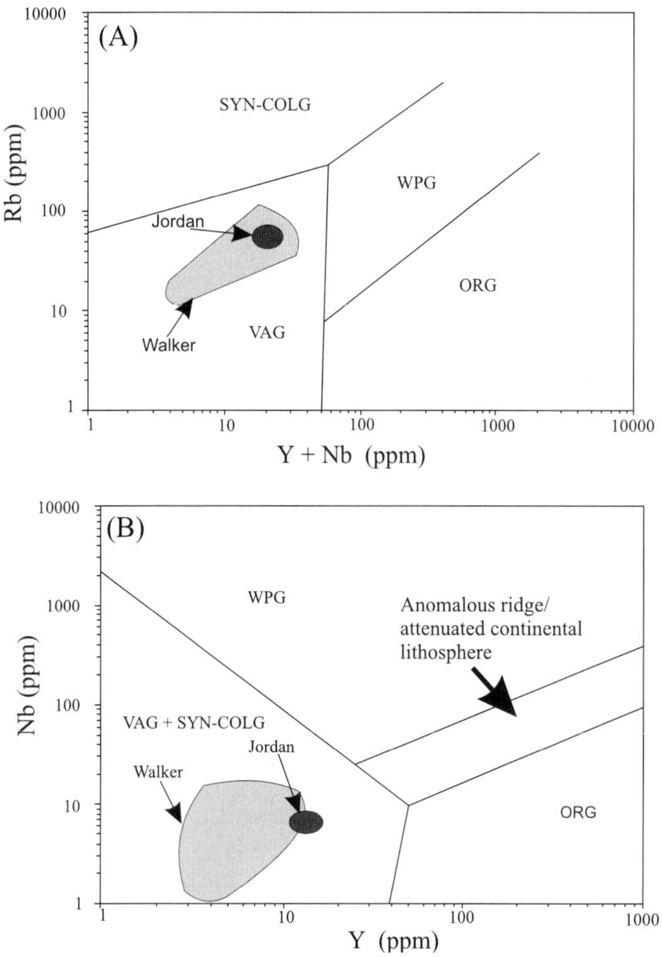

Figure 4. Tectonic affinity diagrams for granitoids from Pearce et al. (1984) with the Jordan River metagranodiorite and the Walker Creek intrusions displayed. A: Rb vs. Y + Nb. B: Nb vs. Y. SYN-COLG—syn-collisional granite, WPG—within plate granite, VAG—volcanic arc granite, ORG—orogenic granite.

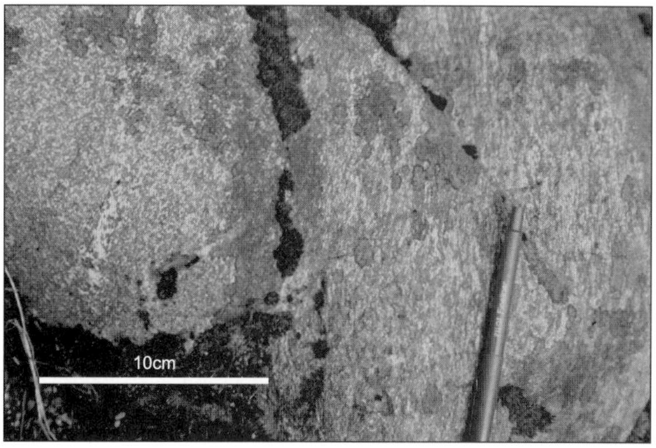

Figure 5. Photograph of gabbroic and schistose textures within the Tripp Creek metabasite.

Figure 6. Photograph of Walker Creek intrusions in outcrop. Only the three largest dikes in the outcrop are outlined. WCI—Walker Creek intrusion.

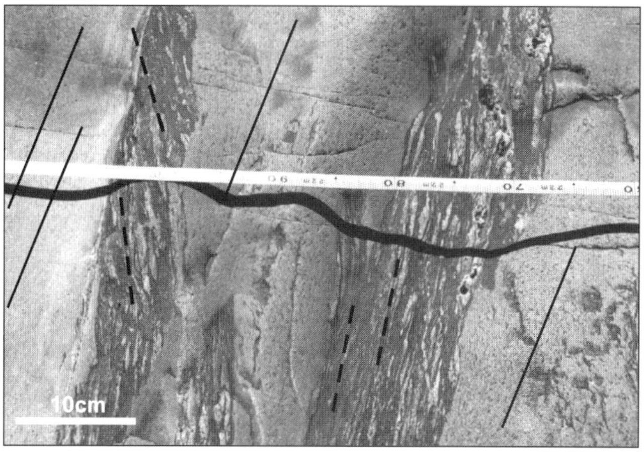

Figure 7. Photograph of xenoliths of Leech River Schist within the Walker Creek intrusions. Solid lines represent the fabric in the intrusions, while the dashed lines represent the fabric in the xenoliths.

orientation of mica, and by a weak preferential orientation of multi-domainal quartz and plagioclase.

Plagioclase feldspar, alkali feldspar, and quartz together constitute up to 90% by volume of the Walker Creek intrusions. Plagioclase (An_5–An_{30}) content in the Walker Creek intrusions contributes between 40% and 60% by volume. Primary alkali-feldspar ranges from 5% to 30%; secondary alkali-feldspar constitutes up to ~15%; and quartz ranges from 15% to 35%. Primary (igneous) alkali-feldspar was recognized optically by the presence of exsolution lamellae as well as textural relationships with other igneous minerals. The primary alkali-feldspars are dispersed throughout the groundmass of many samples and are generally more euhedral than the secondary feldspars. Secondary (alteration) alkali feldspar was recognized optically based on the lack of exsolution lamellae as well as textural relationships with the igneous minerals. Secondary alkali-feldspar commonly occurs in small veinlets that fill fractures in the samples. In thin section, most quartz occurs as multi-domainal masses, which appear to have been single crystals prior to deformation. Quartz is also dispersed throughout the groundmass of most samples.

The main ferro-magnesian silicate in the Walker Creek intrusions is biotite, which makes up approximately 1% to 20%. In some samples, biotite is partially pseudomorphed by chlorite. However, biotite does not appear to be an alteration product of hornblende in any Walker Creek intrusion samples investigated, suggesting that biotite is at least partly igneous in origin, and not solely the result of metamorphism. Biotite is commonly dispersed throughout the groundmass, but in some samples is a micro-phenocryst (~1.5 times the size of the groundmass crystals).

Muscovite (10% to 20% of rock volume) occurs in most of the Walker Creek intrusions. It is primarily a groundmass phase, but also occurs as larger grains, as well as in pressure shadows (discussed above). Groundmass muscovite is commonly preferentially oriented, but coarser-grained muscovite is commonly bent and fractured, with fracture planes slightly discordant to the fabric defined by groundmass muscovite.

Other minerals in the Walker Creek intrusion samples include garnet (<5%), apatite (trace amounts), titanite (trace amounts), and opaques. Garnet occurs as large groundmass crystals (<1 mm) and small phenocryst grains (<2 mm) in several Walker Creek samples. Most of the garnets do not contain inclusions, nor do they have pressure shadows developed around them. This relation suggests that most of the garnet in the Walker Creek intrusions is primary (igneous). Apatite occurs as small (<0.1 mm) nonaligned inclusions in plagioclase crystals. Opaque minerals in the Walker Creek intrusions include magnetite, hematite, and minor chalcopyrite.

Major and trace element analyses and CIPW norms for seven samples (Table 1) support a sedimentary source for the Walker Creek intrusions. All but one are peraluminous, and all but two are sodic. Trace element profiles (Fig. 3C) of the most primitive intrusions (those with highest Mg/Fe, Ca/Na, V, Cr, and Ni) are similar to those of the Jordan River metagranodiorite and the Leech River Schist. The more evolved samples generally have lower trace element abundances.

Petrogenetic modeling carried out by Groome (2000) suggests that the Walker Creek intrusions were derived through (1) partial melting of the Leech River Schist, (2) magma mixing with the Tripp Creek metabasite (or other N-MORB magma), and (3) fractional crystallization. The peraluminous nature of the Walker Creek intrusions, and the abundance of metapelitic xenoliths, point toward the Leech River Schist as a potential source rock (at greater crustal depths). However, melting models using multiple trace elements indicated that the Leech River Schist is an unlikely source for the evolved compositions of the Walker Creek intrusions, and could be the source of the most primitive Walker Creek intrusions only if the schist underwent partial melting in excess of about 70% (high degrees of partial melting would also yield a match with the Jordan River metagranodiorite). At a lower and more reasonable degree of anatexis, e.g., 10%, the resultant liquid has a trace element profile dissimilar to the Walker Creek intrusions. However, mixing of this anatectic liquid with a magma equivalent to the Tripp Creek metabasite yields a composition similar to that of the most primitive Walker Creek intrusions (and the Jordan River metagranodiorite). The Tripp Creek metabasite was chosen as a mixing component in the model because field relations indicate that this unit may have been present as a magma at the time of generation of the Walker Creek intrusions. A generic N-MORB composition provides similar results. Modeling of fractional crystallization from this hybrid magma yielded compositions similar to those of the more evolved Walker Creek intrusions.

GEOCHRONOLOGY OF THE LEECH RIVER COMPLEX

Conventional U-Pb dating of zircon, titanite, and monazite, and $^{40}Ar/^{39}Ar$ dating of muscovite and biotite, provide new constraints on the depositional, magmatic/metamorphic and cooling histories of the Leech River Complex. Prior to these new isotopic ages, the only direct age constraints on the Leech River Complex were a 39 Ma K-Ar cooling date on muscovite of the Walker Creek intrusions (Fairchild and Cowan, 1982), and Rb-Sr data that Fairchild and Cowan (1982) interpreted to indicate a Jura-Cretaceous depositional age.

U-Pb Dating

Detrital Zircons from the Leech River Schist

Metasedimentary staurolite-grade Leech River Schist (sample WG99-ST) was sampled for U-Pb dating of metamorphic titanite or monazite and detrital zircon, if present. Titanite and detrital zircon were recovered, but the sample did not yield monazite. Titanite analyses were attempted, but the material was not suitable for U-Pb dating, due to very low U concentrations (~0.5–1.5 ppm) and high common Pb (>90% of total Pb was nonradiogenic). The sample did yield a small quantity (12 grains) of sufficiently coarse (>100 μm) clear, colorless detrital zircon for single-grain analysis. After strong abrasion, 8 of these grains were selected for analysis. Results are listed in Table 2 and plotted on Figure 8.

Grains A and C are 6% and 13% discordant, and yielded $^{207}Pb/^{206}Pb$ dates of ca. 1389 Ma and 1376 Ma, respectively (Table 2). Due to their significant discordance, these dates are not considered to be reliable age indicators. However, reliably dated detrital zircons that gave similar ages are known in miogeoclinal and pericratonic sedimentary and metasedimentary rocks from southern British Columbia and the western Cordillera of the conterminous United States; grains of this age are not common in the northern Canadian Cordillera (Ross and Parrish, 1991; Gehrels et al., 1995).

The remaining six detrital zircon grains gave Mesozoic Pb/U results. Four are concordant and two are strongly discordant;

TABLE 2. GEOCHRONOLOGY ANALYTICAL DATA FOR ROCKS OF THE LEECH RIVER COMPLEX

U-PB Analytical Data												
Fraction[†]	Wt	U[§]	Pb*[#]	^{206}Pb[††]	Pb[§§]	^{208}Pb[##]	Isotopic ratios (1σ,%)[†††]			Apparent ages (2σ,Ma)[†††]		
	(mg)	(ppm)	(ppm)	^{204}Pb	(pg)	(%)	$^{206}Pb/^{238}U$	$^{207}Pb/^{235}U$	$^{207}Pb/^{206}Pb$	$^{206}Pb/^{238}U$	$^{207}Pb/^{235}U$	$^{207}Pb/^{206}Pb$
WG98-1-1-1C												
M2 m,e,p,12	0.041	1656	75	629	55	84	0.00795 (0.12)	0.0513 (0.37)	0.04675 (0.29)	51.1 (0.1)	50.8 (0.4)	36 (13)
M3 f,t,13	0.042	1662	59	725	48	80	0.00770 (0.22)	0.0503 (0.53)	0.04734 (0.44)	49.4 (0.2)	49.8 (0.5)	67 (21)
M4 ff,e,p,9	0.018	1129	41	664	16	80	0.00798 (0.13)	0.0515 (0.49)	0.04676 (0.43)	51.3 (0.1)	51.0 (0.5)	37 (21)
C c,cl,s,p,1	0.020	347	24	8192	3.6	7.2	0.06777 (0.11)	0.7373 (0.17)	0.07891 (0.08)	422.7 (0.9)	560.8 (1.4)	1169.8 (3.1)
D m,cl,p,3	0.016	335	49	19580	2.5	6.6	0.14621 (0.11)	1.7584 (0.16)	0.08723 (0.08)	879.7 (1.7)	1030.2 (2.1)	1365.5 (2.9)
E f,cl,p,s,5	0.014	183	8.4	2606	2.7	12	0.04325 (0.13)	0.4765 (0.18)	0.07990 (0.13)	273.0 (0.7)	395.7 (1.2)	1194.5 (5.1)
F ff,cl,p,s,15	0.011	301	30	7244	2.8	7.2	0.09778 (0.11)	1.1755 (0.17)	0.08719 (0.09)	601.4 (1.3)	789.2 (1.9)	1364.7 (3.4)
G f,b,p,e,2	0.005	4727	31	3128	3.5	0.2	0.00735 (0.10)	0.0480 (0.20)	0.04736 (0.14)	47.2 (0.1)	47.6 (0.2)	67.4 (6.5)
H ff,b,p,e,20	0.016	10280	69	430	186	0.8	0.00737 (0.16)	0.0482 (0.55)	0.04739 (0.46)	47.4 (0.2)	47.8 (0.5)	69 (22)
WG98-5-4-1C												
M1 c,cl,10	0.075	2224	93	129	776	82	0.00803 (0.53)	0.0511 (1.8)	0.04619 (1.5)	51.6 (0.5)	50.6 (1.8)	8 (71/74)
M3 m,cl,12	0.063	2122	91	126	627	83	0.00790 (0.54)	0.0513 (1.8)	0.04710 (1.5)	50.7 (0.5)	50.8 (1.8)	54 (68/71)
M4 m,cl,11	0.058	2008	85	160	412	83	0.00793 (0.32)	0.0512 (1.3)	0.04679 (1.1)	50.9 (0.3)	50.7 (1.3)	39 (53/55)
A m,b,cl,e,p,5	0.010	3781	28	1696	11	3.2	0.00799 (0.16)	0.0521 (0.34)	0.04729 (0.28)	51.3 (0.2)	51.5 (0.3)	64 (13/14)
B m,b,e,16	0.027	8063	59	1959	55	3.0	0.00783 (0.13)	0.0521 (0.25)	0.04828 (0.17)	50.3 (0.1)	51.6 (0.3)	113.2 (8.2)
C m,b,e,8	0.011	9949	72	1614	34	3.1	0.00775 (0.14)	0.0505 (0.26)	0.04724 (0.17)	49.8 (0.1)	50.0 (0.3)	61.3 (8.0)
D c,p,10	0.027	302	41	5055	14	4.1	0.13774 (0.27)	2.0588 (0.30)	0.10841 (0.08)	831.9 (4.2)	1135.2 (4.0)	1772.8 (3.0)
E c,p,5	0.023	766	17	3833	6.3	11.2	0.02196 (0.14)	0.2191 (0.21)	0.07236 (0.12)	140.0 (0.4)	201.2 (0.8)	996.0 (4.9)
WG98-32-2-1												
A c,cl,p,s,3	0.068	398	9.0	3520	11	8.0	0.02290 (0.23)	0.1567 (0.29)	0.04962 (0.18)	146.0 (0.7)	147.8 (0.8)	177 (8.5)
B c,cl,p,s,8	0.086	198	2.9	171	104	10.4	0.01456 (0.38)	0.1014 (1.2)	0.05052 (0.98)	93.2 (0.7)	98.1 (2.3)	219 (45/46)
C c,cl,s,p,12	0.075	250	10	295	172	6.5	0.03974 (0.24)	0.3965 (0.52)	0.07237 (0.38)	251.2 (1.2)	339.1 (3.0)	996 (15/16)
D m,cl,p,s,4	0.052	274	14	8057	5.6	7.8	0.05048 (0.09)	0.5461 (0.16)	0.07846 (0.08)	317.5 (0.6)	442.5 (1.1)	1158.5 (3.1)
E m,cl,p,e	0.093	701	45	57738	4.6	6.2	0.06505 (0.08)	0.7621 (0.14)	0.08498 (0.07)	406.2 (0.6)	575.2 (1.3)	1315.0 (2.8)
F f,cl,p,e	0.030	405	22	14542	2.7	9.4	0.05211 (0.09)	0.5980 (0.15)	0.08324 (0.08)	327.4 (0.6)	476.0 (1.1)	1274.8 (2.9)
G c,p,s,1	0.005	904	153	11731	2.7	5.4	0.17362 (0.13)	1.7788 (0.18)	0.07431 (0.08)	1032.0 (2.5)	1037.7 (2.4)	1049.8 (3.3)
H c,p,s,1	0.004	148	46	4599	2.4	9.4	0.29367 (0.13)	4.3878 (0.18)	0.10836 (0.09)	1659.9 (3.8)	1710 (3.0)	1772.1 (3.2)
T1 c	0.185	111	2.6	44	708	47	0.01360 (2.2)	0.0877 (7.6)	0.04678 (6.4)	87.1 (3.8)	85 (13)	38 (280/338)
T2 m	0.125	254	4.0	146	218	22	0.01377 (0.46)	0.0926 (1.7)	0.04875 (1.4)	88.2 (0.8)	89.9 (2.9)	136 (65/68)

(continued)

TABLE 2. GEOCHRONOLOGY ANALYTICAL DATA FOR ROCKS OF THE LEECH RIVER COMPLEX (continued)

							U-PB Analytical Data					
							Isotopic ratios (1σ,%)[†††]			Apparent ages (2σ,Ma)[†††]		
Fraction[†]	Wt (mg)	U[§] (ppm)	Pb*[#] (ppm)	$^{206}Pb/^{204}Pb$[††]	Pb[§§] (pg)	^{208}Pb[##] (%)	$^{206}Pb/^{238}U$	$^{207}Pb/^{235}U$	$^{207}Pb/^{206}Pb$	$^{206}Pb/^{238}U$	$^{207}Pb/^{235}U$	$^{207}Pb/^{206}Pb$
WG98-7-1-1C												
A c,cl,p,s,5	0.070	329	8.6	5891	6.2	12	0.02532 (0.14)	0.1739 (0.19)	0.04980 (0.10)	161.2 (0.4)	162.8 (0.6)	185.8 (4.5)
B c,cl,p,s,5	0.055	698	45	28588	5.5	5.7	0.06585 (0.33)	0.7268 (0.36)	0.08005 (0.08)	411.1 (2.7)	554.7 (3.0)	1198.1 (3.3)
C c,cl,p,11	0.040	711	12	8720	3.6	7.4	0.01766 (0.11)	0.1515 (0.16)	0.06224 (0.09)	112.8 (.02)	682.4 (0.4)	682.4 (3.8)
D c,cl,p,e,6	0.052	556	39	14730	8.8	3.6	0.07212 (0.11)	0.8056 (0.17)	0.08101 (0.07)	448.9 (1.0)	600.0 (1.5)	1221.7 (2.9)
E c,p,16	0.074	219	8.0	3121	11.8	12	0.03481 (0.09)	0.3464 (0.17)	0.07218 (0.09)	220.6 (0.4)	302.0 (0.9)	990.9 (3.8)
F m,p,e,20	0.023	378	22	13891	2.3	8.0	0.05799 (0.09)	0.7744 (0.15)	0.09685 (0.07)	363.4 (0.6)	582.3 (1.3)	1564.2 (2.8)
I c,pi,p,e,1	0.009	420	96	1365	37	8.2	0.22350 (0.22)	2.7808 (0.30)	0.09024 (0.13)	1300.3 (5.2)	1350.4 (4.4)	1430.6 (4.9)
J c,pi,p,s,1	0.010	217	57	2565	13	5.9	0.26124 (0.11)	3.6746 (0.18)	0.10201 (0.10)	1496.2 (2.9)	1565.9 (2.9)	1661.1 (3.6)
WG99ST												
A m,cl,co,br,1	0.002	1329	294	27464	1.4	4.5	0.22595 (0.10)	2.7513 (0.16)	0.08831 (0.08)	1313.2 (2.5)	1342.4 (2.4)	1389.3 (2.9)
B m,cl,co,br,1	0.002	63	1.4	136	1.4	11	0.02192 (0.54)	0.2016 (2.1)	0.06671 (2.0)	139.8 (1.5)	186.5 (7.2)	828 (82/86)
C m,cl,co,br,1	0.003	389	80	2450	6.0	5.8	0.20703 (0.12)	2.5033 (0.19)	0.08769 (0.11)	1213.0 (2.8)	1273.0 (2.8)	1375.8 (4.2)
D m,cl,co,e,1	0.002	723	20	469	5.3	11	0.02700 (0.15)	0.1843 (0.57)	0.04951 (0.51)	171.7 (0.5)	171.8 (1.8)	172.2 (23/24)
E m,cl,co,e,1	0.002	136	3.0	87	5.6	7.5	0.02274 (0.44)	0.1546 (3.5)	0.04930 (3.3)	144.9 (1.3)	145.9 (9.6)	162 (147/162)
F m,cl,co,e,1	0.004	1021	29	851	8.9	6.5	0.02916 (0.14)	0.2005 (0.42)	0.04986 (0.35)	185.3 (0.5)	185.5 (1.4)	189 (16/17)
G m,cl,co,e,1	0.002	1280	21	474	5.6	10	0.01612 (0.16)	0.1069 (0.88)	0.04813 (0.83)	103.1 (0.3)	103.2 (1.7)	106 (39/40)
H m,cl,co,st,1	0.002	544	16	368	5.0	17	0.02573 (0.32)	0.2653 (0.75)	0.07478 (0.68)	163.8 (1.0)	238.9 (3.2)	1063 (27/28)

[†] Upper case letter—zircon fraction identifier. All zircon fractions air abraded except WG98-1-1-1C: G and H. Monazites and titanites not abraded. Grain size, intermediate dimension: c—>134mm, m—<134mm and >104mm, f—<104mm and >74mm, ff <74mm. Grain character codes: b—brown, br—broken pieces, cl—clear, co—colorless, e—elongate, p—prismatic, pi—pink, s—stubby, t—tabular. The last numeral represents number of grains analyzed if 20 or less. Zircons nonmagnetic on Franz magnetic separator at field strength of 1.8A and sideslopes of 1°–5°. Monazites and titanites nonmagnetic at 0.6A and magnetic at 1.8A and 20° sideslope. Front slope of 20° for all.

[§] U blank correction of 1pg ± 20%; U fractionation corrections were measured for each run with a double 233U-235U spike (about 0.004/amu).

[#] Radiogenic Pb.

[††] Measured ratio corrected for spike and Pb fractionation of 0.0035/amu ± 20% (Daly collector) which was determined by repeated analysis of NBS Pb 981 standard throughout the course of this study.

[§§] Total common Pb in analysis based on blank isotopic composition.

[##] Radiogenic Pb.

[†††] Corrected for blank Pb (1-4 pg, throughout the course of this study), U (1 pg) and common Pb concentrations based on Stacey Kramer's model Pb at the age of the rock or the 207Pb/206Pb age of the rock.

(continued)

TABLE 2. GEOCHRONOLOGY ANALYTICAL DATA FOR ROCKS OF THE LEECH RIVER COMPLEX (continued)

$^{40}Ar/^{39}Ar$ Analytical Data
WG98 1-1-1C White Mica

Weighted average of J from standards = 0.008484 ± 0.000029

Laser Power (mW)	Cumulative ^{39}Ar	$^{40}Ar/^{39}Ar$ measured	37Ar/39Ar measured	$^{36}Ar/^{39}Ar$ measured	Atmospheric ^{40}Ar (%)	Ca/K	±	Cl/K	±	40*/39K	±	Age (Ma)	± (Ma)
150	0.0013	29.97307	0.03471	0.09346	92.21999	0.06368	0.01904	0.0042	0.00103	2.32973	1.10025	35.31	16.51
300	0.0044	9.2029	0.028	0.02348	75.59119	0.05138	0.0118	0.00102	0.00062	2.23935	0.70233	33.95	10.55
450	0.0126	6.31856	0.016	0.01151	54.07474	0.02936	0.00683	0.00061	0.00014	2.88867	0.22819	43.68	3.41
600	0.0276	7.78988	0.00758	0.0153	58.23978	0.0139	0.00204	0.00052	0.00023	3.2411	0.07793	48.94	1.16
800	0.0628	6.12156	0.00243	0.01079	52.34594	0.00446	0.0009	0.0003	0.00008	2.9035	0.06375	43.9	0.95
1000	0.1446	3.79326	0.00025	0.00258	20.28835	0.00045	0.00058	0.00042	0.00003	3.00079	0.02805	45.35	0.42
1250	0.2611	3.85587	0.00013	0.00282	21.78205	0.00024	0.00024	0.00038	0.00004	2.99353	0.02656	45.25	0.4
1500	0.4083	3.87075	0.00028	0.00291	22.36911	0.00051	0.00031	0.00039	0.00004	2.98262	0.02037	45.08	0.3
2000	0.64	3.6225	0.0003	0.00193	15.89645	0.00056	0.00016	0.0004	0.00003	3.02252	0.01118	45.68	0.17
2500	0.8212	3.38153	0.00013	0.00132	11.67189	0.00024	0.00019	0.00043	0.00002	2.96149	0.011	44.77	0.16
3500	0.9018	3.10283	-0.0002	0.00036	3.48614	-0.00037	0.0005	0.00038	0.00007	2.96696	0.02018	44.85	0.3
4000	0.9153	3.10143	0.00058	0.00111	10.65845	0.00106	0.00167	0.00067	0.00027	2.74522	0.12819	41.54	1.92
8500	1	3.06857	-0.00051	0.00014	1.32106	-0.00094	0.00034	0.00037	0.00003	2.99971	0.01498	45.34	0.22
Integrated		3.78538	0.00058	0.00261	20.54261	0.00106	0.00013	0.00041	0.00001	2.98496	0.00764	45.12	0.19
Plateau	45.2 ± 0.2												
Isochron	45.0 ± 0.3		40/36i =	301.8 ± 6.0									

$^{40}Ar/^{39}Ar$ Analytical Data
WG98-4-6-1C Biotite

Weighted average of J from standards = 0.008484 ± 0.000029

Laser Power (mW)	Cumulative ^{39}Ar	$^{40}Ar/^{39}Ar$ measured	$^{37}Ar/^{39}Ar$ measured	$^{36}Ar/^{39}Ar$ measured	Atmospheric ^{40}Ar (%)	Ca/K	±	Cl/K	±	40*/39K	±	Age (Ma)	± (Ma)
150	0.0074	75.87774	0.02135	0.25531	99.46564	0.03918	0.00539	0.00084	0.00019	0.40532	0.62096	6.19	9.47
300	0.0221	16.47513	0.01108	0.04742	85.20272	0.02033	0.0024	0.0009	0.00013	2.43364	0.17192	36.87	2.58
450	0.069	6.01006	0.00783	0.01135	56.08662	0.01436	0.0007	0.00064	0.00011	2.62663	0.10589	39.76	1.59
600	0.1378	4.80107	0.01099	0.00674	41.7284	0.02016	0.0007	0.00056	0.00007	2.78096	0.0588	42.07	0.88
800	0.2072	4.28869	0.01475	0.00502	34.80232	0.02706	0.00068	0.00052	0.00007	2.77744	0.05226	42.02	0.78
1000	0.2985	3.82709	0.02157	0.00336	26.13574	0.03957	0.0006	0.00056	0.00004	2.80569	0.03841	42.44	0.57
1250	0.4123	4.06778	0.02425	0.00418	30.54542	0.0445	0.00078	0.00049	0.00003	2.80537	0.03497	42.43	0.52
1500	0.525	3.85028	0.022	0.00345	26.62076	0.04037	0.0007	0.00049	0.00005	2.80428	0.03406	42.42	0.51
2000	0.6913	3.60386	0.03814	0.00261	21.47938	0.06998	0.00056	0.0005	0.00003	2.80731	0.01728	42.46	0.26
2500	0.8217	3.62523	0.04806	0.00263	21.4696	0.08818	0.00057	0.00051	0.00004	2.82446	0.03067	42.72	0.46
3500	0.8638	3.70205	0.04145	0.00268	21.48364	0.07605	0.00152	0.00049	0.00011	2.88426	0.07968	43.61	1.19
4000	0.9612	3.74785	0.04861	0.00307	24.27361	0.08919	0.00143	0.00058	0.00006	2.81647	0.02781	42.6	0.42
8500	1	3.21913	0.02964	0.00094	8.61091	0.0544	0.0012	0.00057	0.0001	2.91576	0.0632	44.08	0.94
Integrated		4.67665	0.02991	0.00632	40.15576	0.05489	0.00026	0.00053	0.00002	2.78158	0.01316	42.08	0.24
Plateau	42.5 ± 0.2												
Isochron	43.0 ± 0.2		40/36i =	285.8 ± 2.0									

the latter two are likely to contain old inherited cores. The four concordant grains gave $^{206}Pb/^{238}U$ dates of ca. 185, 172, 145 and 103 Ma, which are interpreted as good estimates of their magmatic ages. The youngest grain, which gave concordant results at 103 Ma (grain G), was an elongate prism interpreted as a detrital zircon of igneous origin. The interpreted age of this grain indicates that the sedimentary protolith of this schist, and probably much of the metasedimentary component of the Leech River Complex, was deposited after ca. 103 Ma. The two older concordant grains give ages typical for the Island Intrusions and Westcoast Crystalline Complex of Wrangellia on Vancouver Island (DeBari et al., 1999), and the younger two concordant grains have interpreted ages common in the southern Coast Plutonic Complex (Friedman and Armstrong, 1995). However, igneous rocks of these ages are known along the length of the Cordillera and are thus not latitudinally significant. Discordant grains B and H appear to contain significant old inheritance and were probably derived from igneous rocks that intruded crust containing components of old zircon. Candidates for such crust include North American continental crust, overlying miogeoclinal sequences, and flanking pericratonic terranes.

Igneous Rocks of the Leech River Complex

The Jordan River metagranodiorite was sampled for U-Pb dating at two localities (Fig. 2). Both samples (WG98-32-2-1C and WG98-7-3-2C) are fine-grained, foliated granodiorites. Zircon recovered from these consists of clear, colorless, euhedral grains and rare, clear, pink grains. In addition, a small

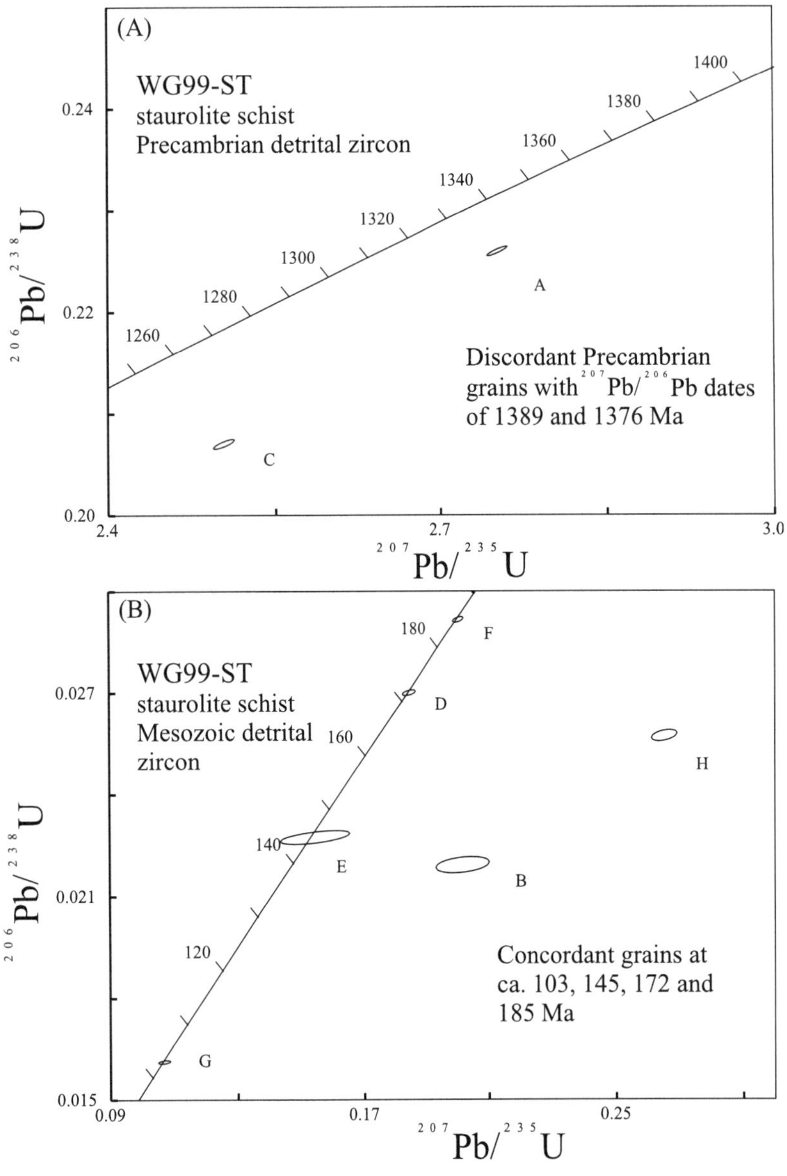

Figure 8. Concordia plots showing U-Pb results for sample WG99-ST. U-Pb analytical data are presented in Table 2 and plotted at the 2 level of uncertainty. A: Detail of Precambrian grains. B: Detail of Mesozoic grains. All fractions are single grain analyses. A and C are stubby, pink zircon grains; all other fractions are colorless stubby fractions.

quantity of clear, yellow titanite was recovered from one of the samples (WG98-32-2-1C).

Discordant results for multi-grain, colorless zircon fractions from both samples indicate the presence of Precambrian to Mesozoic inheritence but do not lead to a precise age estimate for the intrusion (Figs. 9, 10). Single pink grains analyzed from both samples also give discordant results, with $^{207}Pb/^{206}Pb$ dates ranging from 1050 to 1800 Ma. Based on their distinctive physical characteristics and U-Pb systematics, pink zircons are interpreted as Precambrian xenocrysts. Concordant overlapping results were obtained for two multi-grain titanite fractions. Relatively precise fraction T2, with a $^{206}Pb/^{238}U$ date of 88.2 ± 0.8 Ma is interpreted as the time at which the Jordan River metagranodiorite (sample WG98-32-2-1; Fig. 10B) cooled below the closure temperature for titanite, which is estimated to be 660 °C or greater (Frost et al., 2000). Thus, this titanite date is interpreted as the crystallization age for the Jordan River metagranodiorite. Results for discordant zircon fraction B (Fig. 10B) nearly overlap with those of the titanites (Table 2), which supports the interpreted Late Cretaceous age.

The Walker Creek tonalite-trondhjemite-granodiorite suite of intrusions was sampled for U-Pb dating at two localities: one

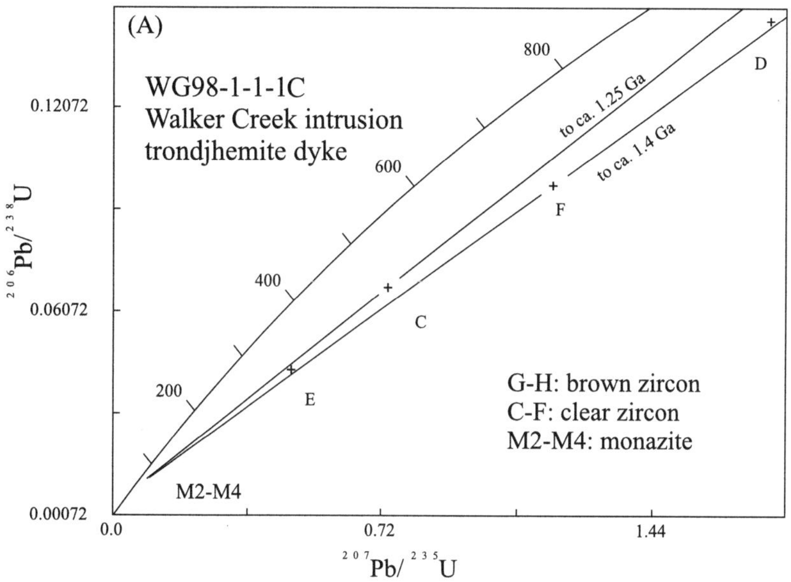

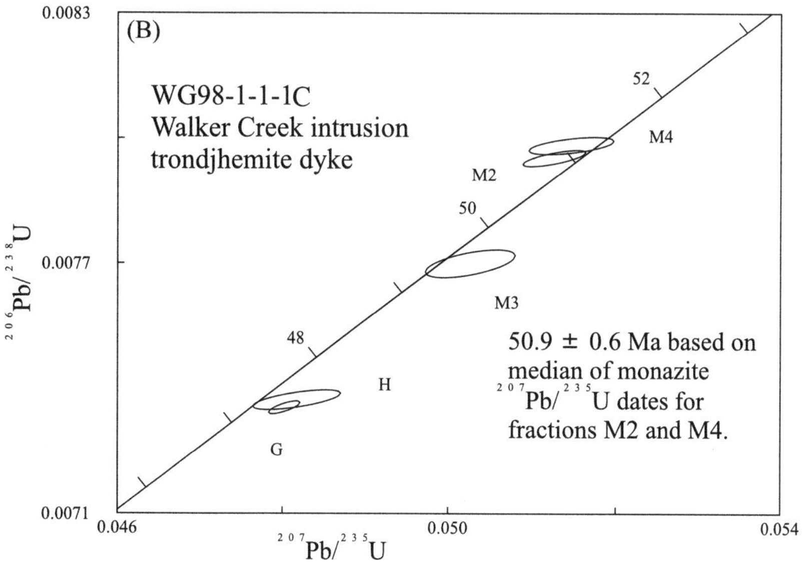

Figure 9. Concordia plots showing U-Pb results for sample WG98-1-1-1C (Walker Creek intrusion). U-Pb analytical data are presented in Table 2 and plotted at the 2 level of uncertainty. A: All fractions. B: Expanded plot showing detail of monazite and brown zircon fractions.

from within a swarm (foliated trondhjemite, WG98-1-1-1C), and the other from an isolated intrusion (weakly foliated tonalite, WG98-5-4-1C; Fig. 2). Monazite and two populations of zircon were recovered and analyzed from each sample. One zircon population consisted of clear, colorless, equant to stubby grains, and the other was made up of brown, elongate prisms. Cores and zoning were not observed in either zircon type. Monazites were clear, pale yellow, nearly inclusion-free, subhedral grains.

Multi-grain clear zircon fractions from both samples yielded discordant results, with $^{207}Pb/^{206}Pb$ dates ranging from 1000 Ma to 1800 Ma, indicating the presence of significant components of old inherited zircon, which appears to occur as cryptic cores (Figs. 11A, 11B; Table 2). Multi-grain brown zircon fractions from both samples contain high concentrations of uranium (~4000–10,000 ppm; Table 2), and give slightly discordant results, with Pb/U dates of 47 Ma to 51 Ma. The slight discordance is attributed to minor inherited zircon, and for most fractions, minor Pb loss (Figs. 11B, 12B). Multi-grain monazite fractions from both samples are concordant to reversely discordant. Interpreted igneous crystallization ages are based on median $^{207}Pb/^{235}U$ dates: 50.9 ± 0.6 Ma for sample WG98-1-1-1C (fractions M2 and M4) and 50.7 ± 1.9 Ma for sample WG98-5-4-1C (fractions M1, M3, and M4) (Figs. 9 and 10 respectively). Discordia lines fit from the interpreted igneous age of each rock, through discordant clear zircon fractions, give upper intercept ages of 1250 Ma to 1800 Ma, which are interpreted as the range of average ages of inherited zircon in the clear grains (Figs. 12A, 12B).

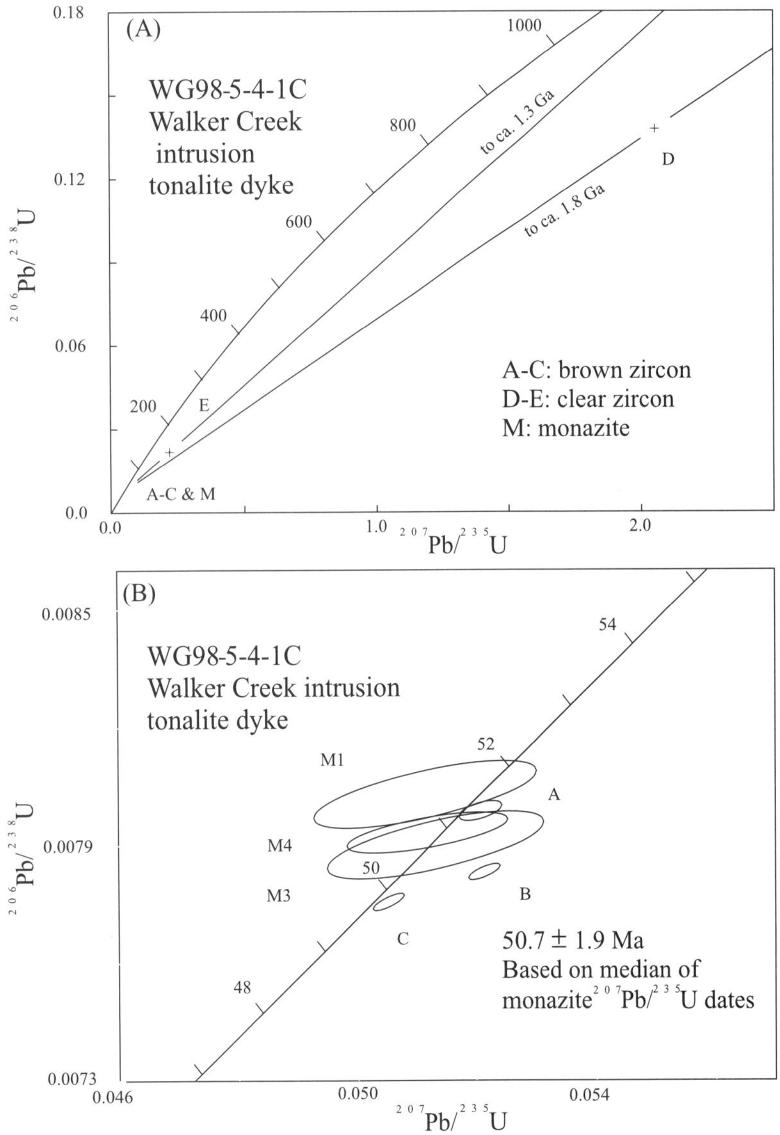

Figure 10. Concordia plots showing U-Pb results for sample WG98-5-4-1C (Walker Creek intrusion). U-Pb analytical data are presented in Table 2 and plotted at the 2 level of uncertainty. A: All fractions. B: Expanded plot showing detail of monazite and brown zircon fractions.

$^{40}Ar/^{39}Ar$ *Cooling Ages*

$^{40}Ar/^{39}Ar$ dating of two mica samples was undertaken in order to delineate a cooling history for the Leech River Complex. Muscovite was sampled from a Walker Creek intrusion (WG98-1-1-1C) (Fig. 13A), and biotite from a sample of amphibolite-facies Leech River Schist (WG98-4-6-1C) (Fig. 13B); the two sample localities are approximately 50 m apart. Plateau ages of 45.2 ± 0.2 Ma and 42.5 ± 0.2 Ma were obtained from the muscovite and biotite samples, respectively. On the basis of Ar-blocking temperatures for micas (Kirschner et al., 1996), the muscovite cooled below approximately 400 °C at ca. 45.2 Ma and the biotite cooled below approximately 330 °C at ca. 42.5 Ma.

METAMORPHISM AND DEFORMATION

Metamorphic grade within the Leech River Complex varies from chlorite zone (greenschist facies) in the north to staurolite-andalusite zone (amphibolite facies) in the south (Fairchild and Cowan, 1982; Rusmore and Cowan, 1985), a gradient that generally correlates with the increasing abundance of intrusive rocks in the southern part of the eastern Leech River Complex (Groome et al., 1999; Groome, 2000). In and near the study area, in the eastern part of the Complex, field and petrographic data indicate that there were at least two periods of metamorphism: a regional greenschist facies event and a localized amphibolite facies event spatially associated with contact aureoles adjacent to

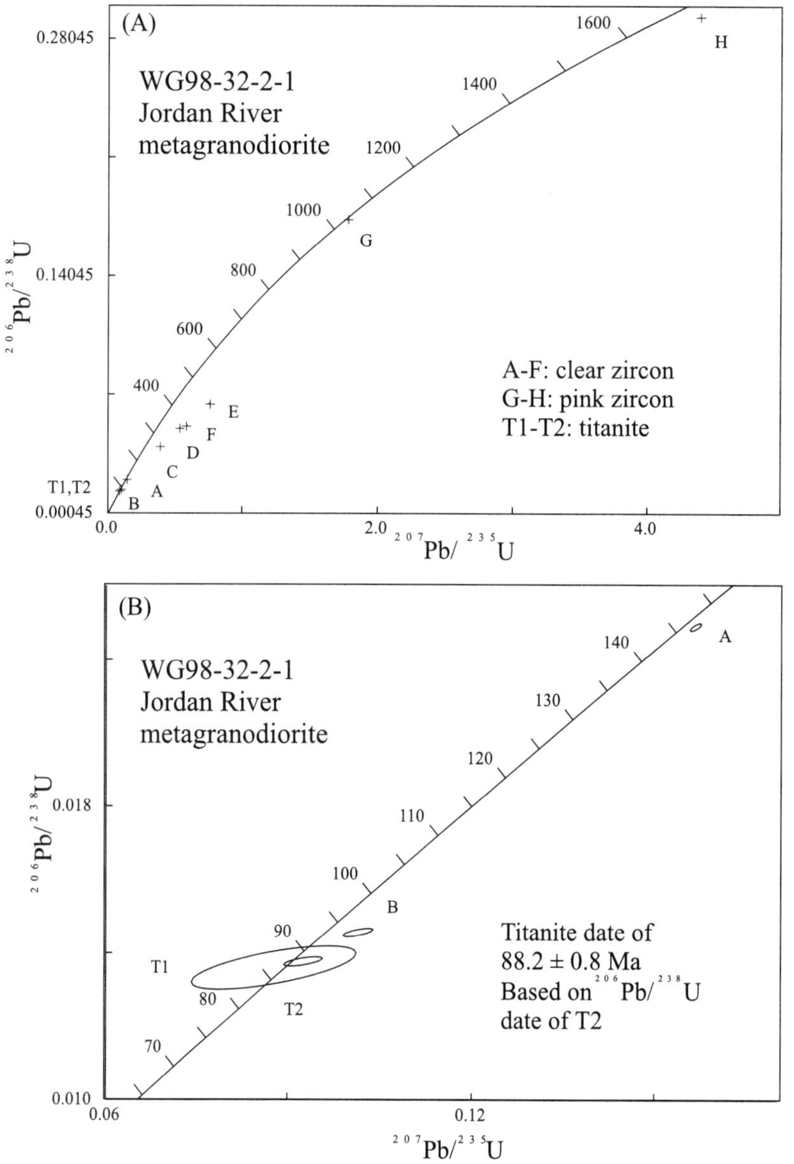

Figure 11. Concordia plots showing U-Pb results for sample WG98-32-2-1 (Jordan River metagranodiorite). U-Pb analytical data are presented in Table 2 and plotted at the 2 level of uncertainty. A: All fractions. B: Expanded plot showing detail of titanite and young zircon fractions.

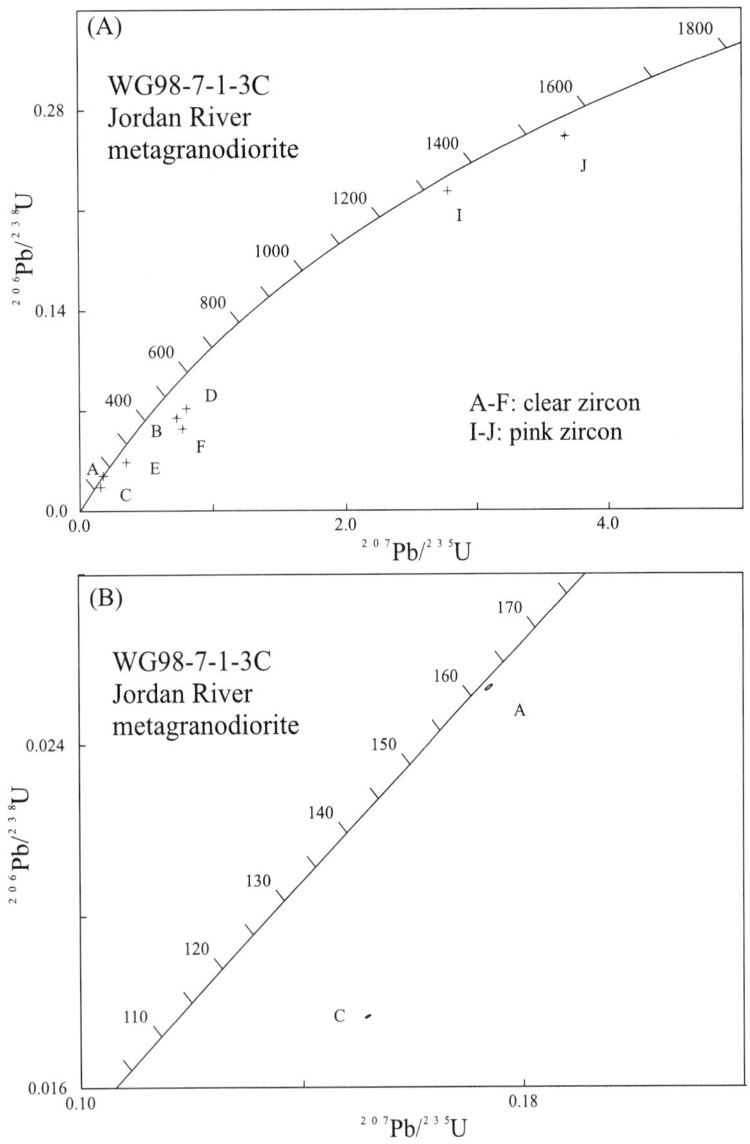

Figure 12. Concordia plots showing U-Pb results for sample WG98-7-1-3C (Jordan River metagranodiorite). U-Pb analytical data are presented in Table 2 and plotted at the 2 level of uncertainty. A: All fractions. B: Expanded plot showing detail of young zircon fractions.

intrusions (Groome, 2000). The greenschist-facies metamorphic assemblage consists primarily of biotite-chlorite-plagioclase-white mica-quartz, but biotite is locally absent in the northern parts of the Complex. The biotite-chlorite-plagioclase-white mica-quartz assemblage corresponds to a temperature range of 400–450 °C, at a pressure of approximately 3 kbar, decreasing to a temperature range of 300 to 400 °C in the absence of biotite, assuming orogenic metamorphism along the kyanite geotherm (Fig. 14) (Bucher and Frey, 1994).

The amphibolite-facies metamorphic assemblage in the Leech River Complex consists of biotite-staurolite-andalusite-plagiolcase-muscovite-quartz, ± garnet, ± cordierite. This assemblage corresponds to a temperature of 500 to 600 °C and a pressure less than the andalusite-kyanite transition, which is approximately 3.5 kbar at 620 °C (Bucher and Frey, 1994). Thus, the transition from greenschist-facies to amphibolite-facies in the study area was apparently the result of an isobaric temperature increase. This, along with the close spatial association with intrusive activity, indicates a contact metamorphic regime. A striking increase in porphyroblast size and abundance occurs at the margins of the Jordan River and Walker Creek intrusions. Staurolite is ubiquitous in the amphibolite facies schists and was used as the main indicator for the amphibolite facies. The presence of garnet and andalusite appears to be strongly controlled by protolith chemistry within the amphibolite facies, as their first occurrences are sporadic near the amphibolite-facies boundary.

Staurolite porphyroblasts range in size from 0.5 mm to 5 mm. They are usually well shaped crystals that commonly have cross-penetration twins. At several localities, staurolite porphyroblasts

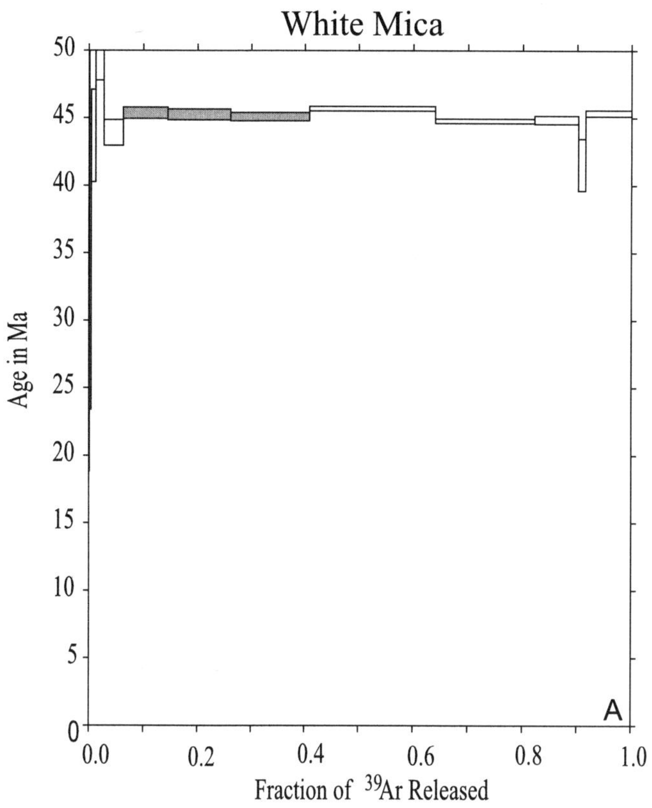

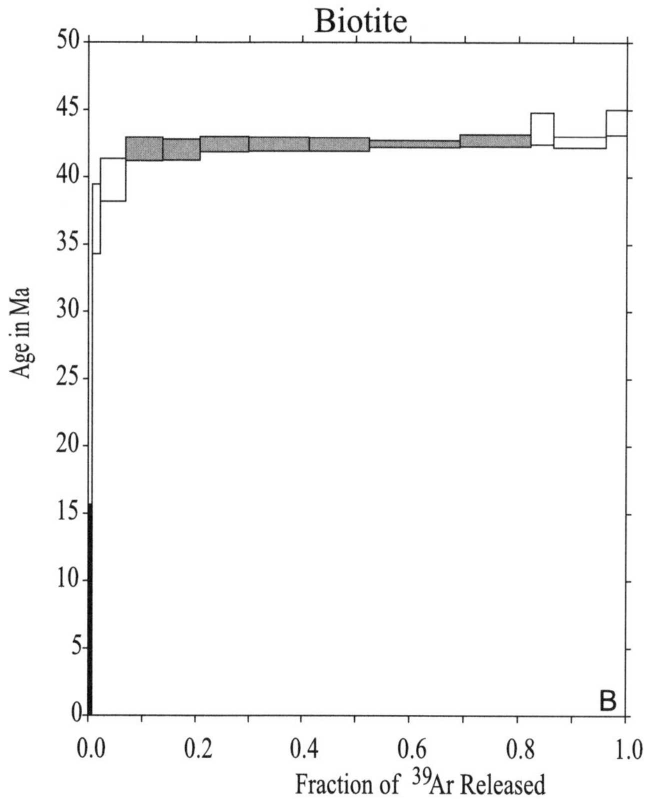

Figure 13. A: $^{40}Ar/^{39}Ar$ spectrum for white mica. Plateau age is 45.2 ± 0.2 Ma. Data are presented in Table 2. The shaded region indicates the plateaus used for the plateau age. B: $^{40}Ar/^{39}Ar$ spectrum for biotite. Plateau age is 42.5 ± 0.2 Ma. Data are presented in Table 2. The shaded region indicates the plateaus used for the plateau age.

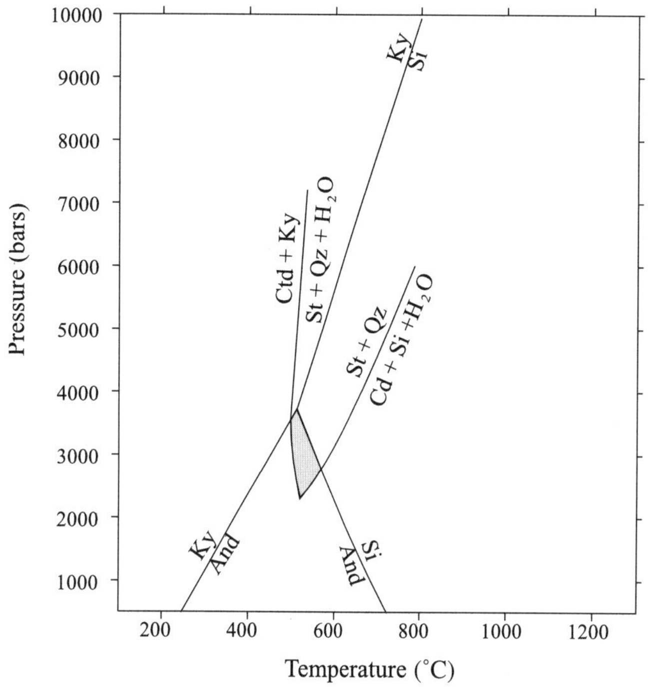

Figure 14. Pressure-temperature conditions of peak metamorphism in the Leech River Schist. Shaded region indicates the *P-T* conditions of peak metamorphism based on geothermometry and field relationships. And—andalusite, Cd—cordierite, Ctd—chloritoid, Ky—kyanite, Qz—quartz, Si—sillimanite, St—staurolite. Reactions are from Bucher and Frey (1994).

have small pressure shadows around them, indicating syn-kinematic growth (probably syn-D1). In thin section, these pressure shadows contain the same mineral assemblage as the ground mass, indicating that these porphyroblasts are not pre-kinematic, as the pressure shadows equilibrated under the same conditions as the rest of the rock (Paterson and Fowler, 1993). At other outcrops, staurolite porphyroblasts are randomly oriented and lack pressure shadows. These relations indicate that at these outcrops staurolite growth was post-kinematic and constrains the amphibolite-facies metamorphism to be syn- to post-kinematic.

The orientations of andalusite porphyroblasts also indicate that amphibolite-facies metamorphism was syn- to post-kinematic (D1). Andalusite porphyroblasts are best observed along the banks of the Jordan River. Where present, these porphyroblasts range in size from 1 cm to more than 15 cm in maximum dimension. The length to width ratios of these porphyroblasts vary from about 1 to 10, with the smaller porphyroblasts being more equant. The highly elongate porphyroblasts are oriented parallel to the schistosity defined by biotite. Locally, some of these porphyroblasts were boudinaged, indicating that deformation had affected them. However, the more equant porphyroblasts crosscut schistosity without the development of pressure shadows. These features indicate that the latter porphyroblasts grew after development of the schistosity. Some andalusite porphyroblasts are pseudomorphed by fine-grained muscovite and biotite, indicating minor retrograde metamorphism. Garnet porphyroblasts range from 0.5 mm to 2 mm in diameter. They are most commonly associated with pressure shadows, indicating mainly syn-kinematic (D1) growth.

The contact nature of the amphibolite-facies metamorphism is supported by mineralogical variations at the margins of dikes and stocks. For example, a small contact aureole was observed in pelitic schist adjacent to a small Walker Creek dike. The visible extent of the contact zone extended less than 35 cm from the dike margin. Within 10 cm of the contact, mm-scale staurolite and garnet are present. The staurolite progressively decreases in abundance away from the dike, and is completely absent in hand specimen beyond 10 cm from the dike. The disappearance of staurolite indicates a drop in grade to the garnet zone, which extends for approximately 25 cm farther from the dike. Beyond 35 cm from the contact, garnet is not visible and the metamorphic grade in the host rock equals the regional middle-greenschist grade in the area. Alongside larger dikes, dike swarms, and stocks, similar increases in metamorphic grade are clearly evident, with broad contact aureoles extending for up to 100 m from intrusive contacts.

Fabric development and deformation of the Leech River Schist were studied by Fairchild (1979), Rusmore (1982) and Groome (2000), but remains imperfectly understood. Fairchild (1979) documented two periods of (post-depositional) deformation in the eastern Leech River Complex. He recognized an early deformational event that produced a regional cleavage in the lower-grade rocks, a foliation in the higher-grade rocks, and transposition of bedding into parallelism with the regional fabric. He also suggested that small, tight to isoclinal folds, including those defined by quartz veins, formed during this deformational this event. Fairchild (1979) recognized a second deformation that produced a localized crenulation cleavage in the lower-grade rocks and large open folds (tens of meters wavelength) in the earlier-formed foliation. These folds are larger near the Leech River fault zone and plunge shallowly to the east. Similar deformational events were reported by Rusmore (1982) in the western portion of the Leech River Complex near Port Renfrew.

Groome (2000) recognized the first and second deformation events described by Fairchild (1979), but also identified a conundrum in the timing of fabric development, igneous intrusion, and formation of contact metamorphic aureoles. Both the Walker Creek and Jordan River intrusions are parallel to the enclosing Leech River Schist, and have contact metamorphic aureoles that apparently record syn- to post-kinematic mineral growth of staurolite and andalusite. In the proximal, coarsest parts of these aureoles, the "syn-kinematic" porphyroblasts are concordant with schistosity, and are locally boudinaged and crosscut by shear bands. The "post-kinematic" porphyroblasts crosscut the schistosity, and apparently postdate the formation of foliation (Fig. 15). However, the ages of the Walker Creek and Jordan River intrusions, and by inference their aureoles, differ by nearly 40 m.y., a relation that requires two separate intervals of foliation development and "post-kinematic" mineral growth. If this is correct, the earlier interval of deformation produced a schistosity that is remarkably similar, both texturally and mineralogically, to the fabric generated by the later one. Additional structural and geochronologic studies are needed to resolve this problem, for instance, direct dating of foliation-defining minerals.

EVOLUTION OF THE LEECH RIVER COMPLEX

The Leech River Complex provides an important link between Wrangellia and other inboard terranes of western North America, and the outboard Paleogene Crescent Terrane, one of the last assemblages to accrete to the continental margin. The main events in the evolution of the complex are indicated in Figure 16. Detrital zircon ages indicate that the Leech River Complex originated as a sedimentary deposit derived from a variety of source areas, including those of Precambrian age or provenance, while whole-rock geochemistry of the metasediments indicate an arc influence for the protolith. The age of earliest sedimentation is not known, but deposition must have occurred after 103 Ma, the interpreted crystallization age of the youngest analyzed detrital zircon grain.

Deposition of the schist protolith ended prior to ca. 88 Ma when the peraluminous Jordan River metagranodiorite was emplaced at approximately 3 kbar. The cause of burial to depths of ca. 10 km is unknown, but may be related to regional west-directed thrusting along the Coast Mountains in early Late

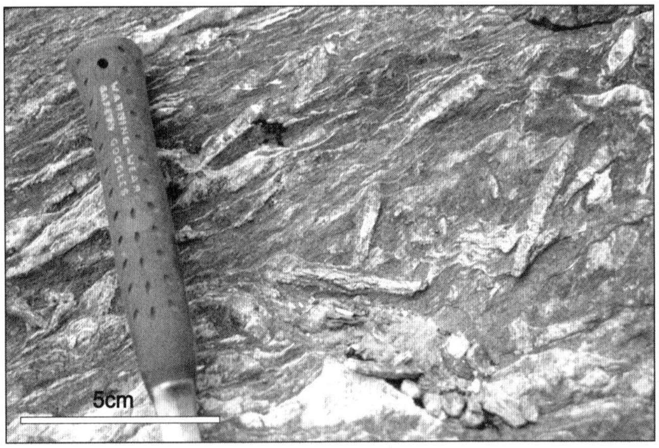

Figure 15. Photograph of randomly oriented andalusite porphyroblasts.

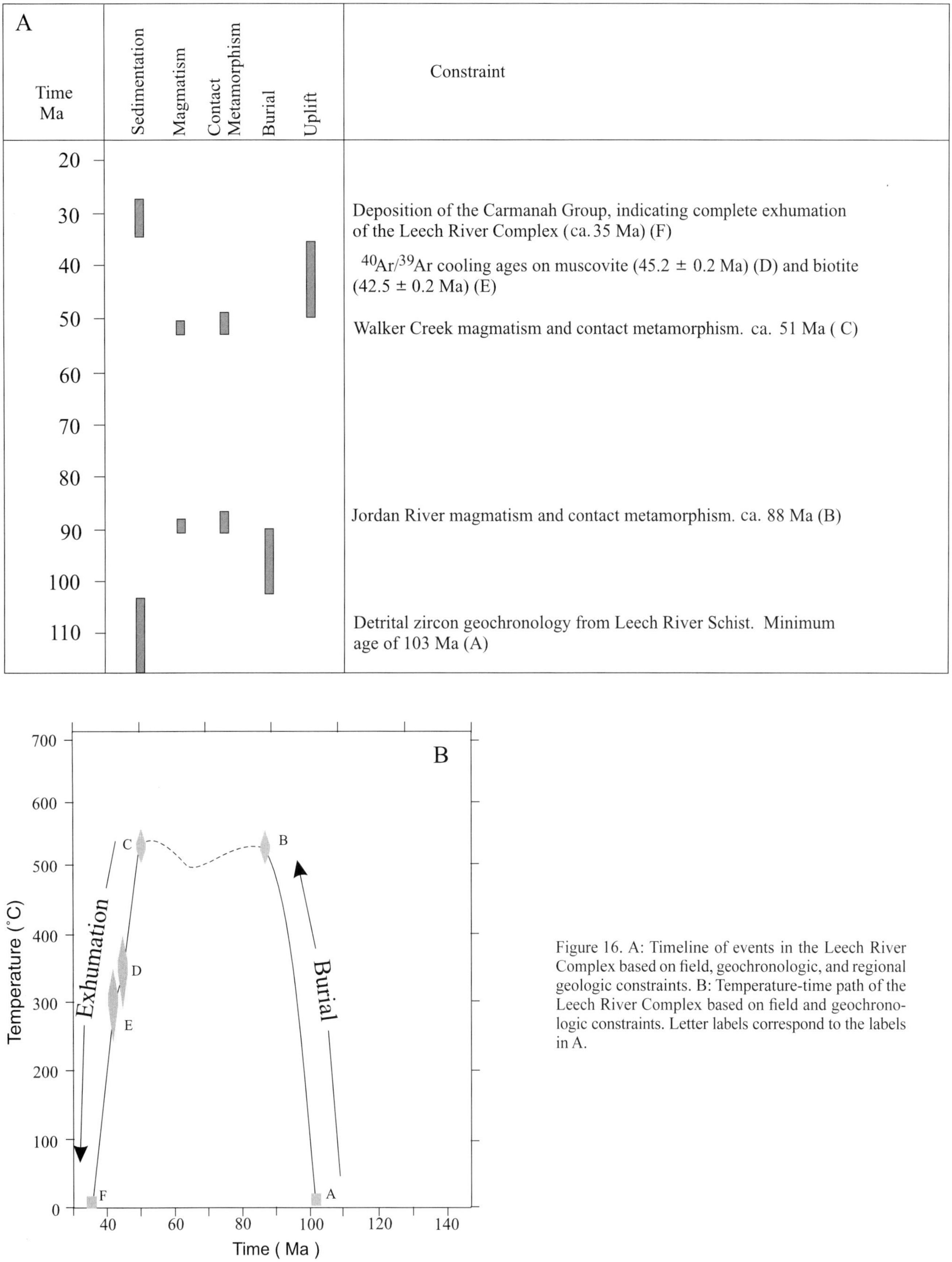

Figure 16. A: Timeline of events in the Leech River Complex based on field, geochronologic, and regional geologic constraints. B: Temperature-time path of the Leech River Complex based on field and geochronologic constraints. Letter labels correspond to the labels in A.

Cretaceous time (Brandon, 1989; Rubin et al., 1990; Journeay and Friedman, 1993). The source of the magmatism is also unknown, but the peraluminous composition of the intrusion suggests that it may have been generated, at least in part, by melting of the Leech River Schist. This forearc magmatism must have been generated by unusually high geothermal conditions, possibly during ridge subduction. The undated Tripp Creek metabasite, with its MORB character, may also have been emplaced at this time (and/or during Walker Creek magmatism), and may have provided much of the heat necessary for anatexis. Deformation, leading to schistosity in the host sediments and foliation in the intrusions, appears to have been broadly synchronous with the magmatism, with peak thermal conditions and porphyroblast growth outlasting the development of foliation.

The peraluminous Walker Creek intrusions were emplaced into the Leech River Schist at ca. 51 Ma, approximately 37 m.y. after crystallization of the Jordan River intrusion. Like the Jordan River intrusions, most of the Walker Creek intrusions are foliated, and imparted on the schist a contact aureole that outlasted deformation. Some of the Walker Creek intrusions are weakly foliated, and a few are non-foliated, indicating that magmatism continued after deformation ended. Geochemical modeling suggests that the Walker Creek intrusions evolved by anatexis or assimilation of the Leech River Schist, mixing with MORB-like melts (possibly magmas of the Tripp Creek metabasite), and fractional crystallization. Subduction of the Kula-Farallon ridge is a plausible cause of high heat flow, MORB-type magmatism, and forearc anatexis.

Rapid cooling of the Leech River Complex followed Walker Creek magmatism. By ca. 45 Ma, muscovite from one of the Walker Creek intrusions had cooled to ~400 °C. By ca. 42.5 Ma, biotite from nearby pelitic schist had cooled to ~330 °C (Fig. 16B). These temperatures may reflect, in part, static cooling of the Complex following cessation of magmatism. However, the rapid decrease in temperature could be linked to exhumation, arguably during a tectonic event. A simple explanation for exhumation is underthrusting of the Metchosin Igneous Complex beneath the Leech River Complex along the Leech River fault (Clowes et al., 1987; Massey, 1986). This fault is exposed as an anastomosing chloritic mylonite zone, with sinistral-oblique sense of displacement, along the contact between the Metchosin and Leech River complexes (Fairchild, 1979; Muller, 1983; Massey, 1986; Groome, 2000). The Leech River Complex was uplifted, denuded, and exposed at the surface by 35 Ma when it was overlain by sediments of the Oligocene Carmanah Group (Fairchild and Cowan, 1982).

THE ROLE OF SLAB WINDOWS

Ridge subduction and slab window formation provide a compelling cause for magmatism in the Leech River Complex because of (1) the inferred forearc position of the Leech River Complex in Cretaceous to Paleogene time, (2) the peraluminous character of the Jordan River and Walker Creek intrusive suites, and (3) the MORB-like composition of the Tripp Creek metabasite. Taken together, these features are compatible with migration of oceanic spreading ridges beneath the Leech River forearc, leading to high heat flow and the injection of MORB-type magmas into the forearc, and concurrent anatexis of the forearc sedimentary and metasedimentary rocks. Known as the "blowtorch" effect (Delong et al., 1979), this model of ridge-induced melting in the forearc is compatible with early models of ridge subduction (e.g., Uyeda and Miyashiro, 1974; Marshak and Karig, 1977), subsequent structural and magmatic models of slab windows (Dickinson and Snyder, 1979; Thorkelson, 1996), and studies of specific Late Cenozoic near-trench igneous complexes in Chile (Kaeding et al., 1990), the Solomon arc (Johnson et al., 1987), California and Mexico (Johnson and O'Neil, 1984; Rogers and Saunders, 1989), Japan (Hibbard and Karig, 1990), and Vancouver Island (Armstrong et al., 1985).

Walker Creek magmatism in the Leech River Complex lies at the southeastern end of a diachronous belt of forearc intrusions and hydrothermal veins that range in age from ca. 61 Ma in southern Alaska to ca. 50 Ma in southeastern Alaska, and apparently record southward migration of the Kula-Farallon ridge along the continental margin (Haeussler et al., 1995). Although the actual history of triple junction migration may have been complex (Sisson and Pavlis, 1993) and complicated by forearc dismemberment and dextral translation (Smart et al., 1996), the overall southward migration apparently ended at or near the Leech River Complex with the emplacement of the Walker Creek intrusions at ca. 51 Ma.

The absence of significant post–Middle Eocene northward translation of the Pacific Rim and Crescent Terranes, as documented in paleomagnetic studies (Irving and Brandon, 1990; Babcock et al., 1992), implies that the present position of the Leech River Complex is close to where it lay at the time of Walker Creek magmatism, in the Middle Eocene. In contrast, paleomagnetic studies on Eocene rocks from southeastern Alaska indicate large dextral displacements (1400 ± 350 km [13 ± 9°]) from the south (Bol et al., 1992), leaving the position of these rocks uncertain at the time of forearc magmatism. Therefore, the Walker Creek intrusions in the Leech River Complex and the broadly coeval Flores volcanics of the Pacific Rim Complex and the Mount Washington and Clayoquot intrusions on Vancouver Island (Massey, 1995) appear to provide the most accurate location of the Kula–Farallon–North America triple junction in early Middle Eocene time (Fig. 17). This position, along Vancouver Island, was predicted by Thorkelson and Taylor (1989) largely on the basis of chemical variations in the Kamloops-Challis arc in south-central British Columbia and the northwestern United States. Geochemical compilation and tectonic modeling by Breitsprecher and Thorkelson (2001) supports the hypothesis that a slab window formed beneath the Pacific Northwest, with the triple junction located off the coast of present-day Vancouver Island at the time of Walker Creek magmatism.

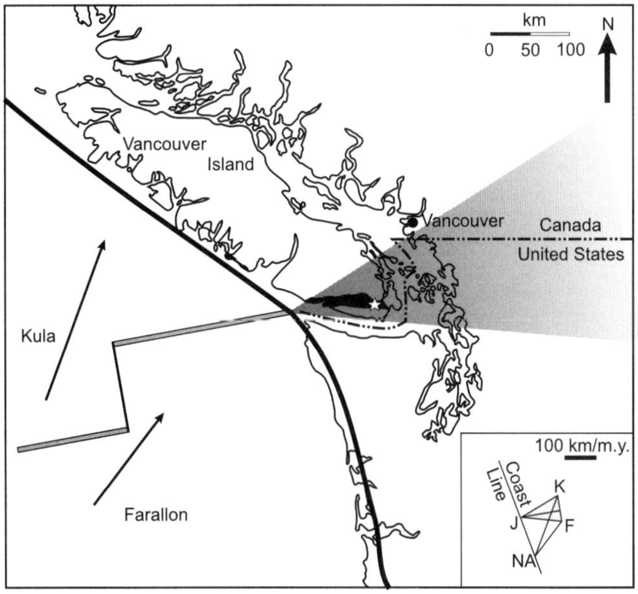

Figure 17. Slab window model for the forearc magmatism in the Leech River Complex at 51 Ma. The Pacific Rim Terrane is indicated by the black shading, and the study area location is indicated by the white star. The heavy dark line represents the inferred location of the North American plate margin in the early Eocene (modified from Dickinson, 1991). Inset: Vector diagram for the Kula (K) and Farallon (F) plates with a fixed North America (NA). J is the position of the triple junction. The shaded region represents the shape of the slab window beneath the North American plate. J-NA vector represents the migration path of the K-F-NA triple junction along the coastline. The orientation of the coastline is taken to be 340°, approximately what it is currently. Vectors are from Babcock et al. (1992) and the principles are from Thorkelson (1996).

CONCLUSIONS

Based on the data and interpretations presented here, a Cretaceous through Eocene history can be deduced for the Leech River Complex. The earliest event in the Leech River Complex's evolution was the deposition of the protolith sediments in a narrow time window in the latest Early Cretaceous and the earliest Late Cretaceous. The timing of sedimentation is constrained between 103 Ma, the age of the youngest detrital zircon in the metasediment, and 88 Ma, the age of Jordan River magmatism.

Between the time of deposition and the time of Jordan River magmatism, the Leech River Complex underwent rapid burial to a depth of approximately 10 km, which is the depth indicated by the metamorphic assemblage in the aureole around the Jordan River metagranodiorite. The speed of this burial, a minimum of 0.6 km/m.y., is indicative of an active tectonic setting. After the burial, it is inferred that the Leech River Complex remained at approximately 10 km depth from the Late Cretaceous through the Middle Eocene. While no direct constraints were found to be dated between 88 Ma and 51 Ma, it can be inferred that the Complex remained at the same *P-T* conditions because the contact metamorphic aureole associated with the Walker Creek intrusions records the same conditions as that around the Jordan River metagranodiorite. The Walker Creek intrusions were dated to be ca. 51 Ma.

After the Walker Creek magmatism, the Leech River Complex underwent rapid exhumation, indicated by the $^{40}Ar/^{39}Ar$ ages from the Leech River Schist. Age determinations on muscovite and biotite indicate that the Leech River Complex underwent rapid cooling shortly after Walker Creek magmatism. Assuming exhumation along the kyanite geotherm, the Complex was exhumed at a minimum rate of 0.8 km/m.y. between ca. 51 Ma and 42.5 Ma, the $^{40}Ar/^{39}Ar$ age from biotite in the schist. The timing of the final exhumation of the Leech River Complex is constrained by the presence of Early Oligocene sedimentary overlap assemblages. For these sediments to be deposited over the Leech River Complex, the Complex must have been completely exhumed. These sediments overlap both the Leech River Complex and Wrangellia, indicating that the Leech River Complex was in its current position relative to Wrangellia by the Early Oligocene.

The timing of these events can be used to constrain the offshore plate configurations along the western margin of North America. The Walker Creek intrusions are interpreted to have formed as a result of the subduction of the Kula-Farallon ridge. Based on paleomagnetic studies of the Eocene Flores Volcanics on Vancouver Island, which indicate that Vancouver Island was in its current location by the Eocene (Irving and Brandon, 1990), the Walker Creek intrusions provide the location of the Kula–Farallon–North America triple junction at ca. 51 Ma.

ACKNOWLEDGMENTS

Funding for this project was provided by Natural Sciences and Engineering Research Council of Canada (NSERC) research grants to DJT and JKM. We would like to thank the critical reviews by Jinny Sisson, Margaret Rusmore and Susan DeBari for insightful and helpful comments and suggestions.

REFERENCES CITED

Armstrong, R.L., Muller, J.E., Harakal, J.E., and Muehlenbachs, K., 1985, The Neogene Alert Bay volcanic belt of northern Vancouver Island, Canada: Descending-plate-edge volcanism in the arc-trench gap: Journal of Volcanology and Geothermal Research, v. 26, p. 75–97.

Babcock, R.S., Burmester, R.F., Engebretson, D.C., and Warnock, A., 1992, A rifted margin for the Crescent basalts and related rocks in the northern Coast Range volcanic province, Washington and British Columbia: Journal of Geophysical Research, v. 97, p. 6799–6821.

Barker, F., Farmer, G.L., Ayuso, R.A., Plafker, G., and Lull, J.S., 1992, The 50 Ma granodiorite of the eastern Gulf of Alaska; melting in an accretionary prism in the forearc: Journal of Geophysical Research, v. 97, p. 6757–6778.

Best, M.G., 1982, Igneous and metamorphic petrology: New York, W.H. Freeman and Company, 630 p.

Bol, A.J., Coe, R.S., Gromme, C.S., and Hillhouse, J.W., 1992, Paleomagnetism of the Resurrection Peninsula, Alaska; implications for the tectonics of southern Alaska and the Kula-Farallon Ridge: Journal of Geophysical Research, v. 97, p. 17,213–17,232.

Brandon, M.T., 1989, Deformational styles in a sequence of olistostromal melanges, Pacific Rim Complex, western Vancouver Island, Canada: Geological Society of America Bulletin, v. 101, p. 1520–1542.

Brandon, M.T., Cowan, D.S., and Vance, J.A., 1988, The Late Cretaceous San Juan thrust system, San Juan Islands, Washington: Boulder, Colorado, Geological Society of America Special Paper 221, 81 p.

Breitsprecher, K., and Thorkelson, D.J., 2001, Spatial coincidence of the Kula-Farallon slab window with trench-distal volcanism of the Eocene Magmatic Belt in the southern Cordillera, *in* Cook, F., and Erdmer, P., compilers, Slave-Northern Cordillera Lithospheric Evolution (SNORCLE) Transect and Cordilleran Tectonics Workshop Meeting (February 22–25), Pacific Geoscience Centre: Lithoprobe Report No. 79, p. 178–182.

Bucher, K and Frey, M., 1994, Petrogenesis of metamorphic rocks (6th edition): Complete revision of Winkler's textbook: Berlin, Springer Verlag, 318 p.

Clowes, R.M., Brandon, M.T., Green, A.G., Yorath, C.J., Sutherland Brown, A., Kanasewich, E.R., and Spencer, C., 1987, LITHOPROBE—Southern Vancouver Island: Cenozoic subduction complex imaged by deep seismic reflections: Canadian Journal of Earth Sciences, v. 24, p. 31–51.

Debari, S.M., Anderson, R.G., and Mortensen, J.K., 1999, Correlations among lower to upper crustal components in an island arc: The Jurassic Bonanza arc, Vancouver Island, Canada: Canadian Journal of Earth Sciences, v. 36, p. 1371–1413.

DeLong, S.E., Schwarz, W.M., and Anderson, R.N., 1979, Thermal effects of ridge subduction: Earth and Planetary Science Letters, v. 44, p. 239–246.

Dickinson, W.R., 1991, Tectonic setting of faulted Tertiary strata associated with the Catalina core complex in southern Arizona: Boulder, Colorado, Geological Society of America Special Paper 264, 106 p.

Dickinson, W.R., and Snyder, W.S., 1979, Geometry of triple junctions related to San Andreas transform: Journal of Geophysical Research, v. 84, p. 561–572.

Duncan, R.A., 1982, A captured island chain in the Coast Range of Oregon and Washington: Journal of Geophysical Research, v. 87, p. 10,827–10,837.

Fairchild, L.H., 1979, The Leech River unit and Leech River fault, southern Vancouver Island, British Columbia [M.S. thesis]: Seattle, University of Washington, 170 p.

Fairchild, L.H., and Cowan, D.S., 1982, Structure, petrology, and tectonic history of the Leech River Complex northwest of Victoria, Vancouver Island: Canadian Journal of Earth Sciences, v. 19, p. 1817–1835.

Forsythe, R.D., and Nelson, E.P., 1985, Geological manifestations of ridge collision: Evidence from the Golfo de Penas-Taitao Basin, southern Chile: Tectonics, v. 4, p. 477–495.

Friedman, R.M., and Armstrong, R.L., 1995, Jurassic and Cretaceous U-Pb geochronometry of the southern Coast Belt, British Columbia, 49°–51° N, *in* Miller, D.M., and Busby, C., eds., Jurassic magmatism and tectonics of the North American Cordillera: Boulder, Colorado, Geological Society of America Special Paper 299, p. 95–139.

Frost, B.R., Chamberlain, K.R., and Schumacher, J.C., 2000, Sphene (titanite); phase relations and role as a geochronometer: Chemical Geology, v. 172, p. 131–148.

Gehrels, G.E., Dickinson, W.R., Ross, G.M., Stewart, J.H., and Howell, D.G., 1995, Detrital zircon reference for Cambrian to Triassic miogeoclinal strata of western North America: Geology, v. 23, p. 831–834.

Gill, J.B., 1981, Orogenic andesites and plate tectonics: Berlin, Springer, 390 p.

Groome, W.G., 2000, Magmatism and metamorphism in the Leech River Complex, southern Vancouver Island, British Columbia, Canada—Implications for Eocene tectonics of the Pacific Northwest [M.S. thesis]: Burnaby, British Columbia, Simon Fraser University, 220 p.

Groome, W.G., Friedman, R.M., Massey, N.W.D., Thorkelson, D.J., Layer, P.W., and Marshall, D., 1999, Evidence from the Leech River Complex for Eocene deformation and accretion on southern Vancouver Island, British Columbia, *in* Evenchick, C.A., et al., eds., Terrane paths '99: Circum-Pacific Terrane Conference, Okanagan Valley, B.C., Canada, September 26–October 1, 1999, Abstracts and Program, p. 39.

Groome, W.G., Thorkelson, D.J., Friedman, R.M., Massey, N.W.D., and Marshall, D.D., 2000, Eocene magmatism in the Leech River Complex: Evidence for spreading ridge subduction and slab window formation: Geological Society of America Abstracts with Programs, v. 32, no. 6, p. A-15.

Haeussler, P.J., Bradley, D., Goldfarb, R., Snee, L., and Taylor, C., 1995, Link between ridge subduction and gold mineralization in southern Alaska: Geology, v. 23, p. 995–998.

Halliday, A.N., Fallick, A.E., Dickin, A.P., Mackenzie, A.B., Stephens, W.E., and Hildreth, W., 1983, The isotopic and chemical evolution of Mount St. Helens: Earth and Planetary Science Letters, v. 63, p. 241–256.

Harris, N.R., Sisson, V.B., Wright, J.E., and Pavlis, T.L., 1996, Evidence for Eocene mafic underplating during forearc intrusive activity, eastern Chugach Mountains, Alaska: Geology, v. 24, p. 263–266.

Hibbard, J.P., and Karig, D.E., 1990, Structural and magmatic responses to spreading ridge subduction: An example from southwest Japan: Tectonics, v. 9, p. 207–230.

Hill, M., Morris, J., and Whelan, J., 1981, Hybrid granodiorites intruding the accretionary prism, Kodiak, Shumagin, and Sanak Islands, southwest Alaska: Journal of Geophysical Research, v. 86, p. 10,569–10,590.

Irving, E., and Brandon, M.T., 1990, Paleomagnetism of the Flores Volcanics, Vancouver Island, in place by Eocene time: Canadian Journal of Earth Sciences, v. 27, p. 811–817.

Johnson, C.M., and O'Neil, J:R., 1984, Triple junction magmatism: A geochemical study of Neogene volcanic rocks in western California: Earth and Planetary Science Letters, v. 71, p. 241–262.

Johnson, R.W., Jaques, A.L., Langmuir, C.H., Perfit, M.R., McColloch, M.T., Staudigel, H., Chapell, B.W., and Taylor, S.R., 1987, Ridge subduction and forearc volcanism: Petrology and geochemistry of rocks dredged from the western Solomon arc and Woodlark Basin, *in* Taylor, B., and Exon, N., eds., Marine geology, geophysics and geochemistry of the Woodlark Basin, Solomon Islands: Houston, Texas, Circum-Pacific Council for Energy and Mineral Resources, Earth Science Series, v. 7, p. 113–154.

Johnston, S.T., and Thorkelson, D.J., 1997, Cocos-Nazca slab window beneath Central America: Earth and Planetary Science Letters, v. 146, p. 465–474.

Journeay, J.M., and Friedman, R.M., 1993, The Coast Belt thrust system: Evidence of Late Cretaceous shortening in southwest British Columbia: Tectonics, v. 12, p. 756–775.

Kaeding, M., Forsythe, R.D., and Nelson, E.P., 1990, Geochemistry of the Taitao ophiolite and near-trench intrusions from the Chile margin triple junction: Journal of South American Earth Sciences, v. 3, p. 161–177.

Kirschner, D.L., Hunziker, J.C., and Cosca, M., 1996, Closure temperature of argon in micas; a review and reevaluation based on Alpine samples: Geological Society of America Abstracts with Programs, v. 28, no. 7, p. 441.

Marshak, R.S., and Karig, D.E., 1977, Triple junctions as a cause for anomalously near-trench igneous activity between the trench and volcanic arc: Geology, v. 5, p. 233–236.

Massey, N.W.D., 1986, Metchosin Igneous Complex, southern Vancouver Island: Ophiolite stratigraphy developed in an emergent island setting: Geology, v. 14, p. 602–605.

Massey, N.W.D., 1995, Geology and mineral resources of the Alberni–Nanaimo Lakes sheet, Vancouver Island 92F/1W, 92F/2E and part of 92F/7E: British Columbia Ministry of Energy, Mines and Petroleum Resources, Geological Survey Branch, Paper 1992-2, 132 p.

Muller, J.E., 1983, Geology, Victoria map area: Geological Survey of Canada Map 1553A, scale 1:100,000.

Paterson, S.R., and Fowler, T.K., 1993, Re-examining pluton emplacement processes: Journal of Structural Geology, v. 15, p. 191–206.

Pearce, J.A., Harris, N.B.W., and Tindle, A.G., 1984, Trace element discrimination diagrams for the tectonic interpretation of granitic rocks: Journal of Petrology, v. 25, p. 956–983.

Read, P.B., Woodsworth, G.J., Greenwood, H.J., Ghent, E.D., and Evenchick, C.A., 1991, Metamorphic map of the Canadian Cordillera: Geological Survey of Canada Map 1714A, scale 1:2,000,000.

Rogers, G., and Saunders, A.D., 1989, Magnesian andesites from Mexico, Chile, and the Aleutian Islands: Implications for magmatism associated with ridge-trench collisions, *in* Crawford, A.J., ed., Boninites and related rocks: London, Unwin Hyman, p. 416–445.

Ross, G.M., and Parrish, R.R., 1991, Detrital zircon geochronology of metasedimentary rocks in the southern Omenica Belt, Canadian Cordillera: Canadian Journal of Earth Sciences, v. 28, p. 1254–1270.

Rubin, C.M., Saleeby, J.B., Cowan, D.S., Brandon, M.T., and McGroder, M.F., 1990, Regionally extensive mid-Cretaceous west-vergent thrust system in the northwestern Cordillera; implications for continent-margin tectonism: Geology, v. 18, p. 276–280.

Rusmore, M.E., 1982, Structure and petrology of pre-Tertiary rocks near Port Renfrew, Vancouver Island, British Columbia [M.S. thesis]: Seattle, University of Washington, 24 p.

Rusmore, M.E., and Cowan, D.S., 1985, Jurassic-Cretaceous rock units along the southern edge of the Wrangellia Terrane on Vancouver Island: Canadian Journal of Earth Sciences, v. 22, p. 1223–1232.

Shouldice, D.H., 1971, Geology of the Western Canadian Continental Shelf: Bulletin of Canadian Petroleum Geology, Canadian Society of Petroleum Geologists, v. 19, p. 405–436.

Sisson, V.B., and Pavlis, T.L., 1993, Geologic consequences of plate reorganization: An example from the Eocene southern Alaska forearc: Geology, v. 21, p. 913–916.

Smart, K.J., Pavlis, T.L., Sisson, V.B., Roeske, S.M., and Snee, L.W., 1996, The Border Ranges fault system in Glacier Bay National Park, Alaska; evidence for major early Cenozoic dextral strike-slip motion: Canadian Journal of Earth Sciences, v. 33, p. 1268–1282.

Sun, S.-S., and McDonough, W.F., 1989, Chemical and isotopic systematics of oceanic basalts: Implications for mantle compositions and processes, *in* Saunders, A.D., and Norry, M.J., eds., Magmatism in the ocean basins: Geological Society [London] Special Publication No. 42, p. 313–345.

Thorkelson, D.J., 1996, Subduction of diverging plates and the principles of slab window formation: Tectonophysics, v. 255, p. 47–63.

Thorkelson, D.J., and Taylor, R.P., 1989, Cordilleran slab windows: Geology, p. 17, p. 833–836.

Uyeda, S., and Miyashiro, A., 1974, Plate tectonics and the Japanese Islands: A synthesis: Geological Society of America Bulletin, v. 85, p. 1159–1170.

Wells, R.E., Engebretson, D.C., Snavely Jr., P.D., and Coe, R.S., 1984, Cenozoic plate motions and the volcano-tectonic evolution of western Oregon and Washington (USA): Tectonics, v. 3, p. 275–294.

Wheeler, J.O., and McFeeley, P., compilers, 1991, Tectonic assemblage map of the Canadian Cordillera and adjacent parts of the United States of America: Geological Survey of Canada Map 1712A, scale 1:2,000,000.

Manuscript Accepted by the Society February 5, 2003

Index

B

C

D

E

N

T

W

X

Y

Z

The following information pertains to Plates 1 and 2 (Chapter 8, Pavlis and Sisson) on the CD-ROM accompanying this volume:

READ ME

The maps contained in this CD-ROM were generated using the commercial drawing program Adobe Illustrator 9.0 for Wintel systems, available in 2001. The encapsulated postscript files can be printed directly to a postscript printer, but the document would be virtually unusable because multiple layers of the document would lie atop one another and be indistinguishable. These maps are best viewed on a computer monitor using a program that can exploit the layered graphics of the eps format. We suggest viewing three general versions of the map to illustrate the geology (toggling the topographic map background on and off to examine topographic versus structural effects): (1) a S1 = S0 map-turn on the form surface of S1-S0 layer plus orientation symbols if desired; (2) a S2 map-turn on the form surface of S2 layer plus symbols if desired; note, S1 and S2 together also show a good picture of the D2 structure; and (3) a similar D3 map (using the S2 form surface layers plus others). Note also that the three major bedrock shading patterns have been filled with transparent shading colors but the igneous bodies and Quaternary cover are opaque. The opaque fill for the Quaternary deposits was done intentionally to emphasize the "cover" characteristics of these units. To show the bedrock structure, turn off the Quaternary layers to view the inferred bedrock structure that extends beneath the cover. We were not able to make the igneous units transparent without introducing odd color combinations. This arose because this map was originally drawn in Adobe Illustrator 8, which did not allow transparent fills and had no easy way of dealing with "islands" in fill patterns. For opaque fills, this was no problem because the opaque fills can be layered to obscure the underlying pattern, but this version, where transparent fills were allowed, produces odd effects. The data should be exported to a drawing program that deals with islands (e.g., Autocad) or a GIS program if the user wishes to fix these display problems.

This file can be easily imported into a GIS program if desired. Georeferencing is straightforward because a UTM grid has been overlain on the topographic base, and most commercial GIS programs available as of this date could easily import these data.